UNIVERSITY OF BATH SCIENCE 16-19

CHEMISTRY

Other titles in the Project

Biology   Martin Rowland
Applied Genetics   Geoff Hayward
Applied Ecology   Geoff Hayward
Micro-organisms and Biotechnology   Jane Taylor
Biochemistry and Molecular Biology   Moira Sheehan

Physics   Robert Hutchings
Telecommunications   John Allen
Medical Physics   Martin Hollins
Energy   David Sang and Robert Hutchings
Nuclear Physics   David Sang

UNIVERSITY OF BATH SCIENCE 16–19

Project Director: J. J. Thompson, CBE

CHEMISTRY

KEN GADD and STEVE GURR

Thomas Nelson and Sons Ltd
Nelson House Mayfield Road
Walton-on-Thames Surrey
KT12 5PL UK

Nelson Blackie
Wester Cleddens Road
Bishopbriggs
Glasgow
G64 2NZ UK

Thomas Nelson (Hong Kong) Ltd
Toppan Building 10/F
22A Westlands Road
Quarry Bay Hong Kong

Thomas Nelson Australia
102 Dodds Street
South Melbourne
Victoria 3205 Australia

Nelson Canada
1120 Birchmount Road
Scarborough Ontario
MIK 5G4 Canada

First published by Thomas Nelson and Sons Ltd 1994

I(T)P Thomas Nelson is an International Thomson Publishing Company.

I(T)P is used under licence.

ISBN 0-17-448236-1
NPN 9 8 7 6 5 4 3 2 1

Contents

The Project: an introduction

The University of Bath Science 16–19 Project, grew out of a reappraisal of how far sixth form science had travelled during a period of unprecedented curriculum reform and an attempt to evaluate future development. Changes were occurring both within the constitution of 16–19 syllabuses themselves and as a result of external pressures from 16+ and below: syllabus redefinition (starting with the common cores), the introduction of AS-level and its academic recognition, the originally optimistic outcome to the Higginson enquiry; new emphasis on skills and processes, and the balance of continuous and final assessment at GCSE level.

This activity offered fertile ground for the School of Education at the University of Bath and a major publisher to join forces with a team of science teachers, drawn from a wide spectrum of educational experience, to create a flexible curriculum model and then develop resources to fit it. This group addressed the task of satisfying these requirements:

- the new syllabus and examination demands of A- and AS-level courses;
- the provision of materials suitable for both the core and options parts of syllabuses;
- the striking of an appropriate balance of opportunities for students to acquire knowledge and understanding, develop skills and concepts, and to appreciate the applications and implications of science;
- the encouragement of a degree of independent learning through highly interactive texts;
- the satisfaction of the needs of a wide ability range of students at this level.

Some of these objectives were easier to achieve than others. Relationships to still evolving syllabuses demand the most rigorous analysis and a sense of vision – and optimism – regarding their eventual destination. Original assumptions about AS-level, for example, as a distinct though complementary sibling to A-level, needed to be revised.

The Project, though, always regarded itself as more than a provider of materials, important as this is, and concerned itself equally with the process of provision – how material can best be written and shaped to meet the requirements of the educational market-place. This aim found expression in two principal forms: the idea of secondment at the University and the extensive trialling of early material in schools and colleges.

Most authors enjoyed a period of secondment from teaching, which not only allowed them to reflect and write more strategically (and, particularly so, in a supportive academic environment) but, equally, to engage with each other in wrestling with the issues in question.

The Project saw in the trialling a crucial test for the acceptance of its ideas and their execution. Over one hundred institutions and one thousand students participated, and responses were invited from teachers and pupils alike. The reactions generally confirmed the soundness of the model and allowed for more scrupulous textual housekeeping, as details of confusion, ambiguity or plain misunderstanding were revised and reordered.

The test of all teaching must be in the quality of the learning, and the proof of these resources will be in the understanding and ease of accessibility which they generate. The Project, ultimately, is both a collection of materials and a message of faith in the science curriculum of the future.

J.J. Thompson

How to use this book

To succeed in any chemistry course you must be able to demonstrate a number of skills. Assessment objectives are useful as a list of specific abilities you are expected to display, and include:

- recall of chemical terms, facts and principles;
- understanding of terms, facts and principles;
- application of chemical knowledge and understanding to new problems;
- interpretation of chemical information;
- conversion of chemical information from one form to another;
- evaluation of chemical information;
- selection and organisation of chemical information;
- communication of your knowledge and understanding;
- a number of practical skills such as accurate observation and recording of data.

A textbook can help you to acquire many of these skills.

Chemistry provides you with a concise summary of the chemical terms, facts and principles which are included in the core of Advanced Level ('A' level) and Advanced Supplementary (AS) chemistry syllabuses of the GCE Examination Boards in the United Kingdom. You should refer to it when consolidating your notes after each chemistry lesson, when preparing answers to assignments and when revising for examinations. If you are studying chemistry in an open-learning scheme, you should use this book to prepare for your tutorials.

The book is divided into six themes, each containing closely-related chapters. You may use it as a course book, working through the chapters of a particular theme in the order in which they are given, visit selected chapters or use the in-text questions and summaries to help your revision.

Although you will probably be following a chemistry course in a school, sixth form college, tertiary college or college of further education, you are unlikely to realise your true potential unless you take an active responsibility for your own learning. Simply reading through a book is too passive a process to develop the important skills you need for an examination. Our book has been designed to help you become actively involved in the learning process: the features designed to achieve this are listed below. Understanding the purpose of each feature will help you gain maximum benefit from the book.

Theme 1

This theme is intended to act as a bridge between GCSE and 'A' level chemistry. It builds upon the knowledge and skills acquired during your GCSE course and prepares you for the 'A' level work to follow.

Learning objectives

These are listed at the beginning of each chapter. You should read them to discover which skills you are expected to acquire in working through the chapter. As you work through a chapter, refer back to the **learning objectives** to check that these are achieved.

Activities

These provide an opportunity for you to develop necessary skills in a more open-ended and problem-solving context. We hope the **activities** will prove challenging but enjoyable.

Questions

In-text questions regularly test your recall and understanding of what you have just read or prepare you for what comes next. These questions are of two types: quick questions (marked with a blue diamond) and longer questions (in blue boxes). Even if you choose not to produce written answers to all these questions, they do provide further evidence of the skills you need and how these may be assessed. Of course, you may wish to revisit many or even all the questions during the course of your revision. Answers to quick questions are given at the end of each chapter. For a selection of the longer questions, answers are provided at the end of the book. For Themes 1 and 2 the answers have been given more fully than for later themes. This has been done purposely to help you to realise what depth is required in answering questions.

Spotlight boxes

The information in these boxes supplements the text by highlighting interesting aspects of chemistry, in a personal, economic, social or technological context.

Tutorial boxes

These offer an opportunity to look at worked solutions to problems, displayed in a step-by-step manner. Our intention here is to encourage you to appreciate the logic involved in each stage in solving a problem.

Examination questions

Whilst there is no substitute for building a collection of examination papers for *your own syllabus*, we have included some recent questions from GCE Advanced Level and Advanced Supplementary examination papers in chemistry. We have tried to ensure that the questions chosen cover a wide range of topic areas. In answering these you will gain valuable experience which will help you to prepare for the examinations.

Key facts

These, as the name suggests, emphasise a term, fact or principle encountered in the text. We hope they will help you to recognise the most important issues covered in a section or paragraph.

Connections

When one topic area is referred to or somehow relates to another dealt with elsewhere in the text, **connections** help you to link different parts of the subject matter. These appear in the margin when useful.

The Database

This book contains a database of information which you may use for general reference during your course and for data for many of the questions asked. This valuable feature is contained in one large section towards the middle of the book. It contains a list of key physical and chemical properties of the elements and many of their compounds. It will be very useful for you when looking for trends in inorganic chemistry and when plotting synthetic routes, particularly in organic chemistry. We hope it will prove a valuable summary of inorganic and organic chemistry for your revision. Explanation of the structure and use of the Database may be found on page 310.

Copies of syllabuses, past examination questions and mark schemes (only for papers from June 1991 onwards) can be obtained from the Publications Departments of the Examinations Board whose syllabus you are following. In the examination question section at the end of the book the examination boards are referred to by the initials of their names, i.e. AEB, JMB, OCSEB, SCE, UCLES, ULSEB (recently renamed ULEAC), UODLE and WJEC.

Acknowledgements

Writing a textbook is a demanding and exacting process. A successful outcome depends on the encouragement and support of family, friends, colleagues and students. There are many people who have contributed either directly or indirectly to the production of this book. We are grateful to everyone who helped.

Thanks are due to the many reviewers who read our early chapters and whose comments helped to shape our approach. We must also thank the students of City of Bath College and Yeovil College, unwitting guinea-pigs for many of our ideas and a great source of inspiration.

For assistance with the photographs we would like to thank Sigi and Christina Wilkinson and the incomparable Steve Cooper.

Finally, we would like to express our personal gratitude to our long-suffering families and to friends who have helped to keep our spirits high during the preparation of the manuscript.

Steve Gurr would like to thank Adele and Connor Crocker, Janis, Bonnie and Alan Gurr, Geoffrey Abbott, Gary Moore and Neil Hankin. Hayden, who must have wondered what all the fuss was about, deserves a special mention.

Ken Gadd would like to thank Barbara, Tim and Mark Gadd and his parents, Terry and Fred, who gave unquestioning support over the years. Finally he acknowledges the skill and good humour of Jessica Walters.

The authors and publishers wish to thank the following for permission to use copyright material:

The Associated Examining Board, Northern Examinations and Assessment Board, Oxford and Cambridge Schools Examination Board, Scottish Examination Board, University of Cambridge Local Examinations Syndicate, The University of London Examinations and Assessment Council, University of Oxford Delegacy of Local Examinations and Welsh Joint Education Committee for questions from past examination papers.

The author and publishers would like to acknowledge with thanks the following photographic sources:

The Advertising Archive *pp* 209, 249, 545; Allsport *pp* 23, 255, 345, 451 (right); Arcaid *pp* 281, 500; Ardea *pp* 400, 486 (lower); Barnaby's Picture Library *pp* 250, 289, 358, 396, 479 (lower); Professor Neil Bartlett, University of California *p* 372; Biofotos/Heather Angel *pp* 52, 148, 382; Anthony Blake Photo Library *p*354 (lower); Boots *p* 201; Bristol Myers Squibb *pp* 207, 210 (right), 506; J Allen Cash Photo Library *pp* 252 (lower left and centre), 511, 573; Cellmark Diagnostics *p* 408; Chemistry in Britain *pp* 17 (right), 264; Chip Clark *p* 354 (upper); Bruce Coleman *pp* 442, 496; Collections *p* 378; R S Components *p* 395; The Environmental Picture Library *p* 451 (lower); Eye Ubiquitous *p* 286; John Frost Newspaper Library *p* 73; Ken Gadd *p* 459 (lower); General Electric Company *p* 373; General Motors Research Laboratories *p* 67; Geoscience Photos *pp* 30, 58 (centre and lower), 63 (left), 94, 298 (left); Alan Godfrey Photography *p* 503; Sally & Richard Greenhill *pp* 122, 141, 252 (lower right), 303, 359, 392 (left), 524; Robert Harding Picture Library *p* 504 (left); Hoffmann La Roche *p* 15 (right); Holt Studios *p* 539; Hutchinson Library *pp* 99, 348 (lower); ICI *pp* 2, 238, 298 (right), 429 (upper), Zeneca 504 (left), 513, 525, 531; The Image Bank *pp* 38, 55 (left), 555; Impact Photos *pp* 42, 375 (upper); Dr Islam, Imperial Cancer Research Institute *pp* 43, 48, 104 (upper); Lawrence Livermore National Laboratory *p* 17 (left); Lockheed Aircraft *p* 547; Jerry Mason *pp* 10 (upper), 58 (upper), 71, 97, 120, 121, 123, 124, 127, 129, 145, 146, 159, 160, 227, 267, 272, 275, 276, 283, 288, 291, 300, 347 (lower), 349, 350, 364, 379, 392, 403, 407, 428, 429 (lower), 445 (upper), 446, 449, 459 (upper), 462 (upper), 474, 476, 479 (upper), 533, 535, 544; Metropolitan Police *pp* 196, Forensic Science Laboratory 178; James L Amos/National Geographic Society *p* 69 (left and right); National Portrait Gallery *p* 15 (left); NASA *pp* 5, 18, 39, 187, 380, 399, 475; The Natural History Museum *p* 65 (upper); NHPA *p* 107; The Associated Octel Company *p* 412 (upper); Swales Parry *p* 84; Perkin Elmer Ltd *pp* 113, 135; Pilkington plc *p* 536; Lord Porter *p* 10 (lower); Punch Publications *p* 75 (left); F H H Roberts *p* 29; A C Rochester *p* 445 (lower); P L Rodriguez *p* 8; The Royal Geographical Society *p* 348 (upper); S H Salter *p* 161; The Science Museum *p* 360; Science Photo Library *pp* 28, 55 (right), 65 (centre), 111, 112, 153, 181, 190, 199, 269, 274, 347 (upper and centre), 412 (lower), 424, 426, 427, 435, 462 (lower), 469, 486 (upper), 490, 493, 499 (left and right), 562, 566, 570; Serengetti *p* 384; T M Seward *p* 6; Shell International *pp* 252 (upper), 505, 517 (upper and lower), 518; Somatogen *p* 421; Gamma/Frank Spooner Pictures *pp* 292, 413; Tony Stone Worldwide *p* 36; Supermarketing *p* 441; Thames Water *p* 157; Tibet Image Bank/Robin Bath *p* 1; TKM *p* 100; R C Treatt *p* 146 (upper); Tripas Associates *pp* 203, 210 (left); Varian *p* 174; Anthony Waltham/Trent University *pp* 63 (right), 431, 453, John Watney *p* 224; C James Webb *pp* 98, 104 (lower), 130, 156, 277, 294, 306, 375 (lower), 561, 567; Professor A R Wellburn, University of Lancaster *p* 572; Janine Wiedel *p* 451 (upper left); Wildlife Matters, John Feltwell *pp* 93, 541; S C R Williams, University of London Chemistry Department *p* 193; Edward Winpenny Photography *p* 532; ZEFA Picture Library *pp* 37, 41, 65 (lower), 74 (left and right), 75 (right), 148, 418;

Every effort has been made to trace all the copyright holders but if any have been inadvertently overlooked the publishers will be pleased to make the necessary arrangement at the first opportunity.

Theme 1

OUR ENVIRONMENT AND THE CHEMIST

The Earth is a vast reservoir of mineral, organic and energy resources. Chemists have learnt how to use these resources to make new substances which benefit human beings, ranging from medicines and fertilisers to plastics, fibres and superconductors.

Chapter 1

CHEMICAL AND ENERGY RESOURCES

LEARNING OBJECTIVES

When you have studied this chapter you should be able to:

1. describe the role of chemistry and chemists in the world;

2. explain that the Earth provides a wealth of chemical and energy resources;

3. describe the Earth's components: the atmosphere, lithosphere, hydrosphere and biosphere;

4. explain the terms elements, compounds and mixtures;

5. explain the terms natural, synthetic and artificial materials;

6. explain that fuels are stores of energy and why we must continue to search for new sources.

1.1 THE EARTH AS A RESOURCE

Fig 1.1 Knowledge, understanding, skills and imagination – the chemist uses all these for the benefit of human beings.

Chemistry and the environment

Chemistry is a remarkable science. At an ever increasing pace, chemists have learned how to take materials which occur naturally – rocks, minerals, organic material (coal, oil, natural gas) – and transform them into materials which benefit human beings. Oil from beneath the Earth's surface has been changed into medicines, plastics, agrochemicals, fibres, detergents, dyes, perfumes and fuels. Rocks, minerals and even the air we breathe have been harnessed by the chemist to make fertilisers, metals and alloys, glass, ceramics, superconductors and other new advanced materials.

Over 10 million different compounds are known, and some 400 000 new ones are reported each year. But this is just the tip of the chemist's iceberg! Such is the growth of our knowledge and understanding of how substances behave, coupled with human beings' imagination and inventiveness, that the days of custom synthesis – making chemical substances with desired properties to order – are almost with us.

The atmosphere around our planet both protects and supports life. The Earth's crust provides beautiful natural surroundings and a wealth of resources, both mineral and organic. The water which covers about 80% of the Earth's surface is essential in the biological processes of life, as well as fulfilling other important roles. Together, the air, the Earth's crust and the waters provide all the raw materials that chemists need. Given an adequate supply of energy, chemists can go about their creative business.

Human beings must manage their environment on Earth and use its natural resources to provide health and comfort for its inhabitants. Looking to the future, we have a duty to protect and preserve this environment and its rich resources for future generations.

CONNECTION: Modelling is examined in Chapter 2.

To make good use of available resources, we need to understand the structure of our world and the changes which take place within it. Structures and changes which interest chemists are often submicroscopic. They are well beyond the scope of the human sense organs. Models are needed to describe and explain some of these unseen objects and processes, and modelling is an essential part of the chemist's tool kit. Models have improved our understanding of the world around us and are bringing the era of 'designer chemistry' ever closer.

The atmosphere, lithosphere, hydrosphere and biosphere

Chemists often find it helpful to categorise and classify chemical substances and types of change. Our first example of this is the classification of the Earth into four components. Between them, they contain the total sum of the Earth's chemical and energy resources.

The atmosphere

An envelope of gases surrounds the Earth. It is mainly a mixture of nitrogen (78.09%) and oxygen (20.95%). Other gases such as argon, carbon dioxide, neon, krypton, xenon and radon are present in much smaller quantities. Water vapour is an important component but its concentration varies. For this reason, percentage composition figures are based on dry air (Table 1.1).

Table 1.1   Elements and compounds in the atmosphere

The composition of the unpolluted atmosphere (expressed as percentage by volume of dry air*)			
nitrogen	78.09	helium	0.00052
oxygen	20.95	krypton	0.00011
argon	0.93	hydrogen	0.00005
carbon dioxide	0.03	xenon	0.000009
neon	0.0018	radon	6×10^{-18}

* Air always contains some water vapour in addition to the gases given in the table; it can vary from 0-1% depending on atmospheric conditions.

The lithosphere

At the centre of the Earth is a solid **core**, made up largely of iron with some nickel. Moving outwards, this core becomes liquid, at temperatures of around 2500–3000°C. Surrounding the core is the **mantle**, solid rock mainly silicates but with a high proportion of iron and magnesium compounds. The outer layer, the **Earth's crust**, is relatively thin but is an invaluable reservoir of resources.

The hydrosphere

More than three-quarters of the Earth's surface is covered by water. The bulk is present in the oceans, some 97%, while much smaller amounts are found in freshwater lakes and rivers (0.6%) and snow and polar ice sheets (2.4%). In total there are 1348 million km^3 of water on the planet. The composition of surface waters depends on whether they are freshwater or salt water, and even then can be extremely variable.

Table 1.2   Elements of the Earth

The composition of the Earth and the Earth's crust (expressed as percentage by mass)		
	Earth	Earth's crust
iron	34.6	5.1
oxygen	29.5	46.4
silicon	15.2	28.0
magnesium	12.7	2.0
nickel	2.4	<0.1
sulphur	1.9	0.10
calcium	1.1	3.5
aluminium	1.1	8.1
sodium	0.57	2.8
chromium	0.26	<0.1
manganese	0.22	<0.1
cobalt	0.13	<0.1
phosphorus	0.10	0.11
potassium	0.07	2.5
titanium	0.05	0.58

Table 1.3   The major compounds found in sea water

Compound	Concentration /g per 1000g H$_2$O	Compound	Concentration /g per 1000g H$_2$O
sodium chloride	27.50	potassium chloride	0.75
magnesium chloride	6.75	calcium carbonate	0.11
magnesium sulphate	5.63	potassium bromide	0.10
calcium sulphate	1.80		

Table 1.4 Elements of the human body

The composition of the human body (expressed as percentage by mass)	
oxygen	65.0
carbon	18.0
hydrogen	10.0
nitrogen	3.0
calcium	1.5
phosphorus	1.0
potassium	0.35
sulphur	0.25
sodium	0.15
chlorine	0.15
magnesium	0.05
iron	0.0004
zinc	0.0004

CONNECTION: The atmosphere, the lithosphere and the hydrosphere are discussed in more detail in Chapters 3, 4 and 5 respectively.

The biosphere

This word is used to describe all the resources bound up in living systems. A plant, an animal or even a microscopic bacterium are all considered to be part of the biosphere. Food provides animals with the chemical and energy resources needed for healthy growth. The biosphere is the source of these foods. It is also the origin of the major fossil fuels – coal, oil and gas.

1 What four elements are found in greatest quantity in the human body?

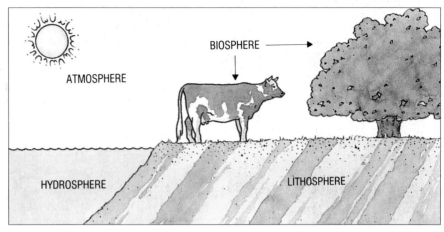

Fig 1.2 The four components of the Earth.

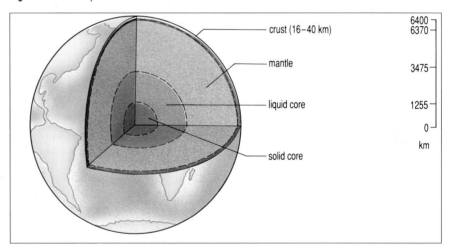

Fig 1.3 The Earth's interior.

An objective view of Earth

It seems certain that a manned space-flight to Mars will take place in the not-too-distant future. Unmanned probes have already investigated the outer planets of our Solar System and, as they travel into space beyond, will take with them evidence of the humans who built them. Information about Earth contained in these probes includes details in diagrammatic form about humans and their planet. The scientists who created these diagrams had to consider exactly what would be of interest to other life forms. Perhaps never before have we needed to be quite so objective about our activities on Earth.

We often take the chemical and energy resources of our planet for granted. It would be interesting and valuable to look at Earth from an entirely objective viewpoint. Perhaps this could only be done by a visitor from another planet.

2 Why do you think it is difficult for us to be entirely objective?

CHEMICAL AND ENERGY RESOURCES

EARTH : AN OUTSIDER'S PRELIMINARY REPORT

Fig 1.4 Third from the Sun. Earth viewed from its satellite, the Moon. It is a unique combination of chemical and energy resources.

Earth is one of nine planets in the Solar System, part of a galaxy known as the Milky Way. The planets orbit a star called the Sun. Earth is the third planet, counting outwards, and occupies an orbit about 93 million miles from the Sun. Earth may be considered to be part of a 'double planet' system since its satellite, the Moon, is larger in size, relative to Earth, than almost any other planetary satellite we have encountered during this mission. However, Earth and Moon are very different in many respects.

One essential difference is that Earth is undergoing rapid change, while the Moon is stable and geologically inert. There is a marked 'surface drift' on Earth. The solid component of the Earth's surface is very mobile, causing mountainous regions and deep troughs. This movement, however, is barely perceptible during the lifetime of an inhabitant of Earth. There is some minor volcanic activity.

The extensive liquid component (the hydrosphere, mainly water with dissolved salts) and gaseous component (the atmosphere, mainly nitrogen and oxygen) are modifying the planet's surface by erosion. We have noticed some significant effects that are causing concern to the inhabitants of the planet:

(a) There has been an increase in temperature of the atmosphere, an effect known as 'global warming'.

(b) There has been a 'thinning' of the protective ozone layer in the upper parts of the atmosphere.

(c) Localised concentrations of chemical compounds have been found in the atmosphere above heavily populated areas of the planet.

---

3 What do you think is meant by the expression 'geologically inert'?

QUESTIONS

1.1 Earth may be classified into four components. List these and explain what features of the Earth belong in each category.

1.2 (a) Elements in Group 0 of the Periodic Table are known as noble gases. Use the Periodic Table at the beginning of the Database to determine the second most common noble gas shown in Table 1.1.

(b) Use Table 1.1 to work out the volume of neon in 1000 cm^3 of dry air.

(c) A sample of air collected in North Africa contained 78.05% nitrogen by volume. A sample collected in Western Europe contained only 77.6% nitrogen. Explain these results.

1.3 (a) What is the percentage of oxygen in the atmosphere and in the Earth's crust? Other than a difference in size, what is the important difference between these figures?

(b) Comment on the difference between the proportion of iron in the Earth as a whole and that present in the Earth's crust.

1.4 Use Table 1.3 to answer the following questions.

(a) What is the total mass of magnesium compounds present in a typical 1000 g sample of sea water?

(b) What mass of sodium chloride could be obtained from 200 g of sea water?

1.5 Why is the concentration of sodium chloride much lower in river water than in sea water?

1.6 (a) Name an element present in small quantities in the human body which forms a large proportion of the Earth's volume.

 (b) Potassium is five times more abundant in the human body than in the Earth as a whole. How do you think this is achieved?

1.7 What are the likely sources of the 'localised concentrations of chemical compounds' described in the outsider's report?

1.2 CLASSIFYING RESOURCES

Chemical resources: elements, compounds and mixtures

An **element** is a substance which cannot be broken down into a simpler chemical substance. About 90 elements occur naturally and nearly 20 more have been made. Each element has different physical and chemical properties to the others and is given a name and a symbol. They are the building blocks from which all chemical substances are made. Elements fall into two broad categories, **metals** (for example, aluminium, iron and copper) and **non-metals** (for example, carbon, oxygen and nitrogen). The division is not sharp, however, and some elements possess properties which make it difficult to classify them as metals or non-metals. Silicon is a good example.

GOING FOR GOLD

The value we place upon an element depends partly upon its usefulness as a resource and partly upon its rarity. Gold, silver and platinum, highly prized as ornamental metals, owe their use as an international currency to their scarcity. The early chemists, or alchemists, dreamt of converting cheap metals into gold but were unsuccessful – gold remains a rare element and a finite resource.

The Earth as a whole contains around 0.29 g of gold for every 1000 kg of rock (0.00000029%). The Earth's crust contains only 0.0004 g of gold per 1000 kg of rock (0.0000000004%). This presents a formidable challenge to would-be extractors of gold and explains the 'gold rush' when nuggets of almost pure gold were found in rivers, streams and underground deposits in the United States in the 1890s. Recently, water from hot springs in New Zealand has been found to precipitate gold at levels of up to 0.03% in scale inside high pressure pipes. This is a potentially valuable source of gold

Fig 1.5 The slight golden colour of the scale lining the inside of the left-hand plate is due to its gold content, precipitated when soluble gold compounds decompose.

4 **Find a list of elements and write the symbols for potassium, phosphorus, strontium and tin.**

Elements combine to form chemical compounds. These may be broken down again by chemical reactions into the elements from which they are made (though this is not always easily done). Water is a compound which contains the elements oxygen and hydrogen. Carbon dioxide can be broken down into carbon and oxygen. Both water and carbon dioxide are compounds.

Compounds have very different properties to those of the elements from which they are formed. For example, sodium chloride is a white crystalline solid which dissolves in water. It is an essential part of a healthy diet. Yet

Fig 1.6 Sodium burns in oxygen to form a white crystalline solid, sodium oxide. The compound has very different properties to the elements from which it is composed.

Fig 1.7 The Periodic Table – arranging elements according to their physical and chemical properties.

sodium is a soft, shiny metal which reacts violently with water, and chlorine is a yellow-green gas which is extremely poisonous.

Any number of elements or compounds may be physically combined to give **mixtures**. For example, air is a mixture made up of nitrogen, oxygen, argon (with other elements in trace quantities), water vapour and carbon dioxide. Sea water and most rocks are mixtures. The elements or compounds in a mixture may be separated by physical methods such as filtration or distillation.

5 **Which of the components of air are elements and which are compounds?**

A mixture has properties which reflect those of its components. For example, things burn in air and they burn in oxygen (but more vigorously). However, carbon dioxide, another component of air, does not support combustion (indeed, it is used in certain fire extinguishers). The reason that air supports combustion is that oxygen is about 700 times more abundant in the atmosphere than carbon dioxide. Clearly the relative amounts of each component affect the overall properties of the mixture.

The relationship between mixtures, compounds and elements may be illustrated by the industrial manufacture of aluminium. Aluminium is found in rocks containing the ore bauxite. The rocks are a mixture of bauxite (impure aluminium oxide) and other substances. The manufacture involves separation and purification of aluminium oxide, a compound. The element aluminium is obtained by electrolysis of aluminium oxide.

The Periodic Table

The elements may be arranged according to their physical and chemical properties. There have been many ways of doing this over the years. The most useful classification is known as the **Periodic Table** (Fig 1.7). The table has vertical columns which contain elements with similar properties to one another. These columns are called **Groups**. The horizontal rows are called **Periods**. The classification of elements using the Periodic Table is one of Chemistry's great success stories, summarising patterns and relationships in a simple form.

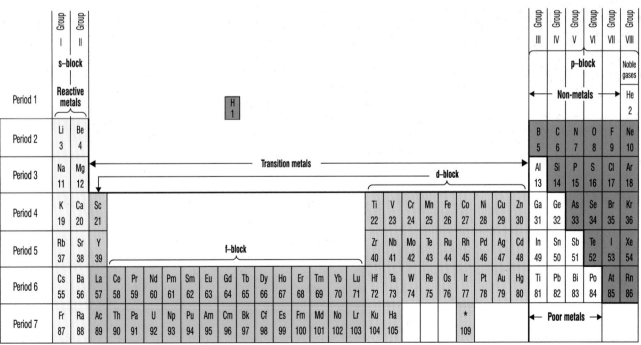

6 **Think about the elements that you know exist as gases at room temperature. Where in the Periodic Table are they found?**

EARTH: AN OUTSIDER'S REPORT ON CHEMICAL RESOURCES

The Solar System under study is approximately 6000 million Earth years old. Radioactive dating using an isotope of potassium confirms that Earth was formed at about that time. Spectroscopic analysis has enabled us to determine the Earth's elemental composition. The bulk of the planet differs from the relatively thin crust. It is from this crust, together with the atmosphere and hydrosphere, that the inhabitants obtain their chemical resources.

The vast quantities of readily accessible silicon ores will be of value to industries requiring semiconductor and advanced materials. High levels of aluminium are important in view of the major role of this element in construction industries. Aluminium is a rare element on other planets but represents more than 8% by mass of the Earth's crust. There are relatively large quantities of potassium compared with the Universe as a whole. Although it is always found in compounds, combined with other elements, it may be valuable to establish a research and development project here. Less valuable metals such as silver and gold are found in trace quantities uncombined with other elements. They may be useful as electrical contacts when working in this oxygen-rich atmosphere. Earth presents major corrosion problems, particularly for articles constructed from iron. This is due to the oxygen-containing atmosphere and the vast quantity of liquid water present. Use of materials based on carbon fibre technology would alleviate the problem of corrosion.

Highly accurate elemental analysis of humans is difficult to obtain. Available data show that humans accumulate certain elements within their bodies in much higher concentrations than are found generally on Earth.

7 **Why do you think that silver and gold are described as 'less valuable metals' in the outsider's report?**

The composition of the Earth (Table 1.2) is different to that of the Universe (Table 1.5). The elements are unevenly distributed throughout the Universe, with some elements being abundant on a planet but others rare.

8 **Where in the Periodic Table is the symbol for the Earth's most abundant element found and where is the symbol for the most abundant element in the universe found?**

Table 1.5   Elements of the Universe

The composition of the Universe (expressed as percentage by mass)	
hydrogen	50.3
helium	47.9
oxygen	1.07
iron	0.61
nitrogen	0.46
carbon	0.14
silicon	0.12
magnesium	0.088
sulphur	0.040
calcium	0.011
aluminium	0.010
sodium	0.0044
chlorine	0.0037
phosphorus	rare
potassium	rare
titanium	rare

PLASTIC TREES?

Can you imagine plastic trees – not just for decoration but which might be planted to turn desert into fertile land?

One of the problems facing farmers in the Third World is that huge areas of land exist as dry desert – useless for growing crops. One possible use for phenolic and polyurethane polymers has been studied by a Spanish inventor, Ibanez Alba. His plastic trees absorb water at night and release it by day. The cool air produced improves the chances of rainfall and, used alongside natural trees, could help to turn deserts green. The 'leaves' are made from phenolic foam and the 'tree-trunk' from fire-resistant polyurethane. Long polyurethane 'roots' anchor the tree against strong winds.

Fig 1.8   The intelligent use of plastics may one day enable crops to be grown in areas which are now deserts. Once natural plant growth is established, deserts may be converted into arable land.

Chemical resources: natural and synthetic materials

A list of all the chemical substances present in the atmosphere, the lithosphere, the hydrosphere and the biosphere would be enormous. Most of these substances are useful to humans but for many purposes they need to be refined and sometimes converted into other, more useful materials.

For example, crude oil (a mixture) is the starting point for a vast range of chemical substances, including medicines, plastics, agrochemicals, fibres, detergents, dyes, perfumes and fuels. Substances which are made by the chemist are said to be **synthetic**. They may occur in nature, and are therefore **naturally-occurring materials**, but are synthesised rather than extracted from natural sources for reasons of convenience or economics. Some synthetic compounds, however, are not found in nature and are said to be **artificial**.

Considerable research and development is concerned with synthesising new compounds and chemically modifying them – all aimed at producing substances to meet specific demands. An example is the development of detergents. Soaps are manufactured from natural oils and fats but suffer from a number of disadvantages, in particular scum formation. Chemists tackled the problem by making synthetic detergents using chemical compounds derived from crude oil. Their first attempts produced effective cleaning materials but the detergents were not biodegradable. This resulted in pollution problems. Chemical modification of the synthetic detergents led to the development of biodegradable products, and this pollution problem was resolved.

Energy resources: using stored energy

Energy is the capacity to do work. Stored reserves of energy are called **fuels**. We are entirely dependent upon ample, reliable and flexible sources of energy. The sources fall into three broad categories:

- non-renewable, finite sources, for example fossil fuels;
- renewable sources, for example biomass (living material from which fuels can be obtained);
- infinite sources, for example solar, wind and tidal energy.

Presently our energy supplies are dominated by fossil fuels and nuclear fuels. Chemists use energy in various forms, such as heat, light, electricity and even sound to bring about chemical reactions.

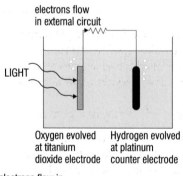

Fig 1.9 Hydrogen will only react with copper(II) oxide to form copper and water if energy is provided in the form of heat. When the Bunsen burner is lit the reaction will take place.

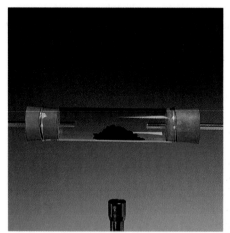

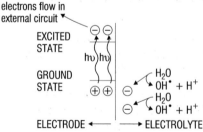

Net process at illuminated electrode is:
$$2H_2O + 2h\nu \rightarrow 2OH^\bullet + 2H^+ + 2\ominus$$
$$2OH^\bullet \rightarrow H_2O + \tfrac{1}{2}O_2$$
$$\overline{H_2O + 2h\nu \rightarrow 2H^+ + 2\ominus + \tfrac{1}{2}O_2}$$

Net process at counter electrode is:
$$2H^+ + 2\ominus \rightarrow H_2$$

9 **List those aspects of modern life styles that consume large amounts of energy.**

Physical changes such as melting and boiling involve energy changes. Water must be heated (provided with energy) to convert it to steam. When steam condenses, energy is released. Similarly, energy changes are associated with all chemical reactions. For example, water may be broken down into hydrogen and oxygen by electrical energy (a process called electrolysis). Conversely, hydrogen and oxygen combine to form water and the change is accompanied by the evolution of energy (the reaction has to be initiated, for example with a lighted match).

Importantly, energy can be changed from one form to another. The chemical energy stored in coal, for example, can be released as heat by burning it. The heat may be used to produce steam from water, which in turn may drive a turbine and generate electricity. The sequence has taken us from stored chemical energy, through heat energy and mechanical energy to, finally, electrical energy.

10 **Why is electricity a particularly convenient form of energy?**

Energy resources: emerging trends

Whatever changes take place in the coming years one thing is certain – our dependence upon ample energy supplies will continue. We will need energy to warm our houses, for transport and for the manufacturing industries. Presently, fossil fuels serve a dual function: an energy source and a source of chemical compounds for the synthetic chemist. But these fossil fuel supplies are finite.

Alternative fuels are being developed, for example methane and ethanol by fermentation of biomass. New methods for the production of electricity in bulk are needed. Investigations into the feasibility of alternative energy sources have been underway for some time and may lead, in due course, to economic power generation based on wind, wave and sun.

Apart from finding new sources of energy we also need to use energy more efficiently. New, more effective insulating materials already contribute to energy conservation. New catalysts, for example allowing chemical reactions to occur at lower temperatures, have improved industrial processes and fuel consumption. Catalysts are also used to reduce the level of pollutants released. Efficient methods of energy storage, such as new electrochemical batteries, may contribute to the growth of electric transport.

More imaginative use of energy from the Sun

The nearness of the Sun (a star) means that its heat, light and other forms of radiation are much more intense than that from other stars. The Sun remains by far our most important source of energy. Millions of years ago large quantities of fossil fuels were laid down beneath the Earth's crust and the oceans. These fuels contain energy trapped by plants. While we continue to use these finite resources, the Sun's energy is still being successfully trapped by plants in a process known as photosynthesis. However, we cannot wait another million years or so for these plants to be converted into fossil fuels. We are not yet using the Sun's energy directly on a large scale. It will be a major advance when we can.

Fig 1.10 Artificial photosynthesis may not be too far away. Already catalysts are being developed which enable water to be split into hydrogen and oxygen by sunlight.

Much energy is reflected into space. The energy trapped by plants is perhaps 1% of that reaching the Earth. Research into photosynthesis has given an improved understanding of this important process and points the way to an entirely chemical process for trapping solar energy – an 'artificial photosynthesis'. There are many approaches to artificial synthesis. Many involve splitting water to give oxygen and hydrogen in a process known as photolysis. Hydrogen is a valuable fuel. It is used, for example, to power the Space Shuttle, and in future it could be an important energy source.

11 **Why is hydrogen often described as a non-polluting fuel?**

12 **What do you think the literal meaning of 'photolysis' is?**

Managing resources

Fossil studies indicate that the 'modern version' of humans has existed for over 30 000 years. In terms of brain development humans are the most advanced species on Earth. Their home, the Earth, is a complex environment and also home to millions of species of plants and other animals. Because of their advanced nature, humans have a great responsibility as 'planetary managers'. The survival of each species depends upon a carefully controlled set of environmental conditions. Our chemical and energy resources must provide for our essential needs but must also take into account the requirements of our descendents and of other life forms.

Problems in managing resources occur where a conflict of interests arises. For example, manufacturing is a necessity yet often produces by-products which contaminate the environment if handled carelessly. Often we must settle for a compromise in which careful monitoring of environmental pollution and restraints on manufacturing industries help to achieve the lowest practical levels of pollutants. To date, 'zero emission options' have proved too expensive.

QUESTIONS

1.8 Iron is converted to rust (hydrated iron(III) oxide) in the presence of water and oxygen. Sea water contains dissolved sodium chloride and other salts and this accelerates the rusting process. Using the above information for illustration, explain what is meant by the following terms:
(a) element; (b) compound; (c) mixture; (d) chemical reaction.

1.9 If iron filings and sand (mainly silicon(IV) oxide) are shaken together a mixture of the two is formed. How could you establish that the two substances were simply mixed and not chemically combined?

1.10 Despite their dangerous properties sodium and chlorine combine to form a white crystalline substance which we readily scatter over food before eating. Explain why it is safe to do this.

1.11 Table 1.5 lists some elements and their relative abundance in the Universe. Contrast this with the abundance of elements on Earth (Table 1.2). What striking differences are there between the abundance of elements in each table?

1.12 In the outsider's report on chemical resources which elements are seen as most desirable/valuable?

1.13 List five synthetic materials you have used today.

1.14 (a) Is the melting of butter on a hot day a chemical or physical change?
(b) How would you describe the change that takes place when an egg is fried?
Justify your answers.

1.15 Explain, in your own words, the meaning of the term photosynthesis.

1.16 In the outsider's report on chemical resources silver and gold are recommended for use as electrical contacts. Why is iron less suitable for this purpose?

1.17 Use the Periodic Table to find two examples of each of the following:
(a) a metal;   (b) a non-metal;   (c) a metalloid.

ACTIVITY

The **Biosphere 2 Project** which took place in the Arizona desert was an attempt to manage a self-contained artificial environment. The project organisers consider the Earth to be Biosphere 1. One day the artificial environment of Biosphere 2 may be duplicated on other distant planets as a working environment for future scientists.

The international team members were sealed inside a giant glass structure for two years along with hundreds of examples of plants and animals. Great care was taken to ensure that the species of animals and plants chosen were compatible. When Biosphere 2 was sealed, the resources available inside had to be constantly recycled to provide the requirements of all species present, including the human beings.

Put yourself in the position of a team member.
(a) List the resources Biosphere 2 would require to sustain itself and explain why these resources require careful management.
(b) What difficulties do you anticipate in the two year project?
(c) How useful do you think this experiment is as a means to test the ability of humans to manage an environment in space or on another planet?

Answers to Quick Questions: Chapter 1

1 Oxygen, carbon, hydrogen and nitrogen.
2 Because we have been brought up in a particular environment and take much for granted. It is hard to think of Earth as 'just another planet'.
3 The nature of rocks and minerals has not changed within the timescale that humans have inhabited the Earth.
4 K, P, Sr, Sn
5 Elements: nitrogen, oxygen, argon; compounds: water, carbon dioxide
6 Top right hand corner of the Periodic Table.
7 Although rare, they play no part in living things and have no value as construction materials.
8 Iron is found in the d-block. Hydrogen, the most abundant element in the Universe, sits alone.
9 Transport, heating.
10 Because it is can be used simply by plugging an electrical appliance into a socket or, in some cases, by putting a battery in an appliance.
11 Because the only product of combustion is harmless water.
12 Splitting by the action of light.

Chapter 2

MODELLING AND MEASURING

LEARNING OBJECTIVES

When you have studied this chapter you should be able to:

1. describe chemical substances as combinations of atoms or ions;

2. describe the structures of atoms and the distribution of mass and charge within atoms;

3. state the characteristics of the subatomic particles;

4. describe the behaviour of the subatomic particles in electric and magnetic fields;

5. represent atoms using their chemical symbols, atomic numbers and mass numbers;

6. explain what is meant by the term isotope;

7. explain why radioisotopes are of importance in medicine and science as analytical tools;

8. describe in outline the origin of the elements;

9. explain what is meant by the term reaction mechanism;

10. outline the kinetic theory of matter in terms of the characteristics and behaviour of solids, liquids and gases;

11. explain what is meant by the term enthalpy change;

12. explain the use of the term amount of substance and its units.

2.1 THE IMPORTANCE OF MODELLING

Modern science is characterised by the collection and analysis of information using sophisticated analytical techniques, computerised data handling and powerful laboratory instruments. Despite this technology the great advances in science still seem to demand that special ingredient – imagination. Central to scientific progress is the extraordinarily creative use by human beings of models to explain the world around us. In recent years the unravelling of the structure of DNA, the development of the electronic computer and progress in drug design have involved the inspired use of models.

Models help to show what something looks like or how something works. They fall into two broad categories: physical models of structures we cannot see and mathematical models of dynamic systems (ones in a state of change). The model may be:

- a picture or construction of something on a smaller scale, such as an astronomer's model of the Solar System;
- a picture or construction of an object on a much larger scale, such as a model of a bacterial cell;
- a model of a process, such as the melting of ice.

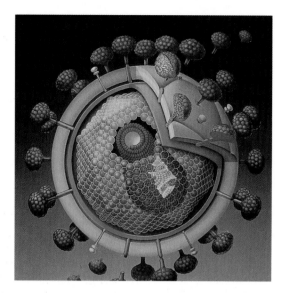

Fig 2.1 A model of the virus responsible for 'acquired immune deficiency syndrome'. Modelling the AIDS virus may one day lead to a cure for this fatal disease

Making sense of the world has often been a controversial struggle, with scientific reasoning sometimes overpowered by political, economic, religious or cultural factors. Sometimes, inaccurate or misleading ideas gain acceptance, only to be thrown out when the weight of evidence against them is overwhelming. This has, in fact, always been the nature of scientific progress. Models are at the heart of this process.

For centuries, a major controversy was concerned with whether matter was **continuous** or **particulate**. Consider the following problem.

Cut a piece of paper in half and two smaller parts remain. Repeat the process and even smaller parts are obtained. Eventually the parts become so small that they cannot be seen with the naked eye. It is possible to continue with a microscope and a tiny knife until even this becomes impossible. An electron microscope and a surgical laser would allow you to work with even smaller pieces but eventually you would have to give up. The important question is this – if it were possible to carry on with the cutting process would you reach a point where the particles could not be divided any further? The continuous theory of matter suggests the paper could be subdivided for ever. The particulate theory of matter states the opposite, that eventually particles would be obtained which could not be broken down further.

1 **Which theory seems more feasible? Give your reasons for this choice.**

Chemical models

Modelling is an invaluable tool and its use has developed through a need to understand and manipulate the material of which our world is composed. Trying to work with chemical compounds with no understanding of their composition is difficult. It is, as early chemists found, a process of trial and error. Good models of matter and change may enable us to predict the chemical or physical behaviour of a substance. This can avoid the time-consuming and costly business of trial and error.

Important chemical models which have been developed to explain structures, properties and events include:

- a model of the atom (the **atomic theory**);
- a model of the structure and behaviour of solids, liquids and gases (the **kinetic theory**);
- many different models to explain how the particles of matter may become rearranged during a chemical change (**reaction mechanisms**).

2 In your own words give an important advantage and one key difficulty associated with the use of models in science.

> **QUESTION**
>
> **2.1** Why are models useful in chemistry? What categories of models are in use?

Fig 2.2a A portrait of Dorothy Hodgkin at work.

DOROTHY HODGKIN AND VITAMIN B₁₂

Vitamin B_{12}, known chemically as cyanocobalamine, was recognised as a vital constituent of a healthy diet many years ago. Establishing the precise spatial arrangements of its 180 atoms, however, took many years of detailed research.

Chemical analysis had shown that the approximate ratio of elements in this substance was $C_{61-64}H_{86-92}O_{14}N_{14}PCo$. It was not until 1957, however, that the total structure was finally announced by Dorothy Hodgkin following years of painstaking work. Using the X-ray diffractometer and the electronic computer (both only recently available at the time) she produced an accurate model of vitamin B_{12}. This fine piece of work earned Dorothy Hodgkin the 1964 Nobel Prize. In 1965 she was admitted to the Order of Merit. Only two other women have received this honour, Florence Nightingale and Margaret Thatcher.

Fig 2.2b Years of careful research revealed the molecular structure of the dark red crystals of vitamin B_{12}.

2.2 THE ATOM: THE FUNDAMENTAL MODEL

The model of the atom is the cornerstone of much chemical thought, but we must not forget that it is only a model. Atoms do not 'look' like the model. Models of the atom have been developed and refined over hundreds of years.

Dalton's atom

The Greeks proposed a particulate model of matter over 3000 years ago, based on logical reasoning alone. The model, describing matter as particulate not continuous, remained essentially unchanged despite numerous experimental investigations.

A key figure in the development of the current model was John Dalton. Central to Dalton's idea was his suggestion that matter consists of tiny, indivisible particles called **atoms**. He also put forward the idea that the chemical elements differ because they consist of different types of atoms. He argued that the elements could not be broken down into any simpler substance by chemical methods.

3 **Was Dalton a believer of the continuous or the particulate theory of matter?**

Dalton's other important idea was that atoms of the same or different elements could combine to form larger structures. Identical atoms combined to give an element. A combination of different types of atoms, in very precise ratios, produced a compound with properties quite unlike those of the individual elements.

Dalton's ideas, taken together, are known as the **atomic theory of matter**. This is a simple example of how models are refined to enable a greater number of observations to be explained. The Greeks had made no distinction between elements and compounds.

The nuclear model of the atom

After Dalton's model was proposed, evidence accumulated which suggested that individual atoms have a substructure. The evidence supported the idea that at the centre of the atom is a **nucleus** in which two different kinds of **subatomic particle, neutrons** and **protons**, are found. They are called **nucleons**. Neutrons are particles with no charge but with a definite, measurable mass. Protons have a positive charge and a mass almost equal to that of a neutron. The positive charge of the nucleus is due to the protons it contains. The other subatomic particles in atoms are the **electrons**. These occupy a region of space outside the nucleus. Electrons have a negative charge and a tiny mass, about 1840th that of a proton. This more sophisticated picture of the atom is known as the **nuclear model of atomic structure**. Its characteristic features are:

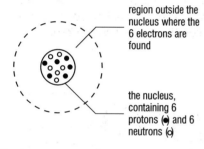

region outside the nucleus where the 6 electrons are found

the nucleus, containing 6 protons (●) and 6 neutrons (○)

Fig 2.3 The carbon atom has a nucleus containing six protons and six neutrons. The six electrons are contained in a region outside the nucleus.

- The number of protons determines which element we are dealing with. Every atom of the same element has the same number of protons.
- In an atom, the number of electrons is the same as the number of protons.
- Electrons held in the atom are bound by their attraction to the nucleus. This type of attraction, between opposite charges is known as an electrostatic force.
- Atoms of the same element may contain different numbers of neutrons. These different atoms are known as isotopes of the element.

4 **Which are the two most abundant elements with odd numbers of protons in their nuclei?**

The test of any model is that it should successfully explain what happens. The nuclear model of atomic structure does this remarkably well. Though refined further, the essential features of the nuclear model remain.

QUESTION

2.2 What are the key differences between Dalton's model of the atom and the nuclear model of the atom?

2.3 In what way, according to Dalton's model, do elements differ from one another?

2.4 Explain what is meant by a subatomic particle.

2.5 Construct a table naming the three subatomic particles, their charge if any and their location in the atom.

2.3 SUBATOMIC STRUCTURE

Unfortunately, we cannot see individual atoms under the microscope, they are far too small. Our 'picture' of the atom has been built up indirectly by careful analysis of experimental work.

The size of atoms

Fig 2.4 Typically, about five million atoms, laid side by side fit into this gap. This number is roughly equal to the population of Finland. Nuclei are much smaller. Around 50 000 million typical nuclei fit into this gap. This number is more than ten times the population of the Earth.

Even the most powerful electron microscopes cannot give us a clear view of an individual atom, though it is possible to see very large groups of atoms using the most modern equipment (Fig 2.5). Just how small atoms are is almost beyond our comprehension. The size varies from element to element, but the radius of a typical atom is about 1×10^{-10} m (one ten thousand millionth of a metre). The nucleus is much smaller, with a radius of about 1×10^{-15} m.

5 **If atoms are so small, how do we measure them?**

CONFIRMING A SUCCESSFUL MODEL

We cannot see atoms, even with the aid of the most powerful microscope. However with the scanning tunnelling microscope (STM), one of the most powerful tools for viewing very small structures, it is possible to 'see' large groups of atoms. The image is built up by computer enhancement.

The development of this technique has allowed confirmation of one of science's most notable successes – the model of DNA developed by James Watson and Francis Crick. Their model, based on data obtained from chemical analysis and X-ray crystallography, consisted of a double helix – two spring-like strands twisted together, held in place by hydrogen bonding.

The first direct image of DNA was recorded late one Friday night in October 1988 by University of California undergraduate Troy Wilson. The 'photograph', in which it is just possible to make out a double helix, is a built-up computer image. However indistinct it may seem, it has helped to confirm Watson and Crick's model for which they won the Nobel Prize in 1962.

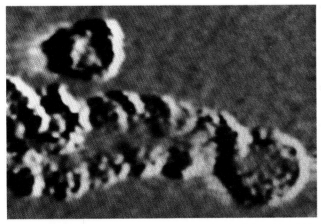

Fig 2.5a Modern technology allows us a glimpse of the shape of larger combinations of atoms. This photograph shows a scanning tunnelling microscope image of DNA.

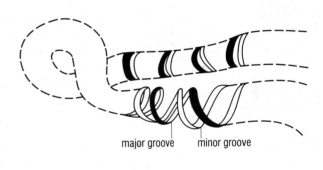

major groove   minor groove

Fig 2.5b A small section of the Watson-Crick model of DNA. The double helix structure was a key feature of the model which helped explain the role of the molecule in inheritance.

Perhaps the most unexpected feature of the atomic model is that the mass is concentrated in the nucleus, which because of its very small size is extremely dense. Another feature of the atomic model is the distribution of charge. Protons in the nucleus make it a region of high positive charge. Electrons are spread out over a much larger volume in a negatively-charged region outside the nucleus. Since the electrons exist in a larger region of space the density of charge is lower. The electrons move rapidly and the volume they occupy is often thought of as a **cloud of negative charge** or a **region of electron density**.

6 **Why do you think the region outside the nucleus is sometimes described as 'mainly empty space'?**

In one of a series of important experiments performed in the period 1907–1911, Rutherford, Geiger and Marsden found that when a beam of positively charged particles (they used helium nuclei, also known as α-particles) is fired at a very thin gold sheet a detector shows that the majority pass through the sheet without any change of direction. However a few are deflected through a small angle and a tiny proportion of the particles (roughly 1 in 10 000) are deflected through a large angle, even through 180°. Rutherford was so surprised by this result that he is said to have compared it to 'firing a 15-inch artillery shell at a piece of tissue paper and having it come back and hit you'.

Imagine you are the presenter of a science programme. Draw a sketch that you could use to explain Rutherford's team's experiment. Hint: Gold atoms may be shown as circles with a simple dot for the nucleus. Outline how you would explain to your audience:

- what this experiment reveals about the structure of the atom and the relative size of its nucleus;
- why Rutherford was so surprised at the results of his team.

THE ORIGIN OF THE ELEMENTS

What is the origin of the universe? Many rival scientific and religious theories have been put forward over the centuries with varying degrees of acceptance. Life on Earth depends upon the properties of the element carbon and the availability of oxygen, hydrogen and nitrogen. All living systems, however, require a much greater number of elements for growth and health. Ninety-one elements occur naturally in the Earth's crust, its atmosphere or its oceans. Where do they all come from?

One model, the 'big-bang' theory, suggests that a massive explosion took place in the universe creating a mixture of hydrogen and helium. The stars, composed almost entirely of hydrogen (75%) and helium (25%), were formed from this mixture. The ratio of hydrogen to helium in stars can be predicted from experiments performed on Earth to compare the stability of hydrogen and helium nuclei. In 1946 Fred Hoyle, a British astrophysicist, suggested that other elements are formed within the stars. Exactly how they were formed remains a mystery, but it is known that elements with an even number of protons are far more common in the universe than elements with an odd number.

We believe now that the nuclei of other elements are formed by a combination or fusion of helium nuclei. This seems reasonable since helium has an atomic number of two and any combination of helium nuclei must therefore have an even atomic number.

Fig 2.6 Stars, of which our own sun is an example, are composed of approximately 75% hydrogen and 25% helium.

Representing atoms

At least 106 different elements are known. In order to describe each of these we need a system which enables us to represent and differentiate each kind of atom. The system has a number of key features:

- Each element is represented by a specific chemical symbol. For example, uranium is represented by the symbol U.

7 **Where would you find the symbols for rhodium, strontium, cerium and selenium in this book, and what are they?**

- Each element has an **atomic number** (represented by the letter Z) which is equal to the number of protons present in the nucleus of an atom of that element. For example, sodium has an atomic number of 11 (Z = 11) and lead has an atomic number of 82 (Z = 82). The number of protons is always the same for a particular element and equals the number of electrons in the atom. In the Periodic Table, elements are arranged in order of increasing atomic number.

8 **Where would you find atomic numbers in this book? Use them to decide how many protons are present in the nucleus of a ruthenium atom.**

- Each isotope has a mass number (represented by the letter *A*) equal to the total number of protons plus neutrons in its nucleus. For example, one chlorine isotope has a mass number of 35 (*A* = 35) as its nucleus contains 17 protons and 18 neutrons.

9 **A second isotope of chlorine has a mass number of 37 (*A* = 37). How many protons and neutrons are present in its nucleus?**

Isotopes of the same element always have the same number of protons (the atomic number) but the number of neutrons will be different. Isotopes of an element, therefore, have different mass numbers.

Representing isotopes

To represent and distinguish a particular isotope of an element, all that is necessary is to show the atomic number, the mass number and the symbol of the element:

$^{A}_{Z}X$ where *A* is the mass number, *Z* is the atomic number and X is the symbol for the element.

For example, one isotope of the element potassium is represented as:

$^{39}_{19}K$

This particular isotope has a mass number of 39. The atomic number of potassium is 19. Since the mass number is equal to the number of protons plus neutrons, the number of neutrons must be 39–19 = 20. We also know the number of electrons in the atom is 19.

10 **Explain how we know that there are 19 electrons in an atom of potassium.**

QUESTIONS

2.6   How many times larger is the radius of a typical atom than that of a typical nucleus?

2.7   The scanning tunnelling electron microscope is a recent addition to the toolkit of physical scientists.
(a) What advantages does it have over conventional microscopes?
(b) How can it help to test the validity of chemical models?

2.8   How would the results of the alpha particle scattering experiment be altered if gold nuclei were negatively charged?

2.9   For each of the following species state (i) the number of protons and (ii) the number of neutrons in the nucleus: $^{20}_{10}Ne$, $^{132}_{54}Xe$, $^{195}_{78}Pt$, $^{79}_{35}Br$.

2.10 Some isotopes of Group IV elements include carbon-12, silicon-28, germanium-73, tin-118 and lead-208.
(a) Use the Periodic Table to represent each of these isotopes with its chemical symbol, atomic number and mass number.
(b) How does the ratio of neutrons to protons change as the Group is descended?

2.11 Three different isotopes of hydrogen exist. Atoms of each contain one proton and one electron. One isotope exists as atoms with no neutrons in the nucleus, one with a single neutron and one with two neutrons. Write suitable symbols for each of the isotopes.

2.12 Uranium is used to fuel nuclear power stations which generate electricity. Natural uranium exists as two common isotopes represented by the symbols $^{235}_{92}U$ and $^{238}_{92}U$. Describe the composition of atoms in each isotope.

2.13 Argon (most abundant isotope $^{40}_{18}Ar$) comes before potassium (most abundant isotope $^{39}_{19}K$) in the Periodic Table. Identify the number of subatomic particles present in each isotope and explain why these elements occur in the order they do.

2.4 PROTONS, NEUTRONS AND ELECTRONS

The masses of atoms

The most sensitive measuring devices cannot measure accurately the masses of individual atoms and so instead a system has been devised for comparing the relative masses of atoms. The mass of an atom of a particular isotope compared to other atoms is known as the **relative isotopic mass** of the atom. By international agreement $^{12}_{6}C$ (carbon-12) is used as a standard and relative isotopic masses are compared on a scale where the relative isotopic mass of $^{12}_{6}C$ is taken as 12.

Relative isotopic mass is defined as:

The ratio of the mass of an atom of a particular isotope to one twelfth of the mass of an atom of the $^{12}_{6}C$ isotope.

Mathematically:

$$\text{relative isotopic mass} = \frac{\text{the mass of one atom of the isotope}}{\text{1/12th of the mass of 1 atom of } ^{12}_{6}C}$$

In nature, elements often exist as a mixture of isotopes. The **relative atomic mass** of an element must take into account the existence of isotopes since atoms of different isotopes have different masses. The relative atomic mass is encountered much more frequently than relative isotopic mass since it is usual to work with the naturally occurring mixtures of isotopes. Relative atomic mass is given the symbol A_r. We can represent the relative atomic mass of, say, nitrogen as $A_r[N] = 14$.

The relative atomic mass of an element, A_r, is the average of the relative isotopic masses of its different isotopes, taking into account the proportions in which they occur.

11 **The relative atomic mass of boron is 10.8 (A_r[B] = 10.8). How can you tell that boron has more than one isotope?**

TUTORIAL

Problem: The element chlorine exists in nature as a mixture of two isotopes, $^{35}_{17}Cl$ (75% abundance) and $^{37}_{17}Cl$ (25% abundance). Calculate the relative atomic mass of chlorine.
Solution: The $^{35}_{17}Cl$ isotope must be given a weighting of 75% and the $^{35}_{17}Cl$ isotope a 25% weighting.

$$\text{the relative atomic mass of chlorine} = \frac{(35 \times 75) + (37 \times 25)}{100} = 35.5$$

Therefore, the relative atomic mass of chlorine is 35.5. This figure takes into account the relative isotopic masses of its two common isotopes chlorine-35 and chlorine-37, and their relative abundances.

Charge and the atom

Although atoms are neutral overall, two of the subatomic particles (protons and electrons) are charged. In many circumstances atoms themselves may acquire a charge to form **ions**. They lose electrons to form positive ions, called **cations**, or gain electrons to form negative ions, called **anions**. For example, formation of a cation:

Na	$\longrightarrow$	Na^+	+	e^-
11 protons		11 protons		0 protons
11 electrons		10 electrons		1 electron
Overall charge = 0		Overall charge = +1		

formation of an anion:

O	+	$2e^-$	$\longrightarrow$	O^{2-}
8 protons		0 protons		8 protons
8 electrons		2 electrons		10 electrons
Overall charge = 0				Overall charge = –2

Atoms are electrically neutral because they always contain equal numbers of protons and electrons. Ions have charges because they contain different numbers of protons and electrons.

12 Calcium forms an ion with a 2+ charge, Ca^{2+}. What are the relative number of electrons and protons in a calcium ion?

We need a convenient way of comparing the charges on particles. Charge is measured in units known as coulombs, but the charges on the subatomic particles and on ions are extremely small compared to one coulomb. The charge on a proton is 1.60206×10^{-19} coulombs. The charge on an electron is -1.60206×10^{-19} coulombs, the same magnitude as that on a proton but with an opposite sign since it is negatively charged. Because of the small sizes of charge chemists often use the concept of relative charge.

All charges in chemistry are considered as multiples of the fundamental unit of charge, 1.60206×10^{-19} coulombs. The relative charge on a proton is taken as +1 and the charge on an electron as –1. Neutrons, as their name suggests, have no charge.

13 Why does the sulphide ion, S^{2-}, have a negative charge?

QUESTIONS

2.14 Define the term relative atomic mass and explain how this concept is used to compare the masses of particles which are far too small to weigh.

2.15 What is the total charge, in coulombs, on one dozen electrons?

2.16 What is the charge in coulombs on a gaseous chromium ion, Cr^{3+}?

2.17 Sand contains the element silicon, combined with oxygen to form silicon(IV) oxide, SiO_2. Silicon has three stable isotopes. The relative mass and percentage abundance of each isotope in a sand sample is given below:

Relative mass	Percentage abundance/%
27.98	92.24
28.98	4.66
29.97	3.10

If this sample of sand were used to make pure silicon calculate:
(a) the A_r value of the silicon produced;
(b) the theoretical mass of silicon which could be produced from 1000 kg of sand.

2.18 Identify the particles W, X, Y and Z in Table 2.1.

Table 2.1

Particle	Subatomic particles present		
	protons	neutrons	electrons
W	12	13	10
X	8	8	8
Y	53	74	54
Z	8	7	10

2.19 A sample of copper in a copper electrode consisted of a mixture of two isotopes, copper-63 (70% abundance) and copper-65 (30% abundance). Calculate the relative atomic mass of copper.

2.20 For the beryllium isotope $^{9}_{4}\mathrm{Be}$ state
   (a) the number of protons in the atom;
   (b) the number of neutrons in the atom;
   (c) the number of nucleons in the nucleus;
   (d) the number of electrons in the neutral atom;
   (e) the number of electrons in the ion Be^{2+}.

2.21 Chlorine exists as molecules containing two chlorine atoms, Cl_2. Two common isotopes of chlorine exist, $^{35}_{17}Cl$ and $^{37}_{17}Cl$. In a sample of chlorine made from sea water what would be the relative masses of the lightest and the heaviest chlorine molecules produced?

2.5 MODELLING THE ARRANGEMENT OF ELECTRONS

A study of the properties of atoms, in particular atomic spectra and ionisation energies, provides evidence that electrons can only possess specific amounts of energy when in an atom. Electrons with the same energy as one another are said to reside in an **energy level**. The different energy levels within an atom are labelled $n = 1$, n = 2, $n = 3$, and so on, where n is called the **principal quantum number**.

Electrons in the $n = 1$ energy level are lowest in energy and are nearest to the nucleus. Electrons in the $n = 2$ energy level are higher in energy, further from the nucleus and easier to remove from the atom. In general, the higher the energy level, the further its electrons are from the nucleus and the easier they are to remove from the atom. The arrangement of electrons in energy levels is known as the atom's **electronic structure** or **electronic configuration**.

KEY FACT

Electrons move rapidly and the volume they occupy is often thought of as a region of electron density. Strictly speaking we should say that an electron in a high energy level is further from the nucleus **on average** than an electron in a lower energy level. Because of its constant movement there will be occasions when it is nearer to the nucleus.

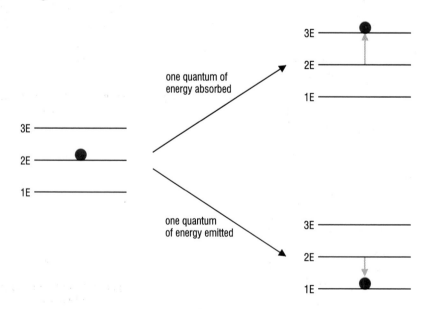

Fig 2.7 An electron can only move to a higher energy level by receiving a 'packet' or **quantum** of energy of fixed size. To return to the lower energy level the electron must emit a quantum of energy of the same size.

Fig 2.8 The colours of fireworks are due to movement of electrons from one energy level to another in atoms. If electrons absorb energy they move to a higher energy level. When they return to lower energy levels they emit radiation of a characteristic frequency, which we see as light of a particular colour.

Each energy level can accommodate only a certain maximum number of electrons. The maximum number of electrons which may be accommodated in a particular energy level is given by $2n^2$, where n = the principal quantum number.

For example, the $n = 2$ energy level can contain a maximum $2 \times 2^2 = 8$ electrons.

14 **How many electrons may be accommodated in the energy levels where n has values of 1, 3 and 4?**

Using the model to describe electronic structures

Although we actually model the electronic structure of a particular type of atom, we often talk about the electronic structure of an element. To model a particular element, electrons are placed in the lowest energy level first until this is filled. Further electrons enter the next lowest energy level. The electronic structures of the first 20 elements are shown in Table 2.2. Alternatively, the arrangement of electrons may be represented by diagrams (Fig 2.9).

Table 2.2

Electronic structures of the first 20 elements

H	1	C	2,4	Na	2,8,1	S	2,8,6
He	2	N	2,5	Mg	2,8,2	Cl	2,8,7
Li	2,1	O	2,6	Al	2,8,3	Ar	2,8,8
Be	2,2	F	2,7	Si	2,8,4	K	2,8,8,1
B	2,3	Ne	2,8	P	2,8,5	Ca	2,8,8,2

15 **What is unusual about the electronic structures of potassium and calcium?**

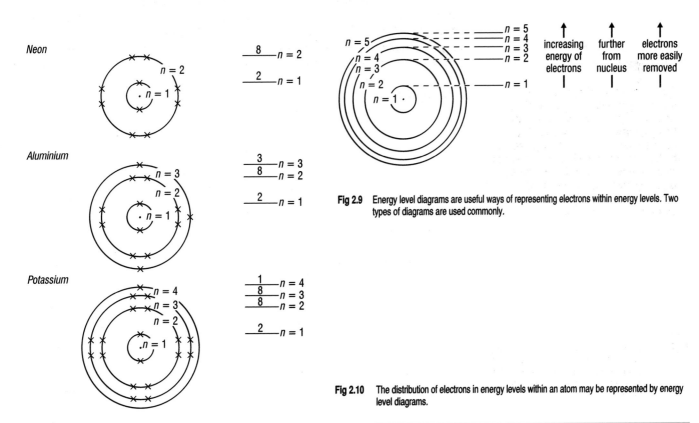

Fig 2.9 Energy level diagrams are useful ways of representing electrons within energy levels. Two types of diagrams are used commonly.

Neon

Aluminium

Potassium

Fig 2.10 The distribution of electrons in energy levels within an atom may be represented by energy level diagrams.

In Table 2.2 we can pick out elements which have identical electronic structures in the outer energy level. Oxygen and sulphur, for example, have six electrons in the outer energy level. All elements which share the same outer electronic structure are found to show very similar chemical behaviour. This characterises elements within a given Group of the Periodic Table.

16 **Can you identify three elements (other than oxygen, sulphur, and selenium) with the same outer electronic structure?**

Electrons in the outermost electron-containing energy level are sometimes referred to as **valency electrons**. When one atom bonds to another it is these electrons which are involved. Their arrangement is always changed by a chemical reaction. Electrons in the inner shells are not involved in bonding and remain unaffected by chemical reactions.

17 **Explain why the chemical behaviour of isotopes of the same element is identical.**

Stable electronic structures

Early chemists had no knowledge of electronic structure and grouped elements together according to their common properties. For example, helium and neon were grouped together because of their unreactive natures. We now attribute this lack of reactivity to their electronic structures. We assume there is some kind of stability associated with the arrangement of electrons found in atoms of the noble gases. Both helium and neon have full outer energy levels and therefore this seems to be a very stable arrangement of electrons. In the Periodic Table, the very unreactive element argon is placed in the same Group as helium and neon. The $n = 3$ energy level can accommodate up to 18 electrons, but in an argon atom it contains only 8 and is therefore incomplete. The arrangement of electrons in an argon atom must be stable, however, since argon is so unreactive.

18 **Where in the Periodic Table are the noble gases to be found?**

There appears to be a stability associated with

- energy levels which contain their maximum number of electrons (2,8,18,32 for $n = 1, 2, 3$ and 4 respectively) and are therefore full;
- energy levels which contain 8 electrons, even though these are not full.

The noble gases all have 8 electrons in their outer energy level. This explains their lack of reactivity.

Further evidence for this pattern is given by the electronic structures of potassium and calcium (Table 2.2). For these two elements the last electrons enter the $n = 4$ energy level even though the $n = 3$ energy level is not full. The Group of elements which includes helium, neon and argon is known as the noble gases. Any ion or combined atom with the same electronic structures as these elements is said to have a **noble gas electronic structure**. Usually, when atoms react, they each achieve this electronic structure.

19 **Predict what the gas argon might be useful for.**

QUESTIONS

2.22 What evidence is there that electrons reside in energy levels?

2.23 A neutral atom has 15 electrons. Without using a Periodic Table explain how the electrons it contains are organised into energy levels.

2.24 In terms of its electronic structure what has a hydrogen atom got in common with an atom of (a) a Group I metal; (b) a Group VII element?

QUESTIONS *continued*

2.25 What are valence electrons?

2.26 What do atoms of Group V elements have in common?

2.27 Hydrogen may exist as individual atoms, H, diatomic molecules, H_2, hydrogen ions, H^+ or as hydride ions, H^-. How many electrons does each of these species have?

2.28 Work out the electronic structures of the elements with atomic numbers 14, 16, 18 and 20.

2.29 Work out the electronic structures of the following ions: Li^+, Be^{2+}, N^{3-}, S^{2-} and P^{3-}.

2.30 Use Table 2.2 to work out which of the following particles is not isoelectronic with the others in the list: O^{2-}, F^-, Ne, Na, Mg^{2+}, Al^{3+}.

2.6 THE NATURE OF RADIOACTIVITY

Stable and unstable nuclei

The model of the atom describes the nucleus as a combination of neutrons and protons. Why are some nuclei unstable and how can we use this to our advantage?

Radioactivity is the result of spontaneous disintegration of some nuclei which have unstable arrangements of protons and neutrons (unstable nuclei). We call this disintegration **radioactive decay**. Many elements exist as more than one isotope. Most isotopes have stable nuclei and do not decay. Those isotopes which are unstable are known as **radioisotopes** or **radioactive isotopes**. Radioisotopes undergo decay continuously until a stable isotope is finally formed. The stable isotope formed is often an isotope of another element.

One important feature of radioactive decay is that its rate does not depend upon physical conditions, such as temperature or pressure, or the presence of other chemical compounds. The stability or instability of a particular nucleus depends upon the ratio of neutrons to protons within the nucleus, the **n/p ratio**. For lighter elements, such as carbon or oxygen, the numbers of neutrons and protons are approximately equal in the stable isotopes. For heavier elements, such as uranium, the number of neutrons must be greater than the number of protons if the isotope is to be stable. Sometimes several unstable isotopes are formed before the final (stable) isotope is reached. This will involve several changes in the structure of the nucleus which are known as **nuclear transformations**. The whole set of transformations is known as a **decay series**.

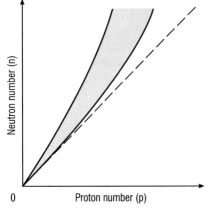

Fig 2.11 The shaded area is known as the **stability band** and all stable isotopes have an n/p ratio in this region. The n/p ratios of unstable isotopes lie outside this region and they will decay until an isotope with an n/p ratio within the shaded region is formed.

FORTUNE FAVOURS THE PREPARED MIND

Often, chance plays a part in research. An unexpected result due to a mistake or an accident can lead to a major breakthrough if the scientist is alert enough to take advantage of such good fortune. As the famous French biochemist Louis Pasteur once said 'fortune favours the prepared mind'. One stroke of good luck fell to another Frenchman, Henri Becquerel, who used crystals of a uranium salt to study X-rays. He accidentally left some of the crystals on a wrapped photographic plate which, by chance, he later developed. Finding a dark spot where the crystals had been he decided that the crystals were emitting a penetrating radiation which had affected the plate. Further experimental work showed that the radiation was quite different to X-rays. Becquerel called the new phenomenon **radioactivity**.

Types of radioactivity

There are three common forms of radioactivity: α-particles, β-particles and γ-rays. These differ in character and in the effect they have on their surroundings.

α-particles

α-particles are helium nuclei, $^4_2He^{2+}$, with a relative mass of 4 and charge of +2. They are deflected in magnetic and electric fields because of their charge. When α-particles pass through a medium, such as air, they collide with atoms of the medium, capturing electrons and causing the atoms to ionise, leaving them with a positive charge. The size of α-particles, and the frequency with which they collide with atoms of the medium, means that they have low penetrating power and are strongly ionising.

20 **How many electons do you think each α-particle can capture?**

β-particles

β-particles are fast moving electrons formed by the breakdown of a neutron into a proton, which remains in the nucleus, and an electron which is ejected from the atom. They are deflected in magnetic and electric fields because of their charge. The charge is smaller than that of α-particles and so β-particles cause less ionisation. The smaller size means that β-particles collide less frequently with atoms of the medium and, therefore, have greater penetrating power than α-particles.

21 **Explain why β-particles are deflected in the opposite direction to α-particles in an electric field.**

γ-rays

γ-rays are not deflected in electric or magnetic fields. They are a form of electromagnetic radiation (or electromagnetic waves) and are part of the electromagnetic spectrum in the same way that infrared, visible, ultraviolet and X-rays are. γ-rays have the highest penetrating power of the three forms of radioactivity. They can penetrate deep inside the human body and they may cause damage to cells and vital organs. There is strong evidence that γ-rays may cause normal cells to mutate and to become cancer cells. γ-decay does not occur on its own and seems to accompany α- and β-particle emission.

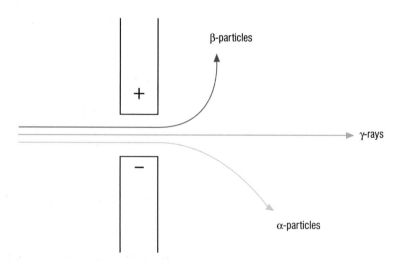

Fig 2.12 The three types of radioactivity differ in how they behave when passed through an electric field.

MODELLING AND MEASURING

22 From which type of radioactivity do workers in the nuclear industry require the greatest level of protection?

23 Why does the decay of a radioactive isotope by α- or β-decay produce an isotope of a different element?

Representing nuclear change with equations

Nuclear changes may be represented by equations. The sum of the mass numbers on the left hand side of the equation must equal that on the right. Similarly the sum of the atomic numbers must also balance on both sides of the equation. In nuclear equations, protons are represented as $_1^1p$, neutrons by $_0^1n$ and electrons by $_{-1}^0e$ The use of nuclear equations may be illustrated by the following two examples:

$$_{91}^{231}Pa \longrightarrow {}_{89}^{227}Ac + {}_2^4He \qquad (\alpha\text{-decay})$$

$$_{89}^{227}Ac \longrightarrow {}_{90}^{227}Th + {}_{-1}^0e \qquad (\beta\text{-decay})$$

Notice that the two equations are balanced for both the mass numbers and atomic numbers.

Using radioisotopes

Radioactivity can be detected using, for example, a Geiger counter, scintillation counter or solid state detector. Therefore, we can both recognise the presence of radioactivity and follow the passage of a radioactive material.

Radioisotopic labels

Sometimes it is helpful to be able to monitor the physical movement of a chemical substance or follow what happens to it when it takes part in a chemical reaction. One extremely useful way of monitoring a chemical substance is to use a **radioisotopic 'label'**. An atom in the compound is replaced by a radioisotope of the same element. The position of the 'label' can then be followed throughout any physical or chemical changes which take place.

Radioisotopic labelling has many useful applications. For example, scientists developing drugs need to know the rate of uptake, the site of action and the final fate of any new drug introduced into the body. This may influence the way in which the drug is given to the patient and the dosage used. Radioisotopes may also be used to measure rates of flow in pipes, the depth of liquids in tanks, leaks in pipes (particularly useful for underground pipes) and wear and tear in machinery.

Radiotherapy

Radiotherapy takes advantage of the damaging effects of γ-rays on cells and uses this to kill cancer cells and so prevent further cancerous growth. The radioisotope cobalt-60 ($_{27}^{60}Co$) has been widely used in the treatment of cancer but its use is accompanied by many unpleasant side effects such as temporary hair loss.

24 How many neutrons are there in the nucleus of an atom of the radioactive cobalt-60 isotope?

Synthesis

Use is made of radioisotopes in the plastics industry. γ-rays are a very high energy form of electromagnetic radiation. They can be used to start or initiate the reaction by which certain polymers are formed.

RADIOISOTOPES AND THE THYROID GLAND

The thyroid gland, situated near the base of the neck, produces hormones which control the rate of metabolic activity in humans. Too little of the hormone and the metabolic rate slows down. In children this can lead to serious impairment of growth and mental disability. Too much and the metabolic rate is too fast resulting in overactivity of the body's organs. The hormones require iodine atoms as an essential part of their structures. For this reason, minute traces of iodide are an essential part of a healthy diet, and in Britain traces of iodide are added to table salt.

The activity of a defective thyroid gland can be monitored by measuring the uptake of iodine. The patient is given a dose of aqueous sodium iodide which has been labelled with a trace quantity of the radiosotope iodine-131. The radioactivity emitted from the thyroid region is monitored over a period of time using a suitable detector and counter.

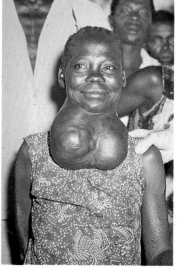

Fig 2.13a These people suffer from an over-active thyroid gland which results in a swelling of the neck.

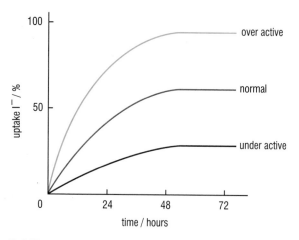

Fig 2.13b Using radioactivity to monitor the uptake of iodine in the thyroid gland.

Dating ancient remains

Carbon dioxide in the atmosphere contains mainly carbon-12 atoms ($^{12}_{6}C$). A very small number, however, contain the heavier radioactive $^{14}_{6}C$ isotope. This is because cosmic rays, mainly from the Sun, cause atoms in the upper atmosphere to disintegrate into smaller fragments. Amongst the fragments are individual neutrons which collide with nitrogen atoms to produce carbon-14 atoms and protons:

$$^{14}_{7}N + ^{1}_{0}n \longrightarrow ^{14}_{6}C + ^{1}_{1}p$$

Fig 2.14 Radiation from the Sun produces carbon-14 in the upper atmosphere.

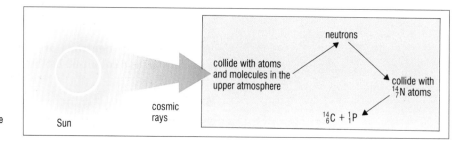

A MAMMOTH FRAUD

In 1889, Hilbourne T. Cresson, an American archaeologist reported finding an engraving of a mammoth on a piece of whelk shell. If genuine and drawn from life this find would have been the oldest example of North American art, at least 10 500 years old and possibly as old as 40 000 years. The Holly Oak pendant, as it became known, appeared to represent an astonishing archaeological find. When finally submitted to radiocarbon analysis in the late 1980s the shell proved to have an approximated age of only 1100 years. It seems likely that Cresson himself engraved the Holly Oak pendant.

Fig 2.15   The Holly Oak pendant was proven to be a fraud by radiocarbon dating; the proportion of $^{14}_{6}C$ was much too high.

Carbon-14 atoms combine with oxygen to form carbon dioxide molecules. Carbon dioxide in the atmosphere, therefore, has a natural level of radioactivity. Carbon dioxide is absorbed by plants during photosynthesis, passing through the food chain as plants are eaten by animals. The proportion of carbon-14 in living systems is constant because they absorb and re-emit carbon-containing compounds continuously. However, once an animal or plant dies the proportion of carbon-14 in its structure decreases. This is because carbon-14 atoms undergo β-decay to give nitrogen:

$$^{14}_{6}C \longrightarrow \, ^{14}_{7}N + \, ^{0}_{-1}e$$

In dead material the carbon-14 atoms are not being replaced and therefore, the older the remains of living things are, the lower the level of radioactivity they show. This principle has proved extremely valuable for dating remains and the technique is known as **carbon dating**. It will only work, however, on remains that were once alive.

QUESTIONS

2.31   How do nuclear changes differ from simple chemical changes?

2.32   Which isotope of lithium, $^{6}_{3}Li$ or $^{9}_{3}Li$ is likely to be more stable? Explain your answer.

2.33   In earlier times alchemists were employed (without success) to make gold from cheaper elements. In a nuclear laboratory gold-198 may now be produced (expensively!) by firing neutrons at $^{198}_{80}Hg$. By writing a nuclear equation for this process work out what other particle must be produced.

2.34   Work out the identity of particle P and explain in words what takes place in the following nuclear transformation:
$^{235}_{92}U + \, ^{1}_{0}n \longrightarrow \, ^{94}_{36}Kr + \, ^{139}_{56}Ba + 3P$

2.35   The Turin Shroud was at one time thought to be around 2000 years old. The level of carbon-14 found in a small sample of the Shroud was, however, much higher than expected. What does this suggest about the Shroud's age? Explain your answer.

2.36   What happens to the n/p ratio of the particles involved in the following nuclear decay processes?
(a) $^{234}_{92}U \longrightarrow \, ^{230}_{90}Th + \alpha$-particle
(b) $^{239}_{92}U \longrightarrow \, ^{239}_{93}Np + \beta$-particle

QUESTIONS *continued*

2.37 What is particle X in the following equation?
$$^{238}_{92}U + ^{2}_{1}H \longrightarrow ^{239}_{92}U + X$$

2.38 When the isotope radium-226 ($^{226}_{88}$Ra) undergoes radioactive decay the process may be described by the following equation:
$$^{226}_{88}Ra \longrightarrow Z + \alpha\text{-particle}$$
What group does element Z belong to?

2.7 MEASURING AMOUNTS OF SUBSTANCES

The need to measure in the laboratory

Chemists need to measure quantities of substances for many purposes. They may need to measure the reactants required for the preparation of a chemical compound and to determine the yield. They may need to measure out compounds needed for quantitative analysis. In the laboratory, the mass or volume of a substance can be easily determined. However, a consequence of the particulate model of matter is that chemists need to be able to measure the number of particles present. We are interested in how many atoms combine to form a compound or rearrange during a chemical reaction.

It is not sufficient to weigh out 10 g of copper, we want to know how many copper atoms are present. It is not sufficient to measure out 50 cm^3 water, we want to know how many hydrogen and oxygen atoms are present. So how can we do this?

To answer these questions we need to return to relative atomic masses. An atom of the carbon-12 isotope, $^{12}_{6}$C, is taken as the standard, with a relative atomic mass of 12.000. The masses of all other atoms are measured relative to an atom of carbon-12. Most elements have more than one isotope and a sample weighed out in the laboratory will contain a mixture of the isotopes in their naturally occurring abundances. The relative atomic mass is an average of the relative isotopic masses, taking account of these relative abundances.

Suppose we weigh out 1 g of carbon-12. There will be a certain number of atoms present. If we weigh out 1 g of magnesium there would be only half as many atoms present because each magnesium atom is about twice the mass of a carbon-12 atom. Therefore, to obtain the same number of atoms as there are in 1 g of carbon-12, we would need to weigh out 2 g of magnesium, and this is the ratio of the relative atomic masses of carbon-12 and magnesium.

Therefore, if we weigh out quantities of elements in the same proportions as their relative atomic masses, they will contain the same number of atoms as one another. For example,

$A_r[\text{C}] = 12; \quad A_r[\text{H}] = 1; \quad A_r[\text{O}] = 16; \quad A_r[\text{Cl}] = 35.5; \quad A_r[\text{Na}] = 23.$

Therefore, the following quantities all contain the same number of atoms as one another:

12 g carbon, 1 g hydrogen, 16 g oxygen, 35.5 g chlorine, 23 g sodium;

as do any masses which are in the ratios 12 : 1 : 16 : 35.5 : 23, such as 1.2 g carbon, 0.1 g hydrogen, 1.6 g oxygen, 3.55 g chlorine, 2.3 g sodium; or
0.24 g carbon, 0.02 g hydrogen, 0.32 g oxygen, 0.71 g chlorine, 0.46 g sodium.

25 **What is the ratio of chlorine atoms in 1.42 g chlorine to carbon atoms 0.24 g carbon?**

CONNECTION: Isotopes and A$_r$, page 20.

KEY FACT

The **empirical formula** of a compound shows the ratio of atoms present in the compound. For example, calcium sulphate has the empirical formula CaSO$_4$. This tells us that calcium, sulphur and oxygen atoms are present in a 1:1:4 ratio. We call this ratio the **stoichiometry** of the compound.

Fig 2.16 The quantities of elements shown all contain the same number of atoms.

The mole: the chemist's unit for counting

We are able to measure relative numbers of particles but we still do not know how many are present in a given quantity of material. The first thing we need is a convenient counting unit. An example will help us understand what is meant by the term counting unit. We often buy eggs in boxes of 'half-a-dozen' or 'a dozen'. A dozen is our counting unit and we use it because it is more convenient than counting individual eggs. We ask for three dozen eggs rather than 36 eggs.

A counting unit may seem to provide little advantage, but as numbers get very large it is increasingly helpful. A ream of paper is 480 sheets. It is far more convenient to ask for 50 reams of paper when ordering it than to ask for 24 000 sheets.

The chemist's counting unit is the **mole** (symbol, **mol**). One mole is 6.0221367×10^{23}. We measure **amounts of substance** using this unit and convert a quantity of substance (measured by mass or volume) into the amount of substance (measured by the number of chemical entities present given as a number of moles). It is essential to say what the entity is that we are counting. For example, 'two moles of oxygen' is too imprecise. We must say whether we mean 'two moles of oxygen atoms', 'two moles of oxygen molecules' or even 'two moles of oxygen cylinders'.

> **26** An oxygen molecule is made up of two oxygen atoms, O_2. How many moles of oxygen atoms are there in five moles of oxygen molecules?

> **KEY FACT**
>
> A **chemical entity** may be an atom, a molecule, an ion or the atoms shown in the empirical formula of a compound.
>
> A **mole** is the amount of a substance which contains the same number of chemical entities as there are atoms in exactly 12.000 g of carbon-12.

SPOTLIGHT

AVOGADRO'S COUNTING UNIT

Imagine covering the surface of our planet with marshmallows – to a depth of one mile! The number required is unimaginably large, in fact 6.02×10^{23}. This would be one mole of marshmallows.

The mole can be thought of as a counting unit. One dozen eggs contains 12 eggs, one mole of particles contains 6.02×10^{23} particles. This number is known as Avogadro's number, after the Italian, Amedeo Avogadro (1776–1856) a professor of physics at Turin University. Avogadro's number is 6.02×10^{23}. Avogadro's constant is 6.02×10^{23} mol^{-1} and is given the symbol L.

The value of Avogadro's constant has been determined experimentally by a variety of methods. The most accurate estimate is $6.022\,136\,7 \times 10^{23}$ mol^{-1}. As the chemists' counting unit, the mole is invaluable. It is far more convenient to say that 6.35 g of copper contains 0.1 moles of copper atoms than to say it contains $6.022\,136\,7 \times 10^{22}$ copper atoms.

Calculating the mass of one mole of chemical entities

Our 'standard' particle is an atom of carbon-12. We have used it to define relative atomic masses by assigning atoms of carbon-12 a relative mass of 12. We can calculate the mass of a mole of any other kind of chemical entity provided we know its mass relative to an atom of carbon-12.

TUTORIAL	**Problem:** What is the mass of 1 mol of sulphur atoms?
	Solution: $A_r[S] = 32$
	Therefore, sulphur atoms are $\frac{32}{12}$ times heavier than atoms of carbon-12
	and so 1 mol of sulphur atoms has a mass of $\frac{32}{12} \times 12\,g = 32\,g$.

Problem: Ethanol exists as molecules with the formula C_2H_5OH. What is the mass of 1 mol of ethanol molecules?

Solution: $A_r[C] = 12$; $A_r[H] = 1$; $A_r[O] = 16$

Therefore, the relative mass of an ethanol molecule, C_2H_5OH, is

$[(2 \times 12) + (6 \times 1) + 16] = 46$

Therefore, ethanol molecules are $\frac{46}{12}$ times heavier than atoms of carbon-12

and so 1 mol of ethanol molecules has a mass of $\frac{46}{12} \times 12$ g = 46 g.

In practice the '12' will always cancel and we can take a short cut and say that if the relative mass of an element or compound is x, then the mass of one mole of the substance will be x g.

27 Sodium hydrogencarbonate, also known as 'baking soda', is used as a raising agent for cakes. If the formula of this compound is $NaHCO_3$, does a 250 g packet contain more or less than one mole of sodium hydrogencarbonate? ($A_r[Na] = 23$; $A_r[H] = 1$; $A_r[C] = 12$; $A_r[O] = 16$)

Molar mass

The mass of one mole of chemical entities is called the molar mass, M. It has the units **g mol^{-1}** ('mass per mole') and the same numerical value as the sum of the relative atomic masses of the atoms present.
Some examples of this are:

1. Copper, Cu
   $A_r[Cu] = 63.5$, therefore the molar mass of copper = 63.5 g mol^{-1}.

2. Carbon dioxide, CO_2
   $A_r[C] = 12$; $A_r[O] = 16$, therefore the molar mass of carbon dioxide
   = $[12 + (2 \times 16)] = 44$ g mol^{-1}.

3. Sodium chloride, NaCl
   $A_r[Na] = 23$; $A_r[Cl] = 35.5$, therefore the molar mass of sodium chloride
   = $[23 + 35.5] = 58.5$ g mol^{-1}.

28 **What is the molar mass of tetrachloromethane, CCl_4?**

The amount of substance is calculated from its mass, m, simply by dividing by the molar mass.

$n = m/M$

The mass of a given amount of substance may be calculated by rearranging the equation to give

$m = nM$.

TUTORIAL

Problem: What amount of substance is present in 22 g of carbon dioxide?

Solution: The molar mass of carbon dioxide is 44 g mol^{-1}.

Using $n = m/M$

Amount of CO_2, $n = \frac{22}{44}$ mol = 0.5 mol.

Problem: What amount of substance is present in 10 g calcium carbonate, $CaCO_3$?

Solution: To determine the molar mass of calcium carbonate:

$A_r[Ca] = 40$; $A_r[C] = 12$; $A_r[O] = 16$

Therefore, molar mass = $[40 + 12 + (3 \times 16)] = 100$ g mol^{-1}

Using $n = m/M$

Amount of $CaCO_3$, $n = \frac{10}{100}$ mol = 0.1 mol.

Problem: Calculate the mass of 0.1 mol copper(II) carbonate, $CuCO_3$.

Solution: To determine the molar mass of copper(II) carbonate:

$A_r[Cu] = 63.5$; $A_r[C] = 12$; $A_r[O] = 16$

Therefore, molar mass = $[63.5 + 12 + (3 \times 16)] = 123.5$ g mol^{-1}

Using $m = nM$

Mass of copper(II) carbonate = $0.1 \times 123.5 = 12.35$ g.

QUESTIONS

2.39 Why is it important to compare the amounts of substances and not simply the masses when considering a chemical reaction?

2.40 Methane is the main component of natural gas. Calculate:
  (a) the number of moles of methane in 2.4 g of methane;
  (b) the mass of 3.5 moles of methane.

2.41 The insecticide Chlordane has the chemical formula $C_{10}H_6Cl_8$. If 2.5 g of the insecticide are used in a spray how many moles does this represent? $A_r[C] = 12$; $A_r[H] = 1$; $A_r[Cl] = 35.5$.

2.42 What mass of copper can be obtained from 1 kg of its ore, copper sulphide, CuS?

2.43 Explain the difference between the terms relative molecular mass, M_r and molar mass, M.

2.44 Diamond is one form of carbon. How many atoms are there in a diamond with a mass of 4 g?

2.45 If a cathedral dome is covered in 90 000 kg of copper sheeting, how many moles of copper is this? ($A_r[Cu] = 63.5$)

2.46 Assuming ice to consist of pure water, how many moles of water would be contained in 1 kg of ice?

2.47 If the compound iodomethane, CH_3I can be bought at £9.00 for 100 g or £9.50 for one mole, which is cheaper? Explain your choice.

2.48 An indigestion tablet contains, as one of its ingredients, 300 mg (0.300 g) of aspirin. Aspirin has the formula $C_9H_8O_4$.
  (a) How many moles of aspirin are contained in one indigestion tablet?
  (b) How many tablets would you need to provide 1 mol of aspirin?

2.49 Calculate the following:

(a) the mass of 0.15 mol of cisplatin, a cancer chemotherapy agent with the formula $Pt(NH_3)_2Cl_2$;

(b) the number of moles of procaine (a dental anaesthetic) contained in 0.005 g of this substance. Procaine has the formula $C_{13}H_{30}N_3O_3$.

$A_r[C] = 12$; $A_r[H] = 1$; $A_r[N] = 14$; $A_r[O] = 16$; $A_r[Pt] = 195$; $A_r[Cl] = 35.5$

Answers to Quick Questions: Chapter 2

1 There is no right or wrong answer but the popularly held view is that matter is particulate.

2 Advantage: enable predictions to be made and short-cuts taken. Difficulty: you may end up believing the model is reality!

3 Particulate theory

4 Hydrogen (atomic number 1), aluminium (atomic number 13)

5 We don't. We measure properties which we believe are dependent on atomic size and make estimates. For example, we can measure electron density maps where X-rays are diffracted by regions of high electron density. By assuming that nuclei are at the centres of very high electron density, distances between atoms and their size can be estimated.

6 Electrons have negligible mass relative to protons and neutrons, yet they occupy by far the greatest part of the volume of the atom.

7 Datasheet 1, Rh, Sr, Ce, Se.

8 Datasheet 1, 44.

9 17 protons, 20 neutrons.

10 Atoms have no charge. Therefore, if there are 19 protons in an atom of potassium, there must be 19 electrons.

11 Because the relative atomic mass is not a whole number.

12 Two more protons than electrons.

13 Because it has two more electrons than protons.

14 2, 18 and 32 respectively.

15 Both have electrons in the $n = 4$ energy level but an incompletely filled $n = 3$ energy level.

16 Li, Na, K all have one electron in their outer energy level. Be, Mg, Ca all have two electrons in their outer energy level.

17 Chemical reactivity depends on the number and positions of electrons in an atom. This is identical for isotopes of the same element.

18 Group O on the far right of the Periodic Table.

19 To provide an inert atmosphere. Some materials react with oxygen and water in the atmosphere. These would not be affected in an argon atmosphere.

20 Two.

21 Because they carry an opposite charge.

22 Gamma radiation.

23 Because both result in a change in the number of protons in a nucleus.

24 33

25 2:1.

26 Ten.

27 One mole $NaHCO_3$ has mass 84 g. Therefore, a 250 g packet of baking powder contains more than 1 mol $NaHCO_3$.

28 154 g mol^{-1}.

Chapter 3

THE ATMOSPHERE, GASES AND MOLECULES

LEARNING OBJECTIVES

When you have studied this chapter you should be able to:

1. describe the regions of the Earth's atmosphere and the gases present;

2. explain atmospheric environmental issues: air pollution, the ozone layer and the greenhouse effect;

3. explain the term 'molecule', the forces holding atoms within a molecule and the nature of bonding pairs and lone pairs of electrons;

4. describe the formation of normal and coordinate covalent bonds using 'dot-and-cross' diagrams;

5. explain the shape of some simple molecules;

6. explain the effect of temperature on the speed of gas molecules;

7. explain the effect of temperature and pressure on the volume of a gas;

8. define and use in calculations: relative molecular mass, molar mass and molar volume.

3.1 THE AIR WE BREATHE

The atmosphere is a mixture of gases unique amongst the planets in our universe. What is more, no matter where you are on the Earth the same gases are present in the same proportions. A deep breath taken in Birmingham, Cardiff, Glasgow or London, will fill your lungs with the same mixture of gases as one taken somewhere in the Sahara desert (give or take a few pollutants).

The composition of the atmosphere is determined by biological, chemical and physical changes that are occurring all the time, changes in which we all play a part. Of course, biological changes are nothing more than the chemical and physical changes associated with living things, from their creation to their death and subsequent decay. The concentrations of gases in the atmosphere have remained almost totally constant for thousands of years, though the atmosphere has evolved and changed over the lifetime of the Earth.

The balance between inputs and outputs

Animals, including ourselves, inspire air and expire carbon dioxide-enriched air. Oxygen from the air enters the blood stream and is carried around weakly bonded to haemoglobin. The oxygen is released at sites in the body where the concentration of oxygen is low and this enables energy-

KEY FACT

Air is a mixture. Nitrogen is the most abundant gas in the atmosphere, accounting for nearly four-fifths of the volume. About one-fifth of the atmosphere is oxygen, with a number of other gases being present but in much smaller amounts (Table 1.1).

Fig 3.1 Variations in atmospheric carbon dioxide concentration through a 24-hour cycle in a forest.

producing oxidation reactions, the process of respiration, to occur. Carbon dioxide is a product of these reactions. Plants convert carbon dioxide and water to carbohydrates, with the aid of sunlight, releasing oxygen into the atmosphere. This is the process of photosynthesis. Maintaining the delicate balance of inputs and outputs, as illustrated by the balance of respiration and photosynthesis and their effect on air composition, is essential for our continued existence on this planet.

1 **During periods of activity people use more energy and therefore require more oxygen. How do you think your output of carbon dioxide varies over a 24-hour period?**

Another example of a balanced system of inputs and outputs is the cycle involving the most abundant gas in the Earth's atmosphere, nitrogen. The nitrogen balance is maintained mainly by a range of complex biological reactions. Bacteria 'fix' atmospheric nitrogen. They convert it into ammonium salts which in turn are converted by bacteria into nitrates. Nitrates are involved in the complex cycle of plant and animal life, ultimately returning nitrogen to the atmosphere (Fig 3.3).

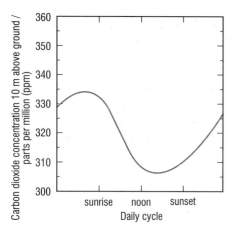

Fig 3.2 Contributing to the oxygen : carbon dioxide balance.

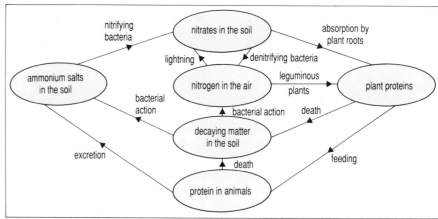

Fig 3.3 The nitrogen cycle.

Industrial uses of air

Air is used extensively in industry:

- it is the raw material for many industrial reactions involving oxidation, for example the Contact Process to manufacture sulphuric acid and the ammonia oxidation process to manufacture nitric acid. Air is also the source of nitrogen for production of ammonia;
- it is used as a coolant in many industrial processes;
- it is fractionally distilled to obtain its constituent gases. These have their own particular uses. For example, vast quantities of oxygen are used in steel works. Nitrogen is used to provide an inert atmosphere for, amongst other applications, food packaging and the manufacture of silicon chips.

QUESTIONS		
	3.1	Crisp packets are often filled with nitrogen rather than air. Why?
	3.2	**(a)** What contribution do the world's rainforests and other plant-life play in maintaining carbon dioxide balance?
		(b) Why do you think the rainforests have been described as 'the lungs of the world'?
	3.3	Account for the variation in carbon dioxide concentration in a forest throughout a 24-hour period, as shown in Fig 3.1?

THE ATMOSPHERE, GASES AND MOLECULES

3.2 ENVIRONMENTAL ISSUES: POLLUTION

Air pollution, the depletion of the ozone layer and the greenhouse effect are major environmental issues which are currently high profile concerns.

There are two types of pollutants: primary and secondary.

- Primary pollutants, for example sulphur dioxide, are the direct product of some activity on Earth and their source is usually readily identified.
- Secondary pollutants, for example sulphuric acid, are often formed by complex chemical reactions of the primary pollutants in the atmosphere.

Foggy, gloomy streets are often used by television and film makers to create the eerie atmosphere of old Victorian London. But in reality the atmosphere was polluted and it was a killer. Throughout the first half of the 20th century smog and air pollution remained a serious problem, culminating tragically in the winter of 1952 when more than 4000 people died in London. Winter fog combined with smoke from coal fires and power stations to produce a lethal smog. Subsequently, the 1956 Clean Air Act brought about a dramatic decrease in the level of air pollution in Britain. In many modern industrial cities, for example Mexico City, smog is still a major problem, but it is a different type of smog. London smogs were characterised by smoke particles and oxides of sulphur. In industrial cities it is the action of the sun's radiation on hydrocarbons and oxides of nitrogen (coming mainly from car exhausts) to produce harmful oxidising compounds, such as ozone, in the air. Chemical reactions initiated by ultraviolet or visible radiation are called **photochemical reactions**.

Fig 3.4   Los Angeles in 1990. Photochemical smogs are common in industrial cities.

The incomplete combustion of fossil fuels, such as coal, oil and natural gas, results in pollution of the atmosphere by carbon monoxide, unburnt hydrocarbons and particles of carbon (soot). In addition, because fossil fuels always contain small amounts of sulphur compounds and nitrogen compounds their combustion produces sulphur dioxide and oxides of nitrogen, which in turn results in acid rain. We can represent the formation of the primary pollutant sulphur dioxide and the secondary pollutant sulphuric acid (present in acid rain) by the following equations:

$[S] + O_2(g) \longrightarrow SO_2(g)$ $[S]$ represents a sulphur-containing material
$SO_2(g) + [O] \longrightarrow SO_3(g)$ $[O]$ represents a suitable source of oxygen
$SO_3(g) + H_2O(l) \longrightarrow H_2SO_4(aq)$

Acidity is measured on the pH scale, which has a range of 1–14. pH 7 is the neutral point. Rain contains dissolved carbon dioxide and is naturally acidic with a pH of about 5.6. Rain which falls in central Europe has a pH of about 4.1 due to the dissolved secondary pollutants, sulphuric acid and nitric acid. In the late 1980s a Los Angeles fog had a pH of 1.7, close to that of the dilute sulphuric acid we use in the laboratory!

Acid rain has had a detrimental affect on vegetation (particularly trees), freshwater fish and buildings. Lakes in Scandinavia and the north-eastern parts of the USA, for example, have become acidified to a level where fish, in particular brown trout and salmon, have disappeared (Fig 3.5).

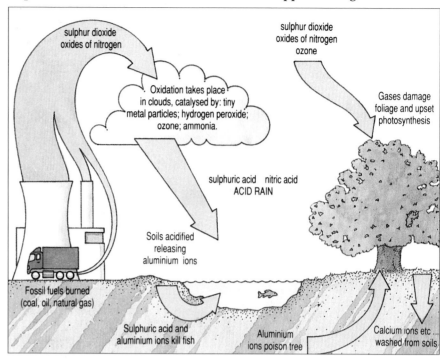

sulphur dioxide
oxides of nitrogen

sulphur dioxide
oxides of nitrogen
ozone

Oxidation takes place in clouds, catalysed by: tiny metal particles; hydrogen peroxide; ozone; ammonia.

Gases damage foliage and upset photosynthesis

sulphuric acid   nitric acid
ACID RAIN

Soils acidified releasing aluminium ions

Fossil fuels burned (coal, oil, natural gas)

Sulphuric acid and aluminium ions kill fish

Aluminium ions poison tree

Calcium ions etc … washed from soils

Fig 3.5 The chain from pollutants to acidified lakes, disappearing fish and dying trees. Understanding the chemistry of the pollution chain may help us to prevent the death of fish and vegetation.

Which substances in the atmosphere may be described as pollutants is, perhaps, a matter for debate. Many of the compounds which we think of as pollutants were present in small quantities even before humans inhabited the Earth. Volcanic eruptions and forest fires were, and still are, sources of 'natural' pollution. Generally speaking, human activity makes a measurable but small contribution to atmospheric pollution. For example, the emission of carbon monoxide due to human activity is about half a billion kilograms each year. This is about one-fifth that from natural sources. In contrast, fifty times as much sulphur dioxide (again, about half a billion kilograms) is produced by human activity compared with natural sources.

2   **Is carbon monoxide a primary or secondary pollutant?**

THIS AIR COMES WITH A HEALTH WARNING!

Air pollution is bad for our health but it is not as easy to quantify the effect as it might seem. However, we can draw some general conclusions. Oxidising agents such as ozone, O_3, and nitrogen dioxide, NO_2, irritate the eyes, nose and throat and make life very uncomfortable, especially for sufferers from respiratory complaints such as asthma or bronchitis. People may also suffer from chest discomfort, coughs and headaches. The problem is made worse during exercise, when people breath more rapidly and take deeper lungfuls of air.

Fig 3.6   Nowadays weather reports also give details of air quality. This cyclist is minimising the risks. What is in the mask?

THE ATMOSPHERE, GASES AND MOLECULES

The ozone layer

The atmosphere protects us from harmful high energy radiation from the Sun. The damaging ultraviolet radiation is absorbed by ozone in the relatively thin layer known as the stratosphere; the ozone layer acts as a radiation filter. Ozone exists as molecules, each of which contains three oxygen atoms and so has the formula O_3. The amount of ozone present in the atmosphere is extremely small but, importantly, it is concentrated in a relatively thin layer some 25–35 km above sea level. It is being badly damaged, however, by certain chemical compounds.

THE CFC TIME BOMB

Chlorofluorocarbons (CFCs) were developed in the 1930s by Thomas Midgely in the USA. They are remarkably inert (for example, they do not burn in air), have a low toxicity and are cheap to manufacture. These properties have led to their use as blowing agents (to make, for example, expanded polystyrene), refrigerants, propellants for aerosols and cleaning solvents. The five most widely used CFCs are:

		ODP			ODP
$CFCl_3$	(code: CFC11)	1.0	$C_2F_4Cl_2$	(code: CFC114)	1.0
CF_2Cl_2	(code: CFC12)	1.0	C_2F_5Cl	(code: CFC115)	0.6
$CF_2ClCFCl_2$	(code: CFC113)	0.8			

The threat of CFCs to the ozone layer was predicted in 1978 by Professor F Sherwood Rowland and his young Mexican postdoctoral assistant Mario Molina. Seven years later, Joe Farman of the British Antarctic Survey and his colleagues discovered a developing 'hole' in the ozone layer over the Antarctic. American NASA satellite measurements have confirmed the presence of this 'hole' and have also shown a general depletion in the ozone concentration over all parts of the globe. Incidentally CFCs are also 'greenhouse gases'; a further environmental problem!

In London, at 9.55pm on 29th June 1990, the environment ministers of 59 nations plus the European Community signed a declaration to completely phase out fully halogenated CFCs by midnight, 31st December 1999. Will we have alternatives in time?

CFCs are produced in large quantities because they have many useful properties, but research has revealed problems associated with their use. In response, chemists are investigating safer alternatives, such as the hydrochlorofluorocarbons (HCFCs) which have ozone depletion potentials (ODPs) in the 0.02–0.10 range and also low Global Warming Potentials (GWPs). One of the most promising is CF_3CHCl_2 (code: HCFCl23). Hydrofluorocarbons have also been studied, with CF_3CH_3F (code: 124a) showing promise. If less harmful chemicals can be produced fast enough we can cut down on the damaging CFCs.

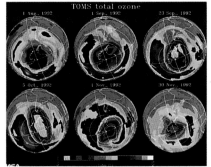

Fig 3.7 The ozone layer over the Antarctic; a computer-generated picture

Chlorofluorocarbons (CFCs) are compounds composed of chlorine, fluorine and carbon atoms. Their lack of reactivity is one reason for the many applications of CFCs, but it also causes problems. CFCs escape into the atmosphere and reside without further reaction until they reach the Earth's protective ozone layer. Here they react with the ozone and destroy it. We can illustrate the problem by considering a typical CFC, dichlorodifluoromethane. In the stratosphere a chlorine atom is split from the molecule:

$$\underset{\underset{Cl}{|}}{\overset{\overset{F}{|}}{Cl-C-F}}(g) \longrightarrow \overset{\overset{F}{|}}{Cl-C-F}(g) + Cl\,(g)$$

Ultraviolet radiation provides the energy to break the C — Cl bond but not the stronger C — F bond. The chlorine atoms react with ozone:

$$Cl(g) + O_3(g) \longrightarrow ClO(g) + O_2(g)$$

and a host of subsequent reactions are possible. Many compounds react with ozone and so reduce its concentration in the stratosphere. The more reactive a compound is towards ozone, the greater is its **ozone depletion potential** (ODP).

The ozone paradox

Ozone can be good or it can be bad, it all depends on where it is! The ozone layer in the stratosphere protects us from harmful solar radiation. It absorbs high energy ultraviolet radiation which would otherwise interact with and damage some of the molecules of life, for example proteins and nucleic acids. We dare not lose our protective shield, which is why there is so much concern about CFCs. Nearer the Earth's surface, ozone is produced by a complex series of photochemical reactions involving the unburnt hydrocarbons and oxides of nitrogen from car exhausts. Ozone is poisonous and low concentrations have an irritant effect on humans. The damaging effect of ozone is put to good use in purifying drinking water; ozone is bubbled into water to destroy harmful bacteria.

The greenhouse effect

Vast amounts of fossil fuels have been burned during the 20th century to obtain warmth and energy. Over the last few decades there has been a steady increase in the concentration of carbon dioxide in the atmosphere, almost certainly due to our use of fossil fuels. For example, the combustion of natural gas, methane, produces carbon dioxide and water.

$$CH_4(g) + 2O_2(g) \longrightarrow CO_2(g) + 2H_2O(g)$$

What is the significance of this increase? Low energy ultraviolet and visible radiation from the Sun, which is not absorbed by the ozone layer, passes through the atmosphere and is absorbed by the surface of our planet. This energy is re-emitted from the Earth's surface as infrared radiation, radiation of longer wavelength. If this heat energy simply escaped the temperature of the Earth's surface would be very much lower, and the Earth could not sustain life forms as we know them. Fortunately certain gases in the atmosphere prevent this energy from escaping by absorbing infrared radiation. This trapping of heat radiating from the Earth's surface is commonly known as the **greenhouse effect**, (Fig 3.8).

One of the most significant greenhouse gases is carbon dioxide. It follows, therefore, that an increase in the carbon dioxide concentration in the atmosphere will cause the temperature of Earth to rise. It is difficult to calculate this rise with accuracy but it has been estimated that doubling the carbon dioxide concentration will cause a 3°C rise in temperature. It is important to prevent all the heat energy from escaping, but a balance must be achieved. Too much trapping leads to global warming and subsequent environmental problems. Many gases have the ability to trap some of the infrared radiation emitted from the Earth's surface, often more efficiently than carbon dioxide. For example, methane is far more effective than carbon dioxide.

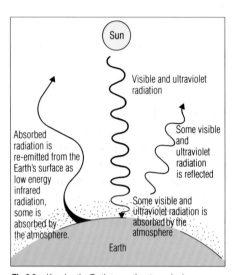

Fig 3.8 Keeping the Earth warm: the atmospheric greenhouse.

QUESTIONS

3.4 Distinguish between primary and secondary pollutants. In which category would you place tobacco smoke?

THE ATMOSPHERE, GASES AND MOLECULES

3.5 Despite its protective function in the upper atmosphere, ozone is potentially harmful. For example it is irritating to lung tissues and attacks rubber, causing cracks in car tyres. How can we reduce the levels of ozone present at ground level in our cities?

3.6 Consider the atmospheres of two nearby planets compared with Earth given in Table 3.1. Explain the trend in surface temperatures for these planets.

Table 3.1

	Venus	Earth	Mars
Distance from the Sun / million km	100	150	225
Atmosphere:			
nitrogen / % by volume	3.5	78.1	2.7
oxygen / % by volume	0.002	20.9	0.13
carbon dioxide / % by volume	96.5	0.03	95.3
Surface temperature / °C	459	15	−50

3.7 The ODP of bromotrifluoromethane, $CBrF_3$, is about ten times that of chlorotrifluoromethane, $CClF_3$.
(a) What does this suggest about the strength of the C—Br bond?
(b) Confirm your ideas by using data from an appropriate Datasheet.

COWS AND THE GREENHOUSE EFFECT

Carbon dioxide is not the only gas in the global carbon cycle. Methane is also important but until recently scientists knew much less about it than carbon dioxide. Although there is little methane in the atmosphere in terms of the carbon it contains, it is increasing faster than carbon dioxide, and each molecule of methane has the same greenhouse effect as 30 molecules of carbon dioxide. Methane comes from agriculture, produced especially in rice paddies and by cattle. It also leaks from pipelines, coal mines and rubbish tips. Bacteria in the soil are another source although some can also break down methane. Controlling emissions of methane and other greenhouse gases may form part of our response to the greenhouse effect, but the most immediate problem concerns the level of carbon dioxide in the atmosphere.

Fig 3.9 A typical cow produces 500,000 cm³ of methane a day. The world population of cattle is estimated at 1300 million. Although methane contributes only slightly to the greenhouse effect, methane is 30 times more effective as a greenhouse gas than carbon dioxide.

ACTIVITY

From *The Independent* 13 September 1990:
Greenhouse climate fear 'unfounded'
The Greenhouse world of the next century will be cloudier, wetter and milder rather than hot and stormy, an American global warming "dissident" told a meeting in London yesterday, writes Nicholas Schoon.

Dr Robert Balling, director of Arizona State University's climatology laboratory, said the evidence of man's impact on the world climate did not yet justify drastic new policies to cut emissions of carbon dioxide, the main greenhouse gas.

"I don't think there is a great world catastrophe about to occur," he said. Dr Balling believed five more years of research were needed before policymakers should begin drawing up options.

He was speaking at a meeting called by the Watt Committee, an energy think tank, financed largely by British Coal. Coal produces more carbon dioxide than other fossil fuels. He said that he largely agreed with the consensus on global warming. But he also believed that since the 1940s sulphur dioxide, a gas produced by burning fossil fuels, has been exerting a growing influence on world climate. Sulphate ions derived from the gas promote the formation of clouds.

Clouds rich in sulphate contain more liquid water, which makes them brighter. As a result they reflect more sunlight back into space, acting as a coolant.

He believes the man-made greenhouse effect is warming the world, but the man-made increase in cloudiness is acting against it.

The reaction of an environmental chemist to this newspaper article was:

"It seems to me we can avoid global warming only if we are prepared to put up with acid rain."

Try to put yourself in the role of the environmental chemist and understand the thinking behind this comment. Taking the part of the chemist, write a letter to the editor of the paper expressing your concerns. Bear in mind the nature of the audience you are addressing.

3.3 MOLECULES

Chemical changes occurring in the atmosphere affect us and our environment. To monitor and control these changes we must understand atmospheric chemistry and gaseous reactions in general. A model of the component particles present in gases is an important start. We will extend this model later to explain how these particles behave, for example, in response to rising temperatures.

With the exception of the noble gases, uncombined atoms rarely occur in nature. One way in which they combine is as discrete assemblies, called molecules. All the molecules of a given compound contain the same atoms, in the same numbers, proportions and spatial arrangement. For example, each carbon dioxide molecule has one carbon atom and two oxygen atoms arranged in a straight line, OCO. We describe it as a linear molecule. Molecules with one carbon atom and one oxygen atom have very different properties to carbon dioxide. They are a different compound, namely the poisonous gas carbon monoxide, CO. Carbon dioxide is a natural component of the atmosphere but excess carbon monoxide is a pollutant.

3 **Why are the molar masses of oxygen and ozone different?**

SPOTLIGHT

Fig 3.10 Aromatherapy being practised in a London Hospice. Can air-borne fragrances affect human moods and behaviour?

MOODS AND MOLECULES

The air around us is essential for breathing. In addition it is a unique transport system. Individual molecules and groups of molecules carried by the air may be detected by animals at levels of less than *1 part per billion*. Such molecules may be hormones which direct an animal to a potential mating partner. They may also direct an animal to a food source or help it to avoid a predator. The ability to detect air-borne molecules may even be inherited. Rats, for example, are frightened by traces of fox odorant even when bred in total isolation.

For human beings scents are rarely a matter of life or death. However, the perfume industry is a multi-million pound industry in which subtle blends of fragrances are created for human enjoyment. Some evidence exists which suggest that certain odours may affect our moods. A steroid molecule found in human sweat has been found to reduce anxiety. The substance, called Osmone 1, has a similar chemical structure to animal musks and to sandalwood oil.

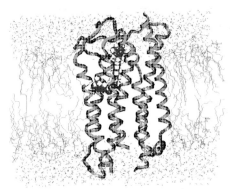

Fig 3.11 Molecules range from assemblies of just two atoms (which may or may not be the same, for example N_2 and CO) to several thousands of atoms, for example enzymes.

Bonds in molecules

Atoms are held together in a molecule by **covalent bonds**. Neighbouring atoms 'share' electrons. To understand this idea of electron sharing it might be helpful to think about a simple example, a molecule of hydrogen, H_2. Each hydrogen atom has a single electron and a single proton. Now, imagine two isolated hydrogen atoms approaching one another (Fig 3.12). The negatively charged electron of each atom will be attracted by the positively charged nuclei of *both* atoms. Electrons under the influence of two nuclei are said to be shared.

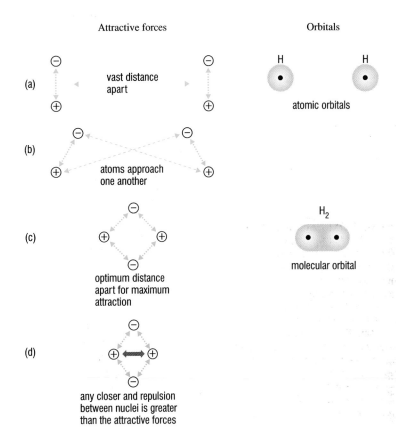

Fig 3.12 As two hydrogen atoms approach one another there is an attractive force. Electrons reside in atomic energy levels prior to bonding, and are under the influence of just one nucleus. After bonding and forming a hydrogen molecule, where they are attracted by two nuclei, they reside in molecular energy levels.

The term covalent bond describes the mutual attraction of two nuclei for the shared electrons. The atoms are bonded by electrostatic attraction between their nuclei and the shared electrons. There is a limit, however, as to how close the atoms can approach one another. At short distances the repulsion between the positively charged nuclei is greater than the attractive forces between electrons and nuclei. The compromise distance between the two nuclei is called the **bond length**.

Work must be done to pull atoms away from one another when bonded in a molecule and so energy is required. This is called the **bond energy** or **bond enthalpy**. Molecules are not rigid assemblies of atoms. They stretch, twist and bend. Also, a covalent bond vibrates, and the distance between the atoms fluctuates about an average value. The bond can only vibrate at certain fixed frequencies, which depend on the energy of the molecule. Absorbed infrared radiation of appropriate energy causes the bond to vibrate more rapidly at a higher frequency. If the energy is great enough, the atoms vibrate so violently that the bond is broken.

CONNECTION: Energy levels for electrons, pages 170–171.

4 **Why are carbon dioxide and methane effective greenhouse gases?**

The strength of a covalent bond is determined by the magnitude of the electrostatic forces between the nuclei and the bonding electrons. It depends on:

- the number of electrons being shared;
- the size of the charge on each nucleus;
- the distance and number of electrons between the two nuclei and the shared electrons.

5 **Why do CFCs release chlorine atoms into the stratosphere and not fluorine atoms?**

We now have a useful model of how atoms combine to form molecules: nuclei of neighbouring atoms have a mutual attraction for shared electrons. However, a closer look at the gases which make up our atmosphere prompts a number of questions which a good model should be able to answer. The questions are relevant to our understanding of covalent bonding in all molecules.

- Why do the noble gases, such as neon and argon, exist as uncombined atoms in the atmosphere?
- Why do oxygen and nitrogen exist as diatomic molecules rather than as uncombined atoms or as larger assemblies?
- Why can oxygen exist as O_2 and O_3 molecules but nitrogen only as N_2?
- Why is the oxygen-oxygen bond in an oxygen molecule weaker than the nitrogen-nitrogen bond in a nitrogen molecule?
- Why is carbon dioxide a linear molecule when water is an angular molecule?

Table 3.2 Gases in the atmosphere

Compound	Formula	M_r	Molecular type and shape	Bond length /nm	Bond energy /kJ mol^{-1}
nitrogen	N_2	28	homonuclear diatomic, linear	0.110	944
oxygen	O_2	32	homonuclear diatomic, linear	0.121	496
argon	Ar	39.9	monatomic	—	—
carbon dioxide	CO_2	44	heteronuclear triatomic, linear	0.122	743
neon	Ne	20.2	monatomic	—	—
water	H_2O	18	heteronuclear triatomic, angular (or non-linear), bond angle 104°	0.096	463 (O — H)

Note. A homonuclear molecule consists of only one type of atom; a heteronuclear molecule consists of more than one type of atom.

To address these questions it is necessary to model more carefully the covalent bonds which bind atoms in molecules. This requires a closer study of the arrangement of electrons in molecules.

Normal covalent bonds

Chlorine exists as diatomic molecules, Cl_2. Each chlorine atom has the electronic structure 2,8,7 and is therefore one electron short of the 2,8,8 electronic structure of an argon atom. By sharing two electrons in a molecule each chlorine atom achieves the electronic structure of argon, having 8 electrons in its outer energy level.

The arrangement of electrons in an atom can be illustrated using circles to represent energy levels with different principal quantum numbers. We can use this model to represent chemical bond formation by **dot-and-cross**

THE ATMOSPHERE, GASES AND MOLECULES

diagrams. The use of dots and crosses does **not** imply two different types of electrons. Their usefulness is in showing the **origin** of electrons in the bond. Some examples are shown in Fig 3.13.

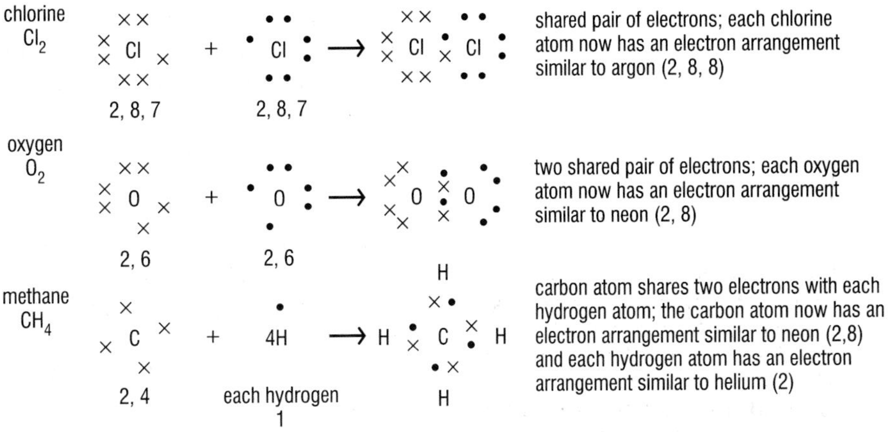

Fig 3.13 Dot and cross diagrams for molecules of chlorine, oxygen and methane.

shared pair of electrons; each chlorine atom now has an electron arrangement similar to argon (2, 8, 8)

two shared pair of electrons; each oxygen atom now has an electron arrangement similar to neon (2, 8)

carbon atom shares two electrons with each hydrogen atom; the carbon atom now has an electron arrangement similar to neon (2,8) and each hydrogen atom has an electron arrangement similar to helium (2)

CONNECTION: Arrangement of electrons in energy levels, and noble gas structures, pages 172 and 175.

The noble gases exist as monatomic molecules and are chemically rather inert. We assume that this is due to their electronic structures. Usually, atoms combine in such a way that they achieve electronic structures similar to those of the noble gases. One or more pairs of electrons may be shared to achieve noble gas electronic structures. Bonds in which two, four and six electrons are shared are called single, double and triple respectively, indicating the number of shared pairs.

6 **Look at the bond diagrams for the compounds in Fig 3.13. Which noble gas electronic structure has each atom?**

7 **Explain the difference in bond energies for O_2 and N_2 (Table 3.2).**

Coordinate bonds

In a normal covalent bond each one of the shared pair of electrons comes from a different atom. In some cases, however, the shared pair comes from the same atom. This is called **coordinate** or **dative covalent bonding**. Examples are the compound formed between boron trifluoride and ammonia (Fig 3.14) and the ammonium ion, NH_4^+, formed between ammonia and a hydrogen ion.

It is important to realise that once a normal, single covalent bond or a coordinate bond has formed there is no way of distinguishing between them. Both are simply two electrons being shared between two nuclei.

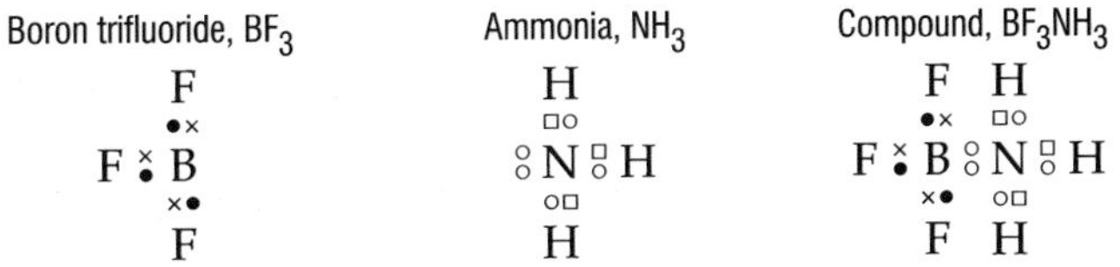

Fig 3.14 Dot and cross diagram for the boron trifluoride/ammonia compound.

The bonding in carbon monoxide cannot be explained by normal covalent bond formation. The bond length and bond energy of carbon monoxide suggest six electrons are being shared, yet the carbon and oxygen atoms cannot provide three electrons each and achieve a noble gas electronic structure. Our model to explain the properties is a diatomic molecule with the atoms linked by a triple bond made up of six electrons, four from the oxygen atom and two from the carbon atom. The bond is composed of two normal covalent bonds and one coordinate bond.

8 Do all of the N—H bonds in the NH_4^+ ion have the same bond strength?

Shorthand bond diagrams

The following 'shorthand' can be used to indicate bonding electrons:

—   two electrons, one from each atom in the bond: **normal covalent bond**,
  — single bond (2 electrons shared)
  = double bond (4 electrons shared)
  ≡ triple bond (6 electrons shared).
→   two electrons, both provided by one of the atoms (the arrow points away from the atom which is donating both electrons): **coordinate bond (or dative covalent bond)**.

Non-bonding electrons may or may not be shown. If they are shown, only dots are used since we do not need to distinguish between the origin of these electrons (Fig 3.15).

9 Draw a shorthand bond diagram for BF_3NH_3 (*Hint:* see Fig 3.14).

Compound	Dot and cross diagram	Shorthand bond diagram	
		With non-bonding electrons	Without non-bonding electrons
Hydrogen chloride, HCl	H ×Cl:	H—Cl:	H—Cl
Nitrogen, N_2	×N : N:	:N≡N:	N≡N
Carbon dioxide, CO_2	O : C : O	O=C=O	O=C=O
Water, H_2O	H × O × H	H—O—H	H—O—H
Hydrogen peroxide, H_2O_2	H × O : O × H	H—O—O—H	H—O—O—H
Carbon monoxide, CO	× C : O:	:C≡O:	C≡O

Fig 3.15   Dot and cross and shorthand bond diagrams.

THE ATMOSPHERE, GASES AND MOLECULES

3.8 Explain what is meant by the term covalent bond.

3.9 The oxonium ion, H_3O^+ is present in aqueous solutions of all acids.
  (a) Show using dot-and-cross diagrams how hydrogen ion capture by a water molecule leads to the formation of this ion.
  (b) Which inert gas has the same electronic structure as the oxygen atom at the centre of the oxonium ion?

3.10 How many lone pairs of electrons are situated on the underlined atoms in the following compounds?
  (a) H<u>Cl</u>; **(b)** $H_2\underline{S}$; **(c)** <u>N</u>H_3; **(d)** $H_2\underline{O}$; **(e)** <u>B</u>F_3.

3.11 Give an example of each of the following bonds and explain the differences in bond strengths.
  (a) single; **(b)** double; **(c)** triple covalent.

3.12 Draw dot-and-cross diagrams for **(a)** water, H_2O; **(b)** ethene, C_2H_4; **(c)** ethyne, C_2H_2.

3.13 Draw shorthand bond diagrams for **(a)** $SiCl_4$; **(b)** H_2S; **(c)** $CF_2ClCFCl_2$.

3.14 **(a)** Draw a dot-and-cross diagram for $AlCl_3$ showing the electrons in the outer energy levels only. .
  (b) Explain why aluminium chloride exists as molecules with the formula Al_2Cl_6 in the gas phase.
  (c) Draw a shorthand bond diagram for a molecule of Al_2Cl_6.

3.15 Write an equation for the photolysis of CF_3Cl in the stratosphere and represent the reaction with a dot-and-cross diagram.

3.16 **(a)** Predict the trend in bond energies for the halogens.
  (b) Find the information in the appropriate Datasheet and comment on your predictions.

3.4 MOLECULAR SHAPES

All molecules have definite shapes. The atoms are arranged in a regular three-dimensional pattern in space, and this arrangement influences the properties of the molecule.

Directional bonds and electron pairs

The covalent bonds which link the atoms in a molecule are said to be **directional**. A comparison of the structures of many molecules shows some particular shapes and bond angles are much more common than others. Shapes based on the tetrahedron and the octahedron, with their respective associated bond angles of 90° and 109.5°, are common. The bond angle 120° is also quite common.

10 **What structure has a shape with bond angles 120°?**

The shapes of molecules can often be explained by applying a simple principle: electron pairs in the outer energy level of an atom arrange themselves so that they are as far away from one another as possible. We call such an arrangement the position of minimum repulsion.

In dot-and-cross diagrams electrons are arranged in pairs. There are two types of electron pairs:

- **bonding electron pairs:** electrons which are shared between two nuclei and which provide the electrostatic 'glue' holding the atoms together in a covalent bond;
- **non-bonding electron pairs**, usually called **lone pairs**: electrons which are associated with just one nucleus.

Fig 3.16 Molecular shapes and electrons. The regions of negative charge associated with bonding pairs and lone pairs in outer energy levels differ slightly in shape. A lone pair is rather squatter (shorter and fatter) than a bonding pair, easy to understand when we realise that bonding electron pairs are shared between two nuclei while a lone pair is attracted by only one nucleus.

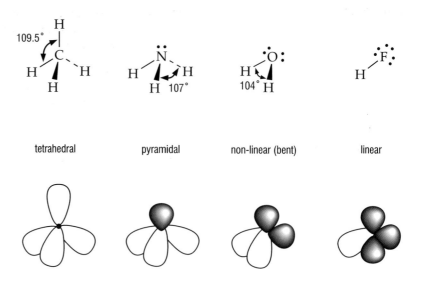

tetrahedral   pyramidal   non-linear (bent)   linear

Electron pairs around a nucleus repel one another, adopting a position of minimum electrostatic repulsion. However, lone pairs are rather squatter than bonding pairs, with a greater electron density near to the nucleus. Consequently, lone pairs exert a stronger repulsive force than do the bonding pairs. The degree of repulsion between electron pairs decreases in the following series:

lone pair/lone pair repulsion > lone pair/bonding pair repulsion > bonding pair/bonding pair repulsion

Some simple molecules

If we look at the dot-and-cross diagrams for methane, CH_4, ammonia, NH_3, and water, H_2O, (Fig 3.16) we see that in each case the eight electrons in the outer energy level are arranged in four electron pairs around the central atom. Methane has a tetrahedral shape. The carbon atom has four pairs of bonding electrons in its outer energy level. The position of minimum repulsion is reached when they point to the corners of a regular tetrahedron. The repulsion between all four electron pairs is the same and the bond angles are 109.5°.

The four electron pairs in ammonia and water also point to the corners of a tetrahedron. However, because of the differences in the repulsion effects of lone pairs and bonding pairs, the bond angles are rather less than the regular tetrahedral angle of 109.5°. The explanation of molecular shape by minimum repulsion between electron pairs is a useful aid for prediction.

11 **How many pairs of electrons around a central atom are present if the angle between them is (a) 120°, (b) 90°?**

The importance of molecular models

Our developing model for molecules is now beginning to help us explain why certain atoms combine to form molecules (generally, to gain noble gas electronic structures), and also to explain the strength of the bonds formed and the shapes of molecules. The reactivity of a molecule depends on the strength of the bonds between its constituent atoms and on its shape. The breaking of covalent bonds is an essential step in all chemical reactions involving molecules and a knowledge and understanding of bond lengths, bond strengths and shape is vital as a means to explain chemical reactivity.

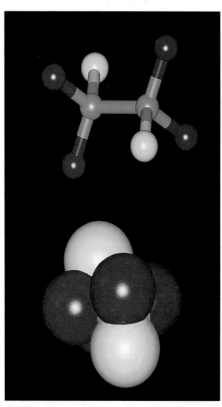

Fig 3.17 Kits are available to make physical models of molecules (both space-filling and exploded). They are very helpful in visualising molecular shapes. Remember, however, that bonds flex and vibrate, and single covalent bonds (but not multiple bonds) can rotate. The models shown in this photograph represent the same molecule – CF_2ClCF_2Cl

3.17 Give an account of the importance of (a) the number and (b) the type of electron pairs arranged around a central atom in dictating the overall shape of a molecule or compound ion.

3.18 Aluminium chloride is a covalent molecule. When in the vapour phase at high temperatures (greater than 180°C) it has the formula $AlCl_3$. Explain why it has bond angles of 120°.

3.19 In proteins the amide group

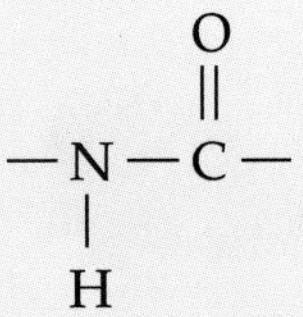

$$-N-C-$$

with O double-bonded to C and H attached to N, is an important linkage between amino acid groups. Explain the shape of the group around **(a)** the nitrogen atom; **(b)** the carbon atom.

3.20 Explain why the $H-N-H$ bond angle in ammonia is slightly greater than the $H-O-H$ bond angle in water.

3.21 Draw diagrams to represent the shapes of the following molecules **(a)** $SiCl_4$; **(b)** CF_3Cl; **(c)** H_2S; **(d)** NF_3; **(e)** BF_3. Write down the expected bond angles.

3.5 MEASURING AMOUNTS OF GASES

To monitor and protect our atmosphere we must be able to measure amounts of gases in the atmosphere. Industrial processes, such as the immensely important manufacture of ammonia and sulphuric acid, often involve reactions between gases and so, again, we must be able to measure gases and the amounts of substances present.

Relative molecular mass and molar mass

Molecules are, of course, far too small to weigh, but we can determine their masses relative to one another. We have met the term relative atomic mass already. The same concept can be applied to molecules. **Relative molecular mass**, M_r, is defined as:

The ratio of the mass of a molecule to one twelfth of the mass of an atom of the $^{12}_6C$ isotope.

Relative molecular mass has no units. It is calculated by adding the relative atomic masses of the atoms present in one molecule. For example:

Nitrogen: $M_r[N_2] = 2 \times A_r[N] = 2 \times 14 = 28$
Methane: $M_r[CH_4] = A_r[C] + (4 \times A_r[H]) = [12 + (4 \times 1)] = 16$
Molar mass, however, has units. For example:
Gaseous nitrogen has a molar mass of 28 g mol^{-1}.
Ethanol (C_2H_5OH) has a molar mass of 46 g mol^{-1}.

Molar volume

A gas is often measured by the volume it occupies rather than its mass. A relationship between the volumes of gases and the number of molecules present was established by Avogadro. He suggested that equal volumes of gases at the same temperature and pressure contain equal numbers of molecules. We can now express **Avogadro's Law** in terms of amounts of substance:

Gases containing equal amounts of substance occupy the same volume at the same temperature and pressure.

CONNECTION: Molar mass is introduced in Chapter 2.

We now know that this law is not entirely accurate, but it is a good approximation and is often helpful in calculations.

The **molar volume**, V_m, of a substance is the volume occupied by one mole of its chemical entities. It has the units **$dm^3 mol^{-1}$**. The molar volume of all gases at room temperature and pressure is about $24 dm^3 mol^{-1}$.

The amount of substance is calculated from its volume, V, simply by dividing by the molar volume both being measured at the same temperature.

$n = V/V_m$

TUTORIAL

Problem: How many molecules are present in $1 dm^3$ of oxygen at room temperature and pressure?

Solution: The molar volume of oxygen at room temperature and pressure = $24 dm^3 mol^{-1}$.
Using $n = V/V_m$

amount of oxygen, $n = \frac{1}{24}$ mol.

Since 1 mol of oxygen gas contains 6×10^{23} O_2 molecules
$\frac{1}{24}$ mol of oxygen gas contains $\frac{1}{24} \times 6 \times 10^{23}$ O_2 molecules

$= 2.5 \times 10^{22}$ O_2 molecules.

QUESTIONS

3.22 Calculate the molar mass of the following substances:
(a) propane, C_3H_8; (b) sulphuric acid, H_2SO_4; (c) benzene, C_6H_6; (d) potassium fluoride, KF; (e) magnesium sulphate -6-water, $MgSO_4.6H_2O$; (f) glucose, $C_6H_{12}O_6$.

3.23 Calculate the volume occupied at room temperature and pressure by 1 g of
(a) hydrogen, H_2; (b) ethane, C_2H_6; (c) chlorine, Cl_2.

3.24 Calculate the mass of $1 dm^3$ of each of the following gases at room temperature and pressure:
(a) helium, He; (b) oxygen, O_2; (c) argon, Ar.

3.25 Assuming that volume measurement is carried out at room temperature and pressure, decide which contains the greater number of atoms, $1 dm^3$ or 1 g of nitrogen, N_2?

3.26 Calculate the amount of substance in (a) 0.032 g SO_2; (b) 0.418 g CF_3Cl; (c) $100 cm^3$ CO (measured at room temperature and pressure).

3.27 Calculate the number of molecules in (a) 0.2 mol SO_2; (b) 30 mol NH_3; (c) 0.005 mol CH_4; (d) 1×10^{-6} mol NO_2.

3.28 Calculate the mass of (a) 0.015 mol NH_3; (b) $2500 dm^3$ CH_4 (measured at room temperature and pressure).

3.29 The composition of dry unpolluted air is 78.09% N_2, 20.95% O_2, 0.93% Ar and 0.03% CO_2. How many molecules of each gas are present in $24 dm^3$ of air at room temperature and pressure.

3.30 Using the data supplied in question 3.29, calculate the average relative molecular mass of dry, unpolluted air.
[Use $A_r[N] = 14$; $A_r[O] = 16$; $A_r[Ar] = 40$; $A_r[C] = 12$]

3.31 Each day an average adult female breathes in about 12 kg of air. The oxygen concentration of the inhaled air is about 21% by volume and of the expired air about 16% by volume. Ignoring the water content of air, calculate (a) the volume of oxygen that an adult female uses daily, and (b) the number of oxygen molecules used each day. Use the value for average relative molecular mass of dry air obtained in question 3.30. 1 mol of any gas occupies $24 dm^3$ at room temperature and pressure.

THE ATMOSPHERE, GASES AND MOLECULES

3.6 MODELLING GASES

The pressure and temperature of the atmosphere varies with altitude (Fig 3.18). Mountain climbing and flying show us the effects of pressure decreases in the atmosphere – the principal one being an insufficient supply of oxygen. If we want to understand and explain the properties of the atmosphere we need a model of gases to go alongside our model for individual molecules.

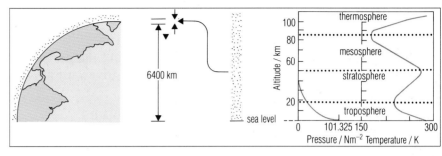

Fig 3.18   The Earth's atmosphere changes with altitude.

We may picture a gas as molecules in constant, chaotic motion. But how fast are they travelling and what kinetic energy do they possess? Are the molecules always moving at the same speed? Do they always have the same energy? The model we use to answer these and other questions about gases is called the **kinetic theory of gases**.

Temperature and speed

The speed at which a molecule moves is constantly changing as a result of collisions with other molecules and any other objects that get in its way. At room temperature and pressure even a volume of gas as small as 1 cm^3 contains some 2×10^{19} molecules. If we had some way of determining the speed of each molecule in the sample at the same moment in time we would observe a characteristic distribution of speeds. This is called the **Maxwell-Boltzmann** distribution (Fig 3.19). The kinetic energy of a molecule is given by $mv^2/2$, where m = mass of the molecule and v = its speed. The energies of a molecule in a gas also display a Maxwell-Boltzmann distribution.

The **root mean square speed**, c_{rms}, is used to describe the speed of gas molecules.

$$c_{rms} = \sqrt{\frac{v_1^2 + v_2^2 + v_3^2 + --- v_n^2}{n}}$$

where n = the number of molecules present in the gas

$v_1, v_2, v_3 --- v_n$ are the speeds of each molecule

We can calculate the effect of temperature on the speed of molecules in a gas from the equation

$$c_{rms} = \sqrt{\frac{3RT}{M}}$$

where R = the gas constant, 8.314 J K^{-1} mol^{-1}; T = temperature / K; M = molar mass / kg mol^{-1} and c_{rms} is the root mean square speed in m s^{-1}.

12   What is the root mean square speed, c_{rms}, of hydrogen molecules at room temperature (298 K)?

KEY FACT

The distribution of speeds of molecules in a gas is described by the Maxwell-Boltzmann equation. It may be represented by a graph, the shape of which depends also upon the temperature of the gas.

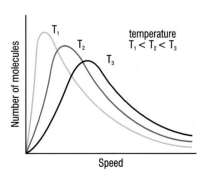

Fig 3.19   The Maxwell-Boltzmann distribution of molecular speeds at different temperatures.

Table 3.3

Gas	Molar mass / kg mol^{-1} (g mol^{-1})	c_{rms} / ms^{-1} (298 K)
water	0.018 (18)	640
nitrogen	0.028 (28)	515
oxygen	0.032 (32)	480
carbon dioxide	0.044 (44)	410

Pressure

A gas exerts a pressure on the surface of any object it comes in contact with. This is because of the constant bombardment of the surface by rapidly and chaotically moving particles. Pressure is the force exerted per unit area of that surface. The pressure will be greater if there are more molecules in a given volume and if they are moving more rapidly. Since the speed of molecules is dependent on the temperature, the pressure a gas exerts will increase with temperature, provided the volume remains the same.

13 **Why should the pressure of a car tyre be measured before the car has been driven any appreciable distance?**

Partial pressures

Air is a mixture of gases. Can we calculate the contribution that each gas makes to the total pressure that air exerts? The contribution of each gas in a mixture is called its **partial pressure**. For a mixture of gases the total pressure exerted is the sum of the partial pressures of the component gases. This is known as **Dalton's Law of Partial Pressures**, first stated by John Dalton in 1801. For a mixture of three gases (a, b and c) the law can be expressed as:

$$p_{total} = p_a + p_b + p_c$$

where p_{total} is the pressure exerted by the gaseous mixture and p_a, p_b, p_c are the partial pressures of the component gases.

The partial pressure of a gas in a mixture is the pressure it would exert if it alone occupied the total volume of the gas at the same temperature. The partial pressure of a gas is given by the general expression:

$$p_a = \frac{n_a}{n_a + n_b + n_c + \ldots} \times p_{total}$$

where n_a is the number of moles of gas a, n_b is the number of moles of gas b, etc.

The ratio of amount of one substance to the total amount of substances present in the mixture is called the **mole fraction**, x.

$$x_a = \frac{n_a}{n_a + n_b + n_c + \ldots}$$

The units for partial pressure are the same as for pressure.

Ideal gases: p, V and T

The relationship between pressure, volume and temperature has been investigated experimentally and can be summarised by the equation:

$$pV = nRT$$

where p = pressure / Pa ($N\,m^{-2}$); V = volume / m^3; n = amount of substance / mol; R = the gas constant, $8.314\,J\,K^{-1}\,mol^{-1}$; T = temperature / K.

There are many reasons for wanting to know the effect of temperature and pressure on the volume of a gas. For example, the Haber process for manufacturing ammonia involves the reaction between hydrogen and nitrogen. Choice of reaction conditions is critical for efficient and economic production of ammonia. These choices can be better made if we have a full understanding of the behaviour of gases. Numerous other important industrial processes require the control of gaseous reactions.

So how reliable is the equation $pV = nRT$? This equation (called the ideal gas equation) evolved from two classical gas laws (Fig 3.21).

THE ATMOSPHERE, GASES AND MOLECULES

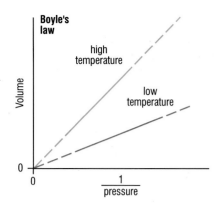

Boyle's law

Volume

high temperature

low temperature

0

0   1
                    pressure

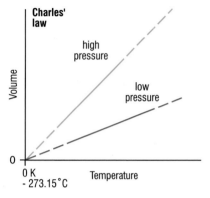

Charles' law

Volume

high pressure

low pressure

0

0 K
− 273.15°C   Temperature

Fig 3.21 Graphs to show Boyle's law and Charles' law.

- Boyle's law: $V \propto 1/p$ where V = volume, p = pressure
- Charles' law: $V \propto T$ where V = volume, T = absolute temperature

We find that no gases obey the equation $pV = nRT$ precisely and departure from it is most evident for dense gases at low temperatures. As the amount of a gas in a container approaches zero, pV/nT approaches a constant value; in other words, the gas obeys the equation $pV = nRT$. For this reason the equation is known as the ideal gas equation. Only ideal (or perfect) gases obey it, and these do not exist! How useful is it, then? Gases at room temperature and pressure behave like ideal gases for practical purposes and the equation $pV = nRT$ can be used to describe them. Given the volume occupied by a gas at known temperature and pressure, we can calculate the amount of substance present. Further, if we know the mass of the gas we can calculate its molar mass.

14 Why does a hot air balloon rise?

Our model of gases is of particles in constant, chaotic motion. However, in our model of an ideal gas we also make two key assumptions:

- there is no attraction between the particles;
- the particles occupy a negligible volume compared with the volume occupied by the gas itself.

Using this model, we can obtain the following mathematical equation to describe the behaviour of a gas:

$$pV = \tfrac{1}{3} Nmc^2_{\text{rms}}$$

where p = pressure, V = volume; N = number of particles; m = mass of each particle; c^2_{rms} = mean square speed of the particles.

The experimental ideal gas equation, $pV = nRT$, can be derived from this theoretical equation. Agreement between the experimental and theoretical equations provides evidence that the kinetic theory model of a gas is a good one. The ideal gas equation does not describe accurately the behaviour of real gases, however, because the assumptions are not accurate. There are forces of attraction between the particles. We call these intermolecular forces. Also, the volume of the particles is not negligible, particularly at high pressures and low temperatures, where there is a high concentration of particles.

QUESTIONS

3.32 The three most common components of air, together with their approximate percentage by volume, are:
nitrogen,   78%;
oxygen,   21%;
argon,   1%.
If on a particular day, atmospheric pressure was 115 kPa, calculate:
(a) the mole fraction of each gas;
(b) the partial pressure of each gas.
(c) Why in practice is the partial pressure of each gas likely to be lower than this?

3.33 Chemical reactions are sometimes performed in sealed tubes. Glass tubes containing the chemical reactants are sealed by melting the end of the tube until it closes over. The tube is then often heated in an oven for several hours to very high temperatures. Thin-walled tubes are considered hazardous. How does the ideal gas equation help to predict this potential hazard?

THE ATMOSPHERE, GASES AND MOLECULES

3.34 The ideal gas equation applies to 'ideal' gases only, yet **real** gases are *not* ideal.

  (a) Describe the key features of the model of an ideal gas and explain how real gases deviate from the ideal.

  (b) Why is the ideal gas equation useful despite its limitations?

3.35 If 100 cm^3 of hydrogen and 50 cm^3 of oxygen are mixed at 25°C and 10^5 Pa, calculate the partial pressure of each gas in the mixture.

3.36 Calculate the mole fraction of oxygen, and its partial pressure, in a sample of dry, unpolluted air at 101 325 Pa (use data in question 3.29 on page 50).

3.37 Calculate the volume occupied by 16 g methane at **(a)** 373 K and 5 × 10^6 Pa; **(b)** 573 K and 5 × 10^6 Pa.

ACTIVITY

Five identical balloons are filled to the same pressure at the same temperature with hydrogen, methane, carbon monoxide and carbon dioxide respectively. They are released simultaneously from a point 10 m above the ground. Predict the sequence of events in the few minutes following the release of the balloons, and describe an experiment to test your predictions.

Answers to Quick Questions: Chapter 3

1 High during the day, especially at times of vigorous activity. Low when sleeping at night.

2 Primary

3 Because they have the molecular formulae, O_2 and O_3 respectively.

4 Because they are good absorbers of infrared radiation.

5 Because the C — F bond is stronger than C — Cl and so the latter is easier to break.

6 Chlorine atoms achieve an argon electronic structure; oxygen atoms and carbon atoms achieve a neon electronic structure; hydrogen atoms achieve a helium electronic structure.

7 Atoms in a nitrogen molecule are held together by a triple covalent bond which is stronger than the double covalent bond which holds atoms together in an oxygen molecule.

8 Although three bonds are normal covalent bonds and the fourth is a coordinate bond, in all four cases two electrons are being shared between a nitrogen atom and a hydrogen atom. All four bonds, therefore, have the same bond strength.

9
```
       F   H
       |   |
   F — B ← N — H
       |   |
       F   H
```

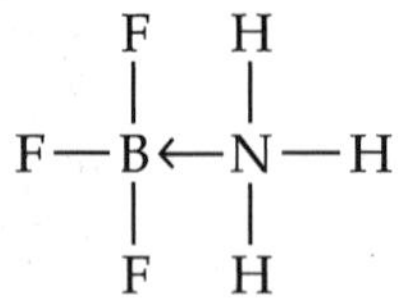

10 Trigonal planar (an equilateral triangle).

11 (a) Three. (b) Six.

12 1980 m s^{-1}

13 Friction between the road and the tyre will cause the temperature of the air inside the tyre to rise. Therefore, the pressure will rise.

14 Because hot air is less dense than cold air.

Chapter 4

THE EARTH'S CRUST

LEARNING OBJECTIVES

When you have studied this chapter you should be able to:

1. describe in broad terms the structure of the Earth;

2. explain the terms mineral and rock;

3. assign oxidation numbers to elements in compounds;

4. name binary inorganic compounds, oxoacids and their salts;

5. describe the structure and bonding present in the four types of solid: molecular, giant molecular, ionic and metallic;

6. explain the properties of solids in terms of their structure and bonding;

7. explain the terms empirical formula, molecular formula and formula unit;

8. calculate empirical formulae from appropriate analytical data.

4.1 ROCKS AND MINERALS

We can only dream of a journey to the centre of the Earth, perhaps as Jules Verne did in his book *Journey to the Centre of the World*. However, earthquake studies have enabled scientists to picture the Earth below our feet – the core and the mantle of our planet (Table 1.2). The thin outer layer of the Earth is called the Earth's crust.

Fig 4.1   The inanimate world of the Earth's crust provides countless examples of great natural beauty, from the immense scale of the Grand Canyon (erosion and rock weathering at its most dramatic) to tiny mineral crystals and gemstones with perfect geometrical shapes and often beautiful colours.

The Earth

The composition of the Earth's crust is very different to that of the Earth as a whole (Table 1.2). Eight elements account for some 98–99% of the crust but many elements commonly found in manufactured articles are not in the list, for example copper, lead, zinc, nickel and tin between them make up less than 0.05% of the Earth's crust. Fortunately, these scarce but very important elements are often concentrated in certain rocks, called **ores**, though these are found only in scattered locations. Even then it often takes the skills of the chemist to concentrate them further and to extract the desired element or compound.

1   **In what types of minerals do we find the most abundant element in the Earth's crust?**

The Earth's crust is the source of elements and compounds essential to life. The physical disintegration of rocks eventually produces soil, fine particles of rock from which the essential elements, combined in chemical compounds, can be dissolved in water and become available to plant life. The crust also supplies chemical and energy resources which have enabled us to make those scientific and technological advances which characterise our society. Fossil fuels, metals, non-metals and chemical substances for a whole range of industrial processes and synthetic products all come from the thin outer layer of our planet.

2   **The names 'stone age', 'iron age' and 'bronze age' reflect human beings' dependence on the Earth's crust. Can you think of a suitable name for our current 'age'?**

SPOTLIGHT

Hardness	Mineral	Others
1	talc, $Mg_3Si_4O_{10}(OH_2)$	1 poly(ethene)
2	gypsum, $CaSO_4.2H_2O$	
3	calcite, $CaCO_3$	
4	fluorite, CaF_2	
5	apatite, $Ca_5(PO_4)_3F$	5 copper coin
6	orthoclase, $KAlSi_3O_8$	5.5 glass window
7	quartz, SiO_2	
8	topaz, $Al_2SiO_4F_2$	
9	corundum, Al_2O_3	
10	diamond, C	

HARD AND OFTEN BORING

The Channel Tunnel is a remarkable feat of engineering, but it would not have been possible without machines which had very hard tips to bore through the rock. Eleven boring machines were used, tipped with double edged picks, and between them they removed 7 million cubic metres of rock in 1305 days. This means that on average, each machine removed about 500 cubic metres of rock per day, an amazing rate of progress.

The Moh's scale can be used to place materials in order of increasing hardness. The rule is simple: a substance below another in the list will scratch it when the two are rubbed together. The scale is from 1 (soft) to 10 (very hard). Carborundum (chemical name silicon carbide, SiC) is an excellent abrasive. Several hundred thousand tons are produced each year. The trivial name carborundum was coined by E.G. Acheson following his accidental synthesis of the compound in 1891. The name comes from carbo(n) and (co)rundum to indicate that SiC has a Moh hardness of 9.5, halfway between diamond (an allotrope of carbon) and corundum (a polymorph of aluminium oxide).

Tooth enamel has a hardness of around 5.5–7.0. It is essential that toothpaste should not damage enamel. For this reason toothpastes should have a hardness of less than 5.5. As a check, a safe toothpaste should not be able to mark glass, which also has a hardness of 5.5.

The difference between rocks and minerals

Minerals are naturally occurring inorganic solids. They are usually compounds but a few are uncombined elements. Elements which occur in the Earth's crust uncombined with other elements are said to be **native** elements. More often, however, elements are found in chemical combination with one another. Each of these minerals has a definite chemical composition and is a pure compound. It may be a simple binary compound, composed of just two elements, or a more complex compound. The two most abundant elements in the Earth's crust are silicon and oxygen, and one of the most common minerals is quartz (chemical name: silicon(IV) oxide, SiO_2). Other examples of minerals are given in Table 4.1. The oldest rocks on Earth are about 3800 million years old; the oldest rocks on the Moon are some 4700 million years old.

Rocks have variable compositions and physical characteristics and usually consist of two or more minerals. There are three types of rock:

- igneous, formed from magma (molten rock), e.g. granite;
- sedimentary, formed by the erosion of rock into smaller fragments and their subsequent cementing together, e.g. sandstone, shale;
- metamorphic, formed when rocks are changed by the action of heat, pressure or certain chemical substances, e.g. marble (from limestone), slate (from shale).

Table 4.1 A classification of minerals according to their chemical composition

Mineral	Examples
Native elements	carbon (C), copper (Cu), gold (Au), silver (Ag), sulphur (S)
Halides	halite (NaCl), fluorite (CaF_2)
Oxides	bauxite (Al_2O_3), hematite (Fe_2O_3), quartz (SiO_2), rutile (TiO_2)
Sulphides	cinnabar (HgS), galena (PbS), marcasite and pyrite (both FeS_2)
Carbonates	calcite ($CaCO_3$), magnesite ($MgCO_3$), siderite ($FeCO_3$), witherite ($BaCO_3$)
Sulphates	anhydrite ($CaSO_4$), barytes ($BaSO_4$)
Silicates*	fosterite (Mg_2SiO_4), spodumene ($LiAlSi_2O_6$), talc ($Mg_3Si_4O_{10}(OH)_2$), muscovite ($KAl_3Si_3O_{10}(OH)_2$), heukandite ($CaAl_2S_{17}O_{18}.6H_2O$)
Molybdates	wulfenite ($PbMoO_4$)
Phosphates	chlorapatite ($3Ca_3P_2O_8.CaCl_2$), fluorapatite ($3Ca_3P_2O_8.CaF_2$)
Tungstates	scheelite ($CaWO_4$)
Vanadates	vanadinite ($Pb_5(VO_4)_3Cl$)

* Vast numbers of silicates are known, all based on the tetrahedral 'SiO_4' unit. About one third of all minerals are silicates, including the main rock-forming minerals.

What we need to work with solids

To work with solids, whether they come from the Earth's crust or are synthetic, chemists need a number of tools. These include

- a system for naming compounds which avoids confusion or ambiguity;
- a system for representing compounds that informs us about their composition and, if necessary, their structure;
- models to describe how atoms pack together in different solids and the type of bonding present, which also explains chemical and physical properties;
- techniques for determining the composition of solids (what elements are present and how much) and for measuring amounts of substance.

4.1 Why are most elements found as ores and not as the free element?

4.2 What is the difference between an inorganic compound and an organic compound?

4.3 Decide which of the following compounds are inorganic or organic:
(a) CH_3COOH; **(b)** CO_2; **(c)** NH_4; **(d)** CCl_4; **(e)** $NaHCO_3$; **(f)** CO; **(g)** KCN; **(h)** $CaCO_3$; **(i)** $HCOOH$.

4.4 List the eight most common elements, in order of decreasing abundance, found in **(a)** the Earth as a whole, **(b)** the Earth's crust. Compare the two lists and comment on them in the light of your knowledge of the Earth's structure.

4.2 NAMING COMPOUNDS

Fig 4.2 Copper(II) sulphate-5-water crystals are blue. When heated gently, water is driven off leaving white anhydrous copper(II) sulphate:

$$CuSO_4.5H_2O(s) \longrightarrow CuSO_4(s) + 5H_2O(l).$$

Fig 4.3 The two minerals calcite and aragonite are chemically identical and share the same systematic name, calcium carbonate ($CaCO_3$). However they have different crystalline forms (they are **polymorphic**). The trivial names provide this information, at least to the trained geologist!

Communication in chemistry is important. Chemical names and symbols must be understood by all who meet and have to use them. But naming chemicals can be confusing. The following names have been used, amongst others, for the well known blue crystals with formula $CuSO_4.5H_2O$:

bluestone
blue vitriol
copper sulphate
cupric sulphate
copper(II) sulphate
hydrated copper(II) sulphate
copper(II) sulphate-5-water
tetraaquacopper(II) tetraoxosulphate(VI)-1-water

The issue is not whether one of these names is correct and the others are wrong. It is that people should be able to share the same understanding of a given name. However, consider what happens when $CuSO_4.5H_2O$ is heated (Fig 4.2). The crystals lose their blue colour and a white solid, $CuSO_4$, remains. The compounds $CuSO_4$ and $CuSO_4.5H_2O$ are different and should not be confused. Both could be called copper(II) sulphate, but to avoid ambiguity it would be better to distinguish between them by using the names copper(II) sulphate and copper(II) sulphate-5-water respectively.

Minerals are usually known by their **trivial** names. These convey little or no information about chemical composition. For example the name siderite provides no clue about the mineral's composition, $FeCO_3$. Many synthetic compounds have also acquired trivial names, for example nylon. Undoubtedly, the systematic naming of compounds, whether naturally occurring or synthetic, would avoid ambiguity and improve communication.

The Commission of the International Union of Pure and Applied Chemistry (IUPAC) have produced a set of recommendations for naming chemical compounds. The important points for inorganic compounds are summarised here, and those for organic compounds later. **Systematic naming** is called **nomenclature**. For example, the IUPAC name for the $CuSO_4.5H_2O$ is tetraaquacopper(II) tetraoxosulphate(VI)-1-water. However, while we want an unambiguous name, we also want one which is manageable. Copper(II) sulphate-5-water is more convenient and more frequently used.

A final word about minerals and trivial names. There are two important minerals containing calcium: gypsum ($CaSO_4.2H_2O$) and anhydrite ($CaSO_4$). The systematic names for these minerals are lengthy, but the abbreviated versions are manageable and still informative: calcium sulphate-2-water and calcium sulphate. However, what about a mineral such as talc, $Mg_3Si_4O_{10}(OH)_2$? Try to imagine the systematic name, or even an abbreviated version. The convenience of using trivial names on occasions is only too apparent. There are other advantages in the occasional use of trivial names (Fig 4.3).

Oxidation numbers

Important in IUPAC nomenclature is the concept of **oxidation number**. The names copper(I) oxide and copper(II) oxide are used for Cu_2O and CuO respectively. In each compound the oxidation number of copper differs and the Roman numeral in the names shows this. An oxidation number is a numerical value assigned to atoms of an element when that element is

- in the free state, e.g. Cl_2;
- present in ionic or covalent compounds, e.g. $NaCl$ or $SiCl_4$;
- present in simple or complex ions, e.g. Cl^- or ClO_4^-.

Unlike **valency**, which concerns the numbers of electrons involved in bonding, oxidation numbers have no physical significance. The oxidation number is assigned by applying a set of simple rules. Although it has certain limitations, the concept is helpful in chemical nomenclature, describing redox reactions and balancing redox equations. Roman numerals, together with 0, are used for oxidation numbers in nomenclature. In algebraic manipulations such as balancing redox equations arabic numerals are used, and these may be positive or negative. We will use the abbreviation $Ox(B)$ for the oxidation number of species B, although there is no agreed convention.

The oxidation number of an atom may be thought of as the charge it would carry if all the other atoms in the species were removed as their common ions. For example, the oxidation number of carbon in tetrachloromethane, CCl_4, is +4. It is obtained by imagining the removal of chlorines as chloride ions, $4 \times Cl^-$, which would leave C^{4+}. Therefore $Ox(C) = +4$.

The rules for assigning oxidation numbers to atoms of an element are:

1. The oxidation number of atoms in an *element* is taken as zero.

2. In combination the oxidation number of some elements is always the same, regardless of whether they are found in ionic or covalent compounds, simple or complex ions:

Element	Oxidation number	Exceptions
fluorine	–1	none
oxygen	–2	F_2O, $Ox(O) = +2$; peroxides, $Ox(O) = -1$
hydrogen	+1	ionic hydrides, $Ox(H) = -1$
Group I	+1	
Group II	+2	

3. The sum of the oxidation numbers in a compound is zero. Examples:

   calcium fluoride, CaF_2
   $Ox(Ca) = +2$, $Ox(F) = -1$
   Therefore,
   $Ox(CaF_2) = +2 + 2(-1) = 0$

   potassium manganate(VII)
   $Ox(K) = +1$, $Ox(Mn) = +7$, $Ox(O) = -2$
   Therefore,
   $Ox(KMnO_4) = +1 + 7 + 4(-2) = 0$

4. The sum of the oxidation numbers in an ion equals the charge on the ion. Examples:

   the chloride ion, Cl^-
   $Ox(Cl) = -1$

   the dichromate(VI) ion, $Cr_2O_7^{2-}$
   $Ox(Cr) = +6$, $Ox(O) = -2$
   Therefore,
   $Ox(Cr_2O_7^{2-}) = 2(+6) + 7(-2) = -2$

Problem: What is the oxidation number of sulphur in Na_2SO_4?

Solution: Use rule 3 above: The sum of the oxidation numbers in a compound is zero.

$Ox(Na) = +1$, $Ox(O) = -2$

Therefore $Ox(Na_2SO_4) = 0 = 2(+1) + Ox(S) + 4(-2)$

$Ox(S) = -2(+1) - 4(-2) = +6$

Problem: What is the oxidation number of chlorine in ClO_4^-?

Solution: Use rule 4: The sum of the oxidation numbers in an ion equals the charge on the ion.

$Ox(O) = -2$
Therefore $Ox(ClO_4^-) = -1 = Ox(Cl) + 4(-2)$
$Ox(Cl) = -1 - 4(-2) = +7$

Binary compounds

A binary compound consists of just two elements. By convention the least electronegative element comes first in both the name and formula of the compound. Electronegativity increases across a period and decreases down a group in the Periodic Table. With a few exceptions, most notably non-metal hydrides such as water, ammonia and methane, all binary compounds end in **-ide**.

The ratio in which atoms combine in a compound with a simple **molecular structure** determines its name. Prefixes are used to indicate how many atoms of each element are present in a simple molecule.
Example:

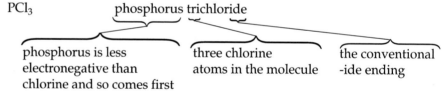

PCl_3 phosphorus trichloride

phosphorus is less electronegative than chlorine and so comes first

three chlorine atoms in the molecule

the conventional -ide ending

Other examples: SO_2, sulphur dioxide; N_2O_4, dinitrogen tetraoxide; BrF_5, bromine pentafluoride; S_2Cl_2, disulphur dichloride.

Oxidation numbers are used in the names of compounds which do *not* have simple molecular structures. Again, the less electronegative element comes first in both name and formula, but this time with its oxidation number indicated by roman numerals in brackets.
Example:

CuO copper(II) oxide

copper is less electronegative than oxygen and so comes first

copper has oxidation number +2

the conventional -ide ending

Other examples: Cu_2O, copper(I) oxide; P_2O_5, phosphorus(V) oxide; P_2O_3, phosphorus(III) oxide; PbS, lead(II) sulphide. The oxidation number is omitted for elements which always display the same value. For example, calcium is always +2 and so the name for its oxide, CaO, is simply calcium oxide rather than calcium(II) oxide.

3   **Does silicon tetrachloride have a molecular structure?**

Oxoacids and their salts

IUPAC-derived names for oxoacids and oxoanions are often cumbersome; abbreviated versions are more commonly used (Table 4.2). Both oxidation numbers and prefixes for numbers of atoms are used in systematic names but not in common names.

Salts are formed when ionisable hydrogen atoms in an acid are replaced by metal or ammonium ions. When all the ionisable hydrogen atoms are replaced, a normal salt is formed, for example sodium sulphate (Na_2SO_4) and ammonium nitrate (NH_4NO_3). Acid salts are formed when the ionisable hydrogen atoms are only partially replaced, for example $NaHSO_4$ sodium hydrogensulphate, K_2HPO_4 dipotassium hydrogenphosphate(V) and KH_2PO_4 potassium dihydrogenphosphate(V).

Table 4.2 Oxoacids and oxoanions

Formula	Systematic name	Common name
H_2SO_4	tetraoxosulphuric(VI) acid	sulphuric acid
SO_4^{2-}	tetraoxosulphate(VI) ion	sulphate ion
H_2SO_3	trioxosulphuric(IV) acid	sulphurous acid
SO_3^{2-}	trioxosulphate(IV) ion	sulphite ion
HNO_3	trioxonitric(V) acid	nitric acid
NO_3^-	trioxonitrate(V) ion	nitrate ion
HNO_2	dioxonitric(III) acid	nitrous acid
NO_2^-	dioxonitrate(III) ion	nitrite ion
CO_3^{2-}	trioxocarbonate(IV) ion	carbonate ion
MnO_4^-	tetraoxomanganate(VII) ion	manganate(VII) ion
$Cr_2O_7^{2-}$	heptaoxodichromate(VI) ion	dichromate(VI) ion
CrO_4^{2-}	tetraoxochromate(VI) ion	chromate(VI) ion
PO_4^{3-}	tetraoxophosphate(V) ion	phosphate(V) ion

4 **What do you think the systematic name for the tungsten-containing mineral, scheelite, is?** *Hint:* Table 4.1.

Hydrates

Many compounds crystallise from aqueous solution as **hydrates** which have a fixed amount of water present. To name a hydrate, the name for the anhydrous compound is followed by a hyphen, an arabic numeral indicating the stoichiometric proportion of water, then another hyphen and the word 'water'. For example, $CuSO_4.5H_2O$ is called copper(II) sulphate-5-water, $NiSO_4.6H_2O$ is called nickel(II) sulphate-6-water and $Na_2CO_3.10H_2O$ is called sodium carbonate-10-water (also known as washing soda crystals). The full stop is used simply to separate the formula of the anhydrous compound from the amount of water present in the hydrated salt.

QUESTIONS

4.5 Work out the oxidation number of the underlined element in each of the following species:
(a) <u>Na</u>$_2$O; (b) Na<u>I</u>; (c) H<u>Br</u>; (d) <u>Cr</u>O$_3$; (e) Na<u>Br</u>O$_3$; (f) H$_2$<u>O</u>$_2$;
(g) <u>I</u>$_2$; (h) <u>Cl</u>$_2$O$_7$; (i) <u>P</u>Cl$_5$; (j) <u>C</u>HCl$_3$; (k) <u>I</u>O$_3^-$; (l) <u>P</u>O$_4^{3-}$.

4.6 Examine the following equation. Explain which substances have changed their oxidation states and give the oxidation state before and after the reaction:

$$2KMnO_4(aq) + 5Na_2C_2O_4(aq) + 16H^+(aq)$$
$$\longrightarrow 2Mn^{2+}(aq) + 2K^+(aq) + 10Na^+(aq) + 10CO_2(g) + 8H_2O(l)$$

4.7 Write chemical formulae for the following compounds:
(a) sodium sulphide; (b) iron(III) oxide; (c) mercury(I) iodide;
(d) copper(I) cyanide; (e) lead(II) bromide; (f) iron(II) oxide.
(g) iodine(VII) fluoride

4.8 Name the following inorganic compounds:
(a) KH_2PO_4; (b) $SrSO_4$; (c) Ag_2CO_3; (d) K_2O;
(e) $MgSO_4.6H_2O$; (f) $CrCl_3$; (g) $Al(OH)_3$; (h) NH_4NO_2;
(i) NH_4NO_3; (j) $KHSO_4$.

4.9 Write down the oxidation number of uranium in the following species: UCl_3, UBr_4, UF_6, UO_2Cl_2, UO^+, UO_2^{2+}, UF_8^{3-}.

4.10 Name the following compounds, all of which exist as simple molecules: BF_3, CS_2, I_2Cl_6, N_2O, XeF_2, XeF_4.

4.11 Name the following compounds, which do not exist as simple molecules: $BiCl_3$, I_2O_5, PbO_2, MoS_2, TiO_2, V_2O_5.

4.3 MODELLING SOLIDS

The arrangement of particles in a solid

We can model molecules and collections of vast numbers of molecules in a gas in order to understand and explain their properties. In the same way, we need to develop models for the solid state, to explain the physical and chemical properties of solids.

Most solids are crystalline with closely packed particles arranged in a regular, repeating array called a **crystal lattice**. The external symmetry of crystals reflects this highly ordered arrangement at the atomic level. Furthermore, the bulk properties of crystalline solids result from this arrangement and the nature of the forces holding the particles together. Some solids do not have a regular repeating array of particles (though they are still closely packed) and they are said to be **amorphous**. An example is the gemstone opal.

Nearly all solids can be placed in one of four groups (Table 4.3):

- molecular solids;
- giant covalent solids (sometimes called giant molecular solids or macro-molecules);
- ionic solids;
- metals.

To understand the properties of these solids we will need to extend our treatment of covalent bonding and then turn our attention to ionic and metallic bonding.

Table 4.3 Classification of solids by their structures and bonding

Type of solid	Structure and bonding	Typical properties
Molecular	covalent bonding between atoms in a molecule and intermolecular bonding between molecules	low melting and boiling points; soft; non-conductors of electricity or heat; insoluble in water; soluble in polar solvents
Giant covalent	covalent bonding between atoms in an extensive array	very high melting and boiling points; hard; non-conductors of heat or electricity; insoluble in water; insoluble in polar solvents
Ionic solids	ionic bonding between ions arranged in a regular three-dimensional lattice	high melting and boiling points; hard; non-conductors of electricity (decomposed by the passage of electricity when molten – electrolysis); soluble in water; insoluble in polar solvents
Metals	metallic bonding between atoms arranged in a regular three-dimensional lattice	high melting and boiling points; hard, malleable and ductile; good conductors of electricity and heat (soild and liquid); insoluble in water; insoluble in polar solvents

Note The close packed particles in a solid may be atoms, molecules or ions.

THE EARTH'S CRUST

Molecular solids

Only two elements occur uncombined in the Earth's crust in appreciable quantities, carbon and sulphur. Sulphur is an example of a typical molecular solid and a consideration of its structure and properties will improve our understanding of molecules. Carbon, as we will see shortly, can occur naturally in two forms. Both have giant molecular structures and, consequently, very different properties to sulphur.

Sulphur

Sulphur has two allotropic forms, rhombic and monoclinic. Both consist of S_8 molecules arranged in a three-dimensional lattice but packed differently in each allotrope. In each molecule, single covalent bonds hold the eight sulphur atoms in a puckered ring (Fig 4.4). Although sulphur has a low melting point (113°C for rhombic sulphur and 119°C for monoclinic sulphur) it is not as volatile as oxygen or nitrogen, which also have molecular structures but are gases. The reason for this difference lies in the relationship between molecular size and the strength of the bonding between molecules – **intermolecular bonding**.

CONNECTION: How the existence of intermolecular forces help us to explain why real gases do not obey the ideal gas equation, Chapter 3.

Fig 4.4 The two allotropes of sulphur – both contain S_8 molecules but packed differently in each allotrope. Monoclinic crystals are shown in the left-hand photograph and rhombic crystals on the right.

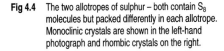

(a)

balance point

(b)
I_2 molecule: centres of positive and negative charge coincide. But electrons are in constant motion:

(c)
Temporary dipole in I_2 molecule: centres of positive and negative charge no longer coincide. Dispersion forces attract neighbouring dipoles:

(d)
$\delta - \quad \delta + \qquad \delta - \quad \delta + \qquad \delta - \quad \delta +$
$|\!-\!| \cdots\cdots |\!-\!| \cdots\cdots |\!-\!|$
The molecules pack together in a three-dimensional lattice:

(e)
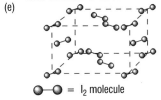

◯—◯ = I_2 molecule

Fig 4.5 The centres of positive and negative charge in a covalent bond may be pictured as analogous to the balance point for a set of barbells. The centre of negative charge is the 'balance point' for the electrons, and the centre of positive charge is the 'balance point' for the two nuclei. Dispersion forces result from temporary separation of these centres of positive and negative charge due to electron movement around the molecules.

> **5** **Rhombic and monoclinic sulphur have different melting points. What does this suggest about the strength of the bonding between the molecules in each allotrope? Explain whether or not you would expect the allotropes to have different boiling points.**

The low melting point of sulphur, resulting from its molecular structure, enables it to be extracted from the Earth's crust by the Frasch process. Steam at 163°C is pumped into the sulphur deposits, causing the sulphur to melt. The molten sulphur is then forced to the surface by the steam where it solidifies on cooling. It is then broken up mechanically.

Halogens and dispersion interactions

A study of the halogens will help us to understand the relationship between molecular size and volatility. Fluorine and chlorine are gases, bromine is a liquid and iodine is a solid, yet all four elements exist as diatomic molecules. In each halogen molecule, two electrons are shared equally between two atoms.

The centres of positive and negative charge for the molecule coincide (Fig 4.5). We say the molecule is **non-polar**. However, electrons are in constant motion and, at any one moment in time, the molecule will possess a **temporary dipole** because the centres of positive and negative charge no longer coincide. The overall effect is a weak electrostatic attraction between molecules due to **dispersion interaction** (sometimes called **London forces**). The larger the molecule and the more electrons there are, the more easily a temporary dipole is formed. It is important to realise, however,

CONNECTION: Intermolecular bonds, pages 81–83.

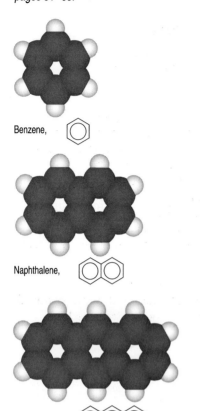

Benzene,

Naphthalene,

Anthracene,

Fig 4.6 The structures of benzene, naphthalene and anthracene. All are planar molecules.

CONNECTION: Bonding in benzene, page 104.

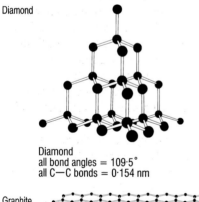

Diamond

Diamond
all bond angles = 109·5°
all C—C bonds = 0·154 nm

Graphite

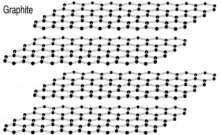

Graphite
all bond angles within a plane = 120°
C—C bonds in plane = 0.142 nm
distance between planes = 0.335 nm

Fig 4.7 The structures of diamond and graphite.

that intermolecular bonds are much weaker than the covalent bonds within molecules.

We can see now why the larger halogens are less volatile. The dispersion forces are strongest between the largest halogen molecules. Therefore iodine is the least volatile halogen. Table 4.4 shows a clear relationship between the relative molecular masses of N_2, O_2, P_4, S_8 and their volatility.

Although we have used elements with molecular structures to illustrate intermolecular bonding, molecular compounds also display increasing intermolecular attraction with increase in size. The related compounds benzene, naphthalene and anthracene are examples (Fig 4.6 and Table 4.4).

Table 4.4 Melting points and boiling points for some molecular substances

	Molecular formula	Relative molecular mass	Melting point /°C	Boiling point /°C
Elements				
nitrogen	N_2	28	–210	–196
oxygen	O_2	32	–218	–183
white phosphorus	P_4	124	44	280
rhombic sulphur	S_8	256	113	445
Compounds				
benzene	C_6H_6	78	5.5	80.1
naphthalene	$C_{10}O_8$	128	80.2	218
anthracene	$C_{14}H_{10}$	178	216	340

6 The tetrahalides of tin have simple tetrahedral molecular structures. The melting points of $SnCl_4$ and $SnBr_4$ are –33.3°C and 31°C, and their respective boiling points are 114°C and 205°C. Predict the melting point and boiling point of SnI_4.

Giant covalent solids

The two allotropes of carbon, diamond and graphite, have quite different properties. Diamond is colourless and transparent, hard (the hardest naturally occurring substance) and is an insulator. It is used in cutting tools, watch-bearings and, of course, as gemstones. In contrast, graphite is black, soft and flaky (it is the 'lead' in pencils) and is less dense than diamond. It is a good conductor of both electricity and heat and is used as a lubricant.

Why do the physical properties of diamond and graphite differ when they are composed of the same atoms and are chemically identical? All is revealed when we look at the structure and bonding of the allotropes (Fig 4.7). The strength of the C—C bond (348 kJ mol^{-1}) and the scaffold-like three-dimensional lattice account for the physical properties of diamond. All the electrons in the outer energy level are involved in normal covalent bonds.

In contrast, graphite consists of hexagonal layers of carbon atoms, rather like chicken-netting. Each carbon atom uses three of its electrons in covalent bond formation with neighbouring atoms. Each of these electrons is **localised** between two nuclei. The remaining electron from each carbon atom is not associated with any one particular pair of carbon atoms. It can move freely from one atom to another throughout the lattice and is said to be **delocalised**. The layers of carbon atoms are held together by weak intermolecular forces and can slip easily over one another – accounting for the lubricating properties of graphite. The atoms are less closely packed in graphite than in diamond and therefore it is less dense (graphite 2.22 g cm^{-3}; diamond 3.51 g cm^{-3}).

7 Explain why it is that graphite conducts electricity but diamond does not?

DIAMONDS ARE FOREVER (ALMOST)

All the diamonds in the world are turning into pencil lead. Fortunately for diamond owners the change is immeasurably slow. However the feasibility of converting one form of diamond into another is put to good use in industry.

Synthetic diamonds (often called industrial diamonds) can be manufactured by subjecting graphite to intense heat and pressure. It was similar conditions deep in the earth's interior that were responsible for the formation of natural diamonds. However, if diamond is heated to over 1000°C it slowly changes into graphite. At 2000°C the change occurs very rapidly and graphite powder is produced as the crystal shatters.

The reason for this instability is that carbon atoms are held together in graphite by marginally stronger bonding (in total) than the carbon atoms are in diamond. This is reflected by the energy change which accompanies the transformation:

$$C(\text{graphite}) \longrightarrow C(\text{diamond}), \Delta H = +1.9 \text{ kJ mol}^{-1}$$

When one mole of graphite is transformed into diamond under standard conditions (298K and 101 325 N m^{-2}); 1.9 kJ of energy are required.

Fig. 4.8 The largest diamond ever found was known as the Cullinan. It measured about 10 cm × 6.5 cm × 5 cm and weighed 621.1 g (3106 carats). This model of it is on display at the Natural History Museum, London.

MOLECULAR FOOTBALLS

For 10 000 years two forms of carbon (diamond and graphite) were known. In 1985 chemical history was made and a third form of carbon was discovered. Professor Harry Kroto and his colleagues vaporised graphite with a laser to give a stable molecule of formula C_{60}, detected by mass spectrometry. It has been suggested, however, that such a molecule is present in soot but had, remarkably, remained undetected. Kroto called the molecule 'buckminsterfullerene' after the architect Buckminster Fuller who designed the Geodesic Dome for the US pavilion at Expo 67 in Montreal. Molecules of C_{60} are popularly known as 'bucky balls'. The proposed structure, a combination of hexagons and pentagons, is similar to a modern football, hence the alternative trivial names 'footballene' and 'soccerene'.

The discovery has opened a whole new field of synthetic chemistry which promises a host of valuable compounds. Engineering on the molecular scale seems possible – but we can only speculate and anticipate. The world of chemistry has been more excited by the discovery of C_{60} and related molecular footballs than, perhaps, any other discovery of the last few decades.

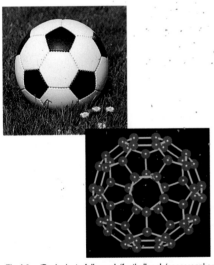

Fig 4.9 'Buckminsterfullerene', 'footballene', 'soccerene' or 'bucky ball'. Whatever you call it, C_{60} is a remarkable molecule.

Another example of a compound with a giant three-dimensional lattice, with atoms held by single covalent bonds, is the most common mineral in the Earth's crust, quartz (silicon(IV) oxide, SiO_2). Its structure is shown in Fig 4.10. As with diamond, the strong covalent bonds linking the three-dimensional network of atoms (average bond energy for Si—O is 374 kJ mol^{-1}) account for the hardness and very high melting point of quartz.

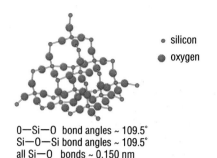

O—Si—O bond angles ~ 109.5°
Si—O—Si bond angles ~ 109.5°
all Si—O bonds ~ 0.150 nm

Fig 4.10 The structure of silicon(IV) oxide.

The sheet silicates, such as talc and muscovite, possess giant two-dimensional lattices similar to graphite but based on the SiO_4 tetrahedron (Fig 4.11). Each plane of silicon and oxygen atoms carries negative charges which are balanced by cations held within the giant structure between the layers. They are therefore more complex in structure than graphite but possess similar cleavage properties; a little talc rubbed between the fingers quickly reveals its lubricating properties.

Ceramics are synthetic materials with giant covalent structures and their strength results from their three-dimensional arrangement of atoms and the strong bonding between them.

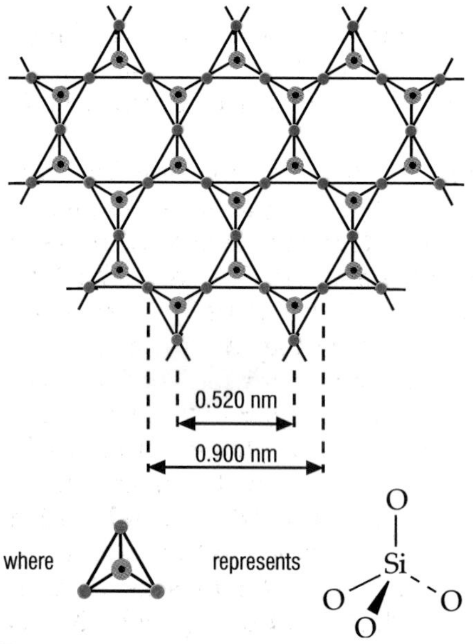

Ionic solids

Low volatility (high melting and boiling points) characterises compounds with giant structures. However, many involatile solids have properties that cannot be explained by a giant covalent structure. They conduct electricity when molten (being simultaneously decomposed), and are often soluble in water. We call them **ionic solids**. They consist of ions packed in a repeating three-dimensional lattice (ionic lattice). Amongst the minerals, halite (NaCl sodium chloride) and fluorite (CaF_2 calcium fluoride) are examples of ionic solids (Fig 4.12).

silicon
oxygen

0.520 nm
0.900 nm

where ▲ represents

$$O$$
$$\begin{array}{c} O \\ Si \\ O \quad O \\ O \end{array}$$

Fig 4.11 The structure of a typical sheet silicate.

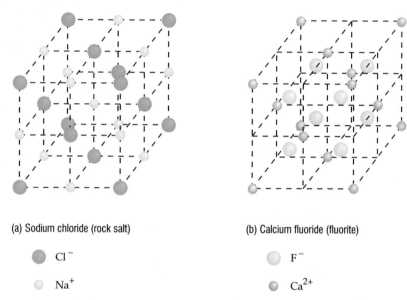

(a) Sodium chloride (rock salt)

Cl^-
Na^+

(b) Calcium fluoride (fluorite)

F^-
Ca^{2+}

The formation of ions can be described using dot and cross diagrams. In the examples below, only the outer electrons are shown for the formation of sodium, calcium, chloride and fluoride ions.

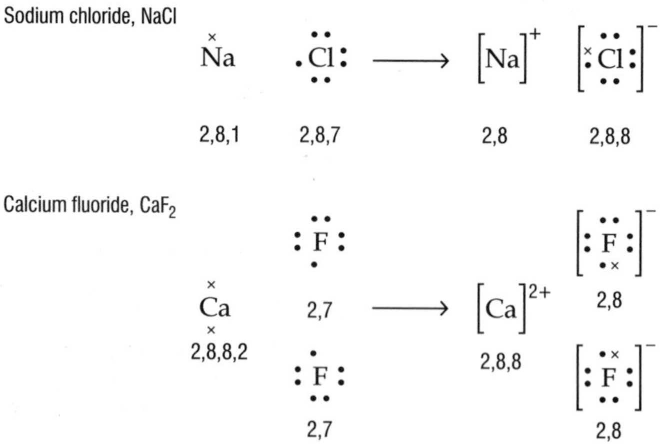

Sodium chloride, NaCl

$$\overset{\times}{Na} \qquad \cdot \overset{\cdot\cdot}{\underset{\cdot\cdot}{Cl}}{:} \longrightarrow \left[Na\right]^+ \quad \left[\overset{\cdot\cdot}{\underset{\cdot\cdot}{:Cl}}{:}\right]^-$$

2,8,1   2,8,7   2,8   2,8,8

Calcium fluoride, CaF_2

$$: \overset{\cdot\cdot}{F} : \qquad \left[: \overset{\cdot\cdot}{F} :\right]^-$$

$$\overset{\times}{\underset{\times}{Ca}} \qquad 2,7 \longrightarrow \left[Ca\right]^{2+} \qquad 2,8$$

2,8,8,2   2,8,8

$$: \overset{\cdot\cdot}{F} : \qquad \left[: \overset{\cdot\cdot}{F} :\right]^-$$

2,7   2,8

Fig 4.12 The structures of sodium chloride and calcium fluoride.

THE EARTH'S CRUST

CRYSTAL STRUCTURE

The overall shape of a crystal provides some information about the likely organisation of particles within the lattice structure. The surface of a cadmium sulphide crystal may be etched with dilute hydrochloric acid solution to reveal six-sided pits under a microscope. Whilst individual ions are too small to be visible the photograph clearly supports the model of a hexagonal crystal lattice for this compound.

Fig 4.13

Positive ions (cations) are formed when an atom loses electrons and negative ions (anions) when an atom gains electrons. Ions of opposite charge are held in the ionic lattice by electrostatic attraction. This is called **ionic** (or **electrovalent) bonding**.

The volatility of an ionic solid depends on the strength of the ionic lattice. The strongest bonding will be exhibited by compounds composed of small, highly charged ions, since the electrostatic attractive force between two ions of charges $z+$ and $z-$, at a distance r apart, is given by $(z+)(z-)/r^2$. The strength of bonding in an ionic solid is measured by its lattice energy.

Anions are larger than the atoms from which they are formed, while cations are smaller than the atoms from which they are formed (Fig 4.14). This can be understood by thinking about the arrangement of electrons in an atom and the nature of the attractive and repulsive forces between sub-atomic particles. Consider the sodium atom. There are 11 protons in the nucleus and 11 electrons arranged in energy levels (2,8,1). A sodium ion forms when the outer electron is lost. The remaining electrons are in energy levels in which they are, on average, closer to the nucleus. Further, the 11+ nucleus is now attracting just 10 electrons. Consequently, Na$^+$ is smaller than Na. In contrast, the chlorine ion (Cl$^-$ 2,8,8) is formed when a chlorine atom gains an electron. The additional electron occupies the same energy level as the outer electrons in a chlorine atom. Repulsion between electrons and a slightly weaker attraction of the 17+ nucleus for the 18 electrons of the chloride ion results in Cl$^-$ being larger than Cl.

	+2	+1	0	–1	–2
Ca	○		◯		
Na		○	◯		
Cl				○	◯
O			○		◯

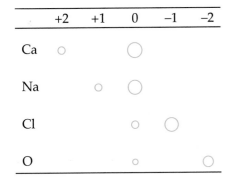

Fig 4.14 The relative sizes of some atoms, cations and anions.

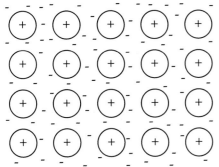

Fig 4.15 Metallic bonding.

8   **Does sodium oxide have a higher lattice energy than magnesium chloride?**

9   **Why would you expect O^{2-} to be larger than F$^-$?**

Metals

In covalent bonds, electrons are shared between a pair of nuclei. We often talk about 'discrete' covalent bonds, bonds in which the bonding electrons are shared between two specific nuclei. In ionic bonding, we imagine all electrons to be associated with one or other of the ions involved. Electrons are said to be **localised** in both types of bonding. The bonding in metals is different.

Metallic bonding can be pictured as positively charged ions embedded in a 'sea' of electrons (Fig 4.15). Sodium will serve to illustrate the idea. The outer electron in each sodium atom is not associated with any one atom or pair of atoms. Instead it is delocalised throughout the structure and free to move under the influence of an electrical field (rather like some of the electrons in graphite). These delocalised electrons act as electrostatic 'glue'.

10   The first three elements in Period 3 are sodium, magnesium and aluminium. What do their respective densities of 0.97 g cm^{-3}, 1.74 g cm^{-3} and 2.70 g cm^{-3} suggest about the way in which atoms pack in these three metals?

QUESTIONS

4.12 Explain what is meant by the term crystal lattice. List the four types of crystal lattice.

4.13 What is an amorphous solid?

4.14 Diamond and graphite are said to be allotropes. Explain what is meant by this term.

4.15 In an examination a student wrote 'the bonding in iodine is weak'. Explain why this answer is misleading and how it may be improved.

4.16 What causes intermolecular bonds and what effect do they have on the melting and boiling points of a substance?

4.17 Graphite is often used for the brushes in electric motors but, being soft, they wear out. Apart from their cost, give two reasons why diamonds would be unsuitable for this purpose.

4.18 Silicon(IV) oxide, SiO_2, is a giant covalent solid with a similar lattice structure to that of diamond. In this structure, however, each silicon atom is covalently bonded to four oxygen atoms. With the aid of a simple sketch, describe the lattice structure of silicon(IV) oxide and explain why its formula is **not** SiO_4.

4.19 Typically, ionic solids have high melting points and boiling points. They are non-conductors in the solid state but conduct readily when molten or when in aqueous solution. How does the model of bonding in an ionic compound explain these observations?

4.20 Place the following ionic compounds in order of increasing lattice energy: K_2O, MgO, CaO, NaF, KF, MgF_2, CaF_2, $NaCl$, KCl, $CaCl_2$.

4.21 Unlike ionic compounds, metals conduct electricity in the solid state. Explain why this is so.

4.22 Draw dot-and-cross diagrams to represent the bonding of each (**a**) sulphur atom in an S_8 molecule; (**b**) phosphorus atom in a P_4 molecule.

4.23 Silicon has the same structure as diamond. Explain why each silicon atom throughout the three-dimensional lattice is surrounded tetrahedrally by four silicon atoms.

4.24 Silicon carbide, SiC, is an extremely hard solid, with a high melting point. It is chemically very stable. Suggest, with reasons, a structure for silicon carbide.

4.25 Draw dot-and-cross diagrams to represent the bonding in (**a**) calcium oxide, CaO; (**b**) magnesium fluoride, MgF_2; (**c**) sodium sulphide, Na_2S.

4.26 The halides of Group I elements have ionic structures. Describe and explain trends in their melting points (Table 4.5).

Table 4.5 Melting points of Group I halides / °C

LiF	867	NaF	988	KF	847
LiCl	607	NaCl	801	KCl	767
LiBr	547	NaBr	747	KBr	727
LiI	447	NaI	647	KI	677

4.27 The Group I elements are metals and each adopts the same arrangement of atoms in a three-dimensional lattice. Explain the trends in the following melting point data: Li, 180°C; Na, 98°C; K, 64°C; Rb, 39°C; Cs, 29°C.

ALUMINIUM FOR VERSATILITY

Despite the fact that it is the most abundant metal and the third most abundant element in the Earth's crust, aluminium has been manufactured commercially only for about 100 years. In this time it has become one of the most versatile metals known. Aluminium is ductile and can be drawn into a wire so that a strand weighing 200 kilograms could encircle the Earth. Aluminium alloys may be as strong as steel and this makes such alloys suitable for aircraft construction. In the home aluminium foil is used for wrapping food whilst aluminium cooking equipment takes advantage of aluminium's excellent thermal conductivity.

The high cost of producing aluminium makes it worthwhile to recycle it. Purification and remoulding is much cheaper than producing more aluminium from the ore, bauxite.

Fig 4.16a Aluminium is used in the construction of electricity cables

Fig 4.16b

Mrs Therese Daubenthaler collecting aluminium drinks cans in Cape Coral, USA. When a plastic bag burst, a deputy sheriff tried to arrest her for littering. "Listen", she said, "instead you should give me an award for cleaning up". The officer helped her pick up the cans.

ACTIVITY

Diamond cutting is an art. The beautiful, sparkling gemstones are produced by cleaving crystals at particular angles.

Construct a molecular model of diamond. By careful inspection, decide at what angles diamond crystals need to be cleaved to produce perfect faces – faces corresponding to layers of carbon atoms.

4.4 MEASURING AMOUNTS OF SOLIDS

Empirical formula

Every compound has a characteristic empirical formula. For example, the mineral bournonite has the empirical formula $PbCuSbS_3$ and a synthetic ceramic superconductor has the empirical formula $YBa_2Cu_3O_7$.

The empirical formula of a compound can be determined by chemical analysis. Imagine an unknown substance has been brought up from a mine or perhaps returned to Earth from space. The first question to answer is whether or not it is pure. If two random samples have the same chemical and physical properties and the same elemental composition, the sample is homogenous. If separation techniques indicate only one substance is present we can then assume it is a pure substance. Once this has been established quantitative analysis of the elements present will enable the empirical formula to be determined.

A sample of the compound is analysed and the masses of each element present determined. The masses are converted into amounts of elements. Now, one mole of any element contains the same number of atoms as one mole of any other element. Therefore, the mole ratio in which atoms combine to form the compound is the ratio in which the atoms combine. This gives the empirical formula.

Problem: The element titanium was discovered in 1791; it was found in the mineral ilmenite. On analysis 0.6068 g of ilmenite is found to contain 0.2232 g iron and 0.1916 g titanium, the remainder of the mineral being oxygen. Calculate the empirical formula of ilmenite, given the following molar masses:

Fe = 55.8 g mol^{-1}; Ti = 47.9 g mol^{-1}; O = 16 g mol^{-1}.

Solution: 0.6068 g ilmenite contains

	simple whole number ratio
$\dfrac{0.2232}{55.8}$ mol iron atoms = 0.0040 mol Fe	1
$\dfrac{0.1916}{47.9}$ mol titanium atoms = 0.0040 mol Ti	1
$\dfrac{[0.6068 - (0.2232 + 0.1916)]}{16}$ mol oxygen atoms = 0.01200 mol O	3

Therefore, empirical formula of ilmenite is $FeTiO_3$.

We can also calculate the mass of an element present in a given mass of compound. For example, the mass of a metal that can be obtained from a known mass of its ore can be calculated.

Problem: Aluminium is obtained by electrolysis of molten aluminium oxide, Al_2O_3. Calculate the mass of aluminium that could be obtained from 500 kg of bauxite, given the following molar masses: Al = 27 g mol^{-1}; O = 16 g mol^{-1}.

Solution: The molar mass of $Al_2O_3 = [(2 \times 27) + (3 \times 16)] = 102$ g mol^{-1}

Fraction by mass of aluminium $= \dfrac{2 \times 27}{102} = 0.53$

Therefore, mass of aluminium in 500 kg bauxite = 500 × 0.53 kg = 265 kg.

In order to take our use of formulae further we need to distinguish between compounds with molecular structures and compounds with giant structures.

Molecular formula and structural formula

Molecular formula

In compounds with molecular structures, the molecular formula gives the number of atoms of each constituent element present in one molecule of the compound. For example, the empirical formula of glucose is CH_2O but the molecular formula is $C_6H_{12}O_6$. Naphthalene, an organic compound used as the starting material in a number of industrial processes, has the empirical formula C_5H_4 and the molecular formula $C_{10}H_8$. The molecular formula does not, however, tell us how the atoms are arranged in the molecule (Figure 4.17).

C₂H₅OH (ethanol)

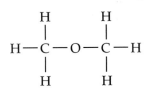

CH_3OCH_3 (methoxymethane)

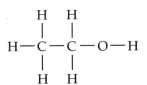

Fig 4.17 Both compounds have the same molecular formula, C_2H_6O, but they have different structural formulae.

Fig 4.18 The pile of sulphur on the left has the same number of sulphur atoms as there are sulphur molecules (S_8) in the pile of sulphur on the right.

CONNECTION: Use of relative atomic masses, pages 20 and 30.

A number of elements exist as molecules (Table 4.6) and these show why the basic unit must be specified when talking about amount of substance. For example:

1 mol of sulphur atoms has mass 32 g but 1 mol of sulphur molecules (S_8) has mass (8×32) g = 256 g.

1 mol of phosphorus atoms has mass 31 g but 1 mol of phosphorus molecules (P_4) has mass (4×31) g = 124 g.

Of course, the concept of molecular formula is not confined to molecular solids. It is used with all molecular compounds, whether, gas, liquid or solid.

Table 4.6 Molecular formulae of some elements

$N_2(g)$	$O_2(g)$	$F_2(g)$
$P_4(s)$	$S_8(s)$	$Cl_2(g)$
		$Br_2(l)$
		$I_2(s)$

Molar mass and formula unit

Compounds with giant structures are characterised by their **formula unit**. This represents the building block of the compound and its chemical composition is given by the empirical formula. Two examples will illustrate the idea. The formula unit of silicon(IV) oxide is SiO_2. As we have already described, it has a giant three-dimensional structure in which each silicon is surrounded by four oxygen atoms and each oxygen atom bridges between two silicon atoms, giving a ratio of silicon to oxygen atoms of 1:2. Sodium chloride is a giant structure of sodium ions and chloride ions; these are in a ratio of 1:1 and the formula unit for sodium chloride is NaCl.

We can measure amounts of compounds with giant structures by using the formula unit. The relative mass of the formula unit can be calculated by adding the relative atomic masses of the atoms present. The molar mass has the same numerical value; it is the mass of one mole of the compound. For example:

one mole SiO_2 has a mass $(28.1 + [2 \times 16])$ g = 60.1 g;

molar mass of SiO_2 = 60.1 g mol⁻¹;

one mole NaCl has a mass $(23 + 35.5)$ g = 58.5 g;

molar mass of NaCl = 58.5 g mol⁻¹.

Structural formula

A structural formula gives some indication of how the atoms are arranged in a particular molecule. The concept of structural formulae is not used for compounds with giant structures, but this does not mean that we do not know about their structures or, indeed, that we are not interested in them. Far from it, as we shall see later.

4.28 Plants need magnesium to make chlorophyll, a green pigment that traps the Sun's energy. The molecular formula of chlorophyll is $C_{55}H_{72}MgN_4O_5$. Calculate the percentage of magnesium present by mass.

4.29 Experiment shows that glucose has a relative molecular mass of 180. Elemental analysis of glucose gives an empirical formula of CH_2O. Calculate the molecular formula of glucose.

4.30 In the solid state phosphorus pentachloride consists of equal numbers of PCl_4^+ ions and PCl_6^- ions. What percentage by mass of a sample of phosphorus pentachloride are the PCl_4^+ ions?

4.31 Calculate the mass of **(a)** 0.02 mol S_8 molecules; **(b)** 0.005 mol P_4 molecules; **(c)** 0.15 mol SiO_2; **(d)** 0.1 mol $CaWO_4$; **(e)** 0.03 mol $CuSO_4.5H_2O$.

4.32 Celestite is a mineral containing strontium. On analysis, 0.3672 g of celestite were found to contain 0.1752 g strontium, 0.0640 g sulphur and 0.1280 g oxygen. Determine the empirical formula of celestite and suggest to what class of minerals it belongs (Table 4.1).

4.33 A 50.2 g sample of enstatite, a silicate mineral, contains 12.15 g magnesium and 14.04 g silicon. The remainder is oxygen. Determine the empirical formula of enstatite.

4.34 The major source of phosphorus is phosphate rock, or more correctly the mineral fluorapatite, $3Ca_3(PO_4)_2.CaF_2$. Calculate the mass of phosphorus that might be obtained from 1000 kg of fluorapatite.

4.35 Ammonium sulphate, $(NH_4)_2SO_4$, is an important fertiliser, providing a valuable source of the essential element nitrogen. Ammonium nitrate, NH_4NO_3, is another source of nitrogen. Calculate the percentage by mass of nitrogen in each fertiliser and state which is the richest supply of nitrogen.

4.36 Calcium sulphate-2-water can be dehydrated by heating at 400°C. Calculate the mass of anhydrous calcium sulphate that would be obtained from 100 g of calcium sulphate-2-water.

Answers to Quick Questions: Chapter 4

1 Silicates
2 Plastics age or polymer age, perhaps.
3 Yes
4 Calcium tetraoxotungstate(VI)
5 Rhombic sulphur has a lower melting point than monoclinic sulphur and so the bonding between its molecules must be weaker. We would expect both allotropes to have the same boiling points because liquid sulphur has the same arrangement of sulphur molecules regardless of their origin.
6 Melting point around 60°C, boiling point around 300°C.
7 Because the delocalised electrons sandwiched between the layers of carbon atoms are relatively free to move.
8 Yes
9 Because the nuclear charge on fluorine is higher than oxygen.
10 Al atoms pack more efficiently than Mg atoms, which in turn pack more efficiently than Na atoms.

Chapter 5

THE WATERS

> ### LEARNING OBJECTIVES
>
> When you have studied this chapter you should be able to:
>
> 1. describe the occurrence of water on Earth and the roles it plays in our environment;
>
> 2. describe how water becomes contaminated and explain how it is purified before return to the environment;
>
> 3. explain the properties of the states of matter – solid, liquid, gas – and how phase changes occur;
>
> 4. explain the concept of dynamic equilibrium;
>
> 5. state that all physical and chemical changes are accompanied by energy changes;
>
> 6. describe and explain the anomalous properties of water;
>
> 7. describe the formation and importance of hydrogen bonds;
>
> 8. explain electronegativity, polar bonds and polar molecules;
>
> 9. explain the different ways in which water can behave towards other compounds;
>
> 10. explain concentration and perform calculations involving concentrations of solutions.

5.1 WATER ON EARTH

Water is the most common substance on our planet and the most widely studied. Water is the one component of the atmosphere that exists naturally in our environment in all three states of matter – solid, liquid and gas (though only about 0.001% of the total amount of water is present as a gas in the atmosphere).

Water has unusual physical properties. With a molar mass of 18 g mol^{-1}, it might be expected to have a much lower melting point and boiling point than it does.

> 1   **The density of water is about 1 g cm^{-3}. What mass of ice and snow is present on the Earth's surface?**

Water is essential to life on Earth:

- It is an essential reaction medium for biological processes. Cells are the smallest living parts of all organisms and vast numbers of reactions take place in these chemical factories, the reaction medium being water.
- It is a reactant in many biological processes, for example photosynthesis.
- It carries dissolved or suspended nutrients to cells in multicellular organisms and carries away the waste products of cell reactions.

Fig 5.1   Water has the most well known of all chemical formulae, H$_2$O. The formula was used in the national campaign advertising the privatisation of the water industry during 1989.

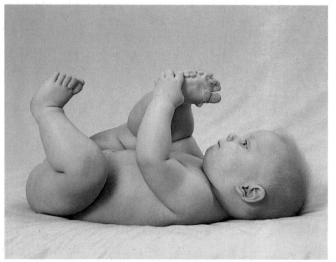

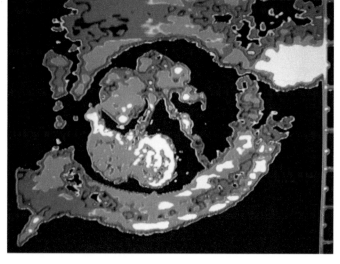

Fig 5.2 About two-thirds of the human body is water in which countless chemical reactions occur. Perhaps even more remarkable is that the human embryo in its first month is about 93 per cent water.

A further example of the importance of water to life is the role played by dissolved oxygen in supporting aquatic life. Fish, for example, take in oxygen which is dissolved in water, through their gills. Equally vital to marine life are dissolved salts resulting from weathering of rocks and outgassing (emissions from volcanoes and hot springs). The solubility of compounds in water with such diverse structures and bonding types as oxygen and sodium chloride (even though the extent of their solubilities varies enormously) shows why water has gained the title 'universal solvent'.

In addition to its involvement in biological processes, water also has two essential non-biological roles on Earth.

The reversible changes of ice into water and water into water vapour, and the accompanying energy changes, provide the Earth with an energy reservoir and one means of distributing the energy.

$$H_2O(s) \underset{\text{releases energy}}{\overset{\text{requires energy}}{\rightleftharpoons}} H_2O(l) \underset{\text{releases energy}}{\overset{\text{requires energy}}{\rightleftharpoons}} H_2O(g)$$

The Sun's radiation provides energy to vaporise water from the oceans. The water vapour which forms is transported by the wind, releasing energy when it condenses to form clouds and fall as rain (or, indeed, snow). Liquid water acts as an energy reservoir, taking energy from the Sun and redistributing it around the Earth (Fig 5.3).

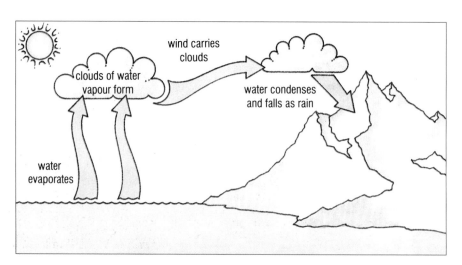

Fig 5.3 The water cycle.

The other non-biological role of water is as an environmental buffer. The oceans, freshwater lakes and rivers contain dissolved atmospheric gases. They act as both providers of these gases and as sinks for them, buffering against dramatic atmospheric changes. For example, carbon dioxide is soluble in water and the oceans contain fifty times more carbon dioxide than does the atmosphere.

2 Why are the oceans not able to buffer the greenhouse effect of methane in the atmosphere?

Water fit to drink

Most of the water on Earth is not drinkable. The oceans account for about 97% of the Earth's water but the dissolved salts make it unsuitable to drink. In fact, drinking sea water can be fatal! In countries where fresh water is in short supply because of low annual rainfall methods must be used to desalinate (remove the dissolved salts) seawater. The two main techniques are **flash distillation** and **electrodialysis**. Freezing, ion exchange and reverse osmosis are also used.

RIVER POLLUTION 1856

Water pollution is not a new phenomenon. Michael Faraday wrote to 'The Times' of 7 July 1855 concerning the River Thames. In his letter sent from the Royal Institution, he said 'The appearance and smell of the water forced themselves at once upon my attention. The whole of the river was an opaque pale brown fluid'. Faraday went on 'The smell was very bad, and common to the whole of the water; it was the same as that which comes up from the gully-holes in the streets; the whole river was for some time a real sewer'.

Sydney Smith, a British clergyman had written some 50 years earlier that 'He who drinks a tumbler of London water has literally in his stomach more animated beings than there are men, women and children on the face of the globe'.

FARADAY GIVING HIS CARD TO FATHER THAMES;
And we hope the Dirty Fellow will consult the learned Professor.

Fig 5.4 The 'Punch' cartoon which followed Faraday's letter to 'The Times'.

Modern freshwater supplies can become polluted through industrial or domestic use or agricultural treatment of the land. In most western countries, organisations such as national river authorities are responsible for monitoring this pollution and cleaning water before it is returned to domestic supplies. Water treatment takes place at sewage works (Fig 5.5); the precise methods used depend on its locality and the likely contaminants.

Fig 5.5 A sewage works. Waste water is a chemical cocktail, discharged into sewers from homes, factories and farms. There are three stages of treatment: primary, secondary and advanced. After the first two stages water is sufficiently clean to be discharged into rivers or the sea. Advanced treatment makes the water suitable for use in domestic drinking supplies.

Sewage treatment

There are three stages of water treatment employed at the sewage works: primary, secondary and advanced.

Primary treatment involves preliminary screening to remove large solids – anything from wooden fencing to toilet paper. This is usually done by mechanical raking. The water is passed to a settlement tank where a very smelly sludge forms.

Secondary treatment. The liquid is drawn off the sludge and pumped into percolating or trickle filters. Bacteria in the filters break down organic material, a process helped by pumping air through the system. Again a sludge forms.

The sludge from primary and secondary treatments may be transferred to a digester tank. Here it is converted into biogas by bacteria which work in anaerobic conditions. Biogas is a mixture of methane and hydrogen which may simply escape into the atmosphere or be used as a fuel. The sludge can be used as a fertiliser. However, if the sludge contains high levels of toxic chemical substances it must be disposed of in some other way, for example in landfill sites.

After secondary treatment, some 10–12 hours after it entered the sewage works, the water may be considered fit to put back into the Earth's natural waters.

Advanced treatment. After secondary treatment water may still contain levels of chemical substances which render it unfit to enter the water system. Nitrates and phosphates, the so-called heavy metals such as lead and mercury (present as soluble salts), and organic compounds not decomposed by the bacteria used in the sewage works, are all difficult to remove.

Advanced treatment removes these poisonous substances, or at least reduces their concentrations to safe levels, and produces water which may be recycled through domestic water supplies or enter rivers and the sea.

QUESTIONS

5.1 (a) Summarise the role of water in supporting life on Earth.
(b) What characteristics of water make it a suitable 'universal solvent'?

5.2 The density of liquid water is about 1 g cm^{-3}. For many practical purposes it is acceptable to assume that aqueous solutions have the same density as water. Why is this assumption a reasonable one?

5.3 Some people believe that the 'greenhouse effect' may 'level off' as carbon dioxide dissolves in the oceans. What information would you need to assess the likelihood of this?

5.4 Using a flow diagram explain how household waste water, which contains sewage, is converted to drinkable water.

5.5 Advanced treatment of water is expensive. What steps could be taken to reduce the cost of water purification?

5.6 Explain why nitrates and phosphates remain in water at the sewage works even after primary and secondary treatment.

5.2 CHANGES OF STATE

States of matter

Water exists in the Earth's atmosphere and on its surface as solid, liquid and gas. These are **states of matter** (also called **phases**). Before we consider the properties of water in detail, we will consider states of matter more broadly. The state in which water exists, just like all other compounds, depends on the temperature and pressure (Fig 5.6).

Perhaps a surprising property of matter is its ability to change state without altering its chemical behaviour. Ice, water and water vapour are chemically identical. The particles present in all three states are water molecules. It is the type and strength of bonding between atoms in molecules which largely determine chemical properties. On the other hand, physical properties depend on the arrangement of water molecules with respect to one another, the strength of bonding between one molecule and another, and the ease with which they can move.

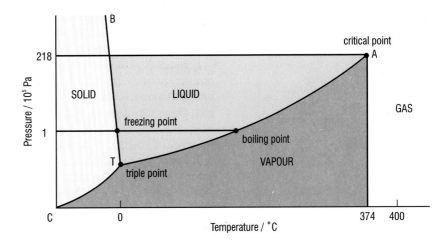

Fig 5.6 Whether water exists as gas, liquid or solid depends on the temperature and pressure. These conditions may be summarised on a **phase diagram**.

The physical properties of all substances depend on the nature, arrangement and freedom of movement of the particles present. It is important, therefore, that scientists and technologists have a good model of matter in its different states. Manufacturing industries, for example, exploit the properties of matter. The ability to control these properties enables improved products to be designed and made – perhaps the bodywork for a high performance car, sports equipment (such as a tennis racket) or a replacement heart valve.

Table 5.1 The characteristics of gases, liquids and solids

	Gases	Liquids	Solids
volume	variable, they expand to occupy the whole of the volume of the container	fixed	fixed
shape	variable, they take the shape of the container	variable, they take the shape of the part of the container they occupy	fixed
relative density	low	high	high
relative effect of pressure	easily compressed (volume decreases)	can be compressed very slightly	cannot be compressed

3 **When water evaporates, 1 cm³ of liquid gives 1 dm³ of vapour. How far apart are water molecules in the vapour phase compared with the liquid phase? Explain the relative effect of pressure on water vapour and liquid water.**

Changing state

Matter can be changed from one state to another. This is called a phase change. The model of matter must be able to explain these changes of state also.

CONNECTION: The arrangement of particles in gases and solids, pages 51 and 62.

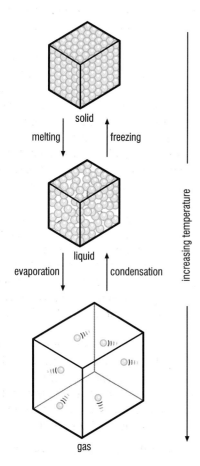

Fig 5.7 Modelling changes of state

Gas particles are in constant, chaotic motion. When a gas is cooled, particles will have, on average, less kinetic energy and move more slowly. The particles begin to attract one another significantly and this begins to restrict their movement. The process of **condensation** begins, a phase change occurs as the gas becomes a liquid. The reverse phase change is called **evaporation**.

In the liquid state, particles are still moving but less rapidly and randomly than in a gas. The attractions between them loosely bind the particles together. Although the liquid as a whole has a fairly random structure, over short distances and for brief periods of time, particles begin to arrange themselves in a slightly ordered way. The particles in a liquid show **short-range order**.

If the liquid is cooled, particles continue to lose kinetic energy as the temperature is lowered. They move even more slowly. The attraction between particles becomes much more difficult to overcome and, eventually, they adopt fixed, regular, three-dimensional arrangements characteristic of solids (Fig 5.7). This phase change is known as **freezing** and the reverse, of course, is **melting**. Particles in a solid show **long-range order**. The particles in a solid are greatly restricted in their movement and simply vibrate about fixed positions. The extent of vibration depends on the temperature of the solid. Theoretically, if a solid were cooled to absolute zero (0K or –273.16°C) the particles would stop vibrating completely.

4 **Explain whether it would be possible to determine that absolute zero had been reached by measurement?**

Dynamic equilibrium

Changes of state are reversible. Water freezes to ice, ice melts to form water. Water boils to form water vapour, which condenses to form water.

Water left in an open beaker slowly evaporates until none remains. However, what happens if the water is placed in a sealed, empty container? Initially water molecules escape from the surface of the liquid into the gas phase. As molecules accumulate in the gas phase they collide with

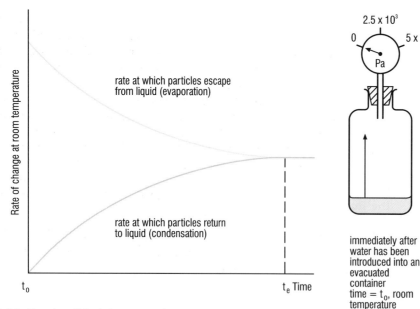

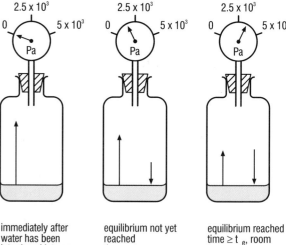

Fig 5.8 Dynamic equilibrium between water and water vapour.

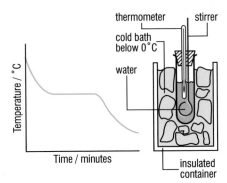

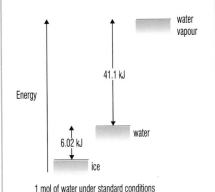

Fig 5.9 A cooling curve for water. In the experiment, water at 60°C is placed in a test-tube fitted with a stopper, a thermometer and a means of stirring during cooling. The test-tube is placed in an ice/sodium nitrate bath (which is well below 0°C). The temperature drops as the water cools to the temperature of the cooling bath. At 0°C the cooling curve flattens out. This is because bonds form as the liquid water solidifies and energy is released in the form of heat. This heat is gained by the water and so the temperature does not drop until all the water has solidified. After this the ice continues to cool until the temperature of the cooling bath is reached.

KEY FACT

Changes which occur with the release of energy in the form of heat are called **exothermic**.

Changes which occur with the absorption of energy in the form of heat are called **endothermic**.

Energy changes accompany both physical and chemical changes and may be represented by an energy diagram. When one mole of water vapour condenses to form liquid water, 41.1 kJ of energy are released. On forming ice the one mole of water releases a further 6.02 kJ of energy in the form of heat. The energy diagram for these phase changes is shown in Fig 5.10.

Fig 5.10 Energy diagram for the phase changes of water.

one another, some gaining energy and others losing it. Molecules with low kinetic energy are unable to overcome the attractive forces of neighbouring molecules and so the gas begins to condense. The water molecules return to the liquid phase. When the rate at which water molecules escape from the liquid equals the rate at which they return, a position of **dynamic equilibrium** is reached. When dynamic equilibrium has been reached the concentration of water molecules in the gas phase is constant. If we had some way of identifying the molecules, however, we would see that they were not necessarily the same molecules. The exchange of water molecules between gas and liquid is continuous. The establishment of equilibrium is shown in Fig 5.8.

Our sealed container is a **closed system**. Neither matter nor energy can enter or leave. Dynamic equilibrium can only be attained in closed systems. A reaction mixture in a highly efficient, stoppered thermos flask is perhaps the closest to a simple closed system possible in the laboratory. Clearly, matter cannot enter or leave the flask since it is sealed. If the flask has a good vacuum and is well silvered, the amount of heat that can enter or leave is minimal.

The pressure exerted by a gas in dynamic equilibrium with a liquid is called the **vapour pressure**. Vapour pressure increases as temperature increases. Molecules gain kinetic energy as the temperature rises, and therefore more of them are able to overcome the attractive intermolecular forces more easily. The temperature at which vapour pressure equals atmospheric pressure is the **boiling point** of the liquid. If we look at the phase diagram for water (Fig 5.6) the variation of boiling point with pressure is given by the line TA.

The solid lines on a phase diagram give those conditions of temperature and pressure, for a closed system, at which two phases would be in dynamic equilibrium. In the phase diagram for water, for example, line TA represents liquid/gas, line TB represents solid/liquid (and therefore shows how melting point varies with pressure), and line TC represents gas/solid. The phase change from solid to gas is called sublimation.

5   What effect does pressure have on the sublimation temperature of ice?

Energy considerations

Phase changes are always accompanied by energy changes. When matter changes from gas to liquid or liquid to solid, energy is released and gained by the surroundings. When matter changes from liquid to gas or from solid to liquid, energy is required and is taken from the surroundings. The energy is usually in the form of heat.

Energy (usually in the form of heat) is required, for example, to change ice to water and water to water vapour, because in each phase change the forces between molecules must be overcome and intermolecular bonds broken. Energy is released due to bond formation when the reverse changes occur – water vapour condenses and ice melts (Fig 5.9). These energy changes characterise the phase changes for all compounds, but those which accompany the phase changes of water for a given mass are higher than almost any other substance.

6   Sketch the cooling curve you would expect when steam at 150°C and above atmospheric pressure is cooled to room temperature without a change in pressure.

5.7 Explain what is meant by the term *phase change*.

5.8 A test tube is half filled with water and corked. Explain why the water present in the test tube may, after a time, be described as 'in dynamic equilibrium'.

5.9 Use Fig 5.6 to:
(a) explain what happens to the boiling point of water as pressure is increased;
(b) predict what happens to a sample of air saturated with water vapour as it is cooled at constant pressure.

5.10 When ammonium chloride is heated strongly in a test tube using a Bunsen burner the solid in the bottom of the test tube gradually disappears. Near the mouth of the tube a white deposit collects which on analysis proves to be ammonium chloride. Explain these observations and discuss the energy changes involved.

5.11 If 1 kg of water vapour is converted to ice under standard conditions, calculate the total amount of energy released.
$A_r[H] = 1$, $A_r[O] = 16$,
$H_2O(g) \longrightarrow H_2O(l)$ $\Delta H = +41.1$ kJ mol^{-1}
$H_2O(l) \longrightarrow H_2O(s)$ $\Delta H = +6.02$ kJ mol^{-1}

5.12 State whether water would be solid, liquid or gas under the following conditions (a) 300°C, 500 Pa; (b) –50°C, 150 000 Pa.

5.13 Why do the energy changes which accompany phase changes provide evidence that the ideal gas equation is only an approximation?

5.14 Heat is released when a gas condenses to form a liquid. The energy changes which accompany the condensation of one mole of water, ethanol (C_2H_5OH) and tetrachloromethane (CCl_4) are –41.1, –43.5 and –30.5 kJ respectively.
(a) Calculate the energy released per gram of each of these compounds.
(b) How much energy is required to change 100 g of water to water vapour at 100°C?

5.3 BONDING WITHIN AND BETWEEN MOLECULES

Bonding within the water molecule

Modern technology has enabled precise measurements of the size and shape of water molecules to be made. In the vapour phase, water is a non-linear (or 'bent') molecule, with 0.095718 nm oxygen-hydrogen bonds and a bond angle of 104.52°.

The oxygen and hydrogen atoms are held together by single covalent bonds in a water molecule. One electron from each atom in the bond is shared. However, the model can be refined. In a perfect covalent bond the electrons are shared equally (Fig 5.11). The centres of positive and negative charge coincide. Perfect covalent bonds only exist in homonuclear molecules such as H_2, Cl_2, P_4 and S_8, since the two atoms in the bond must attract the shared electrons equally. In all other cases one of the two atoms held by the bond has a stronger attraction for the shared electrons than the other. This atom is said to be more electronegative than the other.

Electronegativity

Linus Pauling introduced the term electronegativity in 1932:
'Electronegativity is the power of an atom *in a molecule* to attract electrons to itself.'

It is a qualitative term expressing the relative tendency of an atom to attract a bonding pair of electrons. Strictly speaking, of course, it is the nuclei of atoms held by a covalent bond which attract the shared electrons. Different scales of electronegativity have been proposed by Pauling, R.S. Mulliken, A.L. Allred and E. Rochow, and R.T. Sanderson. The most commonly used is Pauling's, an arbitrary scale from 0 to 4 (4 being most electronegative). The most electronegative element is fluorine, the least electronegative is francium.

Polar molecules

In a water molecule, oxygen has a greater attraction for the shared electrons than hydrogen. The centre of negative charge for each O—H bond is closer to the oxygen atom and each bond is polarised (Fig 5.12). The polarisation of the O—H bonds in the water molecule has a profound effect on the physical properties of water. It also helps to account for the chemical reactivity of water.

The centres of positive and negative charge for the molecule itself do not coincide either. The water molecule is polar and possesses a dipole moment (a measure of the molecule's polarisation). However, molecules that have polar bonds are not necessarily polar themselves. This can be shown by an inspection of the compounds CH_4, CH_3Cl, CH_2Cl_2, $CHCl_3$ and CCl_4 (Fig 5.13). These compounds are a series based on methane, in which the hydrogen atoms in CH_4 are successively replaced by chlorine atoms.

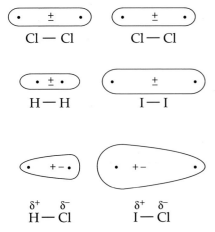

Fig 5.11 In a perfect covalent bond the centres of positive and negative charge coincide. However, in all bonds between unlike atoms, one atom has a stronger attraction for the shared electrons than the other. No longer do the centres of positive and negative charge coincide – the bond is polar.

Fig 5.12 Polarisation of bonds in the water molecule.

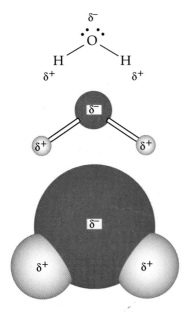

Name	methane	chloro – methane	dichloro – methane	trichloro – methane	Tectrachloro – methane
Molecular formula	CH_4	CH_3Cl	CH_2Cl_2	$CHCl_3$	CCl_4
Dipole moment μ / D	0	1.87	1.54	1.02	0
Shape and bond polarisation					
Direction of dipole					

We see that all the C – Cl bonds are polarised but in tetrachlomethane the nett effect of this is that the centres of posit and negative charge for the molecule still coincide and the molecule possesses no dipole moment.

Dipole – dipole attraction between two chloromethane molecules:

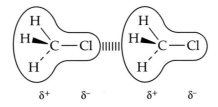

Fig 5.13 Polar bonds and polar molecules.

7 Explain the trend in the dipole moments of the halogenomethanes: CH_3Cl, 1.87 D; CH_3Br, 1.81 D; CH_3I, 1.02 D. (Note: D is the unit of dipole moment)

CONNECTION: Dispersion forces, page 63.

Van der Waals' forces

Attractive forces between molecules result from dipole interaction. This interaction is of two types. **Dispersion forces** occur between all types of molecules (non-polar and polar). The electrons in a molecule are in constant motion. At any one moment the centres of positive and negative charge can be temporarily displaced, and the resultant attraction gives rise to the dispersion forces. **Dipole–dipole** forces occur between polar molecules. Polar molecules have a permanent dipole because their centres of positive and negative charge are permanently displaced.

Together, dispersion and dipole–dipole forces are called **van der Waals' forces**. Intermolecular bonds which form as a result of van der Waals' forces between molecules are much weaker than covalent bonds within molecules.

A special case of dipole–dipole attraction exists for water and a number of other molecules. It is called **hydrogen bonding**.

> ## KEY FACT
>
> All bonding – covalent, ionic, metallic and intermolecular – is the result of electrostatic attraction between unlike charges.

Hydrogen bonding

On the basis of Van der Waals forces alone we might expect water, in view of its small molecular size, to be a gas at room temperature and pressure. There are other unusual properties to explain. For example, most liquids freeze to give a solid which is more dense, yet ice floats on water.

A study of the hydrides of Group IV shows that their volatility (the ease with which they change from liquid to gas) decreases steadily with relative molecular mass. All of these hydrides exist as non-polar, tetrahedral molecules. The molecules have only dispersion forces between them. If the melting points and boiling points of Group VI hydrides followed the same pattern, we might predict that the melting point of water should be about $-90°C$ and its boiling point about $-70°C$ (Fig 5.14). These predictions are a long way from the actual values of $0°C$ and $100°C$. But why?

The low volatility of water suggests that the intermolecular bonding is much stronger than might be expected from molecular size alone. To understand why this is the case, we must look more carefully at the polarised O — H bond. We picture the bond as two nuclei of oxygen and hydrogen atoms embedded in a lobe of shared electron density. But a hydrogen atom is unique because it has no electrons other than the one shared in its covalent bond. In contrast, for example, the oxygen atom in H_2O has two bonding pairs, two lone pairs and two electrons in a lower energy level ($n = 1$).

Oxygen attracts the shared electrons more strongly than hydrogen. As the electron density is reduced, the hydrogen nucleus (simply a proton) becomes increasingly exposed. The positively charged proton will have a powerful attraction for any available negative charge, and the lone pairs on oxygen atoms of neighbouring water molecules provide just this (Fig 5.15). The result is hydrogen bond formation (a special case of dipole–dipole attraction).

The other elements in Group VI are much less electronegative than oxygen, although all are more electronegative than hydrogen. The sulphur atom in H_2S has two bonding pairs, two lone pairs and ten electrons in lower energy levels ($n = 1$ and $n = 2$). The pull of its nucleus on the shared pair of electrons is much less than the pull of oxygen because of this shielding, despite the fact that the nucleus carries a greater charge. Consequently, the S — H bond is less polar and hydrogen bonding does not occur.

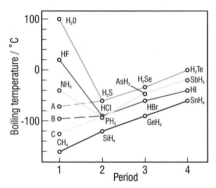

A, B, C, are values predicted for H_2O, HF and NH_3 respectively if no hydrogen bonding occurred

Fig 5.14 The melting points and boiling points of the hydrides of elements in Groups IV, V, VI and VII (extrapolated to give expected values for water if it is assumed that only Van der Waals forces are acting).

Fig 5.15 Hydrogen bonds between water molecules.

8 What forces of attraction are present between molecules of hydrogen sulphide, H_2S?

THE WATERS

KEY FACT

An electron is attracted to the nucleus by electrostatic forces. The higher the energy level in which the electron resides, the weaker is the attraction. There are two reasons:

- it is further from the nucleus on average;
- it is **shielded** from the nucleus by electrons in lower energy levels which are closer to the nucleus.

KEY FACT

For a hydrogen bond to form the following requirements must be met:

- a hydrogen atom must be covalently bonded to a highly electronegative atom, usually oxygen, nitrogen or fluorine;
- the highly electronegative atom must have a lone pair of electrons.

The hydrogen bond is a strong intermolecular bond, some ten times stronger than the other types of bonds between one molecule and another. It also differs from other bonds in that it is directional. For example, the O—H······O linkage is linear. Evidence for this comes from the crystal structure of ice (Fig 5.16). The open structure of ice collapses upon melting and water molecules are packed rather more closely in the liquid state. Ice is slightly less dense than liquid water and this explains why ice floats on water.

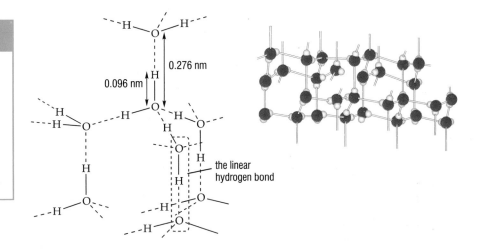

Fig 5.16   The structure of ice: hydrogen bonds hold water molecules in a giant scaffold-like structure. The structure contains cavities which may be occupied by small molecules such as methane. These compounds are called clathrates.

QUESTIONS

5.15 (a) Draw a sketch of the water molecule, showing bonding pairs of electrons as lines and non-bonding pairs as electron clouds. Include the bond lengths and H—O—H bond angle on your sketch.

 (b) Why are the bond angles not exactly those of a regular tetrahedron (you may need to consult section 3.4).

5.16 'Life could not exist on Earth if water molecules did not undergo hydrogen bonding.' What notable effects would the absence of hydrogen bonding have on Earth and how important would this be for life?

5.17 If a positively charged rod is brought close to a stream of trichloromethane running out of a burette, the stream is visibly deflected towards the rod. Explain with the aid of a simple diagram why this deflection takes place.

5.18 At the top of Ben Nevis water boils at 95°C, on Mont Blanc at 83°C and on Mount Everest at 74°C. Explain this trend. Although the technique is not used, how could the height of a mountain be estimated from measurements of the boiling point of water or some other liquid?

5.19 Explain why glass bottles containing frozen water are often found to be cracked and leaking when the water melts.

5.20 Explain why water shows greater anomalous behaviour than the other hydrides of Group VI.

5.21 Hair is made of the protein, keratin. Like all proteins, the basic building blocks (amino acids) are held together in a long molecular chain by peptide links (Fig 5.17). The molecules are held together by electrostatic attraction between ionic groups in the chain and, importantly, by hydrogen bonding. Explain why wet hair loses its shape and temporary waving can be achieved by wetting the hair, shaping it and allowing it to dry.

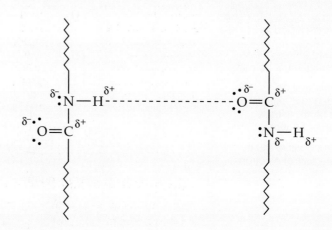

Fig 5.17 Hydrogen bonds between keratin molecules in hair.

ACTIVITY

Icebergs are a spectacular result of hydrogen bonding. Some 5000 form each year from the glaciers and ice shelves of Antarctica. Their total mass is about 10^{15} kg.

The largest iceberg recorded was larger in area than Belgium – it measured 335 km by 97 km, a surface area in excess of 31 000 km². The iceberg was sighted in 1956 in the South Pacific Ocean.

Try to estimate the mass of this iceberg. An initial experiment with an ice cube in water might be helpful, together with the approximation that the density of ice is about 1 g cm⁻³.

SPOTLIGHT

WATER AND SURFACE TENSION

The strong hydrogen bonding between water molecules is responsible for the high **surface tension** of liquid water. This is a force which acts in parallel to the surface of the water and tends to reduce the surface to a minimum. **Surface active agents or surfactants**, such as detergents, lower the surface tension of water by disrupting the hydrogen bonds between water molecules. The formation of foam, a dispersion of a gas in a liquid, may take place at the surface when a surfactant is present.

Fig 5.18 Foam is produced when the surface tension of water is reduced by a surfactant. Here the surfactant is a detergent. Some surfactants produce a persistent foam which is difficult to disperse. Modern biodegradable surfactants are decomposed by microorganisms to harmless products.

5.4 WATER – SOLVENT OR REACTANT?

The composition of sea water has remained essentially constant over a considerable span of geological time. The salts dissolved in it are vital to marine life and also provide us with an abundance of resources.

10 **What is the difference in the relative abundances of sodium and potassium in the Earth's crust and in sea water (see Table 1.2 and 1.3)?**

The major constituents of sea water are ionic compounds (Table 5.2) and we have seen that a characteristic property of most ionic compounds is their solubility in water. But what do we mean by dissolving, and do all compounds that are soluble in water dissolve, or do some react with the water?

Table 5.2 The major constituents in sea water

Compound	Formula	Ions present in aqueous solution
sodium chloride	$NaCl$	$Na^+(aq)$ and $Cl^-(aq)$
magnesium chloride	$MgCl_2$	$Mg^{2+}(aq)$ and $Cl^-(aq)$
magnesium sulphate	$MgSO_4$	$Mg^{2+}(aq)$ and $SO_4^{2-}(aq)$
calcium sulphate	$CaSO_4$	$Ca^{2+}(aq)$ and $SO_4^{2-}(aq)$
potassium chloride	KCl	$K^+(aq)$ and $Cl^-(aq)$
calcium carbonate	$CaCO_3$	$Ca^{2+}(aq)$ and $CO_3^{2-}(aq)$
potassium bromide	KBr	$K^+(aq)$ and $Br^-(aq)$

SPOTLIGHT

SEA, SALT AND SEAWEEDS

The oceans contain an immense reservoir of sodium chloride (and many other salts). In many parts of the world sodium chloride is extracted commercially from the sea. In some countries there are cheaper sources of salt. In Britain, for example, the major source of sodium chloride is the Cheshire salt field. Large as it is, this source is small compared to the world's oceans. It has been calculated that if the sodium chloride contained in the oceans were isolated, dried and compressed in the form of rock salt it would occupy 19 million km^3. This is enough to cover the whole of England, Scotland and Wales to a depth of about 80 km. Ben Nevis, the highest point in the British Isles, is only 1.33 km high!

The principal source of bromine is sea water, which contains about 65 ppm of bromine, in the form of bromide ions. Annually around 3 million kg are extracted. Although the concentration of iodine in sea water is too low for economic recovery, some sea weeds can concentrate it to 0.45% of their dry weight and have been used as a source of iodine.

The physical properties of water can be explained by the structure of water molecules and the bonding within and between them. It is often said that the difference between a physical change and a chemical change is that a physical change is easily reversible. Phase changes take place relatively easily and are said to be physcial changes. What, then, is chemical change? What characterises it and how can we model such change?

We will look now at how four different compounds interact with water: sodium chloride, hydrogen chloride, ammonia and ammonium chloride. This study will increase our understanding of the nature of change (both physical and chemical), accompanying energy changes and the idea of dynamic equilibrium. The examples have been chosen to represent different types of changes which can occur when substances are placed in water.

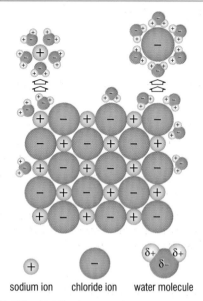

sodium ion · chloride ion · water molecule

Fig 5.19 Modelling the process of sodium chloride dissolving in water.

CONNECTION: The equilibrium between excess solid and a saturated solution is analogous to the dynamic equilibrium between a liquid and a gas, page 78.

These are:
- ionic compounds which dissolve in water as hydrated ions;
- covalent compounds which dissociate into ions when they dissolve in water, forming ions (a process called ionisation). Some compounds fully dissociate, others only partially;
- covalent compounds which react with water (a process called hydrolysis).

Sodium chloride, a water-soluble ionic compound

The temperature of water drops slightly when sodium chloride dissolves in it. The process is weakly endothermic. The solution which forms conducts electricity, indictating the presence of freely moving ions.

$$NaCl(s) + aq \longrightarrow Na^+(aq) + Cl^-(aq)$$

Imagine a crystal of sodium chloride being placed in water (Fig 5.19). A sodium ion on the outside of the crystal is surrounded by water molecules. The negative ends of water molecules are attracted to the positively charged sodium ions. With agitation, a sodium ion is shaken loose from the crystal and is free to move about in the water. The ion is surrounded by bound water molecules and is said to be hydrated. Hydrated sodium ions, $Na^+(aq)$, are present in solution. Similarly chloride ions pass into solution as hydrated chloride ions, $Cl^-(aq)$.

11 **Sodium chloride crystals are cubic. Which part of the crystal is most likely to begin dissolving first?**

When sodium chloride is added to water until no more dissolves and there is an excess of the solid, a **saturated solution** is formed. However, the system is not static. Ions from the solid are constantly being hydrated and passing into solution. At the same rate, hydrated ions shed water molecules and return to the bulk of the solid. The solid and the hydrated ions are in **dynamic equilibrium**.

The dissolving of ionic compounds in water is considered to be a physical change, even though chemical bonds (the electrostatic forces holding ions together in a lattice) are broken and new ones form (the bonds between water molecules and ions).

12 **A number of ionic compounds are insoluble, for example AgCl, $BaSO_4$, and CaF_2. What determines whether or not an ionic compound will dissolve?**

Hydrogen chloride, a covalent compound which fully ionises in water

Hydrogen chloride gas consists of polar hydrogen chloride molecules. It dissolves in non-polar solvents to form solutions which do not conduct electricity. The accompanying energy change is small. In contrast, when hydrogen chloride dissolves in water a solution which conducts electricity forms and the process of dissolving is highly exothermic. The solution, of course, is hydrochloric acid and can be shown to contain hydrated hydrogen ions, $H^+(aq)$, and hydrated chloride ions, $Cl^-(aq)$.

$$HCl(g) + aq \longrightarrow H^+(aq) + Cl^-(aq)$$

We can model the change which occurs when hydrogen chloride dissolves in water by using our knowledge of molecular structure and bonding (Fig 5.20). A proton (the nucleus of the hydrogen atom) is transferred from a hydrogen chloride molecule to a water molecule. It is held by a coordinate bond in the oxonium ion, $H_3O^+(aq)$. Hydrogen chloride is behaving as an acid (a proton donor). Water is behaving as a base (a proton acceptor).

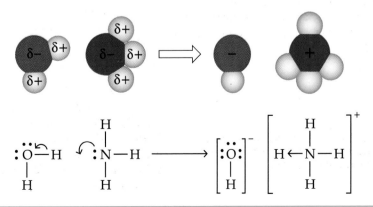

KEY FACT

In aqueous solution, the hydrogen ion (H^+, simply a proton) has one water molecule firmly bonded to it, and we use the formula H_3O^+ to represent this. It is called the oxonium ion. Other water molecules cluster around, because of electrostatic attraction, and hydrate the ion. We represent this as $H_3O^+(aq)$. For simplicity, we use $H^+(aq)$ as our chemical shorthand to show this situation.

13 **Draw a shorthand bond diagram to represent H_3O^+ and suggest a shape for the ion.**

The dissolving of hydrogen chloride in water, at low concentrations, is not appreciably reversible. The process involves breaking a covalent bond to form ions which become hydrated. Whether a physical or chemical change has occurred depends on how one defines physical and chemical change. Most people consider it a physical change.

14 **Energy is released when bonds form. Explain why more energy is released when hydrogen ions are hydrated than when chloride ions are hydrated.** *Hint:* **Think about the relative sizes of the ions.**

Ammonia, a covalent compound which partially ionises in water

Ammonia gas consists of polar ammonia molecules. It dissolves in water to give an alkaline solution. The properties of the solution suggest the presence of hydrated ammonia molecules, $NH_3(aq)$, hydrated ammonium ions, $NH_4^+(aq)$, and hydrated hydroxyl ions, $OH^-(aq)$.

$$NH_3(g) + aq \longrightarrow NH_3(aq)$$
$$NH_3(aq) + H_2O(l) \rightleftharpoons NH_4^+(aq) + OH^-(aq)$$

Ammonia dissolves initially in water as hydrated ammonia molecules. These can react with water molecules to form ammonium ions and hydroxide ions. A dynamic equilibrium is established between ammonia molecules, ammonium ions and hydroxide ions. This reaction has the same characteristics as other dynamic equilibria that we have met, such as phase changes and a solid in the presence of its saturated solution of it.

We can model the ammonia-water system, again by using our understanding of molecular structure (Fig 5.21). Ammonia is soluble in water because of the ease with which ammonia molecules can form hydrogen bonds with water molecules. Once ammonia molecules are in solution, a proton is transferred from a water molecule to an ammonia molecule to

Fig 5.21   Modelling the process of ammonia reacting with water.

THE WATERS

form hydrated hydroxide ions, $OH^-(aq)$, and hydrated ammonium ions, $NH_4^+(aq)$, respectively. The reverse proton transfer also occurs from $NH_4^+(aq)$ to $OH^-(aq)$ to form NH_3 and H_2O. The system is in a state of dynamic equilibrium. In this case water is behaving as an acid and ammonia as a base.

We are left with the dilemma of whether to describe the dissolving of ammonia in water as a physical or chemical change. Perhaps simple dissolving to give aqueous ammonia molecules is a physical change, but the dynamic equilibrium established involves chemical changes.

15 Why is phosphine, PH_3, only sparingly soluble in water?

Aluminium chloride, a covalent compound which reacts with water

Aluminium chloride, $AlCl_3$, is a white solid. When it is added to water an extremely exothermic reaction occurs and hydrogen chloride gas is given off. A white, gelatinous residue of aluminium hydroxide remains.

$$AlCl_3(s) + 3H_2O(l) \longrightarrow Al(OH)_3(s) + 3HCl(g)$$

Aluminium chloride is hydrolysed and the reaction is not reversible. Aluminium-chlorine covalent bonds have been broken and aluminium-oxygen bonds formed. The reaction of aluminium chloride with water has all the characteristics of a chemical change – it is accompanied by a marked energy change and is not reversible.

16 Aluminium chloride is prepared by the direct reaction between chlorine and aluminium. Why is it essential that dry reagents and apparatus are used?

Water – the universal solvent?

The examples given show the versatility of water, ranging from a simple solvent (dissolving ionic compounds) to a reactant (hydrolysing covalent compounds). We have also seen that it is not always easy to classify a process as a physical or chemical change, but then perhaps this is not important. What matters is that we can use our knowledge of structure and bonding to explain and predict the ways in which water will behave towards other substances.

QUESTIONS

5.22 Give the formula of all of the species formed when the following substances dissolve in water:
(a) $CaCl_2$; (b) $Pb(NO_3)_2$; (c) Na_2CrO_4; (d) $KHSO_4$ (e) $Al(NO_3)_3$.

5.23 When a saturated solution of silver chloride is in contact with undissolved solid silver chloride the resulting mixture is described as a dynamic equilibrium. Explain this statement.

5.24 When aqueous ammonia reacts with hydrochloric acid a solution is formed containing ammonium chloride. Explain why the reaction may simply be regarded as one between oxonium ions and hydroxide ions.

5.25 Hydrogen sulphide dissolves in water to give a weakly acidic solution. The process can be summarised by the two equations:

$H_2S(g) + aq \longrightarrow H_2S(aq)$
$H_2S(aq) + H_2O(l) \rightleftharpoons H_3O^+(aq) + HS^-(aq)$

(a) Explain why hydrogen sulphide is far less soluble in water than ammonia. (b) Use shorthand bond diagrams to explain how aqueous hydrogen sulphide molecules react with water.

5.5 MEASURING AMOUNTS OF SUBSTANCE IN SOLUTION

CONNECTION: The formula n = m/M, page 32.

We must be able to measure the amounts of substances in solution, to ensure that we have appropriate quantities for preparing new compounds and to enable accurate quantitative analysis.

Concentration

The concentration, or molarity, of a solution is measured by the number of moles of solute in a known volume of solution. Usually we measure concentrations in moles of solute dissolved in and made up to 1 dm^3 of solution. The units are **mol dm^{-3}**.

TUTORIAL

Problem: What is the concentration of a solution of 0.0585 g of sodium chloride in 500 cm^3 of aqueous solution?

Solution: Mass of sodium chloride in 1000 cm^3 (1 dm^3) = 0.1170 g. The molar mass of sodium chloride is 58.5 g mol^{-1}

Using $n = m/M$,

amount of NaCl in 1 dm^3 = 0.1170/58.5 mol = 0.002 mol.

Therefore, the concentration of sodium chloride in aqueous solution is 0.002 mol dm^{-3}.

Problem: What is the concentration of hydrated sodium ions and hydrated sulphate ions in 0.2 mol dm^{-3} sodium sulphate solution?

Solution: Sodium sulphate dissolves in water to give Na^+(aq) and SO_4^{2-}(aq).

Na_2SO_4(s) + aq $\longrightarrow$ 2Na^+(aq) + SO_4^{2-}(aq)

We know one mole of Na_2SO_4 contains 2 mol Na^+ and 1 mol SO_4^{2-}.

The concentration of hydrated sodium ions in 0.2 mol dm^{-3} aqueous sodium sulphate is 0.4 mol dm^{-3}, and that of hydrated sulphate ions is 0.2 mol dm^{-3}.

Problem: What is the mass of magnesium sulphate present in 50 cm^3 of a 0.05 mol dm^{-3} aqueous solution?

Solution: 0.05 mol $MgSO_4$ in 1 dm^3 (1000 cm^3).

Therefore, $0.05 \times \frac{50}{1000}$ mol $MgSO_4$ in 50 cm^3 = 2.5×10^{-3} mol.

The molar mass of magnesium sulphate is 120.3 g mol^{-1}.

Therefore, 2.5×10^{-3} mol $MgSO_4$ has mass $2.5 \times 10^{-3} \times 120.3$ g = 0.301 g.

Very dilute solutions

The concentrations of pollutants and toxic chemicals in the atmosphere or in water may be extremely low, yet present at harmful levels. Convenient units to measure very low concentrations are parts per million (ppm). Sometimes, for even lower concentrations, parts per billion (ppb) are used.

$$\text{ppm} = \frac{\text{mass of solute /g}}{\text{mass of solution /g}} \times 10^6$$

One ppm is very small. It is the equivalent of a 1 cm^3 pebble in a block of concrete with sides 1 m in length.

Note that in the formula for ppm, the mass of solution and not its volume is used. For concentration, the volume of solution is used.

5.26 Calculate the concentration of solution formed by dissolving the following masses of solute in the stated volume of solution.
  (a) 12 g of sodium hydroxide, NaOH, in 1 dm^3;
  (b) 3.5 g of calcium chloride, CaCl$_2$, in 250 cm^3;
  (c) 17 g of hydrogen bromide, HBr, in 3 dm^3;
  (d) 2.9 g of ethanedioic acid, C$_2$H$_2$O$_4$, in 250 cm^3;
  (e) 5 g of sodium benzene-1,2-dicarboxylate, C$_8$H$_5$NaO$_4$, in 1 dm^3.

5.27 Tartaric acid, C$_4$H$_6$O$_6$, and citric acid, C$_6$H$_8$O$_7$, are often used in wine-making. Describe how you would prepare 4.5 dm^3 of an aqueous solution in which the the concentration of both of these substances is 0.05 mol dm^{-3}.

5.28 Sulphuric acid and potassium hydroxide react together in a mole ratio of 1:2. 100 cm^3 0.1 mol dm^{-3} and 150 cm^3 of 0.15 mol dm^{-3} hydroxide are mixed.
  (a) Write a balanced equation for the reaction which takes place.
  (b) Calculate the amount of potassium hydroxide remaining at the end of the reaction.
  (c) Calculate the concentration of potassium hydroxide at the end of the reaction.
  (d) Name the salt formed.
  (e) Calculate the concentration of the salt formed in the final solution.

5.29 Calculate the concentration of each of the solutions (a) 20 g of ammonium nitrate, NH$_4$NO$_3$, in 4 dm^3 of solution; (b) 1.25 g copper(II) sulphate-5-water, CuSO$_4$.5H$_2$O, in 750 cm^3 of solution.

5.30 Calculate the mass of potassium iodide dissolved in 250 cm^3 of KI(aq) if the concentration is 0.1 mol dm^{-3}.

5.31 How many moles of chloride ions are present in (a) 1 dm^3 of 0.02 mol dm^{-3} HCl(aq); (b) 10 cm^3 of 0.005 mol dm^{-3} CaCl$_2$(aq); (c) 0.373 g KCl dissolved in 100 cm^3 of solution.

5.32 A solution is prepared by dissolving 1.11 g CaCl$_2$ and 1.17 g NaCl in water and making the solution up to 250 cm^3. Calculate the concentration of chloride ions in the solution.

Answers to Quick Questions: Chapter 5

1 In Chapter 1 we are told that there are about 1348 million km^3 of water on Earth. 2.4% is snow and ice = $1.348 \times 10^3 \times 0.024$ km^3
$$= 1.348 \times 10^3 \times 0.024 \times 10^6 \text{ cm}^3$$
$$= 3.235 \times 10^{10} \text{ g}$$

2 Methane is insoluble in water.

3 Ten times further apart in vapour compared with liquid. Therefore, the molecules in the vapour phase can more easily be compressed under increasing external pressure.

4 No, because nothing would move, including the people making the measurements!

5 Sublimation temperature increases with increasing pressure.

6 (See diagram on left)

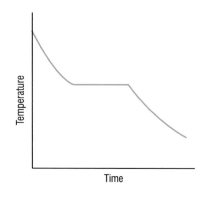

7 Trend in electronegativity: Cl > Br > I. Therefore, CH$_3$Cl is more polar than CH$_3$Br which in turn is more polar than CH$_3$I.

8 Dispersion forces and permanent dipole–dipole.

9 28 (2, 8, 18) and 46 (2, 8, 18, 18)

10 Sodium is greater than seven times more abundant by mass than potassium in the Earth's crust. It is nearly thirty times more abundant in sea water.

11 The corners

12 The strength of the ionic lattice relative to the energy that is released when freed ions are hydrated by water.

13 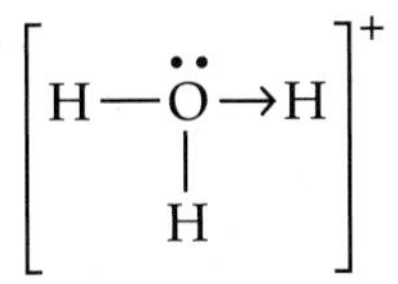
$$\left[\text{H} - \overset{\cdot\cdot}{\underset{|\ \ \text{H}}{\text{O}}} \rightarrow \text{H} \right]^{+}$$ Probable shape: pyramidal

14 Hydrogen ions are extremely small (they are simply protons) and have a very high charge density. Therefore, they attract water molecules strongly. Chloride ions are much larger but still with a single charge (even though it is opposite in sign). The charge density is less and water molecules are less strongly attracted.

15 The electronegativities of phosphorus and hydrogen are similar and so the P — H bonds are not strongly polarised. Hydrogen bonds between PH_3 and H_2O do not form.

16 Because aluminium chloride reacts with water (hydrolyses) easily.

Chapter 6

CARBON COMPOUNDS: TEN MILLION AND RISING

LEARNING OBJECTIVES

When you have studied this chapter you should be able to:

1. summarise the general properties of living systems;

2. explain the terms metabolism and metabolites;

3. explain the formation of fossil fuels;

4. describe the properties of carbon which give rise to the great diversity of its compounds;

5. use some of the language associated with organic chemistry: hydrocarbon, aliphatic, aromatic, alkane, alkene, alkyne, arene, saturated, unsaturated and homologous series;

6. explain the system of nomenclature of organic compounds and use this to name hydrocarbons;

7. recognise and name functional groups in organic compounds.

6.1 THE BIOSPHERE

As far as we know, life is unique to Earth. All life forms appear to require certain essential chemical substances and environmental conditions. For example, plants and animals depend upon the availability of certain key chemical substances, such as oxygen, carbon dioxide and water, and all forms of life are destroyed by extreme heat.

Why is it that rabbits rapidly increase their numbers but rocks do not? We often take the characteristics of plants and animals for granted but what is it about these complex structures that makes them so special? There have been many attempts to define 'life' and 'living systems'. One is 'entities capable of reproduction'. Other characteristic properties of living systems are movement, excretion, growth, nutrition, respiration and response to stimuli.

Cells are the smallest component of all but a few living systems. They are small bags of chemical compounds, bounded by a membrane, and in which a vast number of reactions occur. Cells vary in size, shape and the extent to which they aggregate into larger structures. The purpose of a cell, however, is always the same – to use chemical and energy resources from its environment to survive and to reproduce.

Living systems and energy

Metabolic reactions are the chemical reactions which occur in living organisms. The thousands of chemical substances which take part in these reactions are known as metabolites. Their reactions fall into two classes.

- In **catabolic reactions** complex chemical substances are broken down into simpler ones. Many of these reactions release energy for the life form to use. For example, sugars are broken down into carbon dioxide and water in the human body, releasing energy which can be used to do work.
- In **anabolic reactions** more complex substances are synthesised from simpler compounds, using energy available in the cell. For example, plants use carbon dioxide from the air and water to form sugars – the process of photosynthesis. This is the first step in the creation of complex compounds which, in turn, are used to make larger structures such as leaves and stems. The energy which drives this construction work is provided by sunlight.

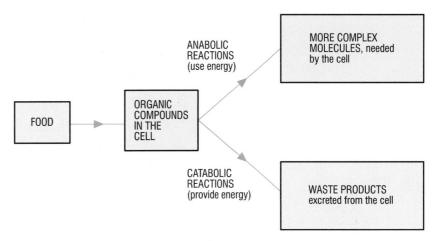

Fig 6.1 Chemical reactions may either provide or use energy in the cell.

1 **Why are the drugs taken illegally by body builders called anabolic steroids?**

Cell metabolism is under the control of biological catalysts called **enzymes**. Enzymes are formed within living cells and control the rate and course of cell reactions. Reactions which occur in a living cell are called metabolic changes and their products are called **metabolites**. Metabolic changes in cells occur in a series of small steps, rather than as one big, possibly explosive, reaction. A series of step-like metabolic reactions is called a **metabolic pathway**.

2 **What advantages might there be in metabolic change taking place in small steps?**

Death and decay: the creation of fossil fuels

Plants may be eaten by animals or they may simply die and decompose by the action of bacteria and fungi. To animals, bacteria or fungi these plant remains are a valuable food, a rich store of chemical compounds built up by the plant when alive. As the plant is decomposed, its components are recycled and become available to other plants and animals.

Under certain conditions, dead plant material may be buried quickly and cannot be eaten by animals or decomposed by bacteria. Instead, heat and pressure change the material into coal. Several grades of coal exist, with increasing carbon content. The higher the carbon content the more efficient the coal is as a fuel.

All the fossil fuels upon which we are so dependent derive from dead plants. Coal, crude oil and natural gas all formed from the action of heat and pressure on dead material buried deep in the Earth's crust.

Fig 6.2 All processes of decay are exothermic. The steam rising from this compost heap shows that this is an example of such a change. Nutrients and humus in the rotting pile are recycled by returning it to the land.

Microscopic sea creatures are believed to have been trapped in rock strata and transformed under heat and pressure to crude oil (also called petroleum) and natural gas. Traces of plant and animal pigments found in crude oil support the theory that petroleum was formed between 100 and 200 million years ago from the remains of marine plankton and algae.

3 **Why cannot petroleum and natural gas provide fossil evidence of their origin?**

Fig 6.3 Evidence that coal is derived from plants comes from many well-preserved fossils of plants laying trapped within coal. It is estimated that peat and coal began to form over 300 million years ago.

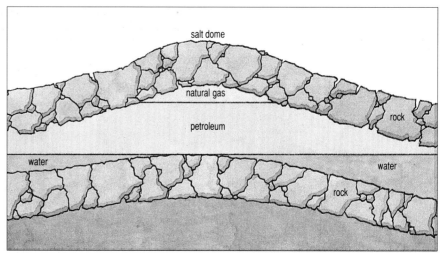

Fig 6.4 Petroleum and natural gas move through porous rocks until trapped by non-porous rocks. Petroleum may be trapped by a 'salt dome' which lines a bulge in non-porous rock. It is often found associated with water and natural gas.

Carbon: the element of life

Chemically, living systems have many similarities. They contain a much higher proportion of certain elements than are present in their surroundings. The element which dominates the chemistry of living systems, however, is carbon. Despite the multitude of different life forms, every living cell includes thousands of compounds which contain carbon. It could be said that the diversity of life depends upon the diversity of carbon chemistry.

During the 19th century the study of carbon compounds became known as **organic chemistry**. At this time all carbon compounds were obtained from plants and animals and were thought to be in some way different from other chemicals. It was thought that a 'vital force', found only within living cells, was needed to make such substances. As more and more of these natural products were isolated from plants and animals the new chemical science grew at a tremendous rate and organic chemistry became a separate study.

In 1828, Friedrich Wöhler, a German chemist, made carbamide, (more commonly known as **urea**). He prepared this organic compound in the laboratory by heating the inorganic compound, ammonium cyanate, NH_4CNO. The vital force theory was exploded. His synthetic carbamide was identical in every respect to 'animal' carbamide. Shortly afterwards, other natural products such as methanol, ethanol and ethanoic acid were synthesized. Despite this, the term organic chemistry has remained. It refers to the chemistry of all carbon compounds (excluding simple substances such as carbon dioxide or simple metal salts containing carbon). In 1848, Leopold Gmelin redefined organic chemistry as 'the chemistry of the compounds of carbon'.

4 **What do you understand by the term 'synthetic carbamide'?**

CARBON COMPOUNDS: TEN MILLION AND RISING

6.1 **(a)** What is meant by the term metabolism?

(b) How is the rate of metabolism controlled within living cells?

6.2 What are fossil fuels and how are they formed?

6.3 What was significant about Wöhler's synthesis of carbamide (urea)?

6.4 A chlorofluorocarbon, $C_2Cl_3F_3$, is an effective degreasing agent and may be used for specialised dry-cleaning. How many different structural formulae can you produce for this molecular formula?

6.2 WHY SO MANY CARBON COMPOUNDS?

More than 90% of all the compounds known contain carbon, yet it is only one of the 91 elements which occur on Earth. So why carbon? What is it about carbon that makes it a component of over seven million different chemical compounds? A simple question but one with several answers.

Catenation and bond energy considerations

Carbon-carbon bonds are not easily broken. Atoms of carbon form strong bonds with each other and can form long chains and rings of varying sizes. The tendency for atoms of an element to link to one another to form a long chain is called catenation. Vast numbers of molecules can form, all of which have backbones of carbon atoms with other atoms, usually hydrogen, bonded to them. These backbones are sometimes called **carbon skeletons**.

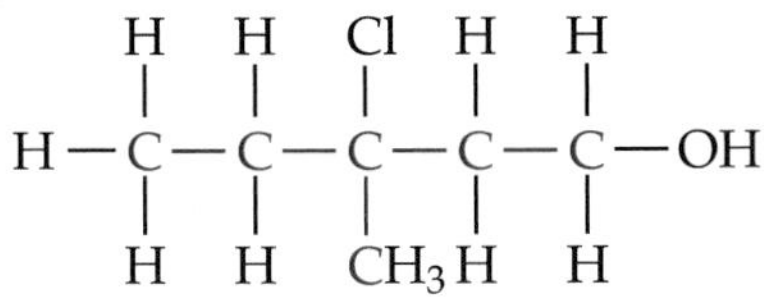

Silicon is in the same group of the periodic table as carbon. The two elements might be expected, therefore, to behave similarly. However, a $C—C$ bond is stronger than a $Si—Si$ bond (347 kJ mol^{-1} and 222 kJ mol^{-1} respectively) because the two carbon atoms are smaller and therefore closer together. Further, the shared electron pair is less shielded.

The $Si—O$ bond energy (368 kJ mol^{-1}) is much higher than that of $Si—Si$. Consequently, in an oxygen-rich atmosphere such as that on Earth $Si—O$ bonds are far more common. Silicon(IV) oxide is, in fact, the most common mineral in the Earth's crust.

> **5** Explain why the silicon-oxygen bond is stronger than the silicon-silicon bond.

A consideration of size and shielding might lead us to expect strong $N—N$, $O—O$ and $F—F$ single bonds. However, all are much weaker than the $C—C$ single bond. We have seen how lone pairs of electrons influence the shape of a molecule. An inspection of the molecules C_2H_6, N_2H_4, H_2O_2 and F_2 shows that only ethane does not have lone-pairs of electrons (Fig 6.5). In all the other three molecules the shortness of the bond (the atoms of Period II are small) means that the lone-pairs on neighbouring atoms are close together. Electrostatic repulsion between them weakens these bonds compared with $C—C$, and so catenation of nitrogen, oxygen and fluorine does not occur.

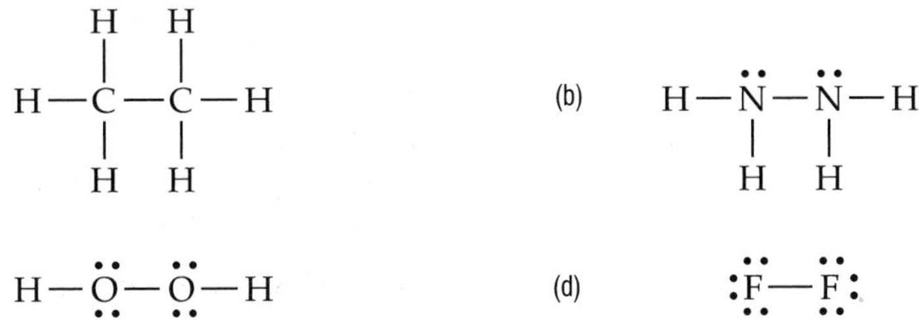

KEY FACT

A carbon skeleton is the backbone of an organic molecule. For example, the carbon skeleton in 3-chloro-3-methylpentan-1-ol is shown here in red:

$$
\begin{array}{ccccc}
H & H & Cl & H & H \\
| & | & | & | & | \\
H—C—&C—&C—&C—&C—OH \\
| & | & | & | & | \\
H & H & CH_3 & H & H \\
\end{array}
$$

Fig 6.5 Shorthand bond diagrams for (a) ethane, C_2H_6; (b) hydrazine, N_2H_4; (c) hydrogen peroxide, H_2O_2; (d) fluorine, F_2.
Bond energies are: $C—C$ 347 kJ mol^{-1}; $N—N$ 159 kJ mol^{-1}; $O—O$ 146 kJ mol^{-1}; $F—F$ 159 kJ mol^{-1}.

KEY FACT

Electron pairs, both bonding and lone-pairs, repel one another. This property may be used to explain the shapes of simple molecules.

Arrangements of carbon atoms: structural isomers

Carbon normally forms four covalent bonds, allowing up to four other atoms to be bonded to a carbon atom. A number of carbon atoms may bond together to give skeletons which are unbranched, branched or have ring structures.

- An **unbranched skeleton** (sometimes called **straight-chain** even though the carbon chain is not straight) consists of carbon atoms linked in a chain with no side chains. For example:

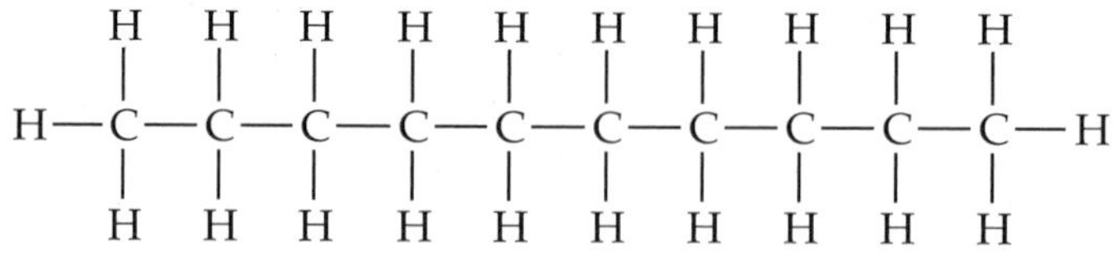

- A **branched skeleton** consists of carbon atoms linked in a chain with other carbon atoms as side chains. For example:

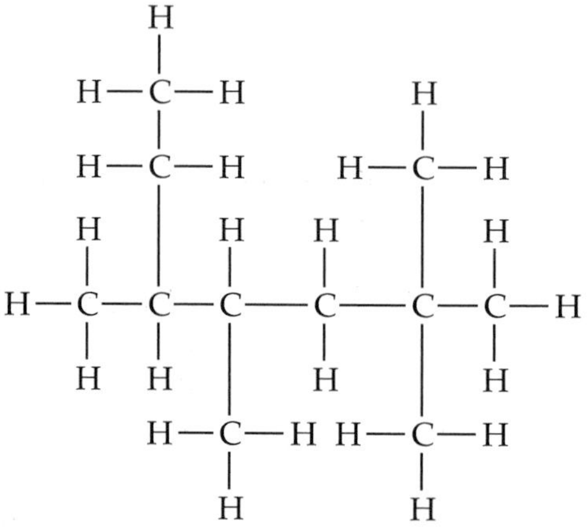

- A ring skeleton consists of carbon atoms arranged in a **cyclic** (or ring) structure. For example:

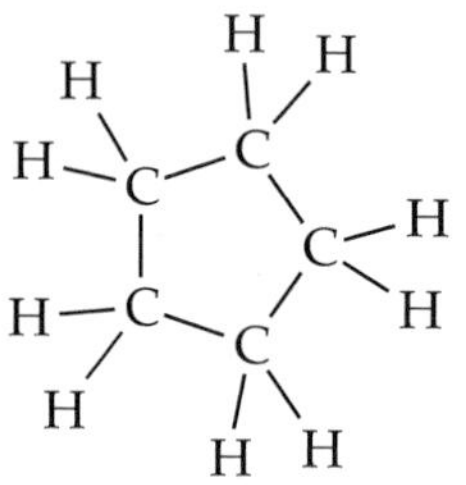

Often two or more molecules have the same molecular formula. For example, three compounds exist with the molecular formula C_5H_{12} (Fig 6.6). These are called **structural isomers**. As the number of carbon atoms in a molecule increases the number of possible structural isomers increases rapidly.

KEY FACT

Structural isomers are compounds with the same molecular formula but different structural formulae. The atoms are linked in different ways in each compound (Fig 6.6).

(a) (b) (c)

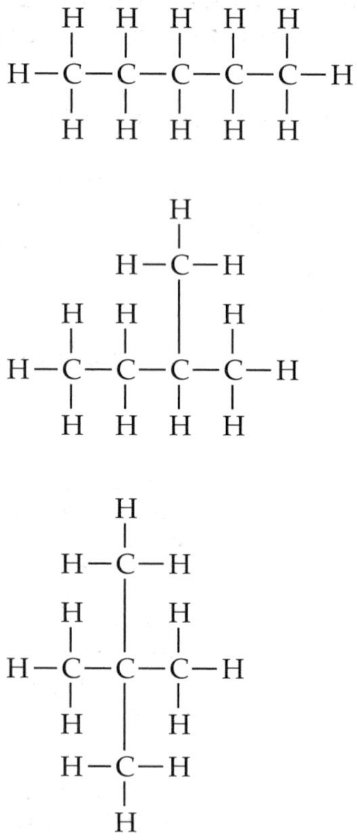

Fig 6.6 The three structural isomers of molecular formula C_5H_{12}.

Fig 6.7 At first sight these two structural formulae may seem different. However, molecular models show that they are the same molecule.

6 Remembering that carbon has a valency of four, draw shorthand bond diagrams for structural isomers with the molecular formula C_6H_{14} (each isomer contains only $C-C$ and $C-H$ bonds).

Carbon atoms, because of their small size, form strong covalent bonds with other atoms and this increases the variety of carbon compounds known. Each carbon atom has four electrons in its outer energy level which are available for bonding. Hydrogen is the most common element, other than carbon, found in organic compounds, though they often contain one or more of the following elements: oxygen, nitrogen and the halogens. Less common are organic compounds containing sulphur or phosphorus, although such molecules are very important biologically.

Another consequence of the small size of carbon atoms is their tendency to form multiple covalent bonds, both with one another and with other small atoms, in particular nitrogen and oxygen. Again, this extends the range of possible molecules which are based on carbon. The following multiple bonds are found in organic compounds:

$$\begin{array}{c} >C=C< \\ -C\equiv C- \\ >C=O \\ >C=N- \\ -C\equiv N \end{array}$$

Functional groups

Particular arrangements of atoms and bonds have chemical properties which are largely independent of the remainder of the molecule. We call these **functional groups**.

The identification of functional groups enables us to make sense of the vastness of organic chemistry, and provides a short cut to learning the chemistry of organic compounds. Compounds with the same functional group are classed together and given a collective name. These are summarised in Table 6.1.

7 To what class of organic compounds does $C_2H_5NH_2$ belong?

Table 6.1 Important classes of organic compounds and the functional groups present

Class of compound	Formula	Functional group	Notes
Alkene	$R^1R^2C=CR^3R^4$	$>C=C<$	
Alkyne	$R^1C\equiv CR^2$	$-C\equiv C-$	
Halogenoalkane	RX (general)	$-X$	X is a halogen atom
primary	R^1CH_2X	$-X$	
secondary	R^1R_2CHX	$-X$	
tertiary	$R^1R^2R^3CX$	$-X$	
Alcohols	ROH (general)	$-OH$	$-OH$ is called a **hydroxyl** group,
primary	R^1CH_2OH	$-OH$	regardless of whether it is part
secondary	R^1R^2CHOH	$-OH$	of an aliphatic compound (an
tertiary	$R^1R^2R^3COH$	$-OH$	alcohol) or an aromatic compound
Phenols	ArOH	$-OH$	(a phenol).
Aldehydes	RCHO*	$-C=O$	
		H	$>C=O$ is called a **carbonyl** group in
Ketones	RR^{1}CO	$>C=O$	both aldehydes and ketones.

Table 6.1 continued

Class of compound	Formula	Functional group	Notes
Carboxylic acids	RCOOH*	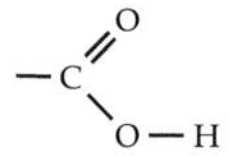	
Esters	RCOOR¹*	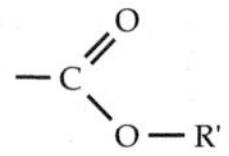	Esters, acid halides, amides and anhydrides are said to be **derivatives** of carboxylic acids. They are obtained by replacing the —OH by an appropriate group.
Acid halides	RCOX	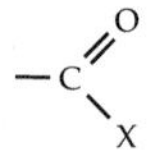	
Acid amides	RCONH₂*	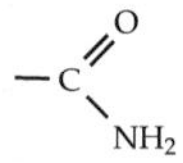	
Acid anhydrides	(RCO)₂O	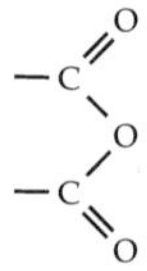	
Amines			
primary	R¹NH₂		
secondary	R¹R²NH		
tertiary	R¹R²R³N		
Nitriles	RCN		
Nitro compounds	RNO₂		

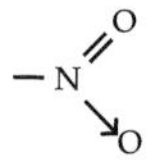

Notes

1. R, R¹, R² etc = hydrocarbon group. These may or may not be the same in a particular given compound.

2. For structures marked *, R may be a hydrogen atom.

SPOTLIGHT

(a)

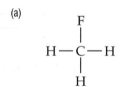

Fig 6.8a Fluoromethane is the simplest possible halogenoalkane. The halogenoalkane functional group is the carbon-halogen bond. The halogen may be F, Cl, Br or I.

(b)

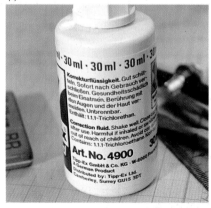

HALOGENOALKANES: MULTI-PURPOSE CHEMICALS

General anaesthetics revolutionised surgery. It is hardly possible to contemplate major operations without their use. An early anaesthetic was trichloromethane, $CHCl_3$, or chloroform as it is still commonly known. However, the lethal dose of chloroform is only slightly higher than the amount needed for anaesthesia and so by the early 1950s chemists had begun the search for the perfect anaesthetic. In 1956, halothane, $CF_3CHBrCl$, was discovered by chemists working for ICI. It seemed ideal – it was non-toxic and non-flammable, and there appeared to be no side-effects. More recently, evidence has been gathering that suggest prolonged exposure may cause liver damage and so the search continues for better anaesthetics.

Some other uses of halogenoalkanes include additives in petrol (1,2-dibromoethane), dry-cleaning (1,1,1-trichloroethane), paint removers (dichloromethane). CFCs are also halogenoalkanes and are used as aerosol propellants and in refrigeration.

Fig 6.8b 1,1,1-Trichloroethane is used as the solvent in some correction fluids.

Fig 6.8c The structural formula of 1,1,1-trichloroethane. The use of locants in the name indicates the position of the three chlorine atoms.

(c)

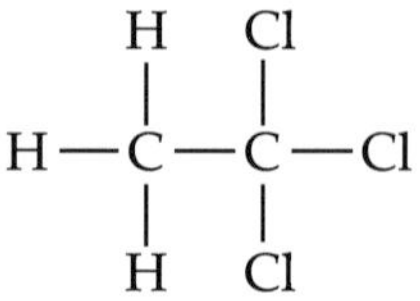

EXPLOITING THE HYDROXYL GROUP

Alcohols are widely used as industrial solvents and fuels. Methylated spirits, for example, is mainly ethanol. Ethanol and propanol are used as disinfectants due to their ability to penentrate the cell walls of bacteria and kill them. *All* alcohols are toxic to human beings if taken in quantity.

Phenol, a benzene ring with one hydrogen atom replaced by a hydroxyl group, is the simplest possible member of the phenol series and was the first surgical antiseptic, introduced by Joseph Lister in 1867. It also has mildly anaesthetic properties. Some substituted phenols (particularly chloro-substituted phenols) are better antiseptics than phenol. For example, 'TCP' (containing 2,4,6-trichlorophenol) and 'Dettol' (containing 2,4,-dichloro-3,5-dimethylphenol) have replaced older medicinal preparations containing phenol itself. Commercial products, such as soaps, deodorants, disinfectant sprays and muscle stimulant sprays contain phenol derivatives. Many drugs, agrochemicals and dyestuffs contain substituted phenols as active ingredients.

Fig 6.9a This Brazilian car runs on ethanol produced by fermentation of sugar-cane. Ethanol and methanol are promising substitutes for petrol and, importantly, are renewable resources.

(b)

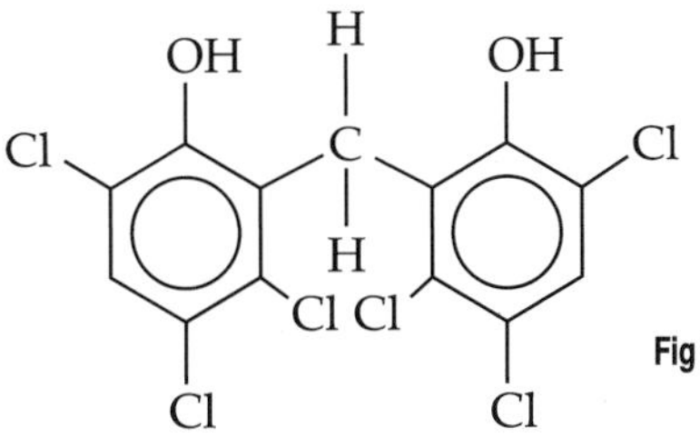

hexachlorophene

Fig 6.9b 4-Chloro-3,5-dimethylphenol is used to combat athletes foot. Hexachlorophene, now banned, was once a widely-used antiseptic, used, for example, in soaps and powders for babies. One batch of baby talc sold in France included, by mistake, a much greater proportion of hexachlorophene than intended. The large number of infant deaths which resulted caused the withdrawal of the product for over-the-counter sales in many western countries.

QUESTIONS

6.5 The huge variety of organic compounds is due to the ability of carbon compounds to catenate.
  (a) Why do the elements nitrogen, oxygen and fluorine not catenate to a significant extent?
  (b) Silicon, in the same Group as carbon might be expected to catenate extensively but does not. Explain why.

6.6 Draw structures for three of the structural isomers with the molecular formula C_7H_{16}.

6.7 Draw all the possible structural isomers for the following hydrocarbons:
  (a) C_4H_{10}; **(b)** C_4H_8; **(c)** C_3H_6.

6.8 Explain why cycloalkanes have the same general formula, C_nH_{2n}, as alkenes.

6.9 Use Table 6.1 to write structural formulae for the simplest possible:
  (a) alkene; **(b)** alkyne; **(c)** primary halogenoalkane; **(d)** primary alcohol; **(e)** secondary alcohol; **(f)** tertiary alcohol; **(g)** aldehyde; **(h)** ketone; **(i)** carboxylic acid; **(j)** primary amine.

6.10 Fig 6.8c shows the structure of 1,1,1- trichloroethane. A structural isomer of this compound is 1,1,2- trichloroethane.
  (a) Give the structural formula of 1,1,2-trichloroethane.
  (b) Explain why 1,2,2-trichloroethane does not exist.

6.3 NAMING HYDROCARBONS

CONNECTION: IUPAC nomenclature for inorganic compounds, page 58.

Organic compounds are so numerous and have such variable structures that any system for naming them must be logical and consistent. Usually, organic compounds are named using the IUPAC system. With the huge growth in the number of organic compounds, a consistent and coherent system such as IUPAC is invaluable. The basis of IUPAC nomenclature is the systematic naming of hydrocarbons.

SPOTLIGHT

CARS: RESOURCE GUZZLERS

A typical car travels only 6 to 8 miles on 1 dm^3 of petrol. That means that a car which covers 15 000 miles in a year uses around 2000 dm^3 of petrol annually. Petrol burns in the engine to give carbon dioxide and water as the main products (although some incomplete combustion occurs, producing carbon monoxide). Our typical car, travelling 15 000 miles a year and consuming around 2000 dm^3 of petrol, releases 4 million dm^3 of carbon dioxide into the atmosphere. Remember the greenhouse effect?

Since their introduction in 1885, the internal combustion engines of cars have been fuelled by petrol, a mixture of hydrocarbons derived from crude oil. These hydrocarbons are a vital source of organic compounds used for the manufacture of plastics, pharmaceuticals, agrochemicals, dyes and many other essential products. Is petrol too valuable to burn?

Fig 6.10 The Ferrari Testarossa is one of the most expensive ways of converting the stored chemical energy of hydrocarbons into mechanical energy.

Alkanes

Alkanes are the simplest of the hydrocarbons. Each carbon atom in an alkane is bonded to four other atoms (either hydrogen atoms or carbon atoms). Since carbon has four electrons in its outer energy level, four is the maximum number of bonds a carbon atom can form. For this reason the alkanes are described as **saturated** compounds. Alkanes are said to be **aliphatic** compounds because they contain only localised covalent bonds (the shared electrons in each bond are associated with two nuclei only).

Alkanes may have unbranched, branched or ring structures. The names of all aliphatic compounds derive from the names of the unbranched alkanes. It is important therefore to know the names of these alkanes. The simplest is methane, CH_4. Some other unbranched alkanes are given in Fig 6.11.

8 **Unbranched alkanes are also called straight chain alkanes. Explain why this term may be misleading.**

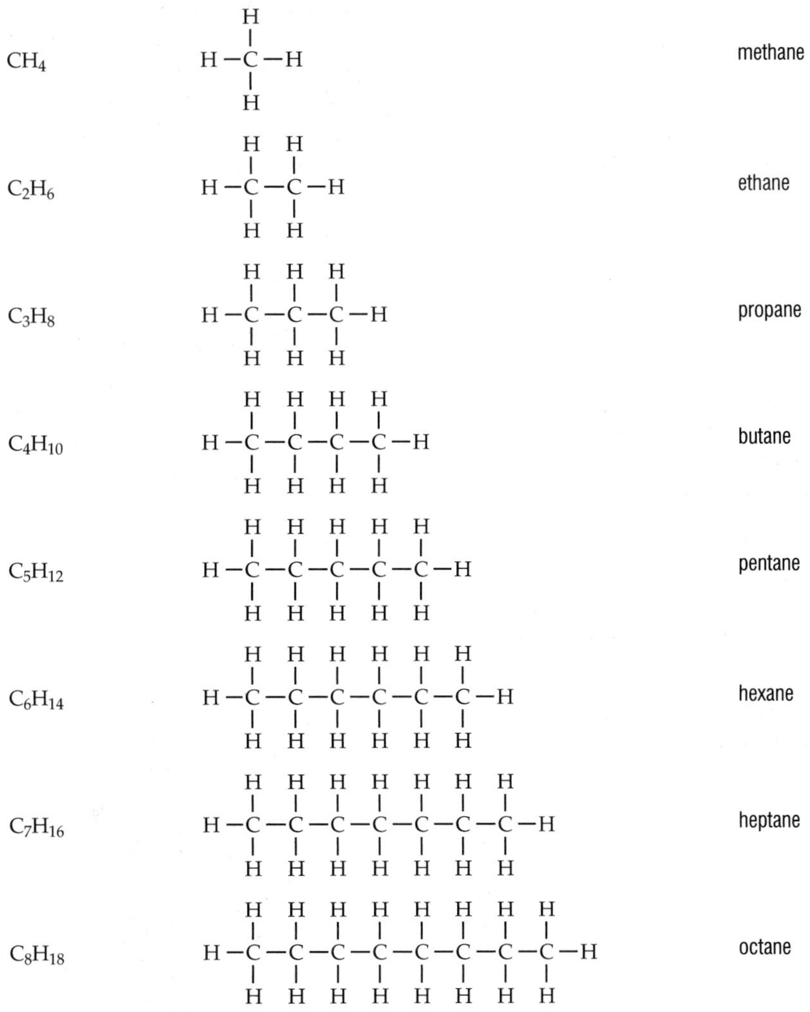

CH_4	H–C–H with H above and H below	methane
C_2H_6	H–C–C–H (each C with H above and below)	ethane
C_3H_8	H–C–C–C–H (each C with H above and below)	propane
C_4H_{10}	H–C–C–C–C–H (each C with H above and below)	butane
C_5H_{12}	H–C–C–C–C–C–H (each C with H above and below)	pentane
C_6H_{14}	H–C–C–C–C–C–C–H (each C with H above and below)	hexane
C_7H_{16}	H–C–C–C–C–C–C–C–H (each C with H above and below)	heptane
C_8H_{18}	H–C–C–C–C–C–C–C–C–H (each C with H above and below)	octane

Fig 6.11   A series of unbranched alkanes, where the chain length is increased by adding a — CH_2— unit as the series is ascended. This is an example of a **homologous series**.

The names of all alkanes end in **ane**. The first part of the name (or **prefix**) shows the carbon skeleton which is present. It is often useful to refer to fragments of alkane structures, which appear as groups in more complicated structures. These fragments are called **alkyl groups** and can be thought of as alkanes from which a hydrogen has been removed. Table 6.2 shows characteristic prefixes for simple alkanes and the alkyl groups derived from them.

KEY FACT

The general molecular formulae for unbranched and branched hydrocarbons is:

Alkanes: C_nH_{2n+2};
Alkenes: C_nH_{2n};
Alkynes: C_nH_{2n-2};
n = an integer.

Cyclic hydrocarbons do not conform to these general formulae.

Table 6.2

No of carbon atoms in alkane chain	Prefix to name	Alkyl group	
1	meth	methyl	— CH_3
2	eth	ethyl	— C_2H_5
3	prop	propyl	— C_3H_7
4	but	butyl	— C_4H_9
5	pent	pentyl	— C_5H_{11}
6	hex	hexyl	— C_6H_{13}
7	hept	heptyl	— C_7H_{15}
8	oct	octyl	— C_8H_{17}

The names of branched alkanes are based upon the unbranched chain present. The name of the compound must tell us

1 the name of the alkane from which the longest unbranched chain derives;
2 the names of substituent alkyl groups;
3 the number of substituents;
4 the positions of substituents.

These four pieces of information are given in reverse order in the name of an alkane. For example the alkane below is called 2,4-dimethylhexane.

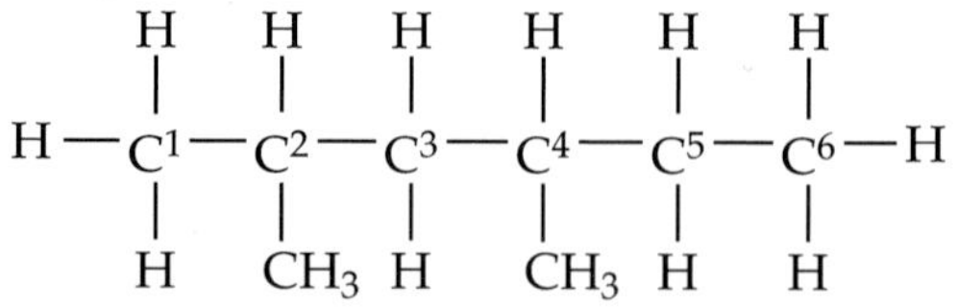

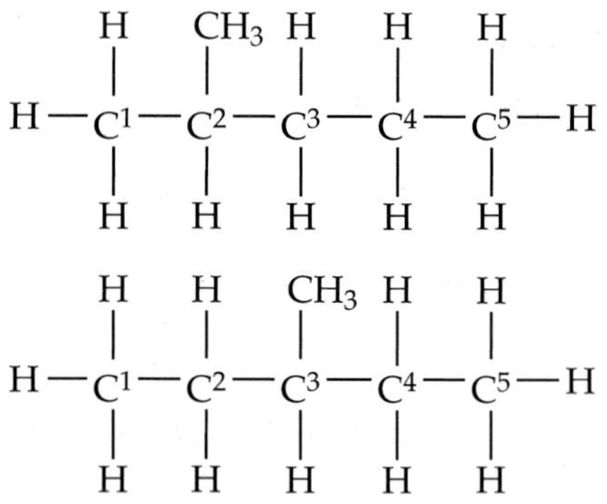

The name is made up as follows:

2,4-dimethylhexane

4 3 2 1

1 The longest unbranched chain has six carbon atoms.
2 The substituents are methyl groups.
3 There are two such groups.
4 They are located on the second and fourth carbon atoms in the chain (they could be considered to be on the third and fifth carbon atoms if we count from the opposite direction, but the rule is that the numbers used are as low as possible). The numbers are also called **locants**.

The method used to name branched alkanes is used to name all organic compounds and so it is essential to understand it.

9 **Why is the first compound in Fig 6.12 called 2-methylpentane and not 4-methylpentane?**

Fig 6.12 Without the use of locants the position of the methyl groups would be uncertain.

TUTORIAL

Problem: Name the compound with structural formula:

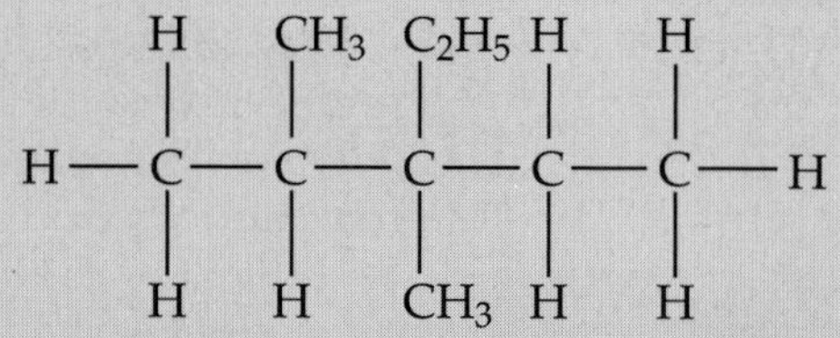

Solution:

Stage 1: The longest unbranched chain is 5 carbon atoms, therefore the name is based on **pentane**.

Stage 2: The substituents are **methyl** groups (two) and **ethyl** groups (one).

Stage 3: The methyl groups are on the **second** and **third** carbon atoms; the ethyl group is on the **third** carbon atom.

Therefore the name must include:

2,3-dimethyl: indicates two methyl groups on carbons 2 and 3

3-ethyl: indicates one ethyl group on carbon 3

and terminate in **pentane**.

Putting things in alphabetical order and making the locants as small as possible, the name for the compound is

2,3-dimethyl-3-ethylpentane

Alkenes and alkynes

Alkenes are aliphatic hydrocarbons which contain a $C=C$ double bond. They are said to be **unsaturated** hydrocarbons since the carbon atoms in the double bond do not have the maximum number (four) of other atoms bonded to them. The name of an alkene always ends in **ene**, indicating the presence of the $C=C$ bond. The first part of the name is based on the name of the alkane of the same chain length and a locant is used to indicate the position of the double bond in the chain (Fig 6.13).

10   Draw a shorthand bond diagram for pent-2-ene.

$$H-\overset{\overset{\displaystyle H}{|}}{C}=\overset{\overset{\displaystyle H}{|}}{C}-\overset{\overset{\displaystyle H}{|}}{\underset{\underset{\displaystyle H}{|}}{C}}-\overset{\overset{\displaystyle H}{|}}{\underset{\underset{\displaystyle H}{|}}{C}}-H \qquad\qquad H-\overset{\overset{\displaystyle H}{|}}{\underset{\underset{\displaystyle H}{|}}{C}}-\overset{\overset{\displaystyle H}{|}}{C}=\overset{\overset{\displaystyle H}{|}}{C}-\overset{\overset{\displaystyle H}{|}}{\underset{\underset{\displaystyle H}{|}}{C}}-H$$

Fig 6.13   These two alkenes are different and their names must distinguish one from the other. The lowest number locant is used in the name to show the position of the first carbon atom of the double bond.

Alkynes are aliphatic, unsaturated hydrocarbons which contain a $C\equiv C$ triple bond. Their names end in **yne** to indicate the presence of the triple bond. As with an alkene, the first part of the name derives from the alkane of the same chain length and a locant is used to indicate the position of the first carbon atom of the triple bond.

11   Why are alkenes examples of aliphatic compounds?

Naming branched alkenes or alkynes is similar to naming branched alkanes. The longest unbranched alkene (or alkyne) and substituent alkyl groups are identified. The same system of locants, prefixes and names is used.

TUTORIAL

Problem: Name the compound with structural formula:

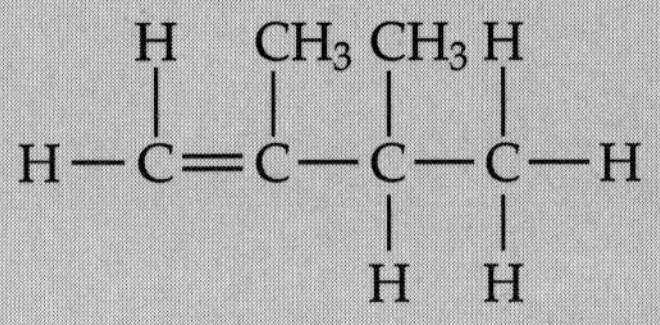

$$H-\overset{\overset{\displaystyle H}{|}}{C}=\overset{\overset{\displaystyle CH_3}{|}}{C}-\overset{\overset{\displaystyle CH_3}{|}}{\underset{\underset{\displaystyle H}{|}}{C}}-\overset{\overset{\displaystyle H}{|}}{\underset{\underset{\displaystyle H}{|}}{C}}-H$$

Solution:
Stage 1: There are four carbon atoms in the longest unbranched chain containing the $C=C$ double bond. The first carbon atom of the double bond is the first in the chain, therefore the name is based on **but-1-ene**.
Stage 2: The substituents are **methyl** groups (two).
Stage 3: The methyl groups are on the **second** and **third** carbon atoms.

Therefore the name must include:

2,3-dimethyl: indicating two methyl groups on carbons 2 and 3 and terminate in **but-1-ene**.

And so the name for the compound is

2,3-dimethylbut-1-ene

---

12   Why is there no substance called 'methene'?

CONNECTION: Explanation of the bonding in benzene, page 251.

Arenes (Aromatic hydrocarbons)

Aromatic hydrocarbons (also known as **arenes**) are unsaturated. They are cyclic compounds with delocalised electrons which are shared by the carbon atoms in the ring. It is the delocalised electrons in the cyclic structure which distinguish aromatic compounds from aliphatic compounds.

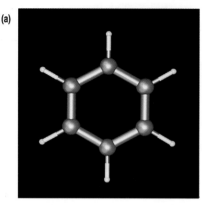

(a)

(b)

or more simply

Fig 6.14 **(a)** shows a photograph of a computer-generated model of benzene showing the six carbon atoms and six hydrogen atoms in the correct relative sizes. **(b)** shows the usual way of representing benzene, with the delocalised system shown as a circle.

The most common aromatic ring is the benzene ring. Benzene itself, C_6H_6, is shown in Fig 6.14, but it is important to realise that the benzene ring is contained in many thousands of organic compounds. The hydrogen atoms in benzene may be substituted by alkyl groups and naming of such substituted benzenes requires identifying what the alkyl groups are and where they are found in the ring. Mono-substituted benzenes (just one hydrogen atom replaced) are straightforward. The name of the alkyl group is used as a prefix to benzene. For example, this substance is called methyl-

Fig 6.15 2,4,6-Trinitromethylbenzene is an explosive, better known as TNT (trinitrotoluene). Locants are used in the name to tell us the positions of four substituents, one methyl (on ring carbon 1) and three nitro groups.

benzene (a methyl group has replaced one hydrogen atom). The pre-IUPAC name for this compound was toluene, and this name is still commonly used.

Where there is more than one substituent, the positions of each substitution must also be indicated, numbering from '1' in alphabetical order, for example:

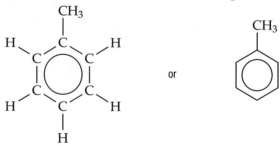

or

is called 1-ethyl-2-methylbenzene.

Fig 6.16 Aromatic compounds have numerous applications, for example as solvents, drugs, polymers, agrochemicals and dyestuffs.

6.11 Give the structural formula of a saturated compound and an unsaturated compound with the molecular formula C_6H_{12}.

6.12 Give the structural formula of two branched-chain isomers of the compound shown in Fig 6.7.

6.13 Petrol constituents are assigned an octane number which reflects aspects of their performance as a fuel for cars. The structural formula of the compound assigned octane number 100 is shown below:

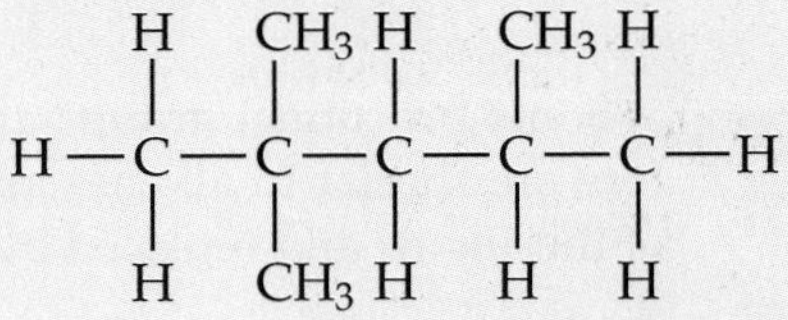

$$
\begin{array}{ccccc}
H & CH_3 & H & CH_3 & H \\
| & | & | & | & | \\
H-C- & C- & C- & C- & C-H \\
| & | & | & | & | \\
H & CH_3 & H & H & H
\end{array}
$$

(a) Give the IUPAC name of the compound.
(b) Explain in your own words why the compound is described as (i) a hydrocarbon; (ii) saturated; (iii) an alkane; (iv) branched; (v) aliphatic.

6.14 Poly(propene) is used for plastic buckets, washing-up bowls and many other household items. It may be made by polymerising a monomer unit known as propene. Give the molecular and structural formulae of propene and explain to which class of hydrocarbons it belongs.

6.15 Rubber is the only natural hydrocarbon polymer. It is formed from a monomer unit with molecular formula C_5H_8, trivial name isoprene. Its IUPAC name is 2-methylbuta-1,3-diene. Give the structural formula of this compound.

6.16 Benzene is an example of an aromatic compound. A dimethylbenzene has two of the hydrogen atoms of benzene replaced by methyl groups. Give the structural formulae and names of all dimethylbenzenes.

6.17 A mono-substituted benzene derivative, molecular formula C_8H_8 may also be considered as a phenyl-substituted alkene. Use this information to work out the structural formula of this compound.

6.4 NAMING COMPOUNDS WITH FUNCTIONAL GROUPS

Functional groups in aliphatic compounds

The naming of compounds containing functional groups follows the same pattern as naming hydrocarbons. We saw that substituent alkyl groups were shown by use of a prefix (for example, methyl) and the $C=C$ or $C\equiv C$ bonds are identified by use of suffixes (ene and yne respectively).

A phenyl group is formed by the loss of a hydrogen atom from benzene:

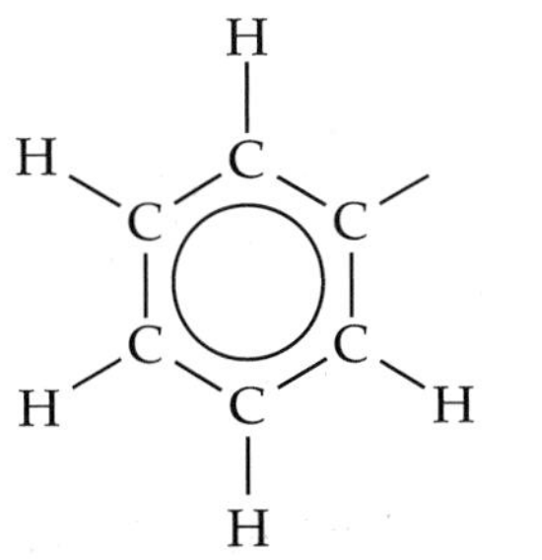

or more simply

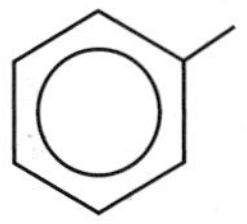

Its presence in a compound is shown by the prefix 'phenyl'. For example, 2-phenylbutane:

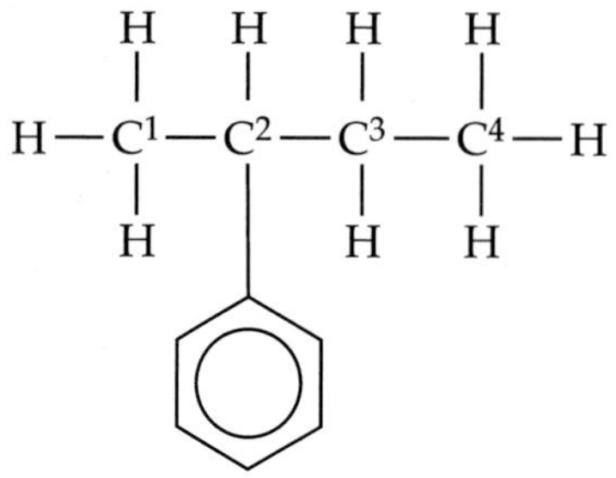

Some functional groups are indicated by the use of a suffix, and therefore are similar to alkenes and alkynes. The 'e' is removed from the name of the alkane and replaced by the appropriate suffix (ending):

alcohols	**ol**
aldehydes	**al**
ketones	**one**
carboxylic acids	**oic acid**
esters	**alkyl...oate**
acid chlorides	**oyl chloride**
amides	**amide**
nitriles	**enitrile**

KEY FACT

Esters are named by treating them as if they were derived from an alkyl group and a group based on a carboxylic acid. The name of the **alkyl** group is placed first, followed by the name for the alkan**oate**.

Other functional groups are shown by the use of a prefix, and therefore naming is similar to alkyl substituents. The prefix is followed by the name of the alkane:

halogenoalkanes	**halo** (fluoro, chloro, bromo, iodo)
amines	**amino**
nitroalkanes	**nitro**

The examples in Table 6.3, where derivatives of propane are listed, will help you to understand the idea.

Table 6.3

Compound	Structural formula	Functional group
propane	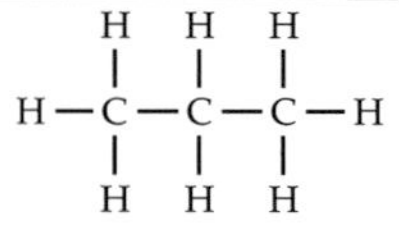	—
propene	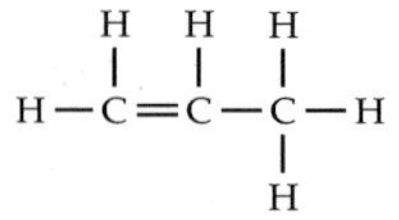	alkene
propyne	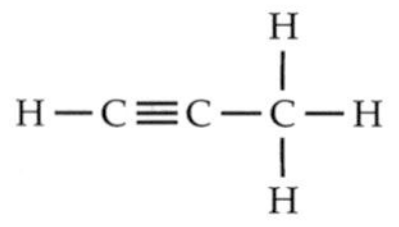	alkyne
1-chloropropane	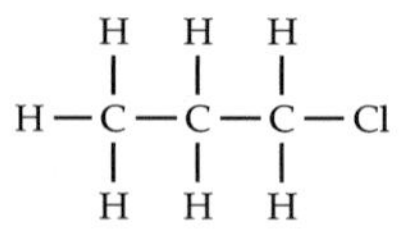	chloro (halogenoalkane)
propan-1-ol	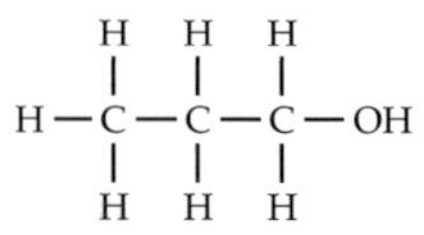	hydroxy (primary alcohol)

Name	Structure	Class
propanal	CH_3-CH_2-CHO	aldehyde
propanone	$CH_3-CO-CH_3$	ketone
propanoic acid	CH_3-CH_2-COOH	carboxylic acid
propanoyl chloride	CH_3-CH_2-COCl	acid chloride
methyl propanoate	$CH_3-CH_2-CO-O-CH_3$	ester
propanamide	$CH_3-CH_2-CO-NH_2$	acid amide
propanenitrile	$CH_3-CH_2-C\equiv N$	nitrile
1-nitropropane	$CH_3-CH_2-CH_2-NO_2$	nitro
1-aminopropane	$CH_3-CH_2-CH_2-NH_2$	amino (primary amine)

Fig 6.17 The irritant present in an ant sting is the smallest possible carboxylic acid, methanoic acid.

CARBOXYLIC ACIDS: THE NATURAL ACIDS

Carboxylic acids are common in living systems and are often produced by bacterial action on food. Vinegar is about 5% ethanoic acid in water produced by bacterial oxidation of ethanol. Rancid butter contains butanoic acid. 2-Hydroxypropanoic acid (lactic acid) is found in sour milk and is also produced in muscles during heavy exercise by the breakdown of carbohydrates, producing the sudden tightening of the muscles known as 'stitch'. Recently, 2-hydroxypropanoic acid has been used to manufacture biodegradable plastics. Another carboxylic acid, 2-ethanoyloxybenzenecarboxylic acid, often called acetylsalicylic acid is perhaps better known as aspirin. Felix Hofmann, working for the Bayer company, first tested aspirin on his arthritic father. After extensive testing Bayer marketed aspirin in 1899.

Problem: Name the compound with structural formula:

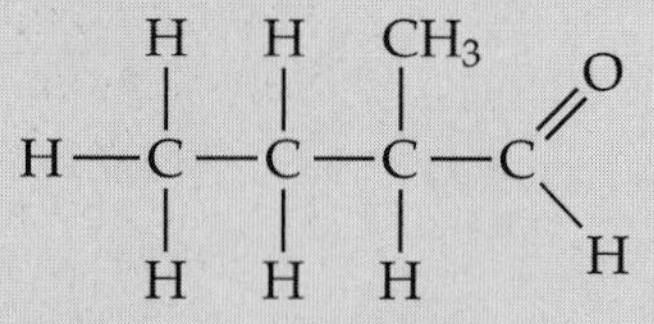

Solution:

Stage 1: The longest unbranched chain contains four carbon atoms, therefore the name is based on **butane**.

Stage 2: The functional group is an **aldehyde** group. Therefore the name is based on **butanal** (butan - 'e' + 'al' = butanal).

Stage 3: There is one substituent group, a **methyl** group, on the **second** carbon atom (counting from the functional group).

Therefore, the name for the compound is

2-methylbutanal

Problem: Name the compound with structural formula:

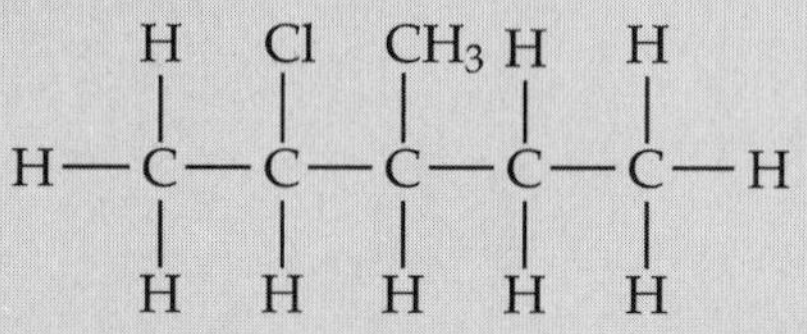

Solution:

Stage 1: The longest unbranched chain contains five carbon atoms, therefore the name is based on **pentane**.

Stage 2: The functional group is a chlorine atom (shown by **chloro**), on the **second** carbon atom.

Stage 3: There is a further substituent, a **methyl** group, on the **third** carbon atom.

Therefore, the name for the compound is

2-chloro-3-methylpentane

Problem: Name the compound with structural formula:

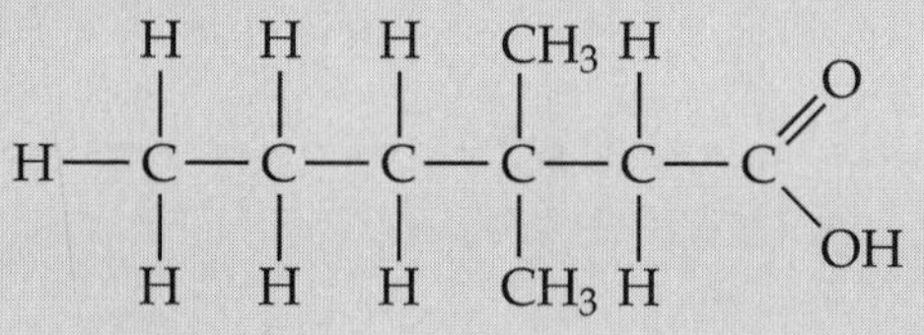

Solution:

Stage 1: The longest unbranched chain contains six carbon atoms, therefore the name is based on **hexane**.

Stage 2: The functional group is a **carboxylic acid**. Therefore the name is based on **hexanoic acid** (hexane – 'e' + 'oic acid' = hexanoic acid).

Stage 3: There are two substituent groups, both **methyl** groups and both on the **third** carbon atom (counting from the functional group).

Therefore, the name for the compound is

3,3-dimethylhexanoic acid

Functional groups in aromatic compounds

We will confine ourselves to compounds which are formed by replacing one or more hydrogen atoms in a benzene molecule. Examples of monosubstituted benzenes are given in Table 6.4. The functional groups they contain are important in organic synthesis. The use of pre-IUPAC names is far more common for aromatic compounds than for aliphatic compounds. The IUPAC names are given in brackets in the table.

Table 6.4 Examples of monosubstituted benzenes

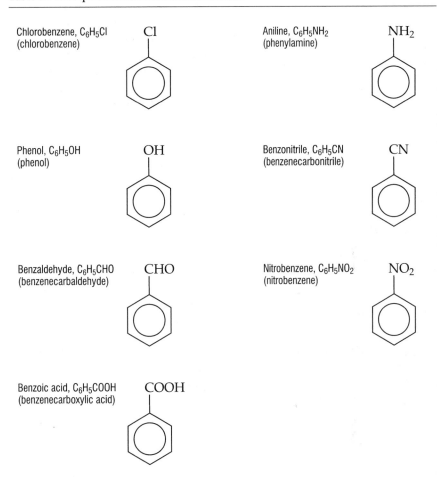

Chlorobenzene, C_6H_5Cl
(chlorobenzene)

Aniline, $C_6H_5NH_2$
(phenylamine)

Phenol, C_6H_5OH
(phenol)

Benzonitrile, C_6H_5CN
(benzenecarbonitrile)

Benzaldehyde, C_6H_5CHO
(benzenecarbaldehyde)

Nitrobenzene, $C_6H_5NO_2$
(nitrobenzene)

Benzoic acid, C_6H_5COOH
(benzenecarboxylic acid)

QUESTIONS

6.18 Name the six simplest primary alcohols.

6.19 **(a)** How many structural isomers containing a benzene ring are there with molecular formula $C_7H_6Br_2$?
(b) Give their names and structural formulae.

6.20 Derive the structural formula of:
(a) butanoic acid; **(b)** butanal; **(c)** butylbenzene;
(d) ethyl butanoate; **(e)** 2-aminobutane; **(f)** butanamide.
Table 6.3 may be helpful. In each case state clearly the class of organic compound to which each belongs.

6.21 'Acrylics' are clothes made from polyacrylonitrile (PAN). The starting material for this polymer is a substance known by the trivial name acrylonitrile. It has the structured formula:

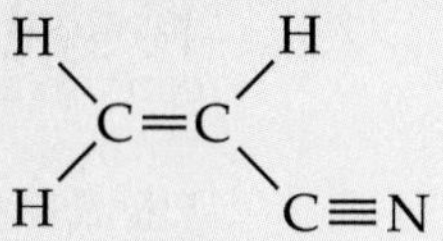

$$\underset{H}{\overset{H}{>}}C=C\underset{C\equiv N}{\overset{H}{<}}$$

(a) What two functional groups does this substance contain.
(b) What would be the IUPAC name for this substance?

6.22 Give the names and structural formulae for structural isomers of C_4H_8O which are (a) aldehydes (b) ketones.

6.23 Draw each of the compounds represented in Table 6.4 circling and naming the functional group it contains.

Answers to Quick Questions: Chapter 6

1 Because they promote the build-up of complex compounds in the body from simpler ones.
2 Chemical reactions can occur in a controlled series of steps rather than one explosive reaction.
3 Because they are liquid and gaseous respectively, imprints of animals and plants cannot be left in them.
4 Carbamide that has been made in the laboratory from simpler starting compounds.
5 Because it is shorter and the shared electrons are shielded less effectively from the nucleus.
6

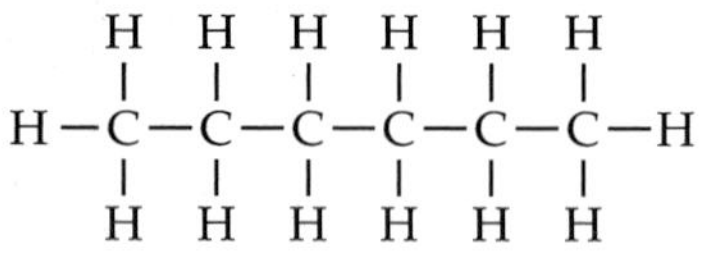

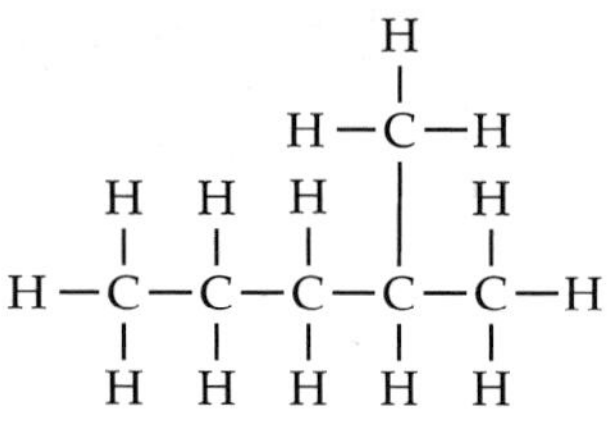

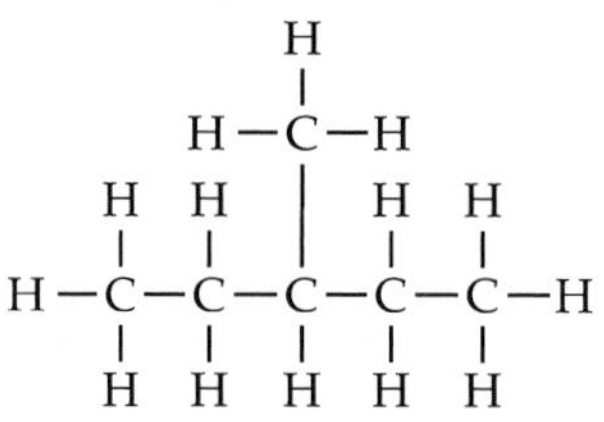

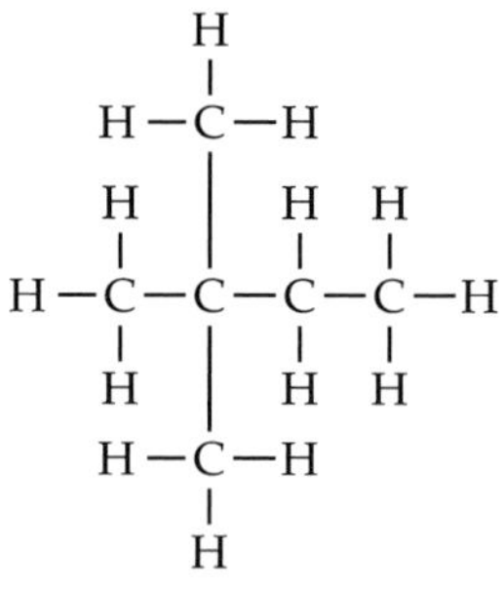

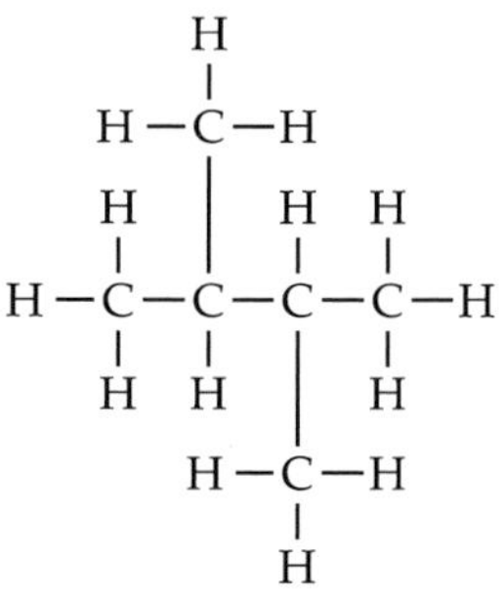

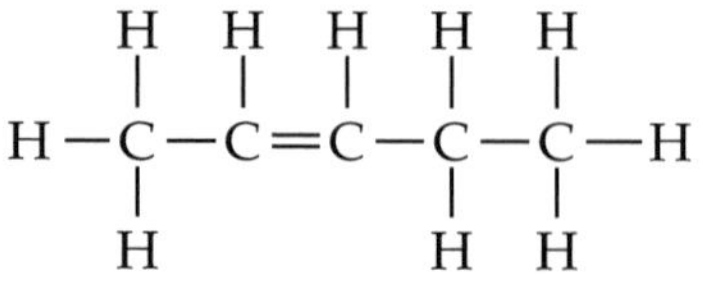

7 The primary amines.
8 The environment about each carbon atom is tetrahedral and so the carbon atoms in the chain are not arranged in a straight line.
9 Because the compound is numbered so that the substituted group is on the lowest numbered carbon atom.
10 (See diagram on left)
11 Because they consist only of discrete covalent bonds, there are no delocalised electrons.
12 Because it has only one carbon atom (as indicated by methene).

Theme 2

ANALYSIS

Identifying and characterising unknown materials; determining the purity of chemical substances and identifying any impurities present; establishing the chemical structure of a substrate. This chemical detective work is a key area of scientific activity.

Chapter 7

ANALYSIS AT THE BENCH

LEARNING OBJECTIVES

When you have studied this chapter you should be able to:

1. explain and give examples of the following types of chemical reactions: acid-base, precipitation, redox, complex formation;

2. explain the underlying chemical principles behind the separation and identification of a mixture containing one or more of the following cations: NH_4^+, Pb^{2+}, Ag^+, Cu^{2+}, Sn^{2+}, Al^{3+}, Cr^{3+} Fe^{3+}, Co^{2+}, Ni^{2+}, Mn^{2+}, Zn^{2+}, Ca^{2+}, Sr^{2+}, Ba^{2+};

3. explain the underlying chemical principles behind the identification of following anions: CO_3^{2-}, HCO_3^-, NO_3^-, Cl^-, Br^-, I^-, SO_4^{2-};

4. write balanced equations for chemical reactions, including ionic equations;

5. describe tests to identify the following classes of organic compounds: alkenes, alcohols, phenols, aldehydes and ketones, carboxylic acids, amines;

6. perform calculations on reacting quantities by using balanced equations;

7. explain the use of volumetric and gravimetric analysis;

8. perform calculations using volumetric and gravimetric analytical data.

7.1 CHEMICAL ANALYSIS

CONNECTION: Separation and purification are discussed in detail in Chapter 8.

The work of the analyst

Analysts identify and characterise unknown materials, both artificial and naturally occurring. They determine the purity of substances and identify impurities in a material. Separation techniques are important to the analytical chemist because, often, samples for analysis are mixtures and the components must be separated in order to identify them.

Analysts work in industrial laboratories, forensic science laboratories, local authority analytical laboratories, hospital pharmacies and so on. Their work is varied. It may lead to the imprisonment of a criminal or an athlete being banned from competing. It may be used to monitor and control pollution and health problems. Analysts also monitor the quality of intermediate compounds and products in a manufacturing process. Often their work helps to develop new and improved materials. Analysis is important, for the analyst's skills make a crucial contribution to many companies and organisations.

Fig 7.1 Accurate titrations play an important part in many quantitative investigations.

Analysis is sometimes performed 'in the field' – anywhere outside the laboratory, where work is carried out on samples where they are found. For example, geologists often take small analytical kits on expeditions. A well thought out kit is portable and allows the rapid identification of rocks and minerals where they are found. Environmentalists use kits to determine the purity of water in rivers and lakes. Soil testing kits, which allow soil pH to be determined and nutrient deficiencies to be identified, are readily available in garden centres.

1   Give another example of an 'on-the-spot' analysis kit.

In the analytical laboratory samples are received for analysis and results reported. The laboratory is judged on the reliability of results and the efficiency with which they are obtained. Efficiency requires a careful balance between accuracy, speed and expense. The analytical chemist often has a choice of techniques for a particular analytical task. Decisions must be made about which to use. For example, the same information might be obtained more quickly by an instrumental technique than by chemical tests. But will it be done often enough to make investment in the instruments worthwhile? What other tasks might the instrument be used for? What is the best solution to the problem?

Laboratory robots are making their impact in analytical chemistry. They need no sleep and can work for 24 hours in a hostile environment. Undoubtedly they will have an increasing influence on analysis in the 21st century. We should remember, however, it is the scientist who designs the tasks for the robot to perform, programs the robot and 'trouble-shoots' when things go wrong.

Fig 7.2   Reagents are automatically measured out and added to the substance under test and the spectrophotometer records the results. The apparatus is programmed to handle dozens of samples without the need for human intervention.

Qualitative and quantitative analysis

Qualitative analysis is about detecting the presence, but not the quantity, of substances. It is concerned with identifying materials, whether they are new compounds or impurities in a sample.

Quantitative analysis is about quantities. It tells us, for example:

- not simply that a drug is impure but the degree of impurity;
- not just that a driver is over the alcohol limit but by how much;
- not just that the local river is polluted by a number of toxic substances but what their concentrations are.

Techniques available for qualitative and quantitative analysis range from those requiring chemical tests using relatively inexpensive glassware to fully automated instrumental techniques. The advantages of instrumental techniques are that they are usually non-destructive (the sample can be recovered after analysis) and very sensitive. Chemical tests are always destructive, but they certainly cost a lot less than modern instruments.

This chapter is concerned with the use of chemical changes for analysis, both qualitative and quantitative.

Equations and reaction quantities

Full equations and ionic equations

We use chemical equations to represent chemical reactions. An equation is shorthand for a great deal of information, given in as little space as possible. It tells us the following about the reactants and products:

- their names;
- their states;
- the relative amounts involved.

CONNECTION: Empirical formulae calculations, pages 69 and 134.

KEY FACT

Quantitative analysis is often the first step in determining the structure of compounds. Elemental analysis, for example, allows us to calculate empirical formulae. Coupled with information about relative molecular mass, this enables the determination of molecular formulae.

There are two types of equations, **full equations** and **ionic equations**. The difference can be illustrated by the reaction between aqueous solutions of silver nitrate and sodium chloride to give a precipitate of silver chloride. The full equation is:

$AgNO_3(aq) + NaCl(aq) \longrightarrow AgCl(s) + NaNO_3(aq)$

However, when ionic compounds dissolve in water they form hydrated ions. Therefore, the reaction might be represented as:

$Ag^+(aq) + NO_3^-(aq) + Na^+(aq) + Cl^-(aq) \longrightarrow AgCl(s) + Na^+(aq) + NO_3^-(aq)$

We can see that $NO_3^-(aq)$ and $Na^+(aq)$ remain unchanged in the reaction – being present in solution before and after the precipitation of silver chloride. They are called spectator ions.

The equation for the reaction simplifies to:

$Ag^+(aq) + Cl^-(aq) \longrightarrow AgCl(s)$

which is the ionic equation for the reaction. There are two advantages of using an ionic equation:

- the equation can be used to represent the formation of silver chloride from any soluble silver salt and soluble chloride.
- the equation is simplified without leaving out any important information.

2 **Express the information provided by the ionic equation above in words.**

Amounts of reactants and products

Information about amounts of reactants and products is vital when we are:

- using chemical reactions for quantitative analysis;
- calculating reacting quantities and yields in a laboratory or industrial preparation.

Simply, the number in front of the formula indicates its mole ratio in the reaction. This is also known as the **stoichiometry**. This is best illustrated by some examples:

$CaCO_3(s) \longrightarrow CaO(s) + CO_2(g)$	In words: one mole of solid calcium carbonate decomposes to give one mole of solid calcium oxide and one mole of gaseous carbon dioxide.
$2C_2H_6(g) + 7O_2(g) \longrightarrow 4CO_2(g) + 6H_2O(l)$	In words: two moles of gaseous ethane react with seven moles of gaseous oxygen to give four moles of gaseous carbon dioxide and six moles of liquid water.
$CaCl_2(s) + aq \longrightarrow Ca^{2+}(aq) + 2Cl^-(aq)$	In words: one mole of solid calcium chloride dissolves in water to give one mole of hydrated calcium ions and two moles of hydrated chloride ions.

KEY FACT

The **stoichiometry** of an equation is the mole ratio of reactants and products in the chemical equation

3 **Write in words all the information contained in the equation:**
$\mathbf{Zn(OH)_2(s) + 2H^+(aq) \longrightarrow Zn^{2+}(aq) + 2H_2O(l)}$

Calculations using balanced equations

Calculations involving reacting quantities may usually be carried out by following this sequence:

1. write the equation for the reaction;
2. convert known quantities of reactants or products into amounts of substance (number of moles);
3. use the equation;
4. convert back from amounts of substance to quantity.

TUTORIAL

Problem: Aluminium chloride, $AlCl_3$, is prepared by the action of dry chlorine on aluminium. Calculate the maximum yield of aluminium chloride that can be obtained from 1.35 g aluminium.
(Molar masses: $Al = 27$ g mol^{-1}; $Cl = 35.5$ g mol^{-1})

Solution:

Stage 1 Equation for the reaction:

$$2Al(s) + 3Cl_2(g) \longrightarrow 2AlCl_3(s)$$

Stage 2 1.35 g aluminium contains $\dfrac{1.35}{27}$ mol Al atoms = 0.05 mol.

Stage 3 From the equation, 2 mol aluminium yields 2 mol aluminium chloride, a 1:1 mole ratio.
Therefore, 0.05 mol Al yields 0.05 mol $AlCl_3$.

Stage 4 Since 1 mol $AlCl_3$ has mass $[27 + (3 \times 35.5)]$ g = 133.5 g;
0.05 mol $AlCl_3$ has mass (0.05×133.5) g = 6.675 g.
Therefore, 1.35 g aluminium can give 6.675 g aluminium chloride.

Problem: Calculate the mass of calcium oxide and the volume of carbon dioxide (measured at room temperature and pressure) produced when 20 g calcium carbonate is decomposed by heating.
(Molar masses: $Ca = 40$ g mol^{-1}; $C = 12$ g mol^{-1}; $O = 16$ g mol^{-1}; molar volume of a gas at room temperature and pressure = 24 dm^3 mol^{-1})

Solution:

Stage 1 Equation for the reaction:

$$CaCO_3(s) \longrightarrow CaO(s) + CO_2(g)$$

Stage 2 One mole of calcium carbonate has a mass of 100 g.

Therefore, 20 g calcium carbonate contains $\dfrac{20}{100}$ mol $CaCO_3$ = 0.2 mol.

Stage 3 From the equation, 1 mol $CaCO_3$ gives 1 mol CaO and 1 mol CO_2.
Therefore, 0.2 mol $CaCO_3$ gives 0.2 mol CaO and 0.2 mol CO_2.

Stage 4 Since 1 mol CaO has a mass of 56 g,
0.2 mol CaO has a mass of (0.2×56) g = 11.6 g.
Since 1 mol CO_2 occupies 24 dm^3 at room temperature and pressure
0.2 mol CO_2 occupies (0.2×24) dm^3 = 4.8 dm^3.
Therefore, 20 g calcium carbonate gives 11.6 g calcium oxide and 4.8 dm^3 carbon dioxide (measured at room temperature and pressure).

Types of chemical reaction

Four important types of chemical reactions are used in analysis. They are:

- acid-base reactions;
- redox reactions;
- complex formation;
- precipitation.

Remember to look out for the reaction types when considering in more detail the analytical methods described in this section.

Acid-base reactions

An acid and a base react to give a salt and water only. These reactions are also called **neutralisation reactions**.

Example 1. $NaOH(aq) + HCl(aq) \longrightarrow NaCl(aq) + H_2O(l)$
(ionic equation: $H^+(aq) + OH^-(aq) \longrightarrow H_2O(l)$)

Example 2. $ZnO(s) + 2HCl(aq) \longrightarrow ZnCl_2(aq) + H_2O(l)$
(ionic equation: $ZnO(s) + 2H^+(aq) \longrightarrow Zn^{2+}(aq) + H_2O(l)$)

Example 3. $Zn(OH)_2(s) + 2HCl(aq) \longrightarrow ZnCl_2(aq) + 2H_2O(l)$
(ionic equation: $Zn(OH)_2(s) + 2H^+(aq) \longrightarrow Zn^{2+}(aq) + 2H_2O(l)$)

Although not strictly acid-base reactions, it is convenient to include the reactions of carbonates and hydrogencarbonates with acids here. For example,

$$CaCO_3(s) + 2HCl(aq) \longrightarrow CaCl_2(aq) + CO_2(g) + H_2O(l)$$

4 **Write an ionic equation for this reaction.**

Redox reactions

An element is said to be **oxidised** if its oxidation number increases as a result of reaction. It is said to be **reduced** if its oxidation number decreases. A redox reaction occurs when one substance is oxidised as another is reduced.

Example 1. $Cl_2(aq) + 2Br^-(aq) \longrightarrow 2Cl^-(aq) + Br_2(aq)$

Example 2. $2Cr^{3+}(aq) + 3H_2O_2(aq) + 2H_2O(l) \longrightarrow 2CrO_4^{2-}(aq) + 10H^+(aq)$

5 **What are the oxidation numbers of manganese and bismuth before and after the following reaction:**

$$2Mn^{2+}(aq) + 5BiO_3^-(aq) + 14H^+(aq) \longrightarrow 5Bi^{3+}(aq) + 2MnO_4^-(aq) + 7H_2O(aq)$$

Complex formation

A complex forms when a **central metal ion** is surrounded by bonded **ligands** (anions or molecules with a lone pair of electrons). Usually the central metal ion is already surrounded by ligands and these are replaced (or substituted) by other ligands. We call this change a **ligand exchange** reaction. For example,

$$[Co(H_2O)_6]^{2+}(aq) + 4Cl^-(aq) \rightleftharpoons CoCl_4^{2-}(aq) + 6H_2O(l)$$

CONNECTION: Neutralisation reactions are dealt with in more detail on page 129.

> ### KEY FACT
>
> The ionic equation for any alkali reacting with any acid is
>
> $H^+(aq) + OH^-(aq) \longrightarrow H_2O(l)$

CONNECTION: Redox reactions described in terms of electron transfer, pages 379–382.

> ### KEY FACT
>
> **Oxidation** may be defined as increase in oxidation number or loss of electrons. **Reduction** may be defined as decrease in oxidation number or gain of electrons.
>
> In a redox reaction, the total increase in oxidation number of atoms and the total decrease in oxidation numbers of other atoms must be equal.

CONNECTION: Complex ions and coordination compounds, section 12.4.

Precipitation

This is the formation of an insoluble compound from two soluble compounds.

Example 1. $Ag^+(aq) + Cl^-(aq) \longrightarrow AgCl(s)$

Example 2. $Ba^{2+}(aq) + SO_4^{2-}(aq) \longrightarrow BaSO_4(s)$

Example 3. $Cu^{2+}(aq) + 2OH^-(aq) \longrightarrow Cu(OH)_2(s)$

The precipitation of metal hydroxides is an example of ligand exchange. Hydroxide ions replace water molecules until a neutral species is formed and the metal hydroxide precipitates:

$$[Fe(H_2O)_6]^{3+}(aq) + 3OH^-(aq) \longrightarrow Fe(H_2O)_3(OH)_3(s) + 3H_2O(l)$$

which we usually simplify by writing the ionic equation:

$$Fe^{3+}(aq) + 3OH^-(aq) \longrightarrow Fe(OH)_3(s)$$

QUESTIONS

7.1 Distinguish between *qualitative* and *quantitative* analysis.

7.2 Why is qualitative analysis of a sample of drinking water usually on its own insufficient to reassure scientists of water quality?

7.3 **(a)** How do instrumental techniques improve *(i)* the accuracy; *(ii)* the efficiency of an analytical chemist's work?
(b) What contribution has automation made to analysis?

7.4 When calcium chloride solution is treated with sodium carbonate solution an instantaneous colourless precipitate of calcium carbonate is observed. For this reaction:
(a) give the full equation;
(b) identify spectator ions;
(c) give the ionic equation.

7.5 Although soluble barium salts are extremely toxic, insoluble barium sulphate is used to investigate the condition of the digestive system. The barium ions show up clearly on X-ray photographs.
(a) Describe how $BaSO_4$ could be prepared from a soluble barium compound such as barium chloride.
(b) Write a full equation for the reaction described in **(a)**.
(c) Identify the spectator ions in this reaction.
(d) What other barium compound could be used in place of the chloride?

7.6 Explain in words precisely what takes place in the reactions described by the equations below:
(a) $2HCl(aq) + Ca(OH)_2(aq) \longrightarrow CaCl_2(aq) + 2H_2O(l)$
(b) $2AlCl_3(s) + 3Na_2CO_3(aq) + 3H_2O(l) \longrightarrow 2Al(OH)_3(s) + 6NaCl(aq) + 3CO_2(g)$
(c) $Pb(NO_3)_2(aq) + MgCl_2(aq) \longrightarrow PbCl_2(s) + Mg(NO_3)_2(aq)$
(d) $FeCl_3(aq) + 3KOH(aq) \longrightarrow Fe(OH)_3(s) + 3KCl(aq)$
(e) $Zn(OH)_2(s) + 2OH^-(aq) \longrightarrow [Zn(OH)_4]^{2-}(aq)$

7.7 Magnesium nitrate, in solution in water, is shown as $Mg(NO_3)_2(aq)$. Explain carefully:
(a) why 'NO_3' is enclosed in brackets;
(b) why there is a '2' *after* the brackets.

7.8 Each of the following compounds is soluble in water. Name each compound and give the amount and formula of each ion produced when one mole of the compound dissolves in water:
(a) KI; (b) Na_2CrO_4; (c) $KMnO_4$; (d) $Cr_2(SO_4)_3$;
(e) $(NH_4)_2SO_4$; (f) $Zn(NO_3)_2$

Relative atomic mass data required for the following questions may be found in the Database.

7.9 A 5 g sample of sulphur reacts with excess fluorine to give sulphur hexafluoride, SF_6. Calculate:
(a) the number of moles of sulphur atoms used;
(b) the mass of sulphur hexafluoride produced;
(c) the mass of sulphur needed to make 100 g of sulphur hexafluoride.

7.10 Nitrogen monoxide reacts with oxygen to give nitrogen dioxide, NO_2. The equation for the reaction is shown below:

$$2NO(g) + O_2(g) \longrightarrow 2NO_2(g)$$

If 142 g of nitrogen monoxide is reacted with 80 g of oxygen, calculate:
(a) which reactant is present in excess;
(b) the mass of product formed;
(c) the mass of the excess reactant left over.

7.11 Lead chromate(VI), $PbCrO_4$, is a yellow pigment present in the paint used for road markings. It precipitates from aqueous solutions of potassium chromate(VI) when hydrated lead ions are added. What is the minimum mass of lead nitrate needed to precipitate all chromate(VI) ions in a solution containing 1.5 g of potassium chromate(VI)? Assume that the solubility of lead chromate(VI) is negligible.

7.12 Anhydrous copper(II) sulphate is a colourless compound which turns blue when wet, a useful test for the presence of water. To make the anhydrous compound, copper(II) sulphate-5-water $CuSO_4.5H_2O$, may be heated.
(a) What mass of the pentahydrate would be needed to prepare 2.5 g of anhydrous copper(II) sulphate?
(b) What mass of water would be liberated in the experiment?

7.13 Baking soda, sodium hydrogencarbonate, may be used as a 'raising agent' in bread and cakes. It is decomposed by heat during baking to form sodium carbonate and carbon dioxide.

$$2NaHCO_3(s) \longrightarrow Na_2CO_3(s) + CO_2(g) + H_2O(g)$$

If 5 g of sodium hydrogencarbonate is used in a cake, calculate:
(a) the mass of carbon dioxide produced;
(b) the volume of carbon dioxide formed, measured at room temperature and pressure;
(c) the mass of sodium carbonate produced.

7.2 QUALITATIVE ANALYSIS

The chemistry behind the identification of cations

Imagine being given an aqueous solution thought to contain one or more of the following cations: NH_4^+, Sn^{2+}, Al^{3+}, Ca^{2+}, Sr^{2+} and Ba^{2+}. How could we determine which were present?

The separation and identification procedure in Table 7.1 has been derived from more complex and sophisticated procedures. The cations have been chosen because:

- chemists often meet these ions and therefore need to understand their chemistry;
- the identification tests illustrate the types of chemical reactions described earlier in this chapter and, therefore, will help us to better understand them.

Reactions are chosen which give selective precipitation of a small number of cations. Further tests are used to identify a specific cation, usually by dissolving the precipitate to give hydrated cations again and examining this solution.

Table 7.1 An analytical procedure to identify: NH_4^+, Pb^{2+}, Ag^+ Cu^{2+}, Sn^{2+}, Al^{3+}, Cr^{3+} Fe^{3+}, Co^{2+}, Ni^{2+}, Mn^{2+}, Zn^{2+}, Ca^{2+}, Sr^{2+}, Ba^{2+} in an aqueous solution.

Step	Procedure	Observation	Inference
	After each step:		**Remember:**
	If no precipitate forms, continue to the next step. If precipitate forms, filter and use filtrate for the next step.		The presence of the cation must be **confirmed** by a suitable test (see text).
STEP 1	Warm the solution with dilute sodium hydroxide.	Evolution of gas, turns universal indicator paper blue	NH_4^+ present
STEP 2	Add dilute hydrochloric acid to the solution and cool thoroughly.	White precipitate	AgCl $PbCl_2$ Ag^+ or Pb^{2+} present
STEP 3	Pass hydrogen sulphide through the weakly acidic solution.	Dark precipitate	CuS (black) SnS (brown) Cu^{2+} or Sn^{2+} present
STEP 4	Boil to remove excess hydrogen sulphide. Cool and add a solution of ammonium chloride in dilute ammonia until alkaline.	Precipitate (white, green or brown)	$Al(OH)_3$ (white) $Cr(OH)_3$ (green) $Fe(OH)_3$ (brown) Al^{3+}, Cr^{3+} or Fe^{3+} present
STEP 5	Add a little more of the solution of ammonium chloride in dilute ammonia and pass hydrogen sulphide.	Precipitate (white, pink or black)	CoS (black) NiS (black) MnS (pale pink) ZnS (white) Co^{2+}, Ni^{2+}, Mn^{2+} or Zn^{2+} present
STEP 6	Boil to remove excess hydrogen sulphide, add ammonium chloride in dilute ammonia, followed by ammonium carbonate solution.	White precipitate	$CaCO_3$ $SrCO_3$ $BaCO_3$ Ca^{2+}, Sr^{2+} or Ba^{2+} present

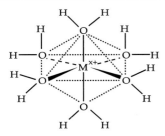

Step 1: Ammonium salts

Ammonium salts release ammonia gas when warmed with dilute sodium hydroxide solution:

$$NH_4^+(aq) + OH^-(aq) \longrightarrow NH_3(g) + H_2O(l)$$

Fig 7.3 The hexaaquametal ion is octahedral.

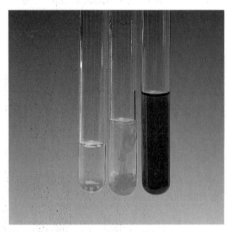

Fig 7.4 The sequence of changes which occur when aqueous ammonia is added to this light blue solution confirms the presence of Cu^{2+}(aq).

Step 2: Silver(I) and lead(II)

AgCl and $PbCl_2$ precipitate upon the addition of dilute hydrochloric acid:

$$Ag^+(aq) + Cl^-(aq) \longrightarrow AgCl(s) \qquad Pb^{2+}(aq) + 2Cl^-(aq) \longrightarrow PbCl_2(s)$$

Confirmation tests: AgCl dissolves in dilute aqueous ammonia as the complex ion $[Ag(NH_3)_2]^+$ (the diamminesilver(I) ion) but $PbCl_2$ is insoluble.

$$AgCl(s) + 2NH_3(aq) \longrightarrow [Ag(NH_3)_2]^+(aq) + Cl^-(aq)$$

6 **What general observation can be made about the solubility of chlorides?**

Step 3: Copper(II) and tin(II)

CuS and SnS precipitate upon bubbling hydrogen sulphide into the solutions:

$$M^{2+}(aq) + H_2S(g) \longrightarrow MS(s) + 2H^+(aq); \text{ where } M = Cu, Sn$$

Hydrogen sulphide is a toxic gas and so must be used with great caution. **Confirmation tests:** Both sulphides dissolve in dilute nitric acid as M^{2+}(aq).

Addition of aqueous ammonia to Cu^{2+}(aq) gives a pale blue precipitate of copper(II) hydroxide. When further ammonia is added, this precipitate dissolves as a mixture of copper(II)/ammine complex ions. The most common one in concentrated aqueous ammonia is $[Cu(H_2O)(NH_3)_5]^{2+}$, which has a deep, royal blue colour. In dilute aqueous ammonia, the most common complex ion formed is $[Cu(H_2O)_2(NH_3)_4]^{2+}$, which is also deep blue in colour.

$$Cu^{2+}(aq) + 2OH^-(aq) \longrightarrow Cu(OH)_2(s)$$
$$Cu(OH)_2(s) + 5NH_3(aq) + H_2O(l) \longrightarrow [Cu(H_2O)(NH_3)_5]^{2+}(aq) + 2OH^-(aq)$$

Addition of sodium hydroxide solution to Sn^{2+}(aq) gives a white precipitate, which dissolves as the complex ion $[Sn(OH)_4]^{2-}$ (the tetrahydroxystannate(II) ion) when further sodium hydroxide is added.

$$Sn^{2+}(aq) + 2OH^-(aq) \longrightarrow Sn(OH)_2(s)$$
$$Sn(OH)_2(s) + 2OH^-(aq) \longrightarrow [Sn(OH)_4]^{2-}(aq)$$

7 **Write an equation for the reaction between copper(II) sulphide and dilute nitric acid.**

Step 4: Aluminium(III), chromium(III) and iron(III)

$Al(OH)_3$, $Cr(OH)_3$ and $Fe(OH)_3$ precipitate if aqueous ammonia is added until the solution becomes weakly alkaline:

$$M^{3+}(aq) + 3OH^-(aq) \longrightarrow M(OH)_3(s); \text{ where } M = Al, Cr, Fe$$

Confirmation tests: All three hydroxides dissolve in dilute hydrochloric acid as M^{3+}(aq).

Addition of sodium hydroxide solution to Al^{3+}(aq) gives a white precipitate of aluminium hydroxide which dissolves as the complex ion $[Al(OH_4]^-$ (the tetrahydroxyaluminate(III) ion) when further alkali is added.

$$Al(OH)_3(s) + OH^-(aq) \longrightarrow [Al(OH)_4]^-(aq)$$

8 **Write an ionic equation for the reaction between aluminium hydroxide and dilute hydrochloric acid.**

Addition of hydrogen peroxide and sodium hydroxide to Cr^{3+}(aq) gives a yellow solution due to chromate(VI) ions, CrO_4^{2-}(aq). The chromium(III) ions have been oxidised.

$$2Cr^{3+}(aq) + 3H_2O_2(aq) + 2H_2O(l) \longrightarrow 2CrO_4^{2-}(aq) + 10H^+(aq)$$

9 What has been reduced in this equation?

Addition of ammonium thiocyanate solution to $Fe^{3+}(aq)$ gives an intense red solution due to the formation of a thiocyanate-iron(III) complex ion by ligand exchange.

$$[Fe(H_2O)_6]^{3+}(aq) + SCN^-(aq) \longrightarrow [Fe(SCN)(H_2O)_5]^{2+}(aq) + H_2O(l)$$

Step 5: Cobalt(II), nickel(II), manganese(II) and zinc(II)

CoS, NiS, MnS and ZnS precipitate when hydrogen sulphide is passed through weakly alkaline solutions:

$$M^{2+}(aq) + H_2S(g) \longrightarrow MS(s) + 2H^+(aq); M = Co, Ni, Mn, Zn$$

Confirmation tests: All four sulphides dissolve in dilute acids as $M^{2+}(aq)$.

Addition of concentrated hydrochloric acid to $Co^{2+}(aq)$ gives a deep blue solution. The complex ion $[CoCl_4]^{2-}$ (the tetrachlorocobaltate(II) ion) forms. Addition of water to this solution gives a pink solution due to the formation of the octahedral complex ion $[Co(H_2O)_6]^{2+}$ (the hexaaquacobalt(II) ion). This is an example of a ligand displacement reaction with a change in coordination number. It differs from the formation of the thiocyanate-iron(III) complex in that the number of ligands which surround the central metal ion changes.

$$[Co(H_2O)_6]^{2+}(aq) + 4Cl^-(aq) \rightleftharpoons [CoCl_4]^{2-}(aq) + 6H_2O(l)$$

10 This reaction takes place with a change in coordination number for Co. What does this mean and what are the coordination numbers before and after reaction?

Addition of sodium hydroxide solution to $Ni^{2+}(aq)$ or $Zn^{2+}(aq)$ gives green and white precipitates respectively of the metal hydroxides. When further sodium hydroxide is added, zinc hydroxide dissolves as the complex ion $[Zn(OH)_4]^{2-}$ (the tetrahydroxyzincate(II) ion) but nickel hydroxide does not dissolve.

$$M^{2+}(aq) + 2OH^-(aq) \longrightarrow M(OH)_2(s); \text{ where } M = Ni, Zn$$
$$Zn(OH)_2(s) + 2OH^-(aq) \longrightarrow [Zn(OH)_4]^{2-}(aq)$$

Addition of concentrated nitric acid and sodium bismuthate (a very powerful oxidising agent) to $Mn^{2+}(aq)$ gives a purple solution containing $MnO_4^-(aq)$. The manganese has been oxidised from the +2 oxidation state to the +7 oxidation state.

$$2Mn^{2+}(aq) + 5BiO_3^-(aq) + 14H^+(aq) \longrightarrow 5Bi^{3+}(aq) + 2MnO_4^-(aq) + 7H_2O(aq)$$

11 Is nickel hydroxide amphoteric?

Step 6: Calcium, strontium and barium

CaCO₃, SrCO₃ and BaCO₃ precipitate when ammonium carbonate is added:

$$M^{2+}(aq) + CO_3^{2-}(aq) \longrightarrow MCO_3(s); \text{ where } M = Ca, Sr, Ba$$

Confirmation tests: All three carbonates react with dilute hydrochloric acid to form $M^{2+}(aq)$.

Addition of potassium chromate(VI) solution to $Ca^{2+}(aq)$ gives no precipitate. However, when it is added to $Sr^{2+}(aq)$ or $Ba^{2+}(aq)$ yellow precipitates form. These can be distinguished since strontium chromate(VI) is insoluble in dilute ethanoic acid, but barium chromate(VI) is soluble.

$$M^{2+}(aq) + CrO_4^{2-}(aq) \longrightarrow MCrO_4(s); \text{ where } M = Sr, Ba$$

Fig 7.5 Addition of concentrated hydrochloric acid to the pink solution changes the colour to an intense blue. Addition of water causes the solution to return to its original pink colour. This sequence confirms the presence of $Co^{2+}(aq)$.

CHEMICALS AND COLOUR CHANGES

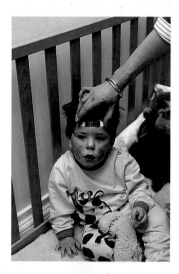

Thermochromic materials change colour with changes in temperature and these may be used as indicators of temperature change. The property can be demonstrated by placing a test tube containing a deep blue aqueous solution of $CoCl_4^{2-}$ in an iced-water bath. It rapidly becomes pale pink. On warming, the deep blue colour returns. Some substances change colour at very precise temperatures. This property has allowed the development of specialised thermometers.

Other chemicals change colour as a result of absorbing moisture. The familiar humidity indicators which are pink for damp weather and blue for dry rely on the dynamic equilibrium between octahedral cobalt(II) and tetrahedral cobalt(II). When the atmosphere is moist pink $[Co(H_2O)_6]^{2+}$ is formed. In a dry atmosphere, blue $CoCl_4^{2-}$ is dominant.

Fig 7.6 A thermionic thermometer being used to measure a baby's temperature

QUESTIONS

7.14 Write equations to represent ligand exchange reactions which take place as excess sodium hydroxide is *slowly* added to a solution of aluminium ions. Assume that $[Al(H_2O)_6]^{3+}$ is the species present initially.

7.15 Write a full equation for the reaction between ammonium sulphate, $(NH_4)_2SO_4$ and sodium hydroxide.

7.16 Explain in terms of complex ion formation how *pure* samples of copper(II) hydroxide and tin(II) hydroxide may be obtained from a mixture of copper(II) sulphate and tin(II) sulphate.

7.17 Give the names and formulae of *three* amphoteric hydroxides.

7.18 Ligand exchange reactions occur by a **stepwise mechanism**. Explain this term with reference to the following overall equation:

$$[Ni(H_2O)_6]^{2+}(aq) + 6NH_3(aq) \longrightarrow [Ni(NH_3)_6]^{2+}(aq) + 6H_2O(l)$$

7.19 Iron(III) chloride and sodium chromate(VI) both dissolve in water to form yellow solutions. Explain with the aid of an equation how you could perform a simple test to distinguish between them.

7.20 A solution containing magnesium chloride and zinc chloride was treated with excess sodium hydroxide solution. A white precipitate was obtained. This was filtered off, yielding a white solid on the filter paper. The filtrate was acidified slowly with dilute hydrochloric acid. At one point the solution became cloudy but rapidly cleared when a few more drops of acid were added. Explain:
(a) what was filtered off from the solution;
(b) why the solution became cloudy;
(c) why it cleared again.

7.21 A sample of copper(II) bromide, $CuBr_2$, was dissolved in aqueous ammonia solution, forming a deep royal blue solution. On evaporation of this solution blue crystals were obtained. On elemental analysis these crystals proved to have the formula $Br_2CuH_{16}N_4O_2$. Suggest an identity for the crystals formed.

7.22 A student dissolved chromium(III) sulphate, $Cr_2(SO_4)_3$, in water and tested the solution with universal indicator paper. To her surprise it was acidic.

(a) Suggest a reason for this result.

(b) State, in terms of complex ions, what would happen to the chromium-containing species as excess concentrated $NaOH(aq)$ is gradually added.

Identifying anions

CONNECTION: Properties of inorganic compounds are summarised on Datasheets 6–15.

Many anions can be identified using simple test-tube reactions.

Carbonates or hydrogencarbonates

These react with cold, dilute acids to give carbon dioxide. For example,

$$Na_2CO_3(s) + 2H^+(aq) \longrightarrow 2Na^+(aq) + CO_2(g) + H_2O(l)$$
$$CaCO_3(s) + 2H^+(aq) \longrightarrow Ca^{2+}(aq) + CO_2(g) + H_2O(l)$$

The gas given off is passed through calcium hydroxide solution (often called limewater). A white precipitate of calcium carbonate confirms carbon dioxide.

$$Ca(OH)_2(aq) + CO_2(g) \longrightarrow CaCO_3(s) + H_2O(l)$$

12 **Write an equation for the reaction of sodium hydrogencarbonate with dilute nitric acid.**

13 **Do you think that carbon dioxide is an acid or a base?**

Nitrates

All nitrates are water soluble. The addition of iron(II) sulphate to a solution of nitrate ions, followed by concentrated sulphuric acid to form a lower layer, gives a brown ring between the two layers. This confirms the presence of nitrate ions (Fig 7.7). The colour is due to the formation of the complex ion $[Fe(H_2O)_5NO]^{2+}(aq)$.

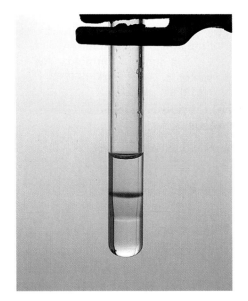

Fig 7.7 The 'brown-ring' test for nitrates.

Halides

Solid halides react with concentrated sulphuric acid to give the hydrogen halide, HX (where X = Cl, Br, I). The reactions of sodium halides illustrate this:

$$NaX(s) + H_2SO_4(l) \longrightarrow NaHSO_4(s) + HX(g)$$

Hydrogen chloride (a colourless gas) does not react further. However, HBr and HI react with concentrated sulphuric acid and this enables us to distinguish the halides. Bromides give a red-brown vapour (bromine) and iodides give a violet vapour (iodine).

$$2HBr(g) + H_2SO_4(l) \longrightarrow Br_2(g) + SO_2(g) + 2H_2O(l)$$
$$8HI(g) + H_2SO_4(l) \longrightarrow 4I_2(g) + H_2S(g) + 4H_2O(l)$$

These reactions show the relative ease of oxidation of the hydrogen halides (Fig 7.8).

CONNECTION: Further properties of Group VII compounds may be found on Datasheet 13.

14 **What are the oxidation numbers of the halogens in the reactions above?**

15 **Hydrogen iodide and sulphuric acid react to produce some sulphur as a minor product. How do you explain this?**

Fig 7.8 The reaction of solid halides with hot, concentrated sulphuric acid. The colours allow us to identify the halide ion, from left to right chloride, bromide and iodide.

Fig 7.9 The characteristic colours of chlorine (left), bromine (centre) and iodine (right) in hexane solution allows us to distinguish between them.

Aqueous solutions of halide ions (except fluorides) give precipitates with aqueous silver nitrate:

$$Ag^+(aq) + X^-(aq) \longrightarrow AgX(s); \text{ where } X = Cl, Br, I$$

AgCl is white and readily soluble in dilute aqueous ammonia, AgBr is a pale cream colour and slightly soluble in excess aqueous ammonia, AgI is pale yellow and insoluble in aqueous ammonia. AgCl darkens easily on exposure to light.

Once the presence of a halide has been established, its identity may be confirmed by making use of the redox reactions:

$$Cl_2(aq) + 2Br^-(aq) \longrightarrow 2Cl^-(aq) + Br_2(aq)$$
$$Cl_2(aq) + 2I^-(aq) \longrightarrow 2Cl^-(aq) + I_2(s)$$

Chlorine is passed through the aqueous solution:

* if the solution remains colourless, chloride ions were present;
* if the solution becomes red-brown, bromide ions were present;
* if the solution becomes violet, iodide ions were present.

The presence of the halogen can be confirmed by shaking the aqueous solution with a suitable organic solvent. The halogens are sparingly soluble in water (with solubility decreasing down the group, chlorine to iodine) but soluble in organic solvents such as hexane. These organic solutions have characteristic colours and extraction of the halogen from aqueous solution can be used to confirm the halogen present (Fig 7.9).

16 **Why are the halogens more soluble in organic solvents than they are in water?**

Sulphates

With a few exceptions, sulphates are water soluble. The addition of barium chloride to a solution of sulphate ions gives an insoluble white precipitate of barium sulphate (Fig 7.10).

$$Ba^{2+}(aq) + SO_4^{2-}(aq) \longrightarrow BaSO_4(s)$$

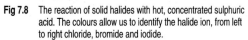

Fig 7.10 The formation of a precipitate of barium sulphate.

7.23 Write balanced equations for the reactions between:
(a) sodium carbonate and sulphuric acid;
(b) calcium carbonate and hydrochloric acid;
(c) magnesium hydrogencarbonate and nitric acid.

7.24 A green crystalline salt was dissolved in water. The solution was divided into 3 portions (i), (ii) and (iii) which were tested as follows:
(i) $NaOH(aq)$ was added (a blue precipitate was obtained);
(ii) $NH_3(aq)$ was added (an initial blue precipitate redissolved to give a royal blue solution);
(iii) $FeSO_4(aq)$ was added, followed by concentrated H_2SO_4 (a brown ring was formed).
Suggest a possible identity for the salt.

7.25 (a) Write an equation for the reaction of $AgNO_3(aq)$ with $NaBr(aq)$.
(b) How may the resulting precipitate be distinguished from that obtained by adding $NaCl(aq)$?
(c) How does use of concentrated H_2SO_4 help to distinguish between *solid* NaCl and NaBr?

7.26 Bromine is obtained industrially by treating acidified sea water with chlorine followed by extraction into an organic solvent. Write an equation for the reaction which produces bromine.

7.27 Barium ethanoate, $Ba(CH_3COO)_2$ is one of few soluble barium salts. On addition of $H_2SO_4(aq)$ to a solution of this compound a white precipitate is obtained.
(a) Explain how this helps to confirm the presence of barium ions in the solution.
(b) Write an equation for the reaction which takes place.

Qualitative organic analysis

It is often useful to be able to identify the functional groups in an organic compound. Along with information about its molecular formula, the identification of functional groups may enable us to work out the structural formula of the compound.

Qualitative analysis of organic compounds is often more complex than the qualitative analysis of cations and anions. This is partly due to the modification of the properties of a functional group by the rest of the molecule. Simple analysis normally identifies the presence of a functional group, but not its environment.

The following are some simple tests which may be used to suggest the presence of a number of important functional groups. As with the cations and anions, selection has been made on the basis of functional groups which are encountered frequently in organic compounds. The list, however, is not exhaustive.

CONNECTION: Other reactions used to identify functional groups are contained in Datasheets 16–23.

Alkenes

Alkenes are insoluble in water. They react with bromine in an organic solvent, such as tetrachloromethane, changing the solution from red-brown to colourless. For example,

$$C_2H_4 + Br_2 \longrightarrow C_2H_4Br_2$$
ethene   1,2-dibromoethane

A NEW KIT FOR BLOOD AND URINE ANALYSIS

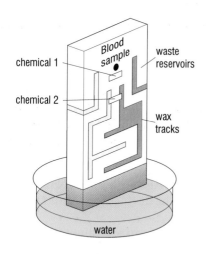

chemical 1

chemical 2

Blood sample

waste reservoirs

wax tracks

water

Fig 7.11 This filter paper analytical kit is simple, cheap and effective means of testing blood and urine samples. It has applications for soil, food and environmental testing and has the advantage that its compact size allows for testing 'in the field'.

An elegant piece of design by a team of British chemists and engineers has enabled samples of blood and urine to be analyzed quickly and cheaply. Potential applications are numerous and include, for example, pregnancy testing and identifying the presence of HIV antibodies.

The method relies on the absorbancy of filter paper. Channels through which water (the eluting solvent) can flow are marked by wax lines a few millimetres apart. Solutions of the appropriate chemical reagents are dropped onto the filter paper at the right places and allowed to dry. The sample of blood or urine is applied in the same way.

When the paper is dipped into a reservoir of water the water moves up the paper by two different routes. As it reaches the reagents it redissolves them and delivers them in the appropriate order to the sample being tested. The reagent to be delivered first is placed in a relatively straight channel. The reagent which must arrive next follows a less direct route. Colour changes take place as each reagent reaches the sample. The technique can be expanded to allow delivery to the sample of several reagents in a precise order.

Alcohols

Alcohols with low molar masses dissolve in water to give a neutral solution. Solubility decreases with increasing molar mass. Alcohols react with phosphorus pentachloride at room temperature to give a chloroalkane and hydrogen chloride. For example, ethanol reacts with PCl_5:

$$C_2H_5OH + PCl_5 \longrightarrow C_2H_5Cl + POCl_3 + HCl$$
ethanol   chloroethane

17 **Name the organic product when propan-2-ol reacts with phosphorus pentachloride.**

Phenols

Phenols are insoluble in water but dissolve in alkali, from which they may be precipitated by the addition of acid. For example,

$$C_6H_5OH(s) + OH^-(aq) \longrightarrow C_6H_5O^-(aq) + H_2O(l)$$
phenol   phenoxide ion

$$C_6H_5O^-(aq) + H^+(aq) \longrightarrow C_6H_5OH(s)$$

Phenols react with a neutral solution of iron(III) chloride to give characteristic intense violet/blue colours.

Carboxylic acids

Carboxylic acids with low molar masses dissolve in water to give an acidic solution. Solubility decreases with increasing molar mass. All carboxylic acids dissolve in alkali and the water-insoluble ones may be precipitated from these solutions by the addition of acid. For example,

$$C_6H_5COOH(s) + OH^-(aq) \longrightarrow C_6H_5COO^-(aq) + H_2O(l)$$
benzoic acid   benzoate ion

$$C_6H_5COO^-(aq) + H^+(aq) \longrightarrow C_6H_5COOH(s)$$

Carboxylic acids react with carbonates and hydrogencarbonates to give carbon dioxide. This reaction distinguishes them from phenols. For example,

$$CH_3COOH(aq) + HCO_3^-(aq) \longrightarrow CH_3COO^-(aq) + H_2O(l) + CO_2(g)$$
ethanoic acid ethanoate ion

18 Write an equation for the reaction between methanoic acid, **HCOOH, and sodium carbonate.**

Aldehydes and ketones

Aldehydes and ketones with low relative molecular mass dissolve in water to give neutral solutions. Solubility decreases with increasing relative molecular mass. Both react with 2,4-dinitrophenylhydrazine to give yellow/orange/red crystalline precipitates because they both contain the carbonyl group, $\underset{/}{\overset{\backslash}{C}}=O$. For example,

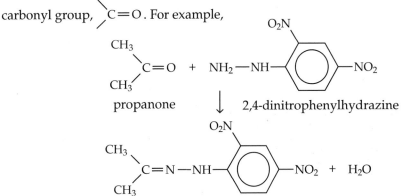

propanone 2,4-dinitrophenylhydrazine

propanone 2,4-dinitrophenylhydrazine

19 Why do you think the solubility of aldehydes and ketones decreases with increasing relative molecular mass?

Aldehydes and ketones may be distinguished by their reaction with either **Fehling's solution** (an alkaline solution of a copper(II) complex) or the diamminesilver(I) ion (the **silver mirror test**). Aldehydes reduce copper(II) to form a red solid, copper(I) oxide, and silver(I) to silver(0). For example,

$$CH_3CHO(aq) + 2Cu^{2+}(aq) + 2H_2O(l) \longrightarrow CH_3COOH(aq) + Cu_2O(s) + 4H^+(aq)$$

$$C_2H_5CHO(aq) + 2Ag^+(aq) + 2H_2O(l) \longrightarrow C_2H_5COOH(aq) + 2Ag(s) + 4H^+(aq)$$

20 Write an equation for the reaction between butanal, C_3H_7CHO, and **Fehling's solution.**

Fig 7.12 This silver mirror is produced when an aldehyde reduces silver from the +1 oxidation state in the diamminesilver(I) ion, $[Ag(NH_3)_2]^+$, to the zero oxidation state in metallic silver, Ag.

Amines

Aliphatic amines dissolve in water to give an alkaline solution. They have a fishy odour. Aromatic amines are insoluble in water but dissolve in acids, from which they may be recovered by the addition of alkali. For example,

$$C_6H_5NH_2(l) + H^+(aq) \longrightarrow C_6H_5NH_3^+(aq)$$
aminobenzene phenylammonium ion
(phenylamine)

$$C_6H_5NH_3^+(aq) + OH^-(aq) \longrightarrow C_6H_5NH_2(l) + H_2O(l)$$

A drop of concentrated hydrochloric acid held in the amine vapour gives a dense white cloud of the chloride (similar to the result obtained with ammonia vapour). For example,

$$C_2H_5NH_2(l) + HCl(g) \longrightarrow C_2H_5NH_3^+Cl^-(s)$$
aminoethane ethylammonium chloride
(ethylamine)

QUESTIONS

7.28 Cyclohexene, C_6H_{10}, reacts rapidly with bromine in 1,1,1-trichloroethane solution to give a colourless solution.
(a) Why is the solution colourless?
(b) Give the structural formula of the organic product;
(c) How could you tell when all the cyclohexene had reacted?

7.29 An organic compound Q, with molecular formula $C_4H_{10}O$, reacted with phosphorus pentachloride. A compound of formula C_4H_9Cl was obtained and fumes of HCl were detected.
(a) Give two possible structural formulae for Q.
(b) Give equations for the reactions of the compounds in **(a)** with PCl_5.

7.30 Some common antiseptics contain 2,4,6-trichlorophenol (TCP). Alkaline solutions of this substance are clear but on addition of HCl(aq) the solution becomes milky.
(a) Give the structural formula of the organic species present in alkaline solution.
(b) Explain what happens when the alkaline solution is acidified with hydrochloric acid.
(c) How could the presence of a phenol be confirmed?

7.31 Write an equation of propanoic acid with potassium carbonate.

7.32 Explain why most aromatic carboxylic acids are insoluble in water yet dissolve readily in sodium hydroxide.

7.33 An organic compound S, molecular formula C_3H_6O, reacted with 2,4-dinitrophenylhydrazine (2,4-DNPH) to give an orange/red precipitate. When warmed with ammoniacal silver nitrate a shiny surface was seen on the inside of the test-tube.
(a) Give the structural formula of S.
(b) Draw the structure of the product formed with 2,4-DNPH.
(c) What caused the shiny surface?

7.34 A water-soluble salt of formula $C_4H_{12}ClN$ gave a fishy odour on warming with dilute NaOH(aq). Warming with dilute HCl(aq) produced no odour.
(a) Explain this observation.
(b) Give one possible structure for the salt.
(c) Why is an aqueous solution of this salt acidic?

7.35 When equimolar quantities of methanoic acid and ethylamine (both liquids) are added to water a colourless solution is formed. Careful evaporation of this solution produced a crystalline solid.
(a) How may this observation be explained?
(b) What is the formula of the solid?

7.3 QUANTITATIVE ANALYSIS

KEY FACT

An indicator is a compound which changes colour close to the end point of a titration. The indicator is chosen so that it changes colour when a drop of excess reagent has been added.

Volumetric analysis

Volumetric analysis is quantitative analysis which uses measurement of volumes. It is sometimes called **titrimetric analysis** because it involves titrations. A general method is as follows:

A measured quantity of material is dissolved and made up to a known volume of solution in a volumetric flask. An aliquot of this is transferred by pipette to a conical flask. After addition of a suitable indicator (if required) the aliquot is titrated with a reagent of known concentration from a burette. Such a solution is called a standard solution. The end-point for the titration (when sufficient reagent has been added) is identified by a change in colour, (Fig 7.14).

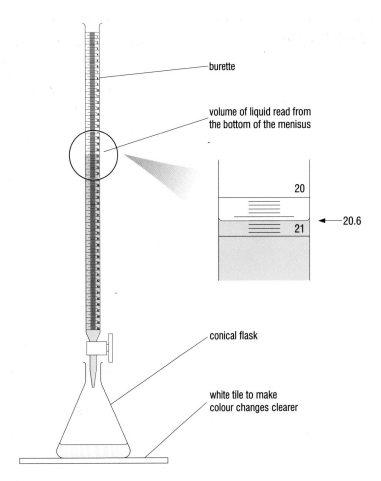

burette

volume of liquid read from the bottom of the menisus

20

21 ← 20.6

conical flask

white tile to make colour changes clearer

Fig 7.13   In a titration, a standard solution is added from a burette to the sample to be analysed. Care must be taken to add the solution very slowly as the end point is approached.

Fig 7.14   In this acid-base titration the indicator used is phenolphthalein. It is colourless in acid solutions and pink in alkaline solutions.

Acid-base titrations

Neutralisation reactions can be used for quantitative volumetric analysis. This type of volumetric analysis is often called an **acid-base titration**, and includes the reactions of carbonates and hydrogencarbonates with acids even though these are not strictly acid-base reactions. We shall look at two typical examples:

Estimating the strength of vinegar. Vinegar is an aqueous solution of ethanoic acid, CH_3COOH, often with a little caramel added to give it a brown colour. The concentration of ethanoic acid may be determined by using an acid-base titration. A known volume of the vinegar to be analysed is diluted with distilled water and titrated against a standard solution of sodium hydroxide, using phenolphthalein as the indicator.

$$CH_3COOH(aq) + NaOH(aq) \longrightarrow CH_3COONa(aq) + H_2O(l)$$

21   How many moles of sodium hydroxide are needed to react completely with 20 cm³ of 1 mol dm⁻³ ethanoic acid solution?

The sodium carbonate content of washing soda crystals. Washing soda crystals, hydrated sodium carbonate, are used to soften water. The Na_2CO_3 content can be determined by dissolving an accurately weighed sample in water and making the solution up to a known volume. Aliquots are titrated with a standard solution of hydrochloric acid, using bromo-cresol green as the indicator. The mass of anhydrous sodium carbonate present in the sample can be determined from the equation:

$$Na_2CO_3(aq) + 2HCl(aq) \longrightarrow 2NaCl(aq) + H_2O(l) + CO_2(g)$$

Calculations for volumetric analysis: acid-base titration

Problem: 40.0 cm³ of dilute sodium hydroxide requires 20.0 cm³ of 0.10 mol dm⁻³ hydrochloric acid for complete reaction. Phenolphthalein is used as indicator. What is the concentration of the sodium hydroxide solution?

Solution:
Stage 1 Equation for the reaction:

$$HCl(aq) + NaOH(aq) \longrightarrow NaCl(aq) + H_2O(l)$$

Stage 2 20.0 cm³ of 0.10 mol dm⁻³ HCl(aq) contains

$$\frac{20}{1000} \times 0.1 \text{ mol HCl} = 2 \times 10^{-3} \text{ mol}$$

Stage 3 From the equation, 1 mol HCl reacts with 1 mol NaOH
Therefore, 2×10^{-3} mol HCl reacts with 2×10^{-3} mol NaOH

Stage 4 Therefore, 40.0 cm³ NaOH(aq) contains 2×10^{-3} mol NaOH

and 1 dm³ NaOH(aq) contains $2 \times 10^{-3} \times \dfrac{1000}{40}$ mol NaOH

$= 0.050$ mol
Therefore, the concentration of the dilute sodium hydroxide solution is 0.050 mol dm⁻³.

Fig 7.15 These and many other substances can be analysed by titration methods. These analyses are carried out for quality control and consumer protection.

Redox titrations

Titrimetric analyses involving oxidation-reduction reactions are known as **redox titrations**. Here are two typical examples:

Manganate(VII) titrations. An aqueous solution of manganate(VII) ions, MnO_4^-(aq), is an intense purple (pink when very dilute). In acid solution MnO_4^-(aq) is reduced to colourless Mn^{2+}(aq). Any substance which reduces MnO_4^- can be determined by titration with standard MnO_4^- solution. No indicator is necessary because of the easily observed change in colour when just sufficient MnO_4^-(aq) has been added to oxidise the substance being determined. The pale pink colour persists in solution.

Iron tablets on sale at a chemist's shop contain iron(II) sulphate. The iron content can be determined by titration with acidified standard MnO_4^- solution.

$$MnO_4^-(aq) + 5Fe^{2+}(aq) + 8H^+(aq) \longrightarrow Mn^{2+}(aq) + 5Fe^{3+}(aq) + 4H_2O(l)$$

22 You are provided with two solutions. Solution A contains 1 mol dm⁻³ MnO_4^-(aq) and solution B contains 1 mol dm⁻³ Fe^{2+}(aq). What volume of solution A must be added to 20 cm³ of solution B to produce a faint but permanent pink colour?

Iodine-thiosulphate titrations. Iodine is produced when a suitable oxidising agent is added to a solution of iodide ions. The liberated iodine can be determined by titration with a standard solution of thiosulphate ions, $S_2O_3^{2-}$(aq). Starch forms an intense blue-black colour with iodine, and the change from black to white is used to indicate the end point of an iodine-thiosulphate titration. The starch solution is usually added when the iodine colour has diminished to a pale straw colour.

$$I_2(aq) + 2S_2O_3^{2-}(aq) \longrightarrow 2I^-(aq) + S_4O_6^{2-}(aq)$$

Note Iodine is not very soluble in water, but does dissolve in aqueous solutions of iodide ions as the complex ion I_3^-(aq). For simplicity, however, we use I_2(aq).

Liquid bleach is an aqueous solution containing sodium chlorate(I), NaOCl(aq). Its concentration can be determined indirectly by an iodine-thiosulphate titration. Chlorate(I) ions react with iodide ions in dilute acidic solution to produce iodine. The concentration of chlorate(I) ions may be determined by titrating the liberated iodine with standard sodium thiosulphate solution.

$$OCl^-(aq) + 2I^-(aq) + 2H^+(aq) \longrightarrow Cl^-(aq) + I_2(aq) + H_2O(l)$$

23 How many moles of $S_2O_3^{2-}(aq)$ are required to react with the iodine produced from an excess of $I^-(aq)$ by 0.15 mol $OCl^-(aq)$?

TUTORIAL

Calculations for volumetric analysis: redox titrations

Problem: 0.312 g of a powdered sample of iron tablets (mainly hydrated iron(II) sulphate) was dissolved in water and a little dilute sulphuric acid added. 20.1 cm^3 of 0.0100 mol dm^{-3} KMnO$_4$(aq) was added to give a pink colour to the solution. Calculate the percentage of iron in the tablets, assuming it was all present as Fe^{2+}.
(Molar mass: Fe = 55.8 g mol^{-1})

Solution:
Stage 1 Equation for the reaction:

$$5Fe^{2+}(aq) + MnO_4^-(aq) + 8H^+(aq) \longrightarrow 5Fe^{3+}(aq) + Mn^{2+}(aq) + 4H_2O(l)$$

Stage 2 20.1 cm^3 of 0.0100 mol dm^{-3} KMnO$_4$(aq) contains

$$\frac{20.1}{1000} \times 0.0100 \text{ mol MnO}_4^-(aq)$$

$$= 2.01 \times 10^{-4} \text{ mol}$$

Stage 3 From the equation, 1 mol MnO$_4^-$(aq) reacts with 5 mol Fe^{2+}(aq). Therefore, 2.01 × 10^{-4} mol MnO$_4^-$(aq) reacts with 1.005 × 10^{-3} mol Fe^{2+}(aq).

Stage 4 We now know 0.312 g of the tablets contained 1.005 × 10^{-3} mol Fe^{2+}. Since 1 mol Fe^{2+} has a mass of 55.8 g:
1.005 × 10^{-3} mol Fe^{2+} has a mass of 1.005 × 10^{-3} × 55.8 g = 0.0561 g.

Therefore, percentage by mass of iron in the iron tablets = $\frac{0.0561}{0.312} \times 100\%$

= 18.0%.

Complexiometric titrations

Finally, we have seen that many reactions involve ligand exchange and the formation of complexes. Again, these may be used for quantitative work and volumetric analyses involving complex formation are known as **complexiometric titrations**. It is beyond the scope of this book to look at these. However, the principles are the same as for all titrimetric analyses.

One example of the use of the technique is in the determination of the hardness of water. The calcium and magnesium ions responsible for the hardness form complexes with EDTA 1,2-bis[bis(carboxymethyl)-amine]ethane (Fig 7.16), and the point at which all the metal ions have complexed is shown by a suitable indicator.

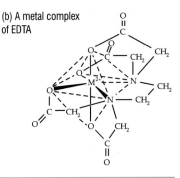

(a) EDTA

(b) A metal complex of EDTA

Fig 7.16 EDTA has six atoms which can bond through their lone-pairs of electrons to a central metal ion. It is the salt of EDTA which complexes with the metal ion.

24 Which six atoms in EDTA (see Fig 7.16) possess lone pairs of electrons which may coordinate to a central metal atom?

7.36 In a titration of hydrochloric acid against standard sodium carbonate solution, aliquots of $Na_2CO_3(aq)$ are transferred by pipette to a titration flask and titrated with $HCl(aq)$ from a burette. How should the following be washed and prepared before use: **(a)** the pipette; **(b)** the burette; **(c)** the titration flask.

7.37 In an acid-base titration the average volume of hydrochloric acid required to titrate a 25 cm^3 portion of 0.049 mol dm^{-3} sodium carbonate solution was 21.70 cm^3.
(a) Calculate the concentration of the acid.
(b) What mass of HCl would the average titre contain?

7.38 Methyl orange is a suitable indicator for titration of aqueous ammonia against dilute sulphuric acid, changing colour at around pH 7. If 25 cm^3 of ammonia solution required 24.3 cm^3 of 0.100 mol dm^{-3} $H_2SO_4(aq)$, calculate:
(a) the concentration of ammonia solution;
(b) the mass of ammonia which must be made up to 1 dm^3 with water to prepare a solution of this concentration.

7.39 A 2 g sample of powdered marble (a form of calcium carbonate, $CaCO_3$), was added to 50 cm^3 of 1 mol dm^{-3} $HCl(aq)$. When the effervescence had completely ceased the solution was titrated against 0.5 mol dm^{-3} $NaOH(aq)$.
(a) What volume of $NaOH(aq)$ would be required if the marble were pure calcium carbonate?
(b) If the titre for this experiment was 20.40 cm^3 calculate the percentage of $CaCO_3$ in the marble.

7.40 Ethanedioate ions, found in the leaves of rhubarb are toxic. In order to assess the toxicity of a rhubarb concentrate, 25 cm^3 samples were titrated against a standard 0.0200 mol dm^{-3} $KMnO_4(aq)$ solution in acid solution at 70°C. The average titre was 19.32 cm^3. Calculate the concentration of the ethanedioate ions in the rhubarb concentrate. The ionic equation for the reaction is:

$$2MnO_4^-(aq) + 16H^+(aq) + 5C_2O_4^-(aq) \longrightarrow 2Mn^{2+}(aq) + 10CO_2(g) + 8H_2O(l)$$

7.41 Hydrogen peroxide, used as a bleach for dyeing hair and in permanent waving, is often bought as a 6% solution. Accurate values for its concentration may be obtained by volumetric analysis using the reaction:

$$2MnO_4^-(aq) + 6H^+(aq) + 5H_2O_2(aq) \longrightarrow 2Mn^{2+}(g) + 5O_2(g) + 8H_2O(l)$$

A solution of $H_2O_2(aq)$ is diluted 40 fold and 25 cm^3 of this solution required 21.32 cm^3 of 0.0200 $KMnO_4(aq)$ for complete reaction in acid solution. Calculate the percentage by mass of H_2O_2 in the original solution.

7.42 **(a)** What is a complexiometric titration?
(b) How many moles of EDTA would be required to form a complex with the calcium ions in 25 cm^3 of 0.1 mol dm^{-3} calcium chloride solution?

Gravimetric analysis

Almost all cations form at least one insoluble compound. Provided an insoluble compound can be precipitated in a pure form and dried without decomposition, it may be used to quantify the amount of cation by gravimetric analysis. One common technique is to precipitate a metal hydroxide, collect it by filtration and heat it strongly to give the metal oxide. For example,

$$Al^{3+}(aq) + 3OH^-(aq) \longrightarrow Al(OH)_3(s)$$
$$2Al(OH)_3(s) \longrightarrow Al_2O_3(s) + 3H_2O(g)$$

25 Why must you be careful not to add an excess of alkali to a solution of aluminium ions when carrying out a gravimetric analysis to determine aluminium as the oxide?

A weighed mass of material is dissolved to form a solution, or a measured volume of an unknown solution is used. Excess of a suitable reagent is added to precipitate a compound formed by reaction with the unknown material. The precipitate is filtered in a dried and pre-weighed sintered glass crucible (Fig 7.17). After washing, it is dried in an oven and left to cool to room temperature in a desiccator. The crucible and precipitate are re-weighed, allowing the mass of precipitate to be calculated.

Fig 7.17 A sintered glass crucible contains a porous glass filter which can collect fine particles of solid. Its advantage over filter paper is that it can be dried at high temperatures, both before and after collection of the precipitate.

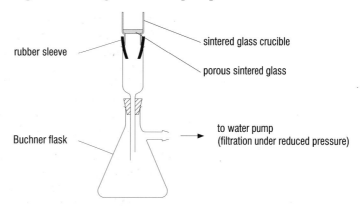

rubber sleeve

sintered glass crucible

porous sintered glass

Buchner flask

to water pump
(filtration under reduced pressure)

Two important examples of anions which are often determined by gravimetric analysis are sulphate ions and chloride ions.

TUTORIAL

Calculations for gravimetric analysis
Problem: When treated with excess barium chloride solution, a sodium sulphate solution precipitates 0.467 g barium sulphate. What mass of sodium sulphate is in the test solution?
(Molar masses: Ba = 137.3 g mol^{-1}; S = 32.1 g mol^{-1}; O = 16.0 g mol^{-1}; Na = 23.0 g mol^{-1})

Solution:
Stage 1 Equation for the reaction:

$$Na_2SO_4(aq) + BaCl_2(aq) \longrightarrow BaSO_4(s) + 2NaCl(aq)$$

Stage 2 1 mol BaSO$_4$ has a mass of (137.3 + 32.1 + [4 × 16.0])) g = 233.4 g.

Therefore, 0.467 g BaSO$_4$ contains $\dfrac{0.467}{233.4}$ mol BaSO$_4$ = 0.00200 mol.

Stage 3 From the equation, 1 mol BaSO$_4$ is formed from 1 mol Na$_2$SO$_4$. Therefore, 0.00200 mol BaSO$_4$ is formed from 0.00200 mol Na$_2$SO$_4$.

Stage 4 1 mol Na$_2$SO$_4$ has a mass of ([2 × 23.0] + 32.1 + [4 × 16.0]) g = 142.1 g

---

Therefore, 0.00200 mol Na_2SO_4 has a mass of (0.00200×142.1) g = 0.284 g.
Therefore, the mass of sodium sulphate in the solution was 0.284 g.

Problem: Calculate the mass of silver iodide that would be precipitated if excess silver nitrate is added to an aqueous solution containing 0.3308 g potassium iodide and 0.2998 g sodium iodide.
(Molar masses: K = 39.1 g mol^{-1}; Na = 23.0 g mol^{-1}; I = 126.9 g mol^{-1}; Ag = 107.9 g mol^{-1})

Solution:
Stage 1 Equation for the reaction:

$$Ag^+(aq) + I^-(aq) \longrightarrow AgI(s)$$

Stage 2 1 mol KI has mass $(39.1 + 126.9)$ g = 166.0 g.
1 mol NaI has mass $(23.0 + 126.9)$ g = 149.9 g.

Therefore, 0.3308 g KI contains $\dfrac{0.3308}{166.0}$ mol KI = 0.00199 mol

and gives 0.00199 mol I^- in solution.

0.2998 g NaI contains $\dfrac{0.2998}{149.9}$ mol NaI = 0.00199 mol.

and gives 0.00199 mol I^- in solution.
Therefore, total amount of I^- in the aqueous solution = 0.00399 mol.

Stage 3 From the equation, 1 mol I^- gives 1 mol AgI
1 mol AgI has mass $(107.9 + 126.9)$ g = 234.8 g.

Stage 4 0.00399 mol AgI has mass (0.00399×234.8) g = 0.937 g.
Therefore, the mass of silver iodide that would be obtained is 0.937 g.

QUESTIONS

7.43 Why is it essential that distilled water be used during gravimetric analysis of chloride ions?

7.44 Copper(II) ions may be determined by their precipitation as copper (II) sulphide, an insoluble copper compound.
   (a) Write an ionic equation for the precipitation of copper(II) sulphide.
   (b) If 0.0184 g of copper(II) sulphide is obtained from 1 dm^3 of solution calculate: *(i)* the mass of copper ions in the solution; *(ii)* the concentration of copper ions.

7.45 Calcium in milk samples is determined by gravimetric analysis. When a 100 cm^3 sample was treated with sodium carbonate solution a white solid was filtered off which, after washing with water and drying, had a mass of 0.24 g. For this sample, calculate:
   (a) the number of moles of Ca^{2+} present;
   (b) the mass of Ca^{2+} present;
   (c) the concentration of Ca^{2+} in the milk sample.

Determining empirical formulae from combustion analysis

The empirical formula of a compound can be obtained if the proportions of each element it contains are determined. The compositions of organic compounds are often determined by **combustion analysis**. The unknown organic compound is burned (or **combusted**) in an excess of oxygen in the presence of a catalyst to ensure rapid and complete reaction. Carbon in the compound is oxidised to carbon dioxide. Hydrogen is oxidised to water.

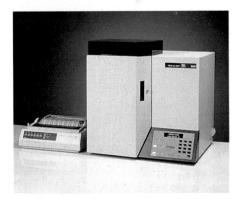

Fig 7.18 A modern CHN analyser. The complete combustion of samples in this apparatus allows the accurate determination of their carbon, hydrogen and nitrogen content. Only very small samples are required – a few milligrams.

CONNECTION: *Calculation of empirical formula, pages 69–70 and 134.*

The amounts of carbon dioxide and water which form are determined. Water is absorbed in pre-weighed tubes containing magnesium chlorate(VII) and carbon dioxide is absorbed in pre-weighed tubes containing potassium hydroxide. The following equation is used to determine the amounts of carbon and hydrogen in the compound.

$$C_xH_y + (x + \frac{y}{4})O_2 \longrightarrow xCO_2 + \frac{y}{2}H_2O$$

For example:

$$CH_4 + 2O_2 \longrightarrow CO_2 + 2H_2O$$

$$C_4H_{10} + 6\frac{1}{2}O_2 \longrightarrow 4CO_2 + 5H_2O$$

26 Write equations for the complete combustion of (a) propane C_3H_8; (b) benzene, C_6H_6.

If the masses of carbon dioxide and water are determined, the masses of carbon and hydrogen in the unknown compound can also be determined. For compounds which contain carbon, hydrogen and oxygen, any difference between the combined masses of carbon and hydrogen and the mass of sample taken is assumed to be due to the oxygen present. Once the masses of elements present in the sample have been determined the empirical formula is determined in the usual way.

27 How might the masses of carbon dioxide and water produced in a combustion analysis be determined?

TUTORIAL

Problem: 0.30 g of colourless liquid, **X**, produced 0.44 g carbon dioxide and 0.18 g water upon complete combustion. Calculate the empirical formula of **X**.
(Molar masses: C = 12 g mol^{-1}; H = 1 g mol^{-1}; O = 16 g mol^{-1})

Solution: Since the molar mass of carbon dioxide = 44 g mol^{-1},

0.44 g CO_2 contains $0.44 \times \dfrac{12}{44}$ g carbon = 0.12 g carbon.

Since molar mass of water = 18 g mol^{-1},

0.18 g H_2O contains $0.18 \times \dfrac{2}{18}$ g hydrogen = 0.02 g hydrogen.

Therefore, mass of carbon + hydrogen = 0.14 g.
Since mass of sample = 0.30 g, mass of oxygen present = (0.30 − 0.14) g = 0.16 g.
Therefore, 0.30 g **X** contains:

	simple whole number ratio
$\dfrac{0.12}{12}$ mol carbon atoms = 0.01 mol C	1
$\dfrac{0.02}{1}$ mol hydrogen atoms = 0.02 mol H	2
$\dfrac{0.16}{16}$ mol oxygen atoms = 0.01 mol O	1

Therefore, empirical formula of **X** is CH_2O.

ANALYSIS AT THE BENCH

7.46 Explain concisely, and in your own words, how combustion analysis is used to establish the formula of an organic compound.

7.47 Butanedione dioxime (dimethylglyoxime) is used in analytical chemistry to identify the presence of nickel ions. It has a relative molecular mass of 116. Combustion analysis (CHN analysis) gave the following results (composition by mass):

C, 41.38%;   H, 6.90%;   N, 24.14%.

The rest of the sample is oxygen. Calculate:
(a) the empirical formula of the compound;
(b) its molecular formula.

7.48 A compound of oxygen and nitrogen is found to contain 30.46% nitrogen and 69.54% oxygen.
(a) Calculate the empirical formula.
(b) Use Datasheet 11 to suggest the two possible identities of the compound.

7.49 Nicotine is present in tobacco leaves. Combustion analysis of a purified sample of nicotine gave the following percentage composition by mass:

C, 74.00%;   H, 8.70%;   N, 17.3%.

(a) Determine the empirical formula of nicotine.
(b) The molar mass of nicotine is approximately 160 g mol^{-1}. Establish the molecular formula.

You are given six aqueous solutions labelled A to F, each containing one of the following solutes (no two solutes are the same): NaOH; K_2CO_3; $BaCl_2$; $Cr_2(SO_4)_3$; $NiCl_2$; $(NH_4)_2SO_4$. Using these solutions together with a known solution of dilute hydrochloric acid and red and blue litmus paper, devise a logical series of tests which allows you to identify the solute in each case. (No other reagent may be used and test-tubes and a Bunsen burner are the only apparatus permitted). Suggest also one confirmatory test for each cation in solution, for which any suitable reagents and apparatus may be used.

Answers to Quick Questions: Chapter 7

1 Pregnancy testing
2 One mole of hydrated silver ions react with 1 mol of hydrated chloride ions, in aqueous solution, to give 1 mol of solid silver chloride.
3 One mole of solid zinc hydroxide reacts with 2 mol of hydrated hydrogen ions, in aqueous solution, to give 1 mol of hydrated zinc ions in solution and 2 mol of water molecules in the liquid phase.
4 $CaCO_3(s) + 2H^+(aq) \longrightarrow Ca^{2+}(aq) + CO_2(g) + H_2O(l)$
5 Before: Mn +2; Bi +5. After: Mn +7; Bi +3.
6 Most chlorides are soluble in water. Two important exceptions are AgCl and $PbCl_2$.
7 $CuS(s) + 2H^+(aq) \longrightarrow Cu^{2+}(aq) + H_2S(g)$
8 $Al(OH)_3(s) + 3H^+(aq) \longrightarrow Al^{3+}(aq) + 3H_2O(l)$
9 Oxygen, from the −1 oxidation state to the −2 oxidation state.
10 The coordination number is the number of atoms attached to the central metal ion. In the reaction, the coordination number of cobalt changes from 6 (six water molecules coordinately bonded through their oxygen atoms) to 4 (four chloride ions coordinately bonded).

11 No

12 $NaHCO_3(aq) + HNO_3(aq) \longrightarrow NaNO_3(aq) + CO_2(g) + H_2O(l)$ or, ionically $HCO_3^-(aq) + H^+(aq) \longrightarrow CO_2(g) + H_2O(l)$

13 Acid

14 HBr $Ox[Br] = -1$; Br_2 $Ox[Br] = 0$.
   HI $Ox[I] = -1$; I_2 $Ox[I] = 0$.

15 Reduction of sulphur in sulphuric acid (+6 oxidation state) to hydrogen sulphide (–2 oxidation state) is not complete. The reduction occurs via sulphur itself (zero oxidation state) and some of this is not reduced further.

16 Because the intermolecular forces between them are dispersion forces. Being non-polar molecules they do not form strong bonds with water molecules and are unable to interupt the strong hydrogen bonding between water molecules.

17 2-chloropropane

18 $Na_2CO_3(s) + 2HCOOH(aq) \longrightarrow 2HCOONa(aq) + CO_2(g) + H_2O(l)$

19 Because the strength of intermolecular bonding increases. The dispersion forces are greater between larger molecules.

20 $C_3H_7CHO(aq) + 2Cu^{2+}(aq) + 2H_2O(l) \longrightarrow C_3H_7COOH(aq) + Cu_2O(s) + 4H^+(aq)$

21 0.02 mol NaOH

22 4 cm^3 of 1 mol dm^{-3} $MnO_4^-(aq)$

23 0.3 mol $S_2O_3^{2-}(aq)$

24 Two nitrogen atoms of the amine groups and four oxygen atoms from the carboxylic acid groups.

25 Because $Al(OH)_3$ is amphoteric and would dissolve in excess alkali.

26 $C_3H_8(g) + 5O_2(g) \longrightarrow 3CO_2(g) + 4H_2O(l)$
   $C_6H_6(l) + 7\frac{1}{2}O_2(g) \longrightarrow 6CO_2(g) + 3H_2O(l)$

27 Carbon dioxide could be absorbed in a pre-weighed sample of an alkali (where it would react to form a carbonate). Water could be absorbed, for example, by a pre-weighed sample of anhydrous calcium chloride, anhydrous magnesium chlorate(VII) or any other material which absorbs water strongly.

Chapter 8

PURIFICATION AND SEPARATION

LEARNING OBJECTIVES

When you have studied this chapter you should be able to:

1. explain why it is important to determine the purity of chemical substances;

2. describe the criteria of purity employed for different purposes and the high standards demanded of the food and drugs industries;

3. identify the main methods used to achieve and establish purity in the laboratory and on an industrial scale;

4. account for the preference of industry for preparative techniques which do not involve a separation stage;

5. describe the use of fractional distillation and use Raoult's law to explain how separation is achieved;

6. describe the use of steam distillation;

7. explain the use of recrystallisation;

8. explain the principles of chromatography, and stationary and mobile phases;

9. describe the use of thin-layer chromatography to monitor the course of chemical reactions and to detect mixtures;

10. describe the technique and explain the principles of gas-liquid chromatography;

11. describe the use of selective precipitation as a separation method;

12. describe the use of ion exchange.

8.1 PURITY

A great number of people are employed in the chemical industries as analytical chemists. A great deal of time, energy and money are spent purifying products (separating them from impurities) and checking that a satisfactory purity has been achieved. Why is purity so important?

The meaning of 'pure'

The description 'pure' suggests good quality and careful control. Pure spring water, a pure-bred Persian cat and the pure genius of a tennis champion all imply desirable and recognised standards. But what does purity mean to the chemist?

A **pure substance** consists of a single element or chemical compound. It may consist of atoms, molecules or a combination of ions but its composition is always fixed. Purity, however, is relative. Absolute purity of any chemical substance has never been achieved, either in the laboratory or in

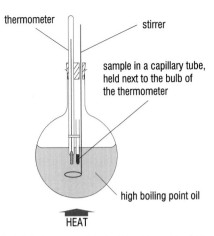

Fig 8.1(a) Apparatus suitable for determining melting points. The melting point is the temperature at which a complete meniscus is observed. A sharp melting point at a temperature in agreement with the literature value is an indication of purity.

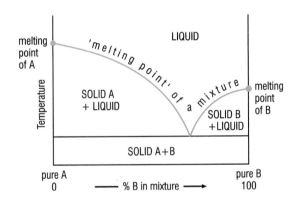

Fig 8.1(b) The melting point of a mixture of two solids depends on its composition. This is shown on a phase diagram. It can be seen that even when the impurity (A) in a compound (B) has a higher melting point, the mixture melts below the melting point of pure B.

CONNECTION: 'Wet chemistry' quantitative analysis, pages 128–133. Instrumental methods, pages 181–197.

nature. A 100% pure sample of a substance could only be produced if we could exclude from it every single atom, molecule or ion except those intended – an impossible task. It makes more sense to examine a product carefully using the best methods available and ensure that it is 'fit for its purpose'. This means analysing a substance with due regard for its likely use. A substance considered to be suitably 'pure' as an additive for a garden fertiliser is unlikely to satisfy the very high standards demanded of medicines or agrochemicals.

1   **Explain why a high purity sample of a compound costs more than one of lower purity.**

Measuring purity

The purity of a sample may be roughly checked in the laboratory by determining its melting point if it is a solid, or boiling point if it is a liquid. Impurities result in a lower melting point or higher boiling point than that of the pure substance. Impurities, therefore, increase the temperature range over which a substance is liquid. The difference between the laboratory result and the literature values gives a first indication of the degree of purity.

Fig 8.2 The boiling point of a substance may be determined by measuring the temperature at which it distils. When a liquid reaches its boiling point its vapours pass through the still-head and condense on the cold surface of the inside of the condenser.

Assessing purity by determining melting or boiling points is usually not accurate enough or appropriate. A quantitative analysis is required. This may involve a 'wet chemistry' method, such as a titration, or the use of an analytical instrument, such as a spectrophotometer.

2   **Explain why ice melts when common salt is sprinkled on it, for example when de-icing roads.**

Controlling standards of purity

Chemical substances are used for a wide range of different purposes. Dependable standards are needed, which producers and their customers can rely on. Any impurities present must be identified and quantified. If an impurity is harmless, no further purification is necessary. For example, a trace of sodium chloride (common salt) in a compound intended for food use would not normally be considered a problem. Detection of a toxic mercury compound in the sample, however, would be cause for alarm. The food and drugs industries operate with very high standards of purity since impurities, if present, may interfere chemically, physically or biologically with the behaviour of a product.

SPOTLIGHT

ENSURING THE PURITY OF MEDICINES

The purity of a drug must be confirmed before it can be dispensed in a hospital. This is usually carried out in the hospital pharmacy department. The standards are laid down in the British Pharmacopeia (BP), and the pharmacist must go through the procedures for identifying and assaying the drug.

Paracetamol

Acetaminophen

$C_8H_9NO_2$ 151.2   103-90-2

Paracetamol is 4'-hydroxyacetanilide. It contains not less than 99.0 per cent and not more than the equivalent of 101.0 per cent of $C_8H_9NO_2$, calculated with reference to the dried substance.

Description A white, crystalline powder; odourless.

Solubility Soluble in 70 parts of *water*, in 7 parts of *ethanol* (*96 per cent*), in 13 parts of *acetone*, in 40 parts of *glycerol* and in 9 parts of *propane-1,2-diol*; also soluble in solutions of alkali hydroxides.

Identification *Test A may be omitted if tests B, C, D and E are carried out. Tests B, C and E may be omitted if tests A and D are carried out.*
A. *The infra-red absorption spectrum*, Appendix II A, exhibits maxima which are only at the same wavelengths as, and have similar relative intensities to, those in the spectrum of *paracetamol EPCRS*.
B. Dissolve 50 mg in 100 ml of *methanol*. To 1 ml of the solution add 0.5 ml of 0.1 M *hydrochloric acid* and dilute to 100 ml with *methanol*. Protect the solution from light and immediately measure the *absorbance* at the maximum at about 249 nm, Appendix II B; A(1 per cent, 1 cm) of the solution at 249 nm, about 880.
C. Boil 0.1 g in 1 ml of *hydrochloric acid* for three minutes, add 10 ml of *water* and cool; no precipitate is formed. Add 0.05 ml of 0.0167 M *potassium dichromate*; a violet colour develops which does not turn red.
D. *Melting point*, 168° to 172°, Appendix V A, Method I.
E. Yields the *reaction* characteristics of acetyl groups, Appendix VI, heating over a naked flame.

Heavy metals Dissolve 1.0 g in sufficient of a mixture of 85 volumes of *acetone* and 15 volumes *water* to produce 20 ml; 12 ml of the solution complies with *limit test B for heavy metals*, Appendix VII (20 ppm). Prepare the standard using *lead standard solution* (*1 ppm Pb*) obtained by diluting *lead standard solution* (*100 ppm Pb*) with the acetone–water mixture.

4-Aminophenol Dissolve 0.50 g in sufficient of a mixture of equal volumes of *methanol* and *water* to produce 10 ml. Add 0.2 ml of *alkaline sodium nitroprusside solution*, mix, and allow to stand for thirty minutes. The solution is not more intensely coloured than 10 ml of a solution prepared at the same time in a similar manner using 0.5 g of *4-amino-phenol-free paracetamol* and incorporating 0.5 ml of a 0.005 per cent w/v solution of *4-aminophenol* in the same solvent mixture.

Related substances Carry out the method for *thin-layer chromatography*, Appendix III A, using *silica gel GF254* as the coating substance and a mixture of 65 volumes of *chloroform*, 25 volumes of *acetone* and 10 volumes of *toluene* as the mobile phase, but allowing the solvent front to ascend 14 cm above the line of application, in an unsaturated chamber. Prepare four solutions as follows. For solution (1), transfer 1.0 g of the substance being examined, finely powdered, to a ground-glass stoppered 15 ml centrifuge tube, add 5 ml of *ether*, shake mechanically for thirty minutes, centrifuge at 1000 rpm for fifteen minutes or until a clear supernatant liquid is obtained, and decant the supernatant liquid. For solution (2), dilute 1 ml of solution (1) to 10 ml with *ethanol* (*96 per cent*). For solution (3), use a 0.005 per cent w/v solution of 4-chloroacetanilide in ethanol)96 per cent). For solution (4), dissolve 0.25 g of 4-chloroacetanilide and 0.1 g of the substance being examined in sufficient ethanol) 96 per cent) to produce 100 ml. Apply separately to the chromatoplate 200 µl of solution (1) and 40 µl of each of the remaining three solutions. After removal of the chromatoplate, dry in a current of warm air and examine under an ultra-violet lamp having a maximum output at about 254 nm. Any spot due to 4-chloroacetanilide in the chromatogram obtained with solution (1) is not more intense than the corresponding spot in the chromatogram obtained with solution (3). Any spot in the chromatogram obtained with solution (2), other than the principal spot corresponding to 4-chloroacetanilide, is not more intense than the spot in the chromatogram obtained with solution (3). The test is valid only if, in the chromagogram obtained with solution (4), the spot corresponding to the chloroacetanilide is clearly separated from the spot corresponding to the substance being examined, the latter having the lower Rf value.

Loss on drying When dried to constant weight at 100° to 105° loses not more than 0.5 per cent of its weight; use 1 g.

Sulphated ash Not more than 0.1 per cent, Appendix IX A, Method II; use g.

Assay Dissolve 0.3 g in a mixture of 10 ml of *water* and 30 ml of M *sulphuric acid*. Boil under reflux for one hour, cool, and dilute to 100 ml with *water*. To 20 ml of the solution add 40 ml of *water*, 40 g of ice, 15 ml of 2M *hydrochloric acid*, 0.1 ml of *ferroin sulphate solution* and titrate with 0.1 M *ammonium cerium(iv) sulphate VS* until a yellow colour is obtained. Repeat the procedure without the substance being examined. Each ml of 0.1 M *ammonium cerium(iv) sulphate VS* is equivalent to 0.007560 g of $C_8H_9NO_2$.

Storage Paracetamol should be kept in a well-closed container, protected from light.

Preparations Pædiatric Paracetamol Elixir
Paracetamol Tablets

Action and Use Analgesic and antipyretic

Usual Dose Range 0.5 to 1 g; up to 4 g daily in divided doses.

Fig 8.3 A page from the British Pharmacopeia for paracetamol. The tests that must be carried out are given but it takes an experienced chemical analyst to perform them reliably.

Both aspirin and paracetamol give relief of mild pain and help to bring down fevers. They are found in OTC pain-relievers (OTC stands for over-the-counter, which means medicines that may be purchased without prescription). In some medicines, aspirin and paracetamol are found in combination with other compounds (for example, caffeine or codeine), and sometimes together. Careful analysis is required to ensure that proportions of drugs in such combined medicines meet the standards of safety and purity laid down in the BP.

Fig 8.4 Aspirin and paracetamol are often taken to relieve headaches.

CONNECTION: More information about pilot plant by-products, pages 509–510.

In the manufacture of a compound, analysis may reveal an impurity which is a by-product of the reaction. Much can often be done to prevent the formation of such impurities or to limit their amounts. Small changes in the operating conditions of a chemical plant (for example, temperature, pressure or solvent) can improve the yield of the desired product. Use of a suitable catalyst may promote formation of one product whilst limiting the formation of another.

Removal of by-products can be difficult and expensive to achieve when working on large scale production of compounds. In industry research chemists spend a great deal of time perfecting processes on a small scale in the laboratory before scaling up for production in a chemical plant. This half-way house between a laboratory preparation and full-scale manufacture is called a **pilot plant**. One of the major considerations at this stage is to develop a synthesis of the compound which gives a product as free from impurities as possible without an expensive separation stage.

3 **What aspects of a laboratory preparation might you want to test in a pilot plant?**

A chemist can suggest suitable operating conditions based on an understanding of the chemistry involved. However, optimum conditions are always determined by trial and error. After such conditions have been established in a pilot plant some fine tuning may be necessary after the process has been scaled up. This is because some factors, such as rate of stirring and heat control, can be altered by scaling up.

Despite all these efforts, impurities invariably remain in a product and will often need to be removed. The techniques used to remove small amounts of impurities are often the same as those suitable for separating mixtures. Techniques such as distillation, recrystallisation and different types of chromatography are common.

QUESTIONS

8.1 (a) Describe, with the aid of a diagram, how melting points are measured in a laboratory.

(b) Explain why the melting point of a sample prepared in a laboratory is usually 'checked against the literature value'.

8.2 (a) What is meant by the term impurity in relation to a chemical sample?

(b) Why do impurities need to be carefully monitored?

PURIFICATION AND SEPARATION

AGENT ORANGE

During the Vietnam War, the USA used a powerful defoliant called Agent Orange to reduce vegetation protecting enemy troops. It was a combination of 2,4-D and 2,4,5-T. These two compounds are herbicides and 2,4-D is found commonly in domestic weedkillers. In the UK, 2,4,5-T has been used extensively to control brambles growing over pathways or railway lines.

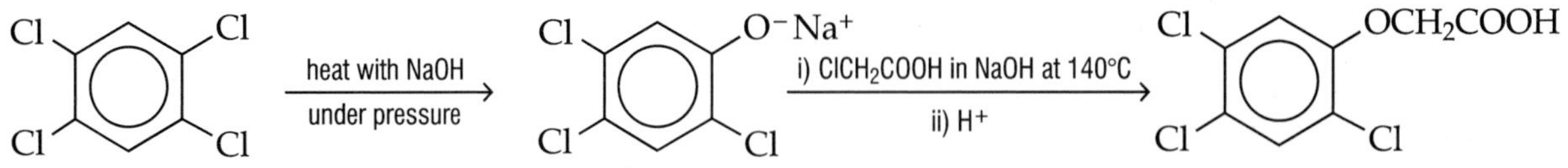

Fig 8.5 2,4,5-T is manufactured from 1,2,4,5-tetrachlorobenzene. The process is simple and cheap.

Cl — Cl — Cl — Cl (1,2,4,5-tetrachlorobenzene) →[heat with NaOH under pressure]→ Cl — O⁻Na⁺ — Cl — Cl →[i) ClCH₂COOH in NaOH at 140°C ii) H⁺]→ Cl — OCH₂COOH — Cl — Cl

Much controversy has surrounded the use of 2,4,5-T since its use in Vietnam was linked to the birth of malformed babies. Tests showed that 2,4,5-T could not be the culprit. However, it was demonstrated that dioxin, a by-product in the manufacture of 2,4,5-T, did cause malformed foetuses. Dioxin forms if the reaction temperature during the synthesis of 2,4,5-T is allowed to rise above 160°C (Fig 8.6). Subsequent research showed that impurities of around 30 ppm of dioxin in 2,4,5-T caused malformations in rat foetuses.

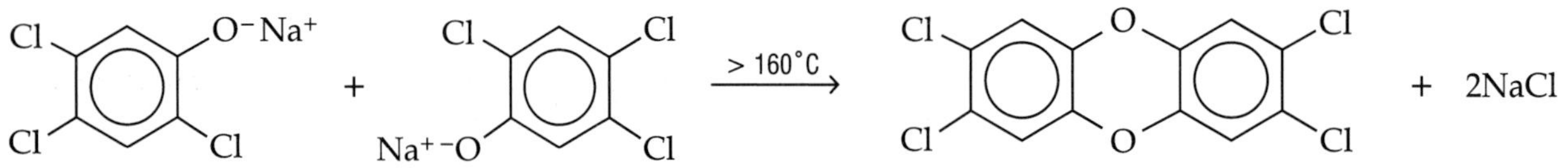

Fig 8.6 If the temperature rises too much during the manufacture of 2,4,5-T the intermediate compound which forms may undergo an alternative reaction, resulting in the formation of the highly toxic dioxin.

Cl — O⁻Na⁺ — Cl — Cl + Cl — Cl — Na⁺O⁻ — Cl →[> 160°C]→ Cl — O — Cl — Cl — O — Cl + 2NaCl

Although no evidence suggests that 2,4,5-T itself is highly toxic, this herbicide, which was so cheap to manufacture has been banned in many countries. The case of 2,4,5-T illustrates the problems of trace quantities of an impurity in a chemical substance.

8.2 DISTILLATION

The purification of solids and liquids is one of the most frequent tasks encountered in the laboratory. Liquid mixtures may usually be purified by distillation. Solids are generally recrystallised from a suitable solvent.

Separating liquid mixtures

Simple distillation

When a vapour and liquid are in dynamic equilibrium, the pressure exerted by the vapour is called its **vapour pressure**. The value will vary with temperature (Fig 8.7). A liquid boils when its vapour pressure equals atmospheric pressure.

Separation of a mixture of a solvent from non-volatile solute may be achieved using **simple distillation**. The equipment for this is quickly and easily set up (Fig 8.2). It is often used in the laboratory to obtain distilled water (dissolved salts remain in the distillation flask) or to recover a solvent from a reaction mixture. Occasionally distillation is carried out under reduced pressure so that the liquid distils below its boiling point. This can be useful for liquids of low volatility.

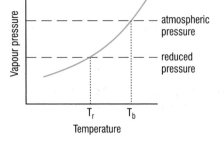

T_b = boiling point at atmospheric pressure

T_r = boiling point at a reduced pressure

Fig 8.7 The variation of vapour pressure with temperature. A liquid boils at the temperature at which its vapour pressure equals atmospheric pressure. Under reduced pressure, the liquid boils at a lower temperature.

Fractional distillation

Fractional distillation is used extensively to separate mixtures of miscible liquids. The efficiency of separation depends on the use of a **fractionating column** between the distillation flask and the stillhead (Fig 8.8). We find by experiment that columns with high internal surface areas (such as those packed with glass rings) allow the most effective separation. Why this is so may be understood if we look at the process of distillation.

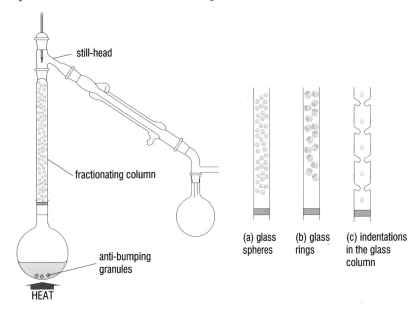

still-head

fractionating column

anti-bumping granules

HEAT

(a) glass spheres

(b) glass rings

(c) indentations in the glass column

Fig 8.8   Fractionating columns come in all shapes and sizes. A high internal surface area for a column creates a greater temperature gradient up its length and gives better separation.

Let us consider a mixture of two miscible liquids, X and Y. If molecules of the two liquids have very similar structures, the intermolecular forces will be almost the same between like (X and X, or Y and Y) and unlike molecules (X and Y). Assuming the difference is negligible, when the two liquids are mixed we would observe:

- no change in temperature;
- no change in volume.

4   **If the combined volume of two liquids is less than the sum of the unmixed volumes, what can be deduced about the intermolecular bonding between unlike molecules?**

Such a mixture is said to be **ideal**. An ideal mixture exerts a vapour pressure to which each component contributes in proportion to its mole fraction. This is expressed in **Raoult's law** which states:

'The partial pressure of a component in a mixture is equal to the vapour pressure of the pure component multiplied by its mole fraction in the mixture.'

For our mixture of X and Y, we can express Raoult's law mathematically:

$$p_X = x_X p^o{}_X \text{ and } p_Y = x_Y p^o{}_Y$$

where $p^o{}_X$ and $p^o{}_Y$ are the vapour pressures of pure X and Y respectively;
p_X and p_Y are the partial pressures of X and Y in the mixture;
x_X and x_Y are the mole fractions of X and Y in the mixture.

$$P = p_X + p_Y$$

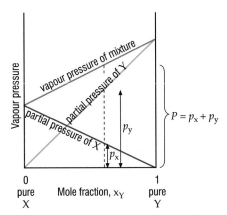

Fig 8.9   Raoult's law may be shown graphically. The line giving the vapour pressure of different composition mixtures is obtained by adding the partial pressures of X and Y for each mixture.

CONNECTION: Partial pressure, mole fraction and Dalton's law of partial pressures are dealt with on page 52.

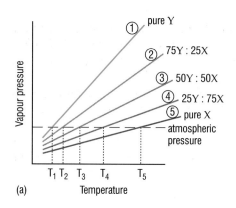

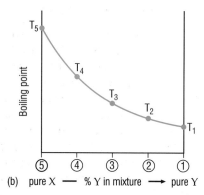

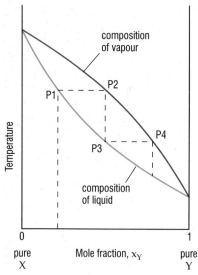

Fig 8.10 Although the vapour pressure/composition relationship is linear (8.10a), a curved line is obtained when boiling point is plotted against composition (8.10b). This is the key to explaining why mixtures can be separated by distillation.

Fig 8.11 The liquid-vapour phase diagram for a mixture of two miscible liquids.

Raoult's law can be used to obtain vapour pressure/composition diagrams at different temperatures and these may be used to draw boiling point/composition diagram (Fig 8.10).

Although no mixture is ideal, some approach ideal behaviour closely, including some whose separation is of great industrial importance. For example:

- liquid air, essentially a mixture of oxygen and nitrogen from which the pure elements may be obtained by fractional distillation under pressure;
- crude oil, which is a mixture of closely related hydrocarbons, particularly alkanes. Separation of these is the basis of the petrochemical industry.

5 **Explain why a mixture of hexane and heptane is more likely to behave as an ideal mixture than a mixture of hexane and hexan-1-ol.**

The boiling point/composition diagram for a non-ideal mixture can be determined experimentally by measuring the boiling point of mixtures of different composition and plotting an appropriate graph. Whether a theoretical diagram or one which has been determined experimentally, either may be used to explain fractional distillation.

When a mixture of X and Y is distilled we find that the composition of the vapour is not the same as that of the boiling liquid. Not surprisingly perhaps, the vapour is richer in the more volatile liquid. Adding a second curve representing the composition of the vapour at the boiling point to Fig 8.11, we obtain a **liquid-vapour phase diagram** (Fig 8.11). The composition of a liquid and vapour at the same temperature is shown by a horizontal line between the liquid and vapour curves. This is called a **tie-line**. In Fig 8.11, line P1-P2 is a tie-line.

We can now use the liquid-vapour phase diagram to predict the composition of the distillate from simple distillation of a particular combination of X and Y. The vertical dotted line shows the composition of a liquid mixture (75% X and 25% Y) before distillation begins. Reading across from point P1 to P2 gives the composition of the vapour. If this vapour condenses, a liquid richer in the more volatile component is formed. When this condensate is boiled, a vapour richer still in the more volatile component (P3 to P4) is produced. The distillation may be repeated over and over again, each time the distillate is richer in the more volatile component.

A series of simple distillations would effect good separation, but it would be extremely time consuming. How can we achieve the same effect as many successive distillations in a single process? This is the purpose of the fractionating column.

6 **Explain why a mixture where the vapour and liquid lines are close would take more successive distillations to effect separation than one for which the lines are further apart.**

The use of a fractionating column enables the equivalent of several simple distillations to be achieved simultaneously. As the vapour rises it condenses in the column. Rising vapours cause this liquid component to evaporate and condense higher up the column. All the time, the mixture in the distillation flask is becoming richer in the less volatile component and the boiling point is slowly rising.

Thus as the vapour ascends the column it becomes richer in the more volatile component. A series of liquid/vapour equilibria have been established in the column and a temperature gradient has been created. The liquid remaining in the distillation flask is the least volatile component. Column efficiency is increased by increasing the internal surface area since this offers more opportunity for rising vapours to meet descending liquids.

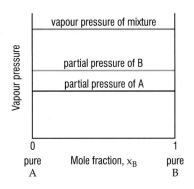

Fig 8.12 The vapour pressure of a mixture of immiscible liquids is independent of its composition. At all compositions, it is the sum of the vapour pressures of the pure components.

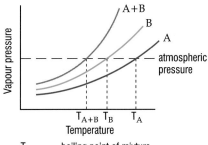

T_{A+B} = boiling point of mixture
T_B = boiling point of pure B
T_A = boiling point of pure A

Fig 8.13 A mixture of immiscible liquids boils below the boiling points of the pure components.

Steam distillation

Mixing a pair of immiscible liquids produces two separate layers, with the less dense liquid forming the top layer. If two such liquids are distilled, the boiling point is lower than that of either of the pure components. For example, water boils at 100°C and benzene at 80°C. A mixture of the two distils over at around 70°C. In contrast, solid impurities cause boiling point elevation. As with mixtures of miscible liquids, we must look at vapour pressure/composition diagrams to explain this.

For a mixture of immiscible liquids the vapour pressure of the mixture is the sum of the vapour pressures of the pure components regardless of the amounts present. For example, in a mixture of two liquids A and B:

$$P = p^o_A + p^o_B$$

where P is the vapour pressure of the mixture;
p^o_A and p^o_B are the vapour pressures of A and B respectively (Fig 8.12).

Remember, a liquid boils when its vapour pressure equals the pressure in the container (normally atmospheric). It follows, therefore, that a mixture of immiscible liquids of any proportions must always boil at a temperature below the boiling points of either of the pure components (Fig 8.13).

Water (as steam) is used to distil many liquid organic compounds at temperatures below their normal boiling point, a process known as **steam distillation** (Fig 8.14). The organic compound to be distilled must, of course, be immiscible with water. The technique is particularly useful for liquids which undergo significant decomposition (often due to oxidation) at their boiling points. Following distillation, the two layers which form are separated using a separating funnel (Fig 8.15).

7 **Why must the rate at which steam is passed be carefully controlled?**
Hint: **the temperature in the distillation flask will be below 100°C.**

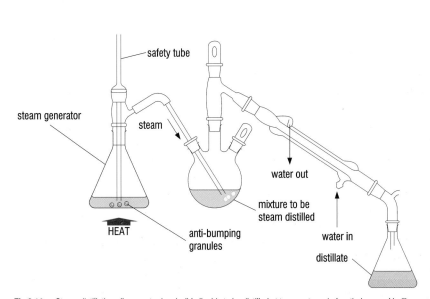

Fig 8.14 Steam distillation allows water-immiscible liquids to be distilled at temperatures below their normal boiling point. The usual procedure is to pass steam continuously through a distillation flask containing a mixture of the compound to be distilled and water.

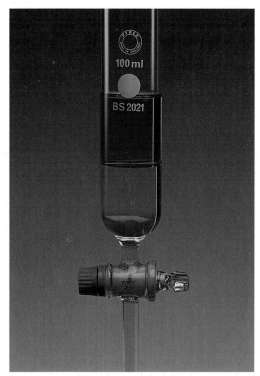

Fig 8.15 A separating funnel is used to separate immiscible liquids.

STEAM DISTILLING ESSENTIAL OILS

Hundreds of natural oils, such as eucalyptus oil, lemongrass oil and tea-tree oil are steam distilled in the perfume and fragrance industries. These oils are often called **essential oils**.

Interest in these essential oils has grown considerably in recent years alongside the growth of herbalism, aromatherapy and alternative medicine. We are only now beginning to understand the value of many of the chemical compounds derived from plants. Many important plants grow only in rainforests, many of which are under threat of destruction.

Fig 8.16 This lemongrass 'still' uses the principles of steam distillation to isolate 0.2-0.4% of lemongrass oil from lemongrass. Lemongrass oil contains 75-80% of citral, a minor component of lemon and orange oils. Used widely for artificial flavours and in perfumery, citral is also a starting material for the synthesis of vitamin A.

Fig 8.17 The natural citral isolated from lemongrass is a mixture of two isomers, often referred to simply as citral a and citral b.

8.3 CRYSTALLISATION AND RECRYSTALLISATION

Solid substances begin to crystallise from solution as the solvent evaporates. Crystals may need to be recrystallised to purify them. The substance is dissolved in a hot solvent, filtered to remove any insoluble impurities, and the filtrate cooled to yield crystals of the substance which can then be separated by filtration (Fig 8.18). Ideally:

- the compound to be purified is soluble in the hot solvent but insoluble in the cold solvent;
- the impurities are either insoluble in both cold and hot solvent or soluble in both cold and hot solvent.

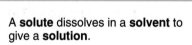

Fig 8.18 The hot solution may be filtered under reduced pressure to increase the filtration rate. As the filtrate cools, crystals of the purified substance form.

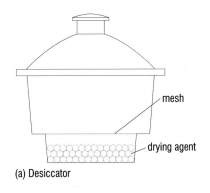

(a) Desiccator

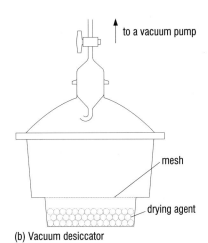

to a vacuum pump

(b) Vacuum desiccator

Fig 8.19 A solvent may be removed from a product by storing it in a **desiccator**. This holds the product and a drying agent which absorbs the solvent. A **vacuum desiccator** is even more effective as the reduced pressure inside encourages rapid evaporation of the solvent.

The recrystallisation solvent should have a suitable boiling point. Too low a boiling point and the solvent will evaporate too quickly during filtration leaving impurities on the surface of the crystals. If the boiling point is too high it may prove difficult to dry the crystals after they have been filtered.

8 **What other properties of the solvent are desirable?**

Water often fulfils many of these requirements. However, most organic compounds do not dissolve in water and require an organic solvent. To find a suitable solvent, small quantities of the crude material are added to each of several solvents in test tubes. Each test tube is warmed and then allowed to cool. Finding a solvent which dissolves the crude material readily when hot, but sparingly when cold is the major challenge of recrystallisation.

9 **What other criterion is important when choosing a solvent for industrial-scale recrystallisation?**

Organic compounds which have been synthesised in the laboratory are often contaminated with small amounts of coloured tars formed as by-products in the reaction. These impurities can be difficult to remove. The problem may usually be overcome by boiling a small quantity of specially treated charcoal with the solution before filtration. The impurity is adsorbed onto the charcoal and removed from solution during filtration.

Water often contaminates reaction products and can be removed by drying them in an oven. In most cases this is the obvious answer. However, it is not desirable to heat some compounds in an oven and, in these cases, water can be removed by recrystallisation from an organic solvent. A small quantity of a **drying agent** is added to the hot solution. It does not dissolve but binds water to itself. Filtration leaves a dry solution from which water-free crystals may be obtained on cooling.

Some common drying agents

Calcium chloride: used in a simple desiccator (Fig 8.19) to absorb water, alcohols, amines or phenols.

Potassium or sodium hydroxides: used to dry liquid amines but also effective in desiccators.

Anhydrous sodium sulphate or anhydrous magnesium sulphate: effective for absorbing water from a range of organic solvents.

Silica gel: an excellent reusable drying agent for removing water from a sample. Can be used in a desiccator. Silica gel is available in self-indicating form. A cobalt salt is used to colour the silica. When anhydrous the salt is blue. In the presence of water, the salt turns pink. Thus a pink colour indicates when the silica gel must be regenerated by heating.

Concentrated sulphuric acid: a dehydrating agent which is particularly useful for drying gases. Can be used in a vacuum desiccator to good effect.

Solvent extraction

Solvent extraction is a useful technique for separating the products of an organic synthesis from an aqueous reaction mixture. A solvent which is immiscible with water is shaken with the aqueous solution. The product distributes itself between the aqueous phase and the organic solvent. The organic layer is separated, using a separating funnel (Fig 8.15), and the product obtained by allowing the solvent to evaporate. The extraction may be repeated a number of times.

Ethoxyethane is often used to separate volatile or thermally unstable products from an aqueous solution. Its low boiling point (34.5°C) enables ethoxyethane to be distilled at a low temperature, avoiding the possibility of thermal decomposition of the desired product.

The distribution of a substance between two immiscible liquids, A and B, is given by the **partition** or **distribution coefficient**. Provided the substance is in the same molecular form in both liquids, the partition coefficient, K, is given by:

$$K = \frac{\text{concentration of substance in liquid A}}{\text{concentration of substance in liquid B}}$$

It is more efficient to use several small portions of the extracting solvent, separating each time and combining the extracts, than to use a large quantity in a single operation.

10 **If a substance is equally soluble in equal quantities of two immiscible solvents, what is the value of the partition coefficient?**

SPOTLIGHT

Fig 8.20a Coffee beans being gathered in Jamaica

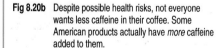

Fig 8.20b Despite possible health risks, not everyone wants less caffeine in their coffee. Some American products actually have *more* caffeine added to them.

THE CAFFEINE-FREE BEAN

The third of the world's population who consume coffee take in between them approximately 120 000 tonnes of the drug per annum. Although there are several biologically active ingredients in coffee beans, it is caffeine which produces the well-known stimulatory effect of a cup of coffee. Although caffeine does not accumulate in the body (it is water-soluble) there are fears that it may affect the health of regular uses through its effect in increasing blood pressure and cholesterol levels.

The German company Kaffee HAG introduced a decaffeinated coffee in the early 1900's. Today many different brands exist along with a number of different methods for removing or reducing the caffeine content of coffee. A common process has been to treat coffee beans with a solvent such as chloromethane which binds to the caffeine. When the beans are steamed, the solvent and the caffeine vaporise together. Another process uses carbon dioxide gas under very high pressure (so that it behaves more like a liquid). This extracts caffeine from the beans without removing many of the desirable soluble solids.

A process preferred by some gourmet coffee drinkers is the Swiss Water Process. This uses hot water to extract all soluble materials from the coffee. The extract is then passed through a charcoal-based filter system which traps the caffeine. It is then returned to extract fresh coffee beans (Fig 8.20(a)). The extract is already saturated with respect to other solids and will dissolve no more of these. It is *not* saturated with caffeine and dissolves further quantities, before being refiltered.

PURIFICATION AND SEPARATION

8.3 Copy the graph from Fig 8.7 and add to it the curve you would expect for a solution of an involatile solute in the same solvent.

8.4 Under what circumstances would fractional distillation be a more appropriate method than simple distillation?

8.5 (a) What is meant by the term homogenous mixture?
    (b) Why do ethanol and water form a homogenous mixture?

8.6 In a sample of dry air (78% N_2, 18% O_2, 1% Ar) at a pressure of 98 000 Pa, calculate:
    (a) the partial pressure of nitrogen;
    (b) the partial pressure (total) of all the other gases in the sample.

8.7 Explain what is meant by the term *ideal mixture*.

8.8 Consider the four mixtures: *(i)* pentane and pent-1-ene; *(ii)* ethanol and ethanoyl chloride; *(iii)* butan-1-ol and water; *(iv)* methyl benzene and water.

Decide whether each mixture consists of the following:

two immiscible liquids;
an ideal mixture;
a non-ideal mixture;
two liquids which react together forming a new compound (use Database).

8.9 Explain, using a liquid-vapour phase diagram, how a 50:50 mixture of heptane (BP; 98°C) and octane (BP; 126°C) could be separated.

8.10 The partition coefficient of a solid in a combination of two immiscible liquids, X and Y, is 0.20. If 5 g of the solid were dissolved in 100 cm³ of X and extracted with 100 cm³ of Y, what mass of solid would remain in X?

8.4 CHROMATOGRAPHY

Chromatography is used extensively in chemical laboratories to

- separate and identify impurities;
- separate mixtures on a large scale;
- monitor the course of a reaction.

Many different chromatography techniques are used and some have developed into specialist, high performance tools capable of identifying trace levels of impurities.

Chromatography refers to a range of closely-related techniques used for separating mixtures. All rely on a **mobile phase** moving through a **stationary phase**. Components in a mixture are attracted to both phases and become distributed between them. It is the relative strength of the attraction of each phase for the components which is important. A component strongly attracted to the stationary phase will be held back. One with a relatively strong attraction to the mobile phase will move along quickly in it. The stationary and mobile phases are carefully chosen to allow components of the separating mixture to move at different speeds and so achieve separation (Fig 8.21).

The rate at which a component moves depends on its equilibrium concentrations in the mobile and stationary phases. The ratio of these concentrations is known as the **distribution coefficient** (symbol, D) of the component.

$$D = \frac{\text{component concentration in mobile phase}}{\text{component concentration in stationary phase}}$$

The substance must be in the same molecular form in both phases for this relationship to apply. Separation may still occur even if it is not, but the equation describing the distribution is more complex. The value of D is a constant for a component distributed between a particular combination of stationary and mobile phases at a certain temperature.

11 **If a compound has a large value for D will it move quickly or slowly through the system?**

12 **Is a compound with a small value for D strongly or weakly attracted to the stationary phase?**

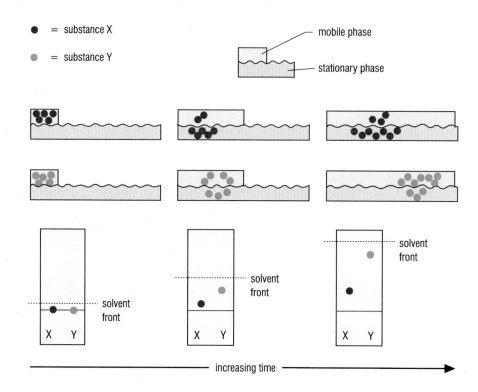

Fig 8.21 These diagrams represent the movement of two substances, X and Y, as a mobile phase containing them moves over a stationary phase.

When a mixture is to be separated by chromatography, time should be spent finding the best mobile and stationary phases. This is often done on a small scale initially to prevent unnecessary waste. The phases are varied until good separation is achieved. To some extent, this is trial and error but an educated guess, based on experience, helps. For example, a polar compound is likely to move quickly when a mobile phase is a polar solvent.

Liquid chromatography

In liquid chromatography, a mobile liquid phase moves through a stationary phase. Thin-layer chromatography, paper chromatography, column chromatography and high performance liquid chromatography (HPLC) are all forms of liquid chromatography. The stationary phase may consist of a number of adsorbent materials including paper, silica (silicon(IV) oxide) and alumina (aluminium oxide).

KEY FACT

A substance is **adsorbed** when it adheres to the surface of a material. It is said to be **absorbed** when it enters pores in the material (rather like water in a sponge).

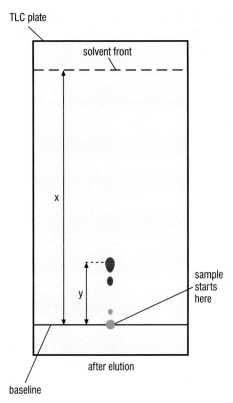

Fig 8.22   Identifying components of a mixture by TLC.

Thin layer chromatography

Thin layer chromatography involves the use of a layer of adsorbent particles (usually fine particles of alumina or silica) as the stationary phase. The layer is usually supported by a glass plate or a thin plastic strip. The combination of adsorbent layer and support is known as a **TLC plate**.

A capillary tube (called an **applicator**) is used to apply a solution of the sample to a baseline (lightly drawn in pencil on the plate). The plate is placed in a suitable jar containing a small amount of a solvent (the mobile phase). As the solvent rises up the plate each component travels a different distance. The distance travelled depends upon the solvent and the adsorbent used. The distances travelled by mixture components on a TLC plate are usually reported as R_f **values**, where R_f is defined as the ratio of the distance travelled by the component to the distance travelled by the solvent front. The values depend on the stationary and mobile phases, so these must be stated.

13 **Why is a pencil preferable to a pen for drawing the base-line?**

14 **For the TLC investigation drawn in Fig 8.22, give the R_f value for the leading component in terms of x and y.**

Components appear as spots on the TLC plate. Coloured components may be identified easily but invisible spots must be exposed. Keeping the plate in iodine vapour for 5–10 minutes is the most usual method used, each component showing up as a brown spot. Alternatively, plates may be pre-treated so that when viewed under ultraviolet light the components show up as pale purple spots.

TLC is most widely used in qualitative analysis to determine whether a substance is pure or to identify the nature of an impurity. If more than one spot is observed then the sample is definitely impure. If one spot is observed then the sample is probably pure. Ideally the plate should be run in more than one solvent and the results compared. TLC is also useful as a means to monitor the course of a reaction.

15 **Why, if only one spot has been observed, is there still some uncertainty about whether a sample is pure?**

16 **Why will repeating the examination with a different solvent help?**

Column chromatography

Column chromatography is essentially scaled-up TLC. In the laboratory, it is used to obtain small quantities of pure compounds (less than 1 g) from a mixture. Larger columns are used in industry where larger quantities of substance are required. Column chromatography uses more adsorbent and solvent than TLC and it can be expensive to run a column. Therefore, TLC is used to find the best adsorbent/solvent combination for use on the larger scale of the column.

High performance liquid chromatography

High performance liquid chromatography (HPLC) is the most up to date form of liquid chromatography. It is used to separate components which are very similar to each other. Very high resolution (separation of components) is possible. High resolution in this context means separating components which have similar distribution coefficients and, therefore, pass through the stationary phase close together. High pressures are used to drive the liquid phase (a solvent or mixture of solvents of very high purity) through a stationary phase consisting of regular micro particles of solid in a stainless steel column. High pressures are required to force the liquid

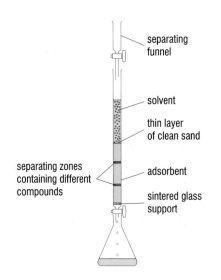

Fig 8.23   In column chromatography, the column is filled with adsorbent and the sample is added in a minimum amount of solvent. Fresh solvent (called the eluent) is run slowly through the column and samples (called fractions) are collected periodically. Their purity is checked by TLC. Consecutive fractions containing the same pure component are combined and evaporated to give the separated material.

phase through the tightly packed column, but the reward is high resolving power. A single spot achieved by a TLC separation has been separated into as many as ten components by HPLC.

HPLC is useful for the analysis of complex mixtures such as body fluids and foods. Most recently it has been used to analyse blood and urine samples of athletes and horses to detect traces of performance drugs.

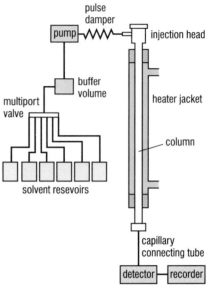

Basic components of a high performance liquid chromatograph

Fig 8.24 In HPLC, particles of adsorbent are tightly packed in the column which is maintained at the desired temperature. The pump forces the solvent or solvents (selected from the reservoirs) through the column at high pressure.

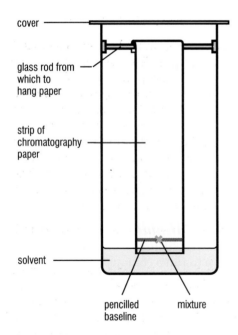

Fig 8.25 Paper chromatography is a cheap and effective way of separating some mixtures. The stationary phase is water, adsorbed onto the cellulose paper, through attraction (by hydrogen-bonding) to the -OH groups of the cellulose.

Paper chromatography

Whereas the other forms of liquid chromatography rely primarily on adsorption, paper chromatography is a simple form of partition chromatography used mainly for qualitative analysis. The procedure is the same as for TLC.

17 Why is it important that the spot itself should not dip into the solvent?

Gas chromatography

In gas chromatography the mobile phase is a gas and the stationary phase is a liquid. It is, therefore, a form of partition chromatography. GLC has developed into one of the most useful techniques for separation and is widely used in industrial preparation of materials as well as a method of analysis. For example, the separation of benzene (bp 80.1°C) and cyclohexane (bp 80.8°C) is almost impossible by fractional distillation, but it can be achieved by GLC. We can use GLC to determine a compound's purity and to identify particular contaminants. It provides both qualitative and quantitative information.

18 Why are benzene and cyclohexane so difficult to separate by conventional means?

A typical GLC set-up is shown in Fig 8.27(a). The mobile phase consists of a chemically inert, high purity gas (called the **carrier gas**). Using a hypodermic needle, the sample is injected into the sample port where it vaporises. It is carried through the column and out into the detector. The time taken for a component of a mixture to pass through the column is known as its **retention time.** The retention time of a substance depends on how its vapour is distributed between the mobile and stationary phases.

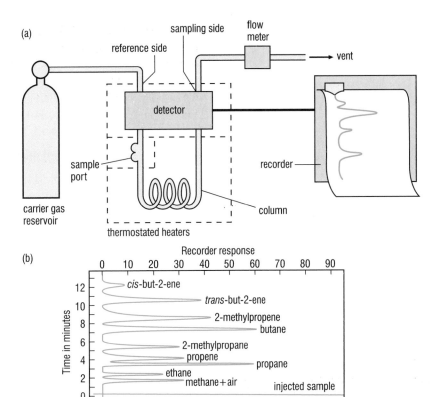

Fig 8.26 In GLC, the stationary phase is packed into a glass or stainless steel column, often coiled to give maximum length in the minimum space. The coil is housed in a thermostatically controlled oven.

Fig 8.27 (a) A GLC arrangement, (b) A GLC chromatogram of a mixture of hydrocarbons. The gas chromatogram plots retention time against peak height. The area under a peak is proportional to the concentration of that component.

19 **Suggest suitable gases for use as carrier gases in GLC.**

Detectors work in different ways but all respond to a component as it leaves the column in a stream of carrier gas. The magnitude of the response depends on the concentration of the component. The flame ionisation detector (FID) is used commonly. The emerging gas is passed through a hydrogen flame. As components arrive, they burn in the flame to produce ions. The presence of ions causes a current to flow between two charged plates in the detector, indicating the arrival of the components. A recorder plots the current that flows against time giving a **gas chromatogram** (Fig 8.27(b)). By using a series of standard solutions, a calibration graph can be drawn which allows the area under a peak to be interpreted in terms of concentration of component in the mixture.

20 **In Fig 8.27(b) can you identify a relationship between chain length and retention time?**

The power of GLC is greatly enhanced if, as it comes off the column, each component is fed directly into a mass spectrometer linked to a computer database. This powerful combination is known as **gas chromatography-mass spectrometry** (GCMS). The total sum of data obtained (retention time, peak areas, relative molecular mass and fragmentation pattern) is compared with the computer database which allows the computer to identify each substance. The most helpful systems will print out a list of the name and percentage of each component in the mixture.

The latest GCMS techniques (using very long, narrow columns called capillaries) allow complex mixtures of trace substances to be identified and quantified. For example, it is possible to identify hundreds of substances in sewage effluents or river water and traces of drugs in urine or blood may be identified with ease.

CONNECTION: Mass spectrometry, page 182.

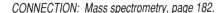

PURIFICATION AND SEPARATION

8.11 On the TLC plate (Figure 8.21) two spots, for substances X and Y, are visible. Using the right hand diagram, determine the R_f values of both components.

8.12 HPLC may be used to separate two isomers of butenedioic acid, $C_4H_4O_4$, successfully but is unable to separate two isomers of 2-hydroxypropanoic acid. Explain why this result may be predicted.

8.13 Why is oxygen unsuitable as a carrier gas for GLC?

8.14 Explain how you could use TLC to distinguish between regular and decaffeinated coffee.

8.5 SEPARATING IONS FROM SOLUTION

Ions can be removed from solution by:

- precipitating them as insoluble compounds;
- passing the solution down an ion exchange column.

Precipitation of ions

In most of the world, water is freely available. It is used in many industrial and domestic processes, for example as a reaction solvent, to transport and concentrate minerals, to carry away human waste of all kinds, to cool industrial plants and to irrigate agricultural land.

However, the excellent solvent properties of water are not without their disadvantages. Chemical substances used in industry and agriculture (particularly fertilisers), together with naturally occurring salts, may all find their way into the waterways. Some of these compounds can cause serious problems for industrial or domestic users. For example, salts responsible for temporary hard water cause damage to industrial and domestic boilers. Nitrate ions, present in most fertilisers, may be a threat to drinking water if the concentration reaches unacceptable levels. We need to monitor and control the quality of water supplies for people, for agriculture and for industry.

Most ionic compounds dissolve appreciably in water yet some are described as insoluble. But 'soluble' and 'insoluble' are relative terms. For example, calcium chloride is described as soluble, 74.5 g dissolves in 100 cm^3 of water at 20°C. Calcium fluoride is regarded as insoluble (only 0.0016 g dissolves in 100 cm^3 of water at 20°C). How, then, would we describe calcium sulphate, whose solubility is 0.21 g per 100 cm^3 of water at 20°C? Clearly, to be unambiguous, we need to give numerical values for solubility. Commonly solubility is given in 'g per 100 g water'.

We can make use of the solubility of certain compounds to effectively remove particular ions from solution. For example, addition of a solution containing excess calcium nitrate to a solution of potassium fluoride will cause precipitation of calcium fluoride.

$$Ca^{2+}(aq) + 2F^-(aq) \longrightarrow CaF_2(s)$$

CONNECTION: Selective precipitation is used extensively in cation analysis, pages 119–121.

Only a very low concentration of fluoride ions remains. The technique of precipitating particular ions from solution is known as **selective precipitation**.

21 **Why is potassium nitrate not precipitated in this reaction?**

The precipitation of insoluble ionic compounds is not restricted to the use of water as solvent. The addition of calcium oxide to molten iron ore during the manufacture of iron leads to precipitation of unwanted silicon(IV) oxide as calcium silicate, or slag. The calcium silicate precipitates as it is insoluble in molten iron:

$$CaO(s) + SiO_2(s) \longrightarrow CaSiO_3(l)$$

SPOTLIGHT

HEAVY METALS IN SHELLFISH

Living systems concentrate certain metals within their bodies by precipitating insoluble metal salts or by converting them to organometallic compounds. The latter are more dangerous due to their ready solubility in fats and consequent rapid movement through food chains.

Shellfish are particularly good at concentrating heavy metals in seawater. Oysters and scallops have been found to concentrate mercury and cadmium to dangerously high levels and should not be eaten from polluted waters. In Australia, oysters found off the New South Wales coast accumulated mercury to over 100 000 times its environmental concentration. Scallops in the same region were able to concentrate cadmium to over 2 000 000 times its environmental concentration.

Selective precipitation may be used to:

- prepare an insoluble compound;
- establish the presence of particular ions in a solution;
- remove unwanted (toxic or troublesome) ions from solution;
- recover economically desirable materials from waste.

Calcium and magnesium ions are responsible for the hardness of water which causes scum to form when soap is used. There are two types of hardness: **permanent** and **temporary**. Permanent hardness remains even after the water has been boiled. It is due to the presence of calcium and magnesium sulphates. Temporary hardness, due to hydrogencarbonates of calcium and magnesium, is removed by boiling. For example:

$$Ca(HCO_3)_2(aq) \longrightarrow CaCO_3(s) + CO_2(g) + H_2O(l)$$

Precipitation may be used to remove both temporary and permanent hardness. Water may be softened for washing purposes by the addition of sodium carbonate, available as washing soda crystals ($Na_2CO_3.10H_2O$). The unwanted calcium ions are precipitated as calcium carbonate:

$$Ca^{2+}(aq) + CO_3^{2-}(aq) \longrightarrow CaCO_3(s).$$

Suitable reagents added to industrial waste can selectively remove ions as insoluble compounds. In addition to preventing environmental pollution this can often recycle valuable chemical substances. Effluent from electroplating works usually contains ions of chromium, nickel, copper or zinc. These may be removed by precipitation, using alkali to precipitate the hydroxides. The waste materials from photographic laboratories are rich in silver and much of this is recovered by precipitation of insoluble silver compounds.

22 **Write an ionic equation for a reaction which could be used to precipitate silver ions from solution.**

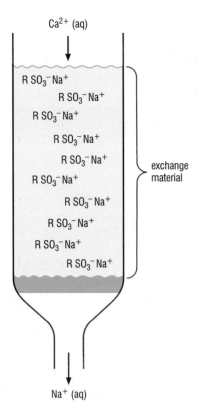

Ca^{2+} (aq)

R SO$_3^-$ Na$^+$
R SO$_3^-$ Na$^+$
R SO$_3^-$ Na$^+$
R SO$_3^-$ Na$^+$
R SO$_3^-$ Na$^+$
R SO$_3^-$ Na$^+$
R SO$_3^-$ Na$^+$
R SO$_3^-$ Na$^+$
R SO$_3^-$ Na$^+$
R SO$_3^-$ Na$^+$

exchange material

Na$^+$ (aq)

Fig 8.28   As the solution containing the unwanted ions passes through the column they are exchanged. The column can be primed before use with ions which are acceptable in the solution. In this exchanger sodium ions are progressively replaced by calcium.

Ion exchange

An alternative method for removing unwanted ions from an aqueous solution is to exchange them for other ions. The process is carried out by passing the aqueous solution through an **ion exchange** material. This is usually a naturally occurring insoluble solid such as a zeolite or a synthetic resin. Zeolites and ion exchange resins have a number of charged sites which have oppositely charged ions associated with them. These ions replace those in the solution as it flows through. Ion exchange materials are available which will replace cations (**cationic exchangers**) or anions (**anionic exchangers**).

Cationic exchangers

These usually contain a negatively charged group bonded covalently to the main structural material. Such groups are sometimes referred to as **fixed anions**. One common group used is the —SO$_3^-$ group. Cations are associated with each group, for example —SO$_3^-$ Na$^+$. The exchanger is placed in a column and the solution to be purified passed through (Fig 8.28).

The operation of the ion exchange material depends upon an equilibrium being established between it and the solution:

$$2RSO_3^- \, Na^+(s) + Ca^{2+}(aq) \rightleftharpoons (RSO_3^-)_2Ca^{2+}(s) + 2Na^+ \, (aq)$$

If a solution with a low concentration of calcium ions is passed down a fresh column the calcium ions are exchanged very efficiently since sodium ions greatly outnumber calcium ions. As the column is used further, however, its efficiency decreases. Sodium ions are increasingly replaced by calcium ions, pushing the position of the above equilibrium to the left. When exhausted, the column must be regenerated by exposing it to a high concentration of sodium ions.

Anion exchangers

An anion exchanger consists of positively-charged groups attached to the solid support. Typical exchange sites are quaternary amine groups – referred to as **fixed cations**. Each positively-charged group has anions associated with it which are available for exchange. The solution containing unwanted anions is passed down the column and the anions exchange. For example, a column primed with chloride ions can be used to remove nitrate ions from solution:

$$RNH_3^+ \, Cl^-(s) + NO_3^-(aq) \rightleftharpoons RNH_3^+ \, NO_3^-(s) + Cl^-(aq)$$

Fig 8.29   Some commercially available 'water filters' contain anion exchange resins, sometimes mixed with charcoal to help to deodorise and purify the water. Such systems purify water for a time but the ion exchange cartridge must be replaced as the resin becomes exhausted.

ALUMINIUM AND PURE WATER

Fig 8.30 Inside the filtration house at a water works. It is essential that the amounts of chemicals added during the purification process are carefully controlled.

To help remove very fine suspended solids from domestic water supplies, a flocculating agent is used. Its purpose, as it settles on the floor of sedimentation tanks is to carry most suspended matter with it. A gelatinous precipitate of aluminium hydroxide is excellent for this purpose and has been used very successfully. It is formed within the tank by precipitation:

$$Al_2(SO_3)_4(aq) + 3Ca(OH)_2(aq) \longrightarrow 2Al(OH)_3(s) + 3CaSO_4(s)$$

Long-term exposure to high aluminium levels from water supplies or from dietary sources has been shown to damage the brain and to induce memory loss. The symptoms are very similar to those of Alzheimer's disease which causes premature senile dementia.

For this reason it is essential that the quantities of aluminium sulphate and calcium hydroxide used in precipitation must be carefully controlled. If too little calcium hydroxide is used, not all the aluminium will be removed from solution. If too much calcium hydroxide is used, aluminium hydroxide, an amphoteric hydroxide, redissolves, again causing high aluminium levels in the drinking water:

$$Al(OH)_3(s) + OH^-(aq) \longrightarrow [Al(OH)_4]^-(aq)$$

In July 1988, 20 tonnes of aluminium sulphate was accidentally dumped into the wrong sedimentation tanks at the Lowermoor treatment works near the Cornish town of Camelford. As a result the level of aluminium ions in the water supply greatly exceeded recommended levels. After the incident some 400 people suffered persistent health problems.

QUESTIONS

8.15 Washing soda crystals have the formula $Na_2CO_3.10H_2O$. A sample of hard water has a concentration of calcium ions of 1×10^{-3} mol dm^{-3}. Calculate the minimum mass of washing soda crystals required to precipitate the calcium ions in a 10 dm^3 sample of the water.
Consult the Database for relative atomic masses.

8.16 (a) Explain what is meant by the terms *cationic exchanger* and *anionic exchanger*. Which of these would you recommend for removing toxic metal ions such as cadmium ions from water?

(b) Give a general equation for the exchange of Cd^{2+} ions using the exchanger you suggested in (a).

(c) Why is it difficult to remove the last traces of cadmium ions from an aqueous sample by this method?

8.17 Some people use domestic 'water purifier' jugs to reduce the level of nitrates in drinking water. Why is it not feasible for water companies to use a similar system on a large scale for tap-water supplies?

Put yourself in the position of leader of a team of analytical chemists working for the Horseracing Forensic Laboratory (HFL). The work of your department involves monitoring the blood and urine of racehorses to identify the presence of traces of performance-enhancing drugs. To expand and improve your analytical capability you hope to purchase *either* a new gas chromatography set-up *or* a high performance liquid chromatography system.

Write a letter to your finance department explaining the advantages of each system. Try to make it clear which system you feel would be more valuable for examination of blood and urine samples thought to contain illegal drugs. Are there any other resources you may wish to purchase in the future to further improve your capability?

Answers to Quick Questions: Chapter 8

1 Because of the costs of extra processing required to purify the compound (for example, by recrystallisation if it is a solid or distillation if it is a liquid).
2 The salt is an 'impurity' and lowers the melting point of ice.
3 The major concerns are whether or not conditions and techniques used on a small scale are still effective when much larger quantities are being used.
4 The intermolecular bonding between unlike molecules is greater than that between like molecules in the separate liquids.
5 Because the type and strength of intermolecular bonding in hexane and in heptane is very similar (dispersion forces, relatively weak) but that in hexan-1-ol (hydrogen bonding, relatively strong) is very different.
6 Because for each distillation there is only a small difference between the composition of the liquid and vapour at a particular temperature.
7 Because water would condense in the distillation flask, increasing the volume of the contents.
8 Non-toxic and non-flammable
9 Cheap
10 One
11 Quickly
12 Strongly
13 Because the graphite (pencil 'lead') is insoluble in the mobile phase and will not interfere with the separation.
14 y/x
15 Under the conditions used, more than one compound might have the same R_f value.
16 Because it is unlikely that two compounds will have the same R_f values for both solvents.
17 Because the mixture would dissolve in the solvent used as the mobile phase and, consequently, travel up the full width of the paper. If more than one spot was being used, the compounds present would simply mix together in the solvent.
18 The molecules have very similar relative molecular masses and structures (both are six membered carbon rings). The intermolecular bonding in each case is very similar (in type and strength).
19 Nitrogen or argon.
20 The greater the chain length, the higher the retention time.
21 Because it is soluble (all nitrates are soluble in water).
22 $Ag^+(aq) + Cl^-(aq) \longrightarrow AgCl(s)$.

Chapter 9

ANALYSIS AND ATOMIC STRUCTURE

LEARNING OBJECTIVES

When you have studied this chapter you should be able to:

1. explain that the interaction of electromagnetic radiation with elements can provide useful information about their structure;

2. explain that atomic emission spectra are due to electronic transitions within the atom;

3. explain the difference between absorption and emission spectra;

4. state how detailed analysis of atomic spectra have helped to develop a model of electronic structure;

5. describe the variations in atomic radii, first ionisation energies, electron affinities and electronegativities observed within a Group and across a Period of the Periodic Table;

6. use successive ionisation energies of an element to provide evidence of its detailed electronic structure;

9.1 ELECTROMAGNETIC RADIATION AND ATOMS

CONNECTION: Analysis by chemical tests is summarised in Chapter 7.

Fig 9.1 A green solution may result from the presence of green Ni^{2+}(aq) but may equally be due to a combination of yellow Fe^{3+}(aq) and blue Cu^{2+}(aq).

Early chemical analysts relied on smell, taste and sight for their detective work. The potential danger rules out taste and smell but a chemist's observational skills remain a vital tool.

Using the human eye to identify the presence of elements in a sample can be a useful first step in qualitative analysis. For example, the colour green may suggest the presence of nickel ions in the sample. However, the eye can be fooled by the colour of a combination of coloured compounds (Fig 9.1). A further limitation is that our eyes are only sensitive to the relatively small range of frequencies corresponding to the visible region of the electromagnetic spectrum.

Nowadays, analysts use a range of instruments to provide information about how different forms of energy interact with different substances. This may be used for:

- qualitative analysis;
- quantitative analysis;
- structural analysis.

To interpret results meaningfully and critically, and to make best use of instruments, a good understanding of both the underlying chemical principles and the equipment is needed. Computers are being used increasingly and so, as in many walks of life, 'computer literacy' is ever more important.

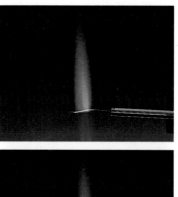

Fig 9.2 Placing a compound which has been wetted with concentrated hydrochloric acid in a Bunsen flame (a flame test) gives the typical flame colours of its elements. This simple test can be used for qualitative analysis, to identify an element. However, to establish its identity beyond doubt, the emitted energy must be analysed by a more accurate device, a spectroscope.
The four photographs on the right suggest the presence of the elements calcium, sodium, potassium and barium reading clockwise from the top left photograph.

Analyses of atomic spectra have helped the development of a refined model of the atom and more refined models of the atom have then increased the value of atomic spectra for use in qualitative, quantitative and structural analysis. Interpretation of spectra and its use for analysis have developed hand in hand.

The Bohr atom

By the early 20th century, James Maxwell had shown that a charged body which travelled on a curved path must radiate energy continuously. If an electron (a 'charged body') in an atom 'orbited' the nucleus in the same way that a planet orbited the sun, then it should lose energy. It would eventually spiral in to the nucleus and the atom would collapse. But this does not happen.

In 1913, Niels Bohr, the Danish physicist, suggested that electrons did not radiate energy providing they possess certain 'allowed' energies. In the Bohr model for the atom, electrons are pictured as revolving around the nucleus in circular paths, called **orbits**. Each orbit has a fixed radius and an electron in the orbit has a fixed amount of energy. Bohr knew of the work of Max Planck and Albert Einstein which suggested strongly that all forms of radiated energy (including light) came in discrete packets known as **quanta**. Bohr proposed that an electron must absorb a single quantum of energy to move from one orbit to another of greater radius.

CONNECTION: The nuclear model of the atom, page 16.

1   **Why would an electron in a large radius orbit possess more energy than one in a smaller radius orbit?**

Electrons occupy energy levels. Each energy level is assigned a principal quantum number and can contain a maximum number of electrons. In this chapter, we will look at experimental evidence which caused the Bohr model to be reconsidered and describe a refined model for the arrangement of electrons in atoms.

KEY FACT

A single discrete packet of energy is called a **quantum** of energy

The electromagnetic spectrum

Light is the visible region of the electromagnetic spectrum. Electromagnetic radiation is a form of energy transported by **waves**. It is characterised by its:

- wavelength (the distance between peaks in a wave);
- frequency (the number of peaks passing a given position every second).

ANALYSIS AND ATOMIC STRUCTURE

The electromagnetic spectrum (Fig 9.3) is continuous but ranges within it have some familiar names, such as **infrared** (long wavelength, low frequency and low energy) and **ultraviolet** (short wavelength, high frequency and high energy). All electromagnetic radiation can be viewed as a means of 'delivering' energy (rather like the waves of the sea 'deliver' energy). The warmth of the Sun's rays, heat from an electric fire and the energy used to cook food in a microwave oven are familiar forms of energy radiated in this way.

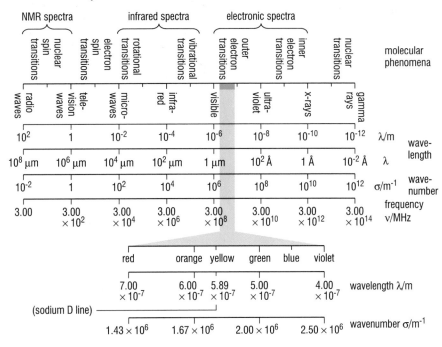

Fig 9.3 The electromagnetic spectrum.

2 **Explain why ultraviolet light and not infrared radiation is responsible for sunburn.**

The idea of energy being transmitted in the form of a wave may be pictured by thinking about the sea. As waves come into shore, the overall movement of water is minimal (a rubber dinghy, for example, rides the waves and drifts only slowly with the tide). The water moves up and down but hardly moves at all in a linear direction. The wave we see a distance from the shore is not made up of the same water by the time it arrives at the shore. The bulk of the water does not move, but the waves are transporting a lot of energy. The damage caused by stormy seas bears witness to this fact. A wave is characterised by the distance between the crests of consecutive waves (wavelength), the height of the waves (amplitude) and the rate at which they reach the shore (frequency). All waves are described in these terms.

3 **Explain why the higher the frequency of a wave, the more energy it is transporting.**

When material is exposed to electromagnetic radiation, it absorbs certain wavelengths and either reflects others or allows them to pass through. For example, red paint absorbs light from the blue region of the visible spectrum and reflects red. A red solution absorbs blue light and allows red to pass through. The absorbed energy may cause electrons to move between energy levels, or bonds in a substance to vibrate more vigorously. The way in which radiation interacts with a chemical substance can provide a great deal of information about the identity and structure of that substance.

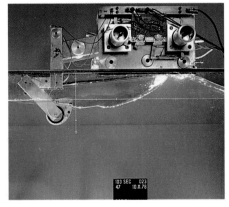

Fig 9.4 The energy stored and transported by waves can be used to do work. This 'bobbing duck' is used to capture wave energy by converting the up and down motion into electrical energy. This is a possible renewable energy source for the future.

CONNECTION: Structural analysis is described in Chapter 10.

Atomic emission spectra

Evidence for the detailed arrangement of electrons in atoms comes from studies of **atomic emission spectra**. The atomic emission spectrum of an element is produced when its atoms are **excited** by giving them energy, for example, by passing a high voltage spark through the gaseous element or by heating the element in a flame. The atoms of elements which have absorbed energy in this way are said to be in an **excited state**. Excited atoms are unstable relative to their **ground state** (normal state) and to regain stability they emit energy in the form of electromagnetic radiation. An **emission spectrum** is produced as they relax (return to their normal state).

The emission spectrum can be seen by using a **spectroscope**, which splits the emitted radiation from an element into different frequencies (Fig 9.5). A prism or diffraction grating (a piece of glass with thousands of parallel lines per centimetre drawn on it) is used to split the radiation coming from an element. The spectrum consists of bright lines on a dark background. It is the 'un-split' emission spectrum of atoms (partly composed of visible light) which is responsible for the attractive colours observed in flame tests (Fig 9.2). Each element produces its own distinctive emission spectrum which reliably identifies the element. We can think of this as an 'elemental fingerprint'.

Fig 9.5 A spectroscope analyses radiation. White light produces a continuous spectrum but each element, when given sufficient energy, produces its own characteristic atomic emission spectrum consisting of discrete lines.

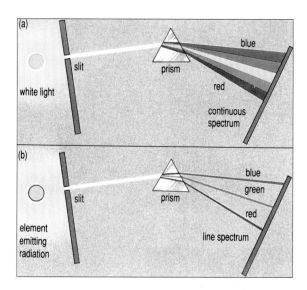

Fig 9.6 The frequencies of the lines in a spectrum allow us to identify the source of the radiation. Atomic emission spectra can be used to identify any element.

4 **Not all elements give coloured flames. Does this mean that their atoms cannot exist in excited states?**

But what is the origin of emission spectra and what can a study of them tell us about the structure of the atom?

The existence of discrete lines in an atomic spectrum, rather than a continuous spectrum, may be explained by a model of the atom in which electrons may possess only certain, specific amounts of energy. Given energy in the form of heat or electricity, an electron absorbs a quantum of energy. It is then promoted from one energy level to a higher one. Radiation of energy corresponding to the difference between the two energy levels is emitted as the electron returns to the lower level (Fig 9.7). If the transition between energy levels produces radiation in the visible region of the electromagnetic spectrum, coloured lines are observed in the emission spectrum. However, suitable detectors may be used to show that line spectra extend beyond the visible region. Higher energy transitions give rise to lines in the ultraviolet range and lower energy transitions to lines in the infrared region.

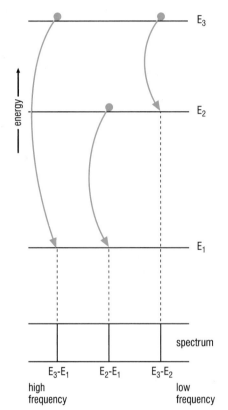

Fig 9.7 The lines in an atomic emission spectrum are due to excited electrons returning to lower energy levels.

5 **Explain in terms of the arrangement of electrons what is meant by the terms 'excited state' and 'ground state' of an atom.**

In the atomic spectrum of an element, series of spectral lines **coalesce**. This means that they get closer and closer together until they merge, and separate lines cannot be distinguished (for example, the barium emission spectrum in Fig 9.6). This suggests that the difference between energy levels gets smaller as they become higher in energy. This is shown by the atomic spectrum of hydrogen (Fig 9.8). There are several groups of lines which coalesce, corresponding to promotion of electrons from lower energy levels to the point where the energy levels converge and the resulting emissions as they return to the lower levels.

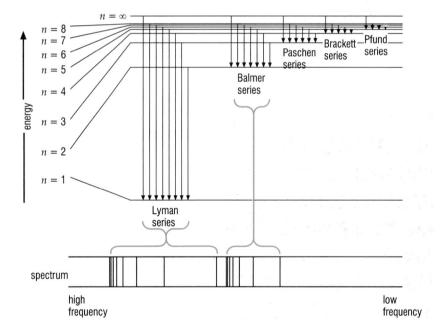

Fig 9.8 The atomic emission spectrum of hydrogen.

In the hydrogen spectrum the **coalescence point** for the Lyman series (Fig 9.8) is the highest observed frequency. It corresponds to the difference in energy between the lowest ($n = 1$) and highest (the **convergence limit**) energy levels. This is the energy released when an electron returns from the convergence limit to the $n = 1$ energy level. Thinking of the process in reverse, if an electron in the $n = 1$ energy level gains this amount of energy, it would escape from the atom and leave a positively charged hydrogen ion. Beyond the convergence limit an electron is independent of the nucleus.

$$H(g) \longrightarrow H^+(g) + e^-$$

ANALYSIS AND ATOMIC STRUCTURE

The amount of energy required to remove one mole of electrons from one mole of gaseous hydrogen atoms is called its **ionisation energy**. Atomic spectra may be used to determine the ionisation energies for gaseous atoms. As we shall see shortly, ionisation energies provide further evidence for the arrangement of electrons in atoms.

6 **What is the important difference between the energy change which produces a line at the convergence limit in the atomic emission spectrum of an element and the first ionisation energy of the same element?**

Spectra and inadequacies of the Bohr model

Careful analysis of atomic emission spectra had shown that each spectral line consisted of a number of very fine lines. The Bohr model, with electrons occupying energy levels within the atom, could explain discrete lines in an atomic emission spectrum – excited electrons emit electromagnetic radiation as they return to the ground state. However, the existence of a fine structure (each spectral line actually being composed of several) could not be explained. A better model was required.

7 **Suggest a simple explanation for the existence of more spectral lines for an element than the Bohr model can explain.**

QUESTIONS

9.1 How does a simple flame test help to establish the identity of metallic elements present in a sample?

9.2 During the early development of the atomic model, James Maxwell argued that atoms should collapse as electrons, continuously losing energy, were drawn to the nucleus. What contribution did Niels Bohr make to the model to suggest why this does not happen?

9.3 When excited atoms relax an atomic emission spectrum is obtained.
   (a) What are the characteristic features of an emission spectrum?
   (b) What does the general appearance of emission spectra tell us about the arrangement of electrons in atoms?
   (c) How would an absorption spectrum differ from an emission spectrum of the same element?

9.4 (a) Explain what happens to a gaseous hydrogen atom when it absorbs sufficient energy to promote its single electron to the $n = \infty$ energy level.
   (b) When the process in *(a)* takes place with *one mole* of hydrogen atoms the energy required is about 1310 kJ. How may this information be used to determine the energy difference between the $n = 1$ energy level and the $n = \infty$ energy level for a single hydrogen atom?

9.5 A 'neon light' takes advantage of the emission spectrum of an element, often neon, argon or krypton, to produce useful light. Explain:
   (a) why such lights require an electrical supply;
   (b) why different elements give rise to different colours.

9.2 THE ARRANGEMENT OF ELECTRONS IN ATOMS

Ionisation energy and electron affinity

Electrons are held in the atom by their electrostatic attraction for the nucleus. Ion formation involves the removal or addition of electrons, both processes involving an energy change. Energy changes which accompany the loss of electrons (to form positive ions) are referred to as **ionisation**

energies. Energy changes which take place when one or more electrons are gained by atoms (forming negative ions) are known as **electron affinities**.

Measurement of ionisation energies, or electron affinities, for a given element yields useful information about its atoms. For example, if a great deal of energy is required to remove an electron from an atom, then that electron must reside in the atom at a point where the attraction of the nucleus is strong. If much energy is released, as an electron is captured by an atom or ion, then this shows that the newly-captured electron must, due to its position, experience a strong electrostatic attraction for the nucleus.

8 Why does it require more energy to remove the second electron from an atom than the first?

To make meaningful comparisons we need to ensure that measurements of ionisation energies or electron affinities are made under standard conditions. Both measurements are made using only gaseous atoms or ions and need to be clearly defined to indicate the species which is gaining or losing electrons. Values are usually quoted for molar quantities of each atom or ion and refer to measurements made at 298K.

Ion formation can be represented by an equation. For example, the first molar ionisation energy of magnesium may be shown as:

$$Mg(g) \longrightarrow Mg^+(g) + e^-; \ \Delta H^{\ominus}_{1st \ IE}(298) = +736 \text{ kJ mol}^{-1}$$

Each of the electrons in an atom may be removed, one by one, and the energies required are known as the successive ionisation energies. For example, the second ionisation energy of magnesium corresponds to the energy required to remove a mole of electrons from one mole of singly-charged gaseous magnesium ions:

$$Mg^+(g) \longrightarrow Mg^{2+}(g) + e^-; \ \Delta H^{\ominus}_{2nd \ IE}(298) = +1450 \text{ kJ mol}^{-1}$$

The third ionisation energy of magnesium is represented as:

$$Mg^{2+}(g) \longrightarrow Mg^{3+}(g) + e^-; \ \Delta H^{\ominus}_{3rd \ IE}(298) = +7740 \text{ kJ mol}^{-1}$$

9 How much energy is required to remove 3 moles of electrons from 1 mole of gaseous magnesium atoms?

The first electron affinity of oxygen is:

$$O(g) + e^- \longrightarrow O^-(g); \ \Delta H^{\ominus}_{1st \ EA}(298) = -141.4 \text{ kJ mol}^{-1}$$

The second electron affinity of oxygen is:

$$O^-(g) + e^- \longrightarrow O^{2-}(g); \ \Delta H^{\ominus}_{2nd \ EA}(298) = +790.8 \text{ kJ mol}^{-1}$$

10 Why is the second electron affinity of oxygen endothermic yet the first electron affinity is exothermic?

Successive ionisation energies

We would expect successive ionisation energies for an element to increase, since each successive electron is removed from an ion of greater positive charge. Therefore, the trend in successive ionisation energies for sodium (Table 9.1) comes as no surprise.

Closer inspection of the ionisation energies of sodium, however, shows that there are two increases in the successive ionisation energies which are much larger than the general trend. The difference between the first and second ionisation energies and between the ninth and tenth ionisation energies are an order of magnitude greater than the preceding difference in each case. It appears, then, that the eleven electrons in a sodium atom are arranged in three groups:

Table 9.1 Successive ionisation energies for sodium / kJ mol^{-1}

1st	496
2nd	4560
3rd	6940
4th	9540
5th	13 400
6th	16 600
7th	20 100
8th	25 500
9th	28 900
10th	141 000
11th	158 700

- one electron is fairly easily removed
- the next eight electrons require more energy but about the same as one another (if we take into account the increasing charge of the remaining ion);
- two electrons which are very difficult to remove.

Using the Bohr model for the atom, these data suggest that electrons in sodium are arranged in the main energy levels as follows:

$n = 3$ energy level contains 1 electron
$n = 2$ energy level contains 8 electrons
$n = 1$ energy level contains 2 electrons

Successive ionisation energies, therefore, provide experimental evidence for our representation of the electronic structure of sodium as 2,8,1. The successive ionisation energies of other elements support the model of an atom with electrons arranged in energy levels, each of which is allowed to contain up to a specified maximum number of electrons. This may be shown graphically by plotting successive ionisation energy against the number of electron removed (Fig 9.9). Conventionally, we use a logarithmic scale.

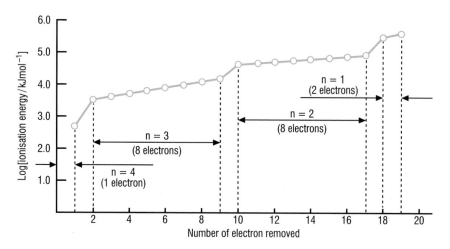

Fig 9.9   Successive ionisation energies provide evidence for the 2,8,8,1 arrangement of electrons in a potassium atom.

But a closer inspection of successive ionisation energies reveals detailed patterns that the Bohr model cannot explain. Suppose we divide a successive ionisation energy by the charge on the ion which remains (IE/charge). This would allow for the increasing charge of the ion from which successive electrons are removed. Now, if all electrons in an atom occupied the same energy level, we might expect all values of IE/charge to be about the same. However, if the electrons occupy different energy levels we might expect a sudden increase when the next electron is removed from a higher energy level.

From experiment, we find that graphs of IE/charge have the same general shape as graphs in which we plot a logarithm of successive ionisation energy against the number of electrons removed.

However, there is one major feature which does not appear on the logarithmic graph – a fine structure for the sections of the graph corresponding to electrons being removed from the main energy levels such as $n = 2$ and $n = 3$.

For example, in the sodium atom there are large increases in IE/charge between the first and second electrons being removed and between the ninth and tenth. This is consistent with the Bohr model. However, the eight electrons in the $n = 2$ energy level are grouped 3-3-2 (Fig 9.10). This observation for the $n = 2$ energy level is confirmed by inspection of data for other elements. Further, we also find that the $n = 3$ energy level has a 3-3-2 pattern for the first eight electrons it can hold.

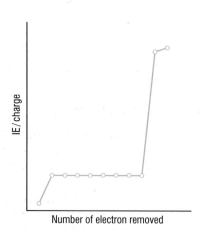

(a) Predicted by the Bohr model for the atom

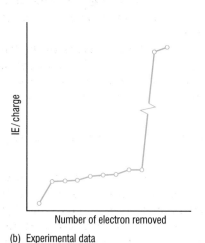

(b) Experimental data

Fig 9.10   Sub-divisions of the $n = 2$ energy level for the sodium atom are suggested by a close analysis of successive ionisation energies, shown in graph(b).

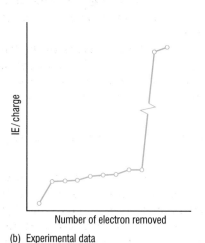

KEY **F**ACT

Successive ionisation energy divided by the charge on the remaining ion may be abbreviated to IE/charge. Such a manipulation allows for the effect of increasing charge on the ion from which an electron is removed. It allows a more useful comparison of successive ionisation energies.

ANALYSIS AND ATOMIC STRUCTURE

Our conclusion is that there are more energy levels in an atom than the Bohr model proposes. At this stage let us remind ourselves of another shortcoming of the Bohr model. It did not allow us to explain the detail of atomic emission spectra, for example there are more lines than expected. Both atomic emission spectra and ionisation energy trends could be explained if more energy levels were available for the electrons to occupy than the Bohr model predicts.

First ionisation energies

A graph of first ionisation energy against atomic number shows a number of characteristic patterns (Fig 9.11). For example, there is a dramatic drop between the last member of a period and the first member of the next period. Also important is the characteristic sawtooth pattern observed for the first two periods of the Periodic Table. Let us consider the first ten elements, hydrogen to neon.

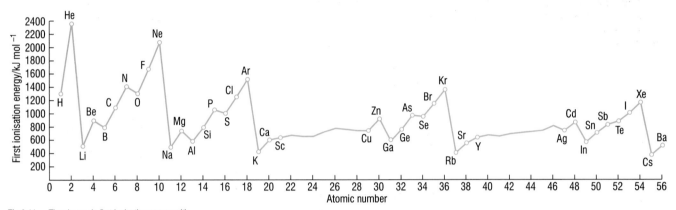

Fig 9.11 The change in first ionisation energy with increasing atomic number.

If the Bohr model were accurate we would predict the following:

- The first ionisation energy of helium would be greater than that of hydrogen since, in each case, the electron is being removed from the $n = 1$ energy level, but the helium nucleus has a 2+ charge compared with the hydrogen nucleus which is 1+.
- The first ionisation energy of lithium would be less than that of helium since the first electron to be removed from a lithium atom (electronic structure: 2,1) comes from the $n = 2$ energy level. Although lithium has a nucleus with a 3+ charge, the electron requires less energy to remove it from a higher energy level than the electron from either hydrogen or helium (both from the $n = 1$ energy level).
- As Period 2 is crossed, the charge on the nucleus increases by 1+ (an extra proton) from one element to the next. The extra electron in each case occupies the $n = 2$ energy level and, therefore, we would predict a smooth increase in first ionisation energy with atomic number for the elements of Period 2.

11 **Why would we expect a sudden drop in first ionisation energy from neon to sodium?**

However, the pattern for Period 2 elements suggest that the eight electrons in the $n = 2$ energy level do not have the same energy as one another (Fig 9.12). Instead, a 2-3-3 pattern is observed (or 3-3-2 if we look from high atomic number to low atomic number). A similar pattern is observed for the first ionisation energies of the Period 3 elements. At this stage we should recall that successive ionisation energy data suggested subdivisions of the $n = 2$ and $n = 3$ energy levels giving rise to a 3-3-2 pattern.

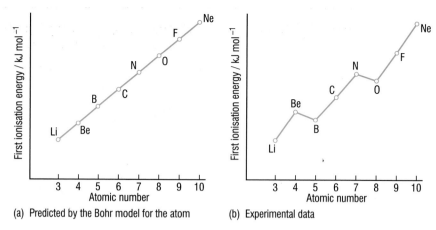

Fig 9.12 The trend in first ionisation energy with increasing atomic number for the Period 2 elements. (a) The graph predicted from the Bohr model of the atom, (b) the experimental graph.

(a) Predicted by the Bohr model for the atom   (b) Experimental data

Clearly the Bohr model of the atom, in particular the arrangement of electrons, needs refining if we are to explain experimental observations of atomic emission spectra and ionisation energies.

QUESTIONS

9.6 **(a)** Explain what is meant by the term ionisation energy;
(b) With the aid of suitable equations, show the process taking place during measurement of *(i)* the first ionisation energy of calcium, *(ii)* the second ionisation energy of lithium.

9.7 **(a)** Define electron affinity;
(b) With the aid of suitable equations, explain the changes taking place during the measurement of the first, second and third electron affinities of nitrogen.

9.8 Explain why:
(a) ionisation energies always have a positive value;
(b) electron affinities may have a positive or a negative value.

9.9 Second ionisation energies are always greater than first ionisation energies. Explain why this is the case.

9.10 The first two ionisation energies for lithium are 519 and 7300 kJ mol^{-1} respectively. The first two ionisation energies for neon (in the same Period as lithium) are 2080 and 3950 kJ mol^{-1}. Explain why:
(a) there is a very large difference between the two values for lithium;
(b) there is a much smaller difference between the two values for neon.

9.11 For aluminium the first four ionisation energies are 577, 1820, 2740 and 11 600 kJ mol^{-1}. Putting yourself in the position of an early modeller of atomic structure, write a short summary explaining what light this information throws on the electron arrangement in an aluminium atom.

9.12 Decide which of the following statements are true, explaining your answer in each case.
(a) Within a Group of the Periodic Table there is a decrease in first ionisation energies;
(b) On crossing a Period in the Periodic Table, from left to right, first ionisation energies decrease;
(c) The first ionisation energy of chlorine is larger than that of fluorine;
(d) The size of the nuclear charge on an atom of an element dictates the size of the first ionisation energy of that element.

9.13 Why does argon have a much higher first ionisation energy (1 520 kJ mol^{-1}) than potassium (418 kJ mol^{-1})?

9.14 Sketch the general shape of the Periodic Table and show on this the trends observed for first ionisation energies.

9.15 Fig 9.11 shows the variation of first ionisation number with atomic number. Explain why:
   (a) there is a general increase in first ionisation energy from lithium to neon;
   (b) *for a given Group* elements in the Period Na to Ar have lower first ionisation energies than those in the Period Li to Ne.

9.3 THE ATOMIC ORBITAL MODEL OF THE ATOM

Wave-like properties of electrons

Following the work of Bohr, and recognition of the weaknesses of his model, much theoretical and experimental work was carried out. It revolved around an intensive study of the nature of electrons, their behaviour and their position within atoms. An important contributor was Louis De Broglie, who in 1925 suggested that electrons showed properties of both particles and waves. Electrons, he argued, had a wavelength. They could transport energy, therefore, in the same way that electromagnetic radiation does.

De Broglie summarised the relationship between the wavelength, λ, mass, m, and velocity, v, of an electron in the **De Broglie equation**:

$$\lambda = \frac{h}{mv} \text{ (where } h \text{ is Planck's constant)}$$

In 1926, Werner Heisenberg stated that it is not possible to measure both the position and the energy of any particle simultaneously, a principle which became known as the **Heisenberg uncertainty principle**. He argued that if we wish to know accurately the energy of an electron we must accept large inaccuracies in determining its position. It seems we can only know either the position or the energy of an electron.

What is possible, however, is a statistical picture of where an electron is likely to be. Erwin Schrödinger, in 1926, derived a **wave equation**, which he used to determine the probability of finding electrons at defined positions within a hydrogen atom. The hydrogen atom was chosen for the calculations because of its simplicity – it contains just one electron. Schrödinger showed this probability in the form of **probability clouds**, mathematically-defined regions of space where there is a likelihood of finding an electron.

Probability clouds are commonly referred to as **atomic orbitals**. Their importance is considerable. The chemical reactions an element and its compounds undergo are determined partly by the number of electrons in atoms of the element, but of equal importance are the atomic orbitals to which its electrons are assigned, their energies and their shapes.

Atomic orbitals

Each main energy level (defined by its principal quantum number, n) consists of one or more atomic orbitals. The atomic orbitals have specific amounts of energy associated with them and are **subdivisions** of the main energy level. We characterise an atomic orbital by its:

KEY FACT

It is important not to confuse **orbits** and **orbitals**. Orbits were a way of describing the path of an electron in the Bohr model of the atom. However, the dual nature of electrons means that this type of orbit is not possible in an atom. Instead we can only describe a volume of space in which we are likely to find an electron of specified energy. This is called an orbital.

- three-dimensional shape;
- orientation in space;
- energy.

An atomic orbital, regardless of its shape, orientation or energy, can accommodate up to two electrons. We shall look at the characteristics of atomic orbitals by considering the main energy levels one by one.

12 **What is the maximum number of electrons which can be accommodated in the first three main energy levels taken together?**

The n = 1 energy level

This consists of a **1s atomic orbital** (usually referred to simply as a 1s orbital) which can accommodate up to two electrons. An electron in a 1s orbital has a specific amount of energy and exists in a spherical region of space around the nucleus. If the electron could be photographed at split-second intervals over a period of time, most of the pictures would catch the electron at a position close to the nucleus. This is another way of saying that the probability of finding an electron near the nucleus is high (Fig 9.13).

Fig 9.13 The shape of a 1s atomic orbital. Strictly speaking, an electron may be found anywhere outside the nucleus. However, most of the time it is close to the nucleus. We can represent an s orbital by enclosing a region which contains about 90% of the electron density.

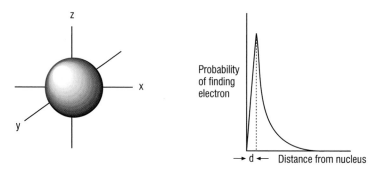

The n = 2 energy level

This consists of four atomic orbitals:
- one 2s orbital;
- three 2p orbitals.

13 **Explain why the four atomic orbitals can accommodate up to eight electrons between them?**

The 2s orbital has the same shape as the 1s orbital but is higher in energy. Electrons in the 2s orbital have more energy and, on average, are further from the nucleus than electrons in a 1s orbital. The three 2p orbitals are **degenerate** (have the same energy as one another) and, in an atom containing more than one electron, are slightly higher in energy than a 2s orbital. They all have the same shape but are distinguished by the different orientation of this shape in space (Fig 9.14).

Fig 9.14 The 2s and three 2p atomic orbitals. Because of its spherical shape, an s orbital has no orientation. However, the double lobed p orbital covers different volumes in space when it is rotated. It turns out that just three orientations are 'allowed', and we define these by reference to a set of x,y,z axes. The three 2p orbitals are labelled $2p_x$, $2p_y$ and $2p_z$. In each case the nucleus is at the origin of the axes.

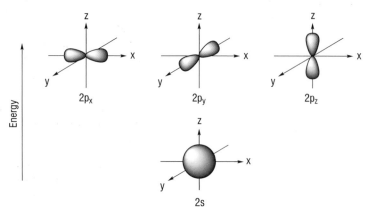

ANALYSIS AND ATOMIC STRUCTURE

The *n* = 3 and *n* = 4 energy levels

The *n* = 3 energy level consists of nine atomic orbitals:

- one 3s orbital;
- three 3p orbitals;
- five 3d orbitals.

The n = 4 energy level consists of sixteen atomic orbitals:

- one 4s orbital;
- three 4p orbitals;
- five 4d orbitals;
- seven 4f orbitals.

The shapes of d orbitals are shown in Fig 9.15. but those of the f orbitals are complex and need not concern us here.

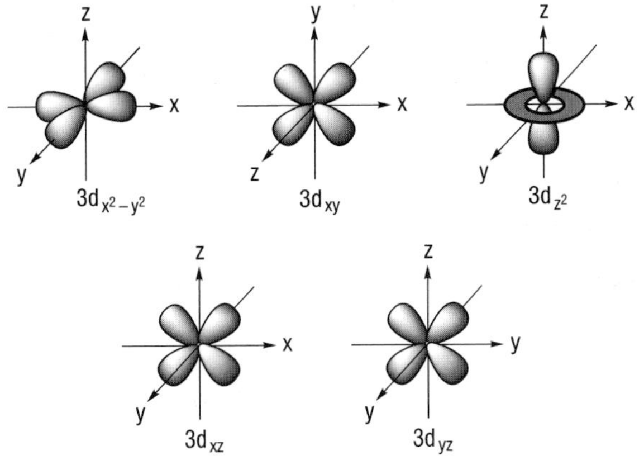

Fig 9.15 The shapes of the five 3d atomic orbitals.

14 **How many electrons can be accommodated in the fourth main energy level?**

An atomic orbital may contain up to two electrons. Experiments show electrons behave as if they were tiny, spinning, bar magnets. When two electrons occupy an atomic orbital they spin in opposite directions and are said to have **paired spins**. The magnetic properties of each are cancelled out. Degenerate orbitals are occupied by single, unpaired electrons before pairing occurs. Single electrons in degenerate orbitals are found to spin in the same direction (have **parallel spins**).

KEY FACT

Two or more energy levels are said to be **degenerate** if they have the same energy as one another.

Defining an electron in an atom

We are familiar with the idea of quanta of energy. Other properties can be quantised, that is have only certain, allowed values. An electron in an atom may be distinguished from any other electron in the same atom by defining four **quantum numbers**:

The **principal quantum number**, *n*, tells us the main energy level from which the orbital containing the electron derived. In the Bohr model this was the only quantum number used.

The **subsidiary quantum number**, *l*, tells us the shape of the orbital. For example:

$l = 1$ indicates an s orbital;
$l = 2$ indicates a p orbital;
$l = 3$ indicates a d orbital.

The **magnetic quantum number, m**, tells us about the orientation in space of the orbital. For example:

when $l = 2$, m may have values $-1, 0, +1$ to show the three allowed orientations of a p orbital;
when $l = 3$, m may have values $-2, -1, 0, +1, +2$ to show the five allowed d orbitals.

The **spin quantum number, s**, tells us about the direction of spin in the orbital. The two spins are differentiated by the use of $+\frac{1}{2}$ and $-\frac{1}{2}$.

Electronic configurations

The atomic orbital model described above was worked out for the hydrogen atom which, since it contains only one electron, is the simplest possible atom. The electron in the hydrogen atom may be found in any of the atomic orbitals. In its ground state (lowest energy state) the electron would be in the 1s orbital. Other one-electron systems such as the helium ion, He$^+$, behave in the same way as hydrogen. In a one-electron system all orbitals within a main energy level (defined by its **n** value) have the same energy.

However, all atoms (other than hydrogen) and most ions are multi-electron systems. Electrons have a negative charge and we would expect some repulsion between them. It is not surprising that multi-electron systems behave differently to single-electron systems. Atomic orbitals which are degenerate in a hydrogen atom lose their degeneracy in a multi-electron system (Fig 9.16).

15 How does the atomic orbital model help to explain the greater number of lines in the atomic emission spectrum of an element than was predicted by the Bohr model?

Fig 9.16 The sub-division of main energy levels for a one-electron system and for multi-electron systems.

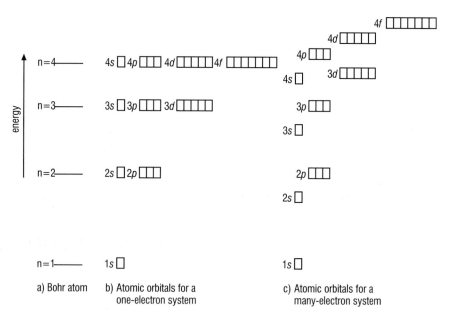

a) Bohr atom   b) Atomic orbitals for a one-electron system   c) Atomic orbitals for a many-electron system

The electronic structure for the ground state of an atom is determined by a simple principle:

electrons are arranged in an atom in such a way that their total energy is kept at a minimum.

To achieve this the following rules are applied when building up the electronic structure of an atom or ion (Fig 9.17). Together these rules are known as the **Aufbau principle**):

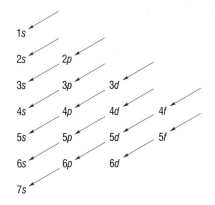

Fig 9.17 This simple diagram shows the order in which orbitals are filled as the electronic structure of an atom is built up. If the arrows are followed the correct sequence for filling orbitals is given.

- the electron is always placed in the lowest energy orbital which is available (in other words, does not already contain two electrons);
- if two similar orbitals are available but one already contains an electron, the electron is placed in the vacant orbital.

16 Why do you think degenerate atomic orbitals are singly occupied initially?

The arrangement of electrons in atomic orbitals may be conveniently represented by the 'electrons-in-boxes' method. A box is used to represent an orbital and single-headed arrows are used to show the presence of an electron, together with its spin. The arrangement of electrons in elements of Period 2 is shown using the electrons-in-boxes method in Fig 9.18.

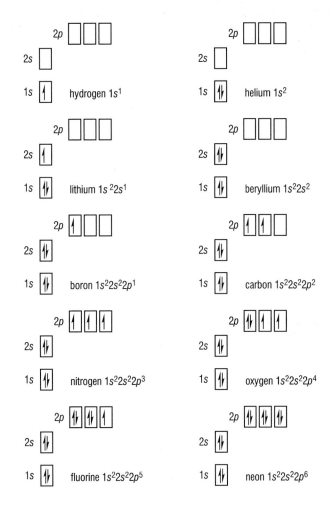

Fig 9.18 Using electrons-in-boxes to show the arrangement of electrons in atomic orbitals for hydrogen, helium and the elements of Period 2.

Chemical shorthand for showing electrons in orbitals

It is useful to have a system which shows at a glance the electronic configuration of atoms and ions. A commonly used system uses small raised numbers (superscripts) to show how many electrons occupy each subshell. For example, silicon atoms have the ground state electronic configuration:
$1s^2 2s^2 2p^6 3s^2 3p^2$

The total number of electrons can be found by adding the superscripts. In this case there are 14 electrons. The system also allows a quick check on the number of electrons in the outer energy level, important since it is their arrangement that dictates chemical properties. A silicon atom has four electrons in the outer energy level. The electronic configuration above is said to be **s,p,d notation**.

KEY FACT

The terms **electronic structure** and **electronic configuration** are both used to mean the arrangement of electrons in atoms and ions. Energy levels may be represented either as main energy levels (for example, fluorine: 2,7) or s,p,d notation (for example, fluorine: $1s^2 2s^2 2p^5$).

ATOMIC ABSORPTION SPECTROSCOPY

The interaction of a element with electromagnetic radiation can be investigated in an atomic absorption spectrometer. With this device the absorption of radiation by a sample at different frequencies can be accurately determined providing the absorbance is **calibrated** (carefully checked against a known standard). The complicated absorption spectrum produced by each element present can be interpreted by a computer linked to a powerful database and can reliably print out a list of the elements present.

Fig 9.19   An atomic absorption spectrometer can rapidly provide multi-element qualitative and quantitative analysis. This may be used, for example, in quality control in the food industry. Modern machines are compact, accurate and reliable.

It is important to be able to interpret the shorthand fully. For example, if we read $2s^2 2p^2$ we understand that there are two paired electrons in the 2s atomic orbital and one unpaired electron in each of two 2p atomic orbitals.

17   **What is the ground state electronic configuration of a potassium ion, K^+?**

QUESTIONS

9.16   The Bohr model of the atom described electrons as particles *'in orbit around a nucleus'*.
(a)   What are the limitations of this simple picture?
(b)   Why is the term *orbit* a poor description of electron movement?

9.17   Explain carefully what is meant by the term *atomic orbital*.

9.18   (a)   Sketch the shapes of all the atomic orbitals in energy level $n = 2$.
(b)   Explain how many electrons may be accommodated in total within this energy level.

9.19   What is meant by the term *degenerate orbitals*?

9.20   A greater number of lines is observed in the atomic emission spectrum of elements than is predicted by the Bohr model. How does the atomic orbital model of the atom help to explain the additional lines?

9.21   (a)   Give the ground state electronic configuration of the element with atomic number 15.
(b)   The same element forms an anion with a charge of 3−. What is its ground state electronic configuration?

9.22   Use Fig 9.17 to work out ground state electronic configurations for the following atoms or ions:
(a) He;   (b) Na^+;   (c) S;   (d) S^{2-};   (e) Ca^{2+};   (f) H^-;   (g) K;   (h) Cl^-;   (i) Br.

9.4   ORBITALS AND IONISATION ENERGIES

Successive ionisation energies

Let us consider again the trends in successive ionisation energies for sodium. We can use the atomic orbital model of the atom to explain the fine detail of the graph shown in Fig 9.10(b).

The electronic structure of sodium is $1s^2 2s^2 2p^6 3s^1$ (Fig 9.20). The first electron to be removed is from the 3s orbital. The next electron is removed from a 2p orbital which, therefore, requires more energy. The third electron must also be removed from a 2p orbital, but now we have a choice – it

KEY FACT

Neon cores:

An electronic structure using s,p,d notation may be simplified by showing the presence of a noble gas 'core'. For example, the full description for a sodium atom is $1s^22s^22p^63s^1$. A neon atom has the electronic structure $1s^22s^22p^6$. Therefore we can write the sodium structure as a neon core plus $3s^1$:

Na [Ne]$3s^1$

This practice may be called '**noble gas notation**'.

could come from the 2p orbital which now contains one electron or a 2p orbital which contains two electrons. If we remember our mental image of an atomic orbital – a region of space which can contain up to two electrons – it is reasonable to suggest that the third electron to be removed is one of the paired electrons. The electron is removed more easily because of its repulsion from the second electron in that orbital. Similarly, the fourth electron removed would be a paired electron.

The next electron to be removed must be an unpaired electron in a 2p orbital. It is not repelled by a second electron and, therefore, slightly more energy is required to remove it. This explains the slight increase in energy needed between the fourth and fifth electron being removed. Once the unpaired electrons in 2p orbitals have been removed, the next one comes from the 2s orbital. This is lower in energy than the 2p orbitals and therefore more energy is required.

The last two electrons to be removed come from the 1s orbital. This derives from the $n = 1$ main energy level and is much lower in energy. Considerably more energy is required to remove these two electrons, and this is shown in the graph of successive ionisation energy divided by charge on remaining ion against number of electron removed (for example, Fig 9.10).

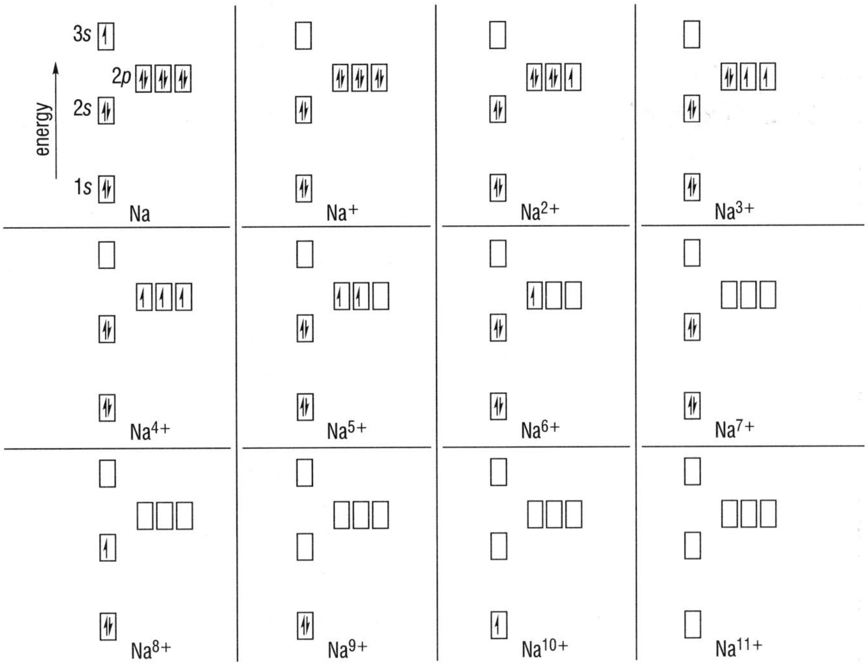

Fig 9.20 The arrangement of electrons as electrons are stripped one by one from a sodium atom. For simplicity the relative energies of the orbitals have been shown to remain the same. This is not the case. For example, a 2p orbital in a Na^+ ion would not have the same energy as a 2p orbital in a Na atom.

First ionisation energies

Fig 9.11 shows a general trend of increasing first ionisation energy across a period, from right to left. But in each period there are two obvious dips which give rise to the sawtooth pattern. In Period 2, the dips occur between beryllium and boron and between nitrogen and oxygen. In Period 3, they are for elements directly below beryllium, boron, nitrogen and oxygen, namely, magnesium, aluminium, phosphorus and sulphur. Clearly the dips repeat in a regular way (they are said to be **periodic**). What is the reason for these reversals of the general trend and the appearance of the characteristic sawtooth pattern?

The atomic orbital model provides an answer. The slight decrease in first ionisation energy from beryllium to boron is due to the slightly higher energy of the 2p orbital (from which the electron is being removed) by comparison to the 2s orbital (from which the beryllium electron is being removed).

18 What is the explanation of the decrease in first ionisation energy from magnesium to aluminium?

The outermost electrons of nitrogen are all unpaired, each occupying a different 2p orbital. Oxygen has one more electron than nitrogen and this is found paired in one of the 2p orbitals. The additional electron is repelled by the electron already occupying this orbital and is more easily removed. This accounts for the slight decrease in first ionisation energy from nitrogen to oxygen.

19 Explain the 'dip' from phosphorus to sulphur seen in Fig 9.11.

The evidence from successive ionisation energies for a given element and trends in first ionisation energies both lend strong support for the refined model for the atom – one in which electrons occupy atomic orbitals. For one more piece of supporting evidence, however, we will revisit the arrangement of electrons in potassium and calcium atoms.

A closer look at the $n = 3$ energy level

The $n = 3$ energy level can contain a maximum of eighteen electrons. We might, therefore, expect that the electrons in a potassium atom would be arranged 2,8,9 or, using our more sophisticated model:

$1s^2 2s^2 2p^6 3s^2 3p^6 3d^1$

However, ionisation energy data shows clearly that the electrons are arranged 2,8,8,1 (Fig 9.9). We have already discussed the stability associated with the presence of just eight electrons in the $n = 3$ energy level. The atomic orbital model offers an explanation. In Fig 9.16 we see that the 4s orbital is lower in energy than the 3d orbitals. Therefore, the 4s orbital fills before any of the 3d orbitals. The correct electronic structure for potassium is:

K $1s^2 2s^2 2p^6 3s^2 3p^6 4s^1$

20 Write down the electronic structure of calcium.

Once the 4s orbital is filled, further electrons are placed in the 3d orbitals. When these are fully occupied the next electrons are placed in the 4p orbitals. The arrangement of electrons in the Period 4 elements is shown in Table 9.2.

Table 9.2 Electronic structures of the Period 4 elements

Element			
Potassium	2,8,8,1	$1s^2 2s^2 2p^6 3s^2 3p^6 4s^1$	$[Ar]4s^1$
Calcium	2,8,8,2	$1s^2 2s^2 2p^6 3s^2 3p^6 4s^2$	$[Ar]4s^2$
Scandium	2,8,9,2	$1s^2 2s^2 2p^6 3s^2 3p^6 3d^1 4s^2$	$[Ar]3d^1 4s^2$
Titanium	2,8,10,2	$1s^2 2s^2 2p^6 3s^2 3p^6 3d^2 4s^2$	$[Ar]3d^2 4s^2$
Vanadium	2,8,11,2	$1s^2 2s^2 2p^6 3s^2 3p^6 3d^3 4s^2$	$[Ar]3d^3 4s^2$
Chromium	2,8,13,1	$1s^2 2s^2 2p^6 3s^2 3p^6 3d^5 4s^1$	$[Ar]3d^5 4s^1$
Manganese	2,8,13,2	$1s^2 2s^2 2p^6 3s^2 3p^6 3d^5 4s^2$	$[Ar]3d^5 4s^2$
Iron	2,8,14,2	$1s^2 2s^2 2p^6 3s^2 3p^6 3d^6 4s^2$	$[Ar]3d^6 4s^2$
Cobalt	2,8,15,2	$1s^2 2s^2 2p^6 3s^2 3p^6 3d^7 4s^2$	$[Ar]3d^7 4s^2$
Nickel	2,8,16,2	$1s^2 2s^2 2p^6 3s^2 3p^6 3d^8 4s^2$	$[Ar]3d^8 4s^2$
Copper	2,8,18,1	$1s^2 2s^2 2p^6 3s^2 3p^6 3d^{10} 4s^1$	$[Ar]3d^{10} 4s^1$
Zinc	2,8,18,2	$1s^2 2s^2 2p^6 3s^2 3p^6 3d^{10} 4s^2$	$[Ar]3d^{10} 4s^2$
Gallium	2,8,18,3	$1s^2 2s^2 2p^6 3s^2 3p^6 3d^{10} 4s^2 4p^1$	$[Ar]3d^{10} 4s^2 4p^1$
Germanium	2,8,18,4	$1s^2 2s^2 2p^6 3s^2 3p^6 3d^{10} 4s^2 4p^2$	$[Ar]3d^{10} 4s^2 4p^2$
Arsenic	2,8,18,5	$1s^2 2s^2 2p^6 3s^2 3p^6 3d^{10} 4s^2 4p^3$	$[Ar]3d^{10} 4s^2 4p^3$
Selenium	2,8,18,6	$1s^2 2s^2 2p^6 3s^2 3p^6 3d^{10} 4s^2 4p^4$	$[Ar]3d^{10} 4s^2 4p^4$
Bromine	2,8,18,7	$1s^2 2s^2 2p^6 3s^2 3p^6 3d^{10} 4s^2 4p^5$	$[Ar]3d^{10} 4s^2 4p^5$
Krypton	2,8,18,8	$1s^2 2s^2 2p^6 3s^2 3p^6 3d^{10} 4s^2 4p^6$	$[Ar]3d^{10} 4s^2 4p^6$

Inspection of Table 9.2. shows two perhaps surprising electronic structures. We might have predicted different electron arrangements for chromium and copper as the orbitals filled according to the Aufbau principle:

Chromium: predicted [Ar]3d^{4}4s^2
           actual   [Ar]3d^{5}4s^1
Copper: predicted [Ar]3d^{9}4s^2
           actual   [Ar]3d^{10}4s^1

If we look at electrons-in-boxes representations for the actual electron arrangements (Fig 9.21), we see that 3d^5 corresponds to a half-filled set of 3d orbitals, that is with each 3d orbital containing a single unpaired electron. We also see that 3d^{10} corresponds to a filled set of 3d orbitals. It seems, therefore, that there is some kind of stability associated with a half filled or completely filled set of orbitals.

Fig 9.21 Electrons-in-boxes diagrams for chromium and copper.

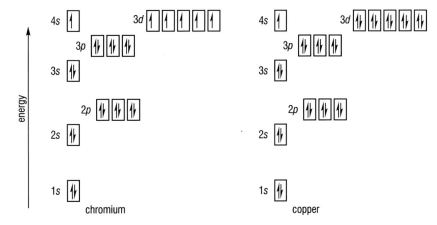

chromium   copper

The electronic configuration of ions

Ions are formed by the loss or gain of electrons by atoms. When cations (positively charged ions) are formed, electrons are removed from the highest electron-containing atomic orbital. For example:

Na 1s^{2}2s^{2}2p^{6}3s^1 or [Ne]3s^1

Na$^+$ 1s^{2}2s^{2}2p^6 or [Ne]

When anions are formed, electrons are added to the next available atomic orbital, applying the Aufbau principle:

O 1s^{2}2s^{2}2p^4 or [He]2s^{2}2p^4

O^{2-} 1s^{2}2s^{2}2p^6 or [Ne]

21 **Give the electronic configurations of fluoride and calcium ions.**

QUESTIONS

9.23 For beryllium (Z = 4) the four successive ionisation energies are: 899, 1 757, 14 849 and 21 006 kJ mol^{-1}. The third successive ionisation energy is very much larger than the second. Explain why the difference between the third and the fourth successive ionisation energies is smaller than that seen between the second and third.

9.24 The first seven successive ionisation energies of an element are: 1314, 3388, 5300, 7469, 10 989, 13 326 and 71 334 kJ mol^{-1}.
(a) Why do the successive ionisation energies increase in size?
(b) To which Group of the Periodic Table is the element likely to belong?

(c) What is the explanation of the large increase in ionisation energy observed for the seventh ionisation energy?

(d) A significant increase in successive ionisation energy is observed for the fifth electron. What is the likely explanation of this?

9.25 In your own words, describe the contribution that a study of atomic spectra and ionisation energies have made to the development of the atomic orbital model of the atom. What do you consider to be the most important developments?

9.26 Examine Fig 9.11 carefully. Write out the electronic configurations of nitrogen and oxygen and use these to explain why the first ionisation energy of oxygen is lower than that of nitrogen, despite the fact that oxygen has a higher nuclear charge.

9.27 On descending a Group in the Periodic Table, nuclear charge increases greatly from one element to the next. Despite this the first ionisation energy decreases. What two factors overcome the effect of increasing nuclear charge?

9.28 For a sulphur atom and a sulphide ion (S^{2-}) give:
(a) the full electronic configuration;
(b) the noble gas notation;
(c) an electrons-in-boxes diagram.

9.29 The electronic configurations of the elements chromium and copper are $[Ar]3d^54s^1$ and $[Ar]3d^{10}4s^1$ respectively. What is unusual about these electronic configurations and why do these elements adopt them?

SPOTLIGHT

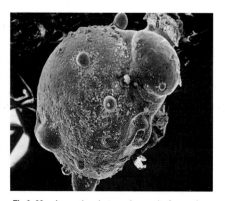

Fig 9. 22 A scanning electron micrograph of a metal globule embedded in a bank note. The elements in this tiny globule can be identified from their X-ray emission spectra, produced by scanning the sample with high energy electrons.

CHEMISTRY AND CRIME

A tiny globule of metal embedded in a bank note but invisible to the eye, enabled forensic scientists to establish a link between the note and safe that had been broken-into using an oxyacetylene torch. A scanning electron microscope is able to detect the globule (Fig 9.22). Its composition is established by changing the mode of action of the electron microscope. A beam of high energy electrons is directed at the globule and radiation in the X-ray region of the electromagnetic spectrum is emitted (Fig 9.23). Each element emits its own characteristic X-ray emission spectrum.

The X-rays are due to energy released when electrons return from a high energy level to a 1s atomic orbital.

Fig 9. 23 X-ray emission spectra of gun shot residues can help to establish the identity of the ammunition used. In this specimen, antimony, barium and lead are clearly present.

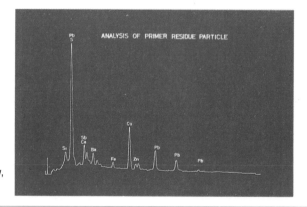

Answers to Quick Questions: Chapter 9

1 Because it has not lost as much energy.
2 Because it has greater energy.
3 Each wave carries a certain amount of energy. The more frequently the waves are transmitted, the more energy they can transport.
4 No. The excitation may require energy greater than that provided by visible light, and so the emitted energy is in the ultraviolet region of the spectrum.
5 In the ground state, all electrons reside in the lowest energy levels available. In an excited state, one or more of the electrons are in higher energy levels.
6 The line at the convergence limit is due to an exothermic process whereas the first ionisation energy is an endothermic process.
7 There are more energy levels than the Bohr model describes.
8 Because the electron (negatively charged) is being removed from an ion which carries a positive charge.
9 9926 kJ mol^{-1}
10 Because, for the 2nd electron affinity, energy must be supplied to overcome repulsion by the negative charge already present.
11 Because the electron is being removed from a higher energy level and, therefore, requires less energy.
12 28
13 Each atomic orbital can accommodate a maximum of two electrons, therefore four atomic orbitals can accommodate up to eight electrons.
14 32
15 Because there are more energy levels within which electrons can reside.
16 Electrons repel one another because of the like charge they carry. Therefore, if two orbitals of the same energy are available to accommodate an in-coming electron, that electron will occupy the vacant orbital rather than the one which already contains an electron.
17 $1s^2 \ 2s^2 2p^6 \ 3s^2 3p^6$
18 The electron being removed from Al is in a 3p atomic orbital, which is higher in energy than the 3s atomic orbital from which the Mg electron is being removed.
19 The electron being removed from the sulphur atom occupies an atomic orbital with another electron, and the two electrons repel one another. However, the electron being removed from the phosphorus atom resides in an atomic orbital alone.
20 $1s^2 2s^2 2p^6 3s^2 3p^6 4s^2$
21 $F^- \ 1s^2 \ 2s^2 2p^6$
   $Ca^{2+} \ 1s^2 2s^2 2p^6 3s^2 3p^6$

Chapter 10

DETERMINING CHEMICAL STRUCTURE

CONNECTION: Designing substances for specific tasks, page 200.

> ## LEARNING OBJECTIVES
>
> When you have studied this chapter you should be able to:
>
> 1. outline the instrumental methods available to chemists to determine chemical structure;
>
> 2. explain the use of X-ray diffraction;
>
> 3. describe how a mass spectrometer works;
>
> 4. explain how mass spectrometry provides evidence for the existence of isotopes;
>
> 5. describe the use of mass spectrometry in compound identification, the determination of relative molecular mass and molecular formula;
>
> 6. explain how mass spectrometry can provide information about the structure of a compound;
>
> 7. calculate the molecular formula of a compound from its empirical formula and relative molecular mass;
>
> 8. explain how nuclear magnetic resonance (NMR) spectrometry assists structure determination
>
> 9. describe the use of infrared spectroscopy and ultraviolet/visible spectroscopy in compound identification and structure determination.

10.1 CHEMICAL ANALYSIS AND STRUCTURE

Structural analysis is concerned with understanding how atoms are arranged in a substance. Since the properties of a compound depend on this arrangement, and on the type of bonds which hold the atoms together, the determination of structure is of great importance. One of the major advances of modern science has been our ability to probe matter and discover how atoms are assembled, in both giant structures and molecular structures. This knowledge has enabled chemists to make better materials, often tailored to meet a specific need.

Chemical tests

Information about the nature of chemical substances can be gained from chemical tests. For example, we can identify the presence of cations and anions in a solid or in solution, and the presence of functional groups in an organic compound, although none of these tests provides information about how all the atoms are arranged. Chemical tests, however, can be a useful first step in determining structure. Qualitative analysis enables the elements present in a compound to be identified, and quantitative analysis is used to show the ratio in which atoms of these elements are present (the empirical formula of the compound).

The usual sequence for structure determination is:

- establish which elements are present;
- determine the empirical formula;
- for molecular compounds, determine the molecular formula;
- determine the structural formula;
- confirmation that the structure has been correctly identified is achieved by synthesising the compound with the proposed structure from known starting compounds. The structure is built in stages, each of which results in a known addition to the structure.

1 **What information is required to determine the molecular formula of a compound whose empirical formula is known?**

Instruments for analysis

Many instruments, often quite sophisticated, are available to provide analytical information. They work by analysing the effect of different forms of energy on matter and may be used to obtain qualitative, quantitative and structural information.

The arrangement of atoms in a substance can be determined by using one or more of the following instrumental techniques:

Mass spectrometry may be used to determine relative atomic masses and relative molecular masses. The more powerful mass spectrometers can be used to obtain molecular formulae. Further, each substance produces a fragmentation pattern in the mass spectrometer. This pattern can be used to identify known substances (a 'fingerprinting' technique) or to provide evidence for the arrangement of atoms in a compound.

Nuclear magnetic resonance spectroscopy may be used to identify substances. Importantly, it is a powerful technique for providing evidence of molecular structure. The environment of atoms such as hydrogen, carbon, fluorine and phosphorus can be probed.

Infrared spectroscopy may be used to identify substances (infrared 'fingerprints' are very useful) and to provide evidence for structural formulae, in particular functional groups present in organic compounds. It is occasionally used for quantitative analysis.

Electronic spectroscopy can provide structural evidence but is more commonly used for quantitative analysis.

X-ray, neutron and electron diffraction can be used to identify compounds. They also offer a means of establishing the structure of a compound.

No one method is definitive but a combination can provide strong evidence for the structure of the molecule. Diffraction techniques, however, provide direct evidence of structure.

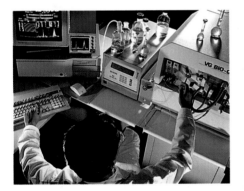

Fig 10.1 A modern mass spectrometer. The real power of this instrument is realised when the mass spectrometer is used in tandem with a chromatograph (usually gas-liquid or high performance liquid chromatograph) and a computer containing a database of known mass spectra.

Diffraction methods and absolute structure determination

The most accurate and powerful technique for structure determination of solids is X-ray diffraction. It provides direct evidence for the arrangement of particles (atoms or ions) in a crystalline solid. The interatomic distances in a solid are similar to the wavelength of X-rays and a crystal acts a three-dimensional diffraction grating. The X-rays are diffracted, though we often talk about X-ray 'reflection' from parallel layers of atoms in the crystal lattice to help us to picture the effect. A structure is determined by trial-and-error in the search for the arrangement of particles which best explains the intensities of 'reflections'. For complicated structures this often requires vast numbers of calculations, best performed by a computer.

2 **Explain why X-ray diffraction cannot be used to determine the structure of an amorphous solid.**

Two other diffraction techniques are used. Electron diffraction may be used for gaseous compounds and gives accurate bond lengths and angles. Neutron diffraction may be used for solids and it is particularly useful for identifying the positions of atoms with small atomic numbers. It is often used after X-ray diffraction has enabled the positions of larger atoms to be determined.

3 **What techniques might be used to determine the structure of a molecular compound which is liquid at room temperature and pressure?**

QUESTIONS

10.1 During the analysis of an unknown chemical substance, explain what information would be obtained by:
(a) qualitative analysis;
(b) quantitative analysis.

10.2 What is the purpose of the following analytical instruments:
(a) a mass spectrometer;
(b) a nuclear magnetic resonance spectrometer;
(c) an infrared spectrometer?

10.3 What is X-ray diffraction used for?

10.2 IN SEARCH OF M_r: THE MASS SPECTROMETER

The empirical formula tells us the simplest ratio of atoms in a compound. To determine the actual *number* of atoms in a molecule of the compound, we need to know the relative molecular mass of the compound.

The mass spectrometer

We cannot weigh individual atoms or molecules, but we can determine their relative masses by **mass spectrometry** (Fig 10.1). Atoms and molecules are ionised and then sorted according to the relative masses of the ions. Relative molecular masses can be determined and, with the more powerful, high resolution mass spectrometers, molecular formulae can be obtained directly.

In a mass spectrometer, a compound is vaporised and its molecules bombarded with high energy electrons. Electrons and molecules collide, the molecules ionise and positive ions are formed. The **molecular ion** is formed when a molecule loses an electron but is otherwise unchanged:

$$M(g) + e^- \longrightarrow M^+(g) + e^- + e^-$$

| gaseous molecule or atom | high energy electron | molecular ion | electron 'knocked out' of the molecule | the bombarding electron, now with lower energy |

Double ionisation does occur, but is much less common:

$$M(g) + e^- \longrightarrow M^{2+}(g) + 2e^- + e^-$$

For simplicity, however, we will only consider the formation of singly charged ions. These are the dominant species produced in a mass spectrometer.

4 **How could double ionisation be minimised in a mass spectrometer?**

The bombarding electrons have sufficient energy to cause cleavage of covalent bonds in molecules. Consequently, most of the molecules **fragment** into smaller, positively charged ions. As we shall see later, the structures of the fragments which form and their relative abundances can tell us a great deal about the compound under investigation.

The ions produced by the bombardment of electrons are accelerated into a magnetic field where they are deflected. For ions with the same charge, deflection in the magnetic field is greatest for ions of lowest mass. Deflection also depends on the size of charge carried by the ion and the mass spectrometer actually determines mass-to-charge ratios. Therefore, ions with high charge are deflected more than ions of the same mass but lower charge. The mass spectrometer records the relative abundance of each particle defined by its mass : charge ratio (m/e).

Results of mass spectrometry are usually interpreted in terms of the relative atomic and relative molecular masses of the fragments produced. The pattern of fragments produced is recorded as a **mass spectrum**. For example, the mass spectrum of methanol (Fig 10.2) shows peaks for ions of relative masses 32, 31, 29 and 28. The particle of relative mass 32 is the molecular ion, CH_3OH^+. The others are fragments of formulae CH_3O^+, CHO^+ and CO^+ respectively.

5 Explain why the particles CH_2^+ and $C_2H_4^{2+}$ cannot be separated in a mass spectrometer.

6 Arrange the following particles in order of decreasing deflection in a magnetic field:

CO^+, CH_2^{2+}, CCl_2^+, CO_2^+, CCl_2^+, CH_2O^+, C^+.

Modern improvements

Mass spectrometers which use fast moving electrons to ionise gaseous molecules are limited to use with small molecules. Extensive fragmentation occurs, with many ions formed and, consequently, complex mass spectra are obtained. The molecules fragment so much that, often, no molecular ion is observed.

Four modern developments of the basic technique for generating ions have greatly enhanced the range of applications of mass spectrometry.

Fast atom bombardment mass spectrometry (FABMS). The solid compound under investigation is ionised by bombarding it with fast moving, high energy atoms of noble gases such as argon. This allows much larger molecules, including sensitive natural products and polymers, to be studied.

Inductively coupled plasma mass spectrometry (ICPMS). Ions are produced by electric discharge at extremely high temperatures (greater than 7000K). This allows material which is not very volatile to be studied. For example, metal ions in a blood sample can be separated and identified (Fig 10.3).

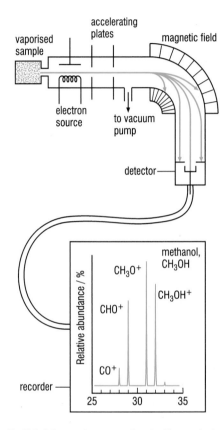

Fig 10.2 A diagram of a mass spectrometer. The vaporised sample is bombarded by high energy electrons and the resulting ions are accelerated. The ions are deflected in a magnetic field, the extent of the deflection being determined by the strength of the field.

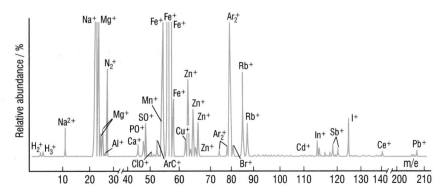

Fig 10.3 The ICP mass spectrum of blood.

7 By inspection of Figure 10.3, which three metals are most abundant in the blood?

Pyrolysis mass spectrometry (PYMS). This is used for complex materials and uses heat to break them into smaller fragments. The fragmentation pattern obtained can be used as a chemical 'fingerprint' to identify the material.

Laser-microprobe mass spectrometry (LAMMS). A laser beam is used to vaporise the surface of the substance to be investigated without also vaporising the substance it rests on. For example, LAMMS can be used to confirm that a pesticide has stuck to a plant following crop spraying because the laser can be directed so that it vaporises only the crystals which are then identified from their mass spectrum (Fig 10.4).

Fig 10.4 Four characteristic fragments in the LAMMS spectrum of crystals found on the leaf of a plant confirm that they are crystals of a pesticide, stuck to the leaf following crop spraying.

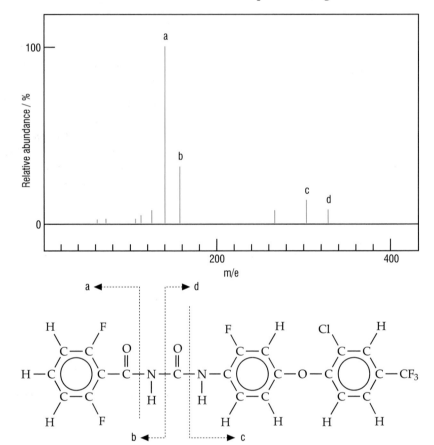

KEY FACT

Isotopes of an element are atoms with the same atomic number (number of protons) but different mass numbers (number of protons and neutrons).

8 From the mass spectrum (Fig 10.4), calculate the relative mass of fragment 'a' and confirm that it corresponds to the molecule fragmenting as suggested in the diagram.

Evidence for isotopes and determination of A_r

Mass spectrometry provides conclusive evidence for the existence of isotopes. The mass spectra of nearly all elements consist of more than one peak. Our explanation is that each peak in the mass spectrum of an element corresponds to an isotope of that element. Elements which have only one natural isotope, for example fluorine, aluminium and iodine, give mass spectra with only one peak. High resolution mass spectrometry allows the determination of relative isotopic masses with great precision – to between five and seven decimal places.

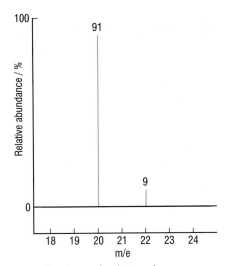

Fig 10.5 The existence of two isotopes of neon are confirmed by its mass spectrum: neon-20 (relative abundance 91) and neon-22 (relative abundance 9)

Fig 10.6 The mass spectrum of chlorine shows two groups of peaks. Those at higher relative mass are due to chlorine molecules containing different isotopes, and those at lower relative mass correspond to chlorine atoms.

Let us look at an example. The mass spectrum of neon contains two peaks due to particles of relative mass 20 and 22 (Figure 10.5). Their relative intensities, 91:9, is a measure of the abundance of each type of particle. The particles are atoms of two isotopes of neon: neon-20 and neon-22.

^{20}Ne: 10 protons, 10 neutrons; relative isotopic mass, 20; relative abundance, 91%;

^{22}Ne : 10 protons, 12 neutrons; relative isotopic mass, 22; relative abundance, 9%;

The relative atomic mass (A_r) of neon can be calculated from the relative abundances of the isotopes.

$$A_r = \frac{(91 \times 20) + (9 \times 22)}{100}$$

$$= 20.18$$

9   **The mass spectrum of magnesium (atomic number 12) has three peaks corresponding to particles of relative masses 24, 25 and 26. Describe the composition of these particles.**

The mass spectrum of chlorine (Fig 10.6) contains five peaks, due to particles of relative mass 35, 37, 70, 72 and 74. Chlorine has two isotopes, $^{35}_{17}$Cl and $^{37}_{17}$Cl. Peaks in the mass spectrum at 35 and 37 are due to atoms of the two isotopes of chlorine, and the peaks at 70, 72 and 74 are due to molecules of chlorine. The composition of the particles corresponding to these peaks are given in Table 10.1.

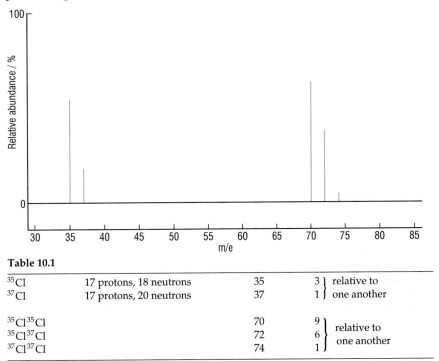

Table 10.1

^{35}Cl	17 protons, 18 neutrons	35	3 } relative to
^{37}Cl	17 protons, 20 neutrons	37	1 } one another
^{35}Cl^{35}Cl		70	9 }
^{35}Cl^{37}Cl		72	6 } relative to one another
^{37}Cl^{37}Cl		74	1 }

10   **Explain why chloromethane, CH$_3$Cl, has two molecular ions of relative masses 50 and 52, and suggest which is the most abundant.**

In low resolution mass spectroscopy (where relative masses are measured to four or five significant figures), all ionised atoms have whole number relative masses since the relative masses of proton and neutrons are both very close to one. However, high resolution mass spectrometry enables very accurate measurements of relative isotopic masses to be made. For example, the relative isotopic masses of $^{35}_{17}$Cl and $^{37}_{17}$Cl have been determined as 34.9688531 and 36.9659034 respectively.

Calculate the accurate relative atomic mass of chlorine and the accurate relative mass of the particle with highest mass observed in the mass spectrum of chlorine gas.

Determining M_r and molecular formulae

The relative molecular mass of a compound can only be determined by mass spectrometry if the molecular ion forms. It is then simply a matter of recording the mass spectrum and determining the relative mass of the molecular ion. The molecular ion of ethanol (Fig 10.7) has a relative mass of 46 which corresponds to a molecular formula of C_2H_6O. For the dye, crystal violet, LAMMS is used because of the low volatility of the compound. The parent ion has a relative mass of 372 (Fig 10.8).

Fig 10.7 The mass spectrum of ethanol, C_2H_5OH. The molecular ion has a relative mass of 46, corresponding to $[C_2H_5OH]^+$.

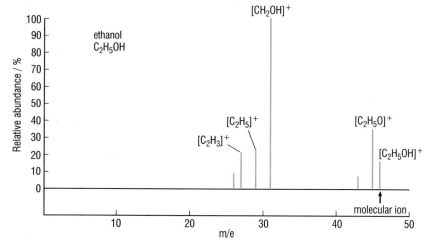

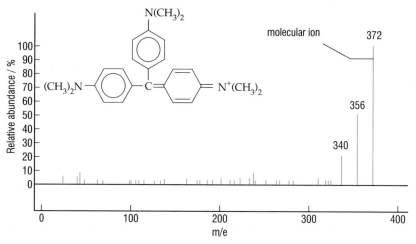

Fig 10.8 The mass spectrum of crystal violet is obtained by laser microphobe mass spectrometry (LAMMS). The molecular ion forms and this enables the relative molecular mass of the compound to be determined.

12 **Use relative atomic masses to calculate the relative molecular mass of crystal violet.**

The molecular formula is either the same as, or a multiple of, the empirical formula. If the relative mass of the empirical formula is divided into the relative molecular mass the scale-up factor is obtained. For example, the empirical formula of benzene is CH and its relative molecular mass is 78. The relative mass of the empirical formula is $(12 + 1) = 13$. Since $(6 \times 13) = 78$, the scale-up factor is 6 and the molecular formula must be $6 \times CH = C_6H_6$.

Problem: The empirical formula of a colourless liquid, X, is found to be CH_2O by combustion analysis. The molecular ion in its mass spectrum has a relative mass of 60. Determine the molecular formula of X.

Solution: The relative mass of each empirical formula unit of X

$$= [12 + (2 \times 1) + 16] = 30$$

$$\frac{\text{Relative molecular mass}}{\text{Relative mass of empirical formula}} = \frac{60}{30} = 2$$

Therefore, the scale-up factor is 2 and the molecular formula is

$2 \times CH_2O = C_2H_4O_2$.

Therefore, the molecular formula of X is $C_2H_4O_2$.

High resolution mass spectrometry

A low resolution spectrometer has two limitations:

- it cannot distinguish between ions of similar relative masses;
- it cannot give very precise measurements.

High resolution spectrometers overcome both these problems. They can, for example, determine relative masses to 9–10 significant figures whereas low resolution spectrometry can only achieve 3–4 significant figures.

High resolution is necessary for accurate work such as the determination of precise molecular formulae. Of course, it is essential to remember that the spectrometer has to be very carefully calibrated since all its measurements are relative to the mass of a singly charged ion formed from an atom of carbon-12. Having found the relative molecular mass accurately the molecular formula can be found.

Databases are available which contain the relative masses of all possible combinations of atoms. A section from a typical database is given in Table 10.2.

Table 10.2

Relative mass	Number of atoms in the molecular ion			
	C	H	N	O
72.057511	4	8	0	1
72.032360	2	4	2	1
72.021127	**3**	**4**	**0**	**2**
72.045107	3	6	1	1
72.068745	3	8	2	0
72.093896	5	12	0	0

13 **What is the molecular formula of a compound whose molecular ion is found to have relative mass 72.021127?**

However, we still do not have information about the structure of the molecule, though closer investigation of the fragmentation pattern may well provide some evidence.

SPOTLIGHT

FIELDWORK ON VENUS

In December 1978 five American and two Russian spacecraft reached Venus. Four of the American probes and both the Russian ones landed on the planet's surface. They had on board a range of instruments – including mass spectrometers, UV spectrometers, gas chromatographs and X-ray spectrometers. Constraints of space and weight meant that the mass spectrometers were only low resolution. A particle of relative mass 64 was identified but the resolution was not good enough to determine whether the particle was SO_2 or S_2. A mass spectrometer capable of higher resolution would have solved the problem.

Accurate relative atomic masses:

^{16}O 15.995; ^{32}S 31.972

Therefore, the relative molecular masses of

$^{32}S^{16}O_2$ would be 63.962; and $^{32}S_2$ would be 63.944.

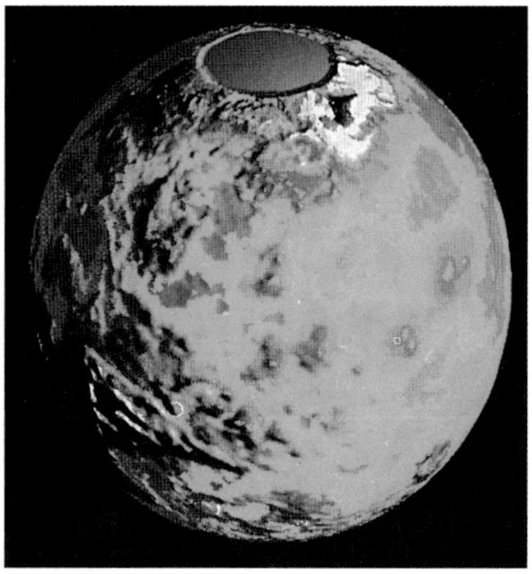

Fig 10.9 A global view of the surface of Venus taken from the Pioneer Orbiter spacecraft

Fig 10.10 Ethanol molecules break up when bombarded with high energy electrons to give a number of fragments. These are some routes by which fragments may form.

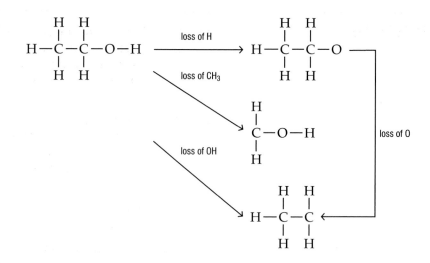

Fragmentation patterns

Structural information is obtained by a study of the fragmentation pattern in a mass spectrum. Looking again at the mass spectrum of ethanol (Fig 10.7), we can identify fragments in the spectrum from their relative masses. For example, an ion appears at relative mass 45, one less than the parent ion. This is due to a hydrogen atom being split from the ethanol molecule, leaving $C_2H_5O^+$. A possible fragmentation path for ethanol is shown in Fig 10.10.

Full analysis of the mass spectra of molecules requires extensive knowledge and much skill. An example is provided by the comparison of mass spectra of two structural isomers, butan-1-ol and 2-methylpropan-2-ol. The way the atoms are arranged in a molecule dictates the way it fragments in the mass spectrometer. Therefore, the two isomers have different yet characteristic spectra (Fig 10.11).

The fragmentation pattern of molecules in a mass spectrometer depends on their structure – the atoms present, how they are arranged in space and the type of bonding between them. The mass spectrometer provides us with the characteristic fragmentation patterns of compounds. The mass spectrum of an unknown compound can be measured and compared with known spectra stored in a computer database and the compound identified.

14 From a consideration of butan-1-ol and 2-methylpropan-2-ol, which kind of carbon skeleton appears to give a more complex fragmentation pattern – unbranched or branched?

> ## KEY FACT
>
> Molecules cleave, or break up, into smaller particles in a mass spectrometer. A compound is characterised by this fragmentation pattern – the relative masses of the fragments which form and their relative abundances.

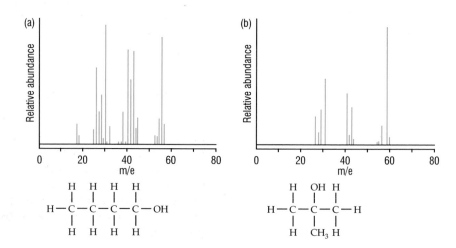

Fig 10.11 The mass spectra of a) butan-1-ol and b) 2-methylpropan-2-ol. Both have the same molecular formula, $C_4H_{10}O$, but they have different structural formulae.

DETERMINING CHEMICAL STRUCTURE

10.4 The relative abundances of the three particles observed in the mass spectrum of magnesium are 78.6, 10.1 and 11.3 (relating to particles of relative mass 24, 25 and 26 respectively). Sketch the expected mass spectrum and calculate the relative atomic mass of magnesium.

10.5 The molecular formula of butan-1-ol and 2-methylpropan-2-ol is $C_4H_{10}O$. Calculate their relative molecular mass and show how, in terms of fragment ions, the two most abundant peaks in each spectrum may be explained.

10.6 Give the m/e ratios of ions which may be observed in a mass spectrum of ethanol. (**Hint:** the mass spectrum of methanol shown in Fig 10.2 may be useful.)

10.7 A hydrocarbon with empirical formula C_2H_4 produces a mass spectrum showing a molecular ion at m/e 56. A fragment ion at m/e 55 is also formed. Suggest:
(a) the likely molecular formula of the hydrocarbon;
(b) a possible structure for the fragment ion.

10.8 An ester of formula $C_4H_8O_2$ is hydrolysed in sodium hydroxide. Following purification and separation of the products a carboxylic acid and an alcohol are obtained. These are analysed separately in a mass spectrometer to give two mass spectra. The alcohol gave a small molecular ion of m/e 46 and a prominent fragment ion at m/e 29. The carboxylic acid gave a mass spectrum with m/e 60 and a prominent fragment ion at m/e 59. Suggest two possible structures for the ester.

10.9 The mass spectrum of a sample of antimony shows peaks at m/e 121 (relative abundance 57%) and at m/e 123 (relative abundance 43%). Calculate the relative atomic mass of antimony.

10.10 **(a)** Why does low resolution mass spectrometry fail to distinguish between the molecular ions of propanone and butane?
(b) How can high resolution mass spectrometry help with this problem?

10.3 NMR: A TOOL FOR STRUCTURAL ANALYSIS

So far we have seen how the molecular formula of a compound can be determined. The next stage of compound characterisation is to determine the arrangement of atoms in a molecule. Our search takes us on to the determination of structural formulae.

No modern industrial or academic research centre is complete without an impressive array of nuclear magnetic resonance (NMR) spectrometers. They often provide the key to unlocking the structural secrets of a molecule.

The NMR spectrometer

Nuclear magnetic resonance spectroscopy is **non-invasive**. Small quantities of the sample are suspended between the poles of a magnet and are recovered unharmed after investigation. Modern instruments are relatively small, robust, highly sensitive and capable of collecting and processing vast amounts of information very quickly. Invaluable information about the arrangement of atoms in molecules can be gained from NMR spectrometers. They are very powerful tools for structural analysis. So how do they work?

Fig 10.12 A modern NMR spectrometer is a powerful tool for structural analysis.

We are familiar with the idea that electrons spin. In the same way, atomic nuclei spin (except those with even atomic number and even mass number). The positively charged spheres behave like small bar magnets. Pictured simply, each spinning nucleus can align with the magnetic field or against it (Fig 10.13). The lower energy state is when the spinning nucleus aligns with the magnetic field. This would be expected since we know that unlike poles of magnets attract one another but like poles repel one another.

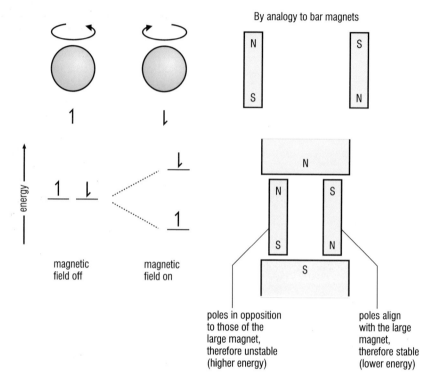

CONNECTION: Electron spin is explained page 171.

Fig 10.13 The spinning charged nucleus behaves as a small bar magnet. In the absence of a magnetic field the two orientations of spin have the same energy, but when a magnetic field is applied the two orientations no longer have the same energy. Given sufficient energy, the nucleus moves from the lower energy state to the higher one.

Given sufficient energy, nuclei of certain isotopes can 'flip' between those two states. In a NMR spectrometer, electromagnetic radiation of fixed energy is passed through the sample while the strength of the magnetic field is varied. The field strength at which the nucleus 'flips' from one state to another can be measured. The behaviour of nuclei in a magnetic field gives us information about their environment and, therefore, the arrangement of atoms in a molecule.

15 Why are electromagnets used in NMR spectrometers and not permanent magnets?

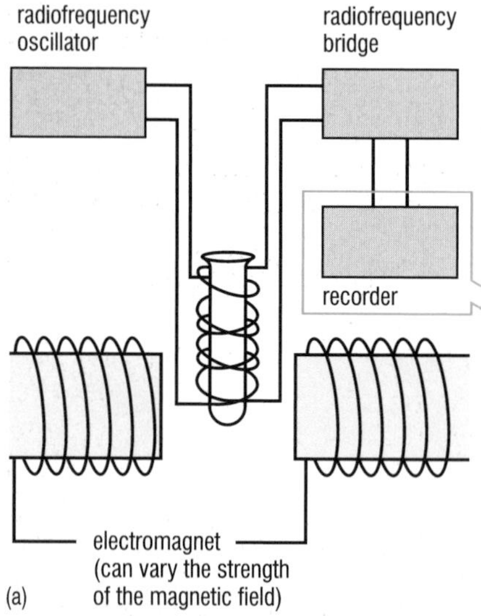

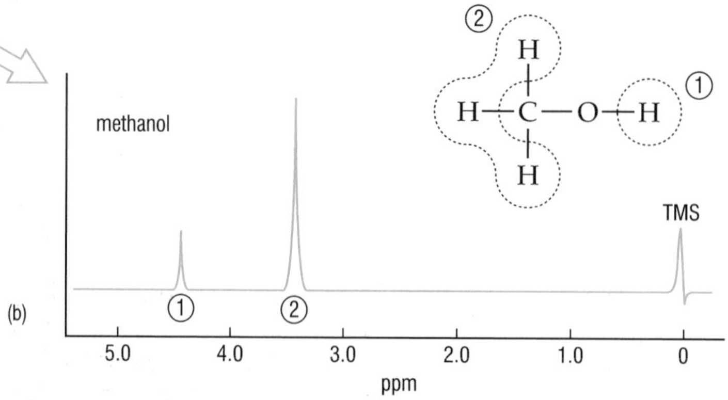

Fig 10.14 (a) A simplified diagram of an NMR spectrometer. The strength of the magnetic field is varied until the spinning nucleus 'flips' from one energy state to another. (b) Low resolution NMR spectrum of methanol, CH_3OH.

DETERMINING CHEMICAL STRUCTURE

The strength of the magnetic field in which a nucleus 'flips' depends on the environment of that nucleus. For example, there are many hydrogen atoms in most organic compounds. We say that these show different **chemical shifts** – in other words they 'flip' at different field stengths. The chemical shift is measured against a standard, the most common being that observed for the protons in tetramethylsilane (TMS), $(CH_3)_4Si$. Hydrogen atoms in particular environments have characteristic chemical shifts. For example, a hydrogen atom in an aliphatic compound has a larger chemical shift than a hydrogen atom in a benzene ring.

CONNECTION: Chemical shifts are given in Datasheet 5.

16 **How can NMR be used to identify the presence of a** $C \equiv C$ **triple bond? (*Hint:* use the Database)**

Focus on structure

What makes NMR spectroscopy so useful is that it can measure the strength of the magnetic field experienced by a nucleus as its environment is altered. Electrons in neighbouring bonds set up their own magnetic field and this influences the magnetic fields experienced by each nucleus. A simple example illustrates this. Methanol has four hydrogen atoms. Three of the hydrogen atoms are in an identical environment, while the fourth (bonded to an oxygen atom) is very different. The NMR spectrum of methanol (Fig 10.14(b)) has two peaks. The large peak is due to the hydrogen atoms attached to the carbon and the smaller peak is due to the hydrogen atom attached to the oxygen.

17 **How many different types of hydrogen atom environment are present in methanol? How does this compare with the area under each peak in the NMR spectrum of methanol?**

TUTORIAL

Problem: Fig 10.15 is the NMR spectrum for a colourless liquid, **X**, with molecular formula C_2H_6O. What is the structural formula of **X**?

Solution: Two structural isomers have the molecular formula C_2H_6O: methoxymethane and ethanol:

(a)
```
      H       H
      |       |
  H — C — O — C — H
      |       |
      H       H
```

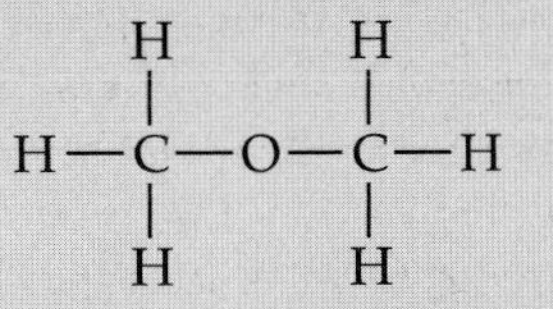

(b)
```
  H   H
  |   |
H—C — C — O — H
  |   |
  H   H
```

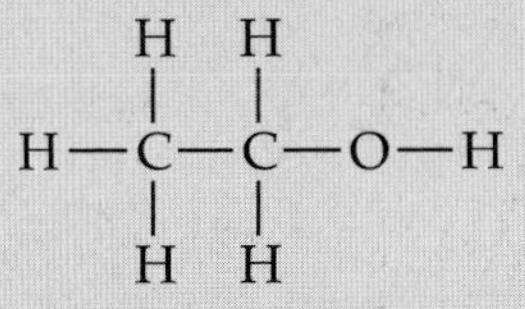

In methoxymethane, the six hydrogen atoms are in an identical environment to one another. Therefore, we predict an NMR spectrum with a single peak.
In ethanol, there are three different environments for the hydrogen atoms:

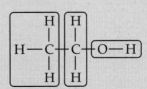

Therefore, we predict a NMR spectrum with three peaks, corresponding to each of the three environments. Further, we predict that the relative areas under the peaks should be in a 3:2:1 ratio. The NMR spectrum of **X** is consistent with our prediction for ethanol and we deduce that **X** is ethanol.

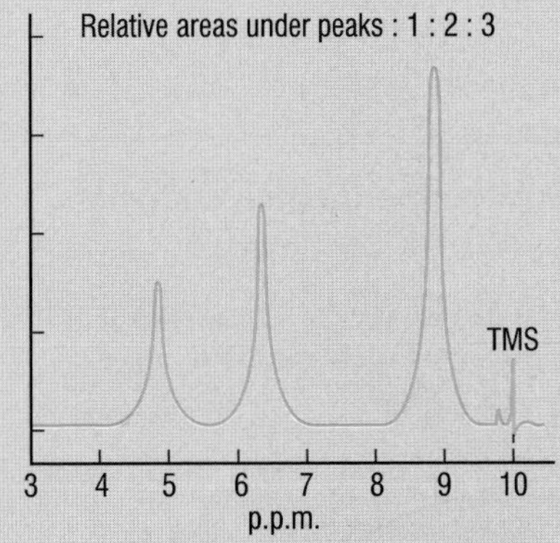

Relative areas under peaks : 1 : 2 : 3

TMS

3   4   5   6   7   8   9   10
p.p.m.

Fig 10.15 Low resolution NMR spectrum of the colourless liquid **X**.

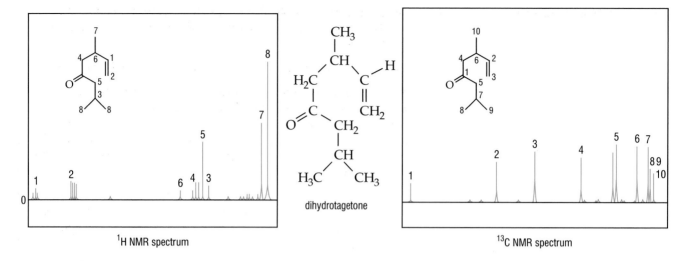

Fig 10.16 Dihydrotagetone, extracted from French marigolds, is used in perfumery. A combination of ^{1}H NMR spectroscopy and ^{13}C NMR spectroscopy can be used to establish its molecular structure.

KEY FACT

NMR spectroscopy of spinning hydrogen atoms is called ^{1}H NMR (also known as **proton magnetic resonance** (PMR)).

High resolution NMR spectroscopy yields more information than low resolution NMR. Neighbouring spinning nuclei affect one another and the strength of the magnetic field they experience. It is beyond the scope of this book to explain this effect and the interpretation of high resolution NMR spectra. It is sufficient to say that determination of the positions, numbers and areas of peaks provides invaluable information for analysing the structure of molecules. An example, however, will help to illustrate the power of the technique.

In dihydrotagetone, there are eight different environments for hydrogen atoms (Fig 10.16). At first, it might appear that sites 7 and 8 are equivalent since both are methyl groups attached to another carbon atom. However, there are subtle differences in these environments, and high resolution NMR allows the two environments to be distinguished.

The examples we have looked at so far are all where NMR spectroscopy has been focused upon the spinning nuclei of hydrogen atoms. However, a number of other nuclei, such as ^{13}C, ^{19}F and ^{31}P, can also be investigated. The use of ^{1}H NMR and ^{13}C NMR in tandem allows the structures of complex molecules to be worked out, since evidence from each technique complements the other. ^{19}F NMR and ^{31}P NMR are used to investigate the environment of fluorine and phosphorus atoms in a molecule.

A recent development is two-dimensional NMR. Combined with computer-aided molecular modelling, two-dimensional NMR can complement X-ray crystallography for determining the structures of large molecules, such as proteins.

QUESTIONS

10.11 Give one advantage of NMR spectroscopy over conventional chemical analysis.

10.12 Fig 10.14(b) shows the NMR spectrum of methanol. Explain why this spectrum has two peaks yet the reference standard used, tetramethylsilane, has only one peak (despite containing twelve hydrogen atoms).

10.13 Predict for a sample of methoxyethane, $CH_3OCH_2CH_3$:
(a) the number of peaks;
(b) the area under each peak.

10.14 (a) What would the NMR spectrum for cyclopropane look like?
(b) Explain why the NMR spectrum of chlorocyclopropane has *three* peaks.

10.15 Describe the principles of operation of a simple NMR spectrometer.

10.16 A compound with molecular formula C_6H_{12} is examined by spectroscopic methods. The mass spectrum shows a clear molecular ion at m/e 84 and although there are a large number of peaks due to fragment ions, there is **no** peak at m/e 15. The NMR spectrum shows only *one* strong peak. Suggest the likely structure of this hydrocarbon.

10.17 Sketch the appearance of the NMR spectrum of methylbenzene using The Database to find the approximate chemical shift values. Indicate on your sketch the number of hydrogen atoms responsible for each peak.

NMR SPECTROSCOPY AND MEDICINE

Body fluids – ranging from urine to blood plasma and sweat to eye fluid – can be studied by high resolution NMR. Spectra obtained can be related to metabolic diseases, organ failure and drug metabolism (Fig 10.17). The medical profession already recognise the advantages of NMR spectroscopy for rapid diagnosis and monitoring of babies.

Fig 10.17 The ^{1}H NMR spectrum of human urine 6 hours after taking a therapeutic dose of paracetamol. Analysis of the spectrum allows the presence of the drug to be identified together with its metabolites and the metabolites usually found in urine.

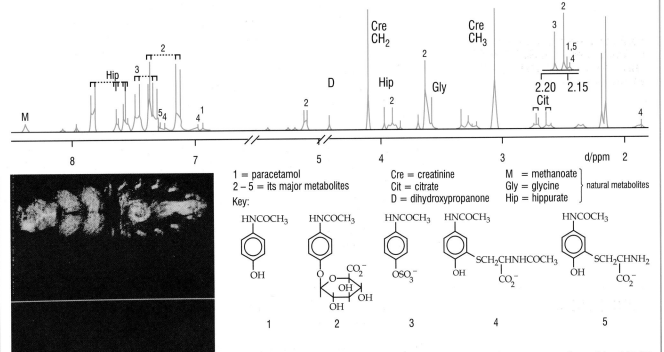

1 = paracetamol
2 – 5 = its major metabolites
Key:

Cre = creatinine
Cit = citrate
D = dihydroxypropanone

M = methanoate
Gly = glycine
Hip = hippurate

} natural metabolites

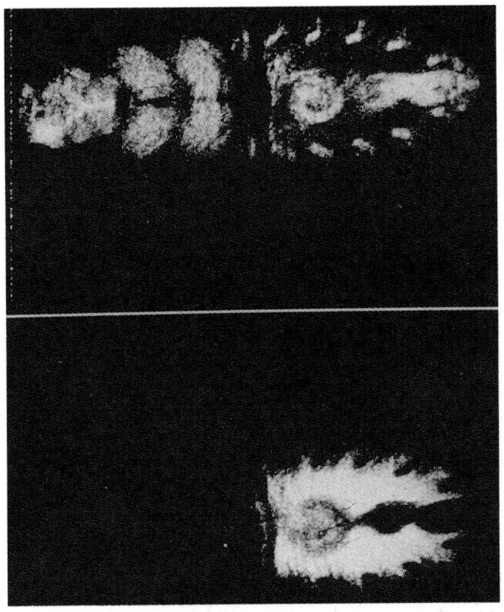

Fig 10.18 Two chemical shift selective NMR images of a live cockroach: a water-only image (top) and a fat-only image (bottom)

Body-scanners used in hospitals make use of images produced by NMR signals. The machines cost around £1 million. Cheaper NMR imaging can be used in the chemistry laboratory to good effect. The growth rate of plant roots can be monitored by NMR imaging without the need to dig up the plant. The fat content of living insects can be determined (Fig 10.18). This may be of use to entymologists and chemists working on lipophilic pesticides.

Although mass spectrometry and NMR spectroscopy are probably the major analytical tools for determination of structure, other techniques are important. We will look briefly at two: infrared spectroscopy and ultraviolet/visible (also called electronic) spectroscopy.

Absorption spectra

Absorption spectra of atoms, molecules and complex ions are often used for quantitative, qualitative and structural analysis. An absorption spectrum is obtained when electromagnetic radiation is passed through a substance and the emerging radiation is split up into a spectrum using a prism or diffraction grating.

An atomic absorption spectrum looks like a continuous spectrum but with certain frequencies missing, showing up as dark lines. If we examine the frequencies of radiation absorbed by atoms as they are excited, we find that they are identical to those emitted as the atoms or ions relax (their emission spectra). The absorption spectra of molecules and complex ions, however, consist of broad bands rather than sharp lines.

Double beam absorption spectrometers are usually used to record infrared and electronic spectra (Figure 10.19). The essential features are the same but the source of electromagnetic radiation and the type of detector used depends on whether infrared or ultraviolet/visible radiation is to be used. For example, an electrically heated rod of rare earth oxides is the source in an infrared spectrometer and the detector is a heat sensor or photoconductivity device. In an ultraviolet/visible spectrometer, the sources are hydrogen discharge lamps (ultraviolet) and tungsten filament lamps (visible) and the detector is a photomultiplier tube.

CONNECTION: Atomic emission spectra, page 162.

KEY FACT

The logarithm of intensity of radiation passing through the reference (I_o) divided by the intensity of radiation passing through the sample (I_s) is called the **absorbance**.

Absorbance, $A = \log \dfrac{(I_o)}{(I_s)}$

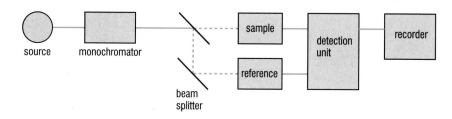

Fig 10.19 The design of a typical double beam absorption spectrometer.

The radiation is split into two beams. One is passed through the sample and the other through a reference (usually the solvent used to dissolve the substance under investigation). The detector measures the absorbance of radiation. An absorption spectrum is a graph of absorbance against frequency.

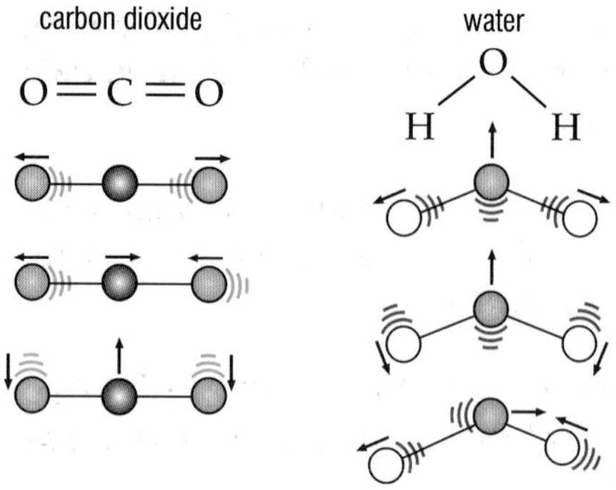

Infrared spectroscopy

The image we may have of molecules from our use of modelling kits, where they are usually portrayed as rigid assemblies of atoms, can be misleading. Bonds in a molecule vibrate, twist and bend (Fig 10.20). But they only do so at certain fixed frequencies. We make use of this property of molecules in infrared spectroscopy.

When infrared radiation is passed through a molecular compound, the molecules absorb radiation with just the right energy to cause the bonds to vibrate, twist or bend more rapidly. The energy required depends in each case on the atoms involved and the nature of the bonds. For example, bonds between light atoms vibrate at higher frequencies (more rapidly) than those between heavy atoms, and multiple covalent bonds vibrate at

carbon dioxide

$O = C = O$

water

Fig 10.20 The typical vibrations of two simple molecules: carbon dioxide and water.

DETERMINING CHEMICAL STRUCTURE

CONNECTION: *Infrared radiation and molecules – Chapter 3.*

KEY FACT

In infrared spectroscopy, the frequency of radiation is often measured as number of waves per centimetre (**cm**$^{-1}$), also called **wavenumber**. Also, transmission rather than absorbance is often recorded, in other words the amount of radiation passing through rather than the amount absorbed. One graph is simply the inverse of the other.

CONNECTION: *Some characteristic infrared absorption frequencies are given on Datasheet 5.*

higher frequencies than single covalent bonds. Frequency of vibration of a bond is also influenced to some extent by other atoms in the molecule.

Place the bonds in order of increasing vibrating frequency:

18 (a) C — Cl, C — Br, C — I, C — H, C — O, C — N;
(b) C = C, C ≡ C, C ≡ N, C = N, C = O.

When the full spectrum of infrared radiation is passed through a compound, an absorption spectrum is obtained. The infrared spectrum may be used to identify the compound, in the same way that a mass spectrum 'fingerprint' can be used for identification.

19 **How could infrared spectroscopy be used to provide evidence that an unknown organic compound is a nitrile? (*Hint:* use the Database)**

The infrared spectrum of an unknown compound can offer clues about its structural formula. This is because certain bonds and functional groups absorb infrared radiation of a characteristic wavenumber. For example, the carbonyl group, $>$C$=$O, absorbs strongly at about 1680 – 1750 cm^{-1}. Therefore, propanone absorbs infrared radiation in this region but ethanol does not (Figure 10.21).

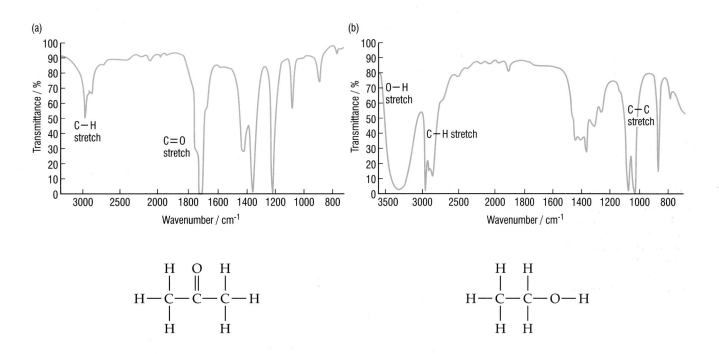

Fig 10.21 Infrared absorption spectra of (a) propanone, and (b) ethanol.

20 **Hex-1-ene and cyclohexane have the same molecular formula, C_6H_{12}. How could infrared spectroscopy be used to differentiate between them? (*Hint:* use the Database)**

Infrared absorption can be used for quantitative analysis but this is not common. One situation in which it is used, however, is the Intoximeter used in most police stations to breathalyse people.

21 **The Intoximeter measures absorbance at 3300 – 3400 cm^{-1}. What causes ethanol to absorb radiation of this wavenumber? (*Hint:* use the Database)**

Electronic spectroscopy

Electrons occupy energy levels within molecules and complex ions in the same way they do for atoms and simple ions. Electrons in atoms occupy atomic orbitals; electrons in molecules occupy molecular orbitals. Ultraviolet (UV) and visible radiation cause the electrons furthest from the nuclei to be promoted to a higher energy level.

ANALYSING THE DRINK-DRIVER

Roadside breathalysers are portable fuel cells which can measure the concentration of alcohol in the breath. However, this is not admissible evidence in a court of law. Such legal evidence must be obtained at the police station using the Intoximeter, which is, in fact, an infrared spectrometer . It measures the absorption of radiation of wavenumber $3300 - 3400$ cm^{-1}. This gives an indirect estimate of the blood alcohol concentration (BAC). A person being prosecuted can ask for a confirmatory test using gas chromatography of a blood sample.

Propanone is found on the breath of people suffering from starvation or diabetes mellitus. Similarly it may be found if people are on a high fat, low carbohydrate diet. Since propanone also absorbs infrared radiation of wavenumber $3300-3400$ cm^{-1} (Fig 10.21), the Intoximeter has a built-in compensation for propanone.

Fig 10.22

The wavelength of radiation of the ultraviolet and visible region of the electromagnetic spectrum is often given, rather than frequency. The usual unit is the nanometre **nm** (10^{-9} m)

CONNECTION: Some characteristic wavelengths of absorption for organic functional groups are given on Datasheet 5.

Fig 10.23 The electronic absorption spectra of (a) propanone – the presence of the C=O group gives rise to a characteristic absorption central at about 275 nm, (b) benzene – the presence of the benzene ring group gives rise to a characteristic absorption centred at about 250 nm.

When the full spectrum of UV/visible radiation is passed through a compound an absorption spectrum is obtained. Because the absorption is due to electronic transitions within a molecule or complex ion, we often call this an **electronic absorption spectrum**. Although not generally as useful as infrared spectroscopy, UV/visible spectroscopy can be used to suggest the presence of certain functional groups in an unknown compound (Fig 10.23). Groupings of atoms in molecules or complex ions which are responsible for the absorption UV/visible radiation are called chromophores.

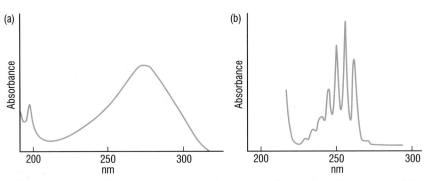

Transition metal ions are often coloured and can be characterised by their absorption spectra. The oxidation state of the metal ion, the nature of molecules or ions bonded to it (the ligands) and the geometrical arrangement of these around it help to determine the colour. For example, the cobalt(II) ion exists in aqueous solution as the octahedral hexaaquacobalt(II) ion, $[Co(H_2O)_6]^{2+}$. Solutions of these ions are pink because visible light from the blue end of the spectrum is absorbed and the remainder

DETERMINING CHEMICAL STRUCTURE

passes through, and that is what we see. Solutions containing the tetrahedral tetrachlorocobaltate(II) ion, $[CoCl_4]^{2-}$, are blue. In this case, red light from the visible spectrum is absorbed and so the solution appears blue (the visible light that is transmitted). The visible spectra of $[Co(H_2O)_6]^{2+}$ and $[CoCl_4]^{2-}$ are characteristic of octahedral and tetrahedral environments for the cobalt(II) ion (Fig 10.24).

Fig 10.24 The electron absorption spectra of a pink solution containing $[Co(H_2O)_6]^{2+}$(aq) and a blue solution containing $[CoCl_4]^{2-}$(aq). The wavelength of absorption and the shape of the absorption bands are characteristic of octahedral and tetrahedral environments of cobalt(II).

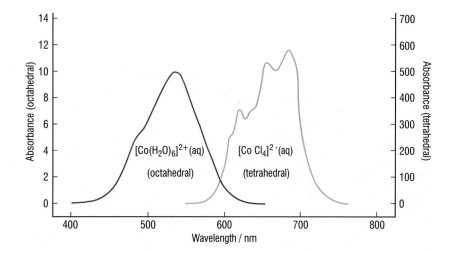

Electronic spectroscopy is a very useful tool in quantitative analysis. This is because, at least for solutions which are reasonably dilute, the intensity of absorption is directly proportional to concentration. For example, the more concentrated a solution of copper(II) sulphate is, the more intensely blue is the solution. The position of the absorption maximum (the wavelength at which absorption is the strongest) does not change, only the amount of radiation that is absorbed.

QUESTIONS

10.18 What is meant by the term *absorption spectroscopy*?

10.19 Explain the origin of energy absorption in the infrared region of the spectrum using ammonia as an example.

10.20 Use the Database to suggest the wavenumbers of the principle infrared absorptions you would expect to see for two isomers of molecular formula C_2H_6O. What features of the spectrum of each compound would enable you to distinguish between them?

10.21 By reference to the Database, discuss the usefulness of:
(a) NMR spectroscopy;
(b) infrared spectroscopy in distinguishing the isomers cyclopropane and propene.

10.22 An aqueous solution of copper(II) sulphate absorbs most strongly at about 800 nm. Explain how a calibration graph may be used to enable the concentration of a solution of copper(II) sulphate to be determined.

10.23 The ultraviolet absorption spectrum of buta-1,3-diene shows an absorption maximum at 220 nm. When bromine is added to a solution of buta-1,3-diene in tetrachloromethane, the absorption maximum disappears. Explain:
(a) the reason for the absorption maximum;
(b) using an equation, the reaction which takes place;
(c) why the absorption maximum disappears.

10.24 When one mole of phenylethene is hydrogenated using a nickel catalyst, one mole of hydrogen molecules is taken up and the ultraviolet spectrum shows an absorption maximum at a different wavelength. Explain:

(a) why one mole of hydrogen molecules reacts;

(b) why the absorption maximum is altered;

(c) why addition of bromine to the product of this reaction produces no further change in the absorption maximum.

ACTIVITY

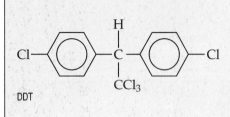

DDT

DDT was used as an insecticide for many years but its use is now banned in most countries. A key reason why it is no longer used is that it persists in the environment and progresses through the food chain, accumulating in the fatty tissues of higher animals.

Imagine you are an analyst given the task of identifying a substance extracted from the liver tissues of seals living in polluted waters. You suspect the sample may be DDT. Explain carefully how you might use mass spectrometry to identify the sample as DDT. Give the mass/charge ratios of the molecular ions you would expect to find in the mass spectrum of DDT. Suggest some likely fragment ions which may be expected to be present. Remember: chlorine exists naturally as a mixture of two common isotopes, ^{35}Cl and ^{37}Cl, in the ratio of 3:1.

How might NMR spectroscopy also be used to identify DDT?

Answers to Quick Questions: Chapter 10

1 Its relative molecular mass or molar mass.

2 Because an amorphous solid does not have a regularly repeating arrangement of particles.

3 Electronic spectroscopy (in particular infrared), NMR spectroscopy and chemical tests for function groups.

4 By ensuring the energy of the bombarding electrons is not too high.

5 Both have the same mass/charge ratio (14:1 in each case).

6 CH^{2+} (6.5) > C^+ (12) > CCl^{2+} (23.75) > CO^+ (28) > CH_2O^+ (30) > CO_2^+ (44) > CCl_2^+ (83) **Note** Mass:charge ratios given in brackets.

7 Sodium, magnesium, iron.   **8** 142

9 Particle of relative mass 24: 12 protons, 12 neutrons
Particle of relative mass 25: 12 protons, 13 neutrons
Particle of relative mass 26: 12 protons, 14 neutrons

10 Chlorine has two isotopes: chlorine-35 and chlorine-37. Therefore, the two molecular ions of CH_3Cl correspond to $CH_3{}^{35}Cl$ (relative mass 50) and $CH_3{}^{37}Cl$ (relative mass 52). The most abundant is $CH_3{}^{35}Cl$.

11 Accurate relative atomic mass of chlorine = 35.4681141
Accurate relative molecular mass of $^{37}Cl_2$ = 73.9318068

12 372 **13** $C_3H_4O_2$ **14** Unbranched.

15 In order that the strength of the magnetic field can be easily changed.

16 The chemical shift of H attached to a $C \equiv C$ triple bond is about 8.0.

17 The relative areas are 3:1. Therefore, there are three hydrogen atoms in one environment and one hydrogen atom in another environment.

18 (a) $C-I < C-Br < C-Cl < C-O < C-N < C-H$
(b) $C=C < C=N < C=O < C \equiv C < C \equiv N$

19 Look for a peak at around $2210 - 2260 \text{ cm}^{-1}$

20 Presence of $C=C$ double bond should give a peak at around $1620 - 1680 \text{ cm}^{-1}$ in the spectrum of hex-1-ene which would not be found in the spectrum of cyclohexane.

21 Absorption due to vibration of the $O-H$ bond.

Theme 3

DESIGN

Designing and making chemical substances with properties to meet a particular purpose requires that the chemist has an understanding of bonding and structure and of the mechanisms of chemical change. A knowledge of how certain types of chemical change can be brought about is equally important.

Chapter 11

DESIGN FOR A PURPOSE

LEARNING OBJECTIVES

When you have studied this chapter you should be able to:

1. explain the term 'research portfolio';

2. explain the terms 'candidate compound' and 'target molecule' and how such compounds are identified;

3. explain what is meant by 'the chemical literature' and describe the problems associated with storing and accessing this;

4. describe the impact of information technology on chemistry;

5. recognise that discovery can be fortuitous or planned and give examples of both;

6. explain the role of mathematical and computer modelling in chemical design;

7. identify functional groups and sub-structures in structural and semi-structural formulae;

8. explain what is meant by 'periodicity' as it applies to Periods within the Periodic Table;

9. explain periodic trends in physical and chemical properties of the elements;

10. use The Database in this book to search for patterns in physical and chemical properties of elements and their compounds;

11. use The Database in this book to choose appropriate reagents and reaction conditions to convert one functional group into another;

12. use synthetic route maps to plot a synthetic route from a starting compound to a target molecule;

13. explain how the length of a carbon skeleton may be altered, ascending and descending an homologous series;

14. explain how a functional group may be 'protected' during reaction;

15. explain the terms 'monomer', 'polymer' and 'polymerisation';

16. explain the difference between 'addition polymerisation' and 'condensation polymerisation' giving examples of each.

11.1 CHEMICAL DATABASES

Research for a purpose

Chemical industries manufacture compounds for particular purposes. Some find their way directly into the marketplace while others are made as intermediates, that is compounds needed in the synthesis of other materials. Often, a new compound must be designed and made for a specific use. To do this, chemists must call upon their chemical knowledge, understanding and imagination.

Market research, a dialogue between a company and consumers, is used to reveal the need for a new material, perhaps a new plastic, agrochemical or perfume fragrance. In the chemical industry and in many academic laboratories, research teams assemble a **research portfolio**, a list of areas to be investigated.

It is not easy to find molecules that do useful things. In 1986 Robert Malpas delivered a lecture to the Royal Society of Chemistry. He said that of the 9 million molecules then known only 100 000 were on sale and only 15 000 of these to the general public. This is an 'application rate' of 1 in 600. For the chemist, the challenge is to identify and make useful compounds.

Imagine being asked to find a compound to fight a disease for which, presently, there is no known cure. How is such a challenge met? It is the business of one of our major chemical industries, the pharmaceutical industry, to solve just this type of problem. Compounds which might have therapeutic properties must be identified. These are the **candidate compounds**, ones which researchers hope may have the required properties and might be considered for more detailed investigation.

PAIN KILLERS BY DESIGN

Rheumatoid arthritis can be a crippling disease. Ibuprofen was developed by the Boots Company to help people suffering from this disease. It doesn't cure rheumatoid arthritis, but it does help alleviate the symptoms – painful, swollen joints.

Ibuprofen was developed by chemists in response to a need. It involved a careful study of candidate compounds and modifications of chemical structure to give a compound with the most suitable properties. One of the candidate compounds was Ibufenac which was available for a short while in the 1960s. However, it was withdrawn in Europe and America because of its side-effects. The first modification was '10499' (Boots' reference number) but it produced rashes in 20% of patients, making it unacceptable. Finally, Ibuprofen was synthesised and, after extensive testing, became available on prescription in 1969. It is now widely available in chemists shops, for example as Nurofen.

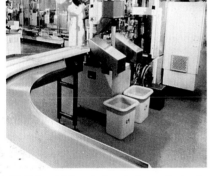

Fig 11.1 **(a)** Ibuprofen in production was developed by chemists to relieve the symptoms of rheumatoid arthritis. The drug is a mild painkiller, an alternative to aspirin and paracetomol.
(b) The structure of Ibuprofen and two of the compounds investigated during the course of its development.

It is difficult to identify candidate compounds worth a further investment of time and effort. Candidate compounds are arranged in a 'league table', with the most likely compounds at the top (to be investigated first) and the less likely (perhaps the more speculative choices) towards the bottom. Chemists use their experience and their understanding of the properties and chemical reactions of substances to plan the programme of work: identifying compounds to be studied and how to make, characterise, and test them. This may involve finding an existing substance (which has not been used for the purpose in mind), perhaps modifying its chemical structure to produce new compounds, or designing a new substance.

New plastics, ceramics, agrochemicals, cosmetics and a host of other everyday, and not so everyday, materials are developed by:
- identifying candidate compounds;
- choosing the most promising ones;
- making;
- testing;
- evaluating.

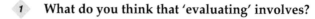 **1 What do you think that 'evaluating' involves?**

Searching the chemical literature

The cornerstone of successful design is access to what is already known and understood. Chemical data are found in what is popularly known as **the chemical literature**. Vast numbers of journals, containing hundreds of thousands of research articles, are published worldwide. In addition, huge numbers of patents are published annually. A research team must know what related work is going on in other laboratories. This requires an extensive study of the literature. In addition to keeping up with work in their own field, researchers need to be aware of broader developments, since lateral thinking can often make significant contributions to research and discovery. However, it is difficult to keep abreast of advancements in key areas of scientific research because of the amount of information available.

Sitting in a library, scouring the journals and textbooks for information takes a great deal of time, time that might be more usefully spent thinking, working in the laboratory or simply relaxing. **Information technology** has come to the rescue of the 'information-swamped' scientist. It allows the immense quantity of chemical literature to be stored, up-dated and accessed easily. Titles and abstracts (short summaries) of research papers are held within a computer database. Search programs are used to identify the relevant journal quickly.

Extensive databases with search facilities enable chemists to find the properties of compounds at the touch of a keyboard. Imagine trying to find a material which could be used as a surface covering for the underside of a ski. Desirable properties might be 'slipperyness', resistance to scratching, ease of adhesion to the ski and, inevitably, cost. Life in the world of chemical design and synthesis is like any other, a compromise. We can use a database containing the properties of all known polymers, for example, to list those which meet our design criteria. Often the ideal material cannot be found and we settle for the best that is available to do the job, an intelligent compromise.

Chemical databases are a means of storing information about the physical and chemical properties of substances. They are designed to allow these properties to be found quickly and, importantly, to allow data to be selected by different criteria and put into different categories.

An example of a computer database is the Beilstein database. It contains some 3.5 million substances. Stereochemical data, in other words full structural information, for each substance is stored and the database has the facility to store up to 400 types of data (such as melting points, boiling points and other physical properties) for each.

2 Imagine you have discovered a new element. What properties would you hope to measure and store in a database?

Databases are also used to store information about reactions, for example, reagents and reaction conditions required to bring about a particular chemical change. Fast, computerised databases allow such information to be accessed quickly and are of enormous help to the chemist designing new compounds and planning ways of making them.

Interrogating a database

Searching a database for information is often known as 'interrogating' the database.

The physical properties of the elements and their compounds can be stored in a database. A sophisticated computer program allows these numerical data to be accessed in several ways. For example, the program might allow elements within a specified melting point range to be listed together with their melting points and in alphabetical order or in order of increasing melting point. Some programs allow the operator to search for the data and present them graphically. They may even search for trends and patterns within the data.

The chemical properties of elements and their compounds can also be accessed in different ways. For example, suppose we want to find out about the chemistry of chlorides of the elements. We might examine each element in the database, its reaction with chlorine and the properties of any chlorides which it forms. Alternatively, or in addition, we might refer to a database on chlorine itself to see if further useful information can be obtained.

Research and development produces an ever increasing amount of data. Much is numerical, for example physical properties, and can be processed by a database management system which can handle only numbers and words. However, the system must also be able to handle chemical structures. These are the international language of chemistry, allowing chemists of all nationalities to communicate. Since the mid-1980s considerable development has occurred in this area. Molecular graphics – the display of three-dimensional chemical structures on the computer screen – are an excellent research tool.

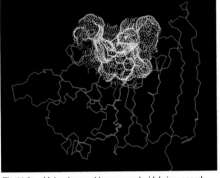

Fig 11.2 Molecular graphics are used widely in research

Searching for substructures

The usual way to interrogate a chemical database of organic compounds is through a **substructure** search. A substructure is part of a molecule – an arrangement of atoms within a molecule, and often contains one or more functional groups. The substructure (or a combination of substructures) to be searched for is specified. We can illustrate the idea by looking at the structures of three commonly used weedkillers – 2,4-D, MCPA and mecoprop.

2,4-dichlorophenoxyethanoic acid (2,4-D) has the structural formula:

OCH$_2$COOH

Cl

Cl

This molecule might be found in a database from a number of different starting points. For example:

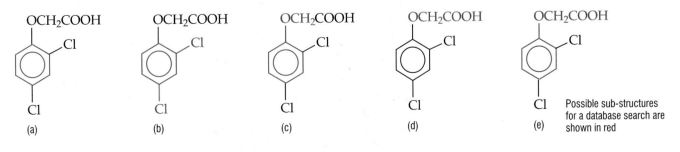

(a) (b) (c) (d) (e) Possible sub-structures for a database search are shown in red

3 If you were trying to find possible candidate compounds which might behave like 2,4-D, what is the disadvantage of searching for substructure (a)?

The structures of the other two commonly used weedkillers are:

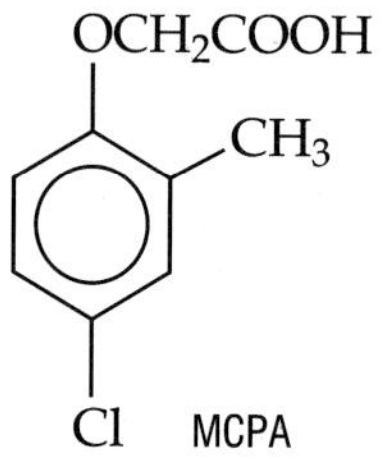

OCH$_2$COOH

CH$_3$

Cl   MCPA

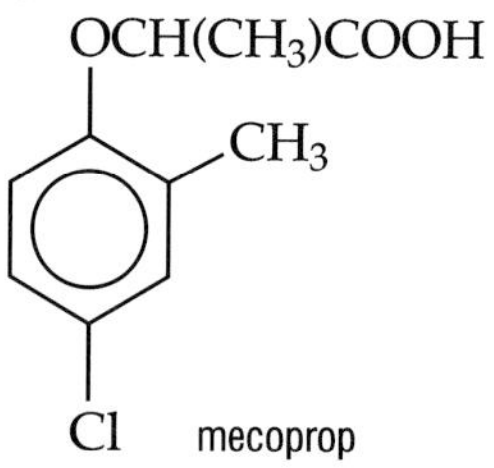

OCH(CH$_3$)COOH

CH$_3$

Cl   mecoprop

4 By looking at the structures of 2,4-D, MCPA and mecoprop, what substructure might you use to search for possible candidate compounds for investigation as weedkillers?

An alternative to the substructure search is the similarity search. This is popular in industry and works on the basis of 'here's an interesting structure – find me those compounds with structures most similar to it'.

This book contains a **database** for your reference. Information is arranged in datasheets. The data recorded are limited for reasons of space but the information included has been chosen to provide essential chemical data relevant to this book. The Database can be searched for patterns and trends. It is for us to develop models which explain observed properties and, importantly, allow us to make predictions.

QUESTIONS

11.1 What kinds of information guide the production of a research portfolio?

11.2 How are candidate compounds identified?

11.3 What is meant by the expression 'substructure search'? What is the simplest substructure containing both the hydroxyl group and the carboxylic acid group in the molecule 2-hydroxypropanoic acid (lactic acid)?

11.4 The structures of some amino acids are given in Fig 11.3.
(a) What substructure is common to these (and indeed all) amino acids?
(b) What functional groups are common to all amino acids?
(c) Write semi-structural formulae for *(i)* alanine, *(ii)* leucine *(iii)* aspartic acid, *(iv)* lysine.

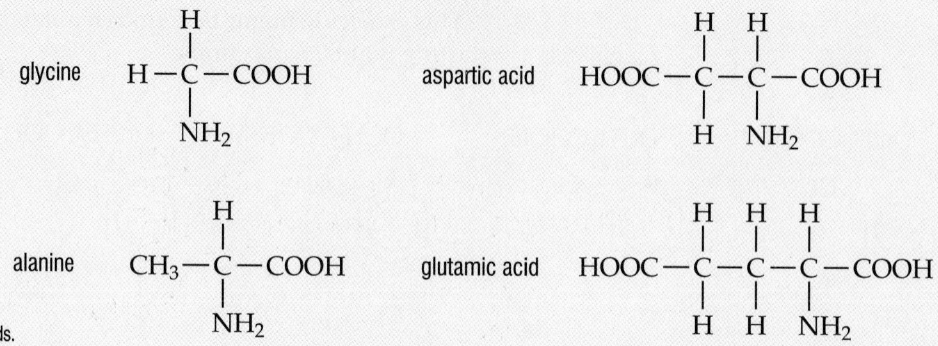

glycine

$$H-\underset{\underset{NH_2}{|}}{\overset{\overset{H}{|}}{C}}-COOH$$

aspartic acid

$$HOOC-\underset{\underset{H}{|}}{\overset{\overset{H}{|}}{C}}-\underset{\underset{NH_2}{|}}{\overset{\overset{H}{|}}{C}}-COOH$$

alanine

$$CH_3-\underset{\underset{NH_2}{|}}{\overset{\overset{H}{|}}{C}}-COOH$$

glutamic acid

$$HOOC-\underset{\underset{H}{|}}{\overset{\overset{H}{|}}{C}}-\underset{\underset{H}{|}}{\overset{\overset{H}{|}}{C}}-\underset{\underset{NH_2}{|}}{\overset{\overset{H}{|}}{C}}-COOH$$

Fig 11.3 The structures of some amino acids.

Fig 11.3 *continued*

valine

CH_3—C—C—COOH

with H, H at top, CH_3, NH_2 at bottom

arginine

$HN=C$—N—C—C—C—C—COOH

with NH_2 and H groups, NH_2 at end

leucine

CH_3—C—C—C—COOH

with H, H, H at top, CH_3, H, NH_2 at bottom

lysine

H_2N—C—C—C—C—C—COOH

with H, H, H, H, H at top, H, H, H, H, NH_2 at bottom

phenylalanine

(benzene ring)—C—C—COOH

with H, H at top, H, NH_2 at bottom

histidine

(imidazole ring) C—C—COOH

with H, H at top, H, NH_2 at bottom

cysteine

HS—C—C—COOH

with H, H at top, H, NH_2 at bottom

H_2N—(benzene ring)—SO_2—NH—(pyridine ring)

H_2N—(benzene ring)—SO_2—NH—(thiazole ring)

H_2N—(benzene ring)—SO_2—NH—(pyrimidine ring)

H_2N—(benzene ring)—SO_2—NH—C(=NH)—NH_2

H_2N—(benzene ring)—SO_2—NH—(isoxazole ring with CH_3)

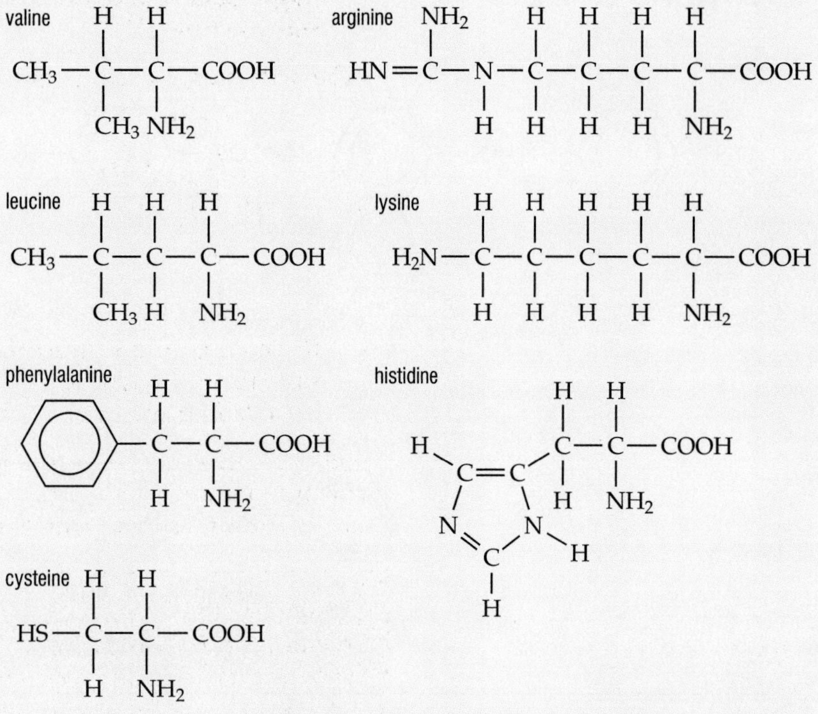

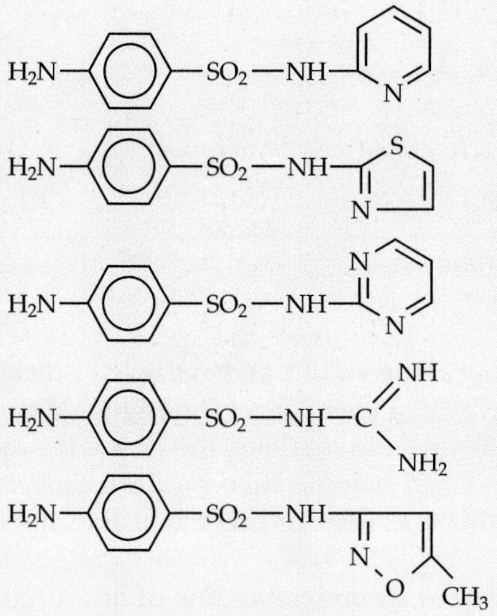

Fig 11.4 The structures of some members of the anti-bacterial drugs, the sulphonamides.

Fig 11.5 The structures of some dyes.

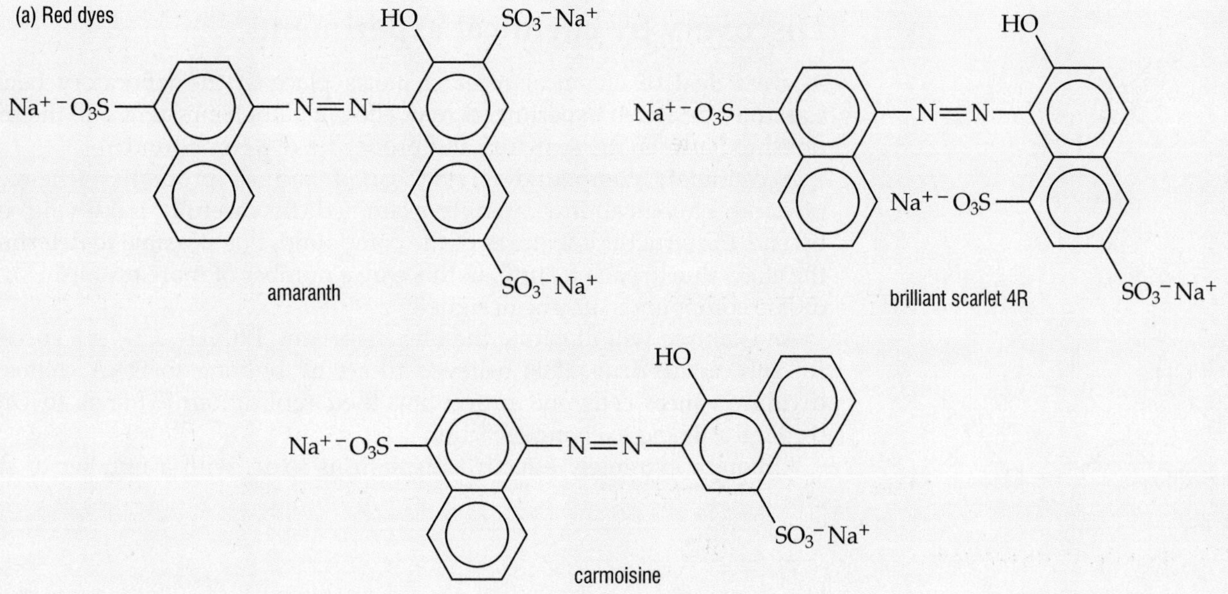

11.5 What qualities would a new antiseptic compound need to possess to achieve success in the marketplace? How would information technology be used to identify suitable candidate compounds for this purpose?

11.6 The structures of some sulphonamide drugs are given in Fig 11.4. What substructure is common to all these molecules?

11.7 The structures of some red dyes and some yellow dyes are given in Figure 11.5.
  (a) What substructure is common to both types of dyes?
  (b) What substructure might you search for in a computer database if you wanted to find candidate molecules which may be (a) red dyes, and (b) yellow dyes?

(a) Red dyes

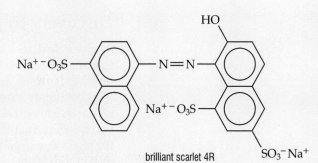

amaranth

brilliant scarlet 4R

carmoisine

(b) Yellow dyes

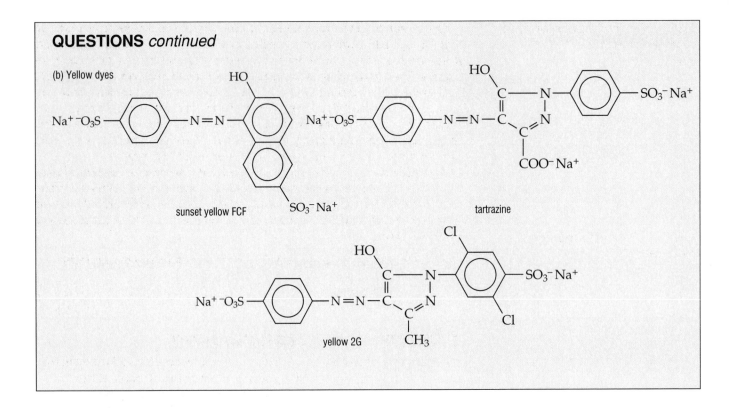

sunset yellow FCF

tartrazine

yellow 2G

11.2 DESIGNING NEW MOLECULES

The properties of known compounds can be stored and retrieved quickly and efficiently using a database. However, a chemist can't design and predict the properties of compounds not yet known. Once the desired properties of a new compound are established it must then be designed and synthesised. How can we be sure that the new product will live up to expectations?

The process of chemical design requires an understanding of how atoms (nuclei and electrons) in a material are put together, and how the structure and bonding determines physical and chemical characteristics. The chemical architect and builder uses this knowledge to design new, and better materials.

Discovery by chemical experiment

A great deal of chemical research takes place at the laboratory bench. Carefully designed experiments may allow a valuable insight into the relationship between the structure and properties of a compound.

A candidate compound is synthesised and its properties (chemical, physical, biological) are carefully examined. By carefully modifying, one by one, the structural features of the compound, it is possible to determine the effect due to each feature. In this way a number of more promising candidate compounds may be designed.

An example will illustrate the idea. Cisplatin, $Pt(NH_3)_2Cl_2$, is a successful anti-cancer drug. It is believed to act by binding to DNA chains in dividing cancer cells and preventing their replication. It bonds to DNA through a ligand exchange reaction.

Although extremely effective, cisplatin is toxic, with a number of side effects.

cisplatin

carboplatin

Fig 11.6   The structures of cisplatin, and carboplatin.

CONNECTION: *Metal complexes and ligand exchange, page 307.*

The research effort was intended to modify the structure of cisplatin to make candidate compounds which can bind to DNA but are less toxic. Researchers knew about the structure of cisplatin and how metal complexes behave. They had a proposed mechanism for its anti-cancer activity. Quite deliberately, chemists set out to make candidate compounds with anti-cancer activity by modifying the structure of a proven therapeutic drug.

Extensive research has taken place since the discovery of cisplatin. This has focused upon identifying and making candidate compounds which are non-toxic yet maintain the beneficial properties of cisplatin.

The work has met with marked success and a second generation of platinum drugs has been developed. Some 2000 compounds which have similar structures to cisplatin were synthesised between 1965 and 1985. One candidate compound which survived the trials and is now in clinical use is carboplatin (Fig 11.6).

CONNECTION: *The way in which cisplatin works, page 275.*

5 **What are the structural similarities of cisplatin and carboplatin?**

SPOTLIGHT

CISPLATIN: LUCK AND HARD WORK

The discovery of cisplatin owes much to good fortune backed up by hard work. In 1961, Barnett Rosenberg set up a Biophysics Department in Michigan State University, USA. As part of his research programme, Rosenberg set out to study the effect of electric fields on bacteria. He exposed bacteria growing in a nutrient solution to an electric field.

The experiments showed that cell division was inhibited when the electric field was applied but that the growth of individual cells was unaffected. In a careful and extensive study, many variables were looked at in an attempt to explain the effect. One by one, variables such as temperature, pH and nutrient concentration were eliminated.

It was postulated that a new compound, formed during the experiment was responsible for the behaviour of the bacteria. Platinum electrodes were used in the apparatus because of their inert character. However, after much meticulous work, Rosenberg's research team found, to their surprise, tiny quantities (less than 10 ppm) of platinum compounds in solution. Clearly the platinum electrodes had been chemically changed. One of the platinum compounds identified in solution was diamminedichloroplatinum(II), or cisplatin, as it is commonly known.

Further research showed that cisplatin alone was responsible for inhibiting bacterial cell division. The research team were imaginative enough to recognise that the application of their discovery would have enormous impact on the treatment of cancer. A fortuitous accident had thus prompted a new research initiative.

The US Food and Drug Administration approved cisplatin for drug use in 1978. This was followed one year later by similar approval in the UK. Work on second generation platinum drugs resulted in approval of carboplatin in the UK in March 1986. The development and production of safe, new drugs takes years of work.

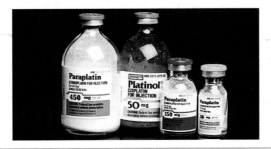

Fig 11.7   Vials of cisplatin and carboplatin.

A second example of discovery by chemical experiments comes from the world of polymers. Nylon is the collective name for polymers with aliphatic hydrocarbon units linked by amide bonds. They are aliphatic polyamides. Three important nylons are manufactured: nylon 6.6, nylon 6 and nylon 6.10, each with slightly different properties. We might expect that the longer the hydrocarbon units in the structure:

- the more flexible the chain will be. The amide bonds are inflexible but the hydrocarbon units are free to flex and twist;
- the weaker the bonding between one polymer molecule and another.

We can modify the properties of nylons by changing the length of the hydrocarbon unit.

6 Explain why we would expect stronger intermolecular bonding between molecules of nylon 6.6 than molecules of nylon 6.10 of the same average relative molecular mass.

CONNECTION: Amide group, page 98.
Polymers, page 236.

KEY FACT

A polyamide contains hydrocarbon sections linked by the amide bonds, — CONH —. The hydrocarbon sections may be aliphatic, aromatic or a mixture.

Table 11.1 Three important nylons

Name	Structural formula	Properties
Nylon 6.6	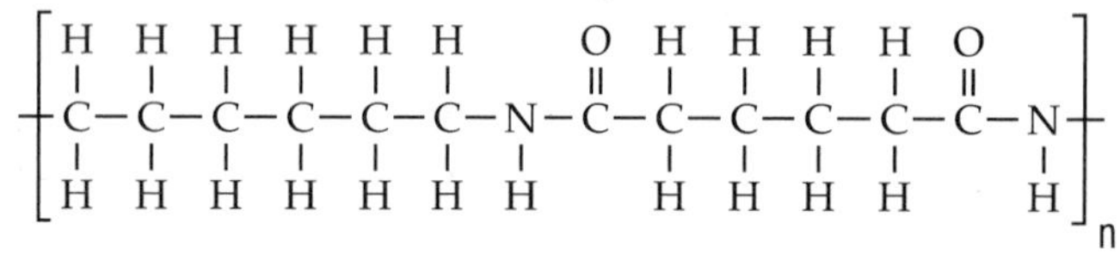	Strong, elastic and low coefficient of friction
Nylon 6.10	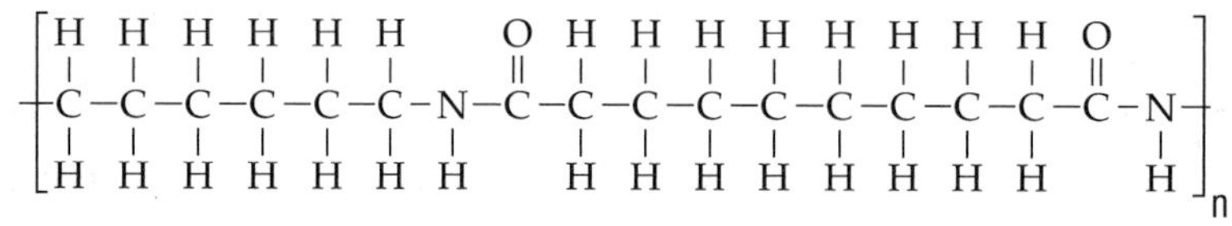	Similar to nylon 6.6 but more flexible
Nylon 6	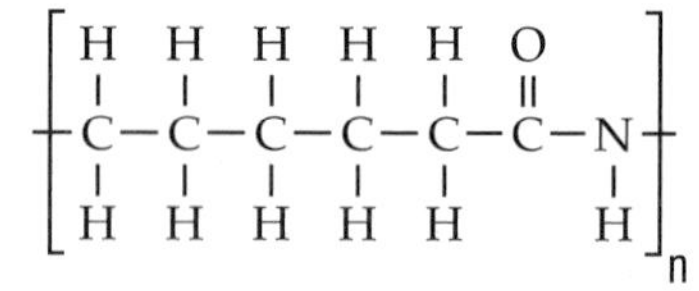	Softer, whiter product than nylon 6.6 or nylon 6.10

A recent development has been to make a polyamide in which the hydrocarbon unit is aromatic rather than aliphatic. This results in a polymer which is inflexible, where the molecules pack together to give rigid, rod-like fibres. These properties are very different to the aliphatic polyamides. One of these aromatic polyamides, Kevlar, is used to make bullet-proof vests, motorcycle helmets and other high performance materials. Importantly, the rigidity of Kevlar was predicted from a knowledge of the structure and bonding of the benzene molecule and related compounds. A new material, therefore, was designed and made for a purpose.

CONNECTION: Structure and bonding in benzene, page 251.

Advanced modelling techniques and design

Mathematical modelling is a powerful tool for the design of new materials. The theoretical properties of a compound are predicted by using mathematical equations to describe particular arrangements of atoms. For example, hardness is an important property of materials in a range of applications. The hardest substance known is diamond, 10 on Moh's scale of relative hardness. The question arises: is the diamond lattice the optimum structure for hardness?

POLYMERS: ONE BY LUCK, ONE BY DESIGN

Discovery by accident has been very common in science, though 'planned' discovery is becoming more common.

Polythene, or more correctly poly(ethene), was discovered during the early 1930s. Two ICI scientists, Reginald Gibson and Eric Fawcett, were investigating chemical reactions which occur at high temperatures and pressures. One particular reaction involved ethene, C_2H_4, and benzaldehyde, C_6H_5CHO. During one experiment the apparatus leaked and the benzaldehyde escaped. When the apparatus was dismantled on 27 March 1933 a waxy, white solid was found. On analysis it was found to have the empirical formula CH_2. Gibson and Fawcett had discovered polythene. As with so many scientific discoveries it was chance rather than design.

In contrast, the discovery of nylon by Wallace Carothers is an example of discovery by design. Carothers used his knowledge of organic chemistry to make polymers. Quite deliberately he set out to join small molecules together to form long polymeric chains. He argued that by putting reactive groups at both ends of a small molecule, such molecules would react with one another to form chains. Initial work on polyesters proved unsuccessful but then Carothers turned to polyamides. Nylon was 'discovered'.

Du Pont patented nylon on 16 February 1938 and the first nylon stockings went on sale in the USA on May 15 1940. In New York alone, four million pairs were sold in a few hours. Carothers never lived to see the impact of his work. He committed suicide two years before the nylon patent was filed by drinking lemon juice containing potassium cyanide. He had suffered from frequent bouts of severe depression.

Fig 11.8 An early advert for nylon stockings

Hardness is a complex concept but mathematical models have been constructed which describe this property. Models have been used to predict the structure of a hypothetical compound, β-C_3N_4, which should be at least as hard as diamond. The predicted structure has a network of CN_4 tetrahedra linked at the corners to form a giant structure (Fig 11.9). If it could be made, and had the anticipated properties, we would be in another classic research situation: a solution looking for a problem – just what might be the possible applications of this new material?

7 Why is diamond so hard?

We can take the modelling of structure one stage further. **Computer aided design (CAD)**, has been used in engineering and construction for some time. Scientists interested in molecular structure are realising the power of this technique in their own field. We can now use a computer to design and construct an image of a molecule on the screen. We specify what atoms we need, together with bond angles and bond lengths. An appropriate software package assembles the molecule and allows us to rotate and move it around the screen so that it can be viewed from different angles. More than this, we can calculate the energies of different spatial arrangements of atoms in a molecule and so predict how one molecule might interact with another.

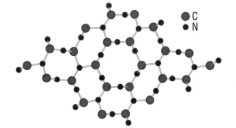

Fig 11.9 The predicted structure of β-C_3N_4. It has been suggested that it might be possible to synthesise it from amorphous carbon nitride, by analogy to the transformation of graphite into diamond. However, even if it could be made, its chemical properties would be central to determining its applicability as an engineering material.

We have to remember, however, that the computer does not do this all by itself. The computer programmer must know about chemical structure – bonding, bond lengths and bond angles. Only then can a program be written with all the necessary information to make CAD possible for chemical compounds. The advantage of the computer, of course, is its speed. Very complicated calculations can be carried out in a fraction of a second. Also, once a complex molecule has been assembled on the screen we can begin altering the structure to produce new molecules. By consulting a database we can predict the likely properties of the new compounds.

What is the significance then of such modelling techniques, for example, to the agrochemical industry? In the search for a new pesticide, the number of candidate compounds to be made, characterised and tested could be reduced drastically. In a typical research project, some ten thousand compounds may have been investigated in the past. Nowadays, the computer can limit this by helping us to eliminate perhaps nine thousand of these where computer modelling has shown that they simply do not have the appropriate shape for an active molecule. The financial saving is enormous.

Fig 11.10   Computer modelling is of great significance in the research for new materials. The computer rapidly calculates spatial arrangements of atoms in a molecule and explores possible interactions. The computer allows us to dismiss some candidate compounds because of their inability to 'fit'. We do not need to go to the expense of synthesising and testing them in the laboratory. This computer-generated model is of interleukin-2, which is used to artificially boost the immune response against certain human cancers

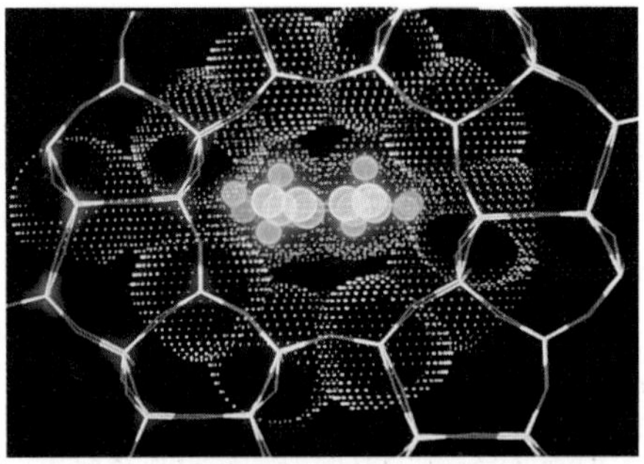

Fig 11.11   The separation of isomers of dimethylbenzene by zeolite ZSM-5 can be modelled by a computer program. The isolation of 1,4-dimethylbenzene from the other isomers is important because of its use in the manufacture of Terylene. Programs can also be used to investigate the structures of solid state catalysts and, an exciting new development, shape-selective catalysts.

Advanced mathematical modelling and computer aided design are important modern techniques. However, scientists have always used models of one kind or another, to help them. One criterion for a successful model is that it enables us to make predictions. Models are not simply about describing objects or events that have already happened. Importantly, they allow us to enter new and exciting fields with some confidence.

Petrol consists of a mixture of hydrocarbons, blended to give a fuel which burns efficiently and causes minimum 'knocking' in the engine. A measure of 'knocking' is the octane number. 2,2,4-Trimethylpentane, $CH_3CH(CH_3)CH_2C(CH_3)_2CH_3$, produces very little 'knock' and has an octane number of 100. The octane numbers of various hydrocarbons are given in Table 11.2.

(a) Draw structural formulae for the hydrocarbons which have the following octane numbers: 24.8, 83.0, 90.9, 102.5, 107.4.

(b) Write a brief account of the relationship between octane number and molecular structure and state the characteristics of a hydrocarbon fuel which would undergo combustion with minimum 'knocking'.

Table 11.2 The octane number of some hydrocarbons

Name of hydrocarbon	Octane number
Octane	−19.0
Heptane	0.0
Hexane	24.8
Cis-oct-2-ene	56.2
Pentane	61.7
2,4-Dimethylhexane	65.2
Trans-oct-4-ene	73.3
Cyclohexane	83.0
Pent-1-ene	90.9
Trans-hex-2-ene	92.7
Butane	93.8
Propane	97.1
But-1-ene	97.4
2,2,4-Trimethylpentane	100.0
Cyclopentane	101.3
Propene	102.5
Benzene	105.8
1,4-Diethylbenzene	106.0
Ethylbenzene	107.4
1,2-Dimethylbenzene	107.4
1,4-Dimethylbenzene	116.4
1,3-Dimethylbenzene	117.5
Methylbenzene	120.1

11.8 The anti-cancer drug cisplatin, discovered by Barnett Rosenberg and his research team (see Spotlight, page 207) is effective at preventing replication of cancer cells.

(a) What led the research team to suspect that a new compound had been created during the experiment?

(b) Why was the research team surprised to find platinum compounds in the nutrient solution?

(c) Why have 'second generation' platinum-based drugs such as carboplatin largely replaced cisplatin?

11.9 Explain:

(a) how the chain length of an aliphatic polyamide may influence its physical properties;

(b) why the aromatic polyamide Kevlar is much less flexible than aliphatic polyamides;

(c) how a chemical database might be used to find materials with similar properties to Kevlar.

11.10 Computer modelling of new chemical compounds is used by the pharmaceutical industry to help identify candidate compounds which may be useful new drugs. Explain:
- **(a)** why this technique is useful in drug design;
- **(b)** why it is often a very *economical* way of shortlisting likely candidate compounds;
- **(c)** what data is needed in order to use computer modelling to identify candidate compounds successfully?

11.11 Mathematical modelling and computer modelling are different but often complementary techniques. Explain briefly what useful information may be obtained through the use of each technique.

11.12 Fig 11.11 shows a computer simulation of 1,4-dimethylbenzene passing through a pore in zeolite ZSM-5.
- **(a)** Draw structural formulae for 1,4-dimethylbenzene and its *aromatic* structural isomers.
- **(b)** Suggest why this zeolite may be used to separate 1,4-dimethylbenzene from the other isomers.

ACTIVITY

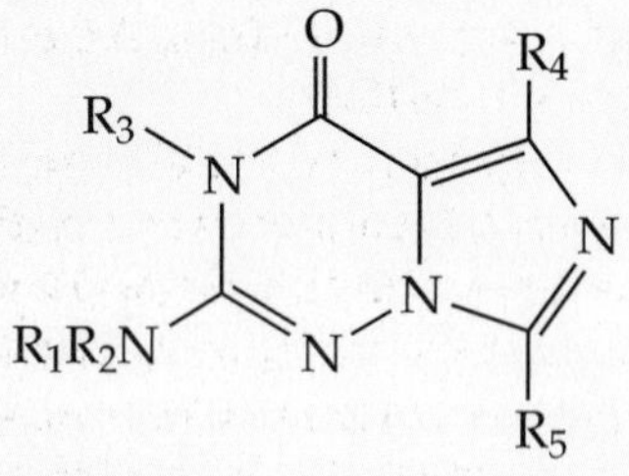

On 19 November 1971, a patent was filed at the Patent Office, London concerning some imidazo-triazine compounds made by Allen and Hanbury's Ltd. The company had good evidence to suggest that these compounds might be effective in the treatment of asthma and bronchitis. The compounds had the general formula shown in Fig 11.12.

According to the patent, R_1, R_2, R_4 and R_5 may be the same or different and each represent a hydrogen atom or a straight or branched chain alkyl (C_{1-6}) or alkenyl (C_{1-6}) radical which may optionally be substituted by one or more aryl groups and R_3 represent hydrogen or a (C_{1-6}) straight or branched chain alkyl group.

In the above description the word 'radical' simply means group. An alkenyl radical is, therefore, an aliphatic group which contains a $C=C$ double bond.

Using this information, write the structural formulae of five different candidate compounds which might be prepared and put forward for biological testing.

Can you suggest a general formula containing a *different* ring system which might be the basis of a new research portfolio. Remember it is likely that it is the fundamental shape of key regions of the molecule which is responsible for its biological activity.

Fig 11.12

11.3 PATTERNS IN THE PERIODIC TABLE

Periodicity: repeating patterns in properties

Early investigations of elements showed that their physical and chemical properties were **Periodic**, that is they repeated in a regular way and are said to display **Periodicity**. The elements were arranged so that ones with similar properties were placed in the same group. This was difficult at first as some elements were unknown. The great achievement of the Russian chemist Dmitri Mendeléev was to leave gaps where it seemed an undiscovered element was needed to complete the pattern. Remarkably, he predicted not only the existence of certain undiscovered elements, but also their properties with considerable accuracy.

The modern Periodic Table consists of a number of Groups and Periods. The main Groups are numbered using Roman numerals I to VIII. Each Period is given a number corresponding to the outer energy level for elements within it. For example, calcium is in Period 4 because its outer electrons are in the $n = 4$ energy level (its electronic structure is $1s^2 2s^2 2p^6 3s^2 3p^6 4s^2$). The Groups and Periods are further organised into blocks.

The blocks of the Periodic Table are named according to the last occupied atomic orbital for the elements in that block.

- Lithium ($1s^2 2s^1$) and potassium ($1s^2 2s^2 2p^6 3s^2 3p^6 4s^1$) are in the **s-block**.
- Nitrogen ($1s^2 2s^2 2p^3$) and silicon ($1s^2 2s^2 2p^6 3s^2 3p^2$) are in the **p-block**.
- Iron ($1s^2 2s^2 2p^6 3s^2 3p^6 3d^6 4s^2$) and silver ($1s^2 2s^2 2p^6 3s^2 3p^6 3d^{10} 4s^2 4p^6 4d^{10} 5s^1$) are in the **d-block** since 3d and 4d are the last atomic orbitals to be added to in building up their respective electronic structures.

KEY FACT

The Periodic Table is not simply a 'list of elements'. It is an invaluable guide to their electronic configurations which allows many of their properties to be *predicted*.

The d-block elements contain the **transition elements**. These are elements which can form at least one stable ion with a partially filled set of d-orbitals.

Trends in physical properties

Several physical properties of elements show clear trends within the Periodic Table. Each depends upon the overall number of electrons in an atom of an element, their arrangement and the charge on the nucleus.

Size of atoms and ions

We generally picture atoms as spheres. A line is often drawn to show the boundary of an atom but in the real world atomic orbitals are not so clearly defined. Tables of atomic radii taken from different sources can show considerable variation and for meaningful comparisons, we must check that data are obtained in the same way.

With simple diatomic molecules of an element, such as Cl_2, the atomic radius is taken as half the distance between the two nuclei (half the bond length). This distance is also known as the **covalent radius**.

In a solid metal half the distance between neighbouring metal atoms is known as the **metallic radius**. For some elements both a covalent radius and a metallic radius may be measured. Which measurement is appropriate depends on the state of the element. The values are different since the atoms are in a very different environment, with different forces acting upon them. For example, half the distance between two sodium atoms in the molecule Na_2 (in the vapour state) is 0.123 nm but half the distance between two sodium atoms in a metallic lattice of sodium is 0.186 nm.

8 How might covalent and metallic radii be measured?

Li
0.152 nm
Na
0.186 nm
K
0.231 nm
Rb
0.244 nm
Cs
0.262 nm

Fig 11.13 Atomic radius data are provided in Datasheet 1. In all Groups of the Periodic Table atomic radius increases with atomic number.

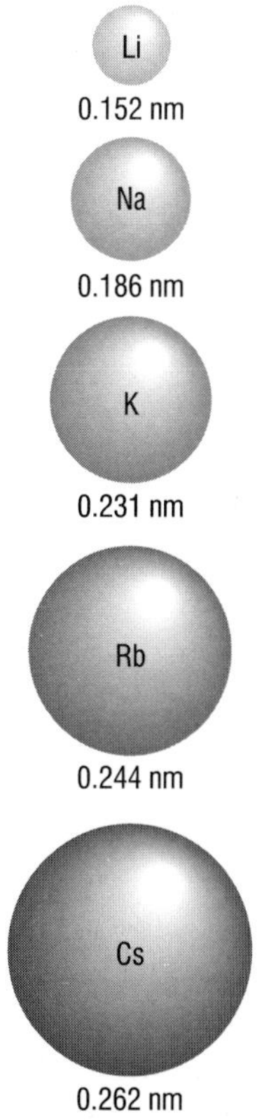

Providing we are consistent in our measurements and compare like with like, we can identify an obvious trend in the atomic radii of elements within any Group of the Periodic Table. The atomic radii of atoms increase down a Group. Within a Group, the outer electrons of a given element occupy a higher energy level than those of the previous element. Electrons in the new energy level are partly shielded from the attraction of the nucleus.

Moving from left to right across a Period in the Periodic Table, atomic radii get smaller. This is because the size of the charge on the nucleus increases by +1 with each element. Although more electrons are added they join the same outer energy level. These electrons are more closely attracted by the increased nuclear charge, shrinking each successive element (Fig 11.13).

9 From the trends shown for the two Periods in Fig 11.14, can you suggest which is the largest element in the Periodic Table?

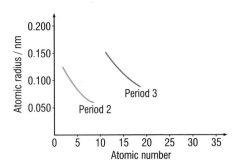

Fig 11.14 From lithium to fluorine the elements show considerable shrinkage due to increasing nuclear charge. The same pattern is observed with the elements of the second Period, from sodium to chlorine (Datasheet 1).

In covalent compounds, we can assign an atomic radius to each of the participant elements providing we can measure the bond length and have a value for the atomic radius of one of the elements involved. For example, the bond length of hydrogen chloride is 0.132 nm (measured by electron diffraction) and the covalent radius of hydrogen is 0.037 nm, then the covalent radius of chlorine is (0.132–0.037) nm = 0.095 nm. The value obtained by halving the bond length of a chlorine molecule is 0.099 nm – close agreement.

In the same way, **ionic radii** must be clearly defined. To do this we make the assumption that the distance which separates two ions in an ionic solid is the sum of the ionic radii of the two ions concerned. If we now assign a value to one ion then all others can be worked out. Usually the radius of the oxide ion, O^{2-} is taken as a standard at 0.140 nm. The radii of similar ions formed by members of a Group increase on descending the Group. The pattern is almost identical to that seen for atomic radii and for the same reasons.

Ionisation energy

The first ionisation energies of elements within a Group of the Periodic Table decrease with increasing atomic number. This may seem surprising at first, since the charge on the nucleus is increasing. There are two reasons for the electron being removed:

- it comes from a higher energy level and, therefore, is further from the nucleus on average;
- it is shielded from the influence of the nucleus by electrons in lower energy levels.

KEY FACT

The first ionisation energy of an element is the energy required to remove one mole of electrons from one mole of gaseous atoms of the element.

Trends in first ionisation energies provided evidence for the arrangement of electrons in atoms.

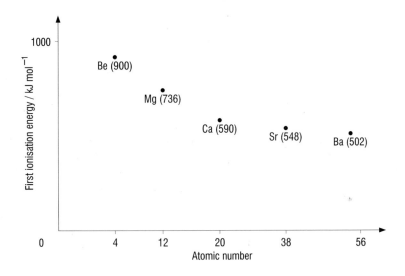

Fig 11.15 First molar ionisation energy decreases with increasing atomic number of elements within a group.

Electron affinity

Descending a Group of the Periodic Table, the electron affinity of the elements tends to become less exothermic. In general, as a Group is descended electrons are accepted less willingly by successive elements. The explanation is that the effect of increasing nuclear charge is overcome by the larger atomic size and by the shielding effect of inner electrons (electrons occupying lower energy levels).

CONNECTION: Electron affinities, Datasheet 1.

10 **Explain why Group VIII elements all show an endothermic first electron affinity.**

DESIGN FOR A PURPOSE

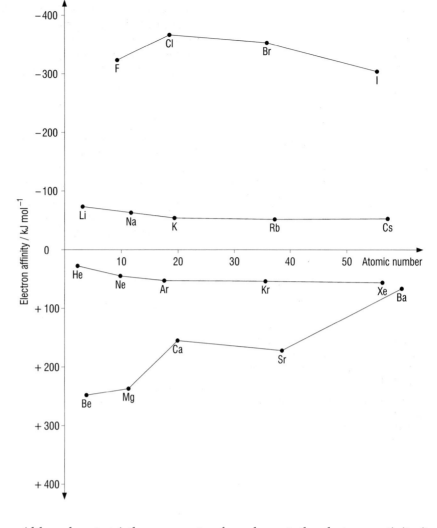

Fig 11.16 Generally, first molar electron affinity decreases with increasing atomic number in a group (Datasheet 1).

Although not strictly a property of an element, the electronegativity its atoms display when combined with other atoms does depend on the element's position in the Periodic Table. The variation of electronegativity with position is similar to that of electron affinity and for the same reasons. The most electronegative elements are found towards the top right-hand corner of the Periodic Table and include fluorine, oxygen, nitrogen and chlorine. The least electronegative elements are found in the bottom-left corner and include francium, caesium and rubidium.

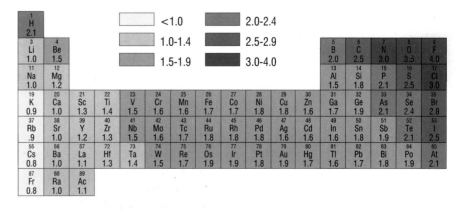

Fig 11.17 Electronegativity increases from left to right across a Period and decreases down a Group.

CONNECTION: Electronegativity values, Datasheet 1.

11 Why are electronegativity values for the Group VIII elements not normally given?

DESIGN FOR A PURPOSE   **215**

In general, the first molar electron affinity of the elements becomes more exothermic from left to right across a Period, reaching a peak with the elements of Group VII. This is due, once again to the effect of increasing nuclear charge attracting electrons more strongly. If an atom of an element towards the right of the Periodic Table gains an electron then the incoming electron enters an orbital relatively close to the nucleus and results in a very stable ion. The gain of an electron here is strongly exothermic, therefore, and energetically favoured. The Group VII elements readily form negative ions.

H −72.8							He +21
Li −59.6	Be +241	B −26.7	C −122	N 0	O −141	F −328	Ne +29
Na −52.9	Mg +230	Al −42.5	Si −134	P −72.0	S −200	Cl −349	Ar +34
K −48.4	Ca +156	Ga −28.9	Ge −119	As −78.2	Se −195	Br −325	Kr +39
Rb −46.9	Sr +167	In −28.9	Sn −107	Sb −103	Te −190	I −295	Xe +40
Cs −45.5	Ba +52.0	Tl −19.3	Pb −35.1	Bi −91.3	Po −183	At −270	Rn +41

Fig 11.18 In general, first molar electron affinity is greater towards the top-right corner of the Periodic Table. Group VIII elements are the exception and do not form negative ions.

12 **The Group VII elements readily form singly-charged anions but not doubly-charged anions. Why is this?**

If atoms of the Group VIII elements were to gain an electron, that electron would enter a new energy level. In this position, relatively far from the nucleus, the electron is not strongly attracted to the nucleus. In these cases and others the enthalpy change for anion formation is endothermic. Put simply, if an element has no affinity for electrons, adding electrons to its atoms will result in the formation of unstable anions and requires an energy input. Such processes tend not to take place.

Trends in chemical properties

The chemical properties of an element are determined by the arrangement of electrons in its atoms. Of particular importance are the electrons in the outer energy level. The members of each Group in the Periodic Table have very similar chemical properties due to the closely-related electronic structures of group members.

For example, Group I elements contain a single electron in the s-orbital of their outer energy level. These elements are metals, all of which react readily, in some cases violently, with water and with more reactive non-metals such as oxygen or the Group VII elements. All elements in Group I form **unipositive ions** (singly-charged positive ions). Their salts are soluble in water and are generally quite stable towards thermal decomposition. Other Groups also show strong similarities amongst their members, making it sensible to study the reactions of Groups as a whole, noting unusual properties of individual members.

KEY FACT

Chemical properties of an element are determined largely by the arrangement of electrons in its outermost energy level.

CONNECTION: Detailed information about the chemical reactions of elements and their compounds is available in Datasheets 6–15.

13 **In which Datasheet can you find information about the chemistry of sulphur and its compounds?**

Period 3 elements and their compounds

Physical properties

In the next chapter we investigate structure and bonding in detail. We can use the elements of Period 3, and of their chlorides and oxides, to show how the Database might be used to obtain evidence for structural types.

The melting points of the Period 3 elements, their chlorides and oxides are found in Datasheet 3. Graphs of these (Fig 11.19) show a trend from high melting points to low melting points, with sharp cut-off points. We interpret these graphs as follows:

Period 3 elements

| Na | Mg | Al | Si | | P | S | Cl | Ar |

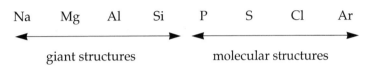

giant structures   molecular structures

Chlorides of elements of Period 3

NaCl   MgCl$_2$ AlCl$_3$*   SiCl$_4$ PCl$_3$ S$_2$Cl$_2$ Cl$_2$
                                                  PCl$_5$**   SCl$_2$

giant structures   molecular structures

* aluminium chloride is molecular (Al$_2$Cl$_6$) in the vapour phase;
** phosphorus pentachloride is molecular in the vapour phase; in the solid state it exists as [PCl$_4$]$^+$[PCl$_6$]$^-$.

Oxides of elements of Period 3

Na$_2$O   MgO   Al$_2$O$_3$ SiO$_2$ P$_2$O$_3$ SO$_2$ Cl$_2$O
                                           P$_2$O$_5$ SO$_3$ Cl$_2$O$_7$

giant structures   molecular structures

CONNECTION: Physical properties of Period 3 elements, Datasheet 3.

14 **The lining of high temperature furnaces are made of an unreactive, refractory (very high melting point) oxide. Which Period 3 oxide is best suited for this use?**

Fig 11.19 Trends in melting points for Period 3 elements, their chlorides and oxides.

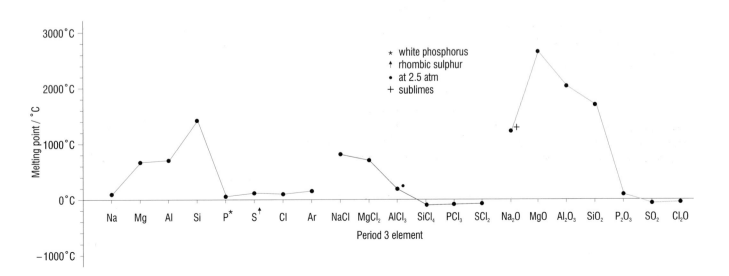

* white phosphorus
* rhombic sulphur
* at 2.5 atm
+ sublimes

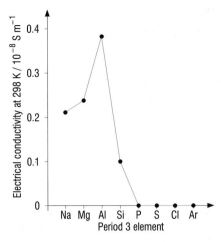

Fig 11.20 Trends in electrical conductivity of Period 3 elements.

CONNECTION: Properties of oxides and chlorides Datasheet 6.

> ### KEY FACT
>
> Oxides of metals are ionic and basic. Oxides of non-metals are covalent and acidic. Oxides which show both basic and acidic properties are said to be **amphoteric**.

CONNECTION: Ionic, covalent and intermediate bond types, page 243.

Evidence for the type of bonding in the Period 3 elements comes from their electrical conductivity. Sodium, magnesium and aluminium have very much higher electrical conductivities than phosphorus, sulphur or chlorine (Fig 11.20). Our interpretation is that the first three elements of Period 3 have metallic bonding, with delocalised electrons relatively free to move through the lattice, enabling an electrical current to pass through. Atoms within molecules of phosphorus, sulphur and chlorine are held together by covalent bonds. Silicon is rather unusual in that the bonding is usually described as covalent, yet its electrical conductivity is about the same as that of iron. Silicon is a semi-conductor and its use has revolutionised the electronics industry.

Chemical properties

Physical properties often lead us to conclusions about the type of structures that oxides and chlorides of the Period 3 elements have. Their chemical properties provide evidence for the type of bonding present in these compounds. The oxides show a trend from basic to acidic character.

Na_2O MgO Al_2O_3 SiO_2 P_2O_3 SO_2 Cl_2O
                                    P_2O_5 SO_3 Cl_2O_7

————————— increasing covalent character ————————➤
————————— increasing acidic character—————————➤

Sodium oxide and magnesium oxide are basic, dissolving in acids to form salts. However, oxides of elements to the right of the Period, for example sulphur dioxide, are acidic, dissolving in water to give acidic solutions. This change is due to the bonding type. Basic oxides tend to be ionic, while acidic oxides are usually covalent.

Across a Period, the elements display decreasing metallic character. This is reflected in the bonding type for the oxides. Metallic oxides are ionic. They have giant structures and are basic, for example,

$$Na_2O(s) + H_2O(l) \longrightarrow 2NaOH(aq)$$

Non-metallic oxides are covalent. They have either giant or molecular structures and are acidic, for example,

$$SO_3(g) + H_2O(l) \longrightarrow H_2SO_4(aq)$$

However, elements towards the centre of the Period form oxides which contain bonds of an intermediate type. The silicon-oxygen bond in silicon(IV) oxide, for example, is about 50% ionic and 50% covalent in character. Aluminium oxide shows both acidic and basic behaviour and is amphoteric.

15   **Write an ionic equation for the reaction of magnesium oxide with an acid.**

The chlorides of Period 3 elements fall into two broad types according to their behaviour in water. The chlorides of sodium and magnesium dissolve to give a neutral solution. However, chlorides of the other Period 3 elements react with water, with varying degrees of vigour, to give acidic solutions. Our explanation is that these chlorides are covalent and hydrolyse in water. In each case hydrogen chloride is formed and dissolves in the solution, making it acidic.

$MgCl_2(s) + aq \longrightarrow Mg^{2+}(aq) + 2Cl^-(aq)$ [dissolves]
$PCl_3(l) + 3H_2O(l) \longrightarrow H_3PO_3(aq) + 3HCl(aq)$ [reacts]

Period 3 chlorides: reactions with water

NaCl	MgCl$_2$	AlCl$_3$	SiCl$_4$	PCl$_3$	S$_2$Cl$_2$	Cl$_2$
				PCl$_5$	SCl$_2$	

————— increasing covalent character —————→
————— increasing tendency to hydrolyse —————→

16 **Using Datasheet 11, describe how phosphorus trichloride might be prepared in the laboratory.**

Transition elements

Transition elements show a 'lull' in periodic trends within a Period. This is illustrated by graphs of first ionisation energy against atomic number (Fig 9.11). The transition elements are all metals and have these characteristics:

- high tensile strength (they can withstand high loads without fracturing);
- ductility (easily drawn into wires);
- malleability (easily beaten into sheets).

They are usually harder than the s-block elements and have higher densities due to their relatively small atomic size. They display all the characteristic properties of metals. Their chemical properties may be summarised as follows:

- transition elements show variable oxidation numbers in the compounds they form;
- they form coloured compounds;
- they form coordination compounds;
- transition elements and their compounds are often good catalysts.

CONNECTION: Electronic structures of transition elements, pages 176–177.

CONNECTION: Oxidation numbers, page 382; coordination compounds, page 266, catalysis, pages 443–446.

Fig 11.21 Copper is a familiar transition element. It shows all the characteristics of a metal and its compounds are brightly coloured. These photographs show a range of copper artefacts and the commercial production of copper(II) sulphate crystals.

Note: To answer some of these questions you will need to find information from the Database and from earlier sections in this book.

11.13 (a) Find values for the ionic radii of the Group I elements, lithium to caesium, and plot a graph of radius against atomic number.
   (b) Explain the trend observed and predict the ionic radius of a francium ion, Fr^+.

11.14 (a) Find values for the first three ionisation energies of Group II elements.
   (b) On the same graph, as you used in question 11.13 show how these values vary with atomic number and explain the trends you observe.

11.15 (a) The boiling points of some iodoalkanes are given in Table 11.3. Use a graphical method to predict the boiling points of 1-iodobutane, $CH_3(CH_2)_3I$. (b) What can you deduce about the strength of intermolecular bonding as the homologous series is descended? Explain the trend.

Table 11.3   Boiling points of iodoalkanes

Iodoalkane	Semi-structural formula	Boiling point / °C
Iodomethane	CH_3I	42.5
Iodoethane	CH_3CH_2I	72.4
1-Iodopropane	$CH_3CH_2CH_2$	102
1-Iodobutane	$CH_3CH_2CH_2CH_2I$	–
1-Iodopentane	$CH_3CH_2CH_2CH_2CH_2I$	155
1-Iodohexane	$CH_3CH_2CH_2CH_2CH_2CH_2I$	181

11.4 REACTIONS OF ORGANIC COMPOUNDS

The Database and organic compounds

The chemistry of a functional group is largely independent of the remainder of the molecule to which it belongs. This is the key to being able to make sense of the vast numbers of organic compounds known and explains the organisation of the datasheets. For example, the chemical reactions of CH_3COOH, CH_3CH_2COOH, $CH_3CH_2CH_2COOH$ and other carboxylic acids are essentially the same, though rates and conditions of reaction may vary. We can predict the type of reactions that a given carboxylic acid will undergo by comparison with other carboxylic acids (Datasheet 21). It is only by experiment, however, that we can confirm its reactions and establish the necessary reagents and conditions.

Reagents and reaction conditions given in the datasheets are generally applicable. However, they may not work for every compound that contains the functional group. A further complication may arise if there is more than one functional group present – a reaction which affects one may bring about unwanted changes in the other group.

To find information about organic compounds in the Database we must to be able to interpret chemical formulae written in a variety of forms. It is essential to be familiar with all types of chemical shorthand used to describe molecules and functional groups, in particular molecular, semi-structural and full structural formulae. It is also important to understand something of the more subtle aspects of molecular structure:

KEY **F**ACT

The chemistry of an organic compound is dictated by the functional groups it contains.

CONNECTION: Molecular formulae, semi-structural formulae, structural formulae and functional groups, pages 71 and 97.

- Firstly, how important is the presence of a —CH_2— group attached to the functional group? The answer is – it all depends. The properties of alcohols and halogenoalkanes are dependent to some extent on the number of hydrogen atoms bonded to the carbon to which they are attached. The properties of other groups, such as carboxylic acids, nitriles, aldehydes and ketones, do not show this same dependence. The chemistry of amines depends on whether they are primary, secondary or tertiary amines, though they do display some common properties.
- Secondly, how important is the carbon skeleton to which a functional group is attached? For example, does it matter whether a functional group is attached to an aliphatic chain or to an aromatic ring? Again, the answer is – it all depends. Often the chemistry is similar but there may be striking differences. For example, alcohols and phenols both contain a hydroxyl group but have different properties.

CONNECTION: Structures of primary, secondary and tertiary halogenoalkanes, Datasheet 18; alcohols, Datasheet 19 and amines Datasheets 20 and 23.

CONNECTION: Alcohols, Datasheet 19; phenols, Datasheet 20.

There are two ways in which the Database might be accessed: through functional groups or through semi-structural formulae.

Accessing through functional groups

The Database contains information about the chemical reactions of different functional groups. To use the Datasheets, the functional group in a compound is identified and the Database consulted to find the relevant Datasheet.

Fig 11.22 A section of Datasheet 22.

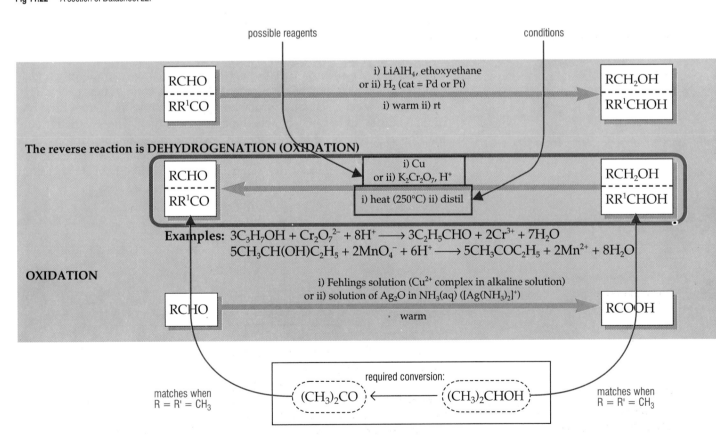

For example, suppose we want to find out about the chemistry of ethanoic acid, CH_3COOH. The compound is a carboxylic acid, RCOOH. Therefore we need to look at Datasheet 21. If we wanted to find out about propanone, CH_3COCH_3, we need to look at Datasheet 22 since propanone is a ketone, R_2CO. Perhaps we wish to know how to convert propanone into propan-2-ol:

$$(CH_3)_2CO \longrightarrow (CH_3)_2CHOH$$

Looking at the appropriate section of Datasheet 22 (Fig 11.22), we can see that this can be achieved either by warming propanone with lithium tetrahydridoaluminate(III) ($LiAlH_4$) in dry ethoxyethane, or by reacting propanone with hydrogen at room temperature over a suitable catalyst such as platinum or palladium.

17 **Which Datasheets contain information about the chemical reactions of (a) propanal, C_2H_5CHO, (b) ethyl ethanoate, $CH_3COOC_2H_5$, (c) ethylamine, $C_2H_5NH_2$?**

Accessing through semi-structural formulae

Datasheets 24 and 25 are synthetic route maps which we will use later to select methods for making organic compounds. In these, general semi-structural formulae are used. For example $RCH_2CH_2CH_2OH$ is used as a general formula for a primary alcohol. To locate a compound, we write its semi-structural formula and match it against the general formulae shown in Datasheet 24 or 25. The map then gives a reference to another Datasheet in which more detailed information can be found.

Suppose we wanted to find out about reaction routes to and from hexan-1-ol, $C_6H_{13}OH$, a primary alcohol. Its semi-structural formula is:

$CH_3CH_2CH_2CH_2CH_2CH_2OH$
($— CH_2OH$ is the primary alcohol functional group)

Looking at the section of Datasheet 24 (Fig 11.23) we can identify a general formula which matches the semi-structural formula of hexan-1-ol. This has been done on a section of Datasheet 24 by circling the parts of the formula common to hexan-1-ol and the general formula. The synthetic route map gives a reference to Datasheet 19 where more detailed information (reagents and reaction conditions) concerning the reactions of hexan-1-ol can be found. Of course, the same Datasheet would have been found through a functional group search of the Database index.

Fig 11.23 A section of Datasheet 24.

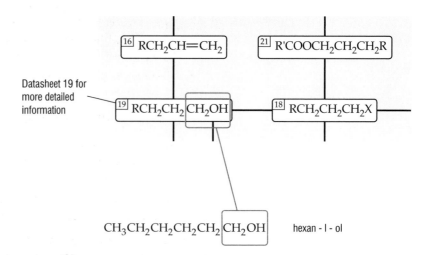

DESIGN FOR A PURPOSE

18 In Fig 11.23, indicate which general formula describes ethyl pentanoate, $C_4H_9COOC_2H_5$. In which Datasheet would you find more information about this compound?

A further example is hexan-2-ol, $C_6H_{13}OH$, a secondary alcohol. Its semi-structural formula is:

$CH_3CH_2CH_2CH_2CH(OH)CH_3$
($-CH(OH)-$ is the secondary alcohol functional group)

Look at the section of Datasheet 24 (Fig 11.24). The general formula which fits hexan-2-ol is easily identified (note that it is different to that for hexan-1-ol).

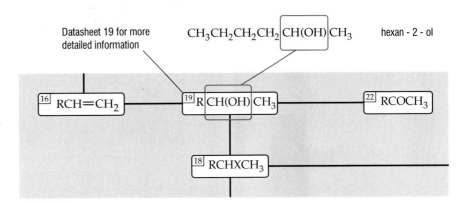

Fig 11.24 A section of Datasheet 24.

19 What is the difference in reaction products when a primary alcohol and a secondary alcohol are oxidised?

Sometimes it requires a bit more thought to identify the most appropriate general formula. For example, ethanoic acid has the semi-structural formula CH_3COOH. Inspection of Datasheet 24 does not reveal a structure which appears to fit this closely. The best matches are RCH_2CH_2COOH and $RCH_2CH_2CH_2COOH$. However, we noted earlier that the carboxylic acid, $-COOH$, displayed similar properties whether it was attached to $-CH_2-$ or not. Therefore, either of the matches in Datasheet 24 could be used. In both cases, Datasheet 21 is referred to and this contains the reactions of carboxylic acids. Similarly, the chemistry of benzoic acid, C_6H_5COOH, will be found in the Datasheet on carboxylic acids since the $-COOH$ group behaves in much the same way whether it is attached to an aliphatic or an aromatic group.

Single step organic conversions

Changing an organic compound into a new compound in a single reaction is called a single step conversion. Datasheets 16–23 can be used to do this by following these stages:

Stage 1. Write semi-structural formulae for the starting compound and the target compound, and identify the functional groups.

Stage 2. Find the appropriate Datasheet and use this to identify the reagents and reaction conditions necessary for the conversion.

SALBUTAMOL

Salbutamol is an important drug used to treat asthmatics. It is available on prescription as Ventolin. Its chemistry can be predicted from an inspection of its molecular formula.

From an inspection of this structure we can identify the following functional groups: a primary alcohol, a secondary alcohol, a secondary amine and a phenol. The chemistry of salbutamol might be expected, therefore, to show the characteristic of these functional groups.

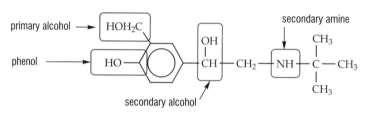

(b) A molecule of salbutamol with the functional groups highlighted.

Fig 11.25 (a) A Ventolin inhaler

The importance of this example is that it shows how the chemistry of quite complex molecules can often be predicted since the chemistry of a functional group is largely independent of the remainder of the molecule in which it is found.

TUTORIAL

Problem. How might propanoic acid be converted into ethyl propanoate?

Solution.
Stage 1. The semi-structural formula of propanoic acid is CH_3CH_2COOH. This is a carboxylic acid, since it contains the —COOH functional group.

The semi-structural formula of ethyl propanoate is $CH_3CH_2COOCH_2CH_3$. This is an ester.

Stage 2. Therefore, information will be found in Datasheet 21 (Carboxylic acids and their derivatives).

From Datasheet 21, we see propanoic acid must be reacted with ethanol, CH_3CH_2OH.

Conditions: reflux propanoic acid and ethanol in the presence of concentrated sulphuric acid, H_2SO_4.

$$CH_3CH_2COOH + CH_3CH_2OH \longrightarrow CH_3CH_2COOCH_2CH_3 + H_2O$$

Note To answer some of these questions you will need to find information from the Database and from earlier sections in this book.

11.16 For each of the following compounds (i) state the class of organic compound to which it belongs, (ii) state which Datasheets should be interrogated to find out more about its chemical properties:

(a) C_2H_5Cl

(b) $C_6H_5COCH_3$

(c) C_3H_7CN

(d) $C_6H_5CH_2CH_2OH$

(e) $C_5H_{11}CHO$

(f) $C_2H_5COOC_3H_7$

(g) $C_6H_5CH_2CN$

(h) C_4H_9COOH

(i) $C_2H_5CH(OH)C_3H_7$

(j) $(CH_3)_2NH$

(k) $C_5H_{11}CONHC_2H_5$

(l) $CH_3CH(COOH)C_5H_{11}$

11.17 For each of the following compounds (i) state the class of organic compound to which it belongs, (ii) write its semi-structural formula, (iii) state which Datasheet should be interrogated to find out more about its chemical properties:

(a) butanal

(b) butanone

(c) but-1-ene

(d) propan-1-ol

(e) ethyl propanoate

(f) ethylamine

(g) ethanenitrile

(h) 1-bromopropane

11.18 Use the Database to find reactions which might be used to distinguish between each of the following pairs of compounds:

(a) C_2H_5Cl and CH_3COCl

(b) $C_6H_5COCH_3$ and C_6H_5COOH

(c) CH_3COCH_3 and C_2H_5CHO

(d) $C_6H_5CH_2CH_2OH$ and $C_6H_5C(CH_3)_2OH$

11.19 Use the Database to describe three reactions of:

(a) propan-1-ol

(b) ethanoic acid

(c) propene

(d) chloroethane

(e) methylamine

(f) ethanal

(g) propanone

(h) benzene

11.20 Use the Database to write structural formulae for the organic products obtained when each of the following takes place:

(a) $CH_3(CH_2)_2COOH$ is treated with PCl_5 at room temperature.

(b) $CH_3(CH_2)_4OH$ is refluxed with acidified potassium dichromate(VI) solution.

(c) $CH_2{=}CH{-}C(CH_3)_2CH{=}CH_2$ is reacted with hydrogen in the presence of a finely divided nickel catalyst.

(d) $CH_3(CH_2)_5CH(CN)CH_3$ is refluxed with dilute sulphuric acid.

(e) C_6H_5OH is treated with ethanoyl chloride, CH_3COCl.

(f) C_6H_5COOH is reacted with ethanol, C_2H_5OH, in the presence of concentrated sulphuric acid.

(g) $C_6H_5CH_2OH$ is treated with PCl_5 at room temperature.

(h) $C_6H_5CH_2CH_2CHO$ is heated with lithium tetrahydridoaluminate(III), $LiAlH_4$, in dry ethoxyethane.

(i) $CH_2(OH)CH_2CH_2COOH$ is treated with PCl_5 at room temperature.

(j) $CH_2(OH)CH_2CH_2CHO$ is refluxed with acidified potassium dichromate(VI).

It is one thing to design a molecule, it is another to make it. The conversion of simple chemicals, with known structures, to new substances, often quite complex, is called **synthesis**. The series of steps required is called a **synthetic route**.

Complex organic compounds can be made by using the vast bank of well-tested reactions and chemical reagents known. The area of 'designer chemistry' is less well advanced in inorganic chemistry though considerable strides have been made in recent years.

Synthetic route maps

It can be very satisfying and rewarding to successfully plot a route to a **target compound** using compounds available in the laboratory and to synthesise it. All the more so if that substance has not been made before. Identifying possible synthetic routes demands a thorough understanding of the chemistry of functional groups. Where more than one route is available, selecting the route to use involves considerations such as costs and availability of starting materials, number of steps, complexity of apparatus needed and ease of manipulation, energy demands, yield, and of course, safety considerations.

Synthesis of a target molecule may involve the transformation of one or more functional groups during a series of steps. The importance of knowing the chemistry of functional groups is only too apparent.

The Database contains two synthetic route maps which summarise some of the major synthetic pathways for organic compounds. They may be used to plot synthetic routes from given starting materials to target compounds. Of course, both maps are limited in the amount of data they contain. A computer database would contain much larger amounts of information. However, for our purposes Datasheets 24 and 25 will be very helpful in their own right and also give an insight into the power of easily accessed databases.

The synthetic route maps and other Datasheets, are used by following the sequence:

Stage 1. Write semi-structural formulae for the starting compound and the target compound. Note the functional groups present.

Stage 2. Match the semi-structural formulae against the general semi-structural formulae in the maps (Datasheets 24 and 25).

Stage 3. Select an appropriate route between starting and target compounds from the synthetic route map.

Stage 4. Use the number in the top left of each box around the compounds to identify the appropriate Datasheet to:
(a) check that the reaction 'goes in the desired direction';
(b) find the reagents and reaction conditions necessary for the conversion to occur.

TUTORIAL
USING THE SYNTHETIC
ROUTE MAPS

The synthesis of butanone, $C_2H_5COCH_3$, from but-1-ene, $C_2H_5CH=CH_2$.

Stage 1. Semi-structural formulae:

Starting compound: but-1-ene, $CH_3CH_2CH=CH_2$ (an alkene, $>C=C<$)
Target compound: butanone, $CH_3CH_2COCH_3$ (a ketone, $>C=O$)

Stage 2. Match against Datasheet 24:

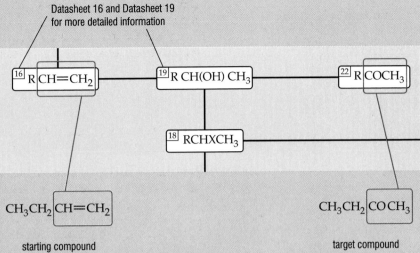

Datasheet 16 and Datasheet 19 for more detailed information

starting compound

target compound

Stage 3. The route between the starting compound and the target compound is a two step conversion.

Stage 4. More detailed information comes from Datasheets 16 and 19.
   Step 1 (Datasheet 16):
   Reagents and conditions: absorb into concentrated sulphuric acid and then dilute with water; both stages at room temperature.
   Step 2 (Datasheet 19):
   Reagents and conditions: potassium dichromate(VI) in dilute sulphuric acid; reflux.

Therefore, synthesis:

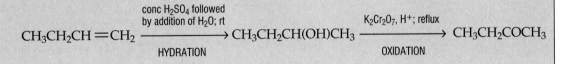

$$CH_3CH_2CH=CH_2 \xrightarrow[\text{HYDRATION}]{\substack{\text{conc } H_2SO_4 \text{ followed} \\ \text{by addition of } H_2O; \text{ rt}}} CH_3CH_2CH(OH)CH_3 \xrightarrow[\text{OXIDATION}]{K_2Cr_2O_7, H^+; \text{ reflux}} CH_3CH_2COCH_3$$

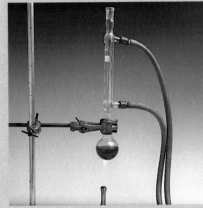

Fig 11.26 Refluxing butan-2-ol with potassium dichromate(VI) in dilute sulphuric acid oxidises the organic compound to butanone. The bright orange colour shows that the oxidation has only just started. Upon completion, the reaction mixture will be green due to $Cr^{3+}(aq)$.

KEY FACT

A **step** refers to a conversion within a synthetic route. The use of the route maps has been broken down into a series of **stages**. Do not get steps and stages confused.

Changing the length of a carbon skeleton

It is often necessary to reduce or increase the length of a carbon skeleton in an organic synthesis. Datasheet 24 is divided into three coloured 'zones'. Crossing from one zone to another means that the length of the carbon skeleton has been increased or decreased by one carbon atom. Two key reactions are involved:

- To increase the length of a carbon chain, an halogenoalkane is made and reacted with potassium cyanide dissolved in ethanol. The halogen atom is replaced by the cyanide to give a nitrile, summarised by the equation:

   $$RX + CN^- \longrightarrow RCN + X^-$$

- To decrease the length of a carbon chain, an acid amide is made and reacted with a solution of bromine in aqueous potassium hydroxide. This is known as a **Hofmann degradation**. The reaction is complex but the net effect is to convert the acid amide to an amine, summarised by the equation:

   $$RCONH_2 + Br_2 + 4KOH \longrightarrow RNH_2 + 2KBr + K_2CO_3 + 2H_2O$$

Increasing length: synthesising propylamine from ethanol

This is an example of a synthesis requiring a carbon skeleton to be
increased in length. The target compound, propylamine, $C_3H_7NH_2$, contains one more carbon atom than the starting compound, ethanol,
C_2H_5OH. This requires crossing from one coloured zone to another in
Datasheet 24.

Stage 1. Semi-structural formulae:
Starting compound: ethanol, CH_3CH_2OH (a primary alcohol, RCH_2OH)
Target compound: propylamine, $CH_3CH_2CH_2NH_2$ (a primary amine,
$RCH_2CH_2NH_2$)

Stage 2. Matching against Datasheet 24:

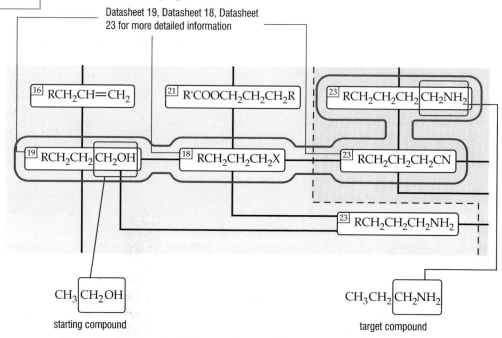

Datasheet 19, Datasheet 18, Datasheet
23 for more detailed information

starting compound

target compound

Stage 3. The route between the starting compound and the target compound is a three step conversion.

Stage 4. More detailed information comes from Datasheets 18, 19 and 23.
Step 1 (Datasheet 19):
Reagents and conditions: phosphorus tribromide, PBr_3; room tem-
perature.
Step 2 (Datasheet 18):
Reagents and conditions: potassium cyanide, KCN, in aqueous
ethanol; reflux.
Step 3 (Datasheet 23):
Reagents and conditions: lithium tetrahydridoaluminate(III) in dry
ethoxyethane; reflux.
Therefore, synthesis:

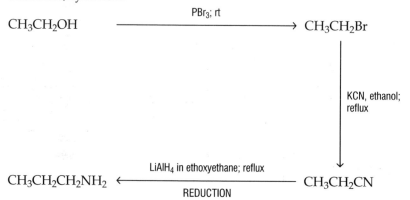

Decreasing length: synthesising pentan-1-ol from hexan-1-ol

This is an example of decreasing the length of a carbon skeleton. It also shows how the synthetic route maps must be used in conjunction with the other Datasheets. From our earlier definition, this is not strictly a synthesis, since we are not building up a molecule. We might more properly call this a multi-step conversion of hexan-1-ol ($C_6H_{13}OH$) to pentan-1-ol ($C_5H_{11}OH$). It requires crossing from one coloured zone to another on Datasheet 24.

Stage 1. Semi-structural formulae:

Starting compound: hexan-1-ol, $CH_3CH_2CH_2CH_2CH_2CH_2OH$ (a primary alcohol, $RCH_2CH_2CH_2OH$)

Target compound: pentan-1-ol, $CH_3CH_2CH_2CH_2CH_2OH$ (a primary alcohol, $RCH_2CH_2CH_2OH$)

Stage 2. Matching against Datasheet 24:

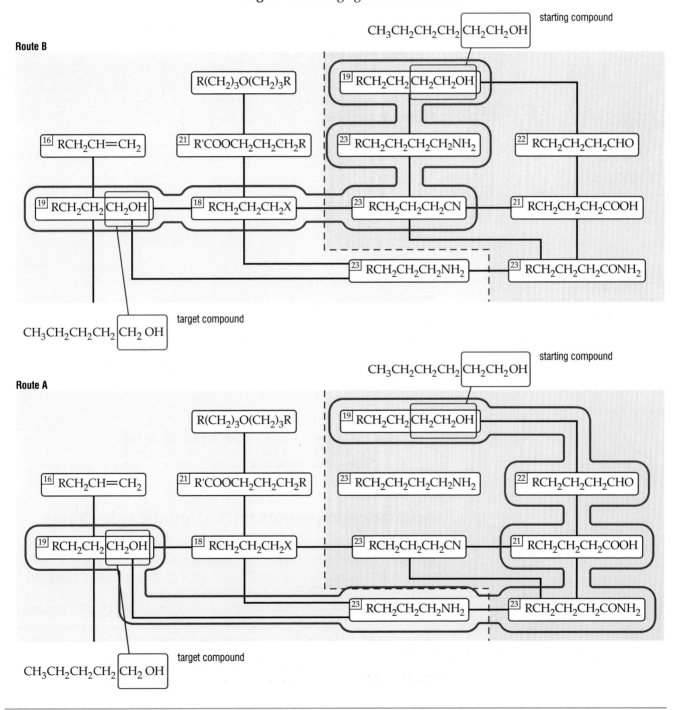

Stage 3. The route map suggests that two possible routes are available, labelled Route A and Route B above. However, inspection of Datasheet 19 shows that a primary alcohol cannot be converted into an amine, even though a primary amine can be converted into a primary alcohol (Datasheet 23). This reinforces the fact that the Datasheets must be used in conjunction with one another.

Route A is chosen, therefore. It is a five step conversion.

Stage 4. More detailed information comes from Datasheets 19, 21 and 23.

Steps 1 and 2 (Datasheet 19):

Reagents and conditions: potassium dichromate(VI), $K_2Cr_2O_7$, in dilute sulphuric acid; reflux.

Step 3 (Datasheet 21):

Reagents and conditions: convert to ammonium salt, $C_5H_{11}COONH_4$, by treatment with aqueous ammonia; heat ammonium salt strongly.

Step 4 (page 227):

Reagents and conditions: bromine, Br_2, in dilute aqueous potassium hydroxide, KOH(aq); warm gently.

Step 5 (Datasheet 23):

Reagents and conditions: sodium nitrite, $NaNO_2$, in cold, dilute hydrochloric acid, HCl(aq); add $NaNO_2$(aq) cold (0–5°C) and then allow to warm to room temperature.

Therefore, synthesis:

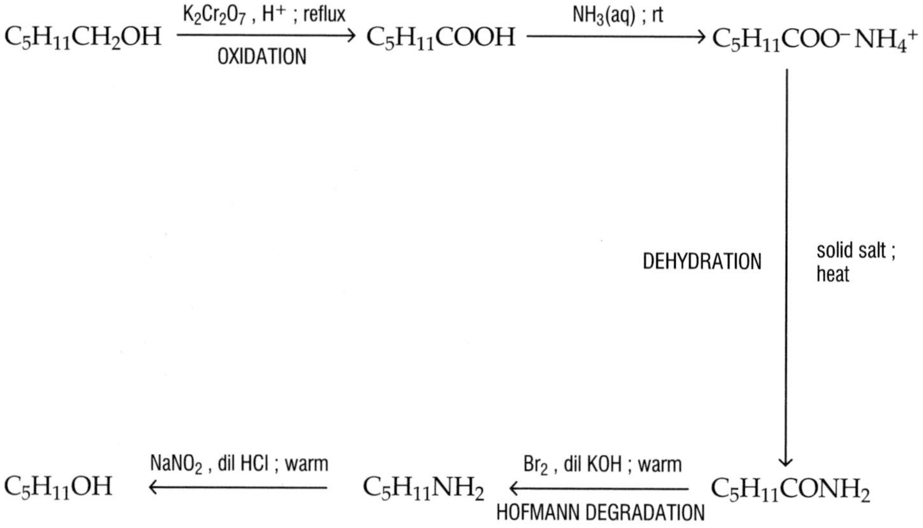

$$C_5H_{11}CH_2OH \xrightarrow[\text{OXIDATION}]{K_2Cr_2O_7 , H^+ ; \text{reflux}} C_5H_{11}COOH \xrightarrow{NH_3(aq) ; \text{rt}} C_5H_{11}COO^- NH_4^+$$

DEHYDRATION — solid salt ; heat

$$C_5H_{11}OH \xleftarrow{NaNO_2 , \text{dil HCl ; warm}} C_5H_{11}NH_2 \xleftarrow[\text{HOFMANN DEGRADATION}]{Br_2 , \text{dil KOH ; warm}} C_5H_{11}CONH_2$$

Note $C_5H_{11}COOH \longrightarrow C_5H_{11}CONH_2$ is shown as a single step on the synthetic route map : inspection of Datasheet 21 shows that it goes via the ammonium salt , $C_5H_{11}COO^- NH_4^+$

Aromatic compounds with aliphatic side chains

Many molecules contain a benzene ring with one or more aliphatic groups attached. To deal with these, it is often necessary to use both of the synthetic route maps, Datasheets 24 and 25. For example, the synthesis of phenymethyl ethanoate from benzene and ethanoic acid

The synthesis of phenylmethylethanoate, $CH_3COOCH_2C_6H_5$, involves a bit of back-tracking. The target compound is an ester. To make it from ethanoic acid, CH_3COOH, the appropriate alcohol must be made, and this is phenylmethanol, $C_6H_5CH_2OH$. Therefore we must first decide how to make $C_6H_5CH_2OH$ from benzene, C_6H_6.

Stage 1. Semi-structural formulae:
Starting compounds: ethanoic acid, CH_3COOH (a carboxylic acid, RCOOH) and benzene

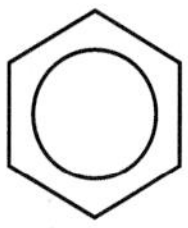

Target compound: phenylmethylethanoate

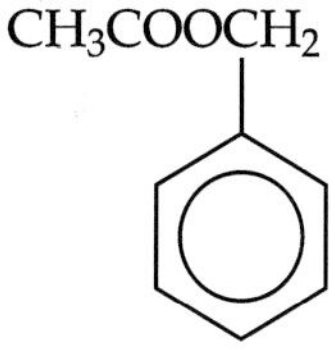

CH_3COOCH_2

Intermediate: phenylmethanol

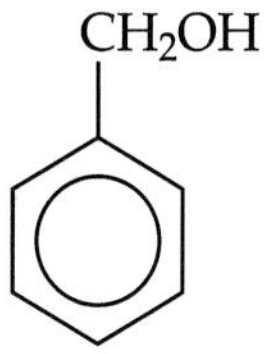

CH_2OH

Stage 2. Matching against Datasheet 25:

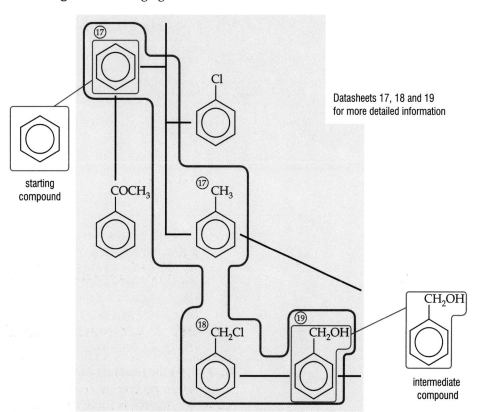

Datasheets 17, 18 and 19 for more detailed information

Stage 3. The route between the starting compound and the intermediate is a three step conversion.

Stage 4. More detailed information comes from Datasheet 17.
    Step 1 (Datasheet 17):
    Reagents and conditions: dilute sodium hydroxide, NaOH(aq); reflux.

Stage 5. To convert the intermediate, phenylmethanol, $C_6H_5CH_2OH$, into the target compound, phenylmethylethanoate, $CH_3COOCH_2C_6H_5$, we need to match against Datasheet 24:

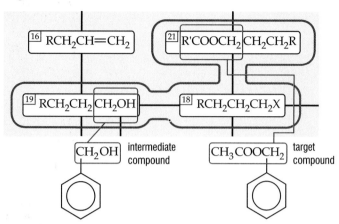

Stage 6. The route between the intermediate and the target compound is a one step conversion.

Stage 7. More detailed information comes from Datasheet 19.

Step 4 (Datasheet 19):

Reagents and conditions: ethanoic acid, CH_3COOH; reflux in the presence of concentrated sulphuric acid, H_2SO_4.

Therefore, synthesis:

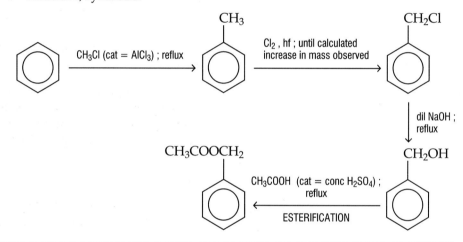

QUESTIONS

11.21 State the reagents and conditions necessary to bring about the following one-step conversions:
   (a) $C_2H_5Cl \longrightarrow C_2H_5CN$
   (b) $C_2H_5OH \longrightarrow CH_3COOC_2H_5$
   (c) $C_2H_5OH \longrightarrow CH_3COOH$
   (d) $CH_3CH(OH)CH_3 \longrightarrow CH_3COCH_3$
   (e) $C_6H_5NO_2 \longrightarrow C_6H_5NH_2$
   (f) $C_6H_5CN \longrightarrow C_6H_5COOH$

11.22 State the reagents and conditions necessary to bring about the following one-step conversions:
   (a) butanoic acid to butan-1-ol
   (b) propanoic acid to propanoyl chloride
   (c) propanenitrile to propanoic acid
   (d) propene to 1,2-dibromopropane
   (e) methylbenzene to benzoic acid
   (f) phenylmethanol to benzaldehyde

11.23 Select synthetic routes for the following conversions, in each case stating the reagents and conditions necessary for each step:
(a) $CH_3CH = CH_2 \longrightarrow C_3H_7CHO$
(b) $CH_3(CH_2)_2OH \longrightarrow CH_3(CH_2)_2COOH$
(c) $CH_3(CH_2)_2COOH \longrightarrow C_2H_5COOH$
(d) $CH_3COCH_3 \longrightarrow CH_3CH(COOH)CH_3$
(e) $C_6H_5CH_3 \longrightarrow C_6H_5COOC_2H_5$
(f) $C_6H_6 \longrightarrow C_6H_5CH(CH_3)COOH$

11.24 Select synthetic routes for the following conversions, in each case stating the reagents and conditions necessary for each step:
(a) butan-1-ol to butanamide
(b) butan-2-ol to 2-aminobutane
(c) 1-iodohexane to heptan-1-ol
(d) hexanoic acid to pentanoic acid
(e) nitrobenzene to benzoic acid
(f) benzene to phenol

11.6 SYNTHESIS OF COMPOUNDS WITH MORE THAN ONE FUNCTIONAL GROUP

The presence of two or more functional groups in a compound can present difficulties when designing synthetic routes. A chemical reagent which brings about a desired conversion of one group may have an unwanted effect on others. There are two important ways of preventing this:

• selection of appropriate reagents and reaction conditions;
• protection of a functional group during reaction.

Aspirin and paracetamol: a case study in reagents and conditions

Aspirin and paracetamol are the two biggest selling pain-relievers available without a prescription. Their syntheses provide an example of how reaction conditions can be chosen to prevent an undesired chemical change from taking place. Important steps in the syntheses of aspirin and paracetamol are:

2-hydroxybenzoic acid + $(CH_3CO)_2O$ → (conc H_2SO_4; warm, ETHANOYLATION) → aspirin + CH_3COOH

4-aminophenol + $(CH_3CO)_2O$ → (warm, ETHANOYLATION) → paracetamol + CH_3COOH

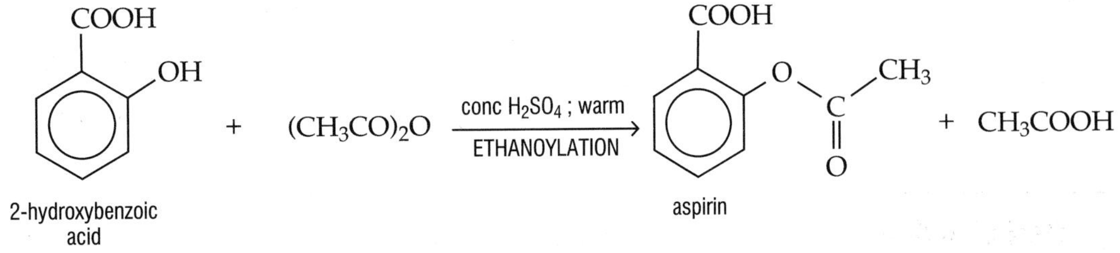

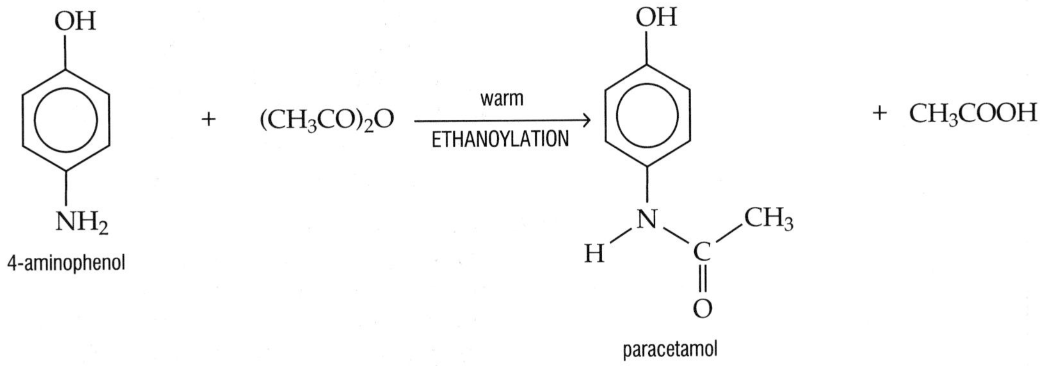

20 Write the structural formula of the product that would have been obtained if ethanoic anhydride had reacted with the phenol group in 4-aminophenol as well as with the amine group.

Why is it that the phenol group reacts with ethanoic anhydride in the synthesis of aspirin but not in the synthesis of paracetamol? A look at the reaction conditions for each reaction provides the answer. The ethanoylation of the phenol group in 2-hydroxybenzoic acid only occurs in the presence of a catalyst – the concentrated sulphuric acid. Ethanoylation of the amine group in 4-aminophenol does not require a catalyst. The amino group, —NH_2, is more reactive than the phenolic —OH group. Reaction conditions can be selected that ensure, in the preparation of paracetamol, that only the amino group is ethanoylated.

21 **Name an alternative reagent which could be used to ethanoylate 2-hydroxybenzoic acid.**

2-Bromophenylamine and 4-bromophenylamine: a case study in protection

Phenylamine reacts readily with bromine water to give 2,4,6-tribromophenylamine.

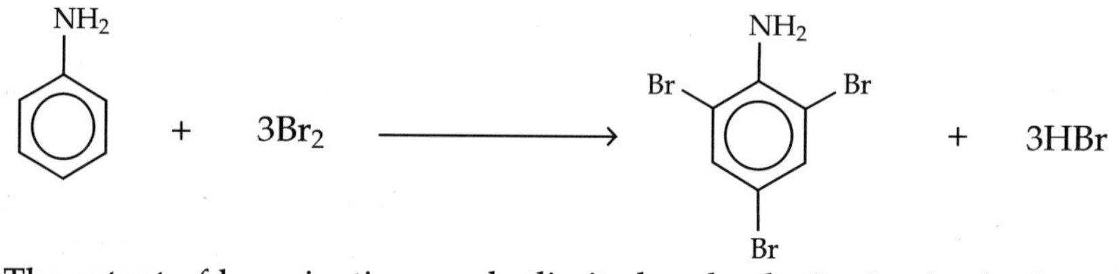

The extent of bromination can be limited and substitution in the 4-position promoted by ethanoylation of the amino group.

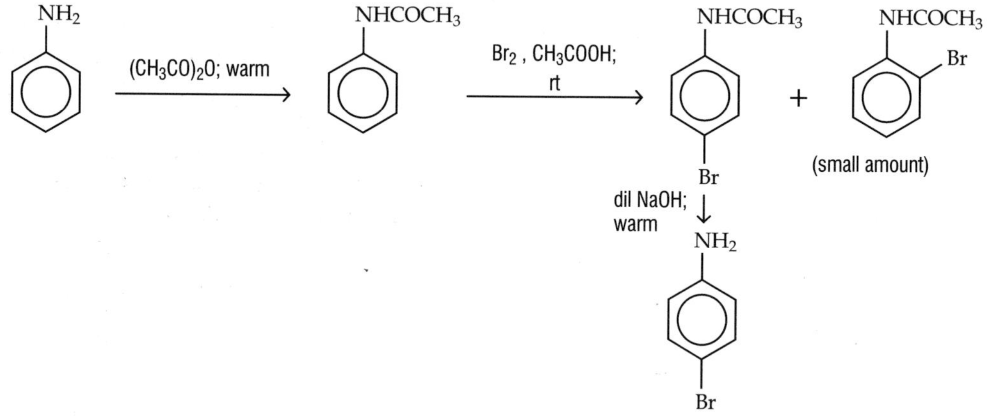

To make 2-bromophenylamine we must first protect the 4-position from attack as well as reducing the activating effect of the —NH_2 group on the ease of substitution in the benzene ring. The —$NHCOCH_3$ is described as a protecting group. This is an example of preventing an undesired chemical change by using a protecting group.

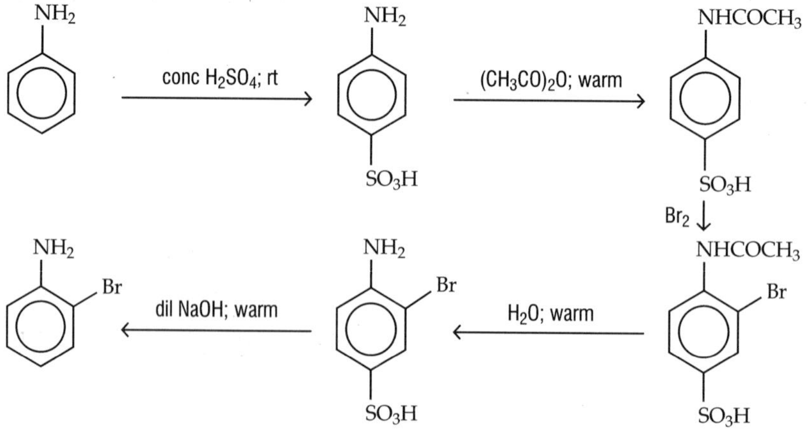

Sulphanilamide: another case study in protection

Sulphonamides are an important group of anti-bacterial drugs. The first one to be discovered was sulphanilamide. It can be synthesised from benzene by the route shown in Fig 11.27. This is an example of temporarily changing a functional group in a compound in order to prevent an undesired chemical change. In a later step, the functional group is reformed.

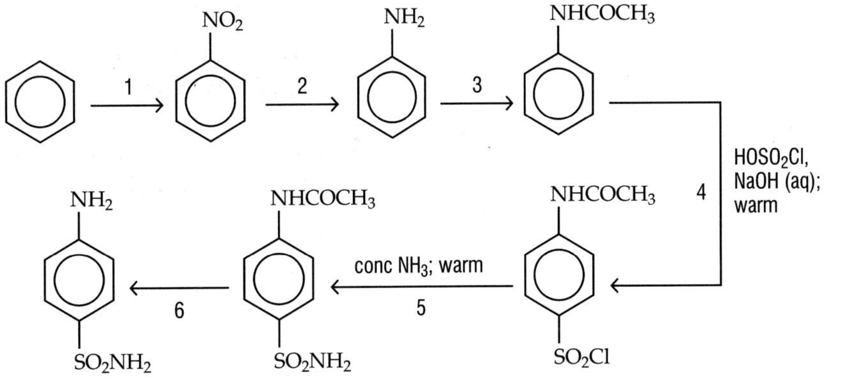

Fig 11.27   The synthesis of sulphanilamide from benzene.

22 **Name the reagents and reaction conditions necessary to bring about the conversions in steps 1,2,3 and 6.**

The reaction sequence shows the formation of an Ar—NH$_2$ group in step 2, followed by its ethanoylation (step 3) and subsequent hydrolysis back to Ar — NH$_2$ in step 6. Why go to all this trouble? Why not make the Ar—NH$_2$ group and then leave it alone? The answer lies in the reaction at step 4. The —SO$_2$Cl group is introduced into the benzene ring by a substitution reaction using chlorosulphonic acid, HOSO$_2$Cl. Changing Ar—NH$_2$ to Ar—NHCOCH$_3$ decreases the tendency for substitution in the benzene ring. Whereas the reaction of phenylamine with chlorosulphonic acid might result in the formation of

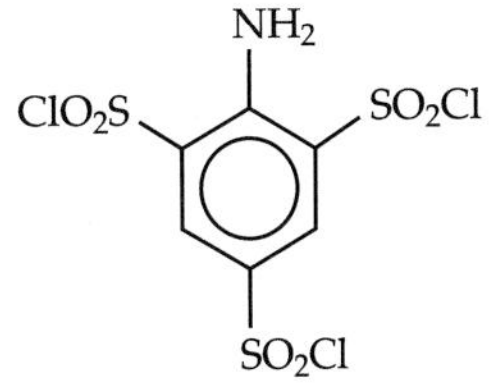

ethanoylation of Ar—NH$_2$ limits substitution to the 4-position (with a very small amount of the 2-substituted product).

QUESTION

11.25 Suggest how the following two conversions might be achieved:

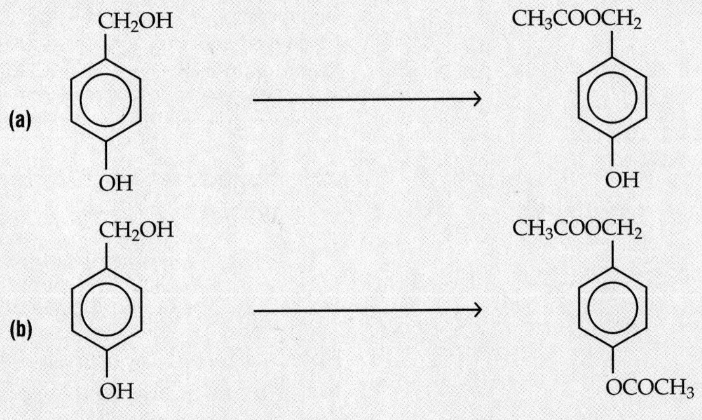

Polymers are giant molecules. They often consist of tens of thousands of atoms joined to one another by covalent bonds. Synthetic fibres and plastics are polymers and are commonplace. It is hard nowadays to imagine our world without these materials.

Giant molecules are made from small molecules. The problem is to find ways of joining small molecules together to form long chains and, if it is desired, to link one chain to another. **Polymerisation** is the process of joining molecules of simple compounds (**monomers**) to form a new compound of much higher relative molecular mass (the **polymer**). Polymer molecules formed from a given reaction mixture vary in size and so we speak about the polymer's **average relative molecular mass**. Some of the polymer molecules will have lower than average values, others will be higher.

There are two types of polymerisation:

- **addition polymerisation**;
- **condensation polymerisation**.

CONNECTION: Addition and condensation reactions, page 284.

Addition polymerisation

Addition polymerisation occurs between molecules which have double or triple covalent bonds. The reactions require high temperatures and pressures unless a catalyst is used.

A very important group of addition polymers are formed from compounds of general formula $CH_2=CHY$ (Table 11.4). The nature of the **substituent group**, Y, determines the properties of the polymer. The properties of the polymer produced can also be influenced by the type of catalyst used. There are two classes of catalysts:

- **ionic catalysts**, the most important ones being Ziegler-Natta catalysts;
- **free-radical catalysts**, such as organic peroxides.

It is a characteristic of this type of polymerisation that no other small molecules are produced at the same time.

> ### KEY FACT
>
> The synthesis of polymers requires knowledge only of those conversions of simple molecules already covered.

$$n \quad \begin{array}{c} H \\ | \\ C \\ | \\ H \end{array} = \begin{array}{c} H \\ | \\ C \\ | \\ Y \end{array} \longrightarrow \begin{bmatrix} H & H \\ | & | \\ C & - & C \\ | & | \\ H & Y \end{bmatrix}_n$$

repeat unit

Table 11.4 Some addition polymers formed from monomers of general formula $CH_2=CHY$

Monomer	Formula	Name of polymer	Common name for polymer
Ethene	$CH_2=CH_2$	poly(ethene)	polythene
Propene	$CH_2=CHCH_3$	poly(propene)	polypropylene
Chloroethene	$CH_2=CHCl$	poly(chloroethene)	PVC
Ethenyl ethanoate	$CH_2=COCOCH_3$	poly(ethenyl ethanoate)	PVA
Propenonitrile	$CH_2=CHCN$	poly(propenonitrile)	polyacrylonitrile
Phenylethene	$CH_2=CHC_6H_5$	poly(phenylethene)	polystyrene

23 **Suggest a structure for the repeat unit when $H-C=C-Y$ polymerises.**

If a single monomer is used, polymerisation gives a **homopolymer**:

—A—A—A—A—A—A—A—A—A—A—A—A—A—A—A—A—A—A—A—A—

But if different monomers (A and B) are mixed a **copolymer** forms. There is not usually a regular sequence of monomers and we say that the polymer is **random**:

—A—A—A—B—A—A—B—B—B—A—B—A—A—A—B—B—B—B—A—A—B—

By varying the ratios of A and B, however, and by varying the rate of addition of each monomer to the reaction mixture, the designer chemist can manipulate the system so that an alternating copolymer is formed:

—A—B—A—B—A—B—A—B—A—B—A—B—A—B—A—B—A—B—A—B—A—

Very occasionally it happens anyway, as when the two monomers do not polymerise individually. A good example is the photoresist polymer, used in the manufacture of microchips, formed between cyclohexene and sulphur dioxide.

Of course, the only reason for manipulating the system to form on alternating copolymer is if we believe the bulk properties of the polymer will be closer to those we want than if random polymerisation had occurred (most commonly, random polymerisation is preferred).

24 **Write shorthand bond diagrams for cyclohexene and sulphur dioxide and draw a section of the ABABABAB copolymer that they form.**

KEY FACT

A homopolymer is made from a single monomer. A copolymer is made from two or more monomers.

Condensation polymerisation

Condensation polymerisation occurs between certain molecules which have more than one functional group. When two monomers join they do so with the elimination of a small molecule. There are two important classes of condensation polymers:

- **polyesters**;
- **polyamides**.

We can find the reagents and reaction conditions required to make esters and amides in Datasheets 21 and 23. For a typical ester (Datasheet 21):

$$\underset{\text{ethanoic}}{CH_3COOH} + \underset{\text{ethanol}}{C_2H_5OH} \xrightarrow{\text{conc } H_2SO_4\text{; reflux}} \underset{\text{ethyl ethanoate}}{CH_3COOC_2H_5} + \underset{\text{water}}{H_2O}$$

The carboxylic acid and the alcohol join with the elimination of a water molecule:

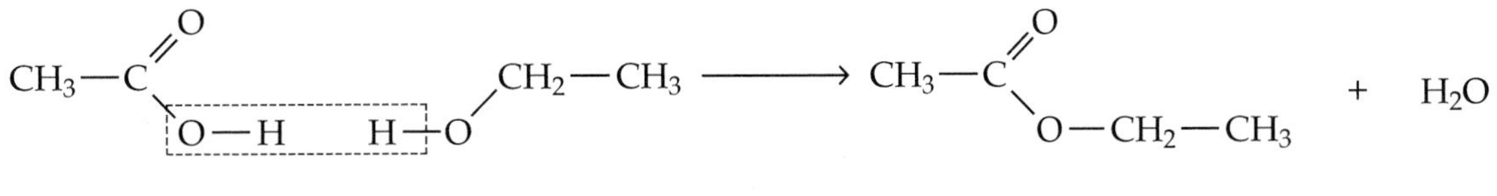

Similarly for the formation of an amide (Datasheet 23):

$$\underset{\text{ethanoyl chloride}}{CH_3COCl} + \underset{\text{ethylamine}}{C_2H_5NH_2} \xrightarrow{\text{rt}} \underset{\text{N-ethylethanamide}}{CH_3CONHC_2H_5} + \underset{\text{hydrogen chloride}}{HCl}$$

The acid chloride and the primary amine join with the elimination of a hydrogen chloride molecule:

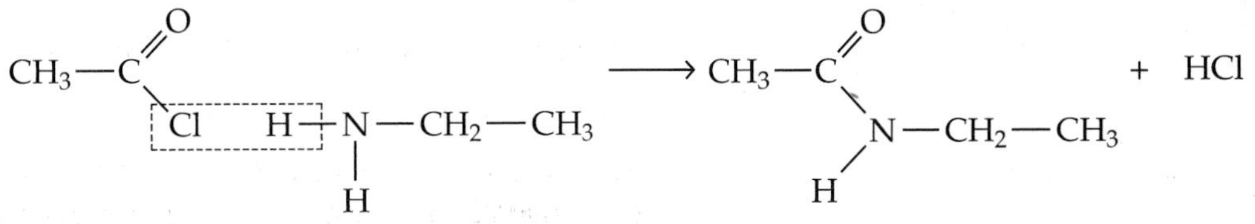

To make condensation polymers we need to take molecules with a reactive functional group at both ends. We may then use the same reaction conditions as those for the preparation of simple esters and amides. This was precisely what Wallace Carothers recognised, and led to his discovery of nylon – by design and not by accident. His thinking is clear in a letter he sent to a colleague, John Johnson, at Cornell University in early 1928:

One of the problems which I am going to start work on has to do with substances of high molecular weight (relative molecular mass). I want to tackle this problem from the synthetic side. One part would be to synthesise compounds of high molecular weight and known constitution. It would seem quite possible to beat Fischer's record of 4,200. It would be a satisfaction to do this, and facilities will soon be available here for studying such substances with the newest and most powerful tools.

Another phase of the problem will be to study the action of substances xAx on yBy where A and B are divalent radicals and x and y are functional groups capable of reacting with each other. Where A and B are quite short, such reactions lead to simple rings of which many have been synthesised by this method. Where they are long, formation of small rings is not possible. Hence reactions must result in either large rings or endless chains. In any event the reactions will lead to the formation of substances with high molecular weights and containing known linkages. For starting materials will be needed as any dibasic fatty acids as can be got, glycols, diamines etc. If you know of any new sources of compounds of these types I should be glad to hear about them!
Wallace Carothers, 14 February 1928

Two important examples illustrate the idea of making condensation polymers from molecules with reactive functional groups at both ends. These are shown in Figs 11.28 and 11.29

Fig 11.28 Terylene is formed by the reaction between benzene-1,4-dicarboxylic acid (a dicarboxylic acid) and ethane-1,2-diol (a dihydric alcohol).

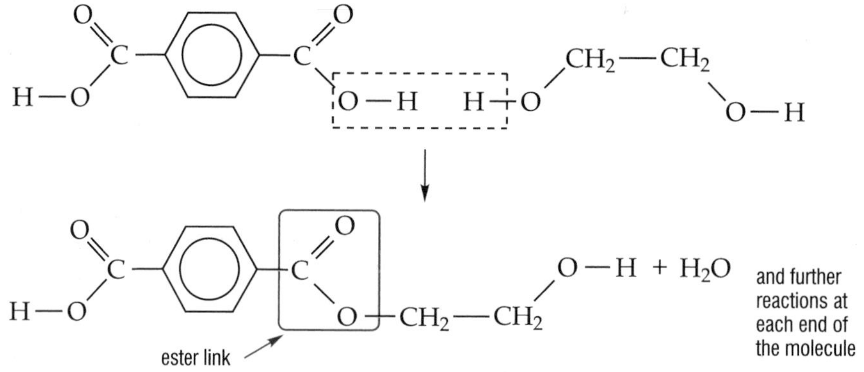

ester link

and further reactions at each end of the molecule

Fig 11.29 Nylon-6,6 is formed by the reaction between hexane-1,6-diamine (a diamine) and hexanedioyl chloride (a diacid chloride).

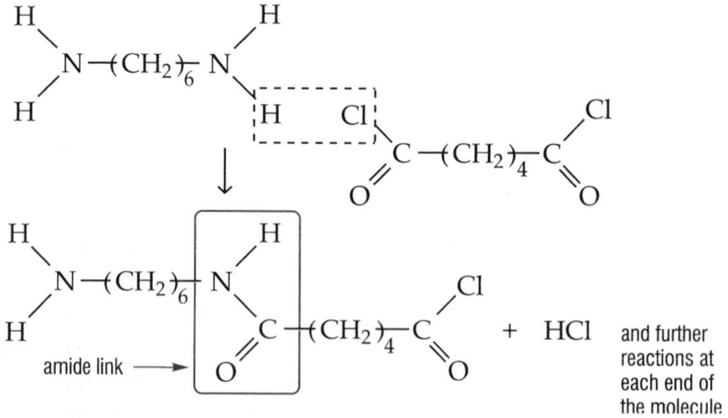

amide link

+ HCl and further reactions at each end of the molecule

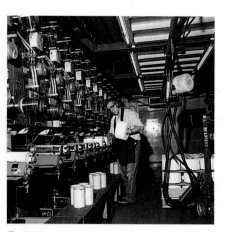

Fig 11.30 Nylon 6.6 being produced in a factory in Gloucester

25 **Would you describe condensation polymers as random copolymers or copolymers with an alternating repeat unit of the type ABABABABA?**

DESIGN FOR A PURPOSE

Homopolymers can be made by selecting a monomer which has two functional groups at opposite ends of the molecule which are able to react with one another. An example is the formation of poly(2-hydroxypropanoic acid), a polymer used to make absorbable surgical sutures (stitches which dissolve in the body fluids and do not need to be removed by the nurse or doctor).

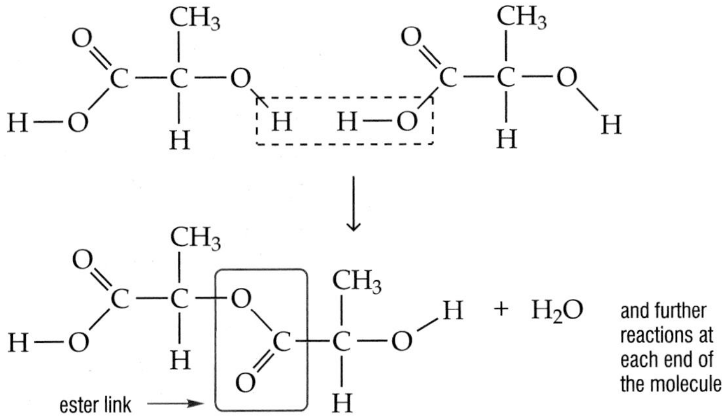

CONNECTION: Hydrolysis of surgical sutures, page 303.

Fig 11.31 Poly(2-hydroxypropanoic acid) is an example of a condensation polymer which is a homopolymer. It is synthesised by joining together molecules of 2-hydroxypropanoic acid in an addition-elimination (condensation) reaction which eliminates water.

26 What would be the relative molecular mass of a poly(2-hydroxypropanoic acid) molecule containing 4 000 monomer units?

QUESTIONS

11.26 How might a polymer containing the following repeat unit be prepared?

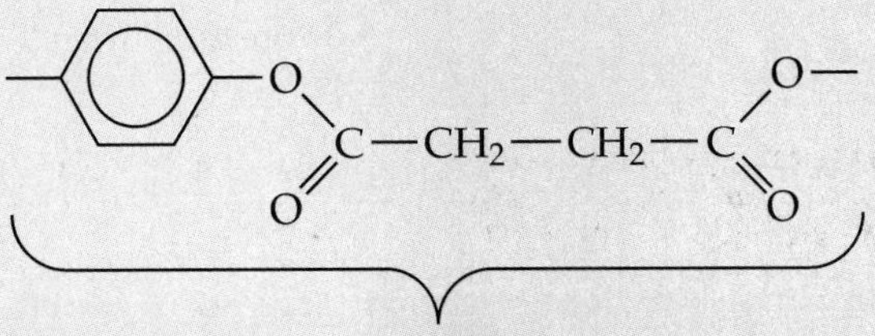

repeat unit

How many repeat units are there in a polymer with this structure of relative molecular mass 20 000?

11.27 A condensation polymer can be prepared by reaction between the following monomers:

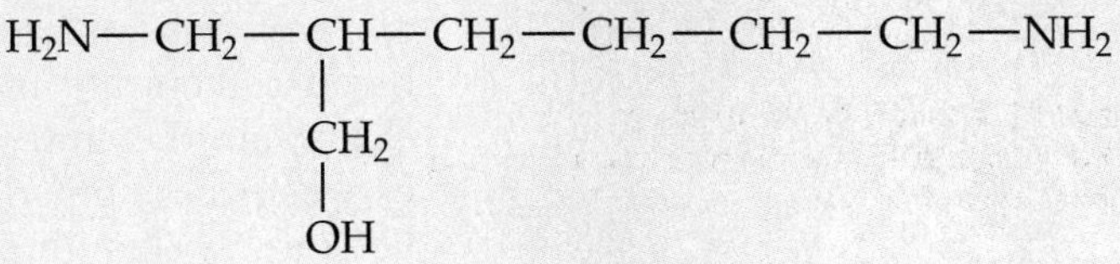

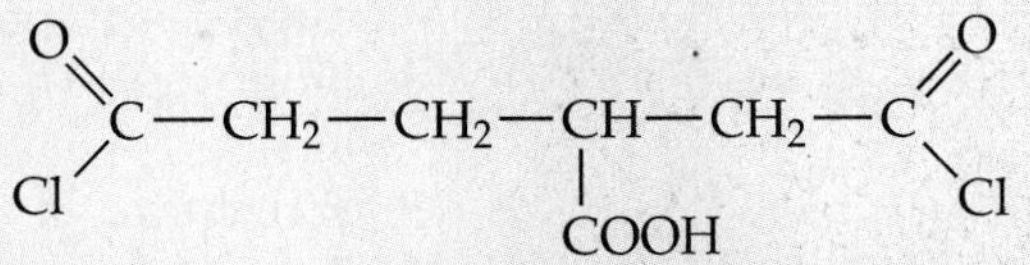

(a) Draw a structural formula for part of the polymer formed between these compounds.
(b) Describe the reaction conditions necessary for polymerisation.
(c) Explain how the polymer molecules might be linked to one another (cross-linked) by further reaction, describing the necessary conditions.
(d) Explain how the choice of reaction conditions allow the extent of polymerisation and the extent of cross-linking to be controlled.

DESIGN FOR A PURPOSE

Proteins are natural polymers made up of amino acids. An amino acid has the general formula:

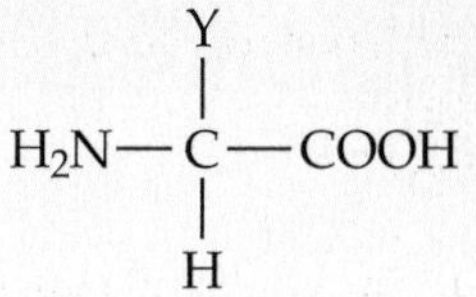

$$H_2N - \underset{\underset{H}{|}}{\overset{\overset{Y}{|}}{C}} - COOH$$

where Y = a variety of substituent groups

Name the type of polymerisation that takes place when proteins are synthesised and draw a structural formula for part of the protein chain.

There are 20 naturally occurring amino acids. Explain why these can give rise to such a huge variety of proteins.

Answers to Quick Questions: Chapter 11

1 Testing whether or not the product has the desired properties.
2 Atomic number, relative atomic mass, atomic emission spectrum, ionisation energy, electron affinity, density, melting point, boiling point, enthalpy of fusion, enthalpy of vaporisation, chemical reactivity towards water, oxygen, acids and alkalis. There may be many others, of course!
3 The search would come up with an enormous number of compounds, many of which would have little in common with 2, 4-D other than a benzene ring.
4

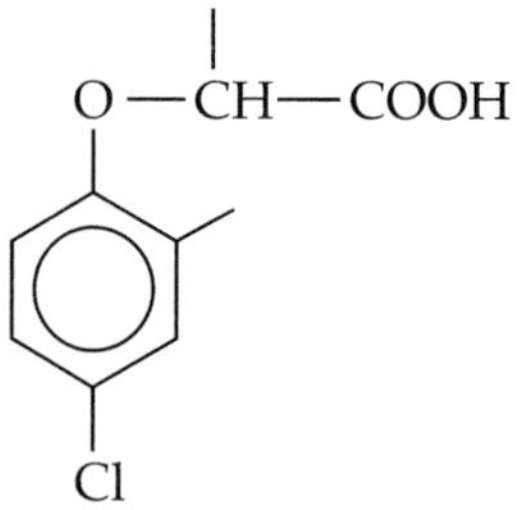

$$O - \underset{|}{CH} - COOH$$

Cl

5 Both contain a platinum(II) ion in a square planar environment, with two ammonia molecules coordinately bonded *cis* to one another.
6 Because there are more amide bonds in a given length of polymer chain and, therefore, more possibilities for hydrogen bond formation between one another.
7 Because the bonding between carbon atoms is equally strong in all directions throughout the three-dimensional lattice.
8 From electron density maps obtained by X-ray diffraction measurements.
9 Francium
10 Because they all have completely filled energy levels and energy is required to place an electron in the next available orbital.
11 Because electronegativity is a measure of the relative attraction of two atoms or groups of atoms for shared electrons. Group VIII elements do not form covalent bonds with other elements (apart from a few compounds formed with fluorine or oxygen).
12 Only one electron is required to achieve a noble gas electronic structure.
13 Datasheet 12
14 Aluminium oxide.

15 $MgO(s) + 2H^+(aq) \longrightarrow Mg^{2+}(aq) + H_2O(l)$

16 Action of dry chlorine on red phosphorus. The apparatus should be flushed with dry chlorine gas to remove air and then the phosphorus warmed until reaction occurs.

17 (a) Datasheet 22

(b) Datasheet 21

(c) Datasheet 23

18 $R^1COOCH_2CH_2CH_2R$; Datasheet 21

19 A primary alcohol gives an aldehyde initially, which is oxidised further to a carboxylic acid. A secondary alcohol is oxidised to a ketone, which cannot be easily oxidised further.

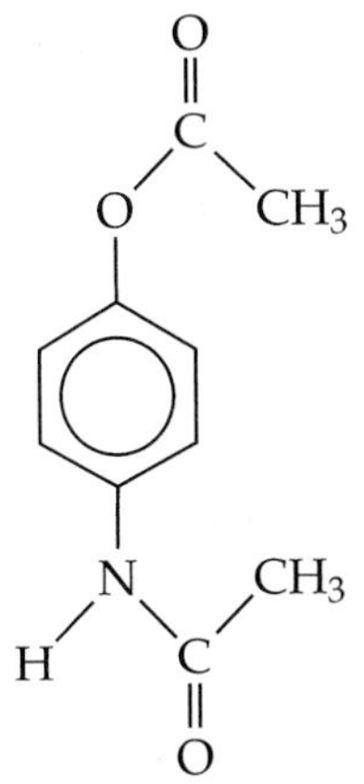

20 (see structural formula on the left.)

21 Ethanoyl chloride

22 Step 1: 50:50 conc HNO_3/H_2SO_4; warm

Step 2: Sn, conc HCl; reflux

Step 3: CH_3COCl; room temperature

Step 6: H_2O (cat = H^+ or OH^-); reflux

23

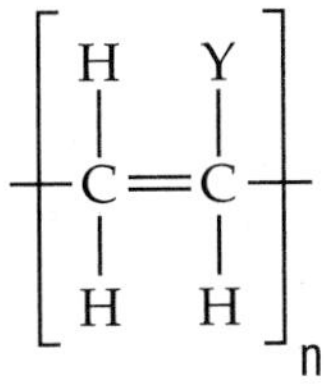

24

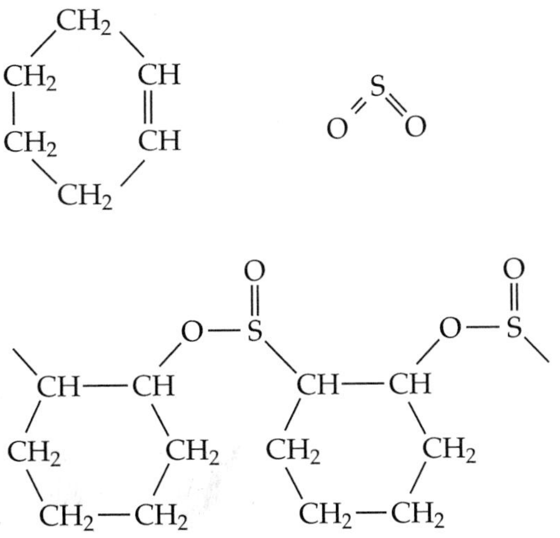

25 Copolymers with a repeating unit of the type ABABABABA.

26 The repeat unit is

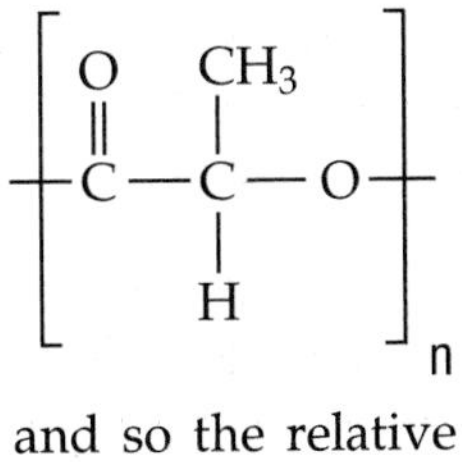

and so the relative molecular mass of a molecule containing 4000 of these units (formed from 4000 monomers) is 288 000.

Chapter 12

BONDING, STRUCTURE AND SHAPE

LEARNING OBJECTIVES

When you have studied this chapter you should be able to:

1. describe evidence for the existence of ionic, covalent and metallic bonds;

2. explain that ionic and covalent bonding are two extreme models of bonding.

3. use dot-and-cross diagrams and electrons-in-boxes to represent ionic and covalent bonding;

4. explain how soaps and detergents work, and how their structures can be designed to give properties which meet particular needs;

5. describe the bonding in benzene in terms of delocalised electrons, and give evidence to support this model;

6. describe the factors which determine the structure of an ionic lattice;

7. explain, with examples, why knowledge of molecular shapes is important;

8. explain the term 'molecular recognition';

9. predict the shape of molecules using electron pair repulsion theory and explain the limitations of this method;

10. explain what is meant by the 'conformation' of a molecule;

11. name simple coordination compounds;

12. explain the terms 'ligand', 'monodentate', 'bidentate', 'polydentate', 'chelate', 'coordination number' and 'coordination geometry';

13. describe the bonding in coordination compounds;

14. explain the terms 'structural isomerism' and 'stereoisomerism';

15. explain why compounds with the same molecular formula but different structural formulae have different properties;

16. give examples of structural isomers of organic compounds and coordination compounds;

17. explain the terms 'geometrical isomerism', 'cis- and trans-isomers' and give examples of the importance of geometrical isomerism;

18. give examples of geometrical isomers of organic compounds and coordination compounds;

19. explain, with examples, the terms 'chiral', 'optical isomerism', 'optical isomers', 'asymmetric carbon centre', 'racemate';

20. give examples of optical isomers of organic compounds and coordination compounds.

12.1 IONIC, COVALENT AND INTERMEDIATE BONDING

CONNECTION: Methods for gathering evidence about structure, page 181.

> ## KEY FACT
>
> Electron density is the concentration of negative charge in a given volume. In a region of high electron density (in an atom, ion or covalent bond) there is a high probability of encountering an electron.

How substances behave, and their possible uses, depend on their chemical make-up, the types of particles present, how these are arranged and the bonding within and between them. In Theme 1, we developed models of structure and bonding which enabled us to explain many properties of matter. These models must be refined and our knowledge of chemical substances extended if we are to become chemical architects and builders.

Evidence for ionic and covalent bonds

The properties of many substances can be explained by ionic and covalent bonding. But what evidence do we have for the existence of these bonds? Properties such as melting points, boiling points, electrical and thermal conduction, and solubilities are all consistent with the models but more direct evidence would be better. X-ray diffraction provides it.

X-rays interact with electrons (not nuclei) in a crystalline solid. A full analysis of the X-ray diffraction patterns allows **electron density** to be calculated. An electron density contour map can be drawn by joining points of equal electron density. The result resembles an Ordinance Survey map – except that the peaks correspond to regions of highest electron density rather than hills and mountains. There are no electrons in the nuclei but, nuclei will be found in the middle of electron density peaks, since electron density will always be highest in the immediate vicinity of a positive nucleus. Electron density maps for substances held together by ionic or covalent bonds are different and an identifying characteristic (Figs 12.1 and 12.2).

◆ 1 **Explain why electron density is highest near to the nucleus.**

Fig 12.1 Electron density maps for (a) sodium chloride and (b) calcium fluoride indicate that the ions can be considered as separate particles in the lattice.

Note The density of electrons is measured as the number of electrons in a volume of 10^{-30} m^3.

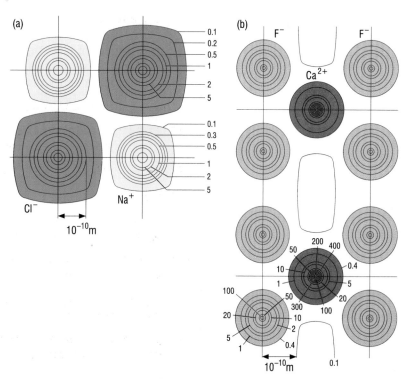

CONNECTION: *Calculations involving lattice energies, pages 370–372.*

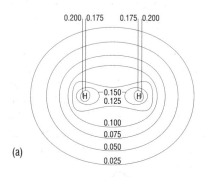

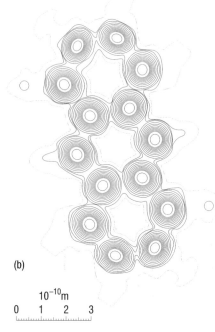

(b)

$$10^{-10} \text{m}$$
0   1   2   3

Fig 12.2   Electron density maps for (a) hydrogen and (b) anthracene indicate a 'sharing' of electron density – discrete spheres of electron density are not observed. **Note** The density of electrons is the number of electrons in a volume of 10^{-30} m^3.

KEY FACT

There are three ways in which we may represent the arrangement of electrons in an atom:

- dot-and-cross diagrams;
- electrons in boxes;
- electron configurations.

Dot-and-cross diagrams describe the arrangement of electrons in the main energy levels ($n = 1$, $n = 2$, $n = 3$ etc.). Electrons-in-boxes, on the other hand, show how the electrons are distributed amongst atomic orbitals. This provides the same information as s, p, d notation.

Further evidence to support our model for ionic bonding comes from the measurement of energy changes. The strength of bonding holding ions together in an ionic solid is measured by the energy change when one mole of the solid is formed from its gaseous ions, its lattice energy:

$$M^+(g) + X^-(g) \longrightarrow MX(s)$$

We cannot measure this energy change directly but we can calculate it from other measurable energy changes using a Born-Haber cycle. This calculated value is called the **experimental lattice energy**. We can also calculate a **theoretical lattice energy** for a compound using a model in which spherical ions, of known size and known magnitude of charge, pack together to form a three-dimensional lattice. A comparison of the experimental lattice energy with the theoretical lattice energy (Table 12.1) shows very close agreement for many compounds which suggests that the model for ionic bonding is a good one.

Table 12.1

Experimental and theoretical lattice energies for sodium and potassium halides		
Compound	Experimental lattice energy / kJ mol^{-1}	Theoretical lattice energy / kJ mol^{-1}
NaF	−902	−891
NaCl	−771	−766
NaBr	−733	−732
NaI	−684	−686
KF	−801	−795
KCl	−701	−690
KBr	−670	−665
KI	−629	−632

Representing ionic bonding

The arrangement of electrons in ions may be represented using main energy levels or atomic orbitals. For example, we may represent the electronic configurations of magnesium ions, Mg^{2+}, and oxide ions, O^{2-}, as follows:

Mg 2,8,2 $\xrightarrow{\text{loss of two electrons}}$ Mg^{2+} 2,8

or

Mg $1s^2 2s^2 2p^6 3s^2$ $\xrightarrow{\text{loss of two electrons}}$ Mg^{2+} $1s^2 2s^2 2p^6$

O 2,6 $\xrightarrow{\text{gain of two electrons}}$ O^{2-} 2,8

or

O $1s^2 2s^2 2p^4$ $\xrightarrow{\text{gain of two electrons}}$ O^{2-} $1s^2 2s^2 2p^6$

2   Use s,p,d notation to describe the arrangement of electrons in Li$^+$, S^{2-}, F$^-$, Ca^{2+} and N^{3-}.

Representing covalent bonding

We have used dot-and-cross diagrams to represent covalent bonding. There are advantages, however, in using electrons-in-boxes for some molecules. In a simple molecule such as methane, CH_4, the arrangement of electrons may be shown by dot-and-cross diagrams or by electrons-in-boxes.

A carbon atom in its ground state has only two unpaired electrons available for bonding. This is shown by our model of the atom which describes the main energy levels as being sub-divided into atomic orbitals. In order to form four single covalent bonds, an electron must be promoted from the 2s orbital to the vacant 2p orbital. This gives four unpaired electrons available for bonding. Promotion of the electron requires energy since it is being moved to an orbital of higher energy. This energy is recovered by the formation of two extra covalent bonds (exothermic).

3   Explain why CH_5 does not exist.

Fig 12.3   The arrangement of electrons in a methane molecule may be represented by (a) a dot-and-cross diagram (showing principal quantum shells only), or (b) electrons-in-boxes (showing the distribution of electrons amongst atomic orbitals).

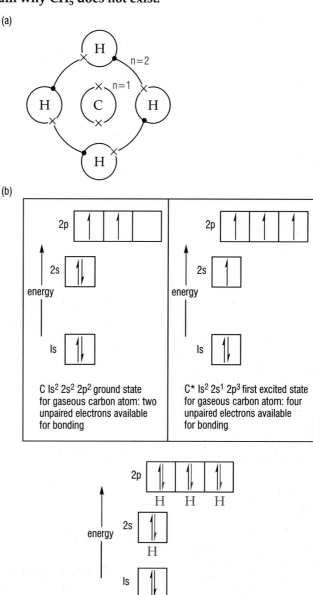

(a)

(b)

C $1s^2 2s^2 2p^2$ ground state for gaseous carbon atom: two unpaired electrons available for bonding

C* $1s^2 2s^1 2p^3$ first excited state for gaseous carbon atom: four unpaired electrons available for bonding

CH_4 : formation of four C — H bonds in methane from carbon in its first excited state

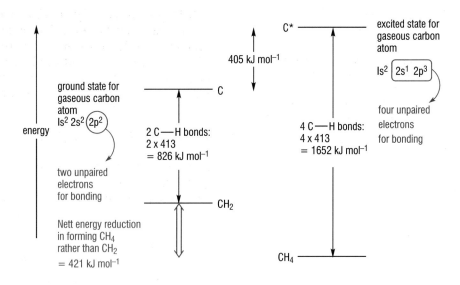

Fig 12.4 It requires energy to change from a carbon atom in its ground state to a carbon atom in its first excited state. However, the energy released from the formation of four C — H bonds in CH_4 is much larger than that released from two C — H bonds in CH_2. The compound CH_2 is unstable and rapidly forms methane, CH_4 and carbon (graphite).

Phosphorus can form two chlorides, PCl_3 and PCl_5, yet nitrogen, which is in the same group as phosphorus (Group V), only forms NCl_3. The use of electrons-in-boxes allows us to explain why. In its ground state, a phosphorus atom has three unpaired electrons in its 3p orbitals. It can form three single covalent bonds, for example in PCl_3. However, promotion of a paired electron from the 3s to a vacant 3d orbital gives five unpaired electrons. In its first excited state, a phosphorus atom can form five single covalent bonds (Fig 12.5).

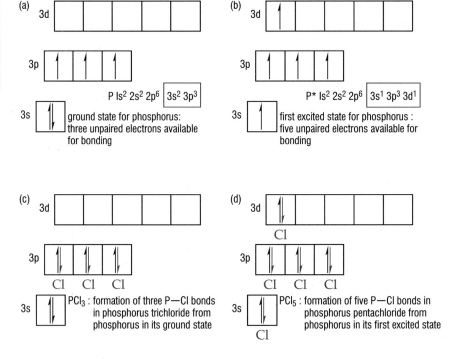

Fig 12.5 The arrangement of electrons in (a) a phosphorus atom in its ground state, (b) a phosphorus atom in its first excited state, (c) phosphorus trichloride, (d) phosphorus pentachloride.

In the case of the chlorides of phosphorus there is a fine balance between the energy required to promote an electron and the energy re-paid by the formation of two extra P — Cl bonds. Both compounds are reasonably stable, though phosphorus pentachloride dissociates at higher temperatures:

$$PCl_5(g) \longrightarrow PCl_3(g) + Cl_2(g)$$

4 **Why is phosphorus pentaiodide unstable?**

BONDING, STRUCTURE AND SHAPE

Nitrogen forms only trivalent compounds, such as NCl_3, because its outer energy level is $n = 2$. This is composed of one 2s and three 2p atomic orbitals. In order to get five unpaired electrons available for bonding, one of the 2s electrons would need a great deal of energy to promote it to the next vacant orbital, the 3s orbital. The promotion energy would not be repaid by the formation of two extra N—Cl bonds.

Intermediate bond types

Ionic and covalent bonds are at the extreme ends of a continuous spectrum of bonding. A more refined model of bonding suggests that all bonds in compounds are intermediate between ionic and covalent. However, some bonds are *predominately ionic* and others are *predominately covalent*.

Consider a bond formed between two atoms, X and Y. Suppose the atoms combine to give an ionic compound of formula X^+Y^-. In our model, the ions are spherical and pack together in a regular, repeating, three-dimensional lattice. It is a characteristic of the perfect ionic bond that there is no electron density directly between the two ions. However, the cation (X^+) will attract the electrons of the anion (Y^-), resulting in some electron density being shared by the two nuclei (Fig 12.6). The anion has been **polarised** by the cation. We say that the bond has acquired some **covalent character**.

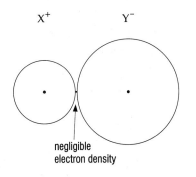

'perfect' ionic bond, electron density maps suggest discrete spheres

negligible electron density

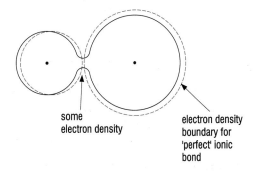

ionic bond acquires some covalent character (sharing of electron density) due to polarisation of the anion, Y^-, by the cation, X^+

some electron density

electron density boundary for 'perfect' ionic bond

Fig 12.6 All ionic bonds have at least some covalent character though it is still convenient for most purposes to consider them simply as ionic.

Table 12.2 Experimental and theoretical lattice energies of some oxides and halides.

Compound	Experimental lattice energy / kJ mol^{-1}	Theoretical lattice energy / kJ mol^{-1}
MgO	−3889	−3929
CaO	−3513	−3477
SrO	−3310	−3205
BaO	−3152	−3042
AgF	−955	−870
AgCl	−905	−770
AgBr	−890	−758
AgI	−876	−736

Evidence for this comes from theoretical and experimental lattice energies. While there was good agreement for the Group I halides, (Table 12.1) discrepancies are observed for some other solids (Table 12.2). An experimental value greater than a theoretical value suggests the bond is stronger than expected. We explain this by proposing that the ionic bond has some additional covalent character.

Taking this idea to its limit, the cation X^+ may attract electrons sufficiently strongly for electron density to be shared equally between the two atoms X and Y. This would be a covalent bond.

Starting with a perfect ionic bond we have seen how the polarisation of an anion by a cation gives increasing covalent character to the bond. Let us now imagine that two atoms, A and B, are held together by a bond with two electrons shared equally between the two nuclei – a covalent bond (Fig 12.7). If B is more electronegative than A, the electrons are no longer shared equally and the bond is polar. The greater the difference in electronegativity is between A and B, the more polar the bond. We say that it has acquired some ionic character. In the extreme the two electrons may be so strongly attracted by one atom compared with the other that an ionic bond forms, A^+B^-.

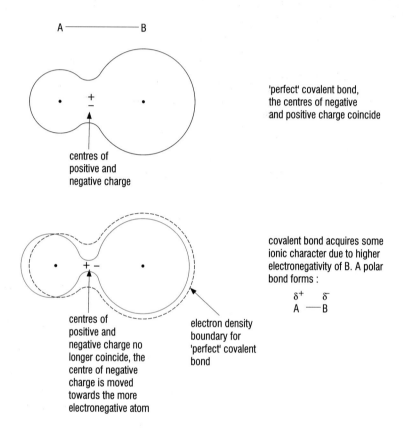

'perfect' covalent bond, the centres of negative and positive charge coincide

centres of positive and negative charge

covalent bond acquires some ionic character due to higher electronegativity of B. A polar bond forms :

$$\overset{\delta^+}{A} - \overset{\delta^-}{B}$$

centres of positive and negative charge no longer coincide, the centre of negative charge is moved towards the more electronegative atom

electron density boundary for 'perfect' covalent bond

Fig 12.7 All covalent bonds have at least some ionic character, unless the two bonded atoms are identical (for example, H_2, Cl_2 or N_2). We often describe a bond simply as 'covalent', however, even if it has some ionic character.

Ionic bonding and covalent bonding may be described by one model with intermediate bond types being common and varying in ionic/covalent character. Indeed, perfect covalent bonds are formed only between two identical atoms and perfect ionic bonds do not exist (though they are approximated very closely). Our model describes the whereabouts of electrons when atoms come together to form bonds. Although it is more correct to talk about predominately ionic bonds and predominately covalent bonds, often we use 'ionic' and 'covalent' for simplicity.

CONNECTION: Polar bonds, page 81.

5   **Using Table 12.2 comment on the likely character of the bonding in silver iodide.**

BONDING, STRUCTURE AND SHAPE

Fig 12.8   A detergent is a compound which will get things clean. It is the collective name for soaps (cleaning agents made from natural products such as animal or plant oils and fats) and synthetic detergents (cleaning agents made from crude oil or coal). The term 'detergent' is commonly used when we really mean a *synthetic* detergent.

Detergents: a case study in using bonding for design

How can research chemists make use of their knowledge and understanding of ionic and covalent bonding, and the relationship between bonding and properties of compounds?

One example is soaps and detergents. These are sold in many forms – from shampoos to washing powders, shower gels to washing up liquid. The action of soaps and detergents can be explained by looking at the bonding present in them. We will also see how this understanding has allowed new, improved products to be designed and made.

Soap is manufactured by heating natural fats and oils with sodium hydroxide. These fats and oils are called triglycerides because they are esters of propan-1,2,3-triol (commonly known as glycerol). The ester groups are hydrolysed by sodium hydroxide.

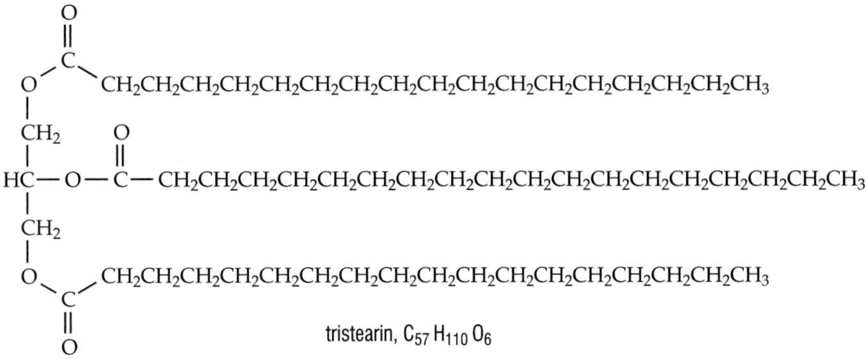

tristearin, $C_{57}H_{110}O_6$

or

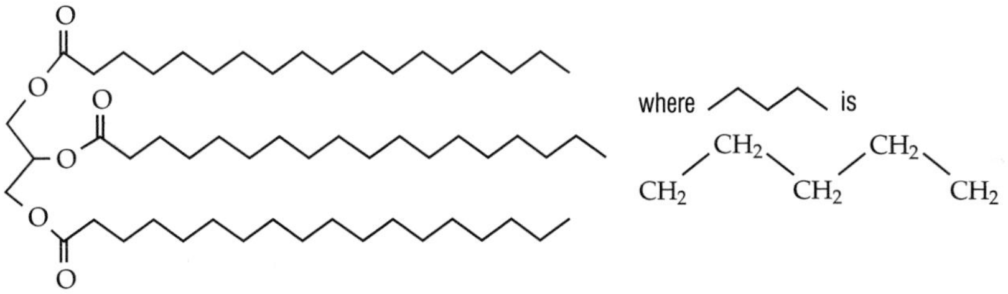

Fig 12.9   A typical triglyceride

where ∧∧∧ is

The sodium salt of the carboxylic acid is a soap. Its cleaning properties are due to its structure and the bonding. The compound has a long hydrocarbon tail of covalently-bonded carbon atoms attached to an ionic head, the sodium carboxylate group, $—COO^-Na^+$. Now, covalent compounds are generally insoluble in water but soluble in non-polar solvents. Ionic compounds are usually water-soluble but insoluble in non-polar solvents. Therefore, the tail of the soap molecule is water-insoluble but dissolves in oil and greases. On the other hand, the head is not attracted by oil and grease but it is strongly attracted to water molecules (Fig 12.10). The action of a soap in removing oil and grease from clothing, hair, dirty dishes and other surfaces can now be explained (Fig 12.11).

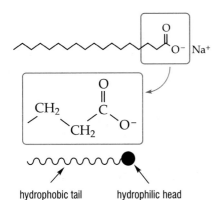

hydrophobic tail   hydrophilic head

Fig 12.10   Sodium octadecanoate, $C_{18}H_{35}O_2Na$, (or, more commonly, sodium stearate) – a typical soap.

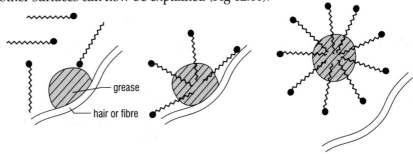

grease

hair or fibre

Fig 12.11   Soap is able to remove oil and grease from a surface because of the combination of covalent and ionic bonds present. One end of the molecule is predominantly covalent in character, the other is predominantly ionic.

BONDING, STRUCTURE AND SHAPE

In hard water areas, an insoluble scum forms when soap is used. This is because calcium and magnesium ions combine with octadecanoate ions to precipitate insoluble salts, represented by the ionic equation:

$$2C_{18}H_{35}O_2^-(aq) + M^{2+}(aq) \longrightarrow (C_{18}H_{35}O_2)_2M(s); \text{ where } M = Ca, Mg$$

6 **Draw the structural formula of 'glycerol'. Identify with a circle and name the functional group it contains.**

Scum is a problem. It means more soap is needed since some reacts with the calcium and magnesium ions. One solution is to remove these ions from solution by adding sodium carbonate ('washing soda crystals'):

$$Ca^{2+}(aq) + Na_2CO_3(aq) \longrightarrow CaCO_3(s) + 2Na^+(aq)$$

Another is to use a water softener, a device through which water is passed to remove the ions responsible for hardness. However, both of these can be expensive and an alternative solution is to synthesise a molecule which resembles a soap but whose calcium and magnesium salts are water-soluble. An example of such a compound is sodium alkylbenzene-sulphonate (Fig 12.12). This is the trivial name, simpler for our purposes here than the full name. The sodium sulphonate head is ionic, $-SO_3^-Na^+$, and therefore hydrophilic. The branched alkane tail is covalently bonded and hydrophobic.

So the problem of scum is solved by the introduction of synthetic detergents. But these give rise to a different problem. Detergents such as sodium alkybenzenesulphonate cannot be broken down by bacteria in sewage works. Consequently, when synthetic detergents first came into widespread use, many rivers began to foam and there was a danger that the detergent might be recycled in our drinking water. Natural soaps did not suffer from this disadvantage and a look at the structures of sodium octadeconoate and sodium alkylbenzenesulphonate reveals why.

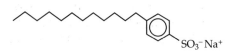

Fig 12.12   Sodium alkylbenzenesulphonate – a synthetic detergent which is not biodegradable.

Fig 12.13   The River Avon in the 1960s when it was polluted by detergent scum and more recently after it had been 'cleaned up'. The designer chemist helped to solve this environmental problem.

7 **What is the difference between the hydrocarbon tail in sodium octadecanoate and in sodium alkylbenzenesulphonate?**

Bacteria readily break down unbranched hydrocarbon chains into smaller molecules. Such a process is called **degradation**. Branched hydrocarbons persist in the environment. Most of the early detergents contained branched hydrocarbons. To the chemical architect this meant designing and building a detergent with an unbranched carbon tail which did not form insoluble salts with calcium and magnesium ions. Biodegradable detergents were born. An example is sodium 3-dodecylbenzenesulphonate (Fig 12.14).

Fig 12.14   Sodium 3-dodecylbenzenesulphonate, $C_{18}H_{29}SO_3Na$, – a biogradable synthetic detergent.

For toothpastes, shampoos and light-duty domestic detergents, sodium alkyl sulphates are often used. Again, they are based on a hydrocarbon tail with an ionic head, $-O-SO_3^-Na^+$. They differ from the other synthetic detergents described above in that they are entirely aliphatic – there is no benzene ring present in the molecule. Bacteria cannot attack the benzene ring, making substances containing this structure non-biodegradable. The nature of the detergent can be modified by changing the cation.

8 **Draw the structural formula of a typical biodegradable sodium alkyl sulphate and try to name it.**

9 **Comment on the similarity between a detergent and an ion exchange resin (page 156).**

12.2 DELOCALISED ELECTRONS IN BONDING

CONNECTION: Metallic bonding, page 67.

Evidence for metallic bonding

Metals have giant structures. They are characterised by their ability to conduct electricity and heat, and properties such as ductility and malleability. The bonding in metals is a form of delocalised bonding. Metallic bonding can be described as a regular lattice of positive ions embedded in a 'sea' of delocalised electrons. Evidence for this model comes from X-ray diffraction maps, as it did for ionic and covalent bonding (Fig 12.15).

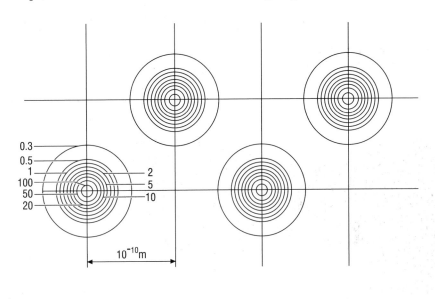

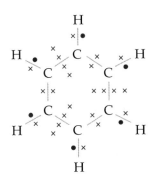

(a) Dot-and-cross bond diagram

Fig 12.15 The electron density map for aluminium is similar to those for ionic compounds but closer inspection shows that the average density of electrons between the nuclei is about 0.21 electrons per 10^{-30} m³. **Note** The density of electrons is the number of electrons in a volume of 10^{-30} m³.

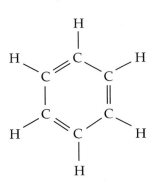

(b) Shorthand bond diagram

Fig 12.16 The alternate single-double bond model for a benzene molecule.

Benzene

In graphite the hexagonal layers of carbon atoms are held together by delocalised electrons. Graphite has a giant structure but bonding involving delocalised electrons is also found in molecular compounds.

One very important example of such a molecular compound is benzene. It has the molecular formula C_6H_6 and has a ring structure. Using dot-and-cross diagrams, we might propose a structure in which the carbon atoms in the ring are held together by alternate single and double covalent bonds. In this way all atoms in the molecule achieve a noble gas electronic configuration (Fig 12.16).

BENZENE AND BLACK GOLD

Aspirin, paracetamol and ibuprofen are safe, tried and tested pain-killers. DDT is an insecticide (now banned) with long-term toxic effects on many living systems. Terylene is used to manufacture clothing. Polystyrene is used for packing and insulation. Kevlar (see Fig 12.20 and Fig 12.21) is used for many purposes due to its remarkable strength.

Remarkably, these materials all contain the benzene ring yet have widely different properties. Even more surprising is the fact that the starting materials for all of these useful substances are derived from crude oil. Its value to human beings as a source of raw materials has earned crude oil the name title *'black gold'*.

Although benzene itself is easily obtained from coal by distillation, it is more economical to convert aliphatic compounds in crude oil to benzene and to substituted benzenes. This is the route to aromatic compounds used industrially.

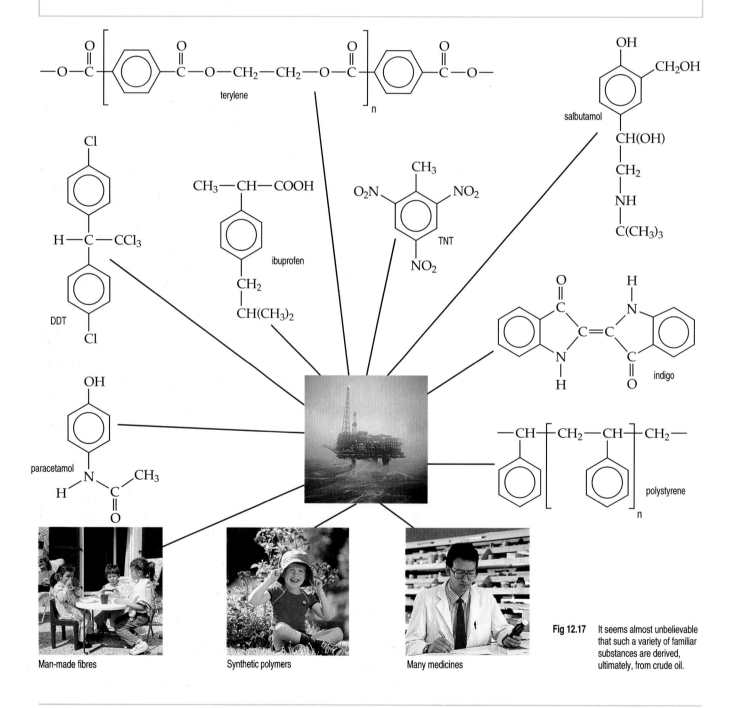

Man-made fibres

Synthetic polymers

Many medicines

Fig 12.17 It seems almost unbelievable that such a variety of familiar substances are derived, ultimately, from crude oil.

BONDING, STRUCTURE AND SHAPE

By experiment we can prove that all the carbon-carbon bonds in a benzene molecule are the same length, 0.139 nm, a value between that for a C—C single bond and a C=C double bond. All the bond angles are identical and the molecule has a symmetrical hexagonal shape. Benzene does not behave chemically as if its molecules contained C=C double bonds. Addition reactions are rare and substitution reactions are the norm.

Many compounds exist in which two neighbouring hydrogen atoms in the benzene ring have been replaced by other atoms or groups. We call these 1,2-disubstituted benzene molecules. One example is 1,2-dichlorobenzene. Only one isomer of this compound exists, yet two would have been expected if the benzene ring contained alternate single and double bonds.

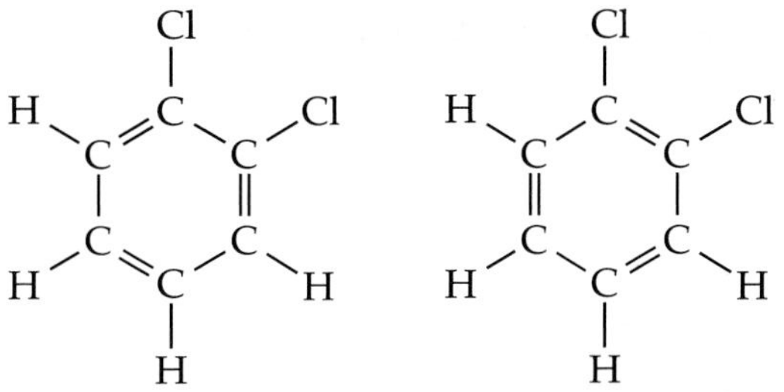

It looks, then, as if our model of the bonding in benzene is inadequate. One further piece of evidence drives home this inadequacy and the need for a refined model.

Hydrogen can be added to C=C double bonds in a reaction known as **hydrogenation**. The addition is accompanied by the release of heat, the reaction is exothermic. When cyclohexene, C_6H_{10}, is hydrogenated to give cyclohexane, C_6H_{12}, the energy released is 120 kJ mol^{-1}.

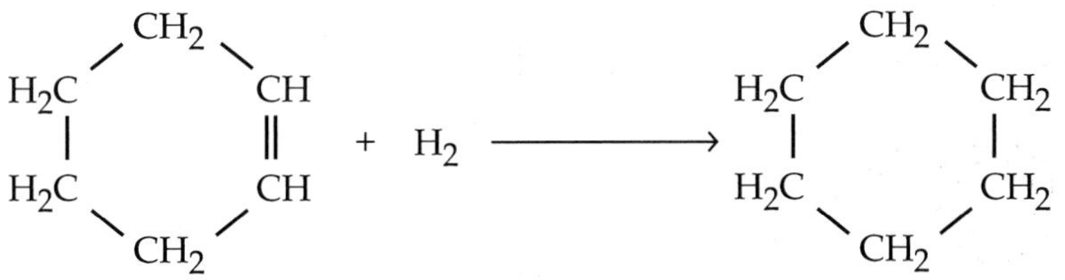

We might predict that the hydrogenation of benzene would release $(3 \times 120) = 360$ kJ mol^{-1} of benzene molecules if there were three C=C double bonds present. However, by experiment we find that the energy released is only 208 kJ mol^{-1}. This suggests that the benzene molecule is more stable than we predicted from our alternate single/double bond model. We say the benzene structure has a stabilisation energy of $(360 - 208) = 152$ kJ mol^{-1} compared to our proposed structure with alternate single and double bonds.

ACTIVITY

Draw a scale diagram of benzene based on diagram (b) in Fig 12.16 using the following data: C—H, 0.109 nm; C—C, 0.154 nm; C=C, 0.134 nm. Use the diagram to work out the distance between carbon atoms which are opposite one another in the ring and the distance between adjacent hydrogen atoms.

Use the Database to write equations for the reaction of 1 mole of this compound with (i) 1 mole of hydrogen chloride, and (ii) 3 moles of bromine.

Use the Database to identify the actual reagents, reaction conditions and products for the reaction of benzene with bromine. Explain in simple terms why this evidence helps to disprove the alternate single/double bond model for benzene.

Fig 12.18 An energy level diagram showing the stabilisation energy of benzene.

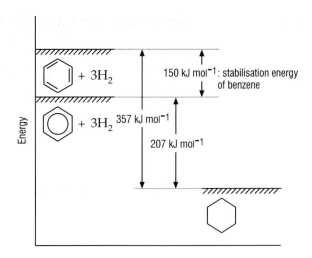

What would this model look like? The answer turns out to be very similar to our model for graphite. We suggest that each carbon atom uses three electrons in normal covalent bond formation with two neighbouring carbon atoms and one hydrogen atom. The remaining electron of each of the six carbon atoms is not associated with any particular pair of carbon atoms. Instead it is delocalised over the six carbon atoms in the benzene ring. Our mental image is of six delocalised electrons forming two dough-nut-shaped rings of electron density above and below the hexagonal plane (Fig 12.18).

10 **Why might we expect our model for benzene to be related to that for graphite?**

Fig 12.19 Our refined model for the bonding in benzene: (a) a three-dimensional representation, with rings of delocalised electron density above and below the plane of the C_6H_6 unit, (b) a two-dimensional representation of the bonding, (c) the commonly used simplified formula for benzene.

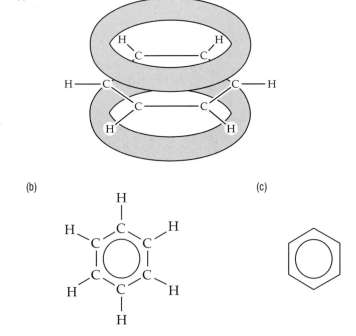

BONDING, STRUCTURE AND SHAPE

CONNECTION: *Nylons, page 208.*

Kevlar: a case study in polymer design

Nylon is commonplace. Clothing, carpets, toothbrush bristles and many other products are made from nylon. It is an aliphatic polymer. Carbon atoms in the polymer chain are held together by single covalent bonds which are free to rotate. Consequently the polymer chains are very flexible.

Fig 12.20 Nylon-6,6 and Kevlar are polyamides, with hydrocarbon units linked by an amide group, —CONH—. Nylon 6.6 is an aliphatic polyamide whereas Kevlar is an aromatic polyamide (also called an aramid).

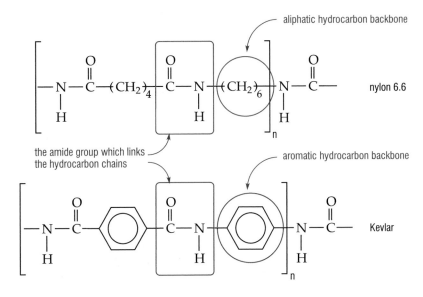

Kevlar is similar to nylon but with benzene rings rather than aliphatic chains linked by the amide group, —CONH—. This difference in molecular structure causes some remarkable changes in the properties of the polymer. These could be predicted through an understanding of the structure and bonding of aliphatic and aromatic compounds. As a consequence, a new material of desired properties could be engineered.

11 **Use the Database to write the structures for the monomers needed to make Kevlar.**

Kevlar is exceptionally strong, five times stronger than steel and ten times stronger than aluminium on a weight for weight basis. It has high heat stability. Kevlar is used extensively in the aerospace and aircraft industries, various car parts (such as tyres, brakes and clutch fittings), ropes and cables, and protective clothing.

The unique properties of Kevlar are due to the delocalised bonding which causes the benzene ring to be inflexible. The delocalisation extends beyond the benzene rings themselves and some electron density is delocalised over the amide groups also. Therefore, Kevlar is far more rigid than nylon and the bonds are stronger throughout the polymer. The high electron density in the chain also results in relatively strong Van der Waals' intermolecular forces between neighbouring polymer molecules.

12 **Nylon is one of the materials used for surgical sutures to stitch wounds. Explain why Kevlar is not likely to be a suitable material for this purpose.**

Fig 12.21 This Formula One racing driver is wearing a crash helmet made from Kevlar. The remarkable strength of Kevlar also accounts for its use in bullet-proof jackets.

QUESTIONS

12.1 Explain what is meant by an electron density map.

12.2 Sketch cross-sections through the 'hills and valleys' of the electron density maps for sodium chloride and hydrogen by taking sections through neighbouring nuclei. Explain why these diagrams offer evidence for ionic and covalent bonding.

BONDING, STRUCTURE AND SHAPE

12.3 The caesium halides show generally good agreement between the theoretical and experimental lattice energies.
  (a) For which of the halides would you expect **(i)** the best agreement; **(ii)** the least agreement?
  (b) Explain your answers to part a.

12.4 In three short sentences, describe the bonding in:
  (a) sodium chloride;
  (b) hydrogen chloride;
  (c) silver chloride.

12.5 Suggest a reason why calcium octadecanoate is insoluble in water although the sodium salt is soluble?

12.6 The general formula for fatty acids is $CH_3(CH_2)_nCOOH$.
  (a) Draw the structural formulae for the sodium salt of the fatty acids found in *(i)* coconut ($n = 10$), and *(ii)* nutmeg ($n = 12$).
  (b) What is the formula for the scum formed when an aqueous solution of a soap based on the compound in (a)*(ii)* is treated with calcium ions?

12.7 Explain why sulphur can exhibit valencies of 2,4, and 6. Give an example of a compound of sulphur for each of these valencies and represent the bonding in each using dot-and-cross diagrams.

12.8 Use electrons-in-boxes to represent the bonding in ICl_3 and IF_7. Explain why ICl_7 is not stable.

12.9 A change in the ionic/covalent character of element-chlorine bonds is shown in the oxides of Period 3 elements. The change reflects the trend of decreasing metallic character of Period 3 elements as atomic number increases. Use the Database to:
  (a) construct a table to summarise the formulae and structural types of the chlorides of Period 3 elements;
  (b) write equations to show how the following chlorides behave towards water and explain how this property provides evidence for the type of bonding present: sodium chloride, magnesium chloride, silicon tetrachloride and phosphorus trichloride.

12.10 Explain why sodium chloride is very soluble in water but silver chloride has a very low solubility.

12.11 Explain, in words only, how a detergent molecule removes oil or grease from clothing.

12.12 An 'invert soap' has a cation containing a long hydrocarbon tail and an anion. Find out about the chemistry of amines from the Database and suggest the structure of a possible 'invert soap' and how it might be synthesised from a suitable alkene.

12.13 Phenylethene, $C_6H_5CH{=}CH_2$ (commonly known as styrene), polymerises to form poly(phenylethene) (or, polystyrene). Use the following enthalpies of hydrogenation to discuss the type of bonding present in phenylethene.
$CH_2{=}CH_2(g) + H_2(g) \longrightarrow CH_3CH_3(g); \Delta H = -157 \text{ kJ mol}^{-1}$
$C_6H_5CH{=}CH_2(g) + H_2(g) \longrightarrow C_6H_5CH_2CH_3(g); \Delta H = -126 \text{ kJ mol}^{-1}$

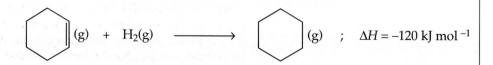

BONDING, STRUCTURE AND SHAPE

QUESTIONS *continued*

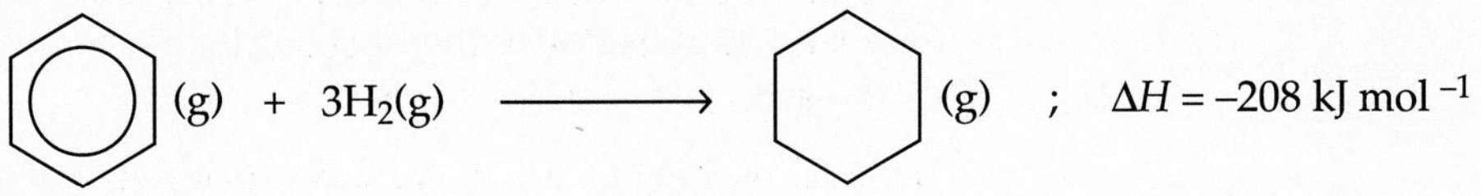
(g) + 3H2(g)
(g) ; ΔH = –208 kJ mol –1

where :

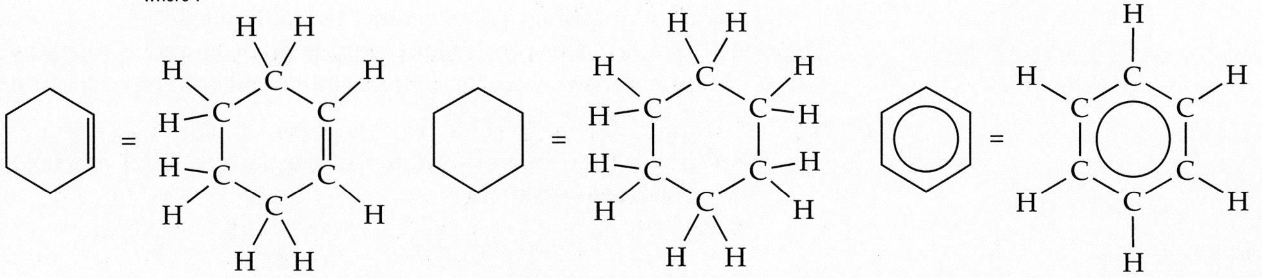

12.14 **(a)** Draw a dot-and-cross diagram to represent the bonding in *(i)* nitric acid, HNO_3, *(ii)* the nitrate ion, NO_3^-.

(b) In nitric acid, two nitrogen-oxygen bonds are approximately the same length but the third is shorter. In the nitrate ion all three nitrogen-oxygen bonds are the same length, and about the same length as the two short bonds in nitric acid. Explain these observations.

12.15 Describe as fully as possible the bonding in solid calcium carbonate.

12.3 PREDICTING STRUCTURE AND SHAPE

Designing molecules or giant structures with a particular purpose in mind would be easier if the arrangement of a group atoms could be predicted in advance, since structure and bonding determine the bulk properties of a substance. Some relatively simple rules allow us to predict structure and shape, at least to a first and often very helpful approximation.

KEY FACT

The bulk properties are those displayed by a substance when present in measurable amounts. They include melting point, hardness, colour, smell etc.

Ionic lattices

Ionic lattices consist of anions and cations packed together in a repeating three-dimensional array held together by electrostatic attraction between oppositely-charged ions. But ions vary in size as does the magnitude of the charge they carry. What determines the pattern in which they pack together and can we predict it?

We know ions pack together in a way that gives the strongest possible lattice structure. Imagine a cation being surrounded by anions. As the coordination number of the cation increases, that is the number of anions surrounding it increases, so the electrostatic attractive forces increase. However, there will also be repulsion between the like-charged anions. A similar situation occurs for each anion surrounded by cations. The compromise solution is easy to understand – coordination numbers will be as high as possible whilst allowing attractive and repulsive forces to balance one another.

For our purposes it is sufficient to illustrate with a simple example, an ionic compound of empirical formula AB. The number of anions that can surround a particular cation may be determined by the ratio of cation radius to anion radius (r^+/r^-). If the anion is smaller than the cation then r^-/r^+ is used. The ratio indicates the likely structure for a salt of given formula. For example,

r^+/r^-	0.155 to	0.23 to	0.41 to	0.73 to higher values
coordination number	3	4	6	8

The two most important AB structures are sodium chloride and caesium chloride (Fig 4.12). The coordination numbers are 6 and 8 respectively. Table 12.3 summarises r^+/r^- for several compounds of empirical formula AB, together with their structures.

13 Why is the coordination number of caesium in CsCl greater than that of sodium in NaCl?

Table 12.3

Compound	r^+/r^-	AB structure
KF	1.00	NaCl
KCl	0.73	NaCl
KBr	0.68	NaCl
KI	0.62	NaCl
RbF	0.92*	NaCl
RbCl	0.82	NaCl (and CsCl)
RbBr	0.76	NaCl
RbI	0.69	NaCl
CsF	0.81*	NaCl
CsCl	0.93	CsCl
CsBr	0.87	CsCl
CsI	0.76	CsCl

* r^-/r^+

CONNECTION: Synthesis of sulphanilamide, page 235.

Shapes of molecules

Molecular shape explains the action of most drugs and agrochemicals

The shapes of molecules have an important influence on their physical properties and chemical reactivity. The ability of one molecule to detect another from its shape and the nature of the functional groups present is known as **molecular recognition**. This recognition is used to great effect in medicine, agrochemicals, fragrances, flavours, catalysts, dyes and many other areas. We will look at two examples – the sulphonamide drugs and pain killers.

Drugs fight bacteria in one of two ways:

- by invading and killing them (said to be **bacteriocidal**);
- by stopping them from growing (called **bacteriostatic drugs**).

Perhaps the most important breakthrough in medicinal chemistry this century was the discovery of the family of antibacterial drugs known as the sulphonamides. The first of these to be discovered was sulphanilamide.

Bacteria and healthy human bodies both require folic acid to grow. The difference is that whilst we obtain folic acid from the food we eat, bacteria make their own. They make it from 4-aminobenzene carboxylic acid, with help from an appropriate enzyme. The 4-aminobenzene carboxylic acid 'fits' snugly into the enzyme which catalyses its synthesis into folic acid.

14 Write structures for the two other compounds required if you wanted to synthesis folic acid from 4-aminobenzene carboxylic acid in the laboratory (use the Database if necessary).

The structure of sulphanilamide is very similar to that of 4-aminobenzene carboxylic acid (Fig 12.22). It acts by occupying the enzyme site into which 4-aminobenzene carboxylic acid fits during the synthesis of folic acid. In this way, bacteria are prevented from synthesising the folic acid they need for growth. The human body remains unaffected. Molecular shape has been the key to the effectiveness of the drug. This is the case for virtually all drugs and agrochemicals.

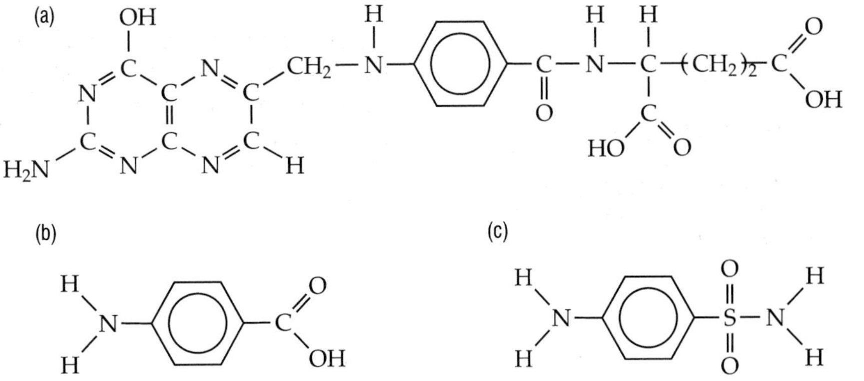

Fig 12.22 The molecular structures of (a) folic acid, (b) 4-aminobenzene carboxylic acid, (c) sulphanilamide. The secret to the drug's action is its similarity in shape and size to 4-aminobenzene carboxylic acid.

15 Is sulphanilamide a bacteriocidal or a bacteriostatic drug?

Aspirin and paracetamol relieve pain because of their molecular shape. Most pain is caused when the body produces a family of chemicals called prostaglandins, for example in response to injury. Prostaglandins are synthesised in the body from fatty acids (aliphatic carboxylic acids with high relative molecular mass) and oxygen. An enzyme is required to hold the molecules in the correct three-dimensional geometry for the oxidation reaction to occur. Pain-relievers such as aspirin and paracetamol work by occupying some of the enzyme sites thus preventing the synthesis of prostaglandins.

Predicting shape

KEY FACT

Enzymes are biological catalysts.

CONNECTION: Page 567

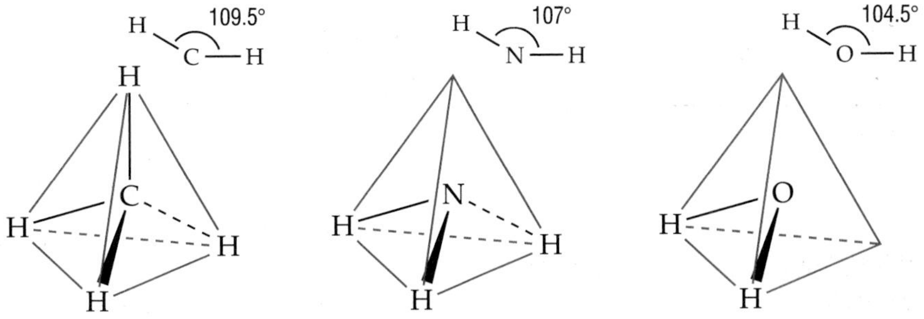

Fig 12.23 The shapes of CH_4, NH_3 and H_2O.

CONNECTION: Shapes of CH_4, NH_3 and H_2O, page 48.

KEY FACT

The following conventions are used to show the spatial arrangement of bonds:

A bond in the plane of the page;

A bond in front of the plane of the page;

A bond behind the plane of the page;

Fig 12.23 shows the shapes of CH_4, NH_3 and H_2O. To summarise the ideas we met earlier:

- Methane, ammonia and water molecules have closely related shapes: CH_4 is tetrahedral, NH_3 is pyramidal and H_2O is bent. All are based on a tetrahedron.
- The molecules CH_4, NH_3, H_2O and HF have four pairs of electrons in the outer energy level of the central atom (C,N,O and F).
- Electron pairs may be bonding pairs (shared between two nuclei) or lone pairs (not involved in bonding within the molecule and only under the influence of one nucleus).
- The molecular shapes can be explained in terms of the electron pairs in the outer energy level getting as far away from one another as possible; pairs of electrons occupy a position of minimum repulsion.
- The relative strengths of electron pair repulsion are: lone pair/lone pair > lone pair/bonding pair > bonding pair/bonding pair

In each of these molecules, four pairs of electrons were involved. The positions adopted by other numbers of electron pairs in order to achieve minimum repulsion are summarised in Fig 12.24.

BONDING, STRUCTURE AND SHAPE

Number of electron pairs	Shape	Description	Bond angles
2		linear	180°
3	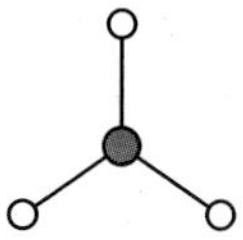	trigonal planar	120°
4	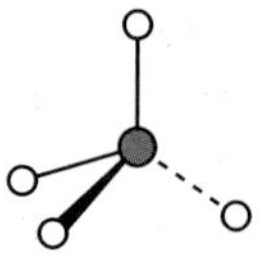	tetrahedral	109.5°
5	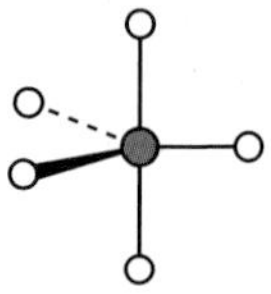	trigonal bipyramidal	120°, 90°
6	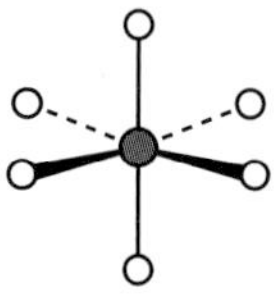	octahedral	90°
7	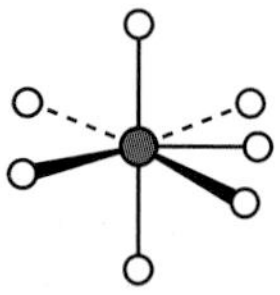	pentagonal bipyramidal	72°, 90°

Fig 12.24 The positions of minimum repulsion for electron pairs.

TUTORIAL

Problem: Predict the molecular shape of phosphorus trichloride using electron pair repulsion theory.

Solution: 1. Write the molecular formula of the molecule: PCl_3
2. Draw a dot-and-cross diagram or electrons-in-boxes diagram to represent the bonding in the molecule.

(a)
$$Cl \overset{\bullet\bullet}{\underset{\bullet\times}{\times}} P \overset{\times}{\bullet} Cl$$
$$Cl$$

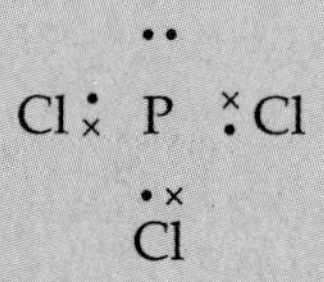

(b)
3p ↑↓ ↑↓ ↑↓
      Cl Cl Cl
3s ↑↓

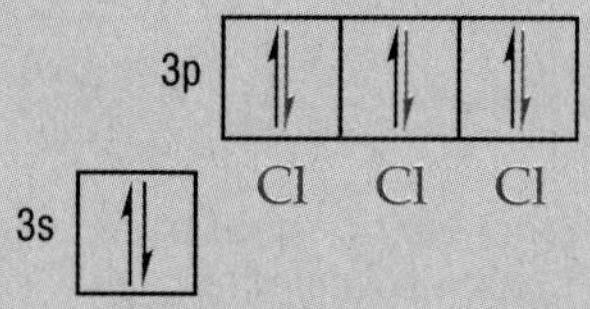

3. Count the number of electron pairs in the outer energy level of the central atom. In PCl_3 there are 4 pairs of electrons – 3 bonding pairs and 1 lone pair.
4. Arrange the pairs of electrons in a position of minimum repulsion.

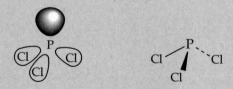

Conclusion: It is predicted that PCl_3 is a pyramidal molecule.

BONDING, STRUCTURE AND SHAPE

The relative repulsion of lone pairs and bonding pairs can be used to explain subtle differences in bond angles from the ideal values given in Fig 12.23. For example, the bond angles in CH_4, NH_3 and H_2O (109.5°, 107° and 104.5° respectively) are explained by the relative strength of electron pair repulsion. The idea can be extended.

Consider water and hydrogen sulphide, two molecules with the same number of bonding pairs and lone pairs as one another. Both are bent molecules; water has a bond angle of 104.5°, while the bond angle in hydrogen sulphide is 98°. Why do the two molecules have different bond angles when both consist of two bonding pairs and two lone pairs of electrons? The concept of electronegativity (the power an atom or group of atoms has to attract electron density) helps us to explain.

Oxygen is more electronegative than sulphur. The oxygen atom has a greater attraction for the bonding pairs in a water molecule than a sulphur atom does in hydrogen sulphide. The electron density of the O—H bonds is distorted more than in the S—H bonds, resulting in a higher concentration of electron density near to the oxygen nucleus in a water molecule. Consequently there is a greater repulsion between the bonding pairs in water than between the bonding pairs in hydrogen sulphide (Fig 12.25).

Fig 12.25 A comparison of molecules of water and hydrogen sulphide.

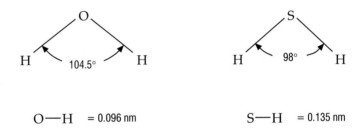

O—H = 0.096 nm   S—H = 0.135 nm

KEY FACT

For the purposes of predicting shape, multiple bonds are treated as if they were single bonds.

16 **The bond angles in ammonia are each 107°. Predict, with an explanation, the bond angles in (a) phosphine, PH_3 and (b) nitrogen trifluoride NF_3.**

TUTORIAL

Problem: Predict the molecular shape of ethene.

Solution: 1. The molecular formula of ethene is C_2H_4.
2. The bonding in C_2H_4 may be represented by

(a)

H   H
   C × C
H   H

(b)

H   H
  \   /
   C === C
  /   \
H   H

3. Around **each** carbon atom there are 2 electron pairs (between the carbon and two hydrogen atoms) and 1 double electron pair (in the double bond between carbon atoms). For the purposes of predicting shape we treat these as 3 pairs of electrons.
4. The position of minimum repulsion for three electron pairs is trigonal planar. Therefore the shape is

H   120°   H
   \   /
120° C === C
   /   \
H   120°   H

Conclusion: Ethene is a planar molecule, with bond angles of approximately 120°.

Problem: Predict the molecular shape of phosphorus trichloride oxide.

Solution: 1. Molecular formula of phosphorus oxotrichloride: $POCl_3$
2. The bonding in $POCl_3$ may be represented by:

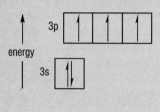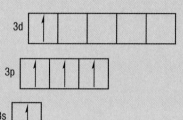

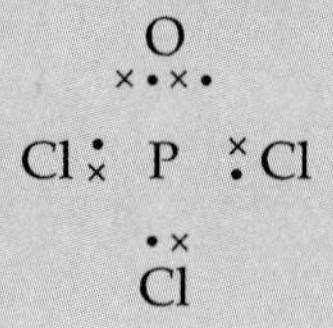

[Ne] core

ground state for phosphorous
(three electrons available for
bonding)

[Ne] core

first excited state for
phosphorus (five electrons
available for bonding)

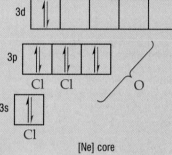

[Ne] core

$POCl_3$

(**Note** It does not matter which of the phosphorus electrons pair off with chlorine or
with oxygen electrons. Remember: electrons-in-boxes is just a way of counting
electrons and explaining valency)

3. In $POCl_3$ there are four pairs of electrons – four bonding pairs (the double
bond consists of two pairs of electrons, but for the purpose of predicting
molecular shape we treat these as if they were a single bonding pair).
4. Arrange the pairs of electrons in a position of minimum repulsion.

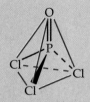

Conclusion: It is predicted that $POCl_3$ is an approximately tetrahedral
molecule.

17 **Predict the Cl—P—Cl bond angles in $POCl_3$.**

BONDING, STRUCTURE AND SHAPE

Larger molecules and bond rotation

The arrangement of atoms in a more complicated molecule may still be predicted by electron pair repulsion theory, but other factors begin to influence shape. Therefore, the theory is limited and its predictions need to be treated with caution. An example of its successful application is poly(ethene), but remember that this is a relatively simple molecule even though it is very large.

Fig 12.26 A section of the polymer chain of poly(ethene), with a dot-and-cross diagram for one of the C—CH$_2$—C units.

Each carbon atom has four bonding electron pairs around it and no lone pairs. Therefore, the prediction is that each carbon is surrounded by two hydrogen atoms and two carbon atoms arranged at the corners of a tetrahedron. We say the central carbon atom is in a tetrahedral environment:

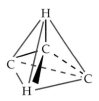

An example of where electron pair repulsion theory lets us down is if we try to predict the shape of the molecule $(CH_3)_3SiOSi(CH_3)_3$. Considering the environment of the central oxygen atom, there are four pairs of electrons, two bonding pairs and two lone pairs. We might sensibly predict a tetrahedral arrangement with a bond angle a little less than 109.5°. However, by experiment the bond angle has been measured as 180°. This indicates the Si—O—Si linkage is linear and not bent as predicted.

To understand the shapes of more complicated molecules it is important to know something about **bond rotation**. Some bonds may rotate but others not. The simple rule is that single covalent bonds are free to rotate but multiple bonds are fixed although, like all covalent bonds, they do vibrate. With this is mind, we can improve our mental image of poly(ethene). While each carbon atom is in a tetrahedral environment, the molecule itself flexes and bends. It can take up many different shapes, called **conformations** (Fig 12.27).

CONNECTION: Infrared spectroscopy and molecular vibrations, page 194.

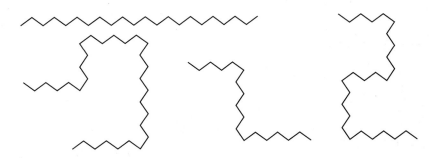

Fig 12.27 A poly(ethene) molecule can adopt various shapes because of the free rotation around C — C single covalent bonds. Chain length, of course, may vary here.

18   **Explain why a sheet of poly(ethene) can be stretched by simply pulling it gently but a sheet of aluminium foil cannot.**

Conformation of molecules

Ethane, C_2H_6, has the structural formula:

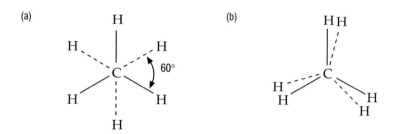

Each carbon atom is in a tetrahedral environment and the C—C bond is free to rotate. As the bond rotates the energy of the molecule changes. The two extreme conformations are

- the staggered form (lowest energy);
- the eclipsed form (highest energy).

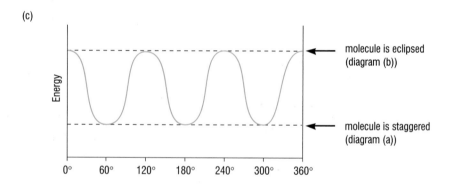

Fig 12.28 (a) The staggered conformation of ethane, (b) the eclipsed conformation of ethane, and (c) the changes in energy of the ethane molecule as the C—C single bond rotates.

Cyclohexane, C_6H_{12}, has the structural formula:

or, more simply,

Again, each carbon atom is in a tetrahedral environment. The C—C bonds are restricted in their rotation by the ring structure. However, the molecule is flexible and can move between two extreme conformations, the **chair** and **boat** forms. The energy changes involved in the transformation from one form to another can be studied with the aid of high powered computers (Fig 12.29).

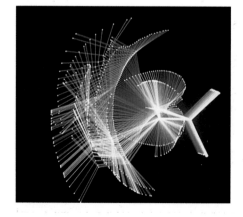

Fig 12.29 A computer graphic of the interconversions of cyclohexane from boat to chair. Each structure is overlaid across three adjacent carbon atoms in the ring. The structures are colour-coded from yellow (lowest energy) through red to blue (highest energy). The chair is the lowest energy form.

BONDING, STRUCTURE AND SHAPE

Note. The use of molecular model kits will help you to understand molecular shape. If possible, you should use them when answering these questions.

12.16 Draw a shorthand bond diagram for hydrogen fluoride, including lone pairs of electrons, and explain why this linear molecule may be considered to derive from a tetrahedral shape.

12.17 The ionic radii of K^+ and Cl^- are 0.133 nm and 0.181 nm respectively. Predict the structure of potassium chloride.

12.18 Predict the shapes of the following molecules and estimate probable bond angles: H_2S, SCl_2, PF_3, PF_5, IF_5, SF_6, XeF_4.

12.19 Predict the shapes of the following molecules and estimate probable bond angles: $SOCl_2$, XeF_2, SO_2, SO_3.

12.20 Predict the shapes and estimate the probable bond angles for NH_4^+ and NH_2^-. Comment on their relationship to the structure of the ammonia molecule, NH_3.

12.21 We can taste and smell only those substances which have particular shapes. Use electron pair repulsion theory to predict, as far as possible, the shapes of the molecules described in the following extracts. You will not be able to draw three-dimensional diagrams for each molecule. Instead you should draw diagrams to represent various parts of the molecules.

(a) Cloves owe their smell largely to the presence of heptan-2-one. This compound accounts for the odours of many fruits and dairy products. Its structural formula is:

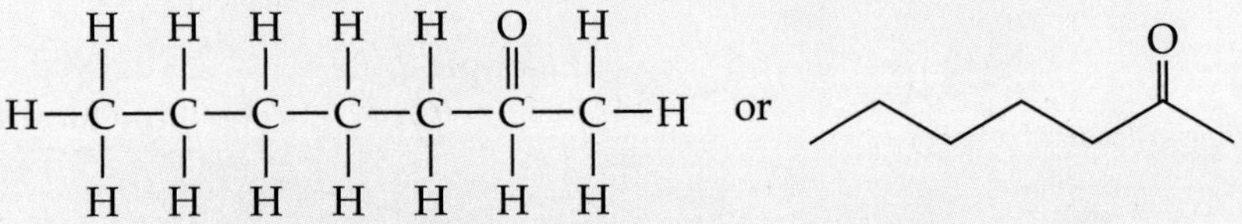

(b) Thiols are similar to alcohols, with the oxygen atom being replaced by a sulphur atom. This subtle change brings about an enormous change in the smell of these compounds – generally speaking thiols have a revolting smell! The compound squirted out by skunks is 3-methylbutan-1-thiol. Its molecular structure is:

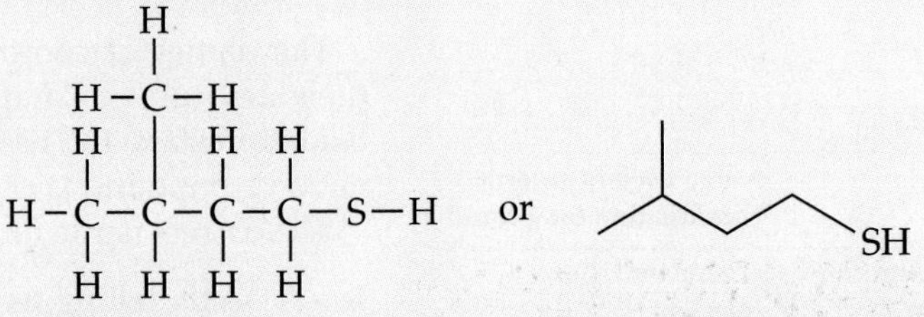

(c) Menthol is extracted from the Japanese peppermint and has a characteristic cool taste. Its structural formula is:

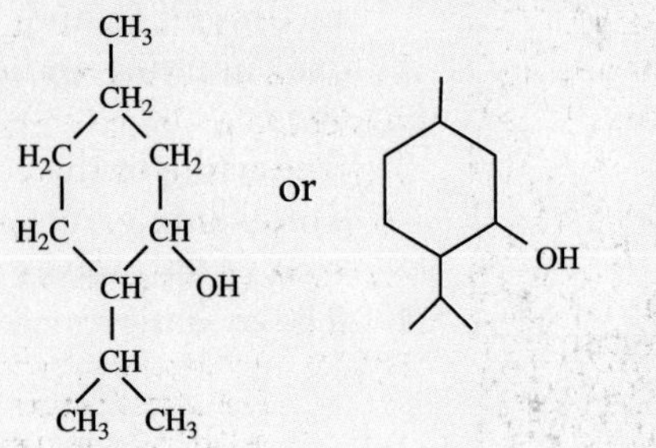

12.4 COORDINATION COMPOUNDS AND THEIR SHAPE

A **coordination compound** consists of a central metal ion or atom surrounded by bonded ligands. Often the cluster is charged and called a **complex ion**. Some clusters carry no charge and the coordination compound is therefore neutral.

The shapes of coordination compounds cannot be predicted as easily as those of simple molecules. Nevertheless, the structure and bonding of coordination compounds determines their reactivity and uses. We have already met an example of a coordination compound in cisplatin, but they are used in a number of important areas, ranging from medicine to analysis, and biochemistry to industrial catalysis.

Nomenclature of coordination compounds

Coordination compounds may be ionic (cationic or anionic depending on whether they carry an overall positive or negative charge) or are neutral. The names of complex ions involve oxidation numbers and the prefixes used in chemical nomenclature. Two examples will illustrate the use of the system of nomenclature for naming coordination compounds.

Example 1

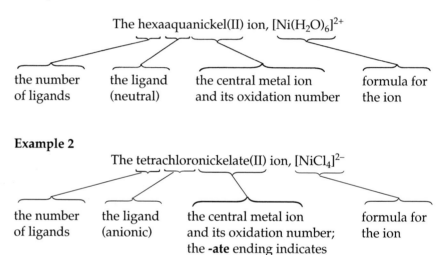

The hexaaquanickel(II) ion, $[Ni(H_2O)_6]^{2+}$

the number of ligands | the ligand (neutral) | the central metal ion and its oxidation number | formula for the ion

Example 2

The tetrachloronickelate(II) ion, $[NiCl_4]^{2-}$

the number of ligands | the ligand (anionic) | the central metal ion and its oxidation number; the **-ate** ending indicates that it is an anionic complex | formula for the ion

The names of coordination compounds do not give information about their structures or shapes. However, they do tell us about the metal and its oxidation state, the nature and number of ligands, and whether the coordination compound is charged or neutral.

The correct name for a coordination compound consists of the following.

1. A prefix telling us how many ligands are present. For example, 'hexa' indicates six.
2. A term telling us what the ligands are. An 'o' ending indicates that it is an anionic ligand, in other words a negatively charged ion. For example, 'ammine' indicates ammonia molecules, 'chloro' indicates chloride ions.
3. The name of the central metal ion if the complex is cationic or neutral. For example, nickel. If the complex is anionic, an appropriate 'ate' version of the metal's name is used (Table 12.4).
4. The oxidation state of the metal in brackets after its name.

Table 12.4

Metal	Name used in anionic coordination compound
Aluminium	aluminate
Titanium	titanate
Vanadium	vanadate
Chromium	chromate
Manganese	manganate
Iron	ferrate
Cobalt	cobaltate
Nickel	nickelate
Copper	cuprate
Zinc	zincate
Tin	stannate
Lead	plumbate

Ligands and bonding

All ligands have at least one atom with a lone pair of electrons which can be used to bond to the central metal ion. A ligand with just one point of attachment is said to be **monodentate**. A ligand with two points of attachment is said to be **bidentate** and to form a **chelate** when it bonds to a central metal ion. Ligands with several points of attachment are said to be **polydentate**.

An ammonia molecule is a monodentate ligand. It bonds to a central metal ion through the lone pair of electrons on the nitrogen atom (Fig 12.30).

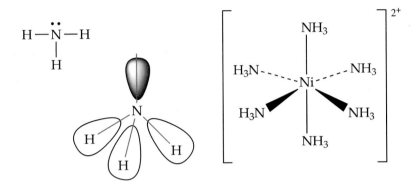

Fig 12.30 The ammonia molecule acts as a monodentate ligand by bonding through the nitrogen atom's lone pair of electrons. The hexaamminenickel(II) ion is typical of the complexes formed by ammonia.

A molecule of 1,2-diaminoethane is a bidentate ligand. It may be thought of as two ammonia molecules linked by a two-carbon atom chain (an ethane skeleton). The molecule has two centres which could bond to a central metal ion – the lone pairs on each of the two nitrogen atoms (Fig 12.31). In complex formation, 1,2-diaminoethane takes the place of two ammonia molecules. 1,2-Diaminoethane can chelate in this way because the C—N and C—C bonds are free to rotate, allowing the molecule to flex and change shape until the two nitrogen atoms are sufficiently close to the metal to form bonds.

19 Explain why 1,2-diaminoethane can replace two ammonia molecules in a coordination compound.

20 How might ethane-1,2-diol bond to a metal ion?

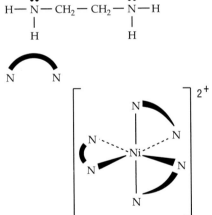

Fig 12.31 1,2-Diaminoethane is a bidentate ligand. It can occupy the positions around a central metal ion that two ammonia molecules would have taken.

Fig 12.32 The successive replacement of pairs of water molecules by 1,2-diaminoethane molecules (ligand exchange) is accompanied by distinctive colour changes. The complex ions: $[Ni(H_2O)_6]^{2+}$, $[Ni(H_2O)_4(en)]^{2+}$, $[Ni(H_2O)_2(en)_2]^{2+}$ and $[Ni(en)_3]^{2+}$ each has its own characteristic colour.

Table 12.5 Some important ligands

	Prefix in nomenclature	Examples of complexes
Monodentate		
H_2O	aqua	$[Cu(H_2O)_6]^{2+}$
NH_3	ammine	$[Co(NH_3)_6]^{3+}$
CO	carbonyl	$Ni(CO)_4$
Cl^-	chloro	$[CuCl_4]^{2-}$, $[TiCl_6]^{2-}$
OH^-	hydroxy	$[Al(OH)_4]^-$, $[Sn(OH)_4]^{2-}$
CN^-	cyano	$[Fe(CN)_6]^{4-}$, $[Fe(CN)_6]^{3-}$
SCN^-	thiocyanato	$[Fe(H_2O)_5(SCN)]^{2+}$
Bidentate		
$NH_2CH_2CH_2NH_2$	1,2-diaminoethane	$[Ni(en)_3]^{2+}$
$C_2O_4^{2-}$	ethanedioato	$[Fe(C_2O_4)_3]^{3-}$
Hexadentate		

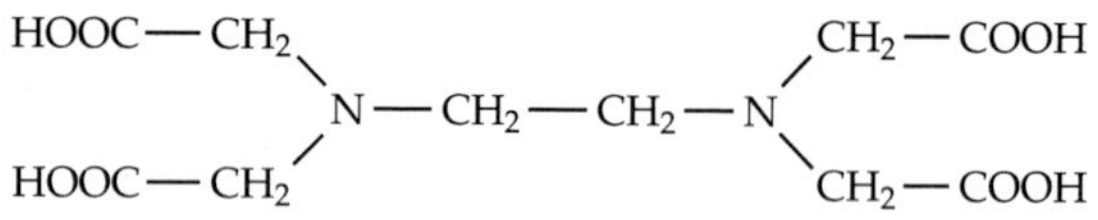

EDTA

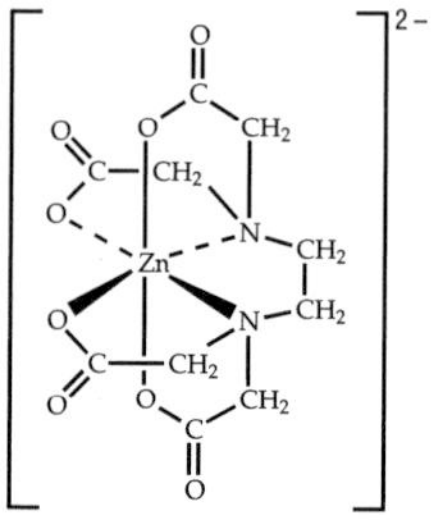

The bond between a ligand and a metal ion may be described as a coordinate bond (hence the name coordination compounds). A lone pair of electrons is donated to a vacant orbital of the metal ion in coordinate bond formation. The nature of the ligand and the charge on the central metal ion, results in the bond having some ionic characteristics.

21 In the complex ions SnF_4^{2-}, $SnCl_4^{2-}$, $SnBr_4^{2-}$ and SnI_4^{2-}, which tin-halogen coordinate bond has the most ionic character?

Shapes of coordination compounds

When we talk about the shapes of coordination compounds we usually confine ourselves to a description of the spatial arrangement of those atoms bonded directly to the central metal ion. The number of atoms bonded to the central metal ion is called its **coordination number** and their arrangement in space is called the **coordination geometry**. The most common shapes of coordination compounds are octahedral and tetrahedral, corresponding to coordination numbers of 6 and 4 respectively.

Table 12.6 The shapes of coordination compounds

Shape		Coordination number	Examples
Linear	$L — M — L$	2	$[Ag(NH_3)_2]^+$, $[Ag(S_2O_3)_2]^{3-}$, $Co[N\{Si(CH_3)_3\}_2]_2$
Trigonal planar (very rare)	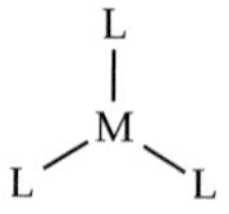	3	$Fe[N\{Si(CH_3)_3\}_2]_3$
Tetrahedral	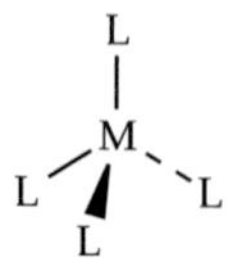	4	$[CoCl_4]^{2-}$, $Ni(CO)_4$, CrO_4^{2-}, MnO_4^-

Note

py = pyridine

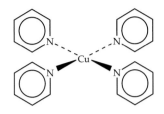

The structure of [Cu(py)₄]²⁺ is $[Cu(py)_4]^{2+}$

acac = pentane-2,4-dione (trivial name, acetylacetone)

The structure of VO(acac)₂ is $VO(acac)_2$

Table 12.6 *continued*

Square planar		4	$[Ni(CN)_4]^{2-}$, $[Cu(py)_4]^{2+}$, $[AuCl_4]^-$, $Pt(NH_3)_2Cl_2$
Trigonal bipyramidal (rare)		5	$VCl_3.2(CH_3)_3N$
Square-based pyramidal (rare)		5	$VO(acac)_2$
Octahedral		6	$[M(H_2O)_6]^{x+}$, $[TiCl_6]^{2-}$, $[Fe(C_2O_4)_3]^{3-}$

HAEMOGLOBIN AND OXYGEN TRANSPORT

Haemoglobin is a molecule vital to our life. It carries oxygen around the body, in the bloodstream, to those places where it is needed for the oxidation reactions which produce energy. The haemoglobin molecule is complex but an essential feature is the presence of four haem groups with iron(II) ions at their centre. Each haem group acts as a tetradentate (four points of attachment) ligand, forming a four-coordinate, planar complex with an iron(II) ion. An oxygen molecule can bond to the iron(II) ion in a position above the plane of the chelate ring. In this way oxygen is transported around the body.

Carbon monoxide is a poisonous gas because it competes with oxygen for the coordination sites in haemoglobin. Carbon monoxide bonds to haemoglobin in the place of oxygen and, therefore, the body becomes 'starved' of oxygen.

Fig 12.33 (a) The photograph shows a sample of normal blood. The change in coordination when carbon monoxide occupies the position normally taken by an oxygen molecule is accompanied by a change in colour and a blood sample would be a much brighter pink. (b) The molecular structure of the haem group.

12.22 Name the following complex ions: $[Cu(H_2O)_6]^{2+}$, $[Cu(NH_3)_6]^{2+}$, $[CoCl_4]^{2-}$, $[MnO_4]^-$, $[MnO_4]^{2-}$, $[Cu(NH_3)_4(H_2O)_2]^{2+}$, $[Ag(NH_3)_2]^+$, $[Al(OH)_4]^-$, $[CuCl_2]^-$.

12.23 Sketch diagrams to represent the shapes of the following complex ions: $[Ni(CN)_4]^{2-}$, $[CoCl_4]^{2-}$, $[Fe(H_2O)_6]^{2+}$, $[Fe(H_2O)_6]^{3+}$, $[MnO_4]^-$, $[Ag(NH_3)_2]^+$, $[CuCl_2]^-$.

12.24 Sodium hexacyanoferrate(II) is often added to domestic salt to improve its pouring qualities. Sketch a diagram to represent the shape of the hexacyanoferrate(II) ion.

12.25 The extraction of gold involves a chemical process to separate it from other metals. The native gold is oxidised by air in the presence of an aqueous solution of cyanide ions. Gold(I) ions combine with cyanide ions to give a complex ion with a coordination number of two. The gold is recovered by treating the aqueous solution of this complex ion with zinc.
   (a) Complete the equation for the oxidation reaction of gold:
   $$4Au(s) + 8CN^-(aq) + O_2(g) + 2H_2O(l) \rightarrow$$
   (b) Name the complex ion formed between gold(I) ions and cyanide ions and suggest a probable shape for it.
   (c) Zinc(II) forms tetracyanozincate(II) ions with cyanide ions. Write an equation for this reaction and explain, in terms of oxidation numbers, what has been oxidised and what has been reduced. Sketch a diagram to show the shape of this tetrahedral complex ion.

12.26 Explain why *trans*-1,2-dichloroethene is non-polar but *cis*-1,2-dichloroethene is polar. How does this account for the difference in boiling points?

12.27 Look at the structures of aminoacids (Fig 11.3). Which is the only amino acid which is not chiral?

12.5 ISOMERISM

CONNECTION: Structural isomers, page 96.

Different arrangements of carbon atoms in molecules with the same molecular formula give rise to structural isomerism. Structural isomers have the same molecular formula but different structural formulae. Other types of isomerism also exist and are important in determining the reactivity of molecules. Whatever the type of isomerism, each isomer can be isolated and purified, and has its own characteristic properties.

Structural isomerism

Organic compounds: structural isomerism of carbon skeletons

Structural isomers have the same molecular formula but different structural formulae. For example, three structural isomers have the molecular formula C_5H_{12}. Their structural formulae are:

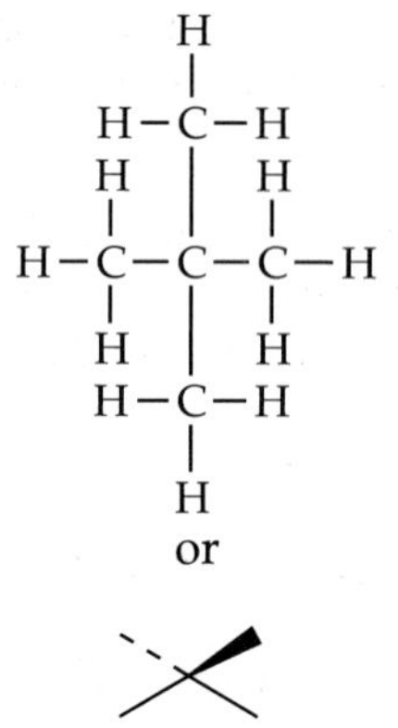

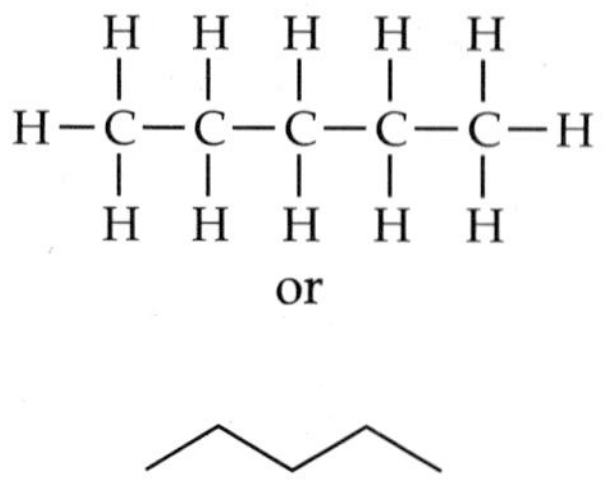

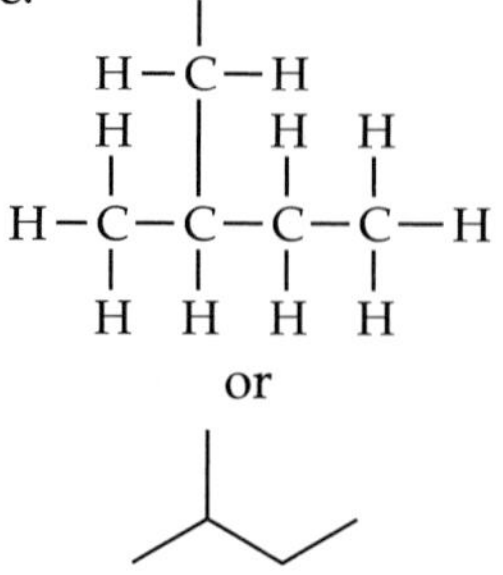

Four structural isomers have the molecular formula C_4H_9OH. Their structural formulae are:

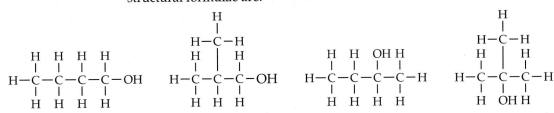

22 Which of the isomers of C_4H_9OH are chemically the most similar?

23 Use The Database to decide how butan-1-ol might be distinguished from 2-methylpropan-2-ol.

Four structural isomers have the molecular formula C_8H_{10}. Their structural formulae are:

CH₂CH₃ ... CH₃ CH₃ ... CH₃ ... CH₃ ... CH₃ ... CH₃

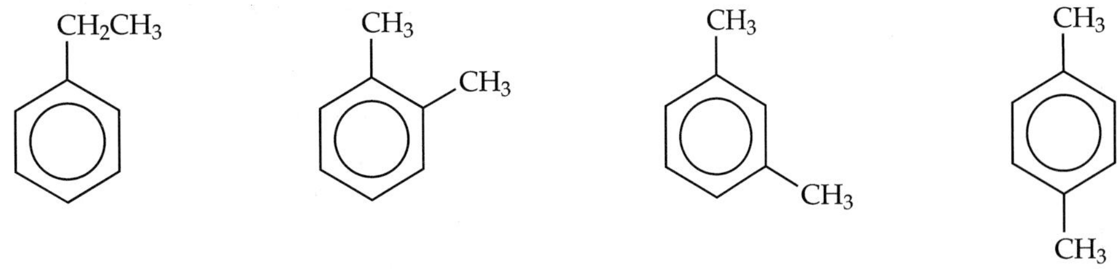

Within each of the sets of structural isomers described above, the compounds have similar, though slightly different, physical properties (such as melting point and boiling point) and undergo very similar chemical reactions. This is because the functional group remains the same in each set.

However, changing the sequence of atoms in a molecule can result in more striking changes in properties. This occurs when the same atoms are linked in different ways so that different functional groups are present. For example, two compounds have the molecular formula C_2H_6O, but one is a primary alcohol and the other is an ether. The structural formulae are:

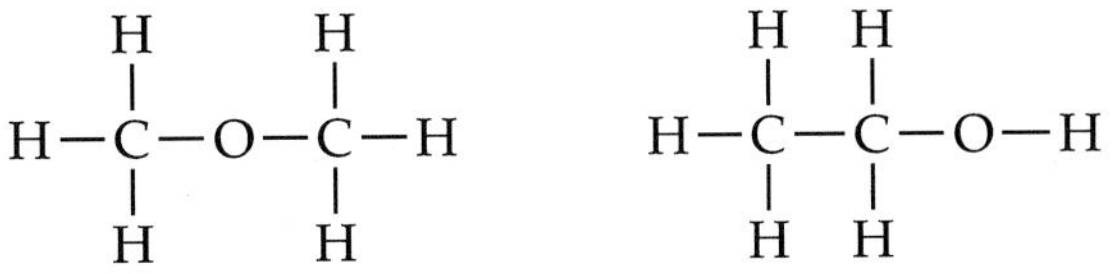

These isomers have different physical properties. Methoxymethane has a boiling point of −24.8°C but ethanol has a boiling point of 78.5°C. The reason is that methoxymethane molecules are attracted to one another by Van der Waals' forces only. In contrast, hydrogen-bonds form between ethanol molecules and these are stronger than Van der Waals' forces. The two compounds also undergo different chemical reactions.

24 Only one of these isomers reacts with carboxylic acids. Which one is it and what is the class of compound formed? Use the database if necessary.

Another example of structural isomers with different functional groups is two compounds which share the molecular formula C_3H_6O. One is a ketone, the other is an aldehyde. Their structural formulae are:

KEY FACT

When deciding upon possible structural formulae for a given molecular formula, a high C:H ratio is a strong clue to the presence of a benzene ring in the structure.

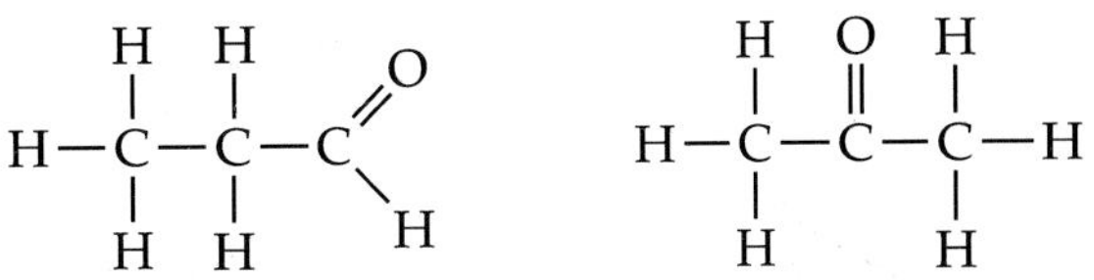

Again, these isomers have different chemical properties to one another because they contain different functional groups. For example, propanal is easily oxidised to a carboxylic acid whereas propanone is quite resistant to oxidation.

25 Select a chemical reaction (using the Database if necessary) that could be used to identify a colourless liquid as either propanal or propanone.

Structural isomerism in coordination compounds

Structural isomerism is not restricted to organic molecules, it is also common in coordination compounds. Four compounds exist with the formula $CoCl_3.6NH_3$, each with its own characteristic colour. They each differ in the position of the chloride ions and ammonia molecules within the compound. They may be in the lattice or coordinately bonded to the central metal ion. The structural isomers are summarised in Table 12.7.

Table 12.7 Different structural isomers with the molecular formula $CoCl_3.6NH_3$.

Structural formula	Colour
$[Co(NH_3)_6]^{3+}$ $3Cl^-$	golden-brown
$[Co(NH_3)_5Cl]^{2+}$ $2Cl^-.NH_3$	purple
$[Co(NH_3)_4Cl_2]^+$ $Cl^-.2NH_3$	two forms: (a) deep violet, (b) green

A similar example is provided by $CrCl_3.6H_2O$. The structural isomers are summarised in Table 12.8.

Table 12.8 Different structural isomers with the molecular formula $CrCl_3.6H_2O$

Structural formula	Colour
$[Cr(H_2O)_6]^{3+}$ $3Cl^-$	violet
$[Cr(H_2O)_5Cl]^{2+}$ $2Cl^-.H_2O$	light green
$[Cr(H_2O)_4Cl_2]^+$ $Cl^-.2H_2O$	dark green

Fig 12.34 The three crystalline compounds which share the molecular formula $CrCl_3.6H_2O$.

The number of chloride ions coordinately bonded in the cobalt(III) and chromium(III) complex ions can be determined by quantitative analysis using silver nitrate solution. When the compounds are dissolved in water, chloride ions not coordinately bonded to the central metal ions enter solution as simple hydrated ions. They react with silver ions to give a precipitate of silver chloride:

$$Ag^+(aq) + Cl^-(aq) \rightarrow AgCl(s)$$

$$[Co(NH_3)_5Cl]^{2+}2Cl^-.NH_3(s) + aq \rightarrow [Co(NH_3)_5Cl]^{2+}(aq) + 2Cl^-(aq) + NH_3(aq)$$

the coordinately bonded Cl^- does not precipitate with $Ag^+(aq)$

the hydrated Cl^- precipitates with $Ag^+(aq)$

BONDING, STRUCTURE AND SHAPE

KEY FACT

Isomers have the same molecular formula but different arrangements of atoms. There are two types of isomerism: **structural isomerism** and **stereoisomerism**.

In **structural isomers**, the atoms join together in a different sequence.

In **stereoisomers**, the atoms are joined in the same sequence but are arranged differently in space. There are two types of stereoisomerism: **geometrical isomerism** and **optical isomerism**.

The chloride ions which are coordinately bonded remain firmly attached to the metal ion even in aqueous solution and do not precipitate as silver chloride when silver nitrate solution is added.

26 If one mole of the violet isomer of $CrCl_3.6H_2O$ were dissolved in water and excess silver nitrate added, how many moles of silver chloride would precipitate?

Looking at Tables 12.7 and 12.8, we see that four isomers of $CoCl_3.6NH_3$ are described but only three of $CrCl_3.6H_2O$. We will explain this shortly by introducing another form of isomerism, geometrical isomerism.

Geometrical isomerism

Geometrical isomers have the same molecular formulae and the same sequencing of atoms but each isomer has a different geometrical arrangement.

Geometrical isomerism in organic compounds

1,2-dichloroethene, $C_2H_2Cl_2$, can exist in two forms, each with its own characteristic physical properties but with the same structural formula, $CHCl{=}CHCl$. Each carbon atom is in a trigonal planar environment giving a molecule which itself is flat. The carbon atoms are linked by a double covalent bond which is not free to rotate. Because of this, two geometrical arrangements are possible:

trans isomer
$$\begin{array}{c} Cl \\ \diagdown \\ \end{array} C{=}C \begin{array}{c} H \\ \diagup \\ \end{array}$$
$$\begin{array}{c} \diagup \\ H \end{array} \qquad \begin{array}{c} \diagdown \\ Cl \end{array}$$

cis isomer
$$\begin{array}{c} Cl \\ \diagdown \\ \end{array} C{=}C \begin{array}{c} Cl \\ \diagup \\ \end{array}$$
$$\begin{array}{c} \diagup \\ H \end{array} \qquad \begin{array}{c} \diagdown \\ H \end{array}$$

Similarly, two geometrical isomers of 1,2-dibromoethene exist. The *trans* isomer is non-polar, with a boiling point of 108°C and a melting point of –7°C; the *cis* isomer is polar, with a boiling point of 110°C and a melting point of –53°C.

27 Draw the displayed formula for a structural isomer of 1,2-dibromoethene and predict its boiling point and melting point in relation to the *trans* and *cis* isomers.

KEY FACT

When the two chlorine atoms are on opposite sides of the molecule we call this the *trans* isomer, and when they are on the same side of the molecule we call this the *cis* isomer.

SPOTLIGHT

HARD AND SOFT MARGARINES

The major fatty acid (a trivial name for high relative molecular mass carboxylic acids) found in many vegetable oils, for example cottonseed and corn oil, is linoleic acid, $C_{18}H_{32}O_2$.

Triglycerides are esters of fatty acids and propane-1,2,3-triol, $CH_2OHCH(OH)CH_2OH$. They are oils. Flexing of the molecules is restricted by the $C{=}C$ double bonds and this limits how closely the molecules can pack together. To convert these oils into margarines which can be spread on bread, for example, the manufacturers bubble hydrogen through the oil in the presence of a finely divided nickel catalyst (very small particles). Hydrogenation occurs, changing the $C{=}C$ double bonds into single bonds. The result is loss in rigidity of the molecules. They can flex and twist more easily and can pack together more closely. The intermolecular bonds are stronger, therefore, and the oil is converted into a fat. Hard margarines result from extensive hydrogenation, while soft margarines are made by limiting the extent of hydrogenation.

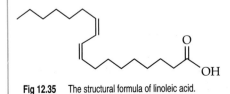

Fig 12.35 The structural formula of linoleic acid.

Sometimes the different geometrical arrangement of atoms results in different chemical as well as physical properties. An example of this is the geometrical isomerism displayed by butenedioic acid, HOOC—CH=CH—COOH. The *trans* and *cis* isomers have different physical properties as would be expected. However, they also behave differently when heated. The *cis* isomer loses water when heated to 160°C, to form an acid anhydride. The *trans* isomer, on the other hand, sublimes at 200°C and does not dehydrate (in fact it converts to the *cis* form at about 250–300°C, indicating that the double bond is partially disrupted and then allows free rotation).

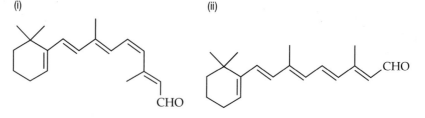

sublimes at 200°C   (very small amount)

GEOMETRICAL ISOMERISM AND VISION

In the rods and cones of the eye's retina, light is absorbed by 11-*cis*-retinal. The alternating single and double bonds of the hydrocarbon chain allow electrons to shift position in the molecule by absorbtion of a small amount of energy. The absorbed light provides sufficient energy for one of the two pairs of electrons in the *cis* double bond to be pulled apart. In consequence the resulting C—C single bond is free to rotate and the molecule suddenly switches to the *trans* form. The split electron pair re-forms to hold the molecule in this new shape. A message is sent to the brain via the optic nerve in response to this dramatic change in shape of the retinal molecule. Once the message has been received the body converts retinal from the *trans* back to the *cis* form, ready to receive and send another message that light has entered the eye.

Fig 12.36 (a) Geometrical isomerism allows the eye to send messages to the brain when light images have been received by the retina. (b) Structural formulae of (i) 11-*cis*-retinal and (ii) 11-*trans*-retinal.

Geometrical isomerism in coordination compounds

Coordination compounds also provide examples of geometrical isomerism. Two compounds exist with the molecular formula $Pt(NH_3)_2Cl_2$. Each has a central platinum(II) ion surrounded by four ligands, two ammonia molecules and two chloride ions.

28 **How do we know that $Pt(NH_3)_2Cl_2$ is not tetrahedral?**

BONDING, STRUCTURE AND SHAPE

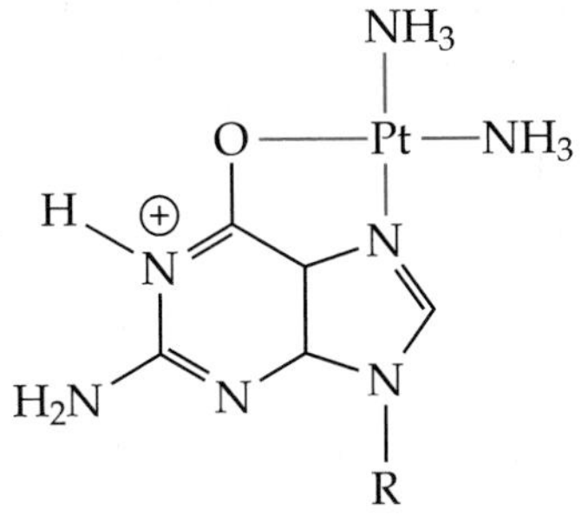

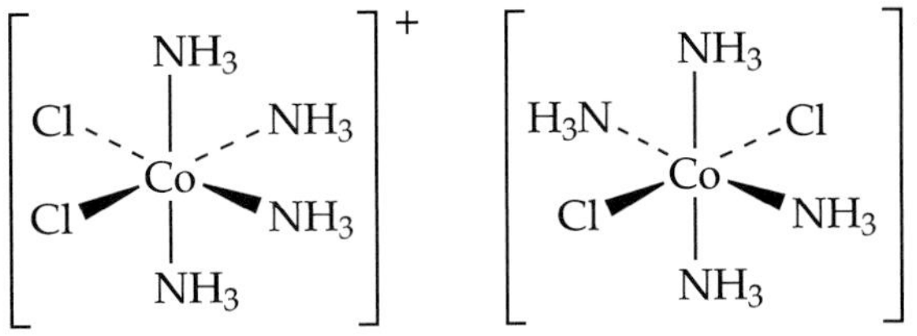

Fig 12.37 One suggestion for the interaction of cisplatin with a section of the DNA molecule.

Both compounds are square planar coordination compounds and so two geometrical isomers exist.

The point of interest is that only the *cis* isomer shows anti-cancer activity. The *trans* compound is inactive. An understanding of ligand exchange reactions and the geometry of coordination compounds can help us to explain this, and is essential in the search for making new and improved platinum anti-cancer drugs. Although the exact mechanism is not known, it has been shown clearly that the platinum(II) ion bonds to sites on the DNA molecule. The two chlorines are replaced by donor atoms from the DNA (one possible way is given in Fig 12.37). This prevents the DNA from replicating. It is clear that in the *trans* compound the two coordination sites occupied by chloride ions are not the correct shape to bond to a section of the DNA molecule. Indeed, a pre-requisite of the second generation anti-cancer platinum compounds is the presence of two *cis* chloride ions bound to the platinum.

Geometrical isomerism is also shown by coordination compounds of general formula Ma_4b_2. We noted earlier that $[Co(NH_3)_4Cl_2]^+Cl^-.2NH_3$ exists in two forms, one which is deep-violet and another which is green. These are *cis* and *trans* isomers:

CONNECTION: Cisplatin, page 206.

We also met the compound $[Cr(H_2O)_4Cl_2]^+Cl^-.2H_2O$ earlier but this time only a single colour was reported for it. The dark green compound is, in fact, a mixture of the *cis* and *trans* isomers.

29 **Sketch diagrams of the *cis* and *trans* isomers of $[Cr(H_2O)_4Cl_2]^+$.**

Optical isomerism

Some molecules can exist as non-superimposable mirror images of one another. Such a pair are called **optical isomers** (sometimes called **enantiomers**). A compound which is optically active and displays optical isomerism is said to be **chiral**. Optical isomers have identical physical properties, except for their effect on plane polarised light. They have identical chemical properties except for their reactions with other optically active compounds.

KEY FACT

Light is a wave which vibrates at right angles to its path through space. This vibration can be limited to a single plane, for example by passing light through a polarising medium such as polarised sun glasses.

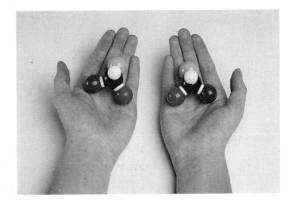

Fig 12.38 A pair of hands demonstrate what we mean by 'non-superimposable' mirror images. The right hand is a mirror image of the left hand. However, a transplanted left hand could not take the place of a right hand. We sometimes call molecules which are non-superimposable mirror images of one another, 'left-handed' and 'right-handed' molecules of the same compound. In this photograph the hands are holding models of such compounds

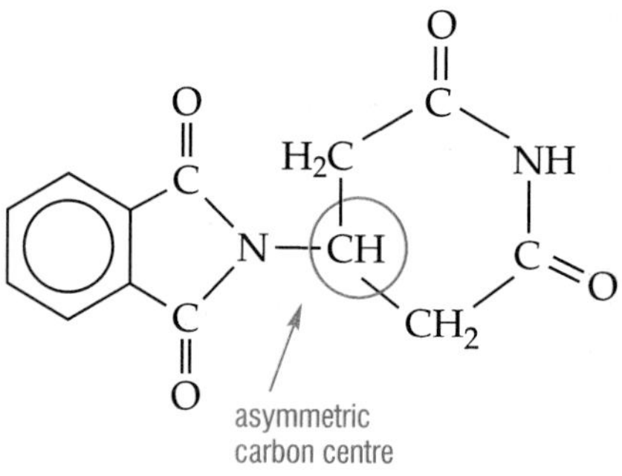

Fig 12.39   The Thalidomide molecule.

asymmetric
carbon centre

Optical isomerism in organic compounds

The minor differences between optical isomers of organic compounds become of vital importance when it comes to reactions with other optically active molecules. This is common in biological reactions. A tragic consequence of this difference was observed during the limited use of the drug Thalidomide.

During the 1960s the use of Thalidomide as a non-addictive sedative for pregnant women led to birth of thousands of badly deformed babies. First prepared in 1953, it was marketed in 1958 without full and extensive trials. The drug was banned in 1961. One optical isomer is an effective sedative which has negligible side-effects. The other affected the growth of the foetus and caused terrible deformities. At the time the two optical isomers were not separated. Had they been, and each tested separately, Thalidomide might still be around as a useful drug for pregnant women. Nowadays the optical isomers of any pharmaceutical compound or agrochemical which is chiral are isolated and tested separately.

Let us look at the Thalidomide molecule (Fig 12.39). Although a fairly complicated molecule with several functional groups, it has one very important feature. Highlighted in Fig 12.39 is an **asymmetric carbon centre**, that is a carbon atom with four different groups attached to it. Because of this the molecule can exist as two non-superimposable mirror images, the two optical isomers. It is a little difficult to see this in a molecule of some complexity and, therefore, it will be helpful to consider something simpler.

Fig 12.40   The idea of mirror image can be seen by comparing one optical isomer of butan-2-ol with the image of the other isomer in a mirror.

Butan-2-ol is a secondary alcohol. It has an asymmetric carbon centre and can exist as two optical isomers:

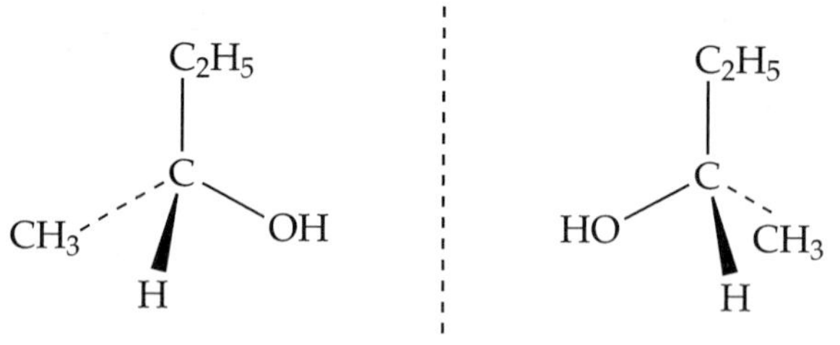

C_2H_5 ... C ... CH_3 — OH — H

(+) isomer   mirror   (−) isomer

30   Why does butan-1-ol not exhibit optical isomerism?

An equimolar mixture of (+) and (−)isomers is called a **racemate** (or sometimes, a **racemic mixture**). It does not rotate plane-polarised light since the effect of the two isomers cancel one another.

Fig 12.41 Chiral compounds are very common in nature. Two examples are found in the food that we eat. Lactic acid (or, more correctly, 2-hydroxy-propanoic acid) is found in foods such as savoury biscuits. Monosodium glutamate (MSG) is a flavour enhancer, very popular in chinese cooking (large quantities can cause sweating and nausea, a complaint known as 'chinese restaurant syndrome').

31 Sketch diagrams of the following molecules and indicate the asymmetric carbon centres: 1-chloroethanol, 2-methylbutanenitrile, 1,2-dichloro-1-fluoroethane.

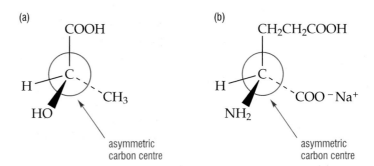

Fig 12.42 (a) 2-Hydroxypropanoic acid (b) Monosodium glutamate

Optical isomerism in coordination compounds

Complex ions also provide examples of optical isomerism. The only criterion, again, is that a molecule or complex ion can exist as two non-superimposable mirror images. A simple example will illustrate. Let us compare $[Ni(NH_3)_6]^{2+}$ with $[Ni(NH_2CH_2CH_2NH_2)_3]^{2+}$. The ammine complex cannot exist in two forms but the 1,2-diaminoethane complex does (Fig 12.43).

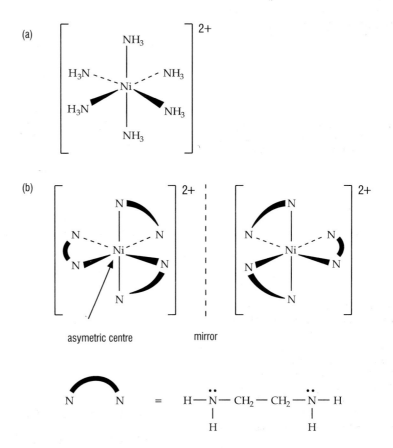

Fig 12.43 (a) $[Ni(NH_3)_6]^{2+}$ (b) two non-superimposable mirror images of $[Ni(NH_2CH_2CH_2NH_2)_3]^{2+}$

32 Which of the geometrical isomers, *cis*-$[Ni(NH_2CH_2CH_2NH_2)_2Cl_2]$ or *trans*-$[Ni(NH_2CH_2CH_2NH_2)_2Cl_2]$, also displays optical isomerism?

12.28 Two esters have the same molecular formula, $C_3H_6O_2$. Write structural formulae for the two esters and name them. Use the Database to write equations for the hydrolysis of each ester, naming the reaction products.

12.29 Write structural formulae for compounds with the molecular formula C_9H_{12} and name them.

12.30 Suggest structural formulae for isomers with the molecular formula $CrCl_3.3NH_2CH_2CH_2NH_2$.

12.31 Write down the structural formulae of all compounds with the molecular formula C_5H_{10}.

12.32 Explain why two forms of 1,2-dichlorohexane exist but only one form of 1,2-dichlorobenzene exists.

12.33 Look at the reaction scheme in Fig 12.44 and describe how *trans* diamminedichloroplatinum(II) could be made in the laboratory.

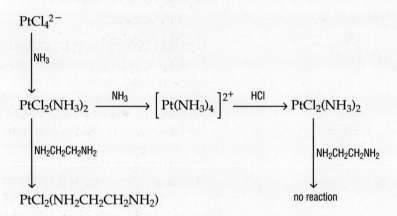

Fig 12.44

Answers to Quick Questions: Chapter 12

1 Because the nucleus has a positive charge and so the electron density is held close by the electrostatic attraction.

2 Li^+ $1s^2$
S^{2-} $1s^2 2s^2 2p^6 3s^2 3p^6$
F^- $1s^2 2s^2 2p^6$
Ca^{2+} $1s^2 2s^2 2p^6 3s^2 3p^6$
N^{3-} $1s^2 2s^2 2p^6$

3 Because five electrons would have to be available for bonding. However, only electrons in the outermost main energy level can be used in bonding and there are only four of these (the other two electrons are in the 1s atomic orbital and so not available for bonding).

4 Because the energy required to promote an electron from the 3s atomic orbital into a vacant 3d atomic orbital is not re-paid by the formation of two extra, relatively weak P–I bonds.

5 Predominately ionic but with some covalent character.

6

$$
\begin{array}{ccc}
OH & OH & OH \\
| & | & | \\
H-C-C-C-H \\
| & | & | \\
H & H & H
\end{array}
$$

Hydroxyl group (the compound is a polyhydric alcohol).

Fig 12.44

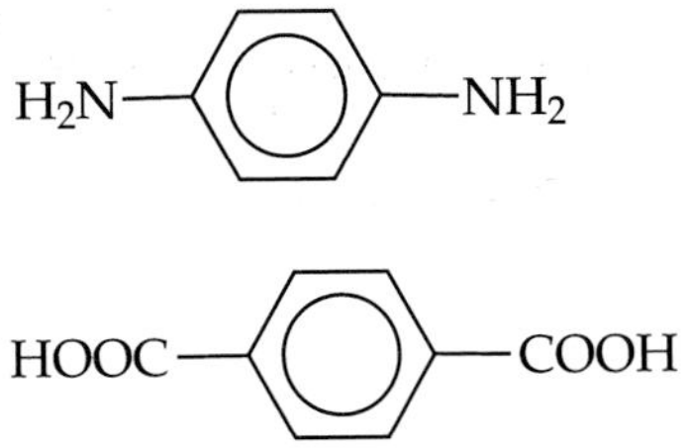

H_2N—(ring)—NH_2

$HOOC$—(ring)—$COOH$

Fig 12.45

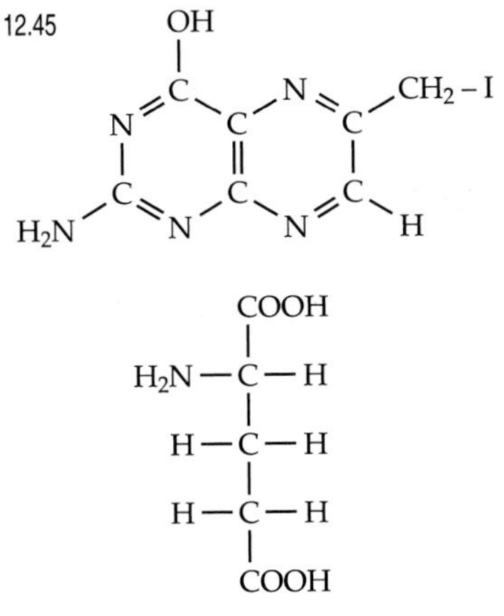

(structure with OH, N, C, CH₂–I, C, H_2N, N, N, C, H)

COOH
|
H_2N—C—H
|
H—C—H
|
H—C—H
|
COOH

Fig 12.46

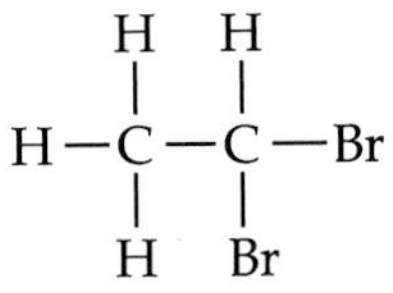

```
     H   H
     |   |
 H — C — C — Br
     |   |
     H   Br
```

Fig 12.47

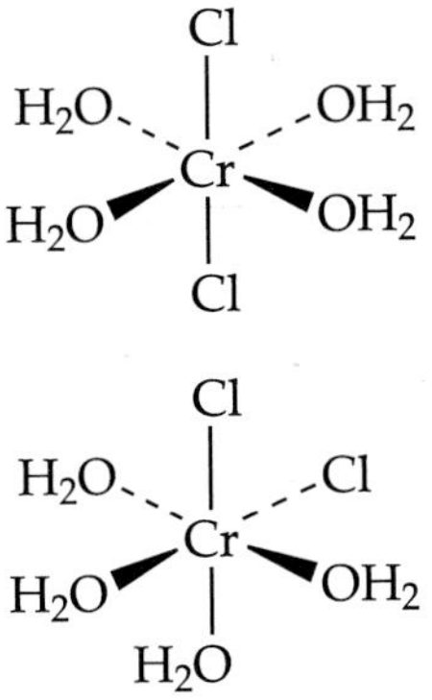

```
         Cl
         |
 H2O .   |   . OH2
      ` Cr ´
 H2O ´   |   ` OH2
         |
         Cl
```

```
         Cl
         |
 H2O .   |   . Cl
      ` Cr ´
 H2O ´   |   ` OH2
         |
         H2O
```

Fig 12.48

```
     H   Cl
     |   |
 H — C — C — OH
     |   |
     H   H
```

```
     H   H   CH3
     |   |   |
 H — C — C — C — CN
     |   |   |
     H   H   H
```

```
     Cl  Cl
     |   |
 H — C — C — H
     |   |
     F   H
```

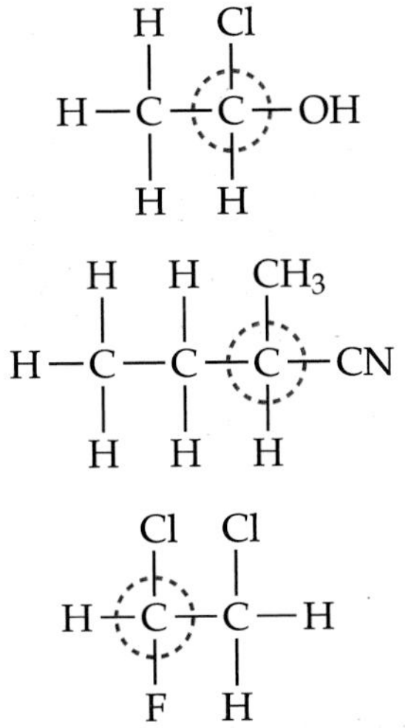

7 The hydrocarbon tail in sodium alkylbenzene sulphonate is branched, but in sodium octadecanoate is unbranched.

8
```
     H   H   H   H   H   H   H   H   H   H        O
     |   |   |   |   |   |   |   |   |   |        ‖
 H — C — C — C — C — C — C — C — C — C — C — O — S — O⁻   Na⁺
     |   |   |   |   |   |   |   |   |   |        ‖
     H   H   H   H   H   H   H   H   H   H        O
```

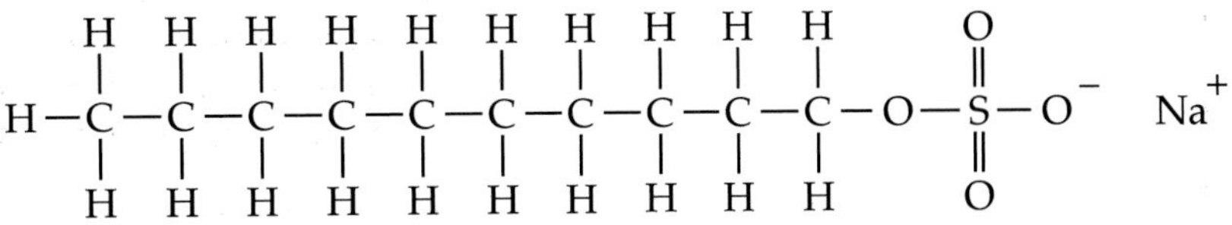

Sodium decyl sulphate

9 Both have a relatively insoluble ion together with a counter ion which can exchange easily with other suitable ions.

10 Because in both compounds, we believe that three electrons from each carbon atom is used for normal covalent bonds formation, while the remaining electron is delocalised.

11 (Fig 12.44)

12 The delocalisation of electrons mean that the backbone is not free to rotate and, therefore, not flexible.

13 Because a caesium ion has a greater ionic radius than a sodium ion.

14 (Fig 12.45)

15 Bacteriostatic.

16 (a) Phosphine. H—P—H bond angle slightly less than 107° because phosphorus is less electronegative than nitrogen and so will not attract the pairs of electrons which surround it as much. Therefore, repulsion will be less.

(b) Nitrogen trifluoride. F—N—F bond angle slightly less than 107° because fluorine is more electronegative than nitrogen and so electron density is pulled away from the central nitrogen atom. Therefore, repulsion will be less.

17 All about 109.5°.

18 In poly(ethene), polymer molecules can adopt numerous non-linear conformations. Pulling a sheet initially 'straightens out' these chains. In aluminium, the atoms are held tightly in a regular, repeating three-dimensional lattice and cannot be easily distorted.

19 Because each 1,2-diaminoethane molecule contains two nitrogen atoms, each with a lone pair of electrons which can be used to form coordinate bonds to a central metal ion.

20 Through the lone pairs of electrons on the oxygen atoms.

21 SnF_4^{2-}

22 Butan-1-ol and 2-methylpropan-2-ol

23 Butan-1-ol is easily oxidised and so will react with a warm acidified solution of dichromate(VI) ions (orange) to give hydrated chromium(III) ions (green). 2-Methylpropan-2-ol will not react with a warm acidified solution of dichromate(VI) ions.

24 Ethanol. Primary alcohol

25 Reaction with Fehling's solution or with a solution containing the diamminesilver(I) ion.

26 Three moles

27 (Fig 12.46)
Higher boiling and melting points because it is a more polar molecule, since both bromine atoms are on the same 'end' of the molecule.

28 Because it can exist as two isomers.

29 (Fig 12.47)

30 Because it cannot exist in two forms which are non-superimposable mirror images. It does not contain an asymmetric carbon centre.

31 (Fig 12.48)

32 cis-[Ni(NH₂CH₂CH₂NH₂)₂Cl₂]

Chapter 13

MECHANISMS OF CHEMICAL CHANGE

LEARNING OBJECTIVES

When you have studied this chapter you should be able to:

1. explain the importance of using models to study physical and chemical changes;

2. explain the terms 'homolytic bond fission' and 'heterolytic bond fission';

3. draw reaction profiles for reactions of known mechanism and explain the terms 'reaction intermediate' and 'transition state';

4. use 'curly arrows' to show electron rearrangements when molecules react;

5. with reference to chemical reactions, explain the terms 'addition', 'substitution', 'elimination' and 'condensation';

6. explain the terms 'electrophile' and 'nucleophile';

7. describe the mechanisms of free-radical addition, nucleophilic addition, electrophilic addition, free-radical substitution, nucleophilic substitution, electrophilic substitution;

8. explain free-radical and electrophilic addition to alkenes;

9. explain nucleophilic addition to carbonyl compounds;

10. explain the formation of addition compounds;

11. explain electrophilic substitution of benzene;

12. explain nucleophilic substitution of halogenoalkanes and alcohols;

13. explain the hydrolysis of covalent chlorides in terms of nucleophilic substitution;

14. explain ligand exchange reactions of coordination compounds in terms of nucleophilic substitution.

13.1 REACTION MECHANISMS

KEY FACT

The model used to describe how a reaction takes place, the way in which bonds are broken and formed, is called the **reaction mechanism**. Whilst this may *sometimes* be predicted for a reaction it can only be confirmed by experiment.

All chemical reactions involve bonds being broken and new ones being formed. The ease or difficulty with which this happens determines the chemical reactivity of a material. An understanding of the principles which affect reactivity is as important to the chemical architect and builder as a knowledge of how to make and characterise the material.

The stability (or instability) of compounds in different environments often determines their suitability for a particular job. We need to know and understand how materials such as fibres, paints, plastics, metals and so on behave in water, oxygen, light and with changes in temperature. The design of biodegradable plastics and detergents depends on an understanding of environmental effects, as does the protection of iron against rusting. In medicine, we need to explain or predict the behaviour of a drug in the environment of the human body. In agriculture, we seek to understand the behaviour of an insecticide in relation to a plant's environment and its action in insect control.

Fig 13.1 We live in a wet and strongly oxidising atmosphere. Materials which remain outside for long periods of time must be resistant to corrosion and to large fluctuations in temperature, from the very low temperatures of winter to much higher summer temperatures.

This chapter explains how covalent bonds in molecular compounds and in compounds with giant structures are broken and formed.

Homolytic and heterolytic bond fission

Our model for a covalent bond is that two atoms share electrons. They are held together by the mutual attraction of the atoms' nuclei for the shared electrons. If both atoms have the same attraction for the electrons as one another, we have a non-polar bond. If one atom has a stronger attraction for the shared electrons than the other (in other words, is more electronegative) then the bond is polar.

Consider a single covalent bond in a molecule. A pair of electrons is shared in a single covalent bond. When the molecule is given sufficient energy, the bond will vibrate with increasing vigour. Eventually the two atoms will break free from one another.

If each atom takes with it a single electron, we call this **homolytic bond fission** and it results in the formation of **free radicals**. The unpaired electron present in a free-radical explains the highly reactive nature of free radicals.

1 **Why are multiple bonds unlikely to undergo bond fission?**

Alternatively, the two shared electrons might remain associated with only one of the atoms when the bond breaks, resulting in the formation of a cation and an anion. We call this **heterolytic bond fission**.

Through our understanding of the nature of covalent bonding, in particular bond polarisation, it is not difficult to predict the conditions which favour each type of bond break. If the two atoms in the bond have a similar attraction for the shared electrons, that is to say there is a very small electronegativity difference between them, it is likely that homolytic bond fission will occur. It is also likely that this will occur in a non-polar solvent. In contrast, a polar bond linking two atoms of different electronegativities is more likely to undergo heterolytic bond fission since the electron density is already distorted. It is more likely this will happen in a polar solvent. The solvent solvates (surrounds and bonds to) any charged particles which form, helping to stabilise them.

CONNECTION: Free radicals and CFCs, page 39.

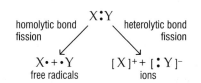

2 What type of bond fission is probable in each of the following molecules: Br_2, HBr, H_2O, CCl_4?

Reaction profiles and intermediate compounds

Energy is required to break bonds and the formation of bonds releases energy. If the formation of new bonds gives out more energy than was required to break the bonds then the reaction is exothermic. If the formation of bonds gives out less energy than that required to break bonds, then it is endothermic.

CONNECTION: *Exothermic and endothermic reactions, page 79.*

We often describe the route of a chemical reaction and the associated energy changes by the use of **reaction profiles**. We represent the relative energies of reactants, products and any intermediate compounds (formed during the course of the reaction) on the reaction profile. The number of intermediate compounds formed determines the number of steps there are in a chemical reaction. If no intermediate forms the reaction is one-step, if one intermediate forms it is two-step, if two intermediates form it is three-step and so on.

Fig 13.2 The reaction profiles for (a) a one-step exothermic reaction (b) a two-step exothermic reaction.

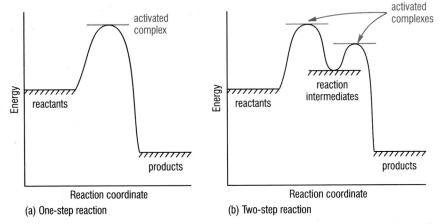

(a) One-step reaction

(b) Two-step reaction

The oxidation of 2,3-dihydroxybutanedioate ions by hydrogen peroxide is catalysed by Co^{2+}. A study of this reaction gives us an example of an intermediate compound. Solutions containing the hexaaquacobalt(II) ion, $[Co(H_2O)_6]^{2+}$, are pink. Hydrogen peroxide oxidises Co^{2+} to Co^{3+}. Although Co^{3+} is not stable in aqueous solution, 2,3-dihydroxybutanedioate ions form a cobalt(III) complex and stabilise the oxidation state. The complex is an intermediate compound (Fig 13.3), observed as a green colour. The coordinated 2,3-dihydroxybutanedioate ions are more easily oxidised to carbon dioxide than those which are not coordinated. As a coordinated ion is oxidised, so another takes its place and is itself then oxidised. This sequence continues until all the 2,3-dihydroxybutanedioate ions have been oxidised. At the end of the reaction cobalt returns to the +2 oxidation state, as the +3 oxidation state can no longer be stabilised. The solution becomes pink again (Fig 13.4).

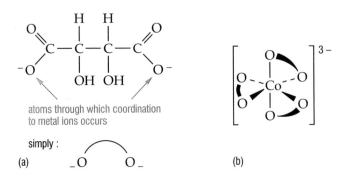

atoms through which coordination to metal ions occurs

simply :

(a)

(b)

Fig 13.3 (a) The 2,3-dihydroxybutanedioate ion;
(b) Structure of the cobalt(III) complex ion.

Fig 13.4 2,3-dihydroxybutanedioate ions are oxidised by hydrogen peroxide. The reaction is catalysed by Co^{2+}. An intermediate compound (a cobalt(III) complex) may be identified by its green colour, which disappears when the reaction is complete.

An **activated complex** is formed for each step in a reaction. This must not be confused with an intermediate compound, which, however short its lifetime, can be isolated and identified. An activated complex, on the other hand, is a species that we speculate might form – a speculation based on experimental evidence and our theories of structure and bonding. It represents our mental image of how reacting species may come together during reaction.

3 How many activated complexes are formed during a two-step reaction?

Representing the movement of electrons

Breaking and forming covalent bonds involves rearranging the electrons in the reactants to give products. To help us follow this rearrangement and to represent the mechanism of change we will need a little more chemical shorthand.

Homolytic bond fission occurs when the shared electrons of a covalent bond separate so that one goes with each of the atoms held by the bond. The convention used to represent the movement of a single electron is a 'curly arrow' with half an arrow head, often called a single-headed arrow:

$$Cl—Cl \longrightarrow Cl\bullet + Cl\bullet$$

Heterolytic bond fission occurs when the shared electrons move together towards just one of the atoms in the bond. The movement of an electron pair is represented by a 'curly arrow' with a full arrow head:

$$H—Cl \longrightarrow [H]^+ + [:Cl]^-$$

An example may be helpful. Pure nitric acid is a molecular compound. It dissolves in water to give an acidic solution and the process is strongly exothermic. The nitric acid ionises, with a proton, H^+, being transferred from the nitric acid molecule to a water molecule. The rearrangement of electrons can be represented using curly-arrows (Fig 13.5). A lone pair of electrons on the oxygen atom of the water molecule is used to form a coordinate bond with a proton captured from the nitric acid. The pair of electrons which was being shared in the O—H bond in HNO_3 is now associated only with the oxygen atom in the resulting nitrate ion.

4 Explain why the oxonium ion, H_3O^+, carries a positive charge and the nitrate ion, NO_3^-, carries a negative charge.

Fig 13.5 The ionisation of nitric acid in water.

Note This is a simple model for nitric acid and the nitrate ion. A more sophisticated one involves extensive delocalisation. However, the use of curly arrows is still very helpful.

Classifying reactions

Changes to molecules which take place during chemical reaction can be classified into three types.

Addition reaction: a small molecule combines with an unsaturated molecule.

For example, $C_2H_4 + Br_2 \longrightarrow C_2H_4Br_2$

Substitution reaction: an atom, or group of atoms, are replaced in a molecule by another atom or group of atoms.

For example, $C_2H_5OH + HBr \longrightarrow C_2H_5Br + H_2O$

Elimination reaction: a small molecule is removed from a larger molecule, leaving an unsaturated group.

For example, $C_2H_5Br + NaOH \longrightarrow C_2H_4 + H_2O + NaBr$

While a reaction may appear more complex, each step can usually be classified under one of these three headings. For example, a **condensation reaction** is said to occur when two molecules join together with the loss of a small molecule. This reaction is strictly an addition reaction followed by elimination.

These three may be extended to include the chemical behaviour of coordination compounds.

> **5** **Ligand exchange is a substitution reaction. Write a definition for a ligand exchange reaction based on the description above.**

The names used to classify different types of reactions are no more than labels for different types of change that can take place. We must understand why and how bonds break and form in a chemical reaction in order to be able to control reactions. For example, synthesising a compound, or ensuring its stability or instability (whichever is required) in its working environment.

QUESTIONS

13.1 Most window frames are made from one of three materials: hardwood, aluminium and PVC (polyvinyl chloride, also known as poly(chloroethene)). Discuss the types of physical and chemical changes that must be withstood by these materials and the advantages and disadvantages of each material.

13.2 Sulphur molecules (S_8 rings) break open when a sample of sulphur is warmed. A viscous liquid is obtained which forms plastic sulphur when poured into cold water.
(a) What type of bond fission has taken place in the S_8 ring?
(b) Represent the bond-breaking using curly arrows.
(c) Suggest how long chains of sulphur atoms then form.

13.3 Sketch reaction profiles for the following types of reaction:
(a) a one-step endothermic reaction;
(b) a three-step exothermic reaction.

In each case state the number of intermediate compounds and transition states that are formed.

13.4 Classify the following reactions as addition, substitution or elimination:
(a) $CH_3CH_2CH_2Br + KCN \longrightarrow CH_3CH_2CH_2CN + KBr$
(b) $CH_3CH_2CH_2CH_2OH \longrightarrow CH_3CH_2CH{=}CH_2 + H_2O$
(c) $CH_3CHO + HCN \longrightarrow CH_3CH(CN)OH$

MECHANISMS OF CHEMICAL CHANGE

QUESTIONS *continued*

(d) $(C_2H_5)_2CO + H_2 \longrightarrow (C_2H_5)_2CHOH$
(e) $CH_3NH_2 + (CH_3CO)_2O \longrightarrow CH_3CONHCH_3 + CH_3COOH$
(f) $C_2H_5CN + 2H_2 \longrightarrow C_2H_5CH_2NH_2$
(g) $C_6H_6 + SO_3 \longrightarrow C_6H_5SO_3H$
(h) $C_6H_5SO_3H + NaOH \longrightarrow C_6H_5OH + NaHSO_3$

13.5 Use the Database to give two examples, other than those above, of:
(a) addition reactions;
(b) substitution reactions.

13.2 ADDITION REACTIONS

Free radical addition

Free radicals are formed during homolytic bond fission. Energy is required to break covalent bonds and, often, ultraviolet light provides this. An example we have already met is the homolytic bond fission of chlorine molecules to give chlorine atoms.

However, some compounds will decompose to give highly reactive radicals simply upon warming. An important group of these compounds is the organic peroxides. Such peroxides are used as initiators in the polymerisation processes in glues such as Araldite and materials such as 'plastic metal'. Free radical addition can be illustrated by the reaction of hydrogen bromide with an alkene in the presence of an organic peroxide (Fig 13.6).

KEY FACT

Free radical reactions involve three steps:

- **initiation** (formation of free radicals from molecules);
- **propagation** (formation of a free radical from reaction between another free radical and a molecule);
- **termination** (formation of a molecule from two free radicals).

initiation

$$C_6H_5-\overset{O}{\overset{\|}{C}}-O-O-\overset{O}{\overset{\|}{C}}-C_6H_5 \xrightarrow{\text{heat (60-80°C)}} 2C_6H_5-\overset{O}{\overset{\|}{C}}-O\bullet$$

propagation

$$C_6H_5-\overset{O}{\overset{\|}{C}}-O\bullet + HBr \longrightarrow C_6H_5COOH + Br\bullet$$

$$CH_3CH{=}CH_2 + Br\bullet \longrightarrow CH_3\dot{C}HCH_2Br$$

$$CH_3\dot{C}HCH_2Br + HBr \longrightarrow CH_3CH_2CH_2Br + Br\bullet$$

termination

$$2Br\bullet \longrightarrow Br_2$$

$$CH_3\dot{C}HCH_2Br + Br\bullet \longrightarrow CH_3CHBrCH_2Br$$

$$2CH_3\dot{C}HCH_2Br \longrightarrow BrCH_2-\underset{CH_3}{\underset{|}{CH}}-\underset{CH_3}{\underset{|}{CH}}-CH_2Br$$

Fig 13.6 The free radical addition of hydrogen bromide to propene, using benzoyl peroxide as the initiator.

6 **Use the convention of curly arrows to show the electron rearrangement that occurs in the first propagation reaction shown in Fig 13.6.**

Several compounds form in the benzoyl peroxide initiated addition of hydrogen bromide to propene. The major product is 1-bromopropane ($CH_3CH_2CH_2Br$). Surprisingly direct reaction between hydrogen bromide and propene, in the absence of a free radical initiator, gives 2-bromopropane ($CH_3CHBrCH_3$) as the major product.

CONNECTION: Direct reaction between hydrogen bromide and propene, page 288.

Fig 13.7 Plastic metal has two components. The mixture can be used to repair minor damage to car-bodies, boats and many other items.

PLASTIC METAL

'Plastic metal' is used to repair small dents and holes in metal structures such as car-body work. Larger repairs need glass fibre and a suitable resin to bind it and set hard. Plastic metal isn't metal at all. It has two components:

- a tube containing phenylethene (commonly known as styrene) and an inorganic bulking agent;
- a tube of 'hardener', which is benzoyl peroxide.

When the contents of the two tubes are mixed, benzoyl peroxide initiates the polymerisation of phenylethene. Plastic metal is, in fact, poly(phenylethene) (more commonly called polystyrene) with an inorganic filling agent.

ACTIVITY

Use the Database together with the information provided, to solve the problem below. In your answer give full named structures for each of the compounds, X, Y, W and Z.

The addition of bromine water to an alkene is complicated but the most important substance which adds across the double bond is bromic(I) acid, HOBr. The bromine atom is the positive pole of this polar compound.

electrophilic attack by bromic(I) acid on an alkene

Ozonolysis is the name used for addition of trioxygen (ozone) across the C=C double bond of an alkene. The product of this reaction is an ozonide which rapidly hydrolyses to give aldehydes or ketones.

unstable ozonide

ozonolysis yields an unstable ozonide which rapidly hydrolyses to yield two carbonyl compounds

An alkene W, of formula C_4H_8, is treated with bromine water to give a compound X. When X is dehydrated by acid catalysis a mixture of two bromoalkenes Y and Z are formed, both having the formula C_4H_7Br. These two compounds may be separated by chromatography. When Y is reacted with trioxygen (ozone) methanal is identified in the reaction mixture. Ozonolysis of Z produced a reaction mixture which contained propanone but no methanal.

Electrophilic addition to aliphatic hydrocarbons

'Electrophile' means 'electron-loving'. An electrophilic addition, therefore, is one which involves initial attack on a region of high electron density. The mechanism of electrophilic addition is best understood by looking at an example. Hydrogen bromide combines with ethene, the simplest alkene, to form 1-bromoethane.

$$C_2H_4 + HBr \longrightarrow C_2H_5Br$$

The hydrogen bromide molecule is polarised since the bromine atom is more electronegative than the hydrogen atom.

$$H^{\delta+}\!\!-\!Br^{\delta-}$$

The C=C double bond in ethene is a region of high electron density because of the four electrons being shared in the covalent bond. The positive end of the HBr molecule is attracted towards this electron density. A new C—H bond forms as the H—Br bond breaks. An intermediate compound, a **carbocation**, is formed, together with a bromide ion. The carbocation reacts rapidly with a bromide ion to form bromoethane. The mechanism is shown in Fig 13.8.

7 What type of bond cleavage has hydrogen bromide undergone?

8 Explain why one of the carbon atoms in the carbocation carries a single positive charge.

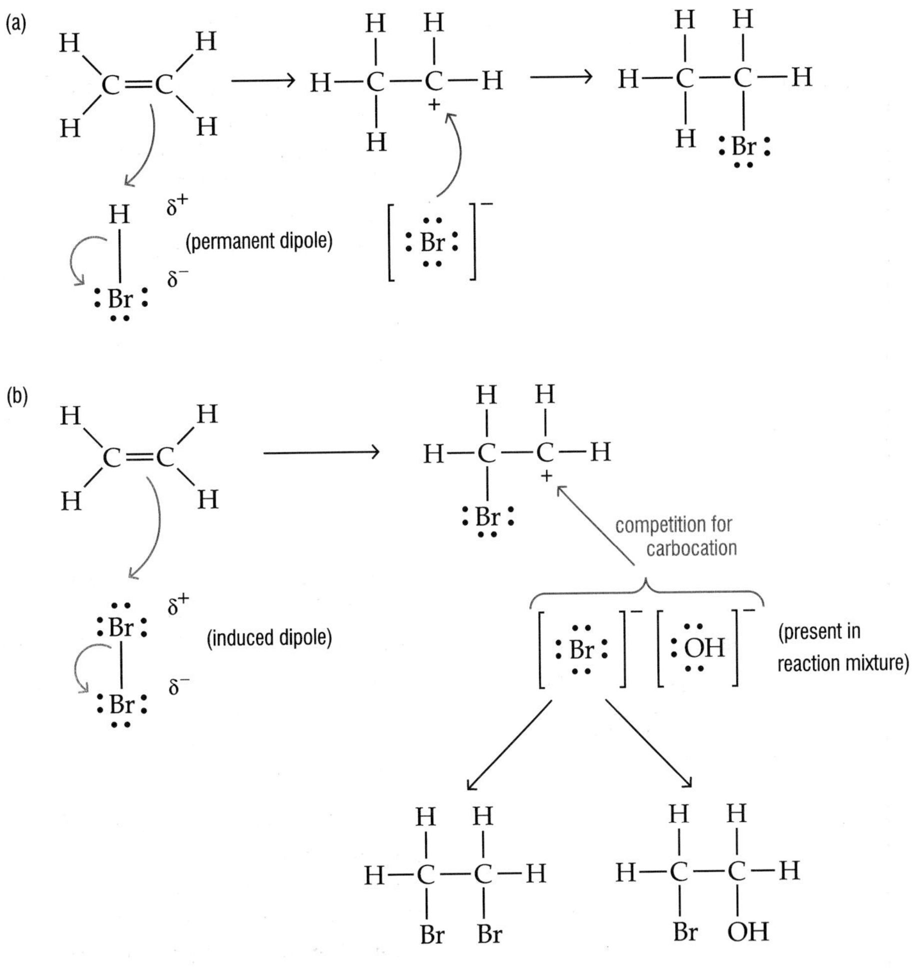

The term 'electron density' describes the probability of finding an electron in a stated volume of space. High electron density means that there is a high chance of finding an electron.

Fig 13.8 (a) Mechanism of the electrophilic addition of hydrogen bromide to ethene, (b) mechanism of the electrophilic addition of bromine to ethene, which takes place when ethene is bubbled through bromine water.

Bromine water can be used to detect the presence of C=C and C≡C bonds in organic compounds. The bromine reacts with compounds containing multiple bonds and, as it is used up, the bromine water becomes colourless. An example,

$$CH_3CH{=}CH_2 + Br_2 \longrightarrow CH_3CHBrCH_2Br$$

However, the bromine molecule is non-polar, so how does it attack a C=C double bond? Our mental image is that, as it approaches the electron density of the C=C double bond, the electron density within the bromine molecule is repelled and a dipole is induced in the bromine molecule. The mechanism of addition is the same as that for hydrogen bromide (Fig 13.8).

MECHANISMS OF CHEMICAL CHANGE   **287**

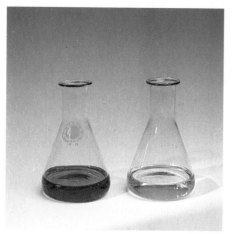

Fig 13.9 Cyclohexene decolorises bromine water because of the presence of a C=C double bond in the hydrocarbon molecule. $C_6H_{12} + Br_2 \longrightarrow C_6H_{12}Br_2$

9 **The bromination of ethene in the presence of water gives a mixture of products, mainly 1,2-dibromoethane and 2-bromoethanol. Why does the formation of this mixture provide evidence for the proposed mechanism of electrophilic addition?**

Molecules which combine with alkenes by this mechanism are said to be **electrophiles**. In summary, we describe the mechanism of the attack of a small molecule on an alkene in this way as **electrophilic addition**.

The problem of asymmetrical alkenes

An interesting problem arises when we add a molecule of the type HX to an asymmetrical alkene. For example, the addition of hydrogen bromide to propene. The attack of hydrogen bromide on propene yields two possible carbocations and, in consequence, we might anticipate that two different reaction products would form, 1-bromopropane and 2-bromopropane (Fig 13.10). We find by experiment, however, that only one of these structural isomers is formed in the reaction, 2-bromopropane. The explanation lies in the relative stability of the two carbocation intermediates. The secondary carbocation is more stable than the primary carbocation.

KEY FACT

An electrophile is an electron pair acceptor. Electrophiles carry a positive charge, either as a cation or the positive pole of a polar molecule.

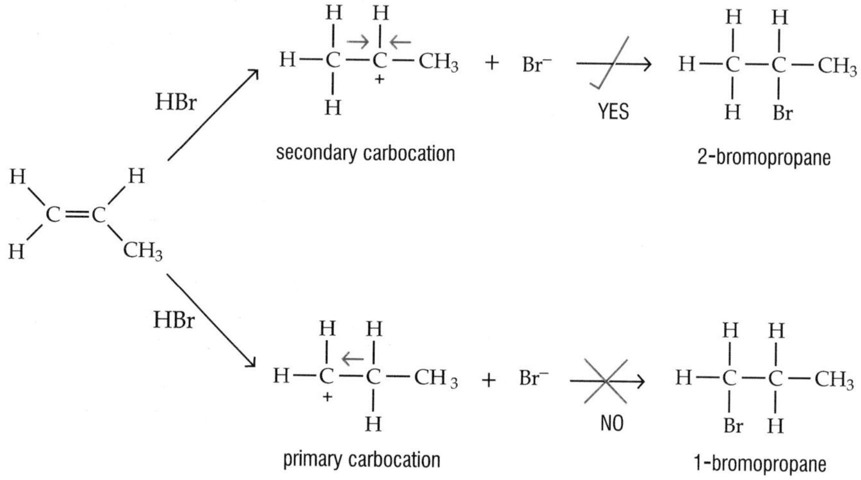

Fig 13.10 The two possible carbocation intermediates which could form during the addition of hydrogen bromide to propene. The secondary carbocation is stabilised relative to the primary carbocation by the positive inductive effect of the methyl groups (⟶).

KEY FACT

An asymmetrical alkene is one with different groups attached to the two carbon atoms of the C=C double bond. It is not symmetrical about the C=C bond.

We have observed this behaviour before in the fragmentation patterns in mass spectrometry. The relative stabilities are due to the positive inductive effect of the methyl groups on the two carbocations. A positive **inductive effect** means that the group has a slight tendency to push electrons away from itself onto a neighbouring atom. Any alkyl group will tend to donate some electron density to the carbon atom carrying the positive charge, and so help to stabilise it. The longer the lifetime of the carbocation, the greater the chance of it reacting with an anion to give a reaction product.

The most stable carbocation is one in which the carbon carrying the positive charge has the greatest number of methyl, or other alkyl groups, attached to it. The addition of hydrogen halides, or other molecules of the type HX, to asymmetrical alkenes is summarised by **Markownikoff's rule**.

Markownikoff's rule: When a molecule, HX, combines with an asymmetrical alkene, the hydrogen atom attaches itself to the carbon that already has the most hydrogen atoms attached to it. The opposite result can sometimes be obtained, for example, during free-radical addition reactions. This is known as anti-Markownikoff addition.

BHC – ONLY ONE ISOMER WORKS

Aromatic hydrocarbons, such as benzene, do not usually undergo addition reactions even though they are unsaturated organic compounds. However, in the presence of intense ultraviolet radiation, chlorine adds to benzene to give 1,2,3,4,5,6-hexachlorohexane (more commonly known as benzene hexachloride). This compound is an important insecticide, marketed as BHC (from the trivial name benzene hexachloride):

Fig 13.11   Farmers can use BHC to control some insects which damage their plants.

BHC can exist, theoretically at least, in eight isomeric forms. However, only one of them is an effective insecticide, the isomer known as gamma-BHC.

Nucleophilic addition to organic compounds

An unsaturated molecule in which a carbon atom is joined to an electronegative atom by a multiple covalent bond usually undergoes **nucleophilic addition**. Good examples are the addition reactions of carbonyl compounds – aldehydes and ketones. The carbonyl group consists of a carbon and an oxygen atom joined by a double covalent bond, $>C{=}O$. Oxygen is more electronegative than carbon and so the carbonyl bond is polarised:

$$>\overset{\delta^+}{C}{=}\overset{\delta^-}{O}$$

Carbonyl compounds undergo nucleophilic addition reactions. A **nucleophile** is an electron pair donor. Nucleophiles are attracted to the carbon atom of the carbonyl group and reaction occurs according to the general equation:

$$>C{=}O \; + \; H{-}X \longrightarrow \; \overset{\displaystyle OH}{\underset{\displaystyle |}{-\overset{|}{C}-X}}$$

10   **Which of the following is not a nucleophile: H_2O, BF_3, NH_3, OH^-?**

Hydrogen cyanide reacts with ethanal to give 2-hydroxypropanenitrile. This can be hydrolysed to give 2-hydroxypropanoic acid, more commonly known as lactic acid. This is a chiral compound and its molecules can exist in two forms which are non-superimposable, mirror images of one another.

CONNECTION: Chiral compounds and optical activity, page 275.

Two experimental observations enable us to propose a mechanism for the addition of hydrogen cyanide to ethanal.

1. The dependence of the reaction rate on reaction conditions. It has been found that addition of hydrogen cyanide to ethanal in acidic solutions is negligible. In neutral solutions, the reaction is measurable but as solutions are made more alkaline, the rate of addition increases very rapidly.
2. The formation of a mixture of two optical isomers in equal concentrations, in other words the reaction product is a racemic mixture.

Our starting point for describing the mechanism of addition is the reaction between hydrogen cyanide and water. Hydrogen cyanide gas dissolves in water to give a solution of hydrogen cyanide molecules. These react with water to give oxonium ions and cyanide ions. A dynamic equilibrium is established between the undissociated molecules and the ions:

$$HCN(g) + aq \longrightarrow HCN(aq)$$
$$HCN(aq) + H_2O(l) \rightleftharpoons H_3O^+(aq) + CN^-(aq)$$

11 Use curly arrows to represent the electron rearrangement when hydrogen cyanide reacts with water.

In acid solutions, the position of equilibrium favours the undissociated molecules but in alkaline solution it favours cyanide ions. Since reaction is negligible in acidic solution, it appears that the cyanide ions rather than hydrogen cyanide molecules attack the ethanal molecule. The cyanide ion is a strong nucleophile; it is a negative ion with a pair of electrons which it can donate in bond formation. A cyanide ion attacks the δ+ carbon of the carbonyl group. The reaction intermediate captures a proton readily, even in alkaline conditions to give the 2-hydroxypropanenitrile product. The mechanism is summarised in Fig 13.12. But why do we get a mixture of the (+) and (−) optical isomers?

CONNECTION: Factors affecting position of equilibrium, pages 460–464.

12 The usual reagent for this reaction is a solution of potassium cyanide in ethanol. Why is it best to avoid alkaline solutions?

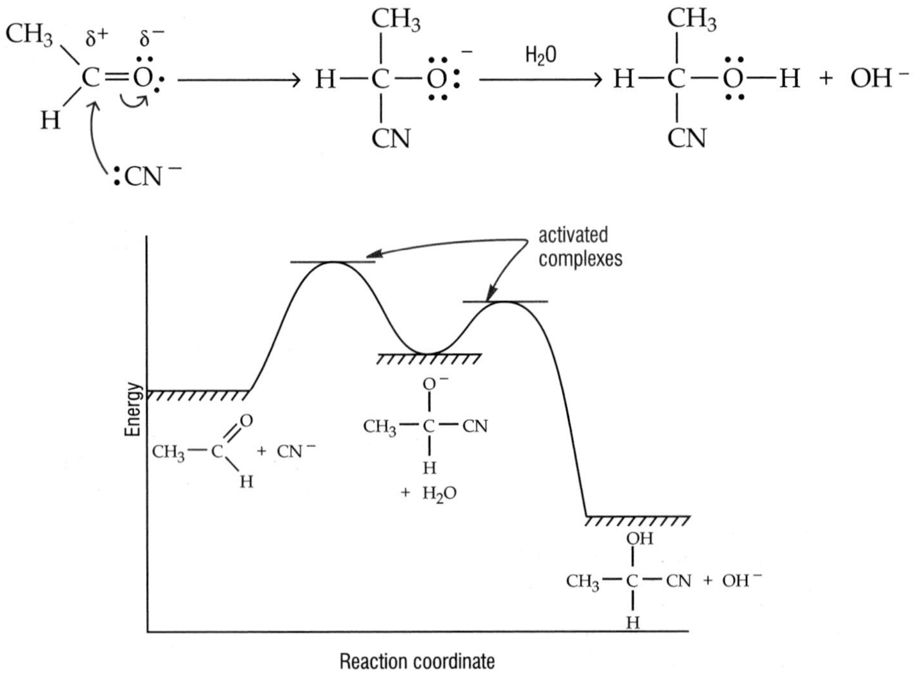

Fig 13.12   The nucleophilic addition of hydrogen cyanide to ethanal.

MECHANISMS OF CHEMICAL CHANGE

Ethanal is a planar molecule. It can be attacked by a cyanide ion from above or below the plane of the molecule (Fig 13.13). Attack from above the plane produces one optical isomer while attack from below the plane produces the other optical isomer. This is a characteristic of the chemical synthesis of chiral compounds. Chemical reagents tend not to be **stereospecific**. Enzymes, Nature's catalysts, always promote the formation of just one optical isomer, they are said to be stereospecific.

13 **Why does the reaction of propanone, $(CH_3)_2CO$, with hydrogen cyanide produce a product which is not optically active?**

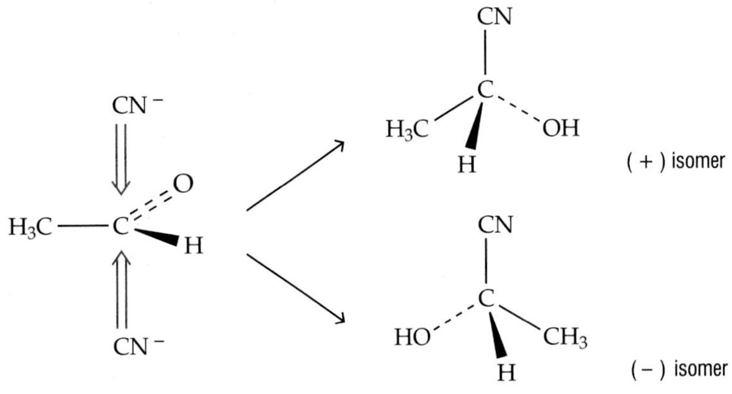

Fig 13.13 Attack of cyanide ions on the ethanal molecule can take place from above or below the plane of the molecule. The result is the formation of an equimolar mixture of (+) and (–) isomers, known as a racemic mixture.

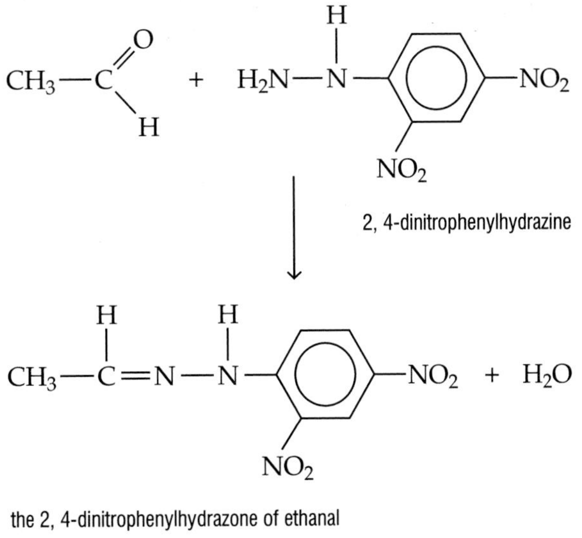

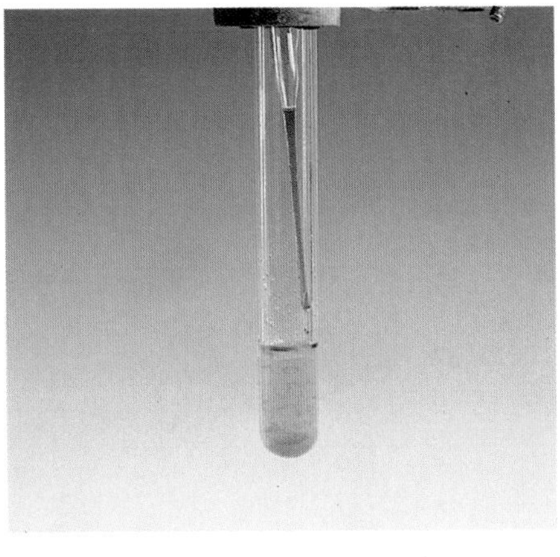

Fig 13.14 An example of using nucleophilic addition reactions in the analytical laboratory is the identification of carbonyl compounds by their reaction with 2,4-dinitrophenylhydrazine to give yellow or orange colour precipitates. The reaction is a condensation reaction – addition followed by elimination.

KEY FACT

The structural features of a nucleophile, a ligand and a base are the same. Each is an electron pair donor. Each contains an electronegative atom with a lone pair of electrons associated with it. Examples are water, ammonia, the hydroxide ion and the cyanide ion.

LACTIC ACID – THE PRICE THE ATHLETE HAS TO PAY

Lactic acid, 2-hydroxypropanoic acid, forms in muscles when there is insufficient oxygen available to metabolise glucose. This happens during strenuous exercise, for example during sprinting. The lactic acid content of the 'sprinting' muscles is much higher than normal at the end of a race. The muscles will have used up the available oxygen and the blood supply cannot replenish the oxygen fast enough, despite the accelerated heart beat and rapid, deep breathing. During the later stages of a race the muscles must release energy anaerobically, that is, without the use of oxygen. Glucose molecules are cleaved into two lactic acid molecules, an energy-producing process which does not require oxygen:

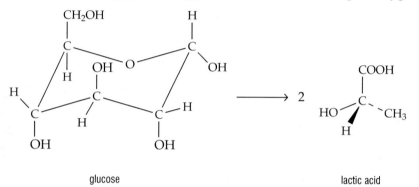

glucose   lactic acid

Fig 13.15 Legs feel like lead at the end of a 400 m sprint due to increasing concentrations of lactic acid in the muscles. Here Marie-Jose Perec is seen winning a gold medal for France in the 1992 Olympics

The athlete is said to have built up an **oxygen debt**. An accumulation of lactic acid will give you the familiar 'stitch' and this will only disappear when the blood can replenish the oxygen supply and the lactic acid is removed by further metabolic reactions which change the lactic acid molecules into others which are easily excreted. It is the build-up of lactic acid in the muscles that make legs feel heavy and weak. Training sessions for top-class sprinters are designed to improve the body's efficiency in metabolising glucose.

Nucleophilic addition to inorganic compounds

Many covalent halides, particularly transition metal halides, react with ligands to form coordination compounds (ionic or neutral). In doing so the central atom increases its coordination number and changes the coordination geometry. These compounds are sometimes called **adducts** or **addition compounds**. The ligand forms a coordinate bond. A lone pair of electrons on the ligand is donated to a vacant orbital on the central atom of the halide. This type of complex formation is another example of nucleophilic addition. The ligand is acting as a nucleophile and is attracted to the central atom of the covalent halide which carries a small positive charge due to the difference in electronegativity between the metal and the halogen.

Fig 13.16 Nucleophilic addition of ligands to a tetrahedral halide. In these examples, the ligand (a halide ion or a neutral molecule) bonds to the central atom, increasing the coordination number and changing the geometry from tetrahedral to octahedral.

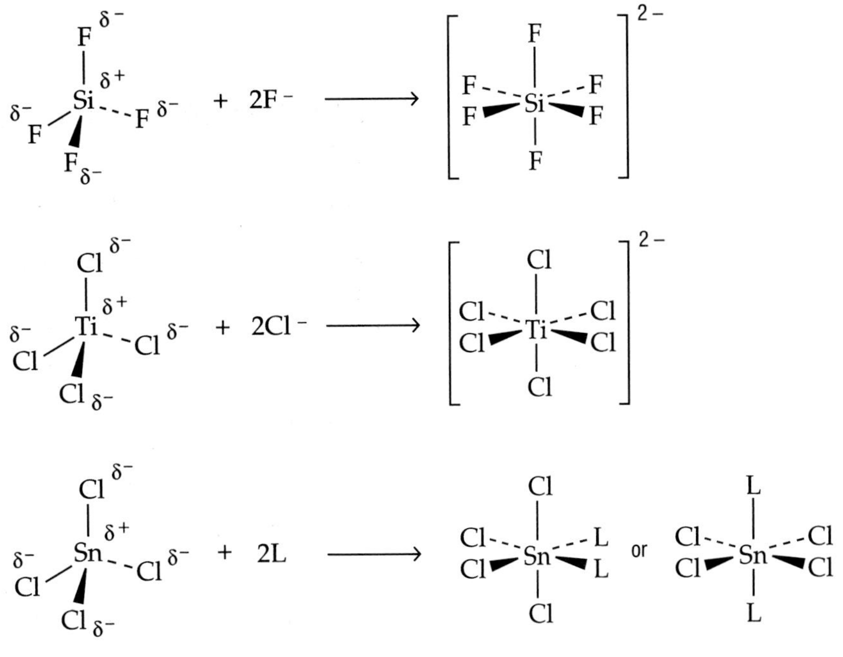

(where L = a ligand and the formation of the *cis* or *trans* addition compounds depends on the nature of this ligand)

For example, silicon tetrachloride ($SiCl_4$) dissolves in concentrated hydrochloric acid to form the hexachlorosilicate(IV) ion, $[SiCl_6]^{2-}$. Silicon has vacant orbitals in its outer energy level ($n = 3$) available for coordinate bond formation (Fig 13.17). A chloride ion has eight electrons (four pairs) in its outer energy level. One of these pairs is donated to a vacant 3d orbital of the silicon. However, tetrachloromethane (CCl_4) does not form an ion of the type $[CCl_6]^{2-}$ because carbon does not have vacant orbitals in the n = 2 energy level which can be used in bonding.

Fig 13.17 In its ground state, a silicon atom has the electronic configuration: $1s^2\,2s^2\,2p^6\,3s^2\,3p^2$. Two unpaired electrons are available for normal covalent bond formation. In its first excited state, a 3s electron is promoted to a vacant 3p orbital, making four electrons available for bonding. The formation of $[SiCl_6]^{2-}$ is possible because the 3d orbitals are available to accept lone pairs of electrons in coordinate bond formation.

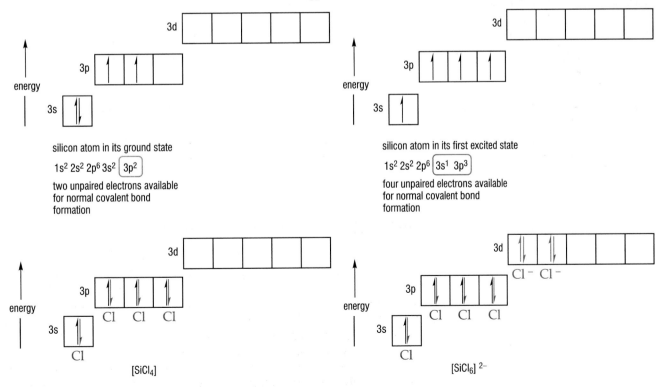

14 Why is it that PCl_6^- exists but PI_6^- does not?

TWO TYPES OF POLYTHENE

Ethene polymerises to form poly(ethene), commonly called polythene, a material that is used for everyday products such as polythene sheets, tubing and containers. But polythene comes in two forms, 'low' and 'high' density. They differ in the amount of branching that occurs during polymerisation and how closely the resulting polymer molecules can pack together. In low density polythene there is extensive branching and, therefore, poor packing, whereas in high density polythene there is far less branching and the polymer chains can pack together more closely.

The predominance of branched polymer chains or of unbranched polymer chains during alkene polymerisation can be controlled by the reaction conditions. Low density poly(ethene) is prepared by using a free radical initiator:

$$R-O-O\cdot \;+\; \underset{H}{\overset{H}{>}}C=C\underset{H}{\overset{H}{<}} \longrightarrow R-O-O-\overset{H}{\underset{H}{C}}-\overset{H}{\underset{H}{C}}\cdot$$

$$R-O-O-\overset{H}{\underset{H}{C}}-\overset{H}{\underset{H}{C}}\cdot \;+\; \underset{H}{\overset{H}{>}}C=C\underset{H}{\overset{H}{<}} \longrightarrow R-O-O-\overset{H}{\underset{H}{C}}-\overset{H}{\underset{H}{C}}-\overset{H}{\underset{H}{C}}-\overset{H}{\underset{H}{C}}\cdot$$

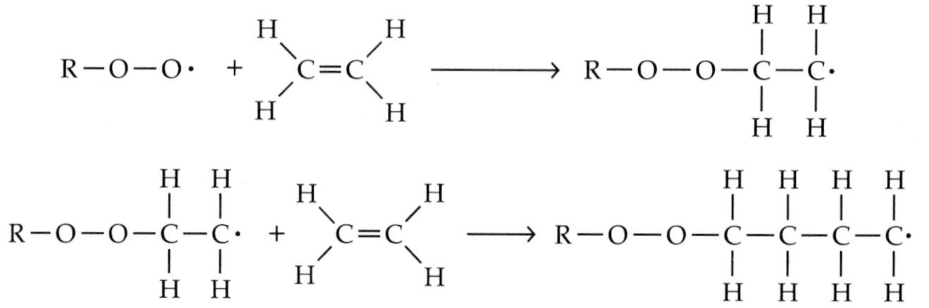

where $R-O-O\cdot$ is a free radical initiator

Ionic catalysts promote the ionic polymerisation of alkenes. Polymerisation gives high density poly(ethene):

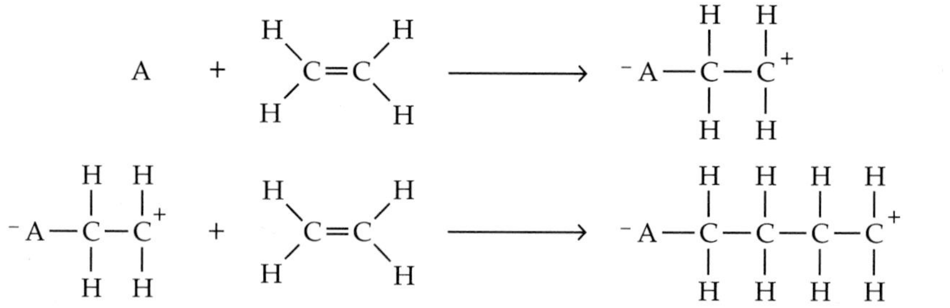

$$A \;+\; \underset{H}{\overset{H}{>}}C=C\underset{H}{\overset{H}{<}} \longrightarrow {}^{-}A-\overset{H}{\underset{H}{C}}-\overset{H}{\underset{H}{C}}{}^{+}$$

$$^{-}A-\overset{H}{\underset{H}{C}}-\overset{H}{\underset{H}{C}}{}^{+} \;+\; \underset{H}{\overset{H}{>}}C=C\underset{H}{\overset{H}{<}} \longrightarrow {}^{-}A-\overset{H}{\underset{H}{C}}-\overset{H}{\underset{H}{C}}-\overset{H}{\underset{H}{C}}-\overset{H}{\underset{H}{C}}{}^{+}$$

where A is a suitable acid

If a free-radical catalyst is used, the polymerisation of an asymmetrical alkene, such as propene, results in a random arrangement of the side groups on the hydrocarbon polymer chain as well as considerable branching. Since the properties of the polymer depend on these two structural features of the chain for many purposes, it is desirable to make polymers where the side groups all point in the same direction and branching is limited. Molecules can then pack together more closely, and, as a result, stronger intermolecular bonds form between chains.

Fig 13.18 The low density form of poly(ethene) has been used for plastic shopping bags and wrapping film. For many purposes, however, a stronger more rigid form is needed – high density polythene. This is used to mould things like plastic milk bottle crates.

Table 13.1 Some properties of low (LDPE) and high density (HDPE) polyethene

	LDPE	HDPE
density / g cm^{-3}	0.920	0.955
percentage crystallinity	55	80
softening point / K	360	400
tensile strength / atm	85-136	204-313

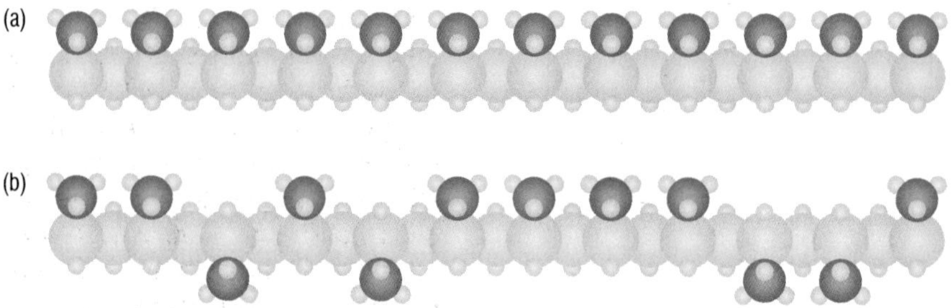

Fig 13.19 **(a)** Isotactic poly(propene), **(b)** atactic poly(propene)

(a)

(b)

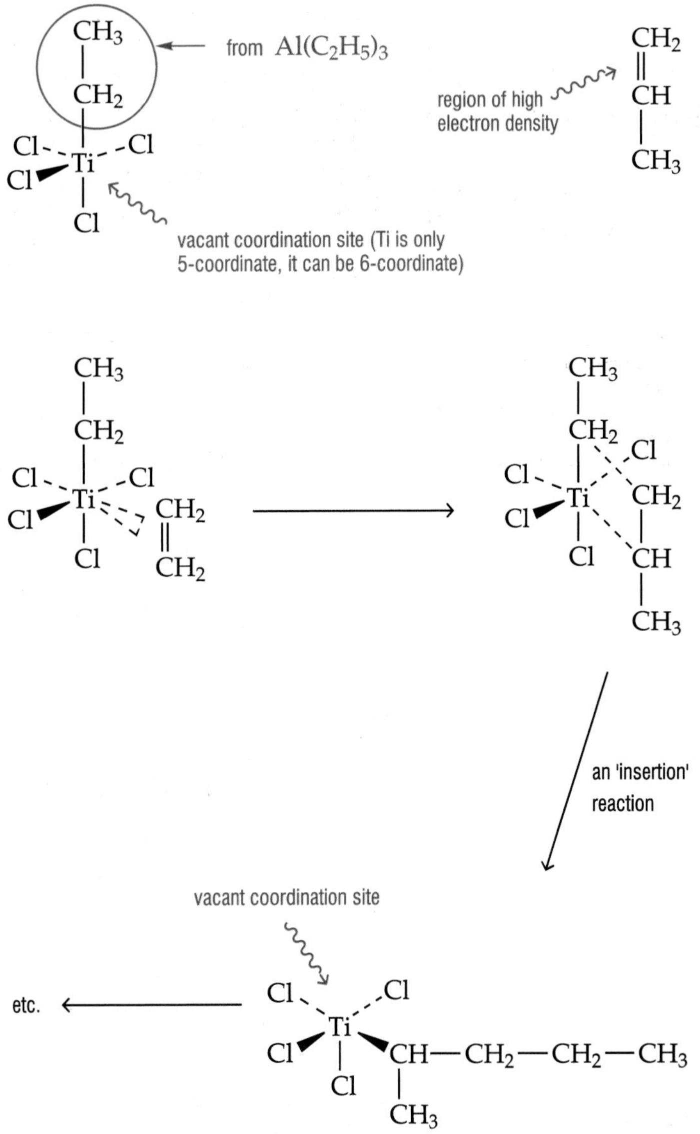

Fig 13.20 The mechanism of polymerisation of propene in the presence of a Ziegler-Natta catalyst.

Ziegler-Natta catalysts promote polymerisation with minimum branching. The catalysts are made usually from titanium(IV) chloride and alkyl aluminium compounds. The solid catalyst consists of titanium ions which are 5-coordinate and which have an alkyl group (from the alkyl aluminium compounds) attached to them. There is a vacant coordination site on the titanium ion to which the alkene can coordinate temporarily. This is nucleophilic addition. Subsequently, an insertion reaction occurs. This is a rearrangement of the alkyl group and the alkene molecule attached to the titanium ion. The alkyl adds to the alkene and the titanium ion returns to 5-coordination. A vacant coordination site is available for another alkene molecule and the reaction sequence continues.

13.6 Suggest initiation, propagation and termination reactions during the changes which occur when the two components of plastic metal are mixed. Explain why the manufacturers recommend that only small amounts of hardener (benzoyl peroxide) should be used.

13.7 For each of the following reactions *(i)* write a structural formula for the product, and *(ii)* suggest a mechanism by showing the rearrangement of electrons:
(a) $CH_2\!\!=\!\!CH\!-\!CH\!\!=\!\!CH_2 + 2HBr$
(b) $C_6H_5CH\!\!=\!\!CH_2 + HCl$
(c) $CH_2\!\!=\!\!CH_2 + H_2O$
(d) $CH_3CH\!\!=\!\!CH_2 + H_2O$

13.8 Ethene can be converted into ethanol by dissolving it in concentrated sulphuric acid and pouring the solution into cold water. Two reactions occur:

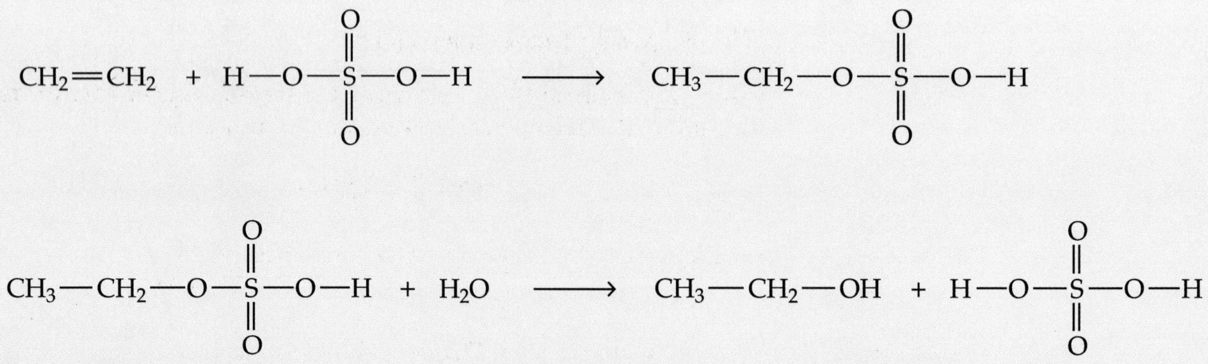

$$CH_2\!\!=\!\!CH_2 \;+\; H\!-\!O\!-\!\overset{\displaystyle O}{\underset{\displaystyle O}{\overset{\|}{\underset{\|}{S}}}}\!-\!O\!-\!H \;\longrightarrow\; CH_3\!-\!CH_2\!-\!O\!-\!\overset{\displaystyle O}{\underset{\displaystyle O}{\overset{\|}{\underset{\|}{S}}}}\!-\!O\!-\!H$$

$$CH_3\!-\!CH_2\!-\!O\!-\!\overset{\displaystyle O}{\underset{\displaystyle O}{\overset{\|}{\underset{\|}{S}}}}\!-\!O\!-\!H \;+\; H_2O \;\longrightarrow\; CH_3\!-\!CH_2\!-\!OH \;+\; H\!-\!O\!-\!\overset{\displaystyle O}{\underset{\displaystyle O}{\overset{\|}{\underset{\|}{S}}}}\!-\!O\!-\!H$$

Propose a mechanism for the reaction between ethene and concentrated sulphuric acid.

13.9 Chlorine reacts with iodine at room temperature to give a dark red-brown liquid. This is iodine monochloride, ICl. Write a structural formula and name the reaction product when ICl is added to:
(a) ethene;
(b) but-1-ene.
In each case, suggest a mechanism, showing the rearrangement of electrons using curly arrows.

13.10 Predict the order of reactivity of the following alkanes towards chlorine, giving the most reactive first. Explain the order you have given.
chloroethene; but-2-ene; 1,1-difluoroethene; tetrafluoroethene; propene; 2,3-dimethylbut-2-ene.

[*Hint;* You may find it helpful to sketch the structural formulae of the molecules]

13.11 Write structural formulae for the products formed when potassium cyanide dissolved in ethanol reacts with the following compounds:
(a) propanal;
(b) pentan-2-one;
(c) cyclohexanone;
(d) pentanal.
Indicate which of the reaction products are chiral compounds.

13.12 Explain why no reaction occurs between aqueous hydrogen cyanide, to which a little alkali has been added, and ethene.

13.13 Draw an electrons-in-boxes diagram to represent the bonding in the following complex ions:
(a) $[SiF_6]^{2-}$;
(b) $[PCl_6]^-$;
(c) $[SnCl_6]^{2-}$;
(d) $[AlCl_4]^-$;
(e) $[AlH_4]^-$.

13.14 Phosphorus pentachloride exists in the vapour phase as PCl_5 molecules. In the solid state, however, it exists as $[PCl_4]^+[PCl_6]^-$. Explain how the atoms must rearrange between two PCl_5 molecules in order to give the pair of ions which exist in the solid state.

13.3 SUBSTITUTION REACTIONS

Free radical substitution

Free radical substitution involves three steps (as in the case of free radical addition). For example, the chlorination of methane:

initiation

$$Cl_2 \xrightarrow{\text{ultra violet}} 2Cl\bullet$$

propagation

$$\begin{cases} Cl\bullet + CH_4 \longrightarrow HCl + \bullet CH_3 \\ \bullet CH_3 + Cl_2 \longrightarrow CH_3Cl + Cl\bullet \end{cases}$$

termination

$$\begin{cases} 2Cl\bullet \longrightarrow Cl_2 \\ 2 \bullet CH_3 \longrightarrow C_2H_6 \\ Cl\bullet + \bullet CH_3 \longrightarrow CH_3Cl \end{cases}$$

Photochlorination is used industrially to make mixtures of halogenoalkanes from petroleum fractions. CFCs are involved in free radical reactions with ozone in the stratosphere.

CONNECTION: Reactions of CFCs, page 39.

15 **Explain why a mixture of chlorinated methanes is produced when methane reacts with chlorine in the presence of ultraviolet radiation.**

Electrophilic substitution in organic compounds

Although benzene is an unsaturated compound, it does not undergo addition reactions readily. This is part of the evidence that leads us to believe benzene does not contain alternate single and double covalent bonds. Rather, we picture a delocalised ring of six electrons around the six carbon atoms.

CONNECTION: Bonding in benzene, page 251.

The most common reactions of benzene involve the substitution of one or more hydrogen atoms in the molecule by other atoms or groups. An example will serve to illustrate the reaction mechanism. We will look at the reaction between nitric acid and benzene to give nitrobenzene. Nitration reactions of this type are very important in the manufacture of explosives, for example TNT and nitroglycerine, and as one of the first stages in the production of many common dyes. The nitration of benzene may be summarised by the equation:

$$C_6H_6 + HNO_3 \longrightarrow C_6H_5NO_2 + H_2O$$

but, as is often the case, the equation disguises the complexity of the reaction mechanisms.

Fig 13.21 Explosives play an important part in our lives, even during periods of peace.

Concentrated nitric acid does not readily attack benzene. However, a mixture of 50:50 concentrated nitric acid and concentrated sulphuric acid reacts with benzene to give nitrobenzene. The two concentrated acids react together to form a solution of the nitronium ion, NO_2^+.

$$HNO_3 + H_2SO_4 \longrightarrow NO_2^+ + HSO_4^- + H_2O$$

16 **Explain what a nitronium ion and a carbon dioxide molecule have in common in terms of number and arrangement of electrons.**

The nitronium ion is a powerful electrophile. It is attracted to the high electron density in the benzene ring and is added to form a carbocation. Two reaction routes are now possible. An anion such as the hydrogen sulphate ion could add to the carbocation, giving a structure with four single covalent bonds and two double covalent bonds. Alternatively, a hydrogen ion could be lost from the carbon to which the nitronium ion has bonded. This second pathway is favoured because it retains the delocalised electron system of a benzene ring which is particularly stable. The overall reaction is that a hydrogen atom in the benzene ring has been replaced by a nitro group. The mechanism is summarised in Fig 13.22.

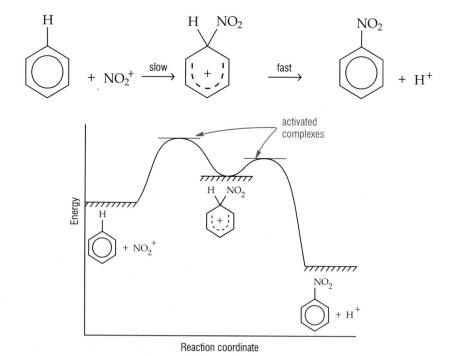

Fig 13.22 The proposed mechanism of nitration of benzene – an example of electrophilic substitution. The two-step reaction pathway, with the first step being the slowest and therefore rate-determining step, is consistent with experimental data.

MECHANISMS OF CHEMICAL CHANGE

Further nitration is possible, though difficult. Reaction conditions are carefully chosen so that only the mono-nitro compound is produced. The nitro group, with its three electronegative atoms, pulls electron density away from the benzene ring, lowering its susceptibility to further electrophilic attack. Again, our understanding of a reaction mechanism enables us to choose reaction conditions which are most likely to give us the desired products.

Electrophilic substitution in coordination compounds

An unusual example of electrophilic substitution comes from coordination chemistry. Although of little importance in terms of applications, it does extend our ideas across a wider range of compounds. Pentane-2,4-dione (acetylacetone, acac) forms coordination compounds with many metal ions. The organic molecule loses a proton and bonds to the metal ion through its two oxygen atoms. Electron density is delocalised around the chelate ring, giving some characteristics similar to benzene. In particular the hydrogen atom attached to the third carbon atom of the coordinated pentane-2,4-dione is susceptible to electrophilic substitution. It can be replaced, for example, by halogen atoms. Delocalisation occurs not just in aromatic molecules and inorganic ions such as CO_3^{2-}, SO_4^{2-} and NO_3^-, but also in some coordination compounds.

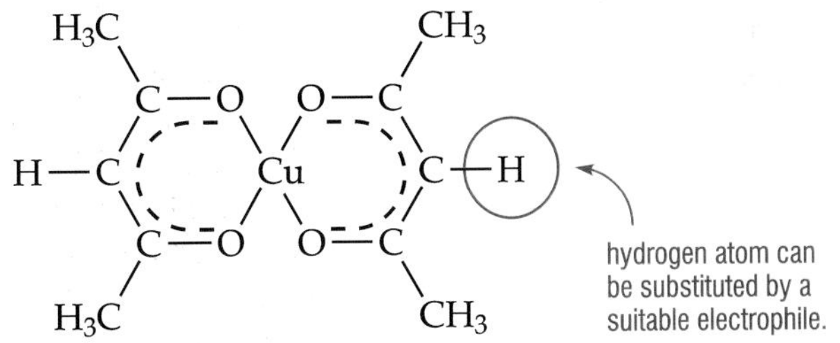

Fig 13.23 Coordination compounds of pentane-2,4-dione display aromatic character, the ring hydrogen atom being susceptible to electrophilic substitution.

hydrogen atom can be substituted by a suitable electrophile.

Nucleophilic substitution in organic compounds

Consider the reaction between a halogenoalkane and an alkali (an aqueous solution containing the hydroxide ion, $OH^-(aq)$). The general equation for substitution is

$$RX + OH^- \longrightarrow ROH + X^-$$

where R = an alkyl group
   X = a halogen atom
   X^- = a halide ion

In the reaction, a carbon-halogen bond is broken and replaced by a carbon-oxygen bond. Because halogens are more electronegative than carbon, the carbon-halogen bond is polarised:

$$C^{\delta+}\!\!-\!X^{\delta-}$$

The carbon atom is susceptible to attack by a nucleophile (in other words, an electron pair donor). A consideration of electronegativities suggests that C—F bonds are most susceptible to attack and that C—I are the least (Table 13.1).

17 Is the breaking of a carbon-halogen bond in nucleophilic substitution homolytic or heterolytic?

Nucleophilic substitution occurs when a molecule is attacked by a nucleophile and an atom or group of atoms in the molecule is replaced by the nucleophile. Such reactions occur in human metabolism, the degradation of a number of polymers, and a host of other chemical changes.

Table 13.1 Electronegativity values (Pauling's scale)

Carbon	2.5
Hydrogen	2.1
Fluorine	4.0
Chlorine	3.0
Bromine	2.8
Iodine	2.5

Fig 13.24 Silver nitrate solution is put into each of three separate test tubes. 1-Chlorobutane added to the first, 1-bromobutane to the second and 1-iodobutane to the third. The test tubes are placed in a hot water bath. Hydrolysis gives halide ions which form silver halides:
$Ag^+(aq) + X^-(aq) \longrightarrow AgX(s)$.
The time taken for the silver halide to appear is a measure of the ease of hydrolysis.

Experiment shows (Fig 13.24) that iodoalkanes hydrolyse the most easily and fluoroalkanes the least easily. Clearly, bond polarity does not explain the trends in reactivity. Instead we must look at the stability of the leaving group and bond strengths. An essential requirement for nucleophilic substitution is that the **leaving group** (still with its bonding pair of electrons) can leave the molecule as a stable species in the solvent used. Reactivity also relates to bond strengths. Ease of hydrolysis of halogenoalkanes reflects the relative strengths of the C—X bonds (Table 13.3).

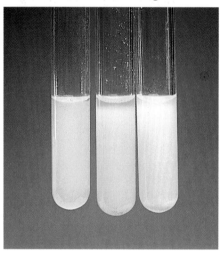

Table 13.3

C—X	Average bond enthalpy / kJ mol⁻¹	
C—F	484	
C—Cl	338	increasing ease of hydrolysis
C—Br	276	
C—I	238	

A simple example shows that nucleophilic substitution does not always take place at the most positively charged centre in a molecule:

A similar reaction occurs in alkali. Hydroxide ions give methanol, CH_3OH, and trimethylamine, $(CH_3)_3N$.

The tetramethylammonium ion, $(CH_3)_4N^+$, is closely related to the ammonium ion, NH_4^+. If nucleophilic substitution occurred at the nitrogen atom carrying the formal positive charge, the leaving group would be CH_3^-. This is a very unstable species. However, substitution at the carbon atom means that $(CH_3)_3N$ is the leaving group, and this is stable.

The mechanism of nucleophilic substitution depends on the type of halogenoalkane involved. Consider a primary halogenoalkane, RCH_2X. Experiments suggest that the hydrolysis with alkali is a one-step mechanism. The rate of reaction depends on the concentration of RCH_2X and on the concentration of OH^-. Also, when a single optical isomer of a chiral

MECHANISMS OF CHEMICAL CHANGE

halogenalkane is used (for example, RCH(F)X where X=Cl, Br, I, **inversion** occurs. The molecule turns inside out (rather like an umbrella caught by the wind) and the optical isomer of the alcohol formed is the mirror image of the halogenoalkane (with OH in place of X, of course). The mechanism is summarised in Fig 13.25. We picture the formation of a transition state in which the carbon-halogen bond stretches to breaking point at the same time as the carbon-oxygen bond is forming.

18 Explain why RCH(F)X could be used to study the effect of hydrolysis on the stereoisomerism of the molecules involved.

Fig 13.25 Nucleophilic substitution: the mechanism of hydrolysis of a primary halogenoalkane.

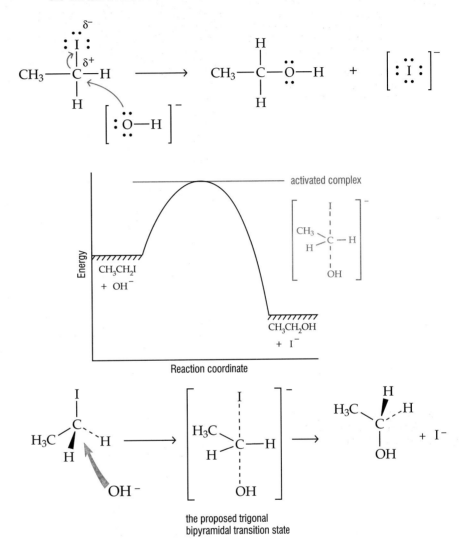

In contrast to primary halogenoalkanes, experiments suggest that hydrolysis of a tertiary halogenoalkane, R_3CX, is a two-step mechanism. The rate depends only on the concentration of the halogenoalkanes. When a single optical isomer of a chiral halogenoalkane, $R^1R^2R^3CX$, is used a racemic mixture of the alcohol is obtained. A different mechanism is operating (Fig 13.26). In the two-step mechanism, it is the first (the formation of a carbocation by breaking the C—X bond) that is the slowest and therefore it determines the overall rate of reaction. Once formed, the carbocation reacts rapidly with OH^- to give the alcohol.

19 Explain why the dependence of rate upon concentration of halogenoalkane rather than upon hydroxide ion concentration supports the mechanism in Fig 13.26.

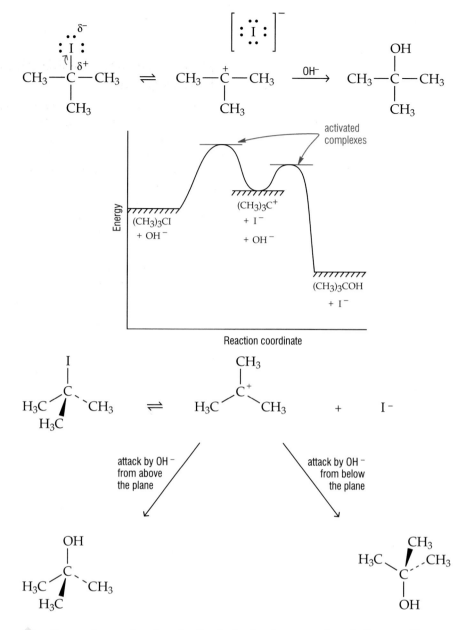

20 **Suggest a mechanism for hydrolysis of a secondary halogenoalkane.**

Acid chlorides hydrolyse very easily, reacting violently even with cold water. The reaction is a nucleophilic substitution, producing a carboxylic acid by replacing the chlorine atom by a hydroxyl group.

$$RCOCl + H_2O \longrightarrow RCOOH + HCl$$

Acid chlorides are different to chloroalkanes in two important respects. Firstly, the carbon atom at which nucleophilic substitution occurs is in a planar environment and therefore more accessible to the nucleophile. Secondly, the presence of an electronegative oxygen atom causes the carbon to carry a slightly higher positive charge and, importantly, pulls some electron density away from the carbon-chlorine bond, therefore weakening it slightly compared to C—Cl in a chloroalkane (Fig 13.27).

the presence of an electronegative oxygen atom increases the attraction of the group bonded to the chlorine, increasing the polarity of the C—Cl bond

Fig 13.27 A comparison of the bond polarities in chloroalkanes and acid chlorides.

SURGICAL SUTURES: ABSORBABLE OR NON-ABSORBABLE?

Stitches in wounds are either removed by the nurse or doctor once the wound has healed (non-absorbable sutures), or they dissolve slowly in the body fluids (absorbable sutures). Repairing internal damage obviously demands the use of absorbable sutures. Chemists have been able to design and synthesise polymers which can be used to make both types of suture. An understanding of the mechanism of amide and ester hydrolysis made this possible.

Polyamides, such as nylon, and aromatic polyesters are used for non-absorbable sutures. This is because they are reasonably resistant to hydrolysis through a nucleophilic substitution mechanism. However, increased susceptibility to hydrolysis (the way in which absorbable sutures 'dissolve') can be achieved by using a polyester made from aliphatic monomer units – an aliphatic polyester.

Fig 13.28 Polymers are used to make surgical sutures. The polymers a re usually polyamides or polyesters, and their resistance to hydrolysis is determined by their precise molecular structure.

(a) Non-absorbable sutures may be made from aliphatic polyamides, eg

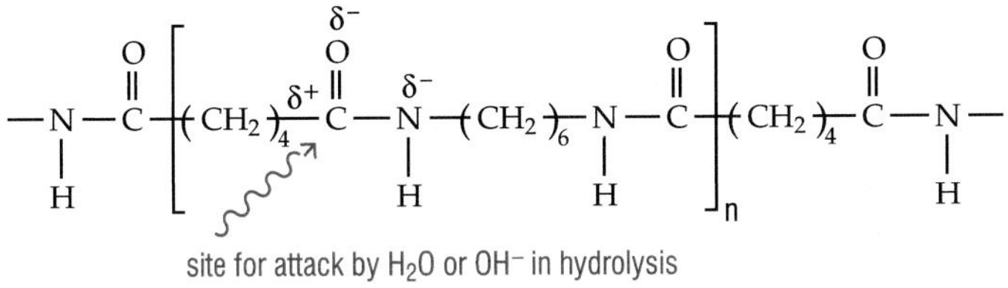

site for attack by H_2O or OH^- in hydrolysis

aromatic polyesters, eg

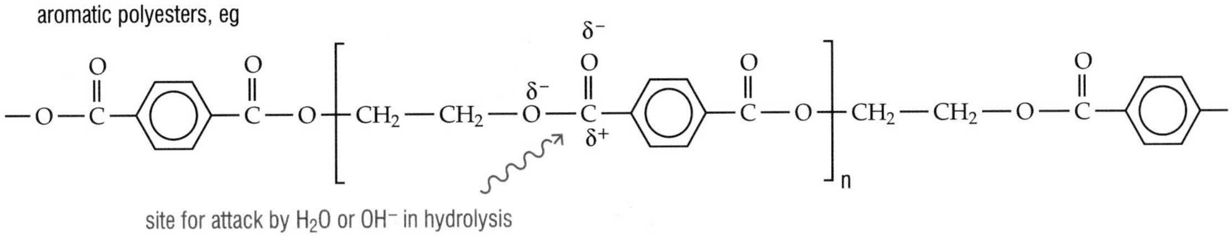

site for attack by H_2O or OH^- in hydrolysis

(b) Absorbable sutures made from aliphatic polyesters, eg

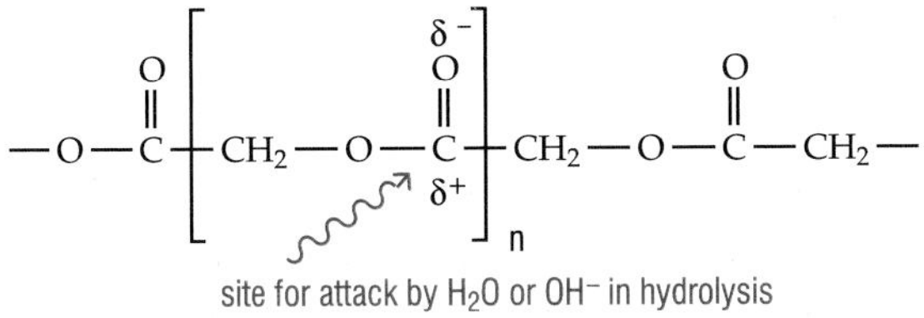

site for attack by H_2O or OH^- in hydrolysis

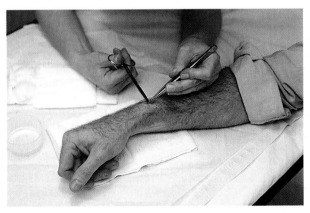

Fig 13.29 The doctor can remove stitches from external wounds but needs to use absorbable sutures for internal repairs.

Hydrolysis of esters and amides

The hydrolysis of both esters and amides is a nucleophilic substitution reaction (Fig 13.30). Esters hydrolyse rather more easily than amides. Aliphatic esters hydrolyse more easily than aromatic esters.

Fig 13.30 A mechanism for the hydrolysis of esters and amides.

(a) Hydrolysis of esters

(b) Hydrolysis of amides

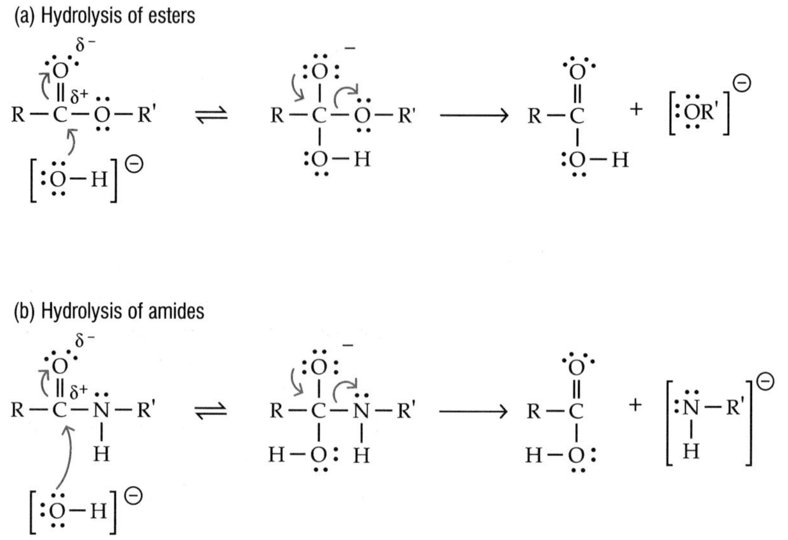

The hydrolysis of covalent halides

Characteristically, non-metals form molecular covalent halides. They differ from ionic halides in their behaviour towards water. Ionic halides tend to dissolve and so form neutral solutions. Covalent halides hydrolyse exothermically to give acidic solutions. Exceptions are the tetrahalides of carbon which do not hydrolyse under any reaction conditions.

To understand the mechanisms of these hydrolysis reactions, the following experimental observations provide a starting point.

* The halides of carbon, CF_4, CCl_4, CBr_4 and CI_4, do not hydrolyse even with hot alkali.
* The halides of silicon, SiF_4, $SiCl_4$, $SiBr_4$ and SiI_4 hydrolyse, the ease of hydrolysis decreasing with increasing relative molecular mass of the halide. SiF_4 and $SiCl_4$ react vigorously with cold water.

Consider the chlorides of carbon and silicon – tetrachloromethane and silicon tetrachloride. Both molecules are tetrahedral, with polarised bonds:

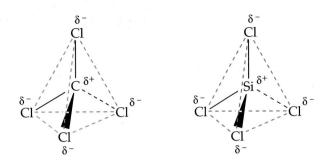

The central atom in both molecules offers a target for attack by a nucleophile. However, reaction will only occur if the nucleophile can get close enough to the central atom and if there is a mechanism by which it can attach itself to form an intermediate compound. A carbon atom is smaller than a silicon atom and, therefore, better protected by the four surrounding chlorine atoms from an approaching nucleophile.

Further, a look at the bonding present reveals a way in which the nucleophile can bond to silicon but not to carbon at the centre of the molecules. The mechanism for the hydrolysis of silicon tetrachloride involves successive replacement (by a similar mechanism) of chlorine atoms by hydroxyl groups leads to $Si(OH)_4$, which rearranges to give hydrated silicon(IV) oxide, $SiO_2.xH_2O$:

Fig 13.31 The mechanism of hydrolysis of silicon tetrachloride.

reaction intermediate

successive replacement to give $Si(OH)_4$ (rearranges to give hydrated silicon (IV) oxide: $SiO_2.xH_2O$

The difference between CCl_4 and $SiCl_4$ is the availability of 3d orbitals in silicon to bond a nucleophile. Silicon may use a vacant 3d orbital to accept a pair of electrons from a nucleophile and form a coordinate bond. For carbon, the outer electrons are in the $n = 2$ energy level and when CCl_4 forms, all $n = 2$ orbitals are occupied. None are vacant to accept a lone pair of electrons in coordinate bond formation (Fig 13.31).

Fig 13.32 An electrons-in-boxes diagram to represent the bonding present in the first intermediate compound formed during the hydrolysis of silicon tetrachloride.

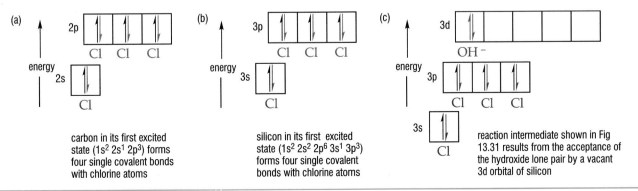

(a) carbon in its first excited state ($1s^2\ 2s^1\ 2p^3$) forms four single covalent bonds with chlorine atoms

(b) silicon in its first excited state ($1s^2\ 2s^2\ 2p^6\ 3s^1\ 3p^3$) forms four single covalent bonds with chlorine atoms

(c) reaction intermediate shown in Fig 13.31 results from the acceptance of the hydroxide lone pair by a vacant 3d orbital of silicon

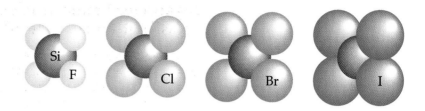

Now consider the other halides of silicon. The protective effect of the halogen atoms explains the trend in ease of hydrolysis observed within the halides of silicon. Even though Si—I bonds are the weakest, the bulky iodine atoms protect the central silicon atom from attack by a nucleophile (Fig 13.33).

21 $SnCl_4$, $SnBr_4$ and SnI_4 exist as tetrahedral molecules. Which is likely to be the most easily hydrolysed?

SPOTLIGHT

Fig 13.34 (a) Silicones are excellent sealants for bathrooms, kitchens and outdoor windows. They remain flexible over a wide temperature range, neither turning runny in hot weather nor brittle in cold weather. They show excellent water-repellency also.

SILICONES

Silicones are polymers based on a —Si—O—Si—O—Si—O— skeleton. Organic polymers are based on a carbon skeleton, —C—C—C—C—C—C—. The silicon 'equivalents' to many naturally occurring organic compounds can be synthesised, but with a silicon-oxygen backbone instead of a carbon skeleton. Silicone oils, greases and rubbers are more expensive than their hydrocarbon counterparts but they have one very important advantage – their physical properties. In particular how easily they flow and the fact that they are much less affected by variations in temperature. The boots worn by Neil Armstrong when he walked on the moon were made from silicone rubber. Conventional rubber (a hydrocarbon polymer) would have become brittle and useless at the very low temperatures on the moon's surface.

Silicones may be prepared by controlled hydrolysis of alkyl silicon chlorides, $R_xSiCl_{(4-x)}$. The polymer-forming reactions are condensation reactions. The first step involves nucleophilic substitution of chlorine by a hydroxyl group, followed by the elimination of a water molecule from two neighbouring Si—OH groups. The result is a Si—O—Si linkage.

$$2(CH_3)_3SiCl \xrightarrow{+2H_2O} 2(CH_3)_3SiOH \xrightarrow{-H_2O} (CH_3)_3Si—O—Si(CH_3)_3$$
hexamethyldisiloxane

$$(CH_3)_3SiCl \xrightarrow{H_2O} (CH_3)_3Si—O—$$ terminal group

$$(CH_3)_2SiCl_2 \xrightarrow{H_2O} \begin{array}{c} CH_3 \\ | \\ —O—Si—O— \\ | \\ CH_3 \end{array}$$ chain-forming group

$$(CH_3)SiCl_3 \xrightarrow{H_2O} \begin{array}{c} CH_3 \\ | \\ —O—Si—O— \\ | \\ O \\ | \end{array}$$ branching and bridging group

(b) Silicone polymers form through a series of condensation reactions.

Ligand exchange: nucleophilic substitution in coordination compounds

Ligand exchange reactions are nucleophilic substitutions. Precise mechanisms are complex but it will be sufficient here to recognise two distinct pathways (Fig 13.35). One involves increasing the coordination number of the central metal ion, and the formation of a 7-coordinate intermediate. The other involves decreasing the coordination number of the central metal ion, and the formation of a 5-coordinate intermediate. These are analogous to the mechanisms proposed for the hydrolysis of primary and tertiary halogenoalkanes respectively.

Fig 13.35 Ligand exchange reactions of coordination compounds are nucleophilic substitutions. There are two possible pathways: (a) via a 5-coordinate intermediate, (b) via a 7-coordinate intermediate.

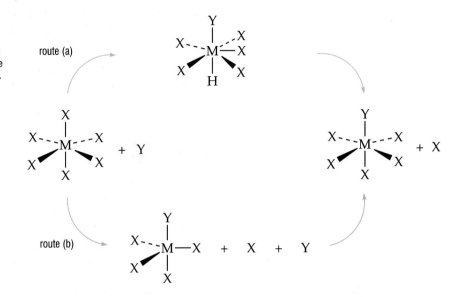

QUESTIONS

13.15 Explain why methane reacts with chlorine only in the presence of ultraviolet light.

13.16 Explain how bromoethane, 1,1-dibromoethane, 1,2-dibromoethane and butane (amongst many other compounds) are formed when bromine reacts with ethane in the presence of light.

13.17 Why is nitrobenzene more difficult to nitrate than benzene?

13.18 For each of the following reactions (i) write a structural formula for the organic product and (ii) suggest a mechanism by showing the rearrangement of electron pairs:
(a) 1-bromopentane and hydroxide ions
(b) ethanoyl chloride and hydroxide ions
(c) ethanoyl chloride and methanol
(d) 1-iodobutane and hydrogensulphide ions (HS⁻)

13.19 Predict the order of hydrolysis of the following: chloroethane; 1-fluorobutane; 2-iodohexane; bromomethane.

13.20 Explain why an ethanol solution of potassium cyanide is used for the conversion of bromoethane to propanenitrile, rather than an alkaline solution of the salt.

13.21 Nitrogen trichloride reacts with water to give NH_3 and $HOCl$. In contrast, phosphorus trichloride reacts with water to give $P(OH)_3$ and HCl.
   (a) Draw shorthand bond diagrams for NCl_3 and PCl_3, and indicate any bond polarisation.
   (b) Explain the different products of hydrolysis formed when these two compounds react with water.

13.22 An important starting compound for the manufacture of silicones is dimethylsilicon dichloride, $(CH_3)_2SiCl_2$. Explain why hydrolysis of this compound involves breaking Si—Cl bonds but not Si—C bonds.

13.23 Suggest two possible mechanisms for the formation of $[Fe(H_2O)_5SCN]^{2+}$, the cause of the intense red solution obtained when SCN^- is added to $[Fe(H_2O)_6]^{3+}$.

Answers to Quick Questions: Chapter 13

1   Because they are much stronger than single covalent bonds.
2   Br_2 homolytic; HBr heterolytic; H_2O heterolytic; CCl_4 homolytic
3   Two
4   The oxonium ions has one more proton than electrons; the nitrate ion has one more electron than protons.
5   A ligand exchange reaction occurs when one or more ligands are substituted in a coordination compound by one or more different ligands.
6

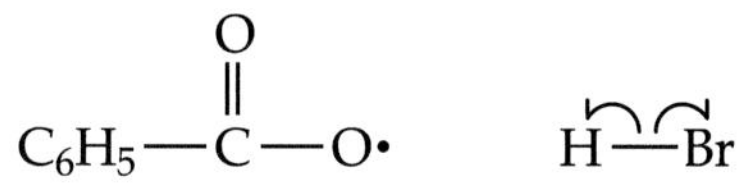

7   Heterolytic
8   Because it has one more proton than electrons.
9   Because the formation of a carbocation could lead to further reaction with any anion present. In water, either $OH^-(aq)$ or H_2O molecules themselves would attack the carbocation.
10   BF_3
11

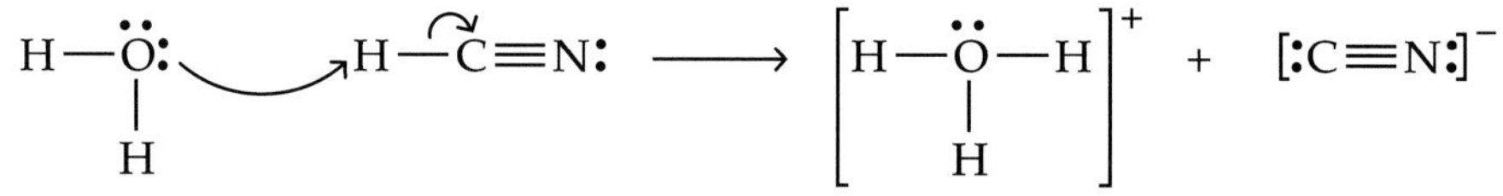

12   Because the hydroxide ion is also a strong nucleophile and so would compete with the cyanide ion.
13   Because the product, 2-hydroxy-2-methylpropanenitrile, does not have an asymmetric carbon centre.
14   Two possible reasons. (i) There is insufficient room around the central phosphorus atom for six iodide ions (which are much larger than the six chloride ions). (ii) PI_5 is unstable because the energy required to promote a 3s electron to a vacant 3d orbital is not repaid by the energy released when two extra P—I bonds form.
15   Because further reactions of CH_3Cl can occur, with hydrogen atoms being replaced by chlorine atoms.

16 They are isoelectronic, that is, have the same number of electrons as one another.
17 Heterolytic
18 Because it is an optically active (chiral) compound.
19 It suggests that the rate is independent of hydroxide ion concentration and, therefore, these are not involved in the rate-determining step.
20 Reaction is likely to proceed by two different mechanisms; one involving two participants in the rate-deterring step (like primary halogenoalkanes) and the other involving one participant (like tertiary halogenoalkanes).
21 $SnCl_4$

THE DATABASE

THE PURPOSE OF THE DATABASE

Data about elements and their compounds is vast. It would be impossible to remember all the available information. The Database contains information about the physical and chemical properties of elements and their compounds. You should refer to it when you need numerical data or information about chemical reactions.

Remember

- Material chosen for inclusion in the Database is that relevant to A- and AS Chemistry syllabuses and other post-GCSE courses.

- Because of this, the Database provides a handy reference for revision. Key chemical reactions (those considered to be most important) are given in the Datasheets.

- The main function of the Database, however, is to help you solve problems without the need to remember lots of facts. Inevitably you will build up such a knowledge through use of the Database.

- The Database is not exhaustive. Much more data could have been included if space permitted.

THE STRUCTURE OF THE DATABASE

The Periodic Table

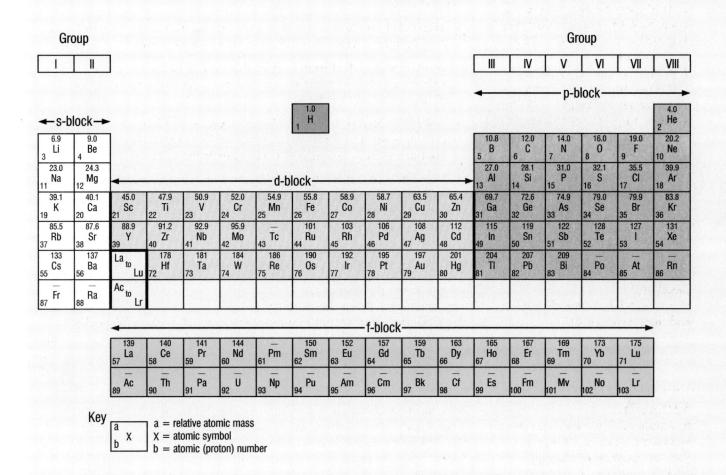

Group

I	II

Group

III	IV	V	VI	VII	VIII

←s-block→

p-block

1.0 H 1

4.0 He 2

d-block

6.9 Li 3	9.0 Be 4

| 10.8 B 5 | 12.0 C 6 | 14.0 N 7 | 16.0 O 8 | 19.0 F 9 | 20.2 Ne 10 |

| 23.0 Na 11 | 24.3 Mg 12 |

| 27.0 Al 13 | 28.1 Si 14 | 31.0 P 15 | 32.1 S 16 | 35.5 Cl 17 | 39.9 Ar 18 |

| 39.1 K 19 | 40.1 Ca 20 | 45.0 Sc 21 | 47.9 Ti 22 | 50.9 V 23 | 52.0 Cr 24 | 54.9 Mn 25 | 55.8 Fe 26 | 58.9 Co 27 | 58.7 Ni 28 | 63.5 Cu 29 | 65.4 Zn 30 | 69.7 Ga 31 | 72.6 Ge 32 | 74.9 As 33 | 79.0 Se 34 | 79.9 Br 35 | 83.8 Kr 36 |

| 85.5 Rb 37 | 87.6 Sr 38 | 88.9 Y 39 | 91.2 Zr 40 | 92.9 Nb 41 | 95.9 Mo 42 | — Tc 43 | 101 Ru 44 | 103 Rh 45 | 106 Pd 46 | 108 Ag 47 | 112 Cd 48 | 115 In 49 | 119 Sn 50 | 122 Sb 51 | 128 Te 52 | 127 I 53 | 131 Xe 54 |

| 133 Cs 55 | 137 Ba 56 | La to Lu | 178 Hf 72 | 181 Ta 73 | 184 W 74 | 186 Re 75 | 190 Os 76 | 192 Ir 77 | 195 Pt 78 | 197 Au 79 | 201 Hg 80 | 204 Tl 81 | 207 Pb 82 | 209 Bi 83 | Po 84 | — At 85 | — Rn 86 |

| — Fr 87 | — Ra 88 | Ac to Lr |

f-block

| 139 La 57 | 140 Ce 58 | 141 Pr 59 | 144 Nd 60 | — Pm 61 | 150 Sm 62 | 152 Eu 63 | 157 Gd 64 | 159 Tb 65 | 163 Dy 66 | 165 Ho 67 | 167 Er 68 | 169 Tm 69 | 173 Yb 70 | 175 Lu 71 |

| — Ac 89 | — Th 90 | — Pa 91 | — U 92 | — Np 93 | — Pu 94 | — Am 95 | — Cm 96 | — Bk 97 | — Cf 98 | — Es 99 | — Fm 100 | — Mv 101 | — No 102 | — Lr 103 |

Key

a
X
b

a = relative atomic mass
X = atomic symbol
b = atomic (proton) number

DATASHEET 1: Atomic properties of the elements

Element	Symbol	Atomic number	Relative atomic mass	Element	Symbol	Atomic number	Relative atomic mass
Actinium	Ac	89	(227)	Neodymium	Nd	60	144.24
Aluminium	Al	13	26.9815	Neon	Ne	10	20.179
Americium	Am	95	(243)	Neptunium	Np	93	(237)
Antimony	Sb	51	121.75	Nickel	Ni	28	58.71
Argon	Ar	18	39.948	Niobium	Nb	41	92.9064
Arsenic (α, grey)	As	33	74.9216	Nitrogen	N	7	14.0067
Astatine	At	85	(210)	Nobelium	No	102	(254)
Barium	Ba	56	137.34	Osmium	Os	76	190.2
Berkelium	Bk	97	(247)	Oxygen	O	8	15.9994
Beryllium	Be	4	9.01218	Palladium	Pd	46	106.4
Bismuth	Bi	83	208.9806	Phosphorus	P	15	30.9738
Boron	B	5	10.81	Platinum	Pt	78	195.09
Bromine	Br	35	79.904	Plutonium	Pu	94	(242)
Cadmium	Cd	48	112.40	Polonium	Po	84	(210)
Caesium	Cs	55	132.9055	Potassium	K	19	39.102
Calcium	Ca	20	40.08	Praseodymium	Pr	59	140.9077
Californium	Cf	98	(251)	Promethium	Pm	61	(147)
Carbon (graphite)	C	6	12.011	Protoactinium	Pa	91	(231)
Cerium	Ce	58	140.12	Radium	Ra	88	(226)
Chlorine	Cl	17	35.453	Radon	Rn	86	(222)
Chromium	Cr	24	51.996	Rhenium	Re	75	186.2
Cobalt	Co	27	58.9322	Rhodium	Rh	45	102.9055
Copper	Cu	29	63.546	Rubidium	Rb	37	85.4678
Curium	Cm	96	(247)	Ruthenium	Ru	44	101.07
Dysprosium	Dy	66	162.50	Samarium	Sm	62	150.4
Einsteinium	Es	99	(254)	Scandium	Sc	21	44.9559
Erbium	Er	68	167.26	Selenium	Se	34	78.96
Europium	Eu	63	151.96	Silicon	Si	14	28.086
Fermium	Fm	100	(253)	Silver	Ag	47	107.868
Fluorine	F	9	18.9984	Sodium	Na	11	22.9898
Francium	Fr	87	(223)	Strontium	Sr	38	87.62
Gadolinium	Gd	64	157.25	Sulphur (α, rhombic)	S	16	32.06
Gallium	Ga	31	69.72	Tantalum	Ta	73	180.9479
Germanium	Ge	32	72.59	Technetium	Tc	43	(99)
Gold	Au	79	196.9665	Tellurium	Te	52	127.60
Hafnium	Hf	72	178.49	Terbium	Tb	65	158.9254
Helium	He	2	4.00260	Thallium	Tl	81	204.37
Holmium	Ho	67	164.9303	Thorium	Th	90	232.0381
Hydrogen	H	1	1.0080	Thulium	Tm	69	168.9342
Indium	In	49	114.82	Tin (white)	Sn	50	118.69
Iodine	I	53	126.9045	Titanium	Ti	22	47.90
Iridium	Ir	77	192.22	Tungsten	W	74	183.85
Iron	Fe	26	55.847	Uranium	U	92	238.029
Krypton	Kr	36	83.80	Vanadium	V	23	50.9414
Lanthanum	La	57	138.9055	Xenon	Xe	54	131.30
Lawrencium	Lr	103	(257)	Yiterbium	Yb	70	173.04
Lead	Pb	82	207.2	Yttrium	Y	39	88.9059
Lithium	Li	3	6.941	Zinc	Zn	30	65.37
Lutetium	Lu	71	174.97	Zirconium	Zr	40	91.22
Magnesium	Mg	25	54.9380				
Manganese	Mn	25	54.9380				
Mendelevium	Md	101	(256)				
Mercury	Hg	80	200.59				
Molybdenum	Mo	42	95.94				

DATASHEET 1 *(continued)*

Element	Atomic radius /nm	Ionic radius /nm	First ionisation energy /kJ mol^{-1}	Electron affinity /kJ mol^{-1}	Electronegativity (Pauling)
H	0.037	0.208	1312.0	−72.8	2.20
He	0.122‡	–	2372.3	+21	–
Li	0.152*	0.060(+1)	513.3	−59.6	0.98
Be	0.112*	0.031(+2)	899.4	+241	1.57
B	0.080	0.020(+3)	800.6	−26.7	2.04
C	0.077	0.015(+4)	1086.2	−121.9	2.55
N	0.074	0.171(−3)	1402.3	0	3.04
O	0.074	0.140(−2)	1313.9	−141.0	3.44
F	0.072	0.136(−1)	1681	−328	3.98
Ne	0.160‡	–	2080.6	+29	–
Na	0.186*	0.095(+1)	495.8	−52.9	0.93
Mg	0.160*	0.065(+2)	737.7	+230	1.31
Al	0.143* 0.125	0.050(+3)	577.4	−42.5	1.61
Si	0.117	0.041(+4)	786.5	−133.6	1.61
P	0.110	0.212(−3)	1011.7	−72.0	2.19
S	0.104	0.184(−2)	999.6	−200.4	2.58
Cl	0.099	0.181(−1)	1251.1	−349.0	3.16
Ar	0.192‡	–	1520.4	+34	–
K	0.231*	0.133(+1)	418.8	−48.4	0.82
Ca	0.197*	0.099(+2)	589.7	+156	1.00
Sc	0.160*	0.081(+3)	631	−18.1	1.36
Ti	0.146*	0.090(+2)	658	−7.6	1.54
V	0.131*	0.074(+3)	650	−50.7	1.63
Cr	0.125*	0.069(+3)	652.7	−64.3	1.66
Mn	0.129*	0.080(+2)	717.4	<0	1.55
Fe	0.126*	0.076(+2) 0.064(+3)	759.3	−15.7	1.83
Co	0.125*	0.078(+2) 0.063(+3)	760.0	−63.8	1.88
Ni	0.124*	0.078(+3)	736.7	−15.6	1.91
Cu	0.128*	0.096(+1) 0.069(+2)	745.4	+118.5	1.90
Zn	0.133*	0.074(+2)	906.4	−9	1.65
Br	0.114	0.195(−1)	1139.9	−324.7	2.96
Rb	0.244*	0.148(+1)	403.0	−46.9	0.82
Sr	0.215*	0.113(+2)	549.5	+167	0.95
Ag	0.144*	0.126(+1)	731.0	−125.7	1.93
I	0.133	0.216(−1)	1008.4	−295	2.66
Xe	0.217‡	–	1170.4	+41	2.6
Cs	0.262*	0.169(+1)	375.7	−45.5	0.79
Ba	0.217*	0.135(+2)	502.8	+52.0	0.89
Pt	0.138*	0.052(+2)	870	−205.3	2.28
Au	0.144*	0.137(+1)	890.1	−222.8	2.54
Pb	0.175*	0.120(+2)	715.5	−35.1	2.33
Fr	0.270*	0.176(+1)	400	−44	0.7

Note All radii are single bond covalent radii unless designated by * (= metallic radius) or ‡ (= van der Waals radius). First ionisation energy data relates to the process: $M(g) \rightarrow M^+(g) + e^-$ for a mole of gaseous atoms, measured under standard conditions. The electron affinity data relates to the process: $M(g) + e^- \rightarrow M^-(g)$ for a mole of gaseous atoms measured under standard conditions. Some electron affinity values are calculated, not measured values. Not all values for a physical property are quoted to the same degree of accuracy. This table reflects the degree of accuracy to which they are known.

DATASHEET 2: Physical properties of the elements

Element	MP /K	BP /K	Density /kg m^3	Electrolytic conductivity /$\mu\Omega^{-1}$ cm^{-1}	$\Delta H^{\circ}_{vap,m}$(298 K) /kJ mol^{-1}	$\Delta H^{\circ}_{fus,m}$(298 K) /kJ mol^{-1}	$\Delta H^{\circ}_{atom,m}$(298 K) /kJ mol^{-1}
H	14	20	0.089	–	0.46	0.12	218
He	1	4	0.179		0.08	0.02	–
Li	454	1620	534	0.108	134.7	4.60	161
Be	1553	3243	1848	0.25	308.8	9.80	321
B	2573	3931	2340	10^{-12}	538.9	22.2	590
C (d)	3820	5100	3513	–	710.9	105.0	
(g)	3640s	–	2260	0.0007			715
N	63	77	1.251	–	5.58	0.72	473
O	55	90	1.429	–	6.82	0.44	248
F	54	85	1.696	–	6.54	5.10	79.1
Ne	24	27	0.900	–	1.74	0.32	–
Na	371	1156	971	0.218	89.0	2.64	109
Mg	922	1363	1738	0.224	128.7	9.04	150
Al	934	2740	2698	0.382	293.72	10.67	314
Si	1683	2628	2329	0.10	383.3	39.6	439
P (P$_4$)	317	553	1820	10^{-17}	51.9	2.51	315
S (α)	386	718	2070	10^{-23}	9.62	1.23	
(β)	392						223
Cl	172	239	3.214	–	20.40	6.41	121
Ar	84	87	1.784	–	6.53	1.21	–
K	337	1047	862	0.143	77.53	2.40	90.0
Ca	1112	1757	1550	0.218	149.95	9.33	193
Sc	1814	3104	2989	0.015	304.8	15.9	340
Ti	1933	3560	4540	0.024	428.9	20.9	469
V	2160	3650	6110	0.04	458.6	17.6	515
Cr	2130	2945	7190	0.078	348.78	15.3	398
Mn	1517	2235	7440	0.054	219.7	14.4	279
Fe	1808	3023	7874	0.10	351.0	14.9	418
Co	1768	3143	8900	0.16	382.4	15.2	427
Ni	1726	3005	8902	0.145	371.8	17.6	431
Cu	1357	2840	8960	0.593	304.6	13.0	339
Zn	693	1180	7133	0.167	115.3	6.67	130
Br	266	332	7.59	10^{-18}	30.0	10.8	112
Rb	312	961	1532	0.080	69.2	2.20	85.8
Sr	1042	1657	2540	0.043	138.91	9.16	164
Ag	1235	2485	10 500	0.62	255.1	11.3	289
I	387	458	4930	10^{-15}	41.67	15.27	107
Xe	161	166	5.90	–	12.65	3.10	–
Cs	302	952	1873	0.053	65.90	2.09	78.7
Ba	1002	1910	3594	0.016	150.9	7.66	176
Pt	2045	4100	21 450	0.094	510.5	19.7	565
Au	1338	3080	19 320	0.42	324.4	12.7	369
Pb	601	2013	11 350	0.046	179.4	5.12	196
Fr	300	950	–	–	–	–	–

Note Data for carbon is given for diamond (d) and for graphite (g)

Standard electrode and redox potentials (at 298 K)

Electrode reaction	$E^\ominus$/V
$Ag^+ + e^- = Ag$	+0.80
$Al^{3+} + 3e^- = Al$	−1.66
$Ba^{2+} + 2e^- = Ba$	−2.90
$\frac{1}{2}Br_2 + e^- = Br^-$	+1.07
$Ca^{2+} + 2e^- = Ca$	−2.87
$\frac{1}{2}Cl_2 + e^- = Cl^-$	+1.36
$HOCl + H^+ + e^- = \frac{1}{2}Cl_2 + H_2O$	+1.64
$Co^{2+} + 2e^- = Co$	−0.28
$Co^{3+} + e^- = Co^{2+}$	+1.82
$[Co(NH_3)_6]^{2+} + 2e^- = Co + 6NH_3$	−0.43
$Cr^{2+} + 2e^- = Cr$	−0.91
$Cr^{3+} + 3e^- = Cr$	−0.74
$Cr^{3+} + e^- = Cr^{2+}$	−0.41
$\frac{1}{2}Cr_2O_7^{2-} + 7H^+ + 3e^- = Cr^{3+} + \frac{7}{2}H_2O$	+1.33
$Cu^+ + e^- = Cu$	+0.52
$Cu^{2+} + 2e^- = Cu$	+0.34
$Cu^{2+} + e^- = Cu^+$	+0.15
$[Cu(NH_3)_4]^{2+} + 2e^- = Cu + 4NH_3$	−0.05
$\frac{1}{2}F_2 + e^- = F^-$	+2.87
$Fe^{2+} + 2e^- = Fe$	−0.44
$Fe^{3+} + 3e^- = Fe$	−0.04
$Fe^{3+} + e^- = Fe^{2+}$	+0.77
$[Fe(CN)_6]^{3-} + e^- = [Fe(CN)_6]^{4-}$	+0.36
$Fe(OH)_3 + e^- = Fe(OH)_2 + OH^-$	−0.56
$H^+ + e^- = \frac{1}{2}H_2$	0.00
$\frac{1}{2}H_2 + e^- = H^-$	−2.25
$\frac{1}{2}I_2 + e^- = I^-$	+0.54
$K^+ + e^- = K$	−2.92
$Li^+ + e^- = Li$	−3.04
$Mg^{2+} + 2e^- = Mg$	−2.38

Electrode reaction	$E^\ominus$/V
$Mn^{2+} + 2e^- = Mn$	−1.18
$Mn^{3+} + e^- = Mn^{2+}$	+1.49
$MnO_2 + 4H^+ + 2e^- = Mn^{2+} + 2H_2O$	+1.23
$MnO_4^- + e^- = MnO_4^{2-}$	+0.56
$MnO_4^- + 4H^+ + 3e^- = MnO_2 + 2H_2O$	+1.67
$MnO_4^- + 8H^+ + 5e^- = Mn^{2+} + 4H_2O$	+1.52
$NO_3^- + 2H^+ + e^- = NO_2 + H_2O$	+0.81
$NO_3^- + 3H^+ + 2e^- = HNO_2 + H_2O$	+0.94
$NO_3^- + 10H^+ + 8e^- = NH_4^+ + 3H_2O$	+0.87
$Na^+ + e^- = Na$	−2.71
$Ni^{2+} + 2e^- = Ni$	−0.25
$[Ni(NH_3)_6]^{2+} + 2e^- = Ni + 6NH_3$	−0.51
$\frac{1}{2}H_2O_2 + H^+ + e^- = H_2O$	+1.77
$\frac{1}{2}O_2 + 2H^+ + 2e^- = H_2O$	+1.23
$\frac{1}{2}O_2 + H_2O + 2e^- = 2OH^-$	+0.40
$O_2 + 2H^+ + 2e^- = H_2O_2$	+0.68
$2H_2O + 2e^- = H_2 + 2OH^-$	−0.83
$Pb^{2+} + 2e^- = Pb$	−0.13
$Pb^{4+} + 2e^- = Pb^{2+}$	+1.69
$PbO_2 + 4H^+ + 2e^- = Pb^{2+} + 2H_2O$	+1.47
$S + 2e^- = S^{2-}$	−0.51
$S + 2H^+ + 2e^- = H_2S$	+0.14
$Sn^{2+} + 2e^- = Sn$	−0.14
$Sn^{4+} + 2e^- = Sn^{2+}$	+0.15
$V^{2+} + 2e^- = V$	−1.20
$V^{3+} + e^- = V^{2+}$	−0.26
$VO^{2+} + 2H^+ + e^- = V^{3+} + H_2O$	+0.34
$VO_2^+ + 2H^+ + e^- = VO^{2+} + H_2O$	+1.00
$Zn^{2+} + 2e^- = Zn$	−0.76

DATASHEET 3: Physical properties of the chlorides, hydrides and oxides of Period 3 elements

Substance	MP /K	BP /K	Substance	MP /K	BP /K
NaCl	1074	1686	H_2S	187	213
$MgCl_2$	981	1685	HCl	159	188
$AlCl_3$	463	–	Na_2O	1548s	–
$SiCl_4$	205	330	MgO	3073	3873
PCl_3	161	349	Al_2O_3	2318	3253
PCl_5	440d	–	SiO_2	1986	2503
S_2Cl_2	193	409	P_2O_3	297	447
SCl_2	191	332d	P_2O_5	856	–
NaH	1073d	–	SO_2	197	263
MgH_2	553s	–	SO_3	306	318
AlH_3	–	–	Cl_2O	253	277
SiH_4	88	162	Cl_2O_7	182	355exp
PH_3	140	185			

s = sublimes; d = decomposes; exp = explodes
No boiling points are quoted for those compounds which decompose or sublime.

DATASHEET 4: Thermochemical data

Note Data on standard molar enthalpy changes of vaporisation, fusion and atomisation are given on Datasheet 2.

Specific heat capacities of selected substances

Substance	Specific heat capacity, c, /kJ kg^{-1} K^{-1}	Substance	Specific heat capacity, c, /kJ kg^{-1} K^{-1}
aluminium	0.902	tetrachloromethane	0.86
carbon (graphite)	0.720	ethane-1,2-diol	2.42
copper	0.385	air (typical value)	1.01
gold	0.128	benzene	1.05
iron	0.451	wood (typical value)	1.76
ethanol	2.46	glass	0.84
water	4.18	cement	0.88
ice	2.06	granite	0.80
ammonia	4.70	stainless steel	0.51

Some average bond enthalpies

Bond	Average bond enthalpy /kJ mol^{-1}	Bond	Average bond enthalpy /kJ mol^{-1}
H—H	432	C≡C	835
H—F	568	C≡N	890
H—Cl	567	Cl—Cl	243
H—Br	364	F—F	159
H—I	297	Br—Br	192
H—O	464	I—I	151
H—S	339	N—H	389
C—F	439	N—N	159
C—Cl	330	N=N	418
C—Br	276	N≡N	946
C—I	238	P—Cl	331
C—C	347	S—Cl	251
C=C	611	Si—H	293
C=O	745	Si—Cl	360
C≡N	615	Si—O	368

DATASHEET 4 *(continued)*

Standard molar enthalpy changes of combustion

Substance	$\Delta H^{\ominus}_{c,m}$(298 K)/kJ mol^{-1}	Substance	$\Delta H^{\ominus}_{c,m}$(298 K)/kJ mol^{-1}	Substance	$\Delta H^{\ominus}_{c,m}$(298 K)/kJ mol^{-1}
C (graphite)	−393.5	C_2H_4	−1410.9	CH_3CH_2OH	−1366.9
C (diamond)	−395.4	C_2H_6	−1559.7	CH_3COOH	−875.3
CH_4	−890.2	C_3H_8	−2219.7	$C_6H_{12}O_6$	−2816.0
CO	−283	C_4H_{10}	−2878.6	H_2	−285.8
C_2H_2	−1300	CH_3OH	−726.3	C_6H_6	−3267.4

Standard molar enthalpies of formation

Substance	$\Delta H^{\ominus}_{f,m}$(298 K)/kJ mol^{-1}	Substance	$\Delta H^{\ominus}_{f,m}$(298 K)/kJ mol^{-1}	Substance	$\Delta H^{\ominus}_{f,m}$(298 K)/kJ mol^{-1}
CaO	−635.5	CO	−110.5	KCl	−436.7
CH_3OH	−238.9	CO_2	−393.5	NaBr	−361.1
CH_3CH_2OH	−277.7	NH_3	−46.1	MgO	−601.8
CH_3COOH	−484.2	$CaCO_3$	−1206.9	KBr	−393.8
CH_4	−74.8	CaO	−635.1	SiO_2	−910.0
C_2H_2	+226.8	$CuSO_4$	−771.4	SO_2	−296.8
CCl_4	−135.4	HCl	−92.3	SO_3	−395.7
C_2H_6	−84.7	HF	−271.1	NO_2	+33.2
C_2H_4	+52.3	H_2O(l)	−285.9	H_2S	−20.6
C_3H_8	−103.8	H_2O(g)	−241.8	NaI	−287.8
C_6H_6	+49.0	NaCl	−411.0	KI	−567.3
$C_6H_{12}O_6$	−1274				

Standard molar enthalpy changes of hydrogenation

Substance	Formula	$\Delta H^{\ominus}_{h,m}$(298 K)/kJ mol^{-1}
ethene	C_2H_4	−132
buta-1, 3-diene	C_4H_6	−239
propene	C_3H_6	−126
but-1-ene	C_4H_8	−127
cis-but-2-ene	C_4H_8	−119.5
trans-but-2-ene	C_4H_8	−115.4
benzene	C_6H_6	−246
phenylethene	$C_6H_5CH\ CH_2$	−126
cyclohexene	C_6H_{10}	−120
cyclohexa-1, 3-diene	C_6H_8	−232

Standard molar enthalpy changes of lattice formation

Substance	$\Delta H^{\ominus}_{lattice,m}$(298 K)/kJ mol^{-1}	Substance	$\Delta H^{\ominus}_{lattice,m}$(298 K)/kJ mol^{-1}
LiF	1012	KBr	689
LiCl	828	KI	645
LiBr	787	AgF	971
LiI	732	AgCl	916
NaF	929	AgBr	903
NaCl	787	AgI	887
NaBr	751	$MgCl_2$	2527
NaI	700	Na_2O	2570
KF	826	MgO	3890
KCl	717	CaO	3461
		$CaCl_2$	2255

DATASHEET 5: Spectroscopic data

Nuclear magnetic resonance: chemical shifts

Type of hydrogen atom	Chemical shift
$CH_3—C—$	9.1
$C—CH_2—C$	8.6
$CH_3—C=C$	8.4
$CH≡C$	8.0
$CH_3—C—$ (C=O)	7.9
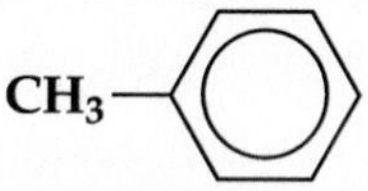 CH_3— (benzene ring)	7.8
$CH_3—O—$	6.7
$C—CH_2—Br$	6.5
$CH_3—O—C—$ (C=O)	6.3
$CH_2=C$	5.3
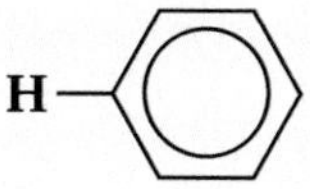 H— (benzene ring)	2.7

Note All chemical shifts are measured relative to tetramethylsilane, $(CH_3)_4Si$.

Ultraviolet/visible absorption for common chromophores in organic molecules

Chromophore	Approximate absorption maxima/nm
$—C—$ (C=O)	188, 279
$—CH=CH_2$	190
$—C—O—$ (C=O)	204
$CH_2=CH—CH=CH_2$	217
$CH_2=CH—C—$ (C=O)	219, 324

DATASHEET 5 *(continued)*

Infrared absorption

Bond	Class of organic compound	Frequency of absorption, wavenumber/cm^{-1}
C—H	alkanes	2850 – 2960
	alkenes	3010 – 3095
	alkynes	3250 – 3300
	arenes	3030 – 3080
C—C		750 – 1100
C=C		1620 – 1680
C≡C		2100 – 2250
C═C	arenes	1500 – 1600
C—O	alcohols, ethers, carboxylic acids, esters	1000 – 1300
C=O	aldehydes, ketones,	1680 – 1750
	carboxylic acids, esters, amides	1640 – 1680
C—N	amines	1180 – 1360
C≡N	nitriles	2210 – 2260
O—H	non-hydrogen-bonded (alcohols, phenols)	3580 – 3670
	hydrogen-bonded (alcohols, phenols)	3230 – 3550
	hydrogen-bonded (carboxylic acids)	2500 – 3000
N—H	non-hydrogen-bonded (amines, amides)	3320 – 3560
	hydrogen-bonded (primary and secondary amines)	3100 – 3400
C—Cl		600 – 800
C—Br		500 – 600
C—I		500

Note All data refers to the stretching frequencies of the bonds.

DATASHEET 6: Inorganic compounds

The following are some common cations and anions which exist in aqueous solution. Many combine to form inorganic compounds.

Cations

+1 Li^+ Na^+ K^+ Cs^+ Rb^+ NH_4^+ Ag^+

+2 Mg^{2+} Ca^{2+} Sr^{2+} Ba^{2+} Sn^{2+} Pb^{2+} Mn^{2+} Fe^{2+} Co^{2+} Ni^{2+} Cu^{2+} Zn^{2+} TiO^{2+} VO^{2+}

+3 Al^{3+} Cr^{3+} Fe^{3+}

Anions

−1 OH^- SH^- F^- Cl^- Br^- I^- NO_3^- NO_2^- HCO_3^- HSO_4^- HSO_3^- $H_2PO_4^-$ ClO^- ClO_3^- ClO_4^- MnO_4^-

−2 O^{2-} S^{2-} CO_3^{2-} SO_3^{2-} HPO_4^{2-} CrO_4^{2-} $Cr_2O_7^{2-}$ MnO_4^{2-} FeO_4^{2-}

−3 PO_4^{3-} VO_4^{3-}

The general reactions shown below apply to those ions given above in **bold**. Information about the others will be found in the relevant Datasheets.

OXIDES

Preparation

Reaction of metal and oxygen; action of heat on hydroxide or carbonate (see below).

Action of water and heat

$M_2O(s)$ Basic oxides, soluble in water and acids:
$$M_2O(s) + H_2O(l) \longrightarrow 2MOH(aq); \quad M_2O(s) + 2H^+(aq) \longrightarrow 2M^+(aq) + H_2O(l)$$

$MO(s)$ Basic oxides, insoluble in water, except CaO, SrO and BaO:
$$MO(s) + H_2O(l) \longrightarrow M(OH)_2(s);$$

Soluble in acids:
$$MO(s) + 2H^+(aq) \longrightarrow M^{2+}(aq) + H_2O(l)$$

SnO, PbO and ZnO are amphoteric, soluble in acids and alkalis:
$$MO(s) + 2OH^-(aq) + H_2O(l) \longrightarrow [M(OH)_4]^{2-}(aq)$$

$M_2O_3(s)$ Amphoteric oxides, insoluble in water, soluble in acids and alkalis:
$$M_2O_3(s) + 6H^+(aq) \longrightarrow 2M^{3+}(aq) + 3H_2O(l)$$
$$M_2O_3(s) + 2OH^-(aq) + 3H_2O(l) \longrightarrow 2[M(OH)_4]^-(aq)$$

HYDROXIDES

Preparation

Soluble hydroxides by reaction of metal and water.
Insoluble hydroxides by addition of $OH^-(aq)$ to soluble salt, $M^{x+}(aq)$:
$$2M(s) + 2H_2O(l) \longrightarrow 2MOH(aq) + H_2(g)$$
$$M^{x+}(aq) + xOH^-(aq) \longrightarrow M(OH)_x(s)$$

Action of water and heat

$MOH(s)$ Soluble in water to give $M^+(aq)$ and $OH^-(aq)$; solutions react with acids:
$$H^+(aq) + OH^-(aq) \longrightarrow H_2O(l)$$

Do not decompose on heating.

$M(OH)_2(s)$ Insoluble in water ($Ba(OH)_2$ slightly soluble, $Ca(OH)_2$ and $Sr(OH)_2$ sparingly soluble).

Soluble in acids:
$$M(OH)_2(s) + 2H^+(aq) \longrightarrow M^{2+}(aq) + 2H_2O(l)$$

Hydroxides of Sn, Pb and Zn also soluble in alkalis:
$$M(OH)_2(s) + 2OH^-(aq) \longrightarrow [M(OH)_4]^{2-}(aq)$$

Decompose on heating:
$$M(OH)_2(s) \longrightarrow MO(s) + H_2O(g)$$

M(OH)$_3$(s) Insoluble in water, soluble in acids and alkalis:
$$M(OH)_3(s) + 3H^+(aq) \longrightarrow M^{3+}(aq) + 3H_2O(l)$$
$$M(OH)_3(s) + OH^-(aq) \longrightarrow [M(OH)_4]^-(aq)$$

Decompose on heating:
$$2M(OH)_3(s) \longrightarrow M_2O_3(s) + 3H_2O(l)$$

CHLORIDES

Preparation

Soluble chlorides by reaction of dilute HCl(aq) and reactive metal, oxide, hydroxide or carbonate (as appropriate):
$$Sn(s) + 2HCl(aq) \longrightarrow SnCl_2(aq) + H_2(g)$$
$$CuO(s) + 2HCl(aq) \longrightarrow CuCl_2(aq) + H_2O(l)$$
$$NaOH(aq) + HCl(aq) \longrightarrow NaCl(aq) + H_2O(l)$$
$$CaCO_3(s) + 2HCl(aq) \longrightarrow CaCl_2(aq) + CO_2(g) + H_2O(l)$$

Chlorides which are easily hydrolysed are prepared by the action of dry chlorine on metal. For example:
$$2Al(s) + 3Cl_2(g) \longrightarrow 2AlCl_3(s)$$

Action of water and heat

MCl(s) Soluble in water to give $M^+(aq)$ and $Cl^-(aq)$.

MCl$_2$(s) Soluble in water to give $M^{2+}(aq)$ and $2Cl^-(aq)$. Some hydrolysis may occur:
$$MCl_2(s) + H_2O(l) \longrightarrow M(OH)Cl(s) + HCl(g)$$

MCl$_3$(s) Hydrolysed:
$$MCl_3(s) + 3H_2O(l) \longrightarrow M(OH)_3(s) + 3HCl(g)$$

Chlorides do not decompose on heating.

CARBONATES AND HYDROGENCARBONATES

Preparation

Reaction of carbon dioxide and alkali:
$$MOH(aq) + CO_2(g) \longrightarrow MHCO_3(aq) \text{ (may be isolated by evaporation)}$$
$$2MHCO_3(aq) + CO_2(g) \longrightarrow M_2CO_3(aq) + H_2O(l)$$
$$M(OH)_2(aq) + CO_2(g) \longrightarrow MCO_3(s) + H_2O(l)$$
$$MCO_3(s) + CO_2(g) + H_2O(l) \longrightarrow M(HCO_3)_2(aq) \text{ (cannot be isolated from solution)}$$

Action of water and heat

M$_2$CO$_3$(s) Soluble in water to give $M^+(aq)$ and $CO_3^{2-}(aq)$.

Carbonates do not decompose on heating but hydrogencarbonates do:
$$2MHCO_3(s) \longrightarrow M_2CO_3(s) + CO_2(g) + H_2O(l)$$

MCO$_3$(s) Generally insoluble in water.

Decompose on heating:
$$MCO_3(s) \longrightarrow MO(s) + CO_2(g)$$

Solutions of hydrogencarbonates decompose on heating:
$$M(HCO_3)_2(aq) \longrightarrow MCO_3(s) + CO_2(g) + H_2O(l)$$

M$_2$(CO$_3$)$_3$(s) Not stable.

NITRATES

Preparation

Reaction of cold, dilute HNO$_3$(aq) on reactive metal, oxide, hydroxide or carbonate (as appropriate). For example:
$$Mg(s) + 2HNO_3(aq) \longrightarrow Mg(NO_3)_2(aq) + H_2(g)$$
$$ZnO(s) + 2HNO_3(aq) \longrightarrow Zn(NO_3)_2(aq) + H_2O(l)$$
$$KOH(aq) + HNO_3(aq) \longrightarrow KNO_3(aq) + H_2O(l)$$
$$NiCO_3(s) + 2HNO_3(aq) \longrightarrow Ni(NO_3)_2(aq) + CO_2(g) + H_2O(l)$$

Action of water and heat

All nitrates are soluble in water to give $M^{x+}(aq)$ and $NO_3^-(aq)$.

MNO_3 decomposes on heating:
$$2MNO_3(s) \longrightarrow 2MNO_2(s) + O_2(g)$$

Other nitrates decompose on heating in other ways. For example:
$$4LiNO_3(s) \longrightarrow 2Li_2O(s) + 4NO_2(g) + O_2(g)$$
$$2Mg(NO_3)_2(s) \longrightarrow 2MgO(s) + 4NO_2(g) + O_2(g)$$
$$2AgNO_3(s) \longrightarrow 2Ag(s) + 2NO_2(g) + O_2(g)$$

SULPHATES AND HYDROGENSULPHATES

Preparation

Soluble sulphates prepared by action of dilute $H_2SO_4(aq)$ on reactive metal, oxide, hydroxide or carbonate (as appropriate). For example:
$$Fe(s) + H_2SO_4(aq) \longrightarrow FeSO_4(aq) + H_2(g)$$
$$CuO(s) + H_2SO_4(aq) \longrightarrow CuSO_4(aq) + H_2O(l)$$
$$2KOH(aq) + H_2SO_4(aq) \longrightarrow K_2SO_4(aq) + 2H_2O(l)$$
$$CoCO_3(s) + H_2SO_4(aq) \longrightarrow CoSO_4(aq) + CO_2(g) + H_2O(l)$$

Hydrogensulphates of Group I metals prepared by action of dilute $H_2SO_4(aq)$ on metal hydroxide in a 1:1 mole ratio:
$$MOH(s) + H_2SO_4(aq) \longrightarrow MHSO_4(aq) + H_2O(l)$$

Insoluble sulphates prepared by precipitation reactions:
$$M^{2+}(aq) + SO_4^{2-}(aq) \longrightarrow MSO_4(s)$$

Action of water and heat

Sulphates are generally soluble in water to give $M^{x+}(aq)$ and $SO_4^{2-}(aq)$. $CaSO_4$ is sparingly soluble; sulphates of Sr, Ba, Pb and Ag are insoluble.

Sulphates do not decompose unless heated very strongly.

DATASHEET 7: Group I elements and their compounds

Element	Electronic configuration
Lithium, Li	[He] $2s^1$
Sodium, Na	[Ne] $3s^1$
Potassium, K	[Ar] $4s^1$
Rubidium, Rb	[Kr] $5s^1$
Caesium, Cs	[Xe] $6s^1$

Group I elements are very reactive metals. Reactivity increases with increasing atomic number. The elements are oxidised easily to form ionic compounds containing M^+. All compounds are white, crystalline solids, unless the anion itself is coloured.

Sodium–caesium

$$2M(s) + H_2(g) \longrightarrow 2MH(s) \qquad [M^+H^-]$$
$$4M(s) + O_2(g) \longrightarrow 2M_2O(s) \qquad [(M^+)_2O^{2-}]$$
$$2M(s) + Cl_2(g) \longrightarrow 2MCl(s) \qquad [M^+Cl^-]$$
$$2M(s) + 2H_2O(l) \longrightarrow 2MOH(aq) + H_2(g) \qquad [M^+(aq) + OH^-(aq)]$$

Salts are soluble in water.

Hydroxides and carbonates do not decompose on heating.

Nitrates give nitrites and oxygen when heated.

Anomolous properties of lithium

Li_2CO_3 is sparingly soluble in water and decomposes on heating:
$$Li_2CO_3(s) \longrightarrow Li_2O(s) + CO_2(g)$$

$LiNO_3$ decomposes on heating:
$$4LiNO_3(s) \longrightarrow 2Li_2O(s) + 4NO_2(g) + O_2(g)$$

DATASHEET 8: Group II elements and their compounds

Element	Electronic configuration
Beryllium, Be	[He] $2s^2$
Magnesium, Mg	[Ne] $3s^2$
Calcium, Ca	[Ar] $4s^2$
Barium, Ba	[Kr] $5s^2$
Strontium, Sr	[Xe] $6s^2$

Group II elements are reactive metals. Reactivity increases with increasing atomic number. The elements are oxidised easily to form ionic compounds containing M^{2+}. All compounds are white, crystalline solids, unless the anion itself is coloured.

Magnesium–strontium

$M(s) + H_2(g) \longrightarrow MH_2(s)$ $\qquad$ $[M^{2+}(H^-)_2]$

$2M(s) + O_2(g) \longrightarrow 2MO(s)$ $\qquad$ $[M^{2+}O^{2-}]$

$M(s) + Cl_2(g) \longrightarrow MCl_2(s)$ $\qquad$ $[M^{2+}(Cl^-)_2]$

$M(s) + 2H_2O(l) \longrightarrow M(OH)_2(aq) + H_2(g)$ $\qquad$ $[M^{2+}(aq) + 2OH^-(aq)]$

except Mg which reacts with steam:

$Mg(s) + H_2O(g) \longrightarrow MgO(s) + H_2(g)$ $\qquad$ $[Mg^{2+}O^{2-}]$

Chlorides and nitrates are soluble in water.

Sulphates, carbonates and hydroxides are sparingly soluble; solubility decreases with increasing atomic number.

Hydroxides and carbonates decompose on heating.

Nitrates give oxides, nitrogen dioxide and oxygen when heated.

Anomolous properties of beryllium

Beryllium is relatively unreactive at room temperature. Salts are soluble in water. Compounds have greater covalent character than other Group II elements.

DATASHEET 9: Aluminium and its compounds

Element	Electronic configuration
Aluminium, Al	[Ne] $3s^2 3p^1$

Aluminium is a shiny white metal of relatively low density. All compounds are white crystalline solids, unless the anion is coloured. Aluminium appears unreactive due to a surface coating of the oxide Al_2O_3. Removal of the oxide layer by immersing aluminium in $HgCl_2(aq)$ reveals the metal's reactivity:

$4Al(s) + 3O_2(g) \longrightarrow 2Al_2O_3(s)$

$2Al(s) + 6H_2O(l) \longrightarrow 2Al(OH)_3(s) + 3H_2(g)$

$2Al(s) + 6H^+(aq) \longrightarrow 2Al^{3+}(aq) + 3H_2(g)$

Aluminium, with or without the thin oxide layer, also reacts:

$2Al(s) + 2OH^-(aq) + 6H_2O(l) \longrightarrow 2[Al(OH)_4]^-(aq) + 3H_2(g)$

$2Al(s) + 3Cl_2(g) \longrightarrow 2AlCl_3(s) \xrightarrow{180°C} Al_2Cl_6(g) \xrightarrow{200°C} 2AlCl_3(g)$

Solutions of $Al^{3+}(aq)$ are acidic:

$[Al(H_2O)_6]^{3+}(aq) \rightleftharpoons [Al(H_2O)_5OH]^{2+}(aq) + H^+(aq) \rightleftharpoons [Al(H_2O)_4(OH)_2]^+(aq) + 2H^+(aq) \rightleftharpoons [Al(H_2O)_3(OH)_3](s) + 3H^+(aq)$

'$AlCl_3(aq)$' contains $\boxed{[Al(H_2O)_6]^{3+}(aq) + 3Cl^-(aq)]} \xrightarrow{\text{evaporate}} AlCl_3.6H_2O(s)$

$2AlCl_3.6H_2O(s) \xrightarrow{\text{heat}} Al_2O_3(s) + 6H^+(aq) + 6Cl^-(aq) + 9H_2O(l)$

DATASHEET 10: Group IV elements and their compounds

Element	Electronic configuration
Carbon, C	[He] $2s^2 2p^2$
Silicon, Si	[Ne] $3s^2 3p^2$
Germanium, Ge	[Ar] $4s^2 4p^2$
Tin, Sn	[Kr] $5s^2 5p^2$
Lead, Pb	[Xe] $6s^2 6p^2$

Group IV elements show a clear trend from non-metallic (carbon) to metallic (lead). Group IV compounds may be divalent or tetravalent. Increasing stability of the divalent state with increasing atomic number is called the inert pair effect. In the following reactions E stands for Group IV elements.

Preparation of the hydrides

$E(s) + 2H_2(g) \longrightarrow EH_4(g)$ [PbH_4 not known]

Methane, CH_4, is formed from graphite and H_2 (Ni catalyst). It is one of thousands of carbon hydrides. Other hydrides may be prepared by reducing ECl_4 in ethoxyethane with $LiAlH_4$. Si_nH_{2n+2} (n = 1–8). Ge_nH_{2n+2} (n = 1–5) and SnH_4 known. Only carbon hydrides are appreciably stable in air.

The hydrides show increasing reducing character on descending the Group. Methane is stable in air and is unaffected by water, acids or alkali. The other hydrides becomes increasingly thermally unstable as the Group is descended. They burn spontaneously in air and hydrolyse in water or alkali.

Preparation of the oxides

$2E(s) + O_2(g) \longrightarrow 2EO$; $E(s) + O_2(g) \longrightarrow EO_2$

CO is formed when carbon is burned in limited oxygen or when T > 750°C otherwise CO_2 is more stable.

The monoxides of Si, Ge and Sn are unstable.

PbO is formed by direct combination at 600–800°C but oxidises to dilead(II) lead(IV) oxide, Pb_3O_4 ('red lead'):
$6PbO(s) + O_2 \longrightarrow 2Pb_3O_4(s)$

The dioxides of C, Si, Ge and Sn are formed by direct combination and at high temperature.
Lead(IV) oxide is formed by the action of dilute nitric acid on Pb_3O_4:

$Pb_3O_4(s) + 4HNO_3(aq) \longrightarrow 2Pb(NO_3)_2(aq) + PbO_2(s) + 2H_2O(l)$

Preparation of the chlorides

Dichlorides: only formed by Ge, Sn and Pb.

$E(s) + Cl_2(g) \longrightarrow ECl_2(s)$ (E = Ge, Sn, Pb)
$E(s) + ECl_4(l) \longrightarrow 2ECl_2(s)$ (E = Ge or Sn)
$Sn(s) + 2HCl(aq) \longrightarrow SnCl_2(aq) + H_2(g)$ (heat Sn with hot, conc HCl(aq))
$Pb^{2+}(aq) + 2Cl^-(aq) \longrightarrow PbCl_2(s)$

$GeCl_2$ and $SnCl_2$ are ionic and dissolve in water to give $E^{2+}(aq)$ ions.

Tetrachlorides: known for all Group IV elements.

$E(s) + 2Cl_2(g) \longrightarrow ECl_4(l)$ (E = Si, Ge, Sn)
$CS_2(l) + 3Cl_2(g) \longrightarrow CCl_4(l) + S_2Cl_2(l)$ (pass Cl_2 into CS_2)
$PbO_2(s) + 4HCl(aq) \longrightarrow PbCl_4(l) + 2H_2O(l)$ (excess conc HCl(aq) at 0°C)

The tetrachlorides are covalent and non-electrolytes. CCl_4 does not hydrolyse in water but other ECl_4 compounds give the dioxide and HCl:
$ECl_4(l) + 2H_2O(l) \longrightarrow EO_2(s) + 4HCl(aq)$ (E = Si, Ge, Sn, Pb)

Thermal stability of the tetrachlorides decreases as the Group is descended. CCl_4, $SiCl_4$ and $GeCl_4$ are stable to high temperatures. $SnCl_4$ decomposes on heating, $PbCl_4$ decomposes spontaneously:
$ECl_4(l) \longrightarrow ECl_2(s) + Cl_2(g)$ (E = Sn, Pb)

Formation of complex ions: $ECl_4(l) + 2Cl^-(aq) \longrightarrow [ECl_6]^{2-}(aq)$ (E = Si, Ge, Sn, Pb)

Reactions of the elements with water

$C(s) + H_2O(g) \longrightarrow CO(g) + H_2(g)$ (1000°C) (mixture known as 'water-gas')
$Si(s) + 2H_2O(g) \longrightarrow SiO_2(s) + 2H_2(g)$ (steam or boiling water)
$2Pb(s) + 2H_2O(l) + O_2(aq) \longrightarrow 2Pb(OH)_2(aq)$ (cold oxygenated water)

Germanium and tin are very unreactive towards water or steam.

Reactions of the oxides

CO is a neutral oxide and a reducing agent:
$2CO(g) + O_2(g) \longrightarrow 2CO_2(g)$

SiO and GeO oxidise rapidly in air to the dioxide. SnO and PbO are amphoteric:
$EO(s) + 2H^+(aq) \longrightarrow E^{2+}(aq) + H_2O(l)$ (E = Sn, Pb)
$EO + OH^-(aq) + H_2O(l) \longrightarrow [E(OH)_3]^-(aq)$ E = Sn, Pb
$EO_2 + 2OH^-(aq) \longrightarrow EO_3^{2-}(aq) + H_2O(l)$ (E = C, Si, Ge))

SnO_2 and PbO_2 react with molten NaOH or KOH to give stannates and plumbates:
$EO_2(s) + 2OH^-(l) \longrightarrow EO_3^{2-}(l) + H_2O(g)$ (E = Sn, Pb)

Acidic character of Group IV oxides decreases with increasing atomic number and increasing oxidation number.

Oxidising power of dioxides increases down the Group. PbO_2 is a strong oxidising agent:

$2PbO_2(s) \xrightarrow{\text{heat}} 2PbO(s) + O_2(g)$

DATASHEET 11: nitrogen and phosphorus, and their compounds

Element	Electronic configuration
Nitrogen, N	[He] $2s^2 2p^3$
Phosphorus, P	[Ne] $3s^2 3p^3$

Nitrogen is unreactive at room temperature due to its strong $N\equiv N$ triple bond and non-polar character.
The commonest allotropes of phosphorus are white and red phosphorus; white phosphorus (P_4) is more reactive than red phosphorus.

Hydrides of nitrogen

$N_2(g) + 3H_2(g) \rightleftharpoons 2NH_3(g)$ (Haber process: 400°C and 200 atm, Fe catalyst)
$2NH_3(g) + NaOCl(aq) \longrightarrow N_2H_4(g) + NaCl(aq) + H_2O(l)$
$4NH_3(g) + 3O_2(g) \longrightarrow 2N_2(g) + 6H_2O(l)$

Oxides of nitrogen

$NH_4NO_3(s) \longrightarrow N_2O(g) + 2H_2O(g)$ (heat)
$3NO(g) \longrightarrow N_2O(g) + NO_2(g)$ (warm at high pressure)
$6NaNO_2(s) + 3H_2SO_4(aq) \longrightarrow 4NO(g) + 2HNO_3(aq) + 2H_2O(l) + 3Na_2SO_4(aq)$
$2NO(g) + O_2(g) \longrightarrow 2NO_2(g)$
$2MOH(aq) + 4NO(g) \longrightarrow 2MNO_2(aq) + N_2O(g) + H_2O(l)$ (M = Group I metal)
$2Pb(NO_3)_2(s) \longrightarrow 4NO_2(g) + 2PbO(s) + O_2(g)$ (heat at 400°C)
$2NO_2(g) \rightleftharpoons N_2O_4(g)$ (N_2O_4 favoured by low temperature and high pressure)
$N_2O_4(g) + H_2O(l) \longrightarrow HNO_2(aq) + HNO_3(aq)$
$4HNO_3(l) + P_4O_{10}(s) \longrightarrow 2N_2O_5(s) + 4HPO_3(l)$ (conc HNO_3 at −10°C)

Oxoacids of nitrogen: nitrous acid, HNO_2 and nitric acid, HNO_3

$$HNO_2(aq) + NaOH(aq) \longrightarrow NaNO_2(aq) + H_2O(l)$$
$$HNO_3(aq) + NaOH(aq) \longrightarrow NaNO_3(aq) + H_2O(l)$$

HNO_2 has not been isolated as a pure compound; aqueous solutions disproportionate:
$$3HNO_2(aq) \longrightarrow HNO_3(aq) + 2NO(g) + H_2O(l)$$

Nitrogen trichloride

$$NH_3(g) + 3Cl_2(g) \longrightarrow NCl_3(l) + 3HCl(g)$$
$$NCl_3(l) + 3H_2O(l) \longrightarrow NH_3(aq) + 3HOCl(aq)$$

Hydrides of phosphorus: phosphine, PH_3 and diphosphine, P_3H_4

$$P_4(s) + 3NaOH(aq) + 3H_2O(l) \longrightarrow 3NaH_2PO_2(aq) + PH_3(g) \quad \text{(hot, conc NaOH)}$$

Oxides of phosphorus

$$P_4(s) + 3O_2(g) \longrightarrow P_4O_6(s)$$
$$P_4O_6(s) + 2O_2(g) \longrightarrow P_4O_{10}(s)$$

Both P_4O_6 and P_4O_{10} are acidic, reacting with water to give phosphonic acid and phosphoric(V) acids respectively:
$$P_4O_6(s) + 6H_2O(l) \longrightarrow 4H_3PO_3(aq)$$
$$P_4O_{10}(s) + 6H_2O(l) \longrightarrow 4H_3PO_4(aq)$$

Chlorides of phosphorus

$$2P(s) + 3Cl_2(g) \longrightarrow 2PCl_3(l)$$
$$PCl_3(l) + Cl_2(g) \longrightarrow PCl_5(s)$$

PCl_5 is ionic in the solid state: $[PCl_4]^+[PCl_6]^-$

$$PCl_3(l) + 3H_2O(l) \longrightarrow H_3PO_3(aq) + 3HCl(aq)$$
$$PCl_5(s) + 4H_2O(l) \longrightarrow H_3PO_4(aq) + 5HCl(aq)$$

Oxoacids of phosphorus

$\begin{array}{ccccc}
H_3PO_3 & \longrightarrow & H_2PO_3^- & + H^+ & \longrightarrow & HPO_3^{2-} & + 2H^+ \\
\text{phosphonic} & & \text{hydrogen-} & & & \text{phosphonate} \\
\text{acid} & & \text{phosphonate ion} & & & \text{ion}
\end{array}$

$\begin{array}{ccccccc}
H_3PO_4 & \longrightarrow & H_2PO_4^- & + H^+ & \longrightarrow & HPO_4^{2-} & + 2H^+ & \longrightarrow & PO_4^{3-} & + 3H^+ \\
\text{phosphoric(V)} & & \text{dihydrogen-} & & & \text{hydrogen-} & & & \text{phosphate(V)} \\
\text{acid} & & \text{phosphate(V) ion} & & & \text{phosphate(V) ion} & & & \text{ion}
\end{array}$

DATASHEET 12: Oxygen and sulphur, and their compounds

Element	Electronic configuration
Oxygen, 6	$[He]\ 2s^2 2p^4$
Sulphur, 16	$[Ne]\ 3s^2 3p^4$

Oxygen, reacts directly with most elements and oxidises many compounds. Generally, metal oxides are basic while non-metal oxides are acidic (see Datasheet 6).

Sulphur is a water-insoluble, non-toxic yellow solid. The two most important allotropes are rhombic and monoclinic sulphur. Both contain S_8 rings. However, other allotropes exist and so sulphur is shown as S and not S_8 in chemical equations. Sulphur has similar reactions to oxygen but usually at higher temperatures.

Sulphides

Group I elements: $2M(s) + S(s) \longrightarrow M_2S(s)$

Group II elements: $M(s) + S(s) \longrightarrow MS(s)$

Group III elements: $2M(s) + 3S(s) \longrightarrow M_2S_3(s)$

Group II and III sulphides: $\begin{cases} 2MS(s) + 3O_2(g) \longrightarrow 2MO(s) + 2SO_2(g) \\ 2M_2S_3(s) + 9O_2(g) \longrightarrow 2M_2O_3(s) + 6SO_2(g) \end{cases}$

Group I, II and III sulphides: $S^{2-}(s) + 2H^+(aq) \longrightarrow H_2S(g)$

H_2S is a weak acid, giving aqueous hydrogensulphide and sulphide ions:
$H_2S(g) + H_2O(l) \rightleftharpoons HS^-(aq) + H_3O^+(aq)$
$HS^-(aq) + H_2O(l) \rightleftharpoons S^{2-}(aq) + H_3O^+(aq)$

Oxides of sulphur

S forms at least thirteen oxides, but only sulphur dioxide, SO_2, and sulphur trioxide, SO_3 are important.
$S(s) + O_2(g) \longrightarrow SO_2(g)$
$SO_3^{2-}(aq) + 2H^+(aq) \longrightarrow H_2O(l) + SO_2(g)$
$Cu(s) + 2H_2SO_4(l) \longrightarrow CuSO_4(aq) + 2H_2O(l) + SO_2(g)$ (hot, conc H_2SO_4)
$Fe_2(SO_4)_3(s) \longrightarrow Fe_2O_3(s) + 3SO_3(g)$ (heat strongly)
$2SO_2(g) + O_2(g) \rightleftharpoons 2SO_3(g)$ (Contact process: catalyst = V_2O_5; 400–450°C; pressure 1–2 Atmospheres)

Oxoacids of sulphur

$\underset{\text{sulphurous acid}}{H_2SO_3} \longrightarrow \underset{\text{hydrogensulphite ion}}{HSO_3^-} + H^+(aq) \longrightarrow \underset{\text{sulphite ion}}{SO_3^{2-}} + 2H^+(aq)$

$\underset{\text{sulphuric acid}}{H_2SO_4} \longrightarrow \underset{\text{hydrogensulphate ion}}{HSO_4^-} + H^+(aq) \longrightarrow \underset{\text{sulphate ion}}{SO_4^{2-}} + 2H^+(aq)$

Chlorides of sulphur

$2S(s) + Cl_2(g) \longrightarrow S_2Cl_2(l)$ (warm)
$S_2Cl_2(l) + Cl_2(g) \longrightarrow 2SCl_2(l)$ (FeCl$_3$ catalyst)
$SCl_2(l) + Cl_2(g) \longrightarrow SCl_4(l)$ (–30°C)

DATASHEET 13: Group VII elements (the halogens) and their compounds

Element	Electronic configuration
Fluorine, F	$[He]\, 2s^2 2p^5$
Chlorine, Cl	$[Ne]\, 3s^2 3p^5$
Bromine, Br	$[Ar]\, 3d^{10} 4s^2 4p^5$
Iodine, I	$[Kr]\, 4d^{10} 5s^2 5p^5$
Astatine, At	$[Xe]\, 4f^{14} 5d^{10} 6s^2 6p^5$

The elements exist as diatomic molecules, X_2. The elements are volatile and reactive. Fluorine, the most reactive element, is a strong oxidising agent and forms compounds with all elements except He, Ne, and Ar.

Preparation

F_2 is produced by electrolysing molten KF and HF.

Cl_2 is made in the laboratory from hot, conc HCl and MnO_2 (industrially by brine electrolysis).

Br_2 and I_2 are obtained by warming KBr or KI respectively with MnO_2 and H_2SO_4.

$$MnO_2(s) + 4H^+(aq) + 2X^-(aq) \longrightarrow Mn^{2+}(aq) + 2H_2O(l) + X_2(g) \quad (X = Cl,\ Br,\ I)$$

Displacement reactions

Show that reactivity and oxidising power of the halogens decreases with increasing atomic number.
$$F_2(g) + 2KCl(s) \longrightarrow 2KF(s) + Cl_2(g)$$
$$Cl_2(aq) + 2Br^-(aq) \longrightarrow 2Cl^-(aq) + Br_2(aq)$$
$$Cl_2(g) + 2I^-(aq) \longrightarrow 2Cl^-(aq) + I_2(aq)$$
$$Br_2(aq) + 2I^-(aq) \longrightarrow 2Br^-(aq) + I_2(aq)$$

Reactions with hydrogen

The elements react with H_2 to give hydrogen halides:
$$H_2(g) + X_2(g,\ l,\ s) \longrightarrow 2HX(g) \quad (F_2\text{: explosive; } Cl_2\text{: explosive when initiated by a spark or UV radiation;}$$
$$Br_2 \text{ and } I_2\text{: at 300°C, with Pt catalyst})$$

$$2HX(g) \longrightarrow H_2(g) + X_2(g,\ l,\ s) \quad (\text{HF and HCl: no reaction; HBr: red hot metal wire; HI: hot glass rod. Stability}$$
$$\text{decreases with increasing molar mass})$$

$$HX(g) + H_2O(l) \longrightarrow H_3O^+(aq) + X^-(aq)$$

Reactions with water

Halogens are sparingly soluble in water. F_2 oxidises water. Other halogens disproportionate.
$$2F_2(g) + 2H_2O(l) \longrightarrow 4HF(g) + O_2(g)$$
$$X_2(aq) + 2H_2O(l) \rightleftharpoons X^-(aq) + H_3O^+(aq) + HOX(aq)$$

Reactions with alkalis

Cl_2, Br_2 and I_2 disproportionate in alkali. In cold, dilute alkali chlorine and bromine give halide and halate(I):
$$X_2(g,\ l,\ s) + 2OH^-(aq) \longrightarrow X^-(aq) + XO^-(aq) + H_2O(l)$$
$$3X_2(g) + 6OH^-(aq) \longrightarrow 5X^-(aq) + XO_3^-(aq) + 3H_2O(l) \quad (Cl_2\text{: hot, conc alkali; } Br_2\text{: warm, dilute alkali; } I_2\text{: room}$$
$$\text{temperature})$$

Halides

Halogens react with most elements to give halides. Generally, metal halides are predominately ionic and non-metals halides are predominately covalent. For a given element, covalent character of halides increases with increasing oxidation state and in the order F<Cl<Br<I.

THE DATABASE

Reactions of ionic halides

$KX(s) + H_2SO_4(l) \longrightarrow KHSO_4(s) + HX(g)$ (X = F, Cl, Br, I)   (hot, conc H_2SO_4)

$2HBr(g) + H_2SO_4(l) \longrightarrow Br_2(l) + SO_2(g) + 2H_2O(l)$ (hot, conc H_2SO_4)

$8HI(g) + H_2SO_4(l) \longrightarrow 4I_2(s) + H_2S(g) + 4H_2O(l)$ (hot, conc H_2SO_4)

$KX(s) + H_3PO_4(l) \longrightarrow KH_2PO_4(s) + HX(g)$ (X = Cl, Br, I; no oxidation to X_2)

The halides are precipitated as insoluble silver salts:

$Ag^+(aq) + X^-(aq) \longrightarrow AgX(s)$

$AgX(s) + 2NH_3(aq) \longrightarrow [Ag(NH_3)_2]^+(aq) + X^-(aq)$ (AgCl soluble in dilute $NH_3(aq)$; AgBr soluble in conc $NH_3(aq)$; AgI insoluble in conc $NH_3(aq)$)

DATASHEET 14: Group VIII elements (the noble gases)

Element	Electronic configuration
Helium, He	$2s^2$
Neon, Ne	[He] $2s^22p^6$
Argon, Ar	[Ne] $3s^23p^6$
Krypton, Kr	[Ar] $3d^{10}4s^24p^6$
Xenon, Xe	[Kr] $4d^{10}5s^25p^6$
Radon, Rn	[Kr] $4d^{10}4f^{14}5s^25p^65d^{10}6s^26p^6$

Group VIII elements (also called the noble gases) are very unreactive due to stable electronic configurations. They are monatomic gases. Helium has the lowest melting point of any substance and, although it is rare on Earth, it is the second most abundant element in the universe. The elements are obtained by fractional distillation of liquid air, followed by adsorption onto charcoal.

Compounds

Few compounds are known. Xenon forms both fluorides and oxides. Krypton forms a fluoride, KrF_2. Radon reacts with fluorine but its radioactivity has prevented detailed studies.

$Xe(g) + F_2(g) \longrightarrow XeF_2(s)$ (in sunlight)

$XeF_2(g) + F_2(g) \longrightarrow XeF_4(s)$ (in sunlight)

$XeF_4(s) + F_2(g) \longrightarrow XeF_6(s)$ (in sunlight and at high pressure)

$XeF_6(s) + Pt(s) \longrightarrow Xe(g) + PtF_6(s)$

XeF_2 is stable in water but XeF_4 and XeF_6 hydrolyse to the explosive xenon trioxide, XeO_3:

$6XeF_4(s) + 12H_2O(l) \longrightarrow 2XeO_3(aq) + 4Xe(g) + 3O_2(g) + 24HF(aq)$

DATASHEET 15: d-block elements and their compounds

Element	Electronic configuration
Scandium, Sc	$[Ar] 3d^1 4s^2$
Titanium, Ti	$[Ar] 3d^2 4s^2$
Vanadium, V	$[Ar] 3d^3 4s^2$
Chromium, Cr	$[Ar] 3d^5 4s^1$
Manganese, Mn	$[Ar] 3d^5 4s^2$
Iron, Fe	$[Ar] 3d^6 4s^2$
Cobalt, Co	$[Ar] 3d^7 4s^2$
Nickel, Ni	$[Ar] 3d^8 4s^2$
Copper, Cu	$[Ar] 3d^{10} 4s^1$
Zinc, Zn	$[Ar] 3d^{10} 4s^2$

1. Aqueous ions, oxides and chlorides

	Sc	Ti	V	Cr	Mn	Fe	Co	Ni	Cu	Zn
Ox No										
+1									Cu_2O $CuCl$	
+2		TiO	VO	Cr^{2+}	Mn^{2+} MnO $MnCl_2$	Fe^{2+} FeO $FeCl_2$	Co^{2+} CoO $CoCl_2$	Ni^{2+} NiO $NiCl_2$	Cu^{2+} CuO $CuCl_2$	Zn^{2+} ZnO $ZnCl_2$
+3	Sc^{3+} Sc_2O_3 $ScCl_3$	Ti^{3+} Ti_2O_3 $TiCl_3$	V^{3+} V_2O_3 VCl_3	Cr^{3+} $CrCl_3$	Mn^{3+}	Fe^{3+} Fe_2O_3 $FeCl_3$	Co^{3+}			
+4		TiO^{2+} TiO_2 $TiCl_4$	VO^{2+} VO_2 VCl_4		MnO_2					
+5			VO_4^{3-} V_2O_5		MnO_4^{3-}					
+6				CrO_4^{2-} $Cr_2O_7^{2-}$ CrO_3	MnO_4^{2-}	FeO_4^{2-}				
+7					MnO_4^- Mn_2O_7					

General trends: Chlorides of d-block elements in low oxidation states tend to be ionic and dissolve in water to give neutral or weakly acidic solutions. For example, $MnCl_2(s)$, $NiCl_2(s)$, $CuCl_2(s)$:

$$NiCl_2(s) + aq \longrightarrow Ni^{2+}(aq) + 2Cl^-(aq)$$

In high oxidation states, they tend to be covalent, molecular compounds which hydrolyse in water. For example, $TiCl_4(l)$, $VCl_4(l)$:

$$TiCl_4(l) + 2H_2O(l) \longrightarrow TiO_2(s) + 4HCl(g)$$

Oxides of d-block elements in low oxidation states tend to be ionic and are basic. For example, $MnO(s)$, $CuO(s)$:

$$CuO(s) + 2H^+(aq) \longrightarrow Cu^{2+}(aq) + H_2O(l)$$

In high oxidation states they tend to be covalent compounds and are acidic. For example, $V_2O_5(s)$, $CrO_3(s)$, $Mn_2O_7(l)$:

$$CrO_3(s) + 2OH^-(aq) \longrightarrow CrO_4^{2-}(aq) + H_2O(l)$$

2. Some common complex ions

$M^{x+}(aq)$. All are octahedral and exist in aqueous solution in equilibrium:

$$[M(H_2O)_6]^{x+}(aq) + H_2O(l) \rightleftharpoons [M(H_2O)_5(OH)]^{(x-1)+}(aq) + H_3O^+(aq)$$

The higher the value of x, the more the equilibrium lies to the right.

$[Ti(H_2O)_6]^{3+}$ violet; $[V(H_2O)_6]^{2+}$ lavender; $[Cr(H_2O)_6]^{2+}$ blue unstable; $[Cr(H_2O)_5OH]^{2+}$ violet;

$[Mn(H_2O)_6]^{2+}$ pale pink; $[Mn(H_2O)_6]^{3+}$ violet unstable;

$[Fe(H_2O)_6]^{2+}$ pale green; $[Fe(H_2O)_6]^{3+}$ purple; $[Fe(H_2O)_5OH]^{2+}$ yellow;

$[Co(H_2O)_6]^{3+}$ orange-brown; $[Co(H_2O)_6]^{2+}$ pink; $Ni(H_2O)_6]^{2+}$ green; $[Cu(H_2O)_6]^{2+}$ blue;

$[Zn(H_2O)_6]^{2+}$ colourless

3. Formation of complex ions by ligand exchange

All ligand exchange reactions occur by a stepwise replacement of ligands. Overall equations for complete replacement of all ligands are shown below.

Coordination number unchanged

$$[M(H_2O)_6]^{x+}(aq) + 6NH_3(aq) \longrightarrow [M(NH_3)_6]^{x+}(aq) + 6H_2O(l)$$

For example: $[Ni(H_2O)_6]^{2+}(aq) + 6NH_3(aq) \longrightarrow [Ni(NH_3)_6]^{2+}(aq) + 6H_2O(l)$
                   green, octahedral   lavender, octahedral

Coordination number changes

$$[M(H_2O)_6]^{x+}(aq) + 4Cl^-(aq) \longrightarrow [MCl_4]^{(x-4)-}(aq) + 6H_2O(l)$$

For example: $[Co(H_2O)_6]^{2+}(aq) + 4Cl^-(aq) \longrightarrow [CoCl_4]^{2-}(aq) + 6H_2O(l)$
                   pink, octahedral   blue, tetrahedral

Other important 6-coordinate complex ions (all octahedral):

$[Cr(NH_3)_6]^{3+}$ yellow;   *trans* $[CrCl_2(H_2O)_4]^+$ dark green; $[Fe(CN)_6]^{4-}$ brown; $[Fe(H_2O)_5SCN]^{2+}$ blood-red;

$[Co(NH_3)_6]^{3+}$ golden-brown; $[Ni(H_2O)_6]^{2+}$ green; $[Cu(NH_3)_4(H_2O)_2]^{2+}$ royal blue.

Other important 4-coordinate complex ions:

$[FeCl_4]^{2-}$ yellow, tetrahedral; $[CoCl_4]^{2-}$ blue, tetrahedral; $[Ni(CN)_4]^{2-}$ orange, square planar;

$[CuCl_4]^{2-}$ yellow, distorted tetrahedral; $[Zn(NH_3)_4]^{2+}$ colourless, tetrahedral.

Important 2-coordinate complex ions (all linear):

$[CuCl_2]^-$ colourless; $[Cu(NH_3)_2]^+$ colourless; $[Ag(NH_3)_2]^+(aq)$ colourless.

4. Some important redox reactions

Vanadium: $VO_2^+(aq)$ may be reduced, for example by Zn/dil HCl(aq):

$VO_2^+(aq) \longrightarrow VO^{2+}(aq) \longrightarrow V^{3+}(aq) \longrightarrow V^{2+}(aq)$
 yellow   blue   green   lavender

Chromium: $CrO_4^{2-}(aq)$ and $Cr_2O_7^{2-}(aq)$ exist in equilibrium; acid solution favours $Cr_2O_7^{2-}(aq)$. $Cr_2O_7^{2-}(aq)$ is a powerful oxidising agent; it may be reduced, for example by Zn/dil HCl(aq):

$2CrO_4^{2-}(aq) + 2H^+(aq) \rightleftharpoons Cr_2O_7^{2-}(aq) + H_2O(l)$
 yellow   orange

$2CrO_7^{2-}(aq) \longrightarrow Cr^{3+}(aq) \longrightarrow Cr^{2+}(aq)$
 orange   green   blue

Manganese: $MnO_4^-(aq)$ is a powerful oxidising agent. In acid solution it is reduced to $Mn^{2+}(aq)$, in strongly alkaline solutions it is reduced to $MnO_4^{2-}(aq)$. In neutral solution a mixture of $MnO_2(s)$ and $Mn^{2+}(aq)$ is obtained.

$MnO_4^-(aq) \longrightarrow MnO_4^{2-}(aq) \longrightarrow MnO_2(s) \longrightarrow Mn^{2+}(aq)$
 purple   green   brown   very pale pink

Guidance notes for Datasheets 16–25

1. Reactions of an organic compound may be found by identifying its functional group(s) and to which class it belongs. Examples are given in section 11.4.

2. Within a Datasheet reactions are described as follows:

   REACTION TYPE (if there is a commonly used name)

   $$\text{REACTANT} \xrightarrow[\text{conditions}]{\text{reagents}} \text{PRODUCT}$$

3. Reagents and reaction conditions are usually applicable. It is not necessarily true that they will work for every compound that contains a given functional group. Also, the list is not exhaustive and other reagents may be suitable.

4. Precise reagents and conditions depend on the actual compound being used. Two aspects are generalised in the Datasheets:

   (a) Reagent concentrations. No details are given other than 'dil' (dilute) and 'conc' (concentrated) for some reagents. Even here, dilute may mean differing concentrations of a reagent for different compounds within the same class.

   (b) Reaction temperature. The following expressions are used:

   For any reaction mixture:

   rt   room temperature

   For liquid reaction mixtures:

   warm   indicates some heating is needed
   reflux   reaction mixture is maintained at its boiling point

   For solid reaction mixtures and reactions which involve passing a gaseous reactant over a solid:

   heat   indicates heating is needed
   heat (x°C) indicates typical temperature

5. The synthetic route maps (Datasheets 24 and 25) are accessed by matching semi-structural formulae. Examples are given in section 11.4. When using the synthetic route maps, remember:

   • they are not exhaustive – there are many, many more reaction pathways available.
   • they use general semi-structural formulae. A compound must be matched against the general formula which best describes it.
   • the lines joining pairs of compounds, do not indicate whether conversion in one or both directions is possible. This information can only be found by looking at Datasheets 16–23, referenced in the synthetic route maps.
   • the two maps compliment one another and, on occasions, it is necessary to use them together.

6. Abbreviations used in the Datasheets in Section III:
R, R^1, R^2, R^3	alkyl groups
Ar	aryl group
H$^+$	acid
OH$^-$	alkali
hf	electromagnetic radiation (usually ultraviolet)
cat	catalyst

DATASHEET 16: Aliphatic hydrocarbons: alkanes and alkenes

Alkanes: saturated hydrocarbons, general formula C_nH_{2n+2}

Alkenes: unsaturated hydrocarbons containing C=C double bond, general formula C_nH_{2n}.

Alkanes and alkenes are **aliphatic** compounds.
In this Datasheet, reactions are given for alkanes and alkenes with the general semi-structural formulae: RCH_2CH_3 and $RCH{=}CH_2$ respectively, (R = alkyl group).
The chemistry of an aliphatic hydrocarbon chain attached to an aromatic group is the same as alkanes and alkenes.

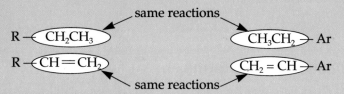

Reactions of the alkanes and alkenes

SUBSTITUTION

RCH_2CH_3 —— X_2 (X = halogen), hf ——→ RCH_2CH_2X, RCH_2CHX_2, RCH_2CX_3, $RCHXCX_3$ etc

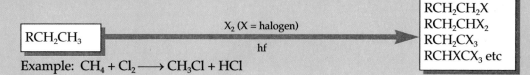

Example: $CH_4 + Cl_2 \longrightarrow CH_3Cl + HCl$

COMBUSTION (OXIDATION)

RCH_2CH_3 —— $O_2(g)$, ignite to initiate reaction ——→ $CO_2 + H_2O$

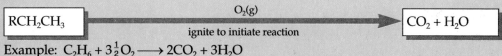

Example: $C_2H_6 + 3\frac{1}{2}O_2 \longrightarrow 2CO_2 + 3H_2O$

HYDROGENATION (REDUCTION)

$RCH{=}CH_2$ —— $H_2(g)$ (cat = Ni, Pd or Pt), 150–300 °C ——→ RCH_2CH_3

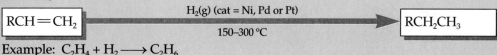

Example: $C_2H_4 + H_2 \longrightarrow C_2H_6$

COMBUSTION (OXIDATION)

$RCH{=}CH_2$ —— $O_2(g)$, ignite to initiate reaction ——→ $CO_2 + H_2O$

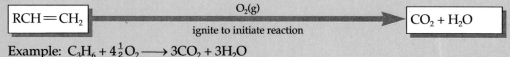

Example: $C_3H_6 + 4\frac{1}{2}O_2 \longrightarrow 3CO_2 + 3H_2O$

ADDITION: HALOGEN

$RCH{=}CH_2$ —— X_2 (X = halogen), rt ——→ $RCHXCH_2X$

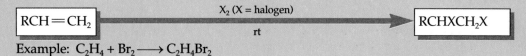

Example: $C_2H_4 + Br_2 \longrightarrow C_2H_4Br_2$

ADDITION: HYDROGEN HALIDE

$RCH{=}CH_2$ —— HX (HX = hydrogen halide), rt ——→ $RCHXCH_3$

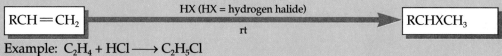

Example: $C_2H_4 + HCl \longrightarrow C_2H_5Cl$

HYDRATION (ADDITION: WATER)

$RCH{=}CH_2$ —— i) conc H_2SO_4, followed by H_2O or ii) $H_2O(g)$ (cat = supported H_3PO_4); i) rt; ii) heat under pressure ——→ $RCH(OH)CH_3$

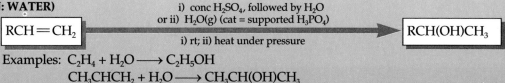

Examples: $C_2H_4 + H_2O \longrightarrow C_2H_5OH$
$CH_3CHCH_2 + H_2O \longrightarrow CH_3CH(OH)CH_3$

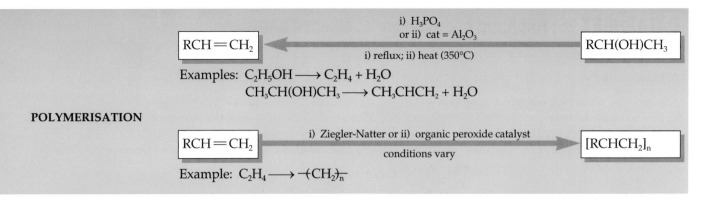

Examples: $C_2H_5OH \longrightarrow C_2H_4 + H_2O$

$CH_3CH(OH)CH_3 \longrightarrow CH_3CHCH_2 + H_2O$

POLYMERISATION

Example: $C_2H_4 \longrightarrow -(CH_2)_n-$

DATASHEET 17: Aromatic hydrocarbons

Arenes: unsaturated hydrocarbons with delocalised ring of electrons, for example, benzene. They are **aromatic** compounds.

Reactions of the arenes

NITRATION + REDUCTION

SULPHONATION

HALOGENATION

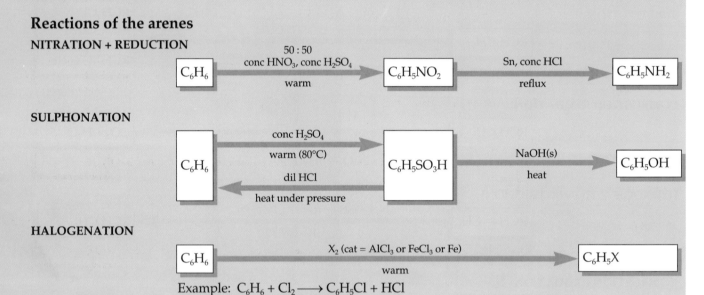

Example: $C_6H_6 + Cl_2 \longrightarrow C_6H_5Cl + HCl$

ALKYLATION

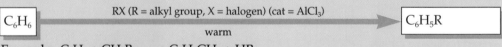

Example: $C_6H_6 + CH_3Br \longrightarrow C_6H_5CH_3 + HBr$

ALKANOYLATION

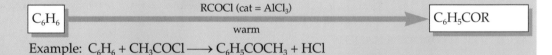

Example: $C_6H_6 + CH_3COCl \longrightarrow C_6H_5COCH_3 + HCl$

OXIDATION

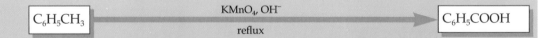

Further substitution in the benzene ring:

2, 4 – directing and activating: —OH, —NH$_2$, —X, —CH$_3$ (X = halogen)

3 – directing and de-activating: —NO$_2$

DATASHEET 18: Halogenoalkanes

General formula: RX (X = halogen)

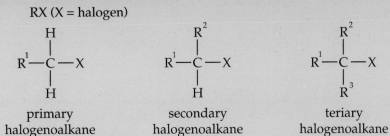

primary halogenoalkane   secondary halogenoalkane   teriary halogenoalkane

R^1, R^2, R^3 = alkyl groups, which may or may not be the same as one another

Reactions of the halogenoalkanes

HYDROLYSIS

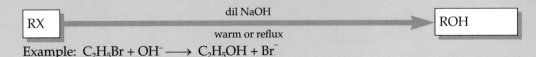

Example: $C_2H_5Br + OH^- \longrightarrow C_2H_5OH + Br^-$

HALOGENATION

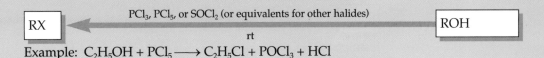

Example: $C_2H_5OH + PCl_5 \longrightarrow C_2H_5Cl + POCl_3 + HCl$

FORMATION OF A NITRILE

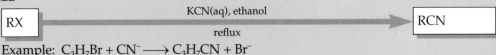

Example: $C_3H_7Br + CN^- \longrightarrow C_3H_7CN + Br^-$

FORMATION OF AN ETHER

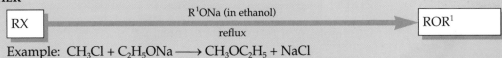

Example: $CH_3Cl + C_2H_5ONa \longrightarrow CH_3OC_2H_5 + NaCl$

Formation of tetraalkyl lead

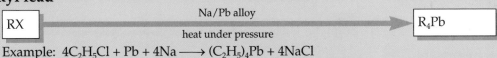

Example: $4C_2H_5Cl + Pb + 4Na \longrightarrow (C_2H_5)_4Pb + 4NaCl$

FORMATION OF AN ALKENE

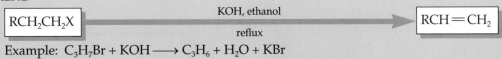

Example: $C_3H_7Br + KOH \longrightarrow C_3H_6 + H_2O + KBr$

Halogenoarenes, ArX, are far more resistant to substitution of —X than the halogenoalkanes, RX.

DATASHEET 19: Alcohols

General formula: ROH

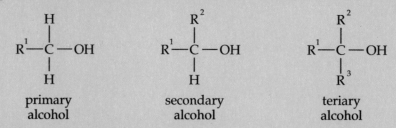

primary alcohol   secondary alcohol   teriary alcohol

R^1, R^2, R^3 = alkyl groups, which may or may not be the same as one another

Reactions of all alcohols

ESTERIFICATION

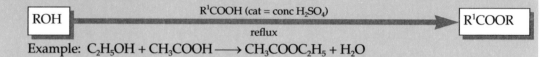

Example: $C_2H_5OH + CH_3COOH \longrightarrow CH_3COOC_2H_5 + H_2O$

HYDROLYSIS

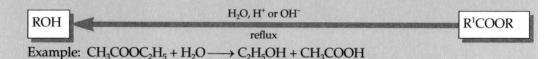

Example: $CH_3COOC_2H_5 + H_2O \longrightarrow C_2H_5OH + CH_3COOH$

HALOGENATION

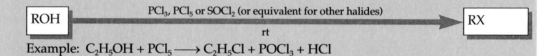

Example: $C_2H_5OH + PCl_5 \longrightarrow C_2H_5Cl + POCl_3 + HCl$

HYDROLYSIS

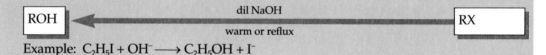

Example: $C_2H_5I + OH^- \longrightarrow C_2H_5OH + I^-$

FORMATION OF AN ALKOXIDE

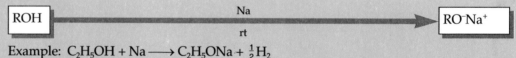

Example: $C_2H_5OH + Na \longrightarrow C_2H_5ONa + \frac{1}{2}H_2$

HYDROLYSIS

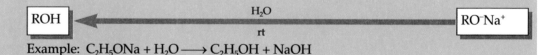

Example: $C_2H_5ONa + H_2O \longrightarrow C_2H_5OH + NaOH$

DEHYDRATION

Dehydration only occurs for alcohols containing the group

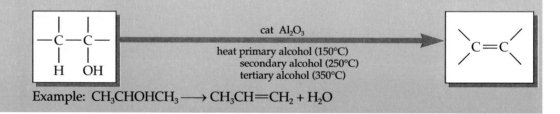

Example: $CH_3CHOHCH_3 \longrightarrow CH_3CH{=}CH_2 + H_2O$

Oxidation reactions

Primary alcohols:

$$R^1CH_2OH \xrightarrow[\text{i) heat (250°C) ii) distil}]{\text{i) Cu or ii) } K_2Cr_2O_7,\ H^+} R^1CHO \xrightarrow[\text{reflux}]{\text{i) } KMnO_4,\ H^+ \text{ or ii) } K_2Cr_2O_7,\ H^+} R^1COOH$$

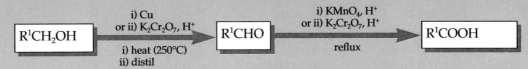

Example: $3C_2H_5OH + Cr_2O_7^{2-} + 8H^+ \longrightarrow 3CH_3CHO + 2Cr^{3+} + 7H_2O$

and further:

$3C_2H_5OH + 2Cr_2O_7^{2-} + 16H^+ \longrightarrow 3CH_3COOH + 4Cr^{3+} + 11H_2O$

The reverse reaction is a REDUCTION

$$R^1CH_2OH \xleftarrow[\text{reflux}]{LiAlH_4,\ \text{ethoxyethane}} R^1COOH$$

Secondary alcohols:

$$R^1R^2CHOH \xrightarrow[\text{reflux}]{\text{i) } KMnO_4,\ H^+ \text{ or ii) } K_2CrO_7,\ H^+} R^1R^2CO$$

further oxidation difficult

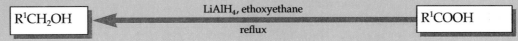

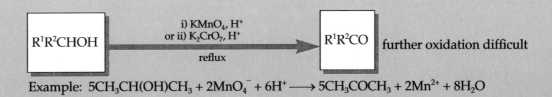

Example: $5CH_3CH(OH)CH_3 + 2MnO_4^- + 6H^+ \longrightarrow 5CH_3COCH_3 + 2Mn^{2+} + 8H_2O$

The reverse reaction is a REDUCTION

$$R^1R^2CHOH \xleftarrow[\text{reflux}]{LiAlH_4,\ \text{ethoxyethane}} R^1R^2CO$$

Tertiary alcohols:

$R^1R^2R^3COH$ oxidation difficult

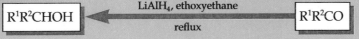

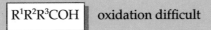

Reactions over copper

Primary alcohols: DEHYDROGENATION (REDUCTION)

$$R^1CH_2OH \xrightarrow[\text{heat (300°C)}]{\text{cat = Cu}} R^1CHO + H_2$$

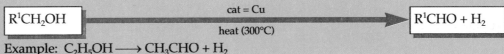

Example: $C_2H_5OH \longrightarrow CH_3CHO + H_2$

Secondary alcohols: DEHYDROGENATION (REDUCTION)

$$R^1R^2COH \xrightarrow[\text{heat (300°C)}]{\text{cat = Cu}} R^1R^2CO + H_2$$

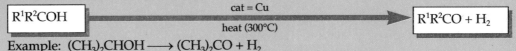

Example: $(CH_3)_2CHOH \longrightarrow (CH_3)_2CO + H_2$

Tertiary alcohols: DEHYDRATION

$$R^1R^2C(OH)CH_2R^3 \xrightarrow[\text{heat (300°C)}]{\text{cat = Cu}} R^1R^2C{=}CHR^3 + H_2O$$

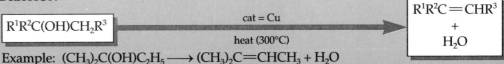

Example: $(CH_3)_2C(OH)C_2H_5 \longrightarrow (CH_3)_2C{=}CHCH_3 + H_2O$

DATASHEET 20: Phenols and aromatic amines

General formulae:

ArOH $ArNH_2$
phenol   aromatic amine

Reactions of phenols and aromatic amines

REACTION WITH NEUTRAL IRON(III) CHLORIDE

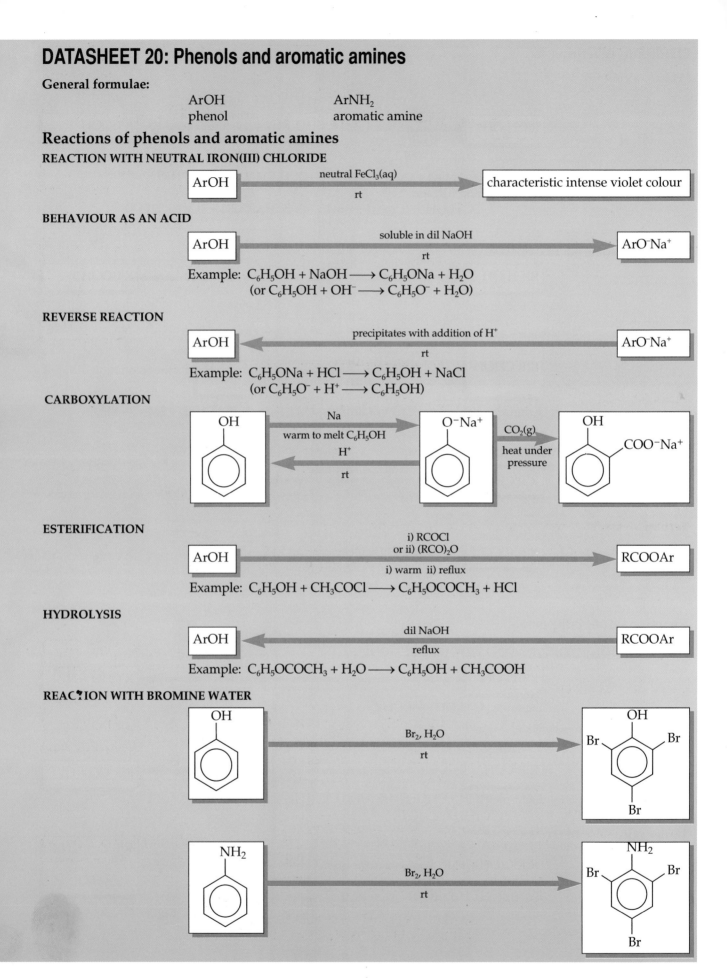

ArOH → (neutral $FeCl_3$(aq), rt) → characteristic intense violet colour

BEHAVIOUR AS AN ACID

ArOH → (soluble in dil NaOH, rt) → ArO^-Na^+

Example: $C_6H_5OH + NaOH \longrightarrow C_6H_5ONa + H_2O$
(or $C_6H_5OH + OH^- \longrightarrow C_6H_5O^- + H_2O$)

REVERSE REACTION

ArOH ← (precipitates with addition of H^+, rt) ← ArO^-Na^+

Example: $C_6H_5ONa + HCl \longrightarrow C_6H_5OH + NaCl$
(or $C_6H_5O^- + H^+ \longrightarrow C_6H_5OH$)

CARBOXYLATION

OH →(Na, warm to melt C_6H_5OH)→ O^-Na^+ →(CO_2(g), heat under pressure)→ OH, COO^-Na^+

(O^-Na^+ ← (H^+, rt) ← OH)

ESTERIFICATION

ArOH → (i) RCOCl or ii) (RCO)$_2$O; i) warm ii) reflux) → RCOOAr

Example: $C_6H_5OH + CH_3COCl \longrightarrow C_6H_5OCOCH_3 + HCl$

HYDROLYSIS

ArOH ← (dil NaOH, reflux) ← RCOOAr

Example: $C_6H_5OCOCH_3 + H_2O \longrightarrow C_6H_5OH + CH_3COOH$

REACTION WITH BROMINE WATER

OH →(Br_2, H_2O, rt)→ 2,4,6-tribromophenol (OH with Br, Br, Br)

NH_2 →(Br_2, H_2O, rt)→ tribromoaniline (NH_2 with Br, Br, Br)

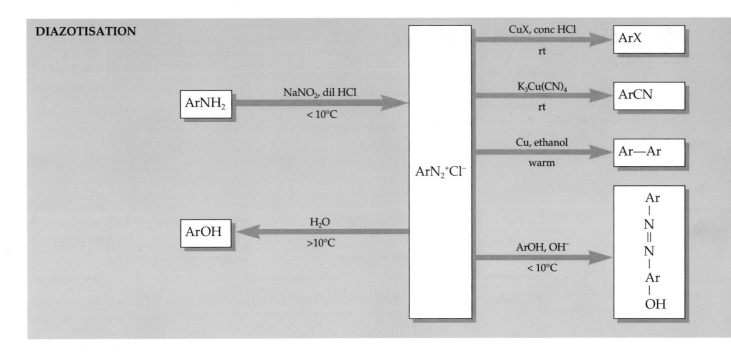

DATASHEET 21: Carboxylic acids and their derivatives

General formulae:

RCOOH	RCOCl	RCOBr	RCOOR^1	RCONH$_2$	(RCO)$_2$O
carboxylic acid	acid chloride	acid bromide	ester	amide	acid anhydride

Behaviour of carboxylic acids as acids (proton donors)

$RCOOH + H_2O \rightleftharpoons RCOO^- + H_3O^+$ $\quad$ $RCOOH + NaOH \longrightarrow RCOO^-Na^+ + H_2O$

$RCOOH + Na \longrightarrow RCOO^-Na^+ + \frac{1}{2}H_2$ $\quad$ $RCOOH + NaHCO_3 \longrightarrow RCOO^-Na^+ + CO_2 + H_2O$

Other reactions

REDUCTION

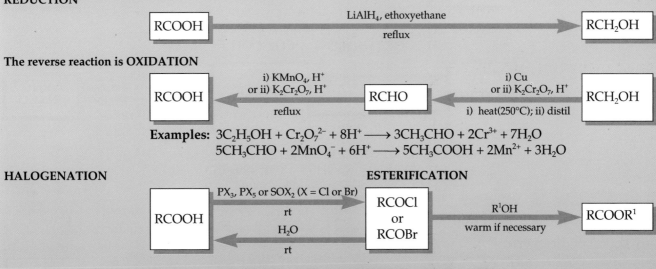

The reverse reaction is OXIDATION

Examples: $3C_2H_5OH + Cr_2O_7^{2-} + 8H^+ \longrightarrow 3CH_3CHO + 2Cr^{3+} + 7H_2O$

$5CH_3CHO + 2MnO_4^- + 6H^+ \longrightarrow 5CH_3COOH + 2Mn^{2+} + 3H_2O$

HALOGENATION $\qquad\qquad\qquad$ **ESTERIFICATION**

Examples: $CH_3COOH + PCl_5 \longrightarrow CH_3COCl + POCl_3 + HCl$

$CH_3COCl + C_2H_5OH \longrightarrow CH_3COOC_2H_5 + HCl$

$CH_3COCl + H_2O \longrightarrow CH_3COOH + HCl$

DEHYDRATION

ESTERIFICATION

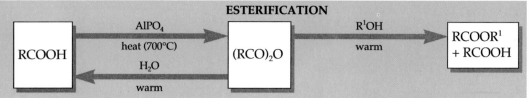

$RCOOH$ $\xrightarrow[\substack{\text{heat (700°C)}}]{\text{AlPO}_4}$ $\xleftarrow[\text{warm}]{\text{H}_2\text{O}}$ $(RCO)_2O$ $\xrightarrow[\text{warm}]{R^1OH}$ $RCOOR^1 + RCOOH$

Examples: $2CH_3COOH \longrightarrow (CH_3CO)_2O + H_2O$
$(CH_3CO)_2O + C_2H_5OH \longrightarrow CH_3COOC_2H_5 + CH_3COOH$
$(CH_3CO)_2O + H_2O \longrightarrow 2CH_3COOH$

FORMATION OF ACID AMIDE

$RCOOH$ $\xrightarrow[\substack{\text{rt} \\ \text{H}^+ \\ \text{rt}}]{\text{NH}_3\text{(aq)}}$ $RCOO^-NH_4^+$ $\xrightarrow[\substack{\text{H}_2\text{O} \\ \text{warm}}]{\text{heat solid salt}}$ $RCONH_2$

Examples: $CH_3COOH + NH_3 \longrightarrow CH_3COO^-NH_4^+$
$CH_3COO^-NH_4^+ \longrightarrow CH_3CONH_2 + H_2O$
$CH_3CONH_2 + H_2O \longrightarrow CH_3COO^-NH_4^+$
$CH_3COO^-NH_4^+ + H^+ \longrightarrow CH_3COOH + NH_4^+$

$RCOOH$ $\xrightarrow[\substack{\text{rt} \\ \text{H}^+ \\ \text{rt}}]{R^1NH_2}$ $RCOO^-R^1NH_3^+$ $\xrightarrow[\substack{\text{H}^+ \\ \text{rt or warm}}]{\text{heat solid salt}}$ $RCONHR^1$

Examples: $CH_3COOH + C_2H_5NH_2 \longrightarrow CH_3COO^-C_2H_5NH_3^+$
$CH_3COO^-C_2H_5NH_3^+ \longrightarrow CH_3CONHC_2H_5 + H_2O$
$CH_3CONHC_2H_5 + H_2O \longrightarrow CH_3COO^-C_2H_5NH_3^+$
$CH_3COO^-C_2H_5NH_3^+ \longrightarrow CH_3COOH + C_2H_5NH_2$

ESTERIFICATION

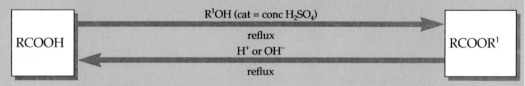

$RCOOH$ $\xrightarrow[\text{reflux}]{R^1OH \text{ (cat = conc } H_2SO_4)}$ $\xleftarrow[\text{reflux}]{\text{H}^+ \text{ or } OH^-}$ $RCOOR^1$

Examples: $CH_3COOH + C_3H_7OH \longrightarrow CH_3COOC_3H_7 + H_2O$
$CH_3COOC_3H_7 + H_2O \longrightarrow CH_3COOH + C_3H_7OH$

$RCOOH$ $\xrightarrow[\substack{\text{rt} \\ \text{H}^+ \\ \text{rt}}]{\text{dil NaOH}}$ $RCOO^-Na^+$ $\xrightarrow[\text{heat}]{\substack{\text{solid salt +} \\ \text{NaOH/Ca(OH)}_2 \text{ (sodalime)}}}$ $RH + Na_2CO_3$

Examples: $CH_3COOH + NaOH \longrightarrow CH_3COO^-Na^+ + H_2O$
$CH_3COO^-Na^+ + NaOH \longrightarrow CH_4 + Na_2CO_3$
$CH_3COO^-Na^+ + HCl \longrightarrow CH_3COOH + NaCl$

An important reaction of amides

HOFMANN DEGRADATION

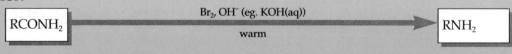

$RCONH_2$ $\xrightarrow[\text{warm}]{\text{Br}_2, OH^- \text{ (eg. KOH(aq))}}$ RNH_2

Example: $CH_3CONH_2 + Br_2 + 4NaOH \longrightarrow CH_3NH_2 + Na_2CO_3 + 2NaBr + 2H_2O$

DATASHEET 22: Aldehydes and ketones

General formulae:

RCHO   RR¹CO

aldehyde   ketone (R, R¹ may or may not be the same)

both classes contain the carbonyl group, $>C{=}O$

Reactions of aldehydes and ketones

HYDROGENATION (REDUCTION)

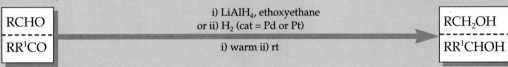

RCHO / RR¹CO →
i) LiAlH₄, ethoxyethane
or ii) H₂ (cat = Pd or Pt)
i) warm ii) rt
→ RCH₂OH / RR¹CHOH

The reverse reaction is DEHYDROGENATION (OXIDATION)

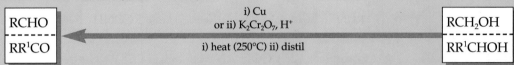

RCHO / RR¹CO ←
i) Cu
or ii) K₂Cr₂O₇, H⁺
i) heat (250°C) ii) distil
← RCH₂OH / RR¹CHOH

Examples: $3C_3H_7OH + Cr_2O_7^{2-} + 8H^+ \longrightarrow 3C_2H_5CHO + 2Cr^{3+} + 7H_2O$

$5CH_3CH(OH)C_2H_5 + 2MnO_4^- + 6H^+ \longrightarrow 5CH_3COC_2H_5 + 2Mn^{2+} + 8H_2O$

OXIDATION

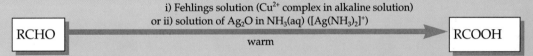

RCHO →
i) Fehlings solution (Cu²⁺ complex in alkaline solution)
or ii) solution of Ag₂O in NH₃(aq) ([Ag(NH₃)₂]⁺)
warm
→ RCOOH

Example: $CH_3CHO + 2[Ag(NH_3)_2]^+ + H_2O \longrightarrow CH_3COOH + 2Ag + 2NH_4^+$

RR¹CO   no reaction with Fehling's solution or with [Ag(NH₃)₂]⁺

ADDITION: CYANIDE IONS

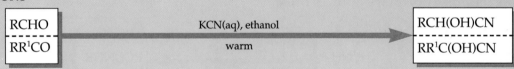

RCHO / RR¹CO →
KCN(aq), ethanol
warm
→ RCH(OH)CN / RR¹C(OH)CN

Examples: $CH_3CHO + HCN \longrightarrow CH_3CH(OH)CN$

$(CH_3)_2CO + HCN \longrightarrow (CH_3)_2 C(OH)CN$

ADDITION: SODIUM HYDROGENSULPHITE

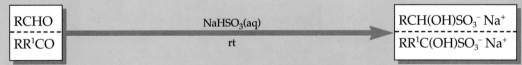

RCHO / RR¹CO →
NaHSO₃(aq)
rt
→ RCH(OH)SO₃⁻ Na⁺ / RR¹C(OH)SO₃⁻ Na⁺

Examples: $CH_3CHO + NaHSO_3 \longrightarrow CH_3CH(OH)SO_3^-Na^+$

$(CH_3)_2CO + NaHSO_3 \longrightarrow (CH_3)_2C(OH)SO_3^-Na^+$

CONDENSATION: WITH Z—NH₂

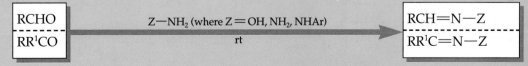

RCHO / RR¹CO →
Z—NH₂ (where Z = OH, NH₂, NHAr)
rt
→ RCH=N—Z / RR¹C=N—Z

Examples: $CH_3CHO + NH_2OH \longrightarrow CH_3CH{=}N{-}OH + H_2O$

$(CH_3)_2CO + NH_2C_6H_5 \longrightarrow (CH_3)_2C{=}N{-}C_6H_5 + H_2O$

DATASHEET 23: Amines and nitriles

General formulae:

RCN	R¹NH₂	R¹R²NH	R¹R²R³N
nitrile	primary amine	secondary amine	tertiary amine

R^1, R^2, R^3 = alkyl groups which may or may not be the same as one another

Reactions of the nitriles

HYDROYLSIS

Examples: $CH_3CN + H_2O \longrightarrow CH_3CONH_2$

$CH_3CONH_2 + H_2O \longrightarrow CH_3COOH + NH_3$

The reverse reaction is DEHYDRATION

Examples: $CH_3COOH + NH_3 \longrightarrow CH_3COO^-NH_4+ \longrightarrow CH_3CONH_2 + H_2O$

$CH_3CONH_2 \longrightarrow CH_3CN + H_2O$

REDUCTION

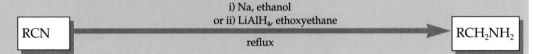

Behaviour of amines as bases

$R^1NH_2 + H_2O \rightleftharpoons R^1NH_3^+ + OH^-$ $R^1NH_2 + HCl \longrightarrow R^1NH_3^+Cl^-$

$R^1R^2NH + H_2O \rightleftharpoons R^1R^1NH_2^+ + OH^-$ $R^1R^2NH + HCl \longrightarrow R^1R^2NH_2^+Cl^-$

$R^1R^2R^3N + H_2O \rightleftharpoons R^1R^2R^3NH^+ + OH^-$ $R^1R^2R^3N + HCl \longrightarrow R^1R^2R^3NH^+Cl^-$

Reactions of primary amines only

ALKANOYLATION

Examples: $CH_3NH_2 + C_2H_5COCl \longrightarrow C_2H_5CONHCH_3 + HCl$

$C_2H_5CONHCH_3 + C_2H_5COCl \longrightarrow (C_2H_5CO)_2NCH_3 + HCl$

The reverse reaction is HYDROLYSIS

Examples: $C_2H_5CONHCH_3 + H_2O \longrightarrow CH_3NH_2 + C_2H_5COOH$

$(C_2H_5CO)_2NCH_3 + H_2O \longrightarrow C_2H_5CONHCH_3 + C_2H_5COOH$

OTHER REACTIONS

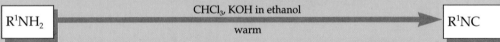

Example: $CH_3NH_2 + CHCl_3 + 3KOH \longrightarrow CH_3NC + 3KCl + 3H_2O$

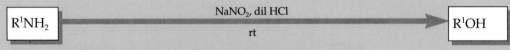

Example: $CH_3NH_2 + HNO_2 \longrightarrow CH_3-N^+\equiv N + OH^- + H_2O$

then $CH_3-N^+\equiv N + H_2O \longrightarrow CH_3OH + N_2 + H^+$

THE DATABASE

DATASHEET 24: Synthetic route map for aliphatic compounds

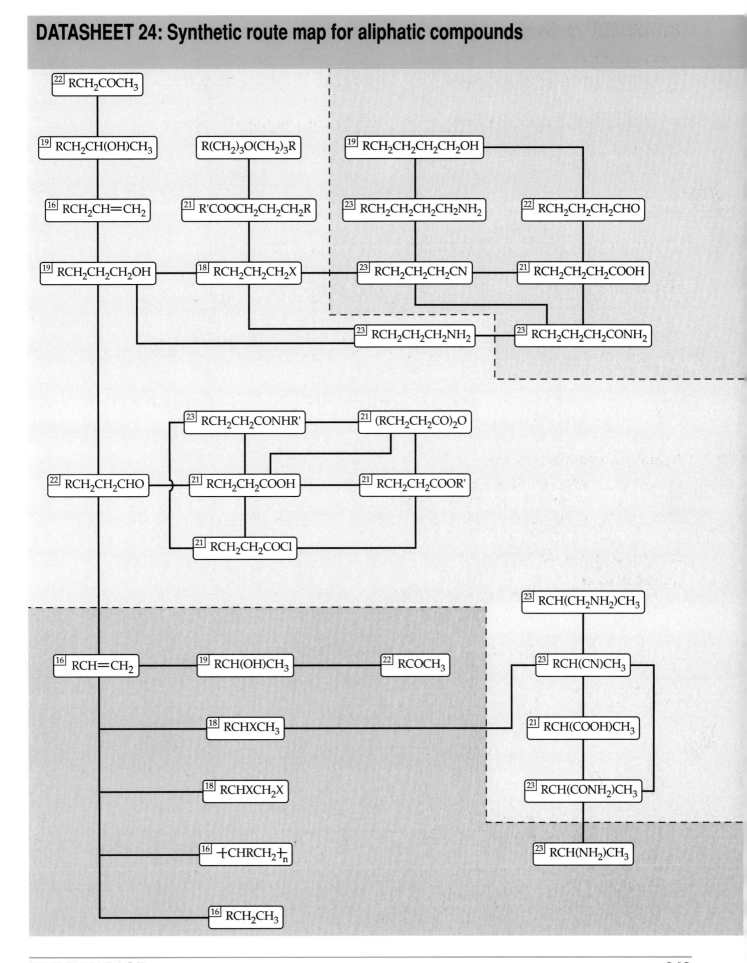

DATASHEET 25: Synthetic route map for aromatic compounds

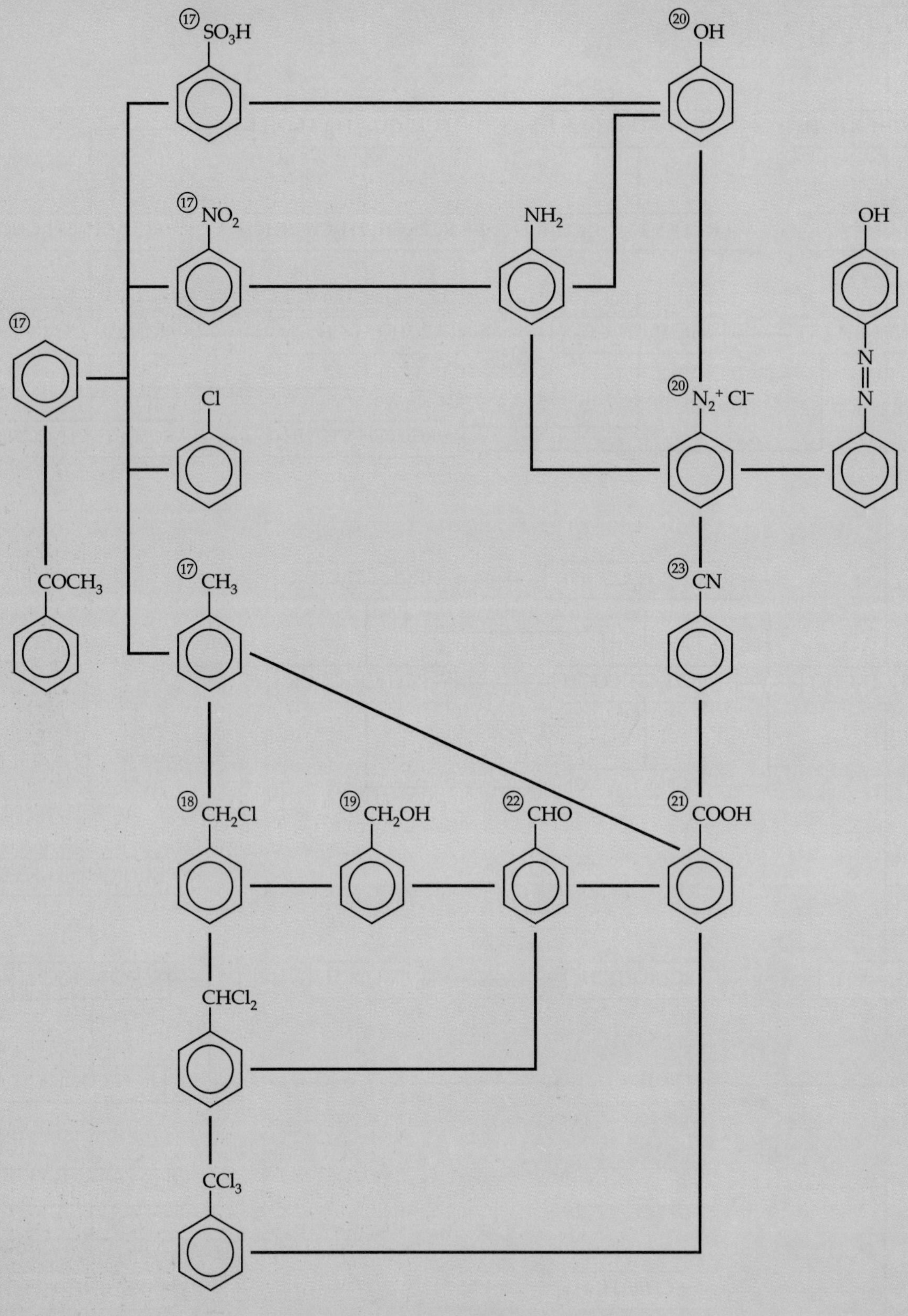

Note Chemistry of Ar—CN and Ar—COOH is essentially the same as aliphatic nitriles and aliphatic carboxylic acids, hence the references to Datasheets 23 and 21.

Theme 4

ENERGY

All chemical reactions, including those in our bodies, are accompanied by energy changes, most commonly loss or gain of heat, when chemical bonds break and new ones form. Electricity is one form of energy which can either be produced from chemical reactions or used to bring about chemical change.

Chapter 14

THERMOCHEMISTRY

LEARNING OBJECTIVES

When you have studied this chapter you should be able to:

1. describe examples of energy transformations;

2. explain the use of the terms endothermic and exothermic;

3. summarise the energy sources available to humans;

4. define the terms 'enthalpy change of reaction' and 'standard conditions';

5. calculate enthalpy changes from experimental measurements;

6. define the terms enthalpy change of formation, combustion, hydration, solution, neutralisation and atomisation;

7. understand and use energy diagrams for chemical reactions;

8. understand that some spontaneous chemical changes are endothermic and that the sign of ΔH does not necessarily determine the direction of chemical change;

9. define lattice enthalpy for simple ionic crystals;

10. state the First Law of Thermodynamics;

11. state Hess's law and explain how this law may be used to find enthalpy changes by an indirect method;

12. construct simple energy cycle diagrams.

14.1 INTRODUCING ENTHALPY

Energy is important. Our modern lifestyles depend heavily on good management of the world's energy resources. It has been estimated that today in Europe, we use 200 times more energy daily than our early ancestors did. We use this energy for many different purposes, including manufacturing, transport, heating, lighting, food production and communication. There is no shortage of energy – it is all around us. Energy comes in many forms and includes heat, light, sound, electricity and chemical energy (the energy stored in a substance through the chemical bonds it contains). All these forms of energy can be converted from one form to another. As managers of this energy, our tasks are to:

- identify useful energy resources;
- convert energy into the most suitable forms to supply our needs;
- harness diffuse, but in a sense infinite, energy sources such as wind, tidal power and solar energy to do useful work;
- develop systems to use energy efficiently and conserve it.

Energy conversion often takes place on a huge scale. A conventional power station changes chemical energy from coal, gas or oil into heat. Nuclear power stations convert into heat the enormous amount of energy produced when uranium-235 atoms are bombarded with neutrons. In both types of power stations, the heat is used to generate steam from water. The steam is used to drive turbines and this produces electricity, a very versatile (and easily transported) form of energy.

> **KEY FACT**
>
> Energy is the capacity of an object or substance to do work. The **First Law of Thermodynamics** (also known as the Law of Conservation of Energy) states that energy cannot be created or destroyed. It can, however, be converted from one form to another.

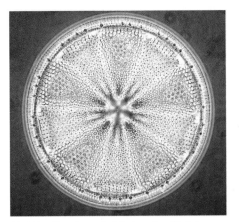

Fig 14.1 Even the smallest plants, such as this diatom, are capable of very efficient energy conversions, converting solar energy to stored carbohydrates. The stored energy can maintain the energy-requiring functions of the cell even when sunlight is not available to the diatom.

1 What energy conversion takes place when a firework is set off?

Plants also convert energy on a massive scale. Plants transform solar energy into stored chemical energy (mainly in the form of carbohydrates) during the day. This stored energy is used for the growth and maintenance of the plant even when sunlight is not available. Animals also take advantage of the stored chemical energy in plants; eating plant materials provides the energy they need for breathing, bodily warmth, movement, growth and repair. Most energy conversions taking place in the natural world are very efficient and waste very little energy.

Fig 14.2 The stored chemical energy contained in a potato may be converted to electrical energy. Here the electricity is being used to run a clock.

However, no energy conversion is 100% efficient. Each conversion involves a loss of energy to the environment and slightly less energy is available to do work than before. Energy not used to do useful work is wasted energy (and energy is expensive). Good 'energy management' involves converting energy as efficiently as possible into the form required *and then conserving* it. For example, a gas central heating boiler converts the stored chemical energy in methane (natural gas) into heat energy to warm the house. To reduce winter heating bills an efficient central heating boiler is important but without good insulation (roof insulation, double glazing, etc) much of the heat may be lost.

2 Conventional light bulbs are only 1% efficient in terms of converting electricity into light. What happens to the other 99% of the electrical energy?

Chemists have a key role to play in the management of energy resources – to make energy available to people safely, at the right time and in the right form. To achieve these aims we need to understand the nature of energy changes taking place in chemical processes. Our understanding of such changes has developed in parallel with our model of the physical world, through careful observation and measurement, supported by creative thinking.

Heat is produced in many reactions yet is absorbed in others. Information about the amount of heat produced or absorbed is of great value. For example, a reaction which releases energy usually causes the temperature of the reaction mixture to rise. This may not, at first, seem that important – but consider the effect of such a temperature rise. The rate of reaction increases which causes further rapid release of energy, heating the reaction mixture further and accelerating the reaction. This spiral of events leads to a 'run-away' reaction which could get out of control and is potentially dangerous. In a modern chemical plant such events are predicted

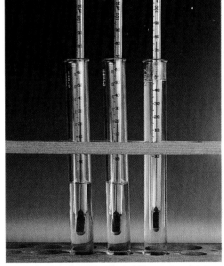

Fig 14.3 Concentrated sulphuric acid generates a lot of heat energy when mixed with water – an exothermic reaction. Unwanted concentrated acid should always be disposed of by pouring it into a large volume of water. Remember: "always do what you oughta, add the acid to the water!".

and therefore avoided. Chemical engineers would, if considering the problem of overheating just described, use information from databases of energy changes to design a plant with cooling systems to control temperature during the course of production.

EARLY CHEMICAL ENGINEERS

Chemical plant management is nothing new. More than 2500 years ago, people in Africa developed early iron-age technology in the region of Lake Victoria. Using only simple and locally available materials, these iron-age smelters were able to generate temperatures of up to 1400°C during the production of iron. Stored chemical energy in charcoal was converted to heat energy to smelt iron(III) oxide (haematite) to the iron needed for tools and weapons. The charcoal required (typically 95 kg for each firing) was obtained by burning the wood from local trees. Analysis of fragments of charcoal from early sites showed that these early technologists were selective in their fuels. They chose trees, such as the thornbush, with its dense wood and high silica content, that burnt slowly. This coupled with a carefully-regulated oxygen supply gave control over the processes.

Fig 14.4   A furnace for smelting iron in the Transvaal. The central shaft was built of clay. The bellows were used to blow air through clay pipes into the base of the furnace to raise the temperature. The photograph was taken in 1890.

Fig 14.5   This tribesman from the Amazon region is converting mechanical energy to heat energy until a self-sustaining oxidation of the chosen fuel begins. The chemical energy of the fuel is converted to heat and light energy. Without mastering this and other energy conversions, would human beings have ventured so far and so successfully from the warm climate of the African continent and the Mediterranean basin?

3   Why is the oxidation of a fuel described as 'self-sustaining' in Fig 14.5?

System and surroundings

Chemical and physical changes take place in many different environments including test tubes, beakers, industrial reactors and living cells. Wherever they take place, if we are to investigate the energy changes which accompany chemical and physical change we must first decide upon the area of interest. It is helpful to define a **system** (the area under investigation) and the **surroundings** (everything outside the system).

During exothermic reactions, stored chemical energy in the system is converted into heat. This heat is transferred to the surroundings. The photograph in Fig 14.3 shows that here the system (sulphuric acid and the

Fig 14.6 Condensation on the outside of the beaker shows that the temperature of the water it contained has dropped as the ammonium nitrate dissolved. The process:

$$NH_4NO_3(s) + aq \longrightarrow NH_4^+(aq) + NO_3^-(aq)$$

must, therefore, be endothermic.

water molecules it reacts with) transfers heat to the surroundings (the aqueous solution, test tube, thermometer and laboratory) during the reaction:

$$H_2SO_4(l) + 2H_2O(l) \longrightarrow 2H_3O^+(aq) + SO_4^{2-}(aq) + HEAT$$

In this reaction, the temperature in the test tube gradually decreases as the heat from the aqueous solution is spread out to the air around it.

During endothermic reactions, the system absorbs heat from the surroundings, which themselves drop in temperature. Fig 14.6 shows such a reaction. Some water molecules interact with the dissolving ions of ammonium nitrate to produce hydrated ions. Together the ions and the water molecules which hydrate them make up the system. The rest of the solution forms part of the surroundings.

4 **Explain the shape of the graph in Fig 14.7(b).**

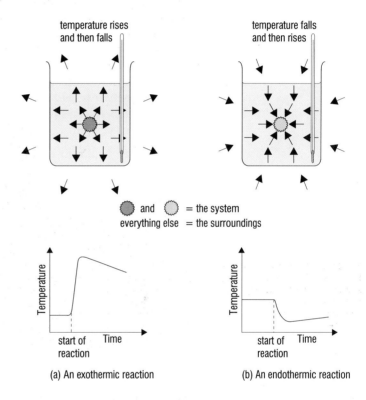

Fig 14.7 In exothermic and endothermic reactions, heat is transferred between system and surroundings.

The internal energy of a system

To study the energy changes involved in a chemical reaction we must look carefully at the energy of the system before and after the reaction. The total energy possessed by a system is called its **internal energy**, U. This is the sum of the kinetic and potential energies of all the particles it contains.

- The kinetic energy of a system is due to the movement through space of its constituent particles and any rotation, vibration and stretching movements they make.
- The potential energy of a system is due to all the electrostatic and nuclear interactions between its particles.

If the sum of kinetic and potential energies is changed then the energy of the system as a whole is changed. A change in internal energy, U, for a system, takes place when:

- heat (given the symbol q) is transferred to or from the system,
- work (given the symbol w) is done by the system or on the system.

Mathematically, we can express this as $U = q + w$.

> **5** Which has the greatest kinetic energy if temperature and pressure are identical: 1 mole of sodium chloride or 1 mole of nitrogen?

Table 14.1 Sign conventions for q and w

The system absorbs heat from the surroundings	$+q$ (q is positive)
The system gives out heat to the surroundings	$-q$ (q is negative)
Work is done on the system by the surroundings	$+w$ (w is positive)
Work is done by the system on the surroundings	$-w$ (w is negative)

Reactions at constant volume

If a reaction takes place in a **closed system** (a sealed flask, for example) then gases cannot escape and 'do work' pushing back the surrounding air. Such a reaction is said to take place at **constant volume**. The only energy transferred between system and surroundings is heat energy.

For a closed system, $w = 0$ and:

$$U = q_v \quad (q = \text{heat transferred at constant volume})$$

This means that for a reaction taking place at constant volume the change in internal energy is equal to the heat energy transferred.

Reactions at constant pressure

Most of the reactions we carry out in the laboratory take place in **open systems**. This term simply means that the reaction is 'open to the atmosphere' and any gases produced can escape. Pressure does not build up in the system and so such a reaction takes place at **constant pressure**. When a gas is formed it occupies space and pushes back the surrounding air. As it does so it is said to 'do work' and uses energy. A reaction may also absorb gases and then work is being done on the system *by the surroundings*. Here the system absorbs energy. Where does this energy come from or go to?

We can easily investigate the amount of work done by or on gaseous systems by examining the changes in pressure and volume which take place. Work may be defined as pV. Here p is the pressure exerted by the system on the surroundings, and V is the change in volume achieved. Consider, as an example, a reaction in which 1.2 g of magnesium (0.05 mol) reacts with 50 cm^3 of 2 mol dm^{-3} hydrochloric acid (0.1 mol):

$$Mg(s) + 2HCl(aq) \longrightarrow MgCl_2(aq) + H_2(g)$$

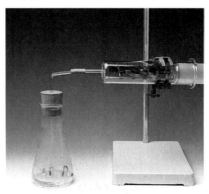

Fig 14.8   Hydrogen released when magnesium and hydrochloric acid react performs work by pushing out the syringe against the pressure of the surroundings.

KEY FACT

Units of energy

The calorie is the unit often used to measure the energy content of foods. One calorie is the quantity of energy needed to raise the temperature of 1 g of water by 1°C (1 K). However, the calorie has been replaced by the SI unit for energy, the **joule**, **J**, which is the preferred unit for the measurement of energy of all kinds. One calorie is equal to 4.184 J. Larger quantities of energy are often measured in kilojoules, kJ, for convenience, 1 kJ = 1 000 J.

The system consists of the magnesium ions and the hydrated hydrogen ions (from the hydrochloric acid). The solvent molecules, container and everything else make up the surroundings.

From the equation we can see that 0.05 mol of magnesium would give 0.05 mol of hydrogen molecules. Since 1 mol of any gas occupies about 24 dm³ at room temperature and pressure, $(0.05 \times 24) = 1.2$ dm³ of hydrogen would be produced. The system would increase in volume by 1.2 dm³ $(1.2 \times 10^{-3}$ m³). The escaping hydrogen must push air out of the way and work done (w) by the expanding system is pV.

If the reaction mixture was at a pressure of 10^5 N m⁻² the work done $= 10^5 \times 1.2 \times 10^{-3}$ J. This means that 120 J of chemical energy has been used to do work. The rest of the energy released by this exothermic reaction is released as heat energy and warms the surroundings.

Despite appearances, this reaction, like most other laboratory reactions, is described as 'a reaction at constant pressure'. The internal pressure created in the system decreases (by pushing the piston out) until the pressure is once again equal to that of the surroundings.

Enthalpy

As we have seen in the example above, the change in internal energy, U, which takes place during a reaction of this kind, at constant pressure, is not only due to heat transferred but also to the work done. The heat transferred during reaction at constant pressure must be less than the heat transferred during a reaction at constant volume. For this reason we must consider these two situations separately.

The heat transferred during a reaction taking place at constant pressure is known as the **enthalpy change** for the reaction. It is the change in enthalpy (the change in 'heat content') which is of interest to the chemist. Enthalpy change, ΔH, is defined as:

$$\Delta H = U + pV.$$

In simple terms this means that the enthalpy change is equal to the change in the internal energy of the system, U, plus any work done by the system pV. We can simplify this to:

$$\Delta H = q_p \qquad (q_p = \text{heat transferred at constant pressure})$$

For a reaction taking place at constant pressure the enthalpy change is equal to the heat transferred.

For the majority of chemical reactions the enthalpy change is the most useful measure of energy change we can determine. However, there are methods for measuring the heat change at either constant volume or constant pressure (Fig 14.9).

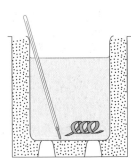 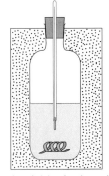

(a) An insulated beaker, open to the atmosphere can be used to measure **enthalpy change**.

(b) A sealed, insulated container (such as a vacuum flask) can be used to measure **internal energy change**.

Fig 14.9 Magnesium reacts with hydrochloric acid to give a solution of magnesium chloride. Hydrogen gas is evolved. The enthalpy change and the internal energy change can be measured using the apparatus shown.

6 For what types of reaction are the enthalpy change and the internal energy change the same?

The enthalpy change of a physical or chemical change taking place at constant pressure may be defined as:

enthalpy change = (enthalpy of final state) − (enthalpy of initial state)

$$\Delta H = H_{final} - H_{initial}$$

Enthalpy changes are due to the reorganisation (breaking or formation) of bonds. Bonds are attractive forces which hold two or more particles together. These particles may be atoms, molecules or ions. Breaking bonds of any kind (ionic, covalent, metallic or intermolecular) requires energy to enable the particles to overcome the attractive force holding them together. If particles become separated by a sufficient distance, such that they no longer influence one another, the bond between them may be regarded as broken.

The formation of bonds releases energy. The breaking of bonds requires energy. The difference between exothermic and endothermic reactions can be explained in simple terms:

- if the energy required to break bonds is less than the energy released when new bonds form, the reaction is exothermic;
- if the energy required to break bonds is greater than the energy released when new bonds form, the reaction is endothermic.

QUESTIONS

14.1 Explain what is meant by the term 'energy conversion' giving a simple example you have seen in the laboratory.

14.2 Explain what energy conversions take place in a car engine.

14.3 State the First Law of Thermodynamics.

14.4 Lead must be heated to 601 K before it melts. Is the melting process an endothermic or an exothermic one?

14.5 When methane burns in a Bunsen burner a chemical reaction takes place.
(a) Write a balanced equation for the reaction taking place.
(b) Identify which is the system and which is the surroundings in this reaction.
(c) Is the system an open or closed system?
(d) Is this reaction exothermic or endothermic?

14.6 Explain carefully the difference between an internal energy change and an enthalpy change.

14.7 In a laboratory experiment 0.12 g of pure tin is added to excess dilute nitric acid at 25°C in a flask connected to a syringe.
(a) Write an equation for the reaction which takes place.
(b) Calculate the increase in volume recorded by the syringe.
(c) Calculate the work done by the system in pushing out the syringe.

The universal gas constant, R = 8.31 J K^{-1} mol^{-1}; molar volume of any gas at s.t.p. = 24 dm^3; assume the pressure is 10^5 N m^{-2}

14.8 A camping stove which runs on 'methylated spirits' oxidises ethanol to provide energy. Make a list of bonds broken and bonds formed during this oxidation.

14.2 ENTHALPY CHANGE AND PHASE CHANGE

KEY FACT

The states of matter, solid, liquid and gas, are called phases. A change of state, for example from solid to liquid or from gas to liquid, is called a **phase change** or **phase transition**.

CONNECTION: States of matter: Section 5.2.

Table 14.2 Specific heat capacities of Period 3 elements

Element	Specific heat capacity /kJ kg^{-1} K^{-1}
Sodium	1230
Magnesium	1030
Aluminium	900
Silicon	711
Phosphorus* (P$_4$)	747
β-Sulphur	732
Chlorine	477
Argon	519

* White phosphorus

KEY FACT

The amount of energy required to raise the temperature of one kilogram of a substance by 1 K is known as its specific heat capacity, c.

A phase change from one state of matter to another involves either breaking bonds (solid to liquid, liquid to gas, or solid to gas) or forming bonds (liquid to solid, gas to liquid, or gas to solid).

Heat and temperature

Temperature and heat are different. The temperature of hot water, does not depend on how much water there is but the amount of heat contained depends on the mass of water present. For example, 50 g of water at 298 K contains twice the heat contained by 25 g of water at the same temperature. The temperature of a substance tells us quantitatively how 'hot' a substance is compared to a reference (0 K). For any substance, the amount of heat it contains depends on its temperature and mass.

Different substances (and different amounts of the same substance) need different amounts of heat energy to bring them to the same temperature, depending on their **specific heat capacities**. The specific heat capacity of a substance, c, is the amount of energy required to raise the temperature of one kilogram of the substance by 1 K. For example, the specific heat capacity of water is 4.184 kJ kg^{-1} K^{-1}.

7 What amount of energy, measured in joules, is required to raise the temperature of 10 g of water by 2.5 K, assuming no loss of energy to the surroundings?

8 Metals are good conductors of heat. Explain why the trend in specific heat capacities of the Period 3 elements supports this statement (Table 14.2).

As we heat a solid, its enthalpy increases. The energy absorbed increases the kinetic energy of its particles. If we were able to see these particles (atoms, molecules or ions) we would observe an increase in their rate of vibration, but their 'average' position would not change. It is a little like hitting a tennis ball attached to a post by an elastic string. Standing in the centre we can hit the ball in all directions around us. If we hit the ball harder it will move faster but its 'average' position is still at the centre. The average speed of such a ball resembles the kinetic energy of particles in a solid lattice.

The temperature of the substance as a whole depends upon the mean kinetic energy of its particles. The amount of energy we must expend to get the ball to move with a certain speed can be compared to the heat energy required to heat a solid to a certain temperature. Using a heavier ball or a stronger elastic would require us to use more effort to get the ball moving at the same average speed. Similarly, the particles in one solid lattice may be larger and heavier than those in another and the strengths of the forces restraining their movement may also vary widely.

Melting and sublimation

There is another way in which the behaviour of a tennis ball 'on elastic' resembles that of a particle in a solid lattice. If we hit the ball hard enough it will have enough momentum to snap the elastic and become independent. If the particles in a solid are given sufficient energy they overcome the forces holding them in their fixed positions. At this point the substance melts, changing from solid to liquid. During a phase change, energy is not used to raise the temperature of the solid but to change the arrangement of particles from that of a solid to that of a liquid or gas. A solid which changes directly into a gas with no liquid phase is said to have sublimed.

Enthalpy changes which accompany phase changes are usually measured per mole of substance. The units are **kJ mol⁻¹**.

The **standard molar enthalpy of fusion,** $\Delta H^{\circ}_{fus,m}$ is the enthalpy change that takes place when one mole of a solid substance is converted to a liquid at its melting point and at 10^5 Pa.

The **standard molar enthalpy of sublimation,** $\Delta H^{\circ}_{sub,m}$ is the enthalpy change that takes place when one mole of a solid substance is converted to a gas at its sublimation temperature and at 10^5 Pa.

Fig 14.10 The melting point of gallium is 30°C. Since human body temperature is 37°C, the heat from a hand is sufficient to change gallium from solid to liquid.

The transition from liquid to solid (freezing) is accompanied by an enthalpy change of the same numerical size but of opposite sign to that for the transition solid to liquid (melting). For example:

$\text{Na(s)} \longrightarrow \text{Na(l)}; \ \Delta H = +2.6 \text{ kJ mol}^{-1}$
$\text{Na(l)} \longrightarrow \text{Na(s)}; \ \Delta H = -2.6 \text{ kJ mol}^{-1}$

9 **Is freezing an exothermic or an endothermic process?**

Fig 14.11 'More ice, please.' Two typical ice-cubes ought to get the temperature down by about 10°C.

A COCA-COLA AND 1.79 MOLES OF ICE PLEASE!

A refreshing iced-drink. Approximately 300 cm³ in volume it contains two ice-cubes with an approximate total volume of 35 cm³. The density of ice is less than that of water (0.92 kg dm⁻³ compared to 1 kg dm⁻³) and their total mass is therefore about 32.2g, approximately 1.79 mol. The specific heat capacity of ice is 2.09 kJ kg⁻¹K⁻¹. If we assume that the temperature of ice is initially –10°C, the energy required to raise its temperature to the melting point of water is $32.2/1000 \times 2.09 \times 10 = 0.67$ kJ. The standard molar enthalpy of fusion for ice is 6.02 kJ mol⁻¹. The energy input required to melt the ice is approximately $1.79 \times 6.02 = 10.78$ kJ.

Further energy is needed to warm the molten ice to 15°C. The specific heat capacity of water is 4.18 kJ kg⁻¹ K⁻¹ so the energy required is $32.2/1000 \times 4.18 \times (15-0) = 2.01$ kJ. The total energy input required is therefore $0.67 + 10.78 + 2.01 = 13.46$ kJ. Since the volume of drink is about 300 cm³, at 25°C, its temperature should fall by about 10°C.

Boiling

When an electric kettle boils, electrical energy is used to raise the temperature of the heating element. Energy is transferred to the water by conduction and convection, raising its temperature. As the temperature rises, so does the average kinetic energy of individual water molecules. Eventually large numbers of water molecules gain sufficient kinetic energy to overcome the attraction they have for the bulk of the liquid and the boiling point of water (100°C at 10^5 Pa) is reached. Once the water has reached 100°C, any further energy supplied is used to change the water from liquid to gas. This energy is sometimes called the **latent heat**. The amount of energy which must be supplied to evaporate a liquid depends upon the size (mass and shape) of its particles and the strength of attractions between them.

The **standard molar enthalpy of vaporisation**, $\Delta H^{\circ}_{vap,m}$ is the enthalpy change which takes place when one mole of a liquid substance is vaporised at its boiling point and at 10^5 Pa.

QUESTIONS

14.9 What is meant by the term specific heat capacity?

14.10 Table 14.2 will be useful in answering the following question. If identical amounts of energy are transferred to samples of aluminium and magnesium of the same mass, which sample would experience the greatest temperature change?

14.11 When an ice-cube melts in a drink the process is endothermic. How does this help to cool the drink?

14.12 An egg is boiled in a saucepan containing 750 cm^3 of water. Calculate the amount of energy which must be supplied to the water (initially at 25°C) to raise its temperature to its boiling point.

14.13 Gallium is a metal with a very low melting point (Fig 14.10). Describe what happens to the gallium metallic lattice when gallium melts.

14.14 Use Datasheet 2 to calculate the enthalpy change which takes place when 64 g of liquid oxygen boils at 90 K and at 10^5 Pa.

14.3 ENTHALPY CHANGES AND CHEMICAL BONDS

Standard conditions

Meaningful comparisons of enthalpy changes associated with chemical reactions can only be made if they are measured under the same conditions. The size of the enthalpy change is affected by:

- amounts of reactants;
- physical states of reactants;
- temperature;
- pressure.

For these reasons, energy changes quoted in the literature are usually measured under **standard conditions**. These are, before and after the change, a pressure of 10^5 Pa and temperature of 298 K (25°C). Under these conditions each substance involved is in its normal physical state. For example:

$$CH_4(g) + 2O_2(g) \longrightarrow CO_2(g) + 2H_2O(l); \quad \Delta H = -890.4 \text{ kJ}$$

The equation tells us that 'one mole of gaseous methane reacts with two moles of gaseous oxygen to give one mole of gaseous carbon dioxide and two moles of liquid water, with the evolution of 890.4 kJ energy'. The complete combustion of two moles of methane would have released 1780.8 kJ, and 0.01 mole would have released 8.904 kJ of energy. The amount of energy released would also be different under other conditions of pressure and temperature.

The enthalpy change under standard conditions is called the **standard enthalpy change**, $\Delta H^{\ominus}(298)$. The superscript $^{\ominus}$ is used to indicate standard pressure, and the figure in brackets is the standard temperature, in Kelvin.

10 How much energy would be released by the complete combustion of 1.6 g methane? (molar mass CH_4 = 16 g mol^{-1})

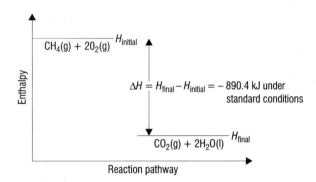

Fig 14.12 The enthalpy content of carbon dioxide and water is less than that of methane and oxygen, and the difference is the amount of energy released when methane combusts. We represent the energy change using an enthalpy profile.

However, it is often difficult or even impossible to measure enthalpy changes under standard conditions and so the standard enthalpy change for such a reaction must be calculated. For example, silicon and chlorine react to form $SiCl_4$ only at high temperatures. In the literature we are likely to find:

$$Si(s) + 2Cl_2(g) \longrightarrow SiCl_4(l); \ \Delta H^{\ominus}(298) = -640 \text{ kJ}$$

This value has been calculated from data obtained from a series of experiments carried out at temperatures at which reaction does occur. A reaction would not actually occur under standard conditions.

Covalent bonds and bond energies

The energy which must be supplied to break a covalent bond is known as the **bond dissociation enthalpy** $\Delta H^{\ominus}_{d,m}$, often called the **bond energy**.

The energy change during the formation of one mole of gaseous bromine molecules from two moles of bromine atoms is −193 kJ mol^{-1}. This is the amount of energy released as one mole of Br—Br bonds form. Conversely, 193 kJ of energy must be supplied to one mole of bromine molecules to cleave one mole of such bonds under standard conditions. For simple diatomic molecules, bond energies may be obtained directly by experiment.

$$Br_2(g) \longrightarrow 2Br(g); \ \Delta H^{\ominus}_{d,m}(298) = +193 \text{ kJ mol}^{-1}$$
$$2Br(g) \longrightarrow Br_2(g); \ \Delta H^{\ominus}(298) = -193 \text{ kJ}$$

Molecules which contain more than two atoms present a more complicated problem. For example, breaking the first O—H bond in a water molecule requires more energy than breaking the second O—H bond:

$$H_2O(g) \longrightarrow H(g) + OH(g); \ \Delta H^{\ominus}_{d,m}(298) = +502 \text{ kJ mol}^{-1}$$
$$OH(g) \longrightarrow H(g) + O(g); \ \Delta H^{\ominus}_{d,m}(298) = +427 \text{ kJ mol}^{-1}$$

THERMOCHEMISTRY

The environment of the O—H bond which is being broken is different in OH compared to that of H_2O. Similarly, the bonds between oxygen and hydrogen in water and in ethanol are in different environments and have different bond energies. In order to compare bonds which exist in different environments, therefore, we use the idea of **average bond energies**. These may be obtained by measuring the bond energy of a particular bond in several different environments and taking an average value.

11 **Why does the same covalent bond have different bond energies in different molecules? Hint: remember what determines the strength of a covalent bond.**

For example, we can calculate the strength of the carbon-chlorine bond in a number of compounds:

CH_3Cl: $\Delta H^{\ominus}[C—Cl] = 335$ kJ mol^{-1}
CH_3CH_2Cl: $\Delta H^{\ominus}[C—Cl] = 342$ kJ mol^{-1}
CCl_4: energy required to break all C—Cl bonds in the compound
$= +1308$ kJ mol^{-1};

therefore, each C—Cl requires $\dfrac{1208}{4} = 327$ kJ mol^{-1}

Extending this to include many other compounds containing the C—Cl bond, we find that the average bond energy for C—Cl is usually given as 330 kJ mol^{-1}.

Ionic bonds and lattice energies

The strong electrostatic interactions present in ionic lattices are difficult to break up and, consequently, compounds with these structures have very high melting points. For example, the melting points of sodium chloride and potassium bromide are 1074 K and 1000 K respectively.

12 **Is the formation of an ionic lattice an exothermic or an endothermic process?**

The **standard molar lattice enthalpy**, $\Delta H^{\ominus}_{lattice,m}$**(298)** is the enthalpy change which occurs when one mole of an ionic lattice is formed from its gaseous ions (from an infinite distance apart in space) under standard conditions. This exothermic process is often called the **lattice energy**. Strictly speaking, a lattice energy is an internal energy change and, therefore, is slightly higher than the corresponding lattice enthalpy. However, the very small difference (about 5 kJ mol^{-1} for M^+X^-) is ignored for most purposes.

The lattice energy may be represented by an equation, for example:

$Na^+(g) + Cl^-(g) \longrightarrow NaCl(s)$; $\Delta H^{\ominus}_{lattice,m}(298) = -787$ kJ mol^{-1}

The size of the lattice energy of an ionic compound influences its properties and the stability of its lattice. Lattice energies cannot be determined by direct measurement and must be obtained indirectly as a result of suitable experimental work. Two main factors dictate the numerical size of lattice energies:

- the charges on the anions and cations of which the lattice is composed;
- the distance between the anions and cations.

Lattices composed of small and highly charged ions have the largest lattice energies. These lattices are particularly stable to heat and there is a very good correlation between the melting points and lattice energies of ionic compounds (Table 14.3). The size of the lattice energy determines other properties of ionic compounds such as their solubility in water.

13 **Which will have the highest melting point: magnesium oxide or magnesium chloride?**

CONNECTION: Bond energies, Datasheet 4.

CONNECTION: Ionic lattices, page 66.

KEY FACT

The **standard molar enthalpy of bond dissociation**, $\Delta H^{\ominus}_{d,m}$**(298)**, (subscript 'd' means dissociation, 'm' means molar) is the enthalpy change which takes place when a mole of specified bonds is cleaved under standard conditions. For simplicity, this is often referred to as the bond energy.

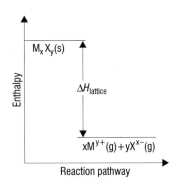

Fig 14.13 An enthalpy profile for the formation of gaseous ions from an ionic lattice:

$M_xX_y(s) \longrightarrow xM^{y+}(g) + yX^{x-}(g)$

CONNECTION: Determination of lattice energies, page 370.

Table 14.3

Ionic compound	Lattice energy / kJ mol^{-1}	Melting point / K
NaF	929	1261
NaCl	787	1074
NaBr	751	1020
NaI	700	920

Covalent solids and energy

Many properties of crystalline solids depend on the strength of the attraction between the particles present. The stronger the attraction the greater the amount of energy needed to disrupt the lattice. The hardness of a solid, its solubility and its melting point depend upon the strength of the attractive forces between its particles. Such properties are of great interest to scientists and engineers who often need to select solids to perform to a certain specification. They fall into two categories:

CONNECTION: Molecular solids, pages 63–64

- simple molecular solids;
- giant molecular solids.

ACTIVITY

The traditional vacuum flask employs a double-walled evacuated glass flask inside a metal container. The walls of the glass flask are silvered to minimise heat loss. Recent alternatives are made from plastic, resulting in fewer breakages.

A flask is to be constructed with an internal surface area of 360 cm^3 and is to be 3 mm in thickness. It could be made of glass or of a number of plastic alternatives. Ignoring heat loss to the surroundings use the following data to decide approximately how much energy would be transferred from hot water to heat a flask made from each of the following materials. Assume that the initial temperature of each flask is 25°C and that it is to contain 500 cm^3 of water at 90°C.

Substance	Density /kg dm^{-3}	Specific heat capacity /kJ kg^{-1} K^{-1}
glass	2.6	0.84
polycarbonate 1.2	1.2	1.2
poly(ethene)	0.92	2.3
poly(phenylethene)	1.06	1.3

Comment critically on your answer. What assumptions have you made? Which material seems most promising? Bear in mind that in the design of a vacuum flask many other factors would have to be considered. What other tests would *you* carry out on the materials before making a decision?

KEY FACT

Crystal lattices are regular, repeating arrays of particles. The particles may be atoms, molecules or ions.

CONNECTION: Intermolecular bonding, pages 80–84.

Simple molecular solids

Simple molecular solids such as solid carbon dioxide, ice or iodine are held together by relatively weak intermolecular bonds. The forces between molecules are easily overcome and simple molecular solids either melt or sublime at low temperatures. The value of solid ice or solid carbon dioxide as refrigerants is that they are able to absorb energy from their surroundings, thus cooling them (Fig 14.14). In the process they undergo endothermic changes of state and produce water and gaseous carbon dioxide respectively.

Fig 14.14 As this ice undergoes a phase change from solid to liquid, it absorbs energy from the fish. The low temperature of the fish is maintained, thus reducing bacterial growth rates and keeping the fish fresh.

THERMOCHEMISTRY

Giant molecular solids

Giant molecular solids such as diamond, graphite or silica are held together by covalent bonds. Such structures are very stable and only melt at very high temperatures. This property can often be put to good use. For example, the resistance of the diamond lattice to thermal and mechanical disruption is reflected in its use for high-performance drilling applications.

Fig 14.15 Some baby powders have a graphite base. Graphite is widely used as a dry lubricant. This is due to (a) its layered structure, which allows layers to slip over one another, and (b) its resistance to thermal disruption. Graphite melts around 4000 K. Therefore, it can also be used as a lubricant in high temperature environments.

QUESTIONS

14.15 What is meant by the term *standard state* for an element?

14.16 In a databook the energy change which takes place when sodium melts is given as $\Delta H^{\ominus}_{fus,m}(298) = 89.0$ kJ mol^{-1}. Explain carefully what is meant by the symbols:
(a) ΔH; (b) fus; (c) m; (d) $^{\ominus}$.

14.17 For hexane, C_6H_{14}; $\Delta H^{\ominus}_{c,m}(298) = -4195$ kJ mol^{-1}. Calculate the enthalpy change which takes place when 1 g of hexane is burnt under standard conditions.

$A_r[C] = 12.01$; $A_r[H] = 1.01$.

14.18 (a) Explain what is meant by the term *standard molar enthalpy change of bond dissociation*.
(b) In what way is this quantity different to an *average bond enthalpy*?

14.19 (a) With reference to potassium iodide, explain what is meant by the term *lattice energy*.
(b) What factors influence the size of the lattice energy of an ionic solid?

14.20 (a) The standard molar enthalpy change of lattice formation, $\Delta H^{\ominus}_{lattice,m}(298)$ for sodium chloride is -787 kJ mol^{-1}. Write an equation (including state symbols) representing the change to which this refers.
(b) Calculate the enthalpy change associated with converting 1 g of solid sodium chloride to gaseous ions under standard conditions.

$A_r[Na] = 22.99$; $A_r[Cl] = 35.45$.

CONNECTION: Dissolving ionic compounds in water, page 372.

KEY FACT

When something dissolves, the system is the dissolving solid and the solvent molecules which directly interact with it. The bulk solvent is treated as the surroundings.

Fig 14.16 This hand-warmer contains supersaturated sodium ethanoate. Disturbing the solution by bending a small metal disc initiates rapid crystallisation of solid from solution. Boiling the bag for six minutes causes all solid to re-dissolve and primes the bag for reuse.

Enthalpy of solution

A solution is a homogeneous phase consisting of more than one substance. One important type of solution is that formed when a solid (known as the **solute**) dissolves in a liquid (the **solvent**). The process of dissolving is difficult to classify either as a physical change or a chemical change. It is useful, however, to discuss, at this point, the enthalpy changes associated with dissolving a solid in a solvent.

Chemists and chemical engineers often need access to data about the solubility of solutes in different solvents. Such data can help, for example, to decide upon the amount of solvent required in a reaction. Too much solvent is wasteful and expensive due to both the cost of solvent and the cost of maintaining it at the reaction temperature. Too little solvent and a product or intermediate may precipitate from solution during a synthesis, blocking pipes and overworking pumps.

The process of dissolving is always accompanied by an enthalpy change. This may be exothermic or endothermic and reflects reorganisation of chemical bonds in the solute and solvent.

The **standard molar enthalpy of solution** is the overall enthalpy change which takes place when one mole of solute dissolves in a solvent to form an infinitely dilute solution under standard conditions. It is given the symbol, $\Delta H^{\ominus}_{sol,m}(298)$.

An infinitely dilute solution is one for which further dilution produces no further enthalpy change. Once sufficient solvent has been added to completely surround every solute particle adding more solvent produces no further enthalpy change because the solvent added does not interact with solute particles.

Why should an enthalpy change occur when a solid dissolves? As a solid dissolves, its lattice must be disrupted. This requires energy and, therefore, is an endothermic process. The dissolving solute breaks existing intermolecular bonds between solvent molecules, also an endothermic process. New bonds are formed between solute and solvent as solvent molecules surround and interact with the solute. This process (called **solvation**) is exothermic. The total enthalpy change associated with the dissolving of a solid in a liquid is the sum of all of these changes.

14 Is the crystallisation of sodium ethanoate (Fig 14.16) exothermic or endothermic?

Sometimes the enthalpy change which takes place as a solution is formed may be relatively small, for example, when iodine dissolves in hexane. Here the only change in bonding is a reorganisation of the weak dispersion forces between non-polar iodine molecules and non-polar hexane molecules. However, when ionic bonds are broken and ions become hydrated, larger enthalpy changes are often involved. For example, when sodium hydroxide dissolves in water:

$$NaOH(s) + aq \longrightarrow Na^+(aq) + OH^-(aq); \quad \Delta H^{\ominus} = -42.7 \text{ kJ mol}^{-1}$$

Enthalpies of formation

We cannot measure the energy content of a substance. However, we can compare the energy content of different compounds because we can measure enthalpy changes. One useful way is to compare their standard molar enthalpies of formation.

The **standard molar enthalpy of formation** of a substance is the overall enthalpy change which takes place when one mole of a compound is formed from its constituent elements under standard conditions. It is given the symbol $\Delta H^{\ominus}_{f,m}(298)$.

Most compounds have negative standard molar enthalpies of formation (energy is released by their formation) although a small number have positive values (energy is absorbed during their formation). The standard molar enthalpy of formation of all elements is taken as zero. However, many elements can exist in allotropes. Unless otherwise stated, the allotrope which is stable under standard conditions is assumed to be involved. For example, both diamond and graphite may be converted to carbon dioxide. The enthalpy change for the formation of carbon dioxide would be different in each case.

$$C(graphite) + O_2(g) \longrightarrow CO_2(g); \ \Delta H^\circ(298K) = -393.5 \ kJ \ mol^{-1}$$
$$C(diamond) + O_2(g) \longrightarrow CO_2(g); \ \Delta H^\circ(298K) = -395.4 \ kJ \ mol^{-1}$$

The standard molar enthalpy of formation for carbon dioxide would assume the use of graphite as a starting material as this is the more stable allotrope. Therefore, $\Delta H^\circ_{f,m}(298)[CO_2(g)] = -393.5 \ kJ \ mol^{-1}$.

15 **Which of the two allotropes of carbon above evolves more heat when one mole of it is burnt?**

Enthalpies of combustion

Data concerned with the enthalpy change which takes place as a fuel is burnt (combusted) is usually provided as a standard molar enthalpy of combustion. The **standard molar enthalpy change of combustion** is the enthalpy change which takes place when one mole of a substance is burnt in excess oxygen under standard conditions. It is given the symbol $\Delta H^\circ_{c,m}(298)$.

Since any standard enthalpy change refers to standard conditions, both reactants and products must be in their standard states. For example, the enthalpy change for the combustion of methane may be given as:

$$CH_4(g) + 2O_2(g) \longrightarrow CO_2(g) + 2H_2O(l); \ \Delta H^\circ(298) = -890.34 \ kJ \ mol^{-1}$$

or

$$CH_4(g) + 2O_2(g) \longrightarrow CO_2(g) + 2H_2O(g); \ \Delta H^\circ(298) = -802.34 \ kJ \ mol^{-1}$$

Both of these enthalpy changes may be measured. However, $\Delta H^\circ_{c,m}(298)[CH_4(g)] = -890.34 \ kJ \ mol^{-1}$.

16 **Why do the two values given above for the combustion of methane differ?**

Knowing the standard molar enthalpy change of combustion of a fuel allows predictions of the energy which can be obtained from it to be made. For example, the amount of fuel required by the Space Shuttle to escape the Earth's gravity has to be carefully calculated. Too little fuel would be a disaster. To carry excess fuel would waste valuable load-carrying capacity. We need to know accurately the amount of energy provided by burning a particular quantity of the fuel.

Enthalpies of neutralisation

When an acid and an alkali are mixed the exothermic reaction which takes place raises the temperature of the solution. Acidic solutions contain hydrated hydrogen ions, $H^+(aq)$ and alkaline solutions contain hydrated hydroxide ions, $OH^-(aq)$. Neutralisation involves reaction between these to yield water:

$$H^+(aq) + OH^-(aq) \longrightarrow H_2O(l)$$

The **standard molar enthalpy of neutralisation** is the enthalpy change, measured under standard conditions, which accompanies the formation of one mole of water during the reaction between an acid and an alkali. It is given the symbol $\Delta H^{\ominus}_{n,m}(298)$.

Since all acids produce hydrated hydrogen ions in aqueous solution and all alkalis give hydrated hydroxide ions, all neutralisation reactions are essentially the same reaction. The spectator ions may change, but the common change is the reaction shown above. For example, the neutralisation of both nitric acid and hydrochloric acid take place with the release of some energy:

$$HNO_3(aq) + NaOH(aq) \longrightarrow NaNO_3(aq) + H_2O(l);$$
$$\Delta H^{\ominus}_{n,m}(298) = -57.1 \text{ kJ mol}^{-1}$$

$$HCl(aq) + NaOH(aq) \longrightarrow NaCl(aq) + H_2O(l);$$
$$\Delta H^{\ominus}_{n,m}(298) = -57.1 \text{ kJ mol}^{-1}$$

In both cases one mole of hydrated hydrogen ions is reacting with one mole of hydrated hydroxide ions. Sodium ions, nitrate ions and chloride ions are spectator ions and play no part in the reaction.

CONNECTION: *Strong and weak acids and bases, pages 474–478.*

17 **Why is the enthalpy change for the neutralisation of one mole of sulphuric acid not the standard molar enthalpy of neutralisation for this reaction?**

Energy and chemical plant design

For comparison purposes enthalpy change data is presented for standard conditions. In reality standard conditions are rarely encountered. Data for standard conditions can be 'corrected' to provide values for different conditions. Without access to such data chemical plant design could only be achieved by an expensive process of 'trial and error'.

For example, the first stage in the **Contact process** for manufacturing sulphuric acid involves spraying molten sulphur into a furnace in the presence of air. An exothermic reaction takes place:

$$S(s) + O_2(g) \longrightarrow SO_2(g); \quad \Delta H^{\ominus}(298) = -296.8 \text{ kJ mol}^{-1}$$

When the reaction mixture is cooled the heat generated is not wasted. Often the heat is removed by a heat exchanger and can make a valuable contribution to the running costs of the chemical plant. The magnitude of the heat change may decide the size and number of heat exchangers required. During planning and designing a sulphuric acid plant, a chemical engineer may be asked to quantify the amount of heat which can be produced.

Although conditions inside the furnace are not standard and quantities far in excess of mole quantities are in use, the chemical engineer can adapt standard molar enthalpy of formation data to predict the energy changes which will take place in the plant. The heat generated by burning sulphur under these conditions can be calculated. If, for example, the heat was used to produce steam to drive a small turbine, with a knowledge of

- the specific heat capacity of water,
- the heat capacity of the furnace,
- the enthalpy of vaporisation of water,

the amount of steam which could be produced under these conditions may be calculated. The calculations are time-consuming, perhaps, but not difficult. They could make a big difference in determining the economic viability of sulphuric acid manufacture and illustrate the value of reliable data.

KEY FACT

Often, the enthalpy change associated with a particular chemical reaction may be described in more than one way. For example, the reaction

$$S(s) + O_2(g) \longrightarrow SO_2(g);$$
$$\Delta H^{\ominus}(298) = -296.8 \text{ kJ mol}^{-1}$$

may be considered as the standard molar enthalpy of combustion of sulphur:
$\Delta H^{\ominus}_{c,m}(298)[S(s)] = -296.8 \text{ kJ mol}^{-1}$

or, the standard molar enthalpy of formation of sulphur dioxide:
$\Delta H^{\ominus}_{f,m}(298)[SO_2(g)] = -296.8 \text{ kJ mol}^{-1}$

14.21 Define standard molar enthalpy change of solution, $\Delta H^\circ_{sol,m}(298)$.

14.22 Explain what happens to potassium ions and nitrate ions as potassium nitrate is stirred into a large excess of water. State which changes are exothermic and which are endothermic.

14.23 For sodium fluoride the lattice energy is –929 kJ mol^{-1}. The total of the standard molar enthalpy changes of hydration for gaseous sodium and fluoride ions is – 927 kJ mol^{-1}.
  (a) If 2 g of sodium fluoride is added to 1 dm^3 of water, calculate the enthalpy change which takes place.
  (b) Would the water increase or decrease in temperature?
  (c) Calculate the change in temperature which takes place given that the specific heat capacity of water is 4.18 kJ kg^{-1} K^{-1} and comment on the size of this change.
    A_r[Na] = 22.99; A_r[F] = 19.00; density of water is 1 kg dm^{-3}

14.24 Explain why standard molar enthalpy changes of combustion for different allotropes of the same element have different values.

14.25 α-sulphur (known as rhombic sulphur) is the most stable form of sulphur at temperatures up to 95.5°C. Above 95.5°C, β-sulphur (known as monoclinic sulphur) is the more stable. Both allotropes consist of S$_8$ rings. Identify, with the reasons, the allotrope with the larger value for $\Delta H_{c,m}(298)$, measured under standard conditions.

14.26 Use Datasheet 4 to determine whether burning 3.2 g of methanol or 2 g of ethanol would produce the larger enthalpy change.

14.27 The standard molar enthalpy change of formation for ammonia is –46.1 kJ mol^{-1}. How much energy is evolved when 1 kg of ammonia is formed from the elements under standard conditions?

A_r[N] = 14.01; A_r[H] = 1.01

14.28 When one mole of nitric acid is neutralised by adding it to excess potassium hydroxide solution, the enthalpy change which takes place is – 57.1 kJ mol^{-1}. When one mole of citric acid is added to excess potassium hydroxide the enthalpy change is around three times the value for nitric acid. What does this imply about the structure of citric acid?

14.5 MEASUREMENT OF ENTHALPY CHANGES

Calorimetry: a direct method

Heat released or absorbed by a reaction can be measured by experiment using a technique called **calorimetry**. The apparatus used is known as a **calorimeter**. A very simple example of a calorimeter which can be used to determine enthalpy changes in aqueous reactions is a polystyrene beaker.

KEY FACT

Calorimetry is the experimental technique used to determine enthalpy changes that accompany phase changes and chemical reactions.

Fig 14.17 A polystyrene beaker fitted with a thermometer has the essential features of a calorimeter – an insulated reaction vessel and a way of measuring temperature. The thermometer may also be used to stir the solution.

18 In the calorimeter, shown in Fig 14.17, which is 'the system' and which is 'the surroundings'?

To calculate an enthalpy change for a reaction taking place in aqueous solution, a number of assumptions are made.

- If the reaction is exothermic, the heat evolved in the reaction is transferred only to the solution.
- If the reaction is endothermic, the heat is absorbed only from the solution.
- The reaction takes place sufficiently rapidly for the maximum temperature to be reached before the reaction mixture begins to return to room temperature.
- The reaction vessel does not gain or lose heat.
- There is no heat exchange with the environment.
- The specific heat capacity of the aqueous solution is taken to be the same as that of water.

Let us use the example of magnesium reacting with dilute hydrochloric acid to illustrate a typical calculation.

$$Mg(s) + 2HCl(aq) \longrightarrow MgCl_2(aq) + H_2(g)$$

In an experiment carried out in a polystyrene beaker, a temperature change from 20.8°C to 26.0°C was measured when 0.500 g magnesium powder was added to 100.0 cm^3 of 2 mol dm^{-3} hydrochloric acid.

We will now assume the following:

- All the heat was transferred to the solution, with no heat loses to the polystyrene beaker (reasonable, since the specific heat capacity of polystyrene is very low) or the environment, and that the maximum temperature was reached before cooling to room temperature occurred.
- The density of the reaction mixture was 1 g cm^{-3} (to allow us to calculate the mass of solution) and its specific heat capacity was 4.2 kJ kg^{-1} K^{-1}.

For the reaction between magnesium and dilute hydrochloric acid described above:

heat released by reaction = heat gained by the solution,
heat gained by the solution = $m \times c \times \Delta T$
where, m = mass of solution (mass of magnesium plus mass of hydrochloric acid) = 100.5 g;
c = specific heat capacity, assumed to be 4.2 kJ kg^{-1} K^{-1} (4.2 Jg^{-1} K^{-1});
ΔT = temperature change = 26.0 – 20.8 = 5.2°C.

Therefore, heat gained by solution = 100.5 × 4.2 × 5.2 J
= 2195 J

and so the heat released by the reaction = 2195 J.

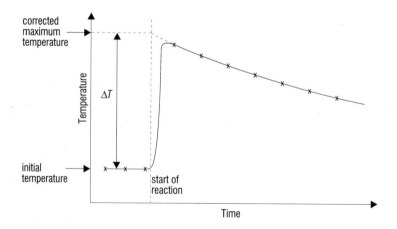

Fig 14.18 A cooling curve can be used to measure the corrected temperature rise for an exothermic reaction. The temperature that the reaction mixture would have reached had it not started to cool to room temperature can be estimated by extrapolation.

Since the molar mass of magnesium = 24.3 g mol^{-1},

the amount of magnesium used in the reaction = $\dfrac{0.500}{24.3}$ = 0.0206 mol.

Therefore, heat released per mole of magnesium = $\dfrac{2195}{0.0206}$ J

$= 106\,700$ J

$= 106.7$ kJ

And so the enthalpy of reaction under the conditions used is:

$Mg(s) + 2HCl(aq) \longrightarrow MgCl_2(aq) + H_2(g);\ \Delta H = -106.7$ kJ mol^{-1}

19 **Does the calorimeter used to determine this enthalpy change operate at constant pressure or constant volume?**

Assessing fuels is an important task in a thermochemical laboratory. In order to be able to compare the usefulness of different fuels we need to be able to measure how much energy is locked inside them. The enthalpy of combustion of a substance provides just such a measure of the value of a fuel.

The flame calorimeter

A flame calorimeter can be used to measure the enthalpy of combustion of a fuel. For example, the enthalpy of combustion of organic liquids can be measured as follows.

A small bottle is partly filled with the sample, and a wick inserted. The bottle is arranged so that when the wick is lit the heat produced is transferred to water held in a container. The temperature rise of this water is recorded. The quantity of fuel used can be found by measuring the difference in the mass of the bottle containing the fuel before and after a period of combustion.

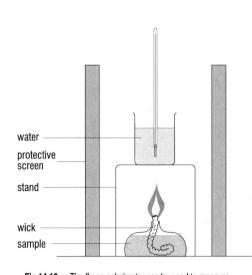

water
protective screen
stand
wick
sample

Fig 14.19 The flame calorimeter can be used to measure the enthalpy changes associated with the combustion of liquid fuels, such as hydrocarbons and alcohols. It is easily set up but less accurate than the bomb calorimeter.

TUTORIAL

Problem: Calculate the enthalpy of combustion of butan-1-ol, using data from the following experiment:

A sample of butan-1-ol (molar mass 74.1 g mol^{-1}) was burnt in a flame calorimeter. The experimental results were as follows:

initial mass of bottle + sample = 19.412 g
final mass of bottle + sample = 18.393 g

initial temperature of water = 21.4°C
final temperature of water = 45.3°C

mass of water in calorimeter = 350 cm^3

Solution:

The specific heat capacity of water is approximately 4.2 kJ kg^{-1} K^{-1} (4.2 J g K^{-1}).

Heat produced during experiment $= m \times c \times \Delta T$
$= 350 \times 4.2 \times (45.3 - 21.4)$ J
$= 35\,133$ J

Mass of butan-1-ol combusted $= (19.412 - 18.393)$ g $= 1.019$ g.

Therefore, amount of butan-1-ol combusted $= \dfrac{1.019}{74.1} = 0.01375$ mol.

Therefore, enthalpy of combustion $= -\dfrac{35\,133}{0.01375}$ J mol^{-1}

$= -2\,555\,000$ J mol^{-1}
$= -2555$ kJ mol^{-1}.

20 **In what way are the conditions non-standard here?**

The bomb calorimeter

Accurate values for the energy changes which accompany combustion can be obtained using a **bomb calorimeter**. A weighed sample of the substance being tested, together with excess compressed oxygen, is sealed into a 'bomb' made of steel. The bomb is surrounded by water which is stirred. Combustion is started by an electrical spark and the sample is completely burnt. Heat released raises the temperature of the bomb and the water. Water temperature is monitored throughout the combustion using a thermometer and the maximum value attained is recorded. By knowing:

- the amount of substance combusted;
- the corrected temperature rise of the water surrounding the bomb (see Fig 14.20);
- the specific heat capacity of water and the quantity in the bomb;
- the heat capacity of the bomb (the amount of heat required to raise the temperature of the whole apparatus by 1 K);

the amount of heat produced by the reaction can be calculated.

21 In a bomb calorimeter, does the reaction take place under conditions of constant pressure or constant volume?

22 What is being determined in a bomb calorimeter – an enthalpy change or an internal energy change?

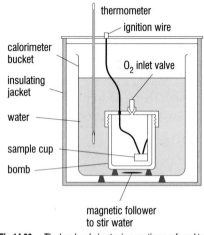

Fig 14.20 The bomb calorimeter is sometimes referred to as a combustion calorimeter. It allows accurate data to be collected.

QUESTIONS

14.29 Describe, with the aid of a simple sketch, how an accurate value for the enthalpy change of combustion may be measured.

14.30 A sample of 1 g of naphthalene, ($C_{10}H_8$), is burned in a constant-volume calorimeter. The water contained within the calorimeter has a mass of 1000 g. The heat capacity of the bomb calorimeter is 1.2 kJ °C^{-1}. During combustion a thermometer shows a temperature rise from 19.84°C to 27.30°C. Calculate the molar enthalpy of combustion for naphthalene.
(The density of water = 1 kg dm^{-3} and the specific heat capacity of water is 4.184 kJ kg^{-1} K^{-1}).

14.31 In a laboratory experiment 1 g of white phosphorus is burned in a crucible. The heat released is used to warm up 200 cm^3 of water in a flask placed above the crucible. The water temperature rises from 24.8°C to 52.0°C.
(a) Calculate a value of the molar enthalpy change of combustion for phosphorus.
(b) Comment on possible sources of error in this experiment.

Hess's law

If no thermochemical data are available for a particular reaction, the enthalpy change can still be predicted by calculation. This indirect method of calculating changes depends upon the First Law of Thermodynamics – energy is neither created nor destroyed. Applying this to a chemical change, we can see that the total enthalpy change which accompanies it will be independent of the route taken. This is **Hess's law** and may be summarised diagrammatically:

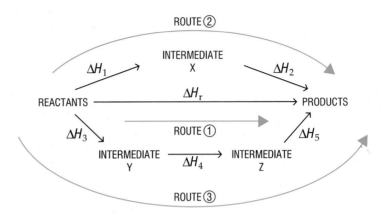

Provided all the enthalpy changes are determined under the same conditions,

enthalpy change for route 1 = sum of enthalpy changes for route 2

= sum of enthalpy changes for route 3

Therefore, $\Delta H_r = \Delta H_1 + \Delta H_2 = \Delta H_3 + \Delta H_4 + \Delta H_5$

Using bond energies

How can we calculate the enthalpy of reaction for the formation of hydrogen fluoride from hydrogen and fluorine?

$$H_2(g) + F_2(g) \longrightarrow 2HF(g)$$

Provided we know the bond energies for H—H, F—F and H—F, we can use Hess's law to calculate ΔH_r.

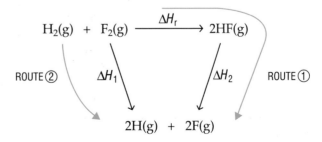

To identify the two routes, simply start from the reactants ($H_2(g)$ and $F_2(g)$) and follow the arrows around. When arrow-heads meet, return to reactants and go around the cycle in the opposite direction until the arrowheads meet again.

From Hess's law, $\Delta H_r + \Delta H_2 = \Delta H_1$

Therefore, $\Delta H_r = \Delta H_1 - \Delta H_2$

ΔH_1 = energy required to break 1 mol H—H bonds + energy required to break 1 mol F—F bonds

ΔH_2 = energy required to break 2 mol H—F bonds

Given:

$\Delta H^{\ominus}_{d,m}(298)[H_2(g)] = + 436 \text{ mol}^{-1}$
$\Delta H^{\ominus}_{d,m}(298)[F_2(g)] = + 151 \text{ mol}^{-1}$
$\Delta H^{\ominus}_{d,m}(298)[HF(g)] = + 568 \text{ mol}^{-1}$
$\Delta H_1 = + 436 + 151 = + 587 \text{ kJ mol}^{-1}$ and $\Delta H_2 = 2 \times (+ 568) = + 1136 \text{ kJ mol}^{-1}$

Therefore, $\Delta H_r = + 587 - (+ 1136) = - 548 \text{ kJ mol}^{-1}$

(Note that standard conditions have been assumed throughout the calculation.)

An important enthalpy change when considering the breaking of covalent bonds is the **standard enthalpy of atomisation**. This is defined as the enthalpy change when one mole of gaseous atoms are formed from an element in its standard state under standard conditions. It is given the symbol $\Delta H^{\ominus}_{atom,m}(298)$ and is always an endothermic change.

23 **Explain why the standard enthalpy of atomisation of chlorine is half the value of the Cl—Cl bond energy.**

TUTORIAL

Problem: Calculate the average bond energy for C—Cl in tetrachloromethane, $CCl_4(g)$ given the following data:

$\Delta H^{\ominus}_{atom,m}(298)[C(g)] = + 715 \text{ kJ mol}^{-1}$
$\Delta H^{\ominus}_{atom,m}(298)[Cl(g)] = + 121 \text{ kJ mol}^{-1}$
$\Delta H^{\ominus}_{f,m}(298)[CCl_4(l)] = - 139 \text{ kJ mol}^{-1}$
$\Delta H^{\ominus}_{v,m}(298)[CCl_4(l)] = + 30 \text{ kJ mol}^{-1}$

Solution: We need to calculate the enthalpy change for the reaction:

$$CCl_4(g) \longrightarrow C(g) + 4Cl(g)$$

The average bond energy for C—Cl in $CCl_4(g)$ is $\frac{1}{4}$ of this enthalpy change.

Constructing a Hess's cycle:

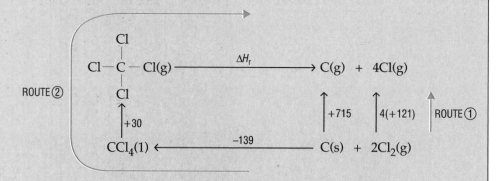

Enthalpy change for route 1 = enthalpy change for route 2

$+715 + 4(+121) = -139 + 30 + \Delta H_r$

Therefore, $\Delta H_r = +715 + 4(+121) + 139 - 30$
$= +1308$ kJ mol^{-1}

Therefore, average bond energy for C—Cl in CCl$_4$(g) $= +1308/4$
$= 327$ kJ mol^{-1}.

Using enthalpies of combustion and enthalpies of formation

It is impossible to measure the enthalpy of formation of methanol directly:

$$C(s) + 2H_2(g) + \tfrac{1}{2}O_2(g) \longrightarrow CH_3OH(l)$$

Instead, we must determine it indirectly from enthalpy changes which can be measured in the laboratory. We know that enthalpies of combustion can be determined accurately using a bomb calorimeter.

Standard enthalpies of combustion are as follows (all kJ mol^{-1}):

$C(s) = -393.5$; $H_2(g) = -285.8$; $CH_3OH(l) = -715.0$

Applying Hess's law enables us to find the enthalpy of formation of methanol:

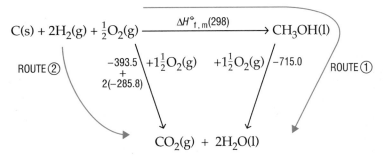

From Hess's law, $\Delta H^{\circ}_{f,m}(298) + (-715.0) = -393.5 + 2(-285.8)$.

Therefore, $\Delta H^{\circ}_{f,m}(298) = -393.5 + 2(-285.8) - (-715.0) = -250.1$ kJ mol^{-1}.

Similarly, we might find it difficult to measure the enthalpy change when ethene and hydrogen bromide react to form bromoethane:

$$C_2H_4(g) + HBr(g) \longrightarrow C_2H_5Br(l)$$

This time we cannot construct a cycle using enthalpies of combustion since neither hydrogen bromide nor bromoethane burn. However, we can use standard enthalpies of formation (all kJ mol^{-1}):

$C_2H_4(g) = +52.3$; $HBr(g) = -36.2$; $C_2H_5Br(l) = -85.4$

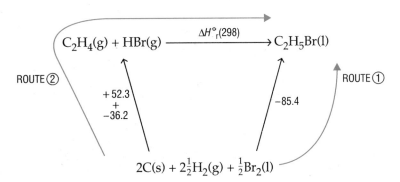

From Hess's law, $-85.4 = \Delta H^\ominus_r(298) + 52.3 + (-36.2)$.
Therefore, $\Delta H^\ominus_r(298) = -85.4 - 52.3 - (-36.2)$
$$= -85.4 - 52.3 - 105.5 = +3.1 \text{ kJ mol}^{-1}$$

24 Explain why it would be impossible to measure the standard enthalpy of formation of sodium sulphate directly.

Born-Haber cycle and lattice energies

CONNECTION: *Lattice energies and the model of ionic bonding, page 357.*

Lattice energies have a major influence on the properties of ionic compounds. A comparison of theoretical values and experimental values provides valuable evidence in support of our model for ionic bonding. Yet lattice energies cannot be determined directly, so how can we determine 'experimental values'?

Hess's law is used to calculate lattice energies from enthalpy changes which can be determined directly. This special case of a Hess cycle is known as a **Born-Haber cycle**. To illustrate its use, consider the Born-Haber cycle for sodium chloride:

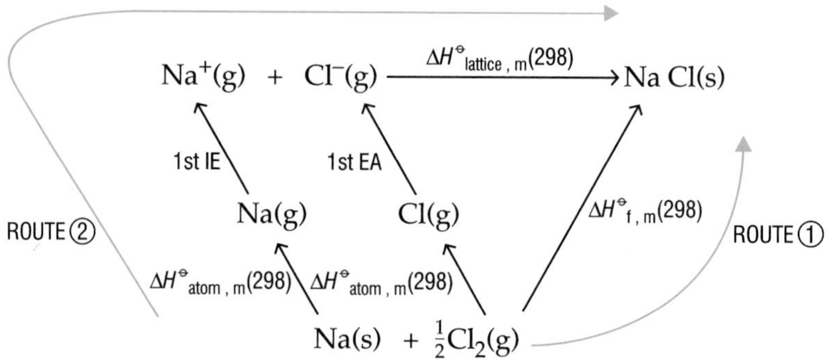

The lattice energy, $\Delta H^\ominus_{latt,m}(298)[\text{NaCl(s)}]$, can be calculated provided we know:

- the standard enthalpy of formation of sodium chloride (-411 kJ mol^{-1});
- the standard enthalpy of atomisation of sodium ($+109 \text{ kJ mol}^{-1}$);
- the standard enthalpy of atomisation of chlorine ($+121 \text{ kJ mol}^{-1}$);
- the first ionisation energy for sodium ($+494 \text{ kJ mol}^{-1}$);
- the first electron affinity for chlorine (-364 kJ mol^{-1}).

From Hess's law, $-411 = +109 + 121 + 494 + (-364) + \Delta H^\ominus_{latt,m}(298)$.
Therefore, $\Delta H^\ominus_{latt,m}(298) = -411 - 109 - 121 - 494 - (-364) = -771 \text{ kJ mol}^{-1}$.

Hypothetical ionic compounds

We can also use theoretical lattice energies to calculate the enthalpies of formation of hypothetical ionic compounds, and use this information to speculate about their likely existence. It provides yet another example of using our chemical knowledge and understanding to design and to predict.

For example, how could we have predicted the most stable chloride of magnesium? The formula of magnesium chloride is $\text{MgCl}_2\text{(s)}$ and we argue that by electron transfer to form Mg^{2+} and Cl^- ions, atoms of both elements achieve a noble gas electronic configuration. But why are MgCl(s) or $\text{MgCl}_3\text{(s)}$ never found in bottles in the chemistry store room? It seems hardly sufficient to say, simply, that in neither of these cases does magnesium attain a noble gas electron arrangement. A much better argument comes from a consideration of energetics.

We can calculate theoretical lattice energies for MgCl(s) and MgCl$_3$(s) by assuming ions of certain radius, packing together in a regular three-dimensional lattice. Based on the perfect ionic model, we find that these theoretical values are:

MgCl(s) = – 750 kJ mol^{-1}, and MgCl$_3$(s) = – 5440 kJ mol^{-1}

We can use Born-Haber cycles to estimate the standard enthalpies of formation of these two hypothetical chlorides of magnesium (Fig 14.21). We find that they are:

$\Delta H^{\circ}_{f,m}$(298)[MgCl(s)] (theoretical) = – 107 kJ mol^{-1}
$\Delta H^{\circ}_{f,m}$(298)[MgCl$_3$(s)] (theoretical) = + 3907 kJ mol^{-1}

MgCl$_3$(s) is clearly unstable with respect to its elements (the estimated standard enthalpy of formation is highly endothermic). This means that MgCl$_3$ ought to decompose rapidly to magnesium and chlorine. It is not surprising, therefore, that the compound has not been prepared. However, MgCl(s) has a negative enthalpy of formation and is energetically stable with respect to decomposition into its elements. So why is it unstable?

The following Hess's cycle provides the answer:

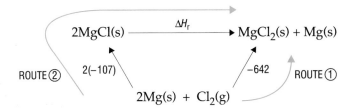

From Hess's law, – 642 = 2(–107) + ΔH°_r.
Therefore, ΔH°_r = – 642 + 2(–107) = – 428 kJ mol^{-1}.

So, although MgCl(s) is stable with respect to decomposition into magnesium and chlorine, it is unstable with respect to decomposition to MgCl$_2$(s) and Mg(s):

2MgCl(s) $\longrightarrow$ MgCl$_2$(s) + Mg(s); ΔH_r = – 428 kJ mol^{-1}

Fig 14.21 Born-Haber cycles for (a) MgCl(s) and (b) MgCl$_3$(s).

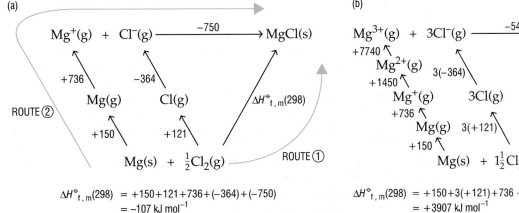

(a)

$\Delta H^{\circ}_{f,m}$(298) = +150+121+736+(–364)+(–750)
= –107 kJ mol^{-1}

(b)

$\Delta H^{\circ}_{f,m}$(298) = +150+3(+121)+736+1450+7740+3(–364)+(–5440)
= +3907 kJ mol^{-1}

25 What is the most important energy consideration which accounts for the fact that $MgCl_3$ does not exist?

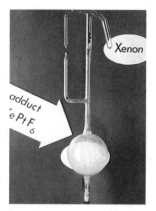

Fig 14.22 The first 'noble gas' compound, $Xe^+PtF_6^-$. Its existence was predicted and its synthesis proved the value of reliable thermochemical data. Since then a number of Group VIII compounds have been made.

XENON – THE 'NOT-SO-NOBLE' GAS

For many years the Group VIII elements, the noble gases, were believed to be completely unreactive. That belief would have continued without some good fortune followed by some very good chemistry. Neil Bartlett, working in California, found that platinum(VI) hexafluoride, PtF_6, changed colour when accidentally exposed to the air. He went on to identify the compound formed as $O_2^+PtF_6^-$ and realised that PtF_6 must have oxidised molecular oxygen.

Bartlett followed this piece of luck by some clear thinking. He realised that, as the first ionisation energy of xenon (1170 kJ mol^{-1}) was very close to that of molecular oxygen (1175 kJ mol^{-1}) it should be possible to make a xenon compound for the first time. His understanding of energetics did not let him down, rewarding him with the first noble gas compound, xenon hexafluoroplatinate(VI), $Xe^+PtF_6^-$. Since then a number of xenon fluorides have been prepared including XeF_2, XeF_4 and XeF_6. They are used successfully as fluorinating agents.

Often the dissolving of solids is an endothermic process. For example,

$NaCl(s) + aq \longrightarrow Na^+(aq) + Cl^-(aq)$; $\Delta H^\ominus = + 3.9$ kJ mol^{-1}.

Yet dissolution occurs spontaneously. The answer lies in the increasing disorder of the system.

Dissolving ionic compounds

From a simple Hess cycle (Fig 14.23), we can see that the enthalpy of solution is determined by the relative amounts of energy
* needed to break up the ionic lattice (the lattice energy);
* released when hydrated ions are formed.

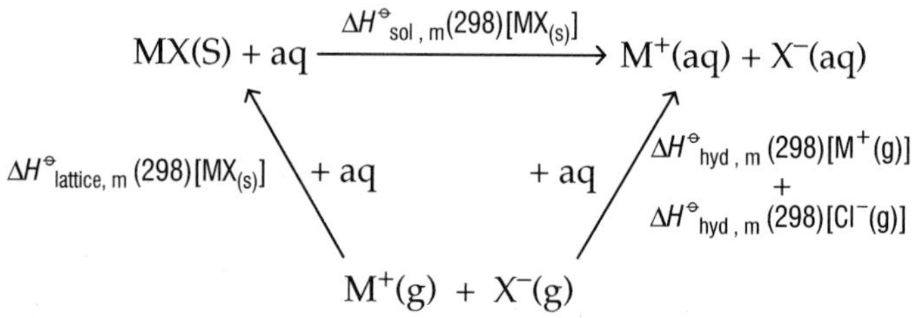

Fig 14.23 The process of a solid dissolving in water does not involve breaking the lattice to give gaseous ions, followed by the hydration of these ions. However, the relative magnitudes of these enthalpy changes help us to explain the size of an enthalpy of solution.

$$MX(S) + aq \xrightarrow{\Delta H^\ominus_{sol, m}(298)[MX_{(s)}]} M^+(aq) + X^-(aq)$$

$\Delta H^\ominus_{lattice, m}(298)[MX_{(s)}] + aq$

$+ aq$ $\Delta H^\ominus_{hyd, m}(298)[M^+(g)] + \Delta H^\ominus_{hyd, m}(298)[Cl^-(g)]$

$$M^+(g) + X^-(g)$$

QUESTIONS

14.32 State Hess's law and give one example of how it may be used to determine an enthalpy change which cannot be measured directly.

14.33 Use the average bond enthalpy data given in the Database to predict the standard enthalpy change of formation of hydrogen bromide.

14.34 Using the enthalpy change of combustion data in the Database:
(a) calculate the standard molar enthalpy change of formation of methanol;
(b) calculate the standard molar enthalpy change of formation of ethanol;
(c) predict $\Delta H^\ominus_{f,m}(298)$ for propanol.

14.35 Use the Database to calculate the lattice energy of calcium chloride, $CaCl_2$, by constructing a Born-Haber cycle. The second ionisation energy of calcium is +1145 kJ mol^{-1}. The standard molar enthalpy change of formation of calcium chloride is −804 kJ mol^{-1}.

14.36 **(a)** Construct a Born-Haber cycle for potassium bromide. Use the cycle and the following data to determine $\Delta H^{\oplus}_{f,m}$ (298) for potassium bromide:

$K(s) \longrightarrow K(g)$	$\Delta H^{\oplus}_{atom,m}(298)$	= +90.0 kJ mol^{-1}
$K(g) \longrightarrow K^+(g)$	$\Delta H^{\oplus}_{1,m}(298)$	= +418.8 kJ mol^{-1}
$\frac{1}{2}Br_2(g) \longrightarrow Br(g)$	$\Delta H^{\oplus}_{atom,m}(298)$	= +112 kJ mol^{-1}
$Br(g) \longrightarrow Br^-(g)$	$\Delta H^{\oplus}_{e,m}(298)$	= − 324.7 kJ mol^{-1}
$K^+(g) + Br^-(g) \longrightarrow KBr(s)$ $\Delta H^{\oplus}_{latt,m}(298)$		= − 689 kJ mol^{-1}

(b) How does your answer to part (a) compare with the value quoted for $\Delta H_{f,m}$(298) in the Database?

14.7 FUELS, ENGINES AND EFFICIENCY

Petrol power

The earliest commercial engines were steam engines. Steam, generated in a boiler by burning coal beneath it, was led through an insulated pipe to push a piston. Connected to a wheel, the piston slid back and forth in a chamber and provided mechanical power. The original steam engines were inefficient. Only 5–10% of the heat energy produced by the burning of coal was transformed into useful mechanical energy. Much of the heat produced escaped to the environment. The coal needed to be physically loaded into a furnace, a messy and labour-intensive process.

Coal is bulky and doesn't 'flow' well. For this reason it is too inconvenient a fuel to be used directly for powering modern transport systems. The universal availability of petroleum products has led to intensive development and refinement of petrol-fuelled internal combustion engines. These engines develop their power through the combustion of petrol inside them. Such is the success of the petrol engine that, worldwide, more than 10 million litres of oil are refined daily to provide petrol for cars and lorries. Petrol consists of a mixture of volatile hydrocarbons obtained by fractionation of petroleum.

The precise mixture of hydrocarbons present in petrol varies according to its grade but it contains alkanes, cycloalkanes and a few aromatic compounds. In general, the hydrocarbons in petrol contain 6–12 carbon atoms and, during fractionation, are collected in the boiling point range 30–180°C This range gives a suitable combination of important physical properties such as volatility and viscosity. The fuel must have a reasonable flow rate, must readily form a satisfactory fuel-air mixture in the carburettor and must be safe enough to allow drivers to serve themselves independently.

Fig 14.24 Steam plays a part in modern generation of electrical power. This modern steam turbine uses high-pressure super-heated steam to drive turbine blades connected to a rotor for electrical power generation. The fuel input is mechanised and automatic. Such turbines convert 30–40 MW of available energy to electrical power.

Combustion of petrol cannot be represented by a single equation as many hydrocarbons are involved. The products of combustion are the familiar products of burning any hydrocarbon; carbon dioxide and water. The equation for the combustion of heptane, a common constituent of a typical petrol, is shown as:

$$C_7H_{16}(g) + 11O_2(g) \longrightarrow 7CO_2(g) + 8H_2O(g)$$

The volume of these gases is greater than that of the hydrocarbon and the oxygen with which it reacts. This increase in volume drives the piston down the cylinder and powers the vehicle.

Petrol additives

Some of the compounds in petrol are far more suitable for petrol engines than others. Additives are often added to improve aspects of performance. Why is this necessary? One major consideration concerns the need for a controlled burning rate within each cylinder. If the mixture of petrol and air burns in an uncontrolled, rapid manner the gases formed by combustion expand suddenly. If this happens the piston within the cylinder is suddenly slapped into movement. This sudden slap is known as 'knocking' and is inefficient and damaging to the engine. Ideally expansion of gases should take place smoothly, increasing engine life and improving the efficiency of conversion from chemical to mechanical energy.

A petrol with a high proportion of straight-chain hydrocarbons tends to produce excessive 'knocking' problems and is unsuitable for use in a modern high-performance engine. A higher proportion of branched-chain alkanes and aromatic compounds gives a smooth controlled expansion of gases and better overall performance. Commercial petrol is given one octane rating despite the fact that it is a complex mixture of hydrocarbons each of which has a different rating. This is achieved by making up mixtures of 2,2,4-trimethylpentane and heptane to give the same amount of 'knock' as the petrol. A petrol with an octane number of 94 has the same 'knock' properties as a mixture of 940 parts of 2,2,4-trimethylpentane and 60 parts of heptane.

The octane rating of petrol can be improved in several ways. One way is to dissolve the lead compounds tetramethyllead and tetraethyllead in petrol in concentrations of between 1–2 g dm^{-3} This can improve the octane number of the mixture by 10 or more. An alternative method for improving the octane number is to add further quantities of branched-chain hydrocarbons to the mixture. Octane numbers of over 100 are obtained by blending in additional quantities of branched-chain hydrocarbons. The latter method has grown rapidly in popularity since the toxicity of lead has long been recognised and its presence in the environment is undesirable. For example, high levels of lead pollution in school playgrounds have been correlated with slow brain development of the children who play there. The switch from leaded to unleaded petrols has already produced lower levels of environmental lead. Analysis of snow accumulating in the North Pole shows decreasing levels of lead pollution and suggests this strategy is successful.

The anti-knock qualities of a petrol are an important property. The availability of new petrol mixtures influences aspects of engine design. At the oil refinery, fractionation patterns change. As demand for a particular boiling point range increases its price rises. It becomes economical to reorganise the way in which fractions are collected. For example, it may become more profitable to collect fractions boiling in the range 40–180°C instead of 30–180°C.

DIESEL IS ECONOMICAL – BUT KEEP IT WARM!

Diesel is the fuel of choice for long-distance transport. Diesel engines are more efficient than petrol engines and give better fuel economy. They produce only around one tenth of the carbon monoxide of a petrol engine but make up for this with smoke, odour and higher noise levels. For domestic transport the petrol engine is still the preferred power source.

Diesel is a fraction of crude oil collected between 250–340°C and contains longer chains of alkanes than petrol. This makes it much less volatile too. One unfortunate consequence is that in cold weather diesel can solidify even whilst the vehicle is in motion, blocking the fuel lines and halting the vehicle.

Fig 14.25 In severe weather desperate lorry drivers have been known to light a series of small fires under the diesel tank and fuel line of their vehicle. Fortunately the temperature at which diesel ignites, its flash-point, is much higher than with petrol.

Diesel fuel typically contains between 14 and 19 carbon atoms. As a result the average chain length of the hydrocarbons in diesel fuel is longer than with petrol. The opportunities for van der Waal's bonding are greater with longer chains and this increases both melting points and boiling points.

Jets and rockets

Jet engines and rockets rely on the principle that 'to every action there is an equal and opposite reaction'. A jet engine of the kind used in airplanes draws air into the front of the engine, compresses this air and passes it into a combustion chamber. Kerosine, often called paraffin, is mainly $C_{12}H_{26}$. It is a fraction of crude oil collected between 150–250°C and is blended to produce aviation fuel. It is injected into the combustion chamber of the jet engine where it reacts with the oxygen in the compressed air. Combustion produces a large increase in the volume of gases and these are expelled through the exhaust and provide the thrust for forward motion.

Rockets need to be able to operate in an environment where oxygen is scarce. They must carry their own oxidiser (liquid oxygen). This oxygen is used to oxidise a fuel or **propellant**. As a result rockets may perform anywhere in space whilst jet engines are restricted to the atmosphere of a planet.

SUGAR-POWER FOR THE FUTURE?

The search for a renewable energy source for powering vehicles continues. We cannot rely on petrol to fuel the cars of the future. In some countries the shortage of petroleum-based fuel has already led to some interesting alternatives. Ethanol made from plants is already mixed with petrol to provide 'gasohol'. Such a mixture is not without its problems. Ethanol, which burns so cleanly in the laboratory to produce carbon dioxide and water, produces ethanal inside a car engine. This is an irritant whose long-term effects are unknown.

One possible alternative for the future is to use carbohydrates directly to provide electrical power. Complete oxidation of a kilogram of sugar provides 16 to 17 million joules of energy. This amount of energy would power a 1 kilowatt electric fire for 5 hours or allow an average car to be driven for 25–30 kilometres.

Fig 14.26 Sugar power: on a full tank of concentrated sugar solution a sugar-powered electric motor could power a vehicle for 1000 kilometres. A microbial fuel cell could oxidise the sugar even more efficiently, producing an efficiency of energy conversion well in excess of the petrol engine.

Energy from the Sun

CONNECTION: Solar energy and plant growth, page 10.

The flow of the Sun's energy through our environment seems relentless. It has been estimated that around 1.7×10^{17} Joules of solar energy reach the Earth *per second*. If converted to electricity, this amount of energy would run 50 million electric fires for every person on earth. For millions of years solar energy has been used by plants to create complex and energy-rich molecules. Human beings and other animals use plants as food to provide basic building materials for growth. Careful oxidation of such molecules during metabolism also provides the energy needed to create complex metabolites. Farming may be viewed as an industry whose basic business is the collection and storage of solar energy. Every 30 g of carbon dioxide trapped by plants (enough to fill two small balloons) stores over 1 kilojoule of energy. However, typical plant crops must be exposed to *over thirty kilojoules* of energy from the Sun to trap 30 g of carbon dioxide.

Although we use the trapped energy of plants in our diet we also use the products of plant decay – oil, coal, gas and peat, as fuels. As these fossil fuels become scarcer and more difficult to extract from the Earth's crust so human interest in methods for trapping solar energy has increased. There are several general methods available for doing this:

- solar-hot water systems;
- photovoltaic cells;
- hydrogen production;
- plant growth.

QUESTIONS

14.37 Coal is a relatively plentiful fossil fuel whilst petrol is obtained from crude oil, a rapidly diminishing resource. Explain why petrol is still the most popular fuel for powering motor vehicles.

14.38 (a) Explain with the aid of an equation, what happens to octane, a constituent of petrol, within a car engine.
(b) Why does this reaction, and other similar reactions, produce a lot of energy?

14.39 Explain why lead is added to some petrols and how high performance petrols are now produced without the need for lead.

14.40 In terms of intermolecular forces, explain why diesel fuel has a higher melting point and lower volatility than petrol.

Answers to Quick Questions: Chapter 14

1 Chemical energy is converted into light, sound and, in some fireworks such as rockets, mechanical energy.
2 It is converted into heat energy.
3 Because once the reaction has started, sufficient energy is produced to keep it going until combustion is complete (assuming an adequate supply of oxygen).
4 As the reaction starts, heat is absorbed from the solution which, therefore, drops in temperature. Slowly, the reaction mixture returns to room temperature as energy is absorbed from the surroundings.
5 One mole of nitrogen
6 One which involves no change in volume.
7 104.6 J
8 Sodium, magnesium and aluminium have relatively high specific heat capacities, a characteristic associated with metals since they are good conductors of heat.

9 Exothermic

10 89.04 kJ

11 Because the strength of the covalent bond depends on the amount of electron density shared by the two atoms involved. This will be influenced by other atoms in the molecule, for example, highly electronegative atoms will tend to attract electron density away from the bond and so weaken it.

12 Exothermic

13 Magnesium oxide

14 Exothermic

15 Diamond

16 Because in one water vapour is formed, while in the other liquid water is formed.

17 Because one mole of H_2SO_4 gives rise to two moles of $H^+(aq)$

18 The system is the reacting components in the beaker. The surroundings are the remainder of the solution, the polystyrene beaker, the air around it and, indeed, the rest of the universe!

19 Constant pressure

20 The rise in temperature due to the fuel burning means that the standard temperature is exceeded.

21 Constant volume

22 Internal energy change

23 The standard enthalpy of atomisation of chlorine refers to the formation of one mole of Cl atoms. When one mole of Cl_2 dissociates (requiring the bond energy), two moles of Cl form.

24 It would be experimentally impossible to form one mole of $Na_2SO_4(s)$ from its constituent gaseous ions, brought together from an infinite distance apart in space.

25 The third ionisation energy is so large.

Chapter 15

REDOX AND ELECTRICITY

LEARNING OBJECTIVES

When you have studied this chapter you should be able to:

1. explain redox in terms of (a) gain or loss of oxygen, (b) electron transfer, (c) changes in oxidation number;

2. write ionic half equations and combine them to give balanced overall equations for redox reactions;

3. balance redox equations by using oxidation numbers;

4. give examples of oxidising and reducing agents;

5. explain redox equilibria and the formation of a Helmholtz double layer;

6. explain the terms half-cell and electrode potential, and describe different types of half-cells;

7. use the correct conventions to represent a half-cell;

8. explain the term standard electrode potentials and how they are measured;

9. explain the terms electrochemical cell and cell e.m.f., and how cell e.m.f. is calculated from electrode potentials;

10. use the correct conventions to represent an electrochemical cell;

11. describe some specialised cells and batteries.

15.1 OXIDATION AND REDUCTION

Fig 15.1 Oxidation reactions supply this person with energy to repair the damage another oxidation reaction has caused to this bridge in Newcastle – rust!

Oxidation reactions may be useful or they may be damaging. Two obvious beneficial oxidations are the burning of fossil fuels to release stored chemical energy as heat and the breakdown of sugars in the body during respiration. However, the oxidation process which causes iron to rust results in only too familiar problems. It is clear that an understanding of oxidation and reduction processes is essential if we are to be able to use them to our advantage and limit the damage they can cause.

A number of definitions of oxidation and reduction are in use. Oxidation and reduction can be defined in terms of:

- gain of oxygen (oxidation) or loss (reduction);
- loss of hydrogen (oxidation) or gain (reduction);
- loss of electrons (oxidation) or gain (reduction);
- increase in oxidation number (oxidation) or decrease (reduction).

The earliest definitions were in terms of gain or loss of oxygen (and, related to this, gain or loss of hydrogen). Hence the term 'oxidation'. A more recent definition is in terms of electron transfer and this can be used with reference to numerous reactions which do not involve oxygen at all.

Most recently, changes in oxidation number have been used. This has the advantage of including all reactions described by the previous two definitions, as well as others which cannot be described either in terms of gain or loss of oxygen or electron transfer. The advantage of the oxidation number definition of oxidation is that we can generalise about an even greater number of reactions. It is another example of a model being refined.

1   Why is the Earth's atmosphere often described as an oxidising atmosphere?

Gain or loss of oxygen

In the processes described below in equation form, the oxygen content of the material being oxidised has increased:

$$CH_4(g) + 2O_2(g) \longrightarrow CO_2(g) + 2H_2O(l)$$
$$C_6H_{12}O_6(s) + 6O_2(g) \longrightarrow 6CO_2(g) + 6H_2O(l)$$
$$4Fe(s) + 3O_2(g) \longrightarrow 2Fe_2O_3(s)$$

Reduction may be considered to be the reverse process – decreasing the oxygen content of a material. Smelting a metal oxide ore to give the metal puts reduction to good use. For example:

$$2ZnO(s) + C(s) \longrightarrow 2Zn(s) + CO_2(g)$$

In terms of gain or loss of oxygen:

Oxidation is the gain of oxygen.
Reduction is the loss of oxygen.

In a similar way, oxidation and reduction can be considered in terms of decreasing and increasing hydrogen content:

$Br_2(g) + H_2(g) \longrightarrow 2HBr(g)$; bromine has gained hydrogen and, therefore, bromine has been reduced.

$H_2S(g) + F_2(g) \longrightarrow S(s) + 2HF(g)$; the hydrogen content of sulphur has decreased and, therefore, sulphur has been oxidised.

Oxidation is the loss of hydrogen.
Reduction is the gain of hydrogen.

2   When hydrogen is passed over hot copper(II) oxide, copper and water are formed. Write an equation for the reaction and explain the change in terms of oxidation and reduction.

Electron transfer and half-reactions

Oxidation and reduction may be explained in terms of electron loss or gain. This is particularly useful when considering electrical energy and chemical reactions (for example, explaining how batteries work). In terms of electron transfer:

Oxidation is the loss of electrons.
Reduction is the gain of electrons.

Consider the reaction between zinc metal and aqueous copper(II) sulphate:

$$Zn(s) + CuSO_4(aq) \longrightarrow ZnSO_4(aq) + Cu(s)$$

We can write an ionic equation for the reaction, omitting the sulphate ions (which are spectator ions):

$$Zn(s) + Cu^{2+}(aq) \longrightarrow Zn^{2+}(aq) + Cu(s)$$

Fig 15.2   The blue colour of copper(II) ions becomes less intense as copper metal is deposited. Zinc metal gradually dissolves, giving colourless zinc ions in solution. Sulphate ions (colourless) are unchanged throughout the reaction.

REDOX AND ROCKET FUELS

Liquid and solid fuels, or propellants, are used in the NASA space shuttle programme. On combustion, these fuels must provide sufficient thrust to place the shuttle into an orbit several hundred miles from the Earth's surface. Over 1 440 000 dm^3 of liquid hydrogen maintained at 70 K and 541 000 dm^3 of liquid oxygen at 90 K are available for powering the shuttle's main engines. Some of the liquid hydrogen is used to cool the engine nozzles during lift-off, the point of maximum power requirement. The low temperatures which must be maintained in order to keep oxygen and hydrogen in liquid form make it difficult for this system to provide all the energy for the shuttle after take-off.

Most of the thrust necessary to place the shuttle in its desired orbit is provided by a solid propellant, in **solid rocket boosters**. The solid propellant is a mixture containing aluminium powder (16% by mass) and ammonium chlorate(VII) (69.83% by mass), NH_4ClO_4. Reaction between these two materials is catalysed by iron(III) oxide. The products of the reaction include aluminium oxide, hydrogen chloride, chlorine, and a mixture of nitrogen oxides.

In space the qualities required of a fuel are dependability, safety and a high energy provision/mass ratio. A compound which satisfies these requirements is hydrazine, N_2H_4. The space shuttle, once in space, oxidises hydrazine to provide energy to power its steering rockets. The early Apollo moon landings used methylhydrazine (a derivative of hydrazine) in the landing rockets, with dinitrogen tetroxide as the oxidising agent.

$$4NH_2NHCH_3(l) + 5N_2O_4(l) \longrightarrow 9N_2(g) + 12H_2O(g) + 4CO_2(g)$$

Such compounds oxidise to produce an enormous volume of gas. The volume is further increased by the high temperature of reaction. As the gases are expelled, forward thrust is achieved.

Fig 15.3 Carrying the first commercial satellite payload in the history of the STS program, and the first four-man crew in a single launch in the history of space flight, Columbia lifts of on schedule on November 11, 1982. The thrust which lifts the space shuttle off its launch pad is the result of a reaction in which hydrogen is oxidised.

The reaction may be considered as a pair of **half-reactions**: an oxidation process and a reduction process. Half-reactions may be represented by **half-equations**.

Oxidation half-equation: $Zn(s) \longrightarrow Zn^{2+}(aq) + 2e^-$; zinc has been oxidised since electrons have been removed.

Reduction half-equation: $Cu^{2+}(aq) + 2e^- \longrightarrow Cu(s)$; copper has been reduced since electrons have been added.

Combining the two half-equations:

$$Zn(s) + Cu^{2+}(aq) + 2e^- \longrightarrow Zn^{2+}(aq) + Cu(s) + 2e^-$$

There are the same number of electrons on either side of the equation and so they cancel. This gives us the ionic equation for the reaction between zinc and copper(II) sulphate solution:

$$Zn(s) + Cu^{2+}(aq) \longrightarrow Zn^{2+}(aq) + Cu(s)$$

The equation is electronically balanced because it has the same total charge on both sides of the equation. Although electrons are not shown, their transfer can be deduced by examining the changes for each substance present. For example, each copper ion converted to copper requires the addition of two electrons.

3 Explain why the blue colour of the solution slowly disappears (Fig 15.2).

Another example, the reaction between magnesium and sulphuric acid, will help to reinforce the idea:

$$Mg(s) + H_2SO_4(aq) \longrightarrow MgSO_4(aq) + H_2(g)$$

The ionic equation for the reaction is:

$$Mg(s) + 2H^+(aq) \longrightarrow Mg^{2+}(aq) + H_2(g)$$

Oxidation half-equation:

$$Mg(s) \longrightarrow Mg^{2+}(aq) + 2e^-$$

magnesium has been oxidised since electrons have been removed.

Reduction half-equation:

$$2H^+(aq) + 2e^- \longrightarrow H_2(g)$$

hydrated hydrogen ions have been reduced since electrons have been added.

Combining the two half-equations:

$$Mg(s) + 2H^+(aq) + 2e^- \longrightarrow Mg^{2+}(aq) + H_2(g) + 2e^-$$

and cancelling the electrons:

$$Mg(s) + 2H^+(aq) \longrightarrow Mg^{2+}(aq) + H_2(g)$$

Electrons do not appear in a balanced redox equation as they must **always** cancel out. Often, combining half-equations involves an additional problem in that a particular oxidation half-equation may involve a different number of electrons to the reduction half-equation with which it is to be combined.

4 In terms of electron transfer, explain why an oxidation reaction is always accompanied by a reduction reaction.

TUTORIAL

Problem: Write a balanced equation for the reaction between potassium manganate(VII) and sodium ethanedioate which takes place in acid solution at about 70°C. The two half-equations are:

reduction half-equation:

$$MnO_4^-(aq) + 8H^+(aq) + 5e^- \longrightarrow Mn^{2+}(aq) + 4H_2O(l)$$

oxidation half-equation: $C_2O_4^{2-}(aq) \longrightarrow 2CO_2(g) + 2e^-$

Solution: The reduction of manganate(VII) involves the addition of five electrons, the oxidation of ethanedioate ions involves the loss of two electrons.
Therefore, $MnO_4^-(aq)$ and $C_2O_4^{2-}(aq)$ must react in a ratio of 2:5 in order that the number of electrons involved in both reduction and oxidation half-equations is the same:

$$2MnO_4^-(aq) + 16H^+(aq) + 10e^- \longrightarrow 2Mn^{2+}(aq) + 8H_2O(l)$$

$$5C_2O_4^{2-}(aq) \longrightarrow 10CO_2(g) + 10e^-$$

Combining these two half-equations, and cancelling electrons, gives the equation for the reaction:

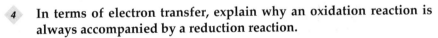

$$2MnO_4^-(aq) + 5C_2O_4^{2-}(aq) + 16H^+(aq) \longrightarrow 2Mn^{2+}(aq) + 10CO_2(g) + 8H_2O(l)$$

PHOTOSYNTHESIS: OXYGEN FORMED BY AN OXIDATION REACTION

In photosynthesis oxygen is produced by an oxidation reaction. This seems to be a paradox, since one of our definitions of oxidation is **gain** of oxygen. However, considering the process in terms of **electron transfer**, we find that it is, nonetheless, an oxidation reaction.

Photosynthesis is a series of chemical changes which bring about the conversion of carbon dioxide and water to carbohydrates and oxygen. The overall reaction may be summarised:

$$6CO_2(g) + 6H_2O(l) \longrightarrow C_6H_{12}O_6(s) + 6O_2(g)$$

The two half reactions are:

1. An oxidation reaction (loss of electrons) which requires the presence of light and chlorophyll (acting as a catalyst):

$$2H_2O(l) \longrightarrow 4H^+(aq) + O_2(g) + 4e^-$$
(the oxidation reaction which produces oxygen)

2. A reduction reaction (gain of electrons) which can occur in the dark:

$$6CO_2(g) + 24H^+(aq) + 24e^- \longrightarrow C_6H_{12}O_6(s) + 6H_2O(l)$$

It is important to realise, however, that these are summary equations for a series of very complex changes. The process of photosynthesis is still not yet fully understood.

Fig 15.4 This oxygenating plant helps maintain a healthy aquatic environment for fish and other pond life by providing oxygen (seen as bubbles on the surface of the leaves).

CONNECTION: Oxidation numbers and redox reactions, page 116.

Oxidation numbers and redox reactions

In terms of oxidation number, Ox:

Oxidation is an increase in oxidation number.
Reduction is a decrease in oxidation number.

The following examples show how reactions covered by earlier definitions of oxidation and reduction may also be described in terms of oxidation number, changes:

1. $Zn(s) + Cu^{2+}(aq) \longrightarrow Zn^{2+}(aq) + Cu(s)$

Reactants: $Zn(s)$: $Ox(Zn) = 0$ $Cu^{2+}(aq)$: $Ox(Cu) = +2$.
Products: $Zn^{2+}(aq)$: $Ox(Zn) = +2$ $Cu(s)$: $Ox(Cu) = 0$.

Zinc has been oxidised (change in oxidation number from 0 to +2) and copper has been reduced (change in oxidation number from +2 to 0).

2. $2MnO_4^-(aq) + 5C_2O_4^{2-}(aq) + 16H^+(aq) \longrightarrow 2Mn^{2+}(aq) + 10CO_2(g) + 8H_2O(l)$

Reactants: $MnO_4^-(aq)$: $Ox(Mn) = +7$; $Ox(O) = -2$
          $C_2O_4^{2-}(aq)$: $Ox(C) = +3$; $Ox(O) = -2$ $H^+(aq)$: $Ox(H) = +1$.
Products: $Mn^{2+}(aq)$: $Ox(Mn) = +2$ $CO_2(g)$: $Ox(C) = +4$; $Ox(O) = -2$
          $H_2O(l)$: $Ox(H) = +1$ $Ox(O) = -2$.

Manganese has been reduced (change in oxidation number from +7 to +2) and carbon has been oxidised (change in oxidation number from +3 to +4). Hydrogen (oxidation number +1 before and after reaction) and oxygen (oxidation number -2 before and after reaction) remain unchanged.

One very powerful use for oxidation numbers is to balance redox reactions. This has one great advantage over using redox half equations in that we need to know only reactants and products, we do not need to remember half equations.

TUTORIAL

Problem: Write a balanced equation for the reaction between manganate(VII) ions, $MnO_4^-(aq)$, and iron(II) ions, $Fe^{2+}(aq)$, in acid solution. The products of reaction are manganese(II) ions, $Mn^{2+}(aq)$, and iron(III) ions, $Fe^{3+}(aq)$.

Solution:
$$MnO_4^-(aq) \quad Fe^{2+}(aq) \longrightarrow Mn^{2+}(aq) \quad Fe^{3+}(aq)$$
$$Ox(Mn) = +7 \quad Ox(Fe) = +2 \qquad Ox(Mn) = +2 \quad Ox(Fe) = +3$$

Changes in oxidation numbers:

Mn +7 to +2 (reduction) = decrease –5
Fe +2 to +3 (oxidation) = increase +1

Since total change in oxidation number must be zero, manganese and iron must react in a 1 $MnO_4^-(aq)$: 5 $Fe^{2+}(aq)$ mole ratio. The balanced equation for the reaction must reflect this.

$$MnO_4^-(aq) + 5Fe^{2+}(aq) \longrightarrow Mn^{2+}(aq) + 5Fe^{3+}(aq)$$

Sufficient water must be added to the right-hand side of the equation to balance the oxygen content:

$$MnO_4^-(aq) + 5Fe^{2+}(aq) \longrightarrow Mn^{2+}(aq) + 5Fe^{3+}(aq) + 4H_2O(l)$$

Finally, sufficient hydrogen ions must be added to the left-hand side of the equation to balance the hydrogen content:

$$MnO_4^-(aq) + 5Fe^{2+}(aq) + 8H^+(aq) \longrightarrow Mn^{2+}(aq) + 5Fe^{3+}(aq) + 4H_2O(l)$$

To confirm that the equation is balanced, the sum of the charges on the ions on the right-hand side of the equation must equal the sum of the charges of the ions on the left-hand side of the equation:

LHS = (1–) + 5(2+) + 8(1+) = 17+; RHS = (2+) + 5(3+) = 17+

Therefore, the equation for the reaction is:

$$MnO_4^-(aq) + 5Fe^{2+}(aq) + 8H^+(aq) \longrightarrow Mn^{2+}(aq) + 5Fe^{3+}(aq) + 4H_2O(l)$$

Common oxidising and reducing agents

Oxidising agents belong to many classes of chemical substances. Their common characteristic is the ability to accept electrons from other substances. Reducing agents are characterised by their ability to donate electrons to other substances. Some common oxidising and reducing agents and their half-equations are shown below. For simplicity the physical states of the substances are omitted:

Oxidising agents ...	undergoing reduction.
Ag^+	$Ag^+ + e^- \longrightarrow Ag$
$Cr_2O_7^{2-}$ (acid solution)	$Cr_2O_7^{2-} + 14H^+ + 6e^- \longrightarrow 2Cr^{3+} + 7H_2O$
MnO_4^- (acid solution)	$MnO_4^- + 8H^+ + 5e^- \longrightarrow Mn^{2+} + 4H_2O$
O_2	$O_2 + 4e^- \longrightarrow 2O^{2-}$
F_2	$F_2 + 2e^- \longrightarrow 2F^-$
Cl_2	$Cl_2 + 2e^- \longrightarrow 2Cl^-$
H_2O_2 (acid solution)	$H_2O_2 + 2H^+ + 2e^- \longrightarrow 2H_2O$
H_2SO_4 (concentrated)	$H_2SO_4 + 2H^+ + 2e^- \longrightarrow 2H_2O + SO_2$
HNO_3	$HNO_3 + H^+ + e^- \longrightarrow NO_2 + H_2O$
	$HNO_3 + 3H^+ + 3e^- \longrightarrow NO + 2H_2O$

Reducing agents	undergoing oxidation
Zn	$Zn \longrightarrow Zn^{2+} + 2e^-$
CO	$CO + O^{2-} \longrightarrow CO_2 + 2e^-$
I^- (acid solution)	$2I^- \longrightarrow I_2 + 2e^-$
Na	$Na \longrightarrow Na^+ + e^-$
Br^-	$2Br^- \longrightarrow Br_2 + 2e^-$
Sn^{2+}	$Sn^{2+} \longrightarrow Sn^{4+} + 2e^-$
$S_2O_3^{2-}$	$2S_2O_3^{2-} \longrightarrow S_4O_6^{2-} + 2e^-$
$C_2O_4^{2-}$	$C_2O_4^{2-} \longrightarrow 2CO_2 + 2e^-$
NH_4^+	$NH_4^+ + 3H_2O \longrightarrow NO_3^- + 10H^+ + 8e^-$
Fe^{2+}	$Fe^{2+} \longrightarrow Fe^{3+} + e^-$

CONNECTION: Redox reactions in qualitative and volumetric analysis, pages 130–132.

SPOTLIGHT

Fig 15.5 Photochromic glasses darken on exposure to strong light and become more transparent when the light intensity falls. Despite initial concern about the slow recovery of darkened glass, new designs promise a much quicker response to low light intensities.

REDUCTION AND PHOTOCHROMIC LENSES

Photochromic lenses darken in sunlight and lighten when light intensity drops. They thus offer a response to changing light intensity and are convenient for purposes such as driving or skiing. At the heart of their operation is an 'intelligent' glass which uses a reduction process to alter its transparency to light.

Trapped in the amorphous material of the glass are very small particles of silver chloride and copper(I) chloride. When sunlight falls on the glass chloride ions are oxidised to chlorine atoms:

$$Cl^- \longrightarrow Cl + e^-$$

The electron is transferred to a silver ion, reducing it to a silver atom. This reflects light and prevents it being transmitted to the eye behind the glass:

$$Ag^+ + e^- \longrightarrow Ag$$

Chlorine atoms are reduced back to chloride ions by copper(I) ions:

$$Cl + Cu^+ \longrightarrow Cl^- + Cu^{2+}$$

In conditions of low light intensity copper(II) ions oxidise silver atoms back to silver ions, restoring the transparency of the glass:

$$Cu^{2+} + Ag \longrightarrow Cu^+ + Ag^+$$

QUESTIONS

15.1 Define oxidation and reduction in terms of the transfer of:
(i) oxygen; (ii) hydrogen; (iii) electrons;
giving examples to illustrate your answer in each case.

15.2 Using the example:

$$2Ag^+(aq) + Cu(s) \longrightarrow Cu^{2+}(aq) + 2Ag(s)$$

explain what is meant by the term redox reaction.

15.3 Identify in each of the following reactions which species undergoes oxidation and which undergoes reduction. Explain (i) which substance behaves as an oxidising agent and (ii) which substance behaves as a reducing agent in each reaction.
(a) $3Fe(s) + 4H_2O(g) \longrightarrow Fe_3O_4(s) + 4H_2(g)$
(b) $2Mg(s) + O_2(g) \longrightarrow 2MgO(s)$
(c) $2AgNO_3(aq) + Fe(s) \longrightarrow Fe(NO_3)_2(aq) + 2Ag(s)$
(d) $Fe_2O_3(s) + 3C(s) \longrightarrow 2Fe(s) + 3CO(g)$

15.4 Explain what is meant by the term oxidation number using examples from the chemistry of chlorine to illustrate your answer.

15.5 Identify the oxidation state of the metal(s) in each of the following compounds or ions:
(a) $KMnO_4$; (b) K_2CrO_4; (c) $Na_2Cr_2O_7$; (d) $AgCl$; (e) Fe_2O_3;
(f) $[Cu(NH_3)_4(H_2O)_2]^{2+}$; (g) MnO_2; (h) Na_2O; (i) Pb_3O_4

15.6 Combine the following pairs of half-equations to obtain balanced redox equations:
(a) $MnO_4^-(aq) + 8H^+(aq) + 5e^- \longrightarrow Mn^{2+}(aq) + 4H_2O(l)$
$Fe^{2+}(aq) \longrightarrow Fe^{3+}(aq) + e^-$
(b) $Cr_2O_7^{2-}(aq) + 14H^+(aq) + 6e^- \longrightarrow 2Cr^{3+}(aq) + 7H_2O(l)$
$Cu(s) \longrightarrow Cu^{2+}(aq) + 2e^-$
(c) $2Br^-(aq) \longrightarrow Br_2(aq) + 2e^-$
$Cl_2(aq) + 2e^- \longrightarrow 2Cl^-(aq)$
(d) $MnO_4^-(aq) + 8H^+(aq) + 5e^- \longrightarrow Mn^{2+}(aq) + 4H_2O(l)$
$2I^-(aq) \longrightarrow I_2(aq) + 2e^-$
(e) $MnO_4^-(aq) + 8H^+(aq) + 5e^- \longrightarrow Mn^{2+}(aq) + 4H_2O(l)$
$C_2O_4^{2-}(aq) \longrightarrow 2CO_2(g) + 2e^-$
(f) $Sn^{2+}(aq) \longrightarrow Sn^{4+}(aq) + 2e^-$
$Cr_2O_7^{2-}(aq) + 14H^+(aq) + 6e^- \longrightarrow 2Cr^{3+}(aq) + 7H_2O(l)$
(g) $Cu^{2+}(aq) + e^- \longrightarrow Cu^+(aq)$
$2I^-(aq) \longrightarrow I_2(aq) + 2e^-$

15.2 ELECTRODE POTENTIALS

Chemical energy stored in fossil fuels is converted into electrical energy at a power station. However, there is a demand for portable sources of electricity – **batteries**. By careful design we can produce highly specialised batteries which use chemical reactions to provide a reliable source of electricity. All batteries depend on redox reactions to provide electrical energy. Before we consider these processes in more detail, we need to understand the nature of redox reactions.

Redox equilibria

A **redox equilibrium** is established when, in an electrochemical process, the rate of electron gain is equal to the rate of electron loss. A redox equilibrium is one which exists between two chemically-related species which are in different oxidation states. For example, in aqueous solution a redox equilibrium may exist between hydrated iron(II) ions and iron(III) ions

$Fe^{2+}(aq) \rightleftharpoons Fe^{3+}(aq) + e^-$

To convert chemical energy into electrical energy in electrochemical cells, we must learn to manipulate such equilibria.

A redox equilibrium is also established when a metal rod is brought into contact with an aqueous solution containing ions of the same metal. Two opposing reactions begin. One involves the metal atoms of the rod entering solution as hydrated metal ions and leaving electrons behind on the surface of the rod. The metal atoms are oxidised:

$M(s) \longrightarrow M^{n+}(aq) + ne^-$ (loss of electrons; increase in oxidation number)

The other reaction (the reverse reaction) involves the hydrated metal ions in solution receiving electrons from the metal rod and being deposited as metal atoms on the surface of the rod. The ions are being reduced:

$M^{n+}(aq) + ne^- \longrightarrow M(s)$ (gain of electrons; decrease in oxidation number)

As with all reversible reactions, an equilibrium is established when the rate of the forward and the reverse reactions is equal. The equilibrium which exists between a metal in contact with a solution of one of its salts is an example of a redox equilibrium. This equilibrium may be represented by the equation:

$$M(s) \rightleftharpoons M^{n+}(aq) + ne^-$$

$M(s)$ and $M^{n+}(aq)$ are the two chemically-related species;
$M(s)$ is the reduced species; $M^{n+}(aq)$ is the oxidised species.

CONNECTION: Equilibrium, page 449.

5 **What would happen to the redox equilibrium if the concentration of the metal salt was increased?**

For some metals the dominant reaction involves oxidation of the metal rod to give ions and so the position of the equilibrium is towards the right in the above equation. This results in a surplus of electrons left on the metal rod, leaving it with a negative charge on the surface. There will also be more positively charged ions in solution than before and the solution will develop an overall positive charge. A double layer develops at the interface between rod and solution, with electrons on the surface of the metal and cations in solution in the immediate vicinity of the metal, attracted to its surface. This is called the **Helmholtz double layer** (Fig 15.6).

For example, when a zinc rod is placed in 1 mol dm^{-3} zinc sulphate solution:

$$Zn(s) \rightleftharpoons Zn^{2+}(aq) + 2e^-$$

the overall change is a net dissolving of the zinc, leaving a negative charge on the zinc rod. The mass of the zinc rod decreases (though only *very* slightly).

6 **What are the two components of the Helmholtz double layer which forms here?**

For other metals, reduction of metal ions and so the deposit of metal atoms dominates. Electrons are provided by atoms of the metal rod which, itself, acquires an overall positive charge as a result. The solution will be left with fewer positive ions than before and will acquire an overall negative charge. For example, when a copper rod is placed in 1 mol dm^{-3} copper(II) sulphate solution:

$$Cu(s) \rightleftharpoons Cu^{2+}(aq) + 2e^-$$

the dominant process is deposition of copper ions as copper atoms and so the rod acquires a positive charge. The mass of the copper rod increases *very* slightly.

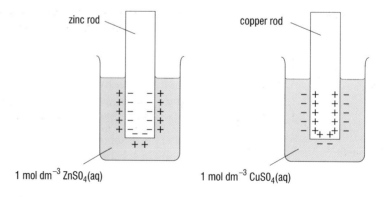

Fig 15.6 A charge separation develops when metal rods are placed in aqueous solutions of their salts. A Helmholtz double layer is produced at the metal surface.

REDOX AND ELECTRICITY

Half-cells and electrode potentials

In both of the situations described above, the solution and rod develop electrical charges which are opposite. A **potential difference** is said to exist between the rod and the solution because of this difference in charge, or **charge separation**. The rod placed in solution is referred to as an **electrode**.

- For the zinc electrode, the potential difference is described as negative because the electrode acquires a negative charge relative to the solution.
- The copper electrode is said to have a positive potential relative to the solution of copper(II) sulphate into which it is placed.

7 **Why does the potential difference between a metal and a solution of one of its salts depend on the position of equilibrium?**

Constituents of a redox equilibrium, taken together, make a **half-cell**. A half-cell equilibrium can be shown by a half-equation which indicates the nature of the chemical species involved in the redox equilibrium and the number of electrons transferred. By convention, it is expressed as a reduction (addition of electrons) reaction:

Oxidised species + ne^- → Reduced species

For example,

$$Cr_2O_7^{2-}(aq) + 14H^+(aq) + 6e^- \longrightarrow 2Cr^{3+}(aq) + 7H_2O(l)$$

$$MnO_4^-(aq) + 8H^+(aq) + 5e^- \longrightarrow Mn^{2+}(aq) + 4H_2O(l)$$

$$Cl_2(g) + 2e^- \longrightarrow 2Cl^-(aq)$$

$$Fe^{3+}(aq) + e^- \longrightarrow Fe^{2+}(aq)$$

Half-cells of any type may be combined to produce an electrochemical cell. There are three common types of half-cell:

1. The metal/metal ion half-cell. In a metal/metal ion half-cell a metal is in contact with a solution of one of its salts. For example, the $Zn^{2+}(aq)/Zn(s)$ and $Cu^{2+}(aq)/Cu(s)$ half-cells.

2. The non-metal/non-metal ion half-cell. In a non-metal/non-metal ion half cell an equilibrium is established between a non-metallic element and an ion derived from that element in solution. The non-metal is usually gaseous and bubbled through a solution of ions. The redox equilibrium is established at the surface of an inert metal conductor such as platinum. For example, the chlorine half-cell in which $Cl_2(g)$ is in equilibrium with $Cl^-(aq)$ (Fig 15.7).

3. The ion/ion half-cell. In an ion/ion half-cell an inert electrode (again, platinum is commonly used) is in contact with a solution containing two types of ion derived from the same element but each in a different oxidation state. For example, the $Fe^{2+}(aq)/Fe^{3+}(aq)$ half-cell (Fig 15.8).

8 **What advantage does a 'platinised' electrode have over a platinum electrode which has not been treated in this way?**

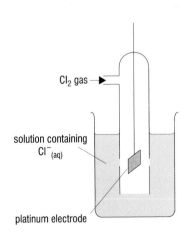

Cl₂ gas →

solution containing
Cl⁻(aq)

platinum electrode

Fig 15.7 The chlorine half-cell contains a 'platinised' platinum electrode – a platinum electrode on which a further black spongy surface layer of platinum has been deposited. The platinised platinum has many porous cavities and is extremely inert.

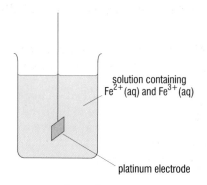

Fig 15.8   In the Fe^{2+}(aq)/Fe^{3+}(aq) half-cell, two oxidation states of the element are in the same phase. Platinum is used as an inert electrode.

solution containing Fe^{2+}(aq) and Fe^{3+}(aq)

platinum electrode

CONNECTION: Standard electrode potentials, Datasheet 2.

Representing half-cells

By convention half cells are usually represented in a standard way. This convention shows a conjugate redox pair. This term means the two substances which take part in a redox equilibrium are related by a loss or gain of electrons. Any phase boundaries which may exist in the half-cell are also shown. The oxidised species is shown first followed by the reduced species and the phase boundary is shown by a single vertical line. The example of the zinc half-cell indicates how the convention is applied:

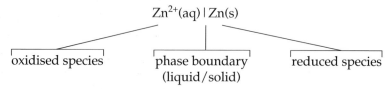

$$Zn^{2+}(aq) \mid Zn(s)$$

oxidised species   phase boundary (liquid/solid)   reduced species

Where an inert electrode, such as a platinum electrode, is used in the half-cell, this is shown next to the more reduced species of the conjugate redox pair. For example, the hydrogen half-cell:

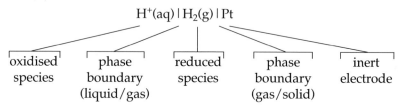

$$H^+(aq) \mid H_2(g) \mid Pt$$

oxidised species | phase boundary (liquid/gas) | reduced species | phase boundary (gas/solid) | inert electrode

These conventions are used in the representation of electrochemical cells formed by combining half-cells.

9 **Represent the half-cell for $Cl_2(g) + 2e^- \longrightarrow 2Cl^-$(aq) (platinum electrode).**

Standard electrode potentials

The electrode potential of a half-cell is the potential difference which arises between a solution and an electrode when a redox equilibrium has been established. As a consequence electrode potentials are often called **redox potentials**. Use of the term emphasises the reduction-oxidation origin of electrode potentials. If standard conditions are employed then we refer to the **standard electrode potentials** (also known as **standard redox potentials**) of a half-cell. Standard electrode potentials, measured under standard conditions, are given the symbol $E^{\ominus}(298)$. By convention, they are reduction potentials because they refer to a reduction process.

For example, the standard electrode potential, $E^{\ominus}(298)$, of a Zn^{2+}(aq)/Zn(s) half-cell is -0.76 V and this can be expressed as:

Zn^{2+}(aq) $\mid$ Zn(s); $E^{\ominus}(298) = -0.76$ V

A redox potential, therefore, is a measure of the tendency for reduction of the more oxidised species to take place and the standard redox potentials allow a direct numerical comparison of the relative strength of oxidising and reducing agents to be made.

The standard electrode potential for Zn^{2+}(aq)/Zn(s) is negative and suggests that zinc ions show less tendency than hydrogen ions (H^+(aq) $\mid$ H_2(g); $E^{\ominus}(298) = 0$ V) to undergo reduction. For the reduction of copper ions to copper metal ($Cu^{2+}(aq) + 2e^- \longrightarrow Cu(s)$), $E^{\ominus}(298) = +0.34$ V. This shows that copper ions are more likely than hydrogen ions to undergo reduction.

In general, the more positive the value of the standard electrode potential of a half-cell the more likely that reduction will take place in that half-cell.

10 What relationship, if any, might you expect to find for metals between their ionisation energies and their standard electrode potentials?

Standard conditions

Experiment shows that the position of equilibrium and, therefore, size of the electrode potential developed by a half-cell depends upon:

- the particular half-cell involved;
- the concentration of any ions in solution which are involved in the redox process;
- the pressure of any gaseous components;
- the temperature of the solution.

In order to compare the potential difference developed by half-cells we need to make our measurements under standard conditions. These are usually taken as:

- concentration of 1 mol dm^{-3} for participating ions;
- temperature of 298 K;
- pressure of 101 325 N m^{-2} (1 atmosphere).

The e.m.f. of a cell

The **electromotive force (e.m.f.)** of a cell is the maximum potential difference produced by a combination of half-cells when zero current is passing. The units are volts. A combination of half-cells is called an **electrochemical cell** and the cell voltage generated in such a cell is measured with a high resistance voltmeter since this limits the passage of a current and, therefore, allows the cell e.m.f. to be determined.

Under standard conditions, the maximum voltage produced is called the **standard cell e.m.f.** and is given by the standard electrode potential of the half-cell which is undergoing reduction minus the standard electrode potential of the half-cell which is undergoing oxidation:

$E_{cell} = E_{reduction} - E_{oxidation}$ (hence the term: **redox**)

This relationship is true regardless of whether or not standard states are used. However, if the standard cell e.m.f. is required then the half-cells must be in their standard states.

Consider the cell made by combining

- a $Zn^{2+}(aq)/Zn(s)$ half cell; $E^{\circ}(298) = -0.76$ V, and
- a $Cu^{2+}(aq)/Cu(s)$ half-cell; $E^{\circ}(298) = +0.34$ V.

Since the copper electrode has the more positive value of the two standard electrode potentials of the half-cells it will undergo reduction. Therefore, the standard cell e.m.f. is

$+ 0.34 - (-0.76)$ V
$= 1.10$ V.

11 Use the Database to calculate the standard cell e.m.f. for a cell in which the reduction reaction is $Ag^{+}(aq) + e^{-} \longrightarrow Ag(s)$, and the oxidation reaction is $Cu(s) \longrightarrow Cu^{2+}(aq) + 2e^{-}$.

> ### KEY FACT
>
> An electrochemical cell converts chemical energy directly into electrical energy. It provides a source of direct electrical current, delivered at known voltage usually in a convenient, portable form.

> ### KEY FACT
>
> An electrochemical cell produces an **electromotive force (e.m.f.)** as a result of the chemical reactions which occur at the electrodes. It has the units volts. One or more cells in series is called a **battery**.

The voltage produced by an electrochemical cell remains constant throughout its lifetime. However, the time over which a cell can produce electricity depends on how hard it is made to work. The more energy it is required to supply in a given time period, the more quickly the reacting compounds are used up. For example, the batteries powering a transistor radio are used up more quickly if the volume is turned up high.

We can understand this by looking at the unit we call the volt:

1 volt = 1 J A^{-1} s^{-1} (A = ampere, the unit of current corresponding to charge flowing per unit time)

The greater the energy demands (the higher the number of joules required), the shorter will be the lifetime at a given current.

Measuring electrode potentials

The standard hydrogen electrode

It is impossible to measure directly the individual electrode potential of a single half-cell. To measure the potential, a voltmeter would need to be connected to the electrode and to the solution used. If, however, a wire from the voltmeter was dipped into the solution this would represent another electrode and would alter the electrode potential. Therefore, the potential difference developed by a particular half-cell must be compared with the potential difference developed by a **standard** or **reference half-cell**.

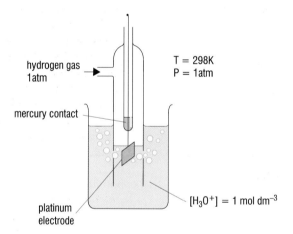

Fig 15.9 A standard hydrogen electrode. A platinised platinum electrode is immersed in a solution of 1 mol dm^{-3} hydrated hydrogen ions, $H^+(aq)$, maintained at 298 K. Hydrogen is passed at a pressure of 1 atmosphere.

The hydrogen half-cell (Fig 15.9) has been universally adopted as a reference half-cell and its electrode potential under standard conditions is taken as zero.

$H^+(aq) + e^- \longrightarrow \frac{1}{2} H_2(g)$; $E^{\ominus}(298) = 0.00$ V

It is called the **standard hydrogen electrode** and is a **primary reference electrode** since all electrode potentials are measured by reference to the electrode potential of this half-cell.

The hydrogen half-cell can be used to measure the standard electrode potential of another half-cell, provided standard conditions are used for both electrodes. The two half-cells are combined to form an electrochemical cell (Fig 15.10). A **salt bridge**, containing a concentrated solution of a strong electrolyte such as potassium nitrate, is used to complete the circuit. This allows the passage of ions from one solution to the other without the two solutions mixing.

12 Why is it important to prevent ions from the two solutions mixing?

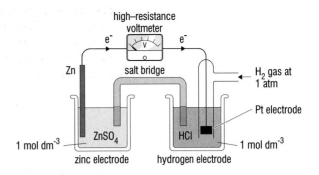

Fig 15.10 If the hydrogen half-cell is combined with a second half-cell to form an electrochemical cell then the cell e.m.f. is equal to the electrode potential of the second half-cell.

The cell e.m.f. being measured is equal to the electrode potential of the other half-cell, since the electrode potential of the standard hydrogen half-cell is taken as zero. Whether the electrode potential is positive or negative depends upon whether the electrode of the half-cell being measured is positively or negatively charged when compared with the standard hydrogen electrode.

Secondary reference electrodes

The standard hydrogen electrode is not convenient to use. For example, it is difficult to maintain a stream of hydrogen gas at exactly 1 atmosphere pressure. Therefore, the standard hydrogen electrode is often abandoned in favour of a more reliable and convenient **secondary reference electrode**. This is an electrode whose electrode potential has been measured using a standard hydrogen electrode. It is used to determine other electrode potentials by measuring the e.m.f. of a cell composed of the secondary reference electrode and the half-cell whose electrode potential is to be determined.

The **calomel electrode** is often used as a secondary reference electrode because it is both reliable and convenient. It consists of mercury metal in contact with a paste made of mercury and mercury(I) chloride ('calomel') (Fig 15.11). The calomel electrode is calibrated by setting it up in a cell with the standard hydrogen electrode. The electrode potential of a standard calomel electrode is + 0.244 V.

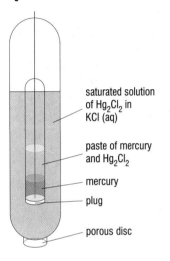

Fig 15.11 The calomel electrode.

saturated solution of Hg_2Cl_2 in KCl (aq)

paste of mercury and Hg_2Cl_2

mercury

plug

porous disc

QUESTIONS

15.7 Explain what is meant by the term *redox equilibrium*.

15.8 A piece of magnesium ribbon is dipped into a solution containing magnesium chloride.
(a) Explain, with the aid of an equation, the nature of the redox equilibrium which is established.
(b) Predict the sign of the charge on the magnesium ribbon once equilibrium is established.
(c) Explain the origin of the potential difference established between solid magnesium and the solution.

15.9 Why does a Helmholtz double layer form when a metal is placed in contact with one of its salts?

15.10 (a) Explain carefully what is meant by the term *standard electrode potential*.
(b) Under what conditions is a standard electrode potential normally measured?
(c) Describe with the aid of suitable diagrams how the standard electrode potential of the silver/silver ion half-cell could be measured.

15.11 (a) Give a labelled diagram of the hydrogen half-cell.
(b) Explain what is meant by *primary reference electrode*.

15.3 CELLS AND BATTERIES

We have seen how two half-cells are combined in order to determine the electrode potential of one, provided the value for the other is known (either a standard hydrogen electrode or a secondary reference electrode). Half-cells are also combined to generate electrical energy. Clever and imaginative combinations enable batteries to be designed and constructed for specialised use.

Many different types of commercial batteries are available, the design of each depending on its potential uses. When designing a battery, decisions must be taken about the relative importance of:

- low mass
- reliability
- low cost of materials
- low production costs
- toxicity (environmental/health considerations of manufacture)
- disposal/recycling potential
- efficiency at likely operating temperature
- efficiency at possible extremes of operating temperature

13 **Give an example of a battery where small size and low mass are the most important considerations.**

Each electrochemical cell consists of a combination of two half-cells one of which has a tendency towards reduction. This involves a net decrease in the number of electrons at its electrode. The other half-cell favours oxidation at its electrode which accumulates electrons. If the two half-cells are combined in an electrical circuit, electrons flow from the electrode with the highest density of electrons to the other. This flow of electrons is an electrical current.

Most commercial batteries provide reliable and consistent conversion over a considerable period of time. Continued research by electrochemical industries has enabled powerful batteries to be manufactured which occupy less space and have a lower mass than ever before.

These characteristics of smaller size and lower mass have great appeal for industry since many uses demand them. For example, pocket calculators use a small, light, yet very efficient battery which provides a small but very reliable voltage over many hours. The demands placed upon a car battery, however, are quite different. Car batteries need to store electrical energy when the electrical demand on the car is low, but must be able to make additional electrical energy available when demand is high. Small variations in the output of the battery are less important here.

Commercial cells are often defined as **primary cells** (which are non-rechargeable as they are unable to regenerate the cell components) or **secondary (storage) cells** (which are capable of accumulating charge by reversing the cell reaction). Secondary cells are rechargeable and include car batteries.

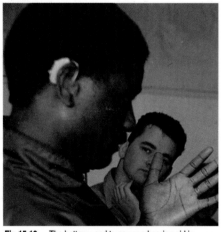

Fig 15.12 The battery used to power a hearing aid is reliable, virtually leakproof and remarkably long-lived.

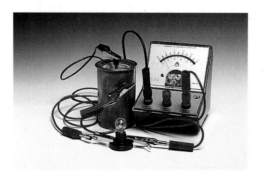

Fig 15.13 The Daniell cell is one of the earliest and simplest electrochemical cells. It is a combination of a zinc/zinc ion half-cell and a copper/copper ion half-cell. Electrons flow from the zinc electrode (the anode) to the copper electrode (the cathode).

Representing cells

Electrochemical cells are shown in circuit diagrams by the symbol:

The electrical current (the flow of electrons) flows from the negative (shorter) pole to the positive (longer) pole.

Cells are usually represented by a **cell diagram** or **cell statement**. This shows the combination of half-cells used in the cell. Using this convention the half-cell in which oxidation takes place (anode) is shown on the left-hand side in the cell diagram. The other half-cell is where reduction occurs (cathode) and is on the right-hand side. Each half-cell is written so that the dominant chemical process taking place in each half-cell may be read left-to-right in the cell diagram. A salt bridge is represented by two broken vertical lines. Phase boundaries are shown by a single vertical line.

The conventions for cell diagrams for electrochemical cells are as follows:

- the half-cell with the more positive electrode potential is placed on the right-hand side;
- the half-cell on the left is written as an oxidation and the one on the right as a reduction;
- if an inert electrode is used, this is placed next to the more reduced species in the correct half-cell.

For the Daniell cell (Fig 15.13) the cell diagram is:

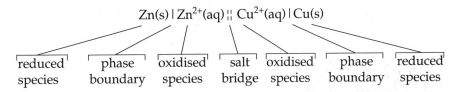

This implies that the reaction taking place in the zinc/zinc ion half-cell is an oxidation:

$$Zn(s) \longrightarrow Zn^{2+}(aq) + 2e^-$$

and the zinc half-cell is therefore the anode of the electrochemical cell. The copper/copper ion half-cell undergoes reduction and this half-cell must be the cathode:

$$Cu^{2+}(aq) + 2e^- \longrightarrow Cu(s)$$

The direction of change can be determined by examination of the cell diagram which shows that the flow of electrons in this cell is from the zinc to the copper electrode around the external circuit.

14　**Represent an electrochemical cell consisting of $Fe^{2+}(aq)/Fe(s)$ and $Ag^+(aq)/Ag(s)$.**

If the members of a half-cell are in the same phase they are separated only by a comma (,). An inert electrode is shown next to the more reduced species in the half-cell in which it is used. For example,

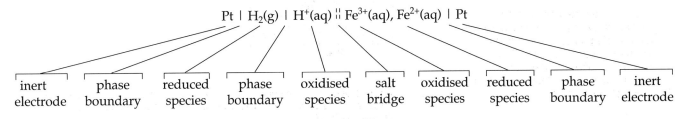

> ## KEY FACT
>
> Once we have drawn a cell diagram, the e.m.f. of the cell can be calculated from
>
> $$E_{cell} = E_{rhs} - E_{lhs}$$
>
> where E_{cell} is the e.m.f. produced by the cell, E_{rhs} is the electrode potential of the right-hand half-cell and E_{lhs} the electrode potential of the left-hand half-cell.

Problem: When the $Fe^{3+}(aq)/Fe^{2+}(aq)$ and $Sn^{4+}(aq)/Sn^{2+}(aq)$ half-cells are combined:

(a) What cell reaction takes place?
(b) Draw the cell diagram.
(c) Calculate the cell e.m.f.

The standard electrode potentials are:

$Fe^{3+}(aq)/Fe^{2+}(aq)$; $E^{\ominus}(298) = +0.77$ V
$Sn^{4+}(aq)/Sn^{2+}(aq)$; $E^{\ominus}(298) = +0.15$ V

Assume that each species is present in its standard state.

Solution:
(a) The $Fe^{3+}(aq)/Fe^{2+}(aq)$ half-cell has the more positive electrode potential. Therefore, it will undergo reduction and the $Sn^{4+}(aq)/Sn^{2+}(aq)$ half-cell will undergo oxidation. The cell reaction is:
$2Fe^{3+}(aq) + Sn^{2+}(aq) \longrightarrow 2Fe^{2+}(aq) + Sn^{4+}(aq)$
(b) The cell diagram should place the $Fe^{3+}(aq)/Fe^{2+}(aq)$ half-cell on the right-hand side and the $Sn^{4+}(aq)/Sn^{2+}(aq)$ half-cell on the left-hand side:
$Sn^{2+}(aq), Sn^{4+}(aq) \mathbin{\vdots\vdots} Fe^{3+}(aq), Fe^{2+}(aq)$
(c) The cell e.m.f. is given by
$E^{\ominus}(298)_{cell} = E^{\ominus}(298)_{rhs} - E^{\ominus}(298)_{lhs}$

Therefore, $E^{\ominus}(298)_{cell} = +0.77 - (+0.15) = +0.62$ V

15.12 For each of the following pairs of half-cells
    (a) Decide in each case what will be the cell reaction;
    (b) Write a cell diagram; **(c)** Calculate $E(298)_{cell}$.
    (i) $Cu^{2+}(aq)/Cu(s)$; $E^{\ominus}(298) = 0.34$ V
          $Ag^{+}(aq)/Ag(s)$; $E^{\ominus}(298) = 0.80$ V
    (ii) $Al^{3+}(aq)/Al(s)$; $E^{\ominus}(298) = -1.66$ V
          $Fe^{2+}(aq)/Fe(s)$; $E^{\ominus}(298) = -0.44$ V
    (iii) $Br_2(aq)/Br^{-}(aq)$; $E^{\ominus}(298) = 1.07$ V
          $Pb^{2+}(aq)/Pb(s)$; $E^{\ominus}(298) = -0.13$ V

15.13 **(a)** Draw a simple labelled sketch showing the main features of the Daniell cell.
    (b) Explain at which electrode reduction would take place.
    (c) Write a cell diagram for this cell.
    (d) Explain why the cell e.m.f. is measured using a high resistance voltmeter.

15.14 Explain the use of a salt bridge in an electrochemical cell.

15.15 **(a)** Use the standard electrode potentials shown in the Database to predict what would happen if magnesium ribbon is dipped into a solution of lead(II) nitrate.
    (b) Write the two half-cell reactions for the reaction taking place.
    (c) Give an ionic equation for the reaction.

15.16 **(a)** Use the Database to determine which two of the following half-cells could be combined *in aqueous solution* to produce a cell with the greatest cell e.m.f.
    (i) $Ni^{2+}(aq)/Ni(s)$;   *(ii)* $Fe^{3+}(aq)/Fe^{2+}(aq)$;   *(iii)* $I_2(s)/I^{-}(aq)$;
    (iv) $F_2(g)/F^{-}(aq)$;   *(v)* $Sn^{4+}(aq)/Sn^{2+}(aq)$
    (b) For the half-cell combination cell you have chosen in (a), write a cell diagram for this combination, and calculate the cell e.m.f.

15.4 SPECIALISED CELLS AND BATTERIES

The dry cell

The modern dry cell, or Leclanché cell, is an adaptation of a cell developed by Georges Leclanché, a 19th century French chemist. The dry cell produces a maximum of 1.5 V and is used in numerous gadgets such as radios, flashlights and toys. Dry cells are virtually leakproof because the main components are solids or pastes which are sealed effectively from the external environment. They are relatively inexpensive but their voltage tends to drop quite quickly and they cannot be reused.

The zinc anode usually encloses the cell (Fig 15.14) and the cathode is a central graphite rod immersed in a moist paste (the electrolyte) consisting of a mixture of ammonium chloride, zinc chloride and manganese(IV) oxide.

The half-cell undergoing oxidation is:

$$Zn(s) \longrightarrow Zn^{2+}(aq) + 2e^-$$

The reactions occurring in the other half-cell are more complex but they may be simplified to:

$$2NH_4^+(aq) + 2e^- \longrightarrow 2NH_3(g) + H_2(g)$$
$$H_2(g) + 2MnO_2(s) \longrightarrow Mn_2O_3(s) + H_2O(l)$$

When a high current is demanded from the battery (when it is required to do hard work), hydrogen cannot react quickly enough, the reaction slows down and the cell voltage tends to drop.

15 What is the purpose of the manganese(IV) oxide in this half-cell?

16 What is the oxidation state of the manganese-containing product?

The alkaline battery

A more advanced (and more expensive) form of the dry cell is the **alkaline battery** which produces a typical voltage of slightly above 1.5 V. The reaction at the anode is again the oxidation of zinc, but under alkaline conditions to give zinc oxide

$$Zn(s) + 2OH^-(aq) \longrightarrow ZnO(s) + H_2O(l) + 2e^-$$

At the cathode, manganese(IV) oxide is reduced to give manganese(III) oxide. Hydroxide ions are regenerated:

$$H_2O(l) + 2MnO_2(s) + 2e^- \longrightarrow Mn_2O_3(s) + 2OH^-(aq)$$

The alkaline battery does not produce hydrogen or, indeed, any other gases. For this reason, it does not suffer the voltage drop of the common dry cell when worked hard.

17 Write an overall equation for the reaction taking place in an alkaline battery.

The lead-acid accumulator

With the current emphasis on fuel economy, a car battery must not add too much to the weight of the vehicle but must provide enough power to start the car even in the coldest weather. The lead-acid accumulator (or lead storage cell) is a secondary cell developed originally by a French physicist, Gaston Planté in 1859. It is cheap and easy to manufacture. Cars and lorries use a battery of six such cells joined in series. Each contains a lead anode and a cathode made of lead coated with lead(IV) oxide.

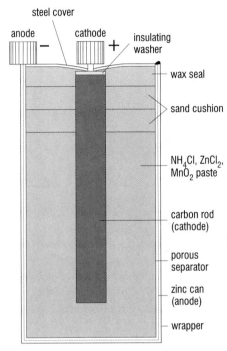

steel cover
anode cathode insulating washer
wax seal
sand cushion
NH_4Cl, $ZnCl_2$, MnO_2 paste
carbon rod (cathode)
porous separator
zinc can (anode)
wrapper

Fig 15.14 The dry cell is a valuable, portable, source of electricity and is a subject of continued research and development. In the United States and Europe about five Leclanché cells are used annually per person.

Fig 15.15 The dry cell is capable of delivering 1.5 V. The longer lasting versions are usually alkaline batteries and are more expensive.

The electrolyte is approximately 30% aqueous sulphuric acid. Both lead and lead(IV) oxide react with the acid to give a surface coating of lead(II) sulphate on the electrodes. The half-equations for each half-cell are:

$$Pb(s) \rightleftharpoons Pb^{2+}(aq) + 2e^-$$
$$PbO_2(s) + 4H^+(aq) + 2e^- \rightleftharpoons Pb^{2+}(aq) + 2H_2O(l)$$

A storage cell must be designed carefully if it is to operate efficiently. It must be able to convert electrical energy to chemical energy when the vehicle is running and then to supply this stored energy on demand when the vehicle is not running (for example, to start the engine). The lead accumulator design (Fig 15.16) ensures that these criteria are met. When the battery is being used to supply electrical energy while the engine is not running, the two half-reactions proceed in the forward direction (left to right in the above equations).

When the engine is running, the alternator charges the battery by reversing the two half-reactions (right to left in the above equations). The overall processes taking place may be represented by a single equilibrium equation:

$$Pb(s) + PbO_2(s) + 2H_2SO_4(aq) \underset{charging}{\overset{discharging}{\rightleftharpoons}} 2PbSO_4(s) + 2H_2O(l)$$

18 Is lead(IV) oxide oxidised or reduced whilst the battery is providing power?

19 What happens to the oxidation state of Pb^{2+} ions when the car engine is running but with low electrical demand?

When the battery is discharging, a deposit of lead(II) sulphate builds up on both electrodes. Due to a gradual leakage of current from a battery, even when not in use, there is a net usage of sulphuric acid. The concentration of sulphuric acid, often measured by its **specific density**, may be increased by the addition of a few drops of a 'battery regenerator'. This is just a more concentrated solution of sulphuric acid.

High performance cells

The mercury battery

Mercury cells are tiny cells used for specialised medical applications and in advanced electronics. The mercury cell is more expensive than an ordinary dry cell but it maintains a reliable voltage of 1.35 V for 95% of its life. Although not particularly light, the mercury cell gives excellent performance for its volume and is used, for example, in heart pacemakers and digital watches.

The mercury cell (Fig 15.18) is encased in a stainless steel case with a zinc/mercury amalgam (a solution of zinc in liquid mercury) as the anode and a strongly alkaline electrolyte paste containing potassium hydroxide, zinc hydroxide and mercury(II) oxide. The paste is the cathode, separated from the zinc by a paper divider. The paper allows migration of ions but maintains separation of the two electrodes.

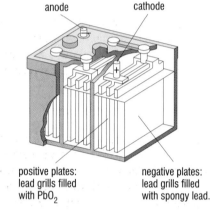

anode   cathode

positive plates: negative plates:
lead grills filled   lead grills filled
with PbO_2 with spongy lead.

Fig 15.16 A single lead accumulator cell of the type used in a typical car battery. In a typical car battery six of these cells are arranged in series and the battery is capable of delivering 12 V.

Fig 15.17 Despite years of neglect, the six-cell lead accumulator battery from this car continues to provide a satisfactory power source with the minimum of maintenance.

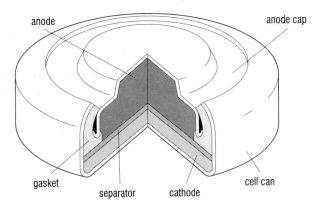

Fig 15.18 The mercury battery has an advantage over a conventional dry cell battery in that its electrolyte composition does not change during its lifetime. As E_{cell} depends upon electrolyte composition, the cell voltage in the mercury cell is very reliable.

The cell reactions are as follows:

Anode: $Zn(s) + 2OH^-(aq) \longrightarrow ZnO(s) + H_2O(l) + 2e^-$

Cathode: $HgO(s) + H_2O(l) + 2e^- \longrightarrow Hg(l) + 2OH^-(aq)$

Overall cell reaction:

$Zn(s) + HgO(s) \longrightarrow ZnO(s) + Hg(l)$

Despite the success of the mercury cell, there is growing concern over the careless disposal of used cells. The cells are not reusable and throwing them away not only wastes a valuable resource (although the mercury could be recycled) but allows toxic mercury and mercury compounds to enter the environment.

Metal/air batteries

One of the most promising developments in battery technology is the aluminium-air battery. Aluminium readily undergoes oxidation, and weight-for-weight this is one of the best providers of electrical energy. Also the battery is rechargeable. Under the right conditions the aluminium-air cell is capable of producing a maximum voltage of 2.7 V. Oxygen in the air is used as the oxidising agent for the cell. Since oxygen is constantly available the mass of an oxidising agent is saved, giving the high power/mass ratio.

High purity aluminium is used for the anode. Air is bubbled constantly through a cathode made of an inert porous metal. The electrolyte is usually either sodium chloride or sodium hydroxide. The half-reactions are:

Anode: $Al(s) + 3OH^-(aq) \longrightarrow Al(OH)_3(s) + 3e^-$

Cathode: $O_2(g) + 2H_2O(l) + 4e^- \longrightarrow 4OH^-(aq)$

Overall cell reaction:

$4Al(s) + 3O_2(g) + 6H_2O(l) \longrightarrow 4Al(OH)_3(s)$

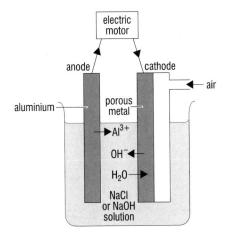

Fig 15.19 The aluminium-air battery is light, rechargeable, efficient and largely recyclable. It is ideally suited to powering electric vehicles including powered golf caddies and invalid carriages.

Recharging the aluminium-air battery involves replacing the aluminium anode with fresh aluminium, topping up with water and cleaning out the precipitate of aluminium hydroxide (which can be recycled). The efficiency of the cell can be reduced by oxide layers building up on the surface of the aluminium but this can be controlled by using an alloy of aluminium with very low levels (less than 0.05%) of tin or gallium.

Metal-air batteries using zinc are well-known. Alloying with a small amount of mercury helps reduce corrosion of the zinc anode. The zinc-air battery has a good power/mass ratio and has an overall energy capacity almost twice that of a mercury cell. It is used for hearing aid batteries.

20 Write down the cathode and anode half-reactions for the zinc-air cell.

Solid state batteries

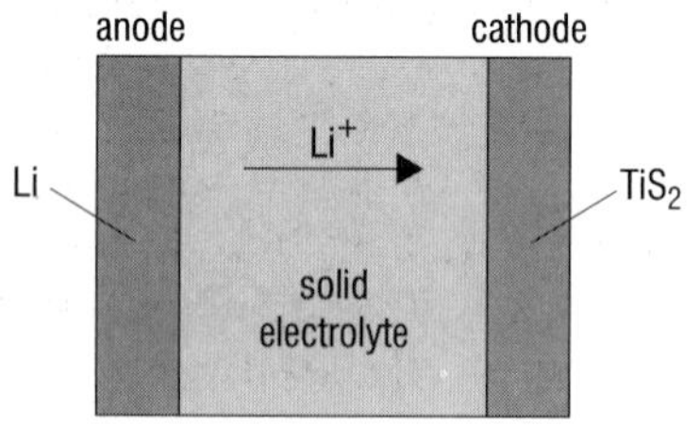

Fig 15.20 The solid-state lithium cell.

The discovery and refinement of solid electrolytes has enabled solid-state batteries to be developed. These cells use lithium metal as the anode and (usually) a cathode made of titanium(IV) sulphide. The electrolyte is lithium iodide supported in a solid, polymeric material. This gives good conductance at room temperatures and removes the need for an aqueous electrolyte. These batteries provide voltages of up to 3.0 V per cell and have a good shelf-life. With further development they may prove to be the 'batteries of the future'.

21 When lithium ($A_r = 6.9$) is oxidised, what mass of lithium is needed to produce one mole of electrons?

22 Why should a lithium battery be expected to provide high voltages?

ACTIVITY

The 1987 Pentax World Solar Challenge race was won by *Sunraycer*, a solar-powered car which completed the 1867 mile race from Darwin to Adelaide in 44 hours 54 minutes, an average speed of 41.6 miles per hour. The car was one of 23 solar vehicles competing in a race which became a showcase for design and alternative technology. *Sunraycer* used two types of solar cells, the first manufactured from pure silicon (1500 of these) and the second a gallium-arsenide cell (8000 of these). The cells drove an electric motor. In strong sunlight the solar cells were able to charge up 68 1.5 V silver-zinc batteries, capable of storing 3 kilowatt hours. In poor sunlight the car was driven by these batteries. They were not ideal in that they may be damaged if allowed to discharge completely, are very expensive and can only be recharged about 15 times. The cells worked well within the context of a 44 hour race but were not ideal for long-term use.

Placing yourself in the position of the designer of an electric car *for urban use*, prepare a design brief for batteries which would power the vehicle. Try to consider all the qualities that the ideal batteries would have and to anticipate some of the compromises you may have to make. You will need to review the qualities of cells described in this chapter in preparing your design brief.

REDOX AND ELECTRICITY

Fuel cells

The combustion of fossil fuel and its conversion to electrical energy is an inefficient process as much heat is lost to the surroundings. Even the most modern power plants rarely achieve a conversion of chemical energy to electricity which exceeds 45% efficiency. Combustion reactions are redox reactions in which a fuel (a reducing agent) burns in air (with oxygen as oxidising agent). This is desirable when the fuel is being burnt to produce heat. However, it is useful to be able to oxidise the fuel electrochemically and avoid unwanted heat losses.

Fuel cells are electrochemical cells which convert the chemical energy of a fuel directly into electrical energy. They differ from typical electrochemical cells in that they receive a constant supply of reactants from which to produce a constant electrical current. Fuel cells may use a wide variety of fuels. The fuel undergoes oxidation at the anode.

One of the most important fuel cells is the **hydrogen-oxygen fuel cell**. The cell uses graphite electrodes impregnated with an electrocatalyst, a metal/metal oxide combination (often nickel/nickel(II) oxide). This catalyses the breakdown of hydrogen and oxygen molecules to individual atoms. Both electrodes are porous. Hydrogen is bubbled into the anode and oxygen into the cathode. Under pressure, the gases diffuse separately through the electrodes into a warm potassium hydroxide electrolyte.

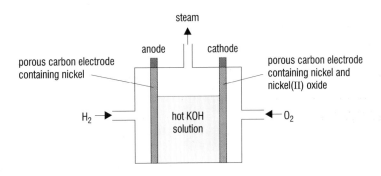

Fig 15.21   A hydrogen-oxygen fuel cell

In spacecraft, such as the Space Shuttle, electrical power is required for a wide variety of purposes. These include life-support systems, communications, control devices, environmental regulators, propulsion control and for scientific experiments. The electrical needs must be met by a combination of solar cells and dependable electrochemical cells. The relative importance of these two sources depends upon the orbit used by the spacecraft. Low Earth-orbit depends more heavily upon storage cells and fuel cells. The main fuel cell employed is the hydrogen-oxygen cell.

Fig 15.22   The hydrogen-oxygen fuel cell has worked well on many space missions where it acts as a reliable source of pollution-free on-board power and provides essential drinking water for astronauts. This photograph shows one of the three fuel cells that make up the generating system. Each fuel cell is capable of generating 12 kilowatts at peak and 7 kilowatts average power.

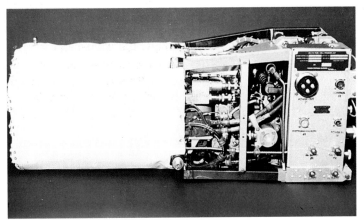

In a hydrogen-oxygen cell (Fig 15.21), electrons are removed from hydrogen and passed around an external circuit to oxygen. The overall reaction involves the formation of water.

Anode: $2H_2(g) + 4OH^-(aq) \longrightarrow 4H_2O(l) + 4e^-$; $E^{\ominus} = -0.83$ V

Cathode: $O_2(g) + 2H_2O(l) + 4e^- \longrightarrow 4OH^-(aq)$; $E^{\ominus} = +0.40$ V

Overall cell reaction:

$2H_2(g) + O_2(g) \longrightarrow 2H_2O(l)$

23 What is the voltage generated by the hydrogen-oxygen fuel cell operating under standard conditions?

24 Examine the operating conditions for the hydrogen-oxygen fuel cell described above. Give two reasons why the voltage you have calculated in the previous question is unlikely to be the actual cell e.m.f.

ELECTRIC EELS: MOBILE FUEL CELLS

The electric eel, *electrophorus electricus* is, in effect, a battery of specialised fuel cells contained within a living system. The eel's food is its fuel. Part of this fuel is oxidised in an 'electric organ', producing an electrical current. Each of the fuel cells making up this organ contributes about 1.5 V to the overall potential difference measured between the head (the cathode) and the tail (the anode) of the eel. Although electron carriers are involved at the cellular level, electrons are ultimately passed to oxygen. A one metre eel produces an average voltage of over 300 volts – larger than the UK mains supply!

Fig 15.23 The electric eel uses fuel cells to produce electricity.

Hydrocarbons and alcohols have been used successfully in fuel cells. When butane is fed into a suitable fuel cell the following half-reactions take place.

Anode: $C_4H_{10}(g) + 8H_2O(l) \longrightarrow 4CO_2(g) + 26H^+(aq) + 26e^-$

Cathode: $13O_2(g) + 52H^+(aq) + 52e^- \longrightarrow 26H_2O(l)$

Overall cell reaction:

$2C_4H_{10}(g) + 13O_2(g) \longrightarrow 8CO_2(g) + 10H_2O(l)$

The overall reaction taking place is identical to that which takes place when butane burns in air. The conversion of chemical energy into electrical energy is 70% efficient, an impressive value.

Fuel cells are efficient, pollution-free and silent in use but are not yet in use on a large scale for energy production despite continued research and development. A major difficulty lies in finding a supply of sufficiently pure fuels. Impure fuels poison the electrocatalysts without which the efficiency of a fuel cell is greatly reduced.

15.17 What would be the major design considerations for a camera battery suitable for use in very cold conditions?

15.18 Explain with the aid of a cell diagram how a lead-acid accumulator (a 'car battery') operates.

15.19 Hydrazine could be used in a fuel cell in alkaline conditions. The standard reduction potentials are given below. (Note both are written as reductions):

$$N_2(g) + 4H_2O(l) + 4e^- \longrightarrow N_2H_4(aq) + 4OH^-(aq); \quad E^\ominus = -1.15 \text{ V}$$
$$O_2(g) + 2H_2O(l) + 4e^- \longrightarrow 4OH^-(aq); \quad E^\ominus = +0.40 \text{ V}$$

(a) In which of these half-cells will reduction be the dominant process taking place?
(b) Calculate the cell e.m.f. for a cell established under standard conditions;
(c) Write a balanced equation for the overall cell reaction which would take place;
(d) Comment on the feasibility of the cell;
(e) If the cell is used, what substances must be supplied to it?
(f) Why could this cell be regarded as a 'clean cell'?

Answers to Quick Questions: Chapter 15

1 Because it contains about 20% by volume of oxygen.
2 $CuO(s) + H_2(g) \longrightarrow Cu(s) + H_2O(l)$
   Copper(II) oxide is reduced because oxygen has been removed. The hydrogen has been oxidised because oxygen has been added.
3 The blue colour is due to $Cu^{2+}(aq)$. As zinc displaces copper from solution, these ions are replaced by colourless $Zn^{2+}(aq)$.
4 For oxidation to occur (loss of electrons), another species must gain these electrons and so be reduced.
5 The position would move to favour the reduced species, the metal.
6 $Zn^{2+}(aq)$ in solution near the electrode and electrons on the surface of the electrode.
7 Because the higher the metal ion concentration, the greater the potential difference.
8 It has a much higher surface area.
9 $Cl_2(g) \mid Cl^-(aq) \mid Pt$
10 One might expect that low ionisation energies are accompanied by high standard electrode potentials – but it is not as simple as this!
11 $+ 0.46$ V
12 Reaction may occur, for example, precipitation.
13 Heart pacemaker, wristwatch.
14 $Fe(s) \mid Fe^{2+}(aq) \mid\mid Ag^+(aq) \mid Ag(s)$
15 It is an oxidising agent.
16 +3
17 $Zn(s) + 2MnO_2(s) \longrightarrow ZnO(s) + Mn_2O_3(s)$
18 Reduced.
19 It disproportionates to oxidation state of 0 in Pb and +4 in PbO_2.
20 **Anode:** $Zn(s) + 2OH^-(aq) \longrightarrow Zn(OH)_2(s) + 2e^-$
    Cathode: $O_2(g) + 2H_2O(l) + 4e^- + 4OH^-(aq)$
21 6.9 g
22 Because lithium has a very high standard electrode potential.
23 1.23 V
24 It does not operate at standard temperature or pressure.

Chapter 16

ELECTROCHEMISTRY IN USE

LEARNING OBJECTIVES

When you have studied this chapter you should be able to:

1. explain the order of elements in the electrochemical series and use the series to predict redox displacement reactions;

2. describe the principles of bromine and iodine extraction;

3. describe chemical reduction and electrolytic reduction as the general methods for extracting metals;

4. explain electrolysis and the terms electrolytic cell, electrolyte, anode and cathode;

5. write equations for redox reactions which occur during electrolysis;

6. predict the products of electrolysis of aqueous solutions of metal salts;

7. describe Faraday's laws of electrolysis and explain the principles that underpin them;

8. explain the size of the charge carried by electrons and ions, using the units coulombs and Faradays;

9. perform calculations involving amounts of substances in electrolysis;

10. describe the manufacture of chlorine and sodium hydroxide (part of the chloralkali industry), and the manufacture of fluorine;

11. describe the electrolytic processes used to extract aluminium and other reactive metals;

12. describe the use of electrolysis for purifying metals

16.1 THE ELECTRO-CHEMICAL SERIES

How do we decide whether an element is resistant to corrosion? How can we predict whether a metal is likely to be found in nature in an uncombined state or as a salt? A useful way of comparing elements is to examine the electrochemical series. Elements with large negative standard electrode potentials are placed at the top of the series. These elements oxidise and form positive ions readily and, for this reason, do not occur uncombined in nature. They are very reactive metals. Elements with large positive values are placed at the bottom of the series. These are either metals which do not oxidise readily and, therefore, are often found in nature in their metallic state or non-metals such as the halogens.

CONNECTION: Occurrence of elements in the Earth's crust, page 3.

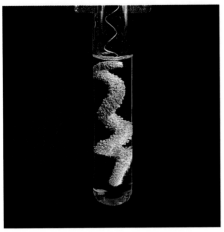

Fig 16.1 Silver crystals are growing on this piece of copper as silver is displaced from an aqueous solution of silver nitrate by the copper. We would predict this reaction, since the standard electrode potentials for copper and silver are + 0.34 V and + 0.80 V respectively.

Metal displacement reactions

If a copper wire is placed in aqueous silver nitrate, beautiful crystals of silver 'grow' on its surface (Fig 16.1). The copper is said to have displaced silver from a solution of silver ions. A redox reaction occurs in which electrons are transferred from copper to silver ions:

Oxidation half equation: $Cu(s) \longrightarrow Cu^{2+}(aq) + 2e^-$

Reduction half equation: $Ag^+(aq) + e^- \longrightarrow Ag(s)$

Overall equation: $Cu(s) + 2Ag^+(aq) \longrightarrow Cu^{2+}(aq) + 2Ag(s)$

Electron transfer occurs because silver ions have a greater affinity for electrons than do copper(II) ions, i.e. they are more easily reduced. This is reflected by their standard electrode potentials (which, remember, are reduction potentials):

$Ag^+(aq)/Ag(s)$; $E^{\ominus}(298) = + 0.80$ V

$Cu^{2+}(aq)/Cu(s)$; $E^{\ominus}(298) = + 0.34$ V

The electrochemical series allows us to predict which metals will displace other metals from their salts. An element with a more negative electrode potential will usually displace an element with a more positive electrode potential. The overall reaction is described as a **redox displacement reaction**.

1 **What would you observe if some zinc powder was added to a blue solution of copper(II) sulphate?**

KEY FACT

A metal will displace metals below it in the electrochemical series from solutions of their salts.

TUTORIAL

Problem: Predict what would happen if a zinc rod was dipped in each of the following solutions in turn, being left in each one for a few minutes before being removed, washed with water and placed in the next solution.

Solution 1: 1 mol dm^{-3} $Pb(NO_3)_2(aq)$

Solution 2: 1 mol dm^{-3} $Mg(NO_3)_2(aq)$

Solution 3: 1 mol dm^{-3} $Cu(NO_3)_2(aq)$

Standard electrode potentials given in Table 16.1.

Solution: When the zinc is placed in solution 1, zinc will displace lead from solution because zinc has the more negative standard electrode potential. Therefore, the rod will become coated with lead and zinc(II) ions will pass into solution:

$Zn(s) + Pb^{2+}(aq) \longrightarrow Zn^{2+}(aq) + Pb(s)$

When the lead coated rod is placed in solution 2, no reaction will occur because lead has a less negative standard electrode potential than does magnesium.

When the lead coated rod is placed in solution 3, lead will displace copper from solution because lead has the more negative standard electrode potential. Therefore, the rod will become coated with copper and lead(II) ions will pass into solution:

$Pb(s) + Cu^{2+}(aq) \longrightarrow Pb^{2+}(aq) + Cu(s)$

CONNECTION: A fuller list of electrode potentials is given in Datasheet 2.

Table 16.1 An electrochemical series of elements

Metals	Non-metals	Half-cell	E (298)/V
Potassium		$K^+(aq)/K(s)$	−2.92
Calcium		$Ca^{2+}(aq)/Ca(s)$	−2.87
Sodium		$Na^+(aq)/Na(s)$	−2.71
Magnesium		$Mg^{2+}(aq)/Mg(s)$	−2.38
Aluminum		$Al^{3+}(aq)/Al(s)$	−1.66
Zinc		$Zn^{2+}(aq)/Zn(s)$	−0.76
Iron		$Fe^{2+}(aq)/Fe(s)$	−0.44
Tin		$Sn^{2+}(aq)/Sn(s)$	−0.14
Lead		$Pb^{2+}(aq)/Pb(s)$	−0.13
	Hydrogen	$H^+(aq)/H_2(g)$	0.00
Copper		$Cu^{2+}(aq)/Cu(s)$	+0.34
	Iodine	$I_2(s)/I^-(aq)$	+0.54
Silver		$Ag^+(aq)/Ag(s)$	+0.80
	Bromine	$Br_2(l)/Br^-(aq)$	+1.07
	Chlorine	$Cl_2(g)/Cl^-(aq)$	+1.36
Gold		$Au^+(aq)/Au(s)$	+1.68

CONNECTION: Using the halogen displacement reactions in analysis, page 124

Halogen displacement reactions

Redox displacement reactions are used in the manufacture of bromine and iodine. Bromine is obtained from sea water, where it is present as bromide ions. Concentrations are low, however, and 20 000 tonnes of sea water produces just 1 tonne of bromine. The most economical sources of bromides are brine deposits such as the Arkansas brines, the Michigan brines and the Dead Sea. In these brines, bromide ions are present at around one thousand times their concentration in sea water.

Chlorine is used to oxidise bromide ions to bromine:

$$2Br^-(aq) + Cl_2(g) \longrightarrow Br_2(aq) + 2Cl^-(aq)$$

This redox reaction comprises two half reactions:

Oxidation half equation: $2Br^-(aq) \longrightarrow Br_2(aq) + 2e^-$
Reduction half equation: $Cl_2(g) + 2e^- \longrightarrow 2Cl^-(aq)$

As with metal displacement reactions, this may be predicted since the standard electrode potential for $Cl_2(g)/Cl^-(aq)$ (+ 1.36 V) is less negative than $Br_2(aq)/Br^-(aq)$ (+ 1.07 V).

Although bromine is not very soluble in water, at these low concentrations it does dissolve. To prevent hydrolysis of the bromine, the pH of the sea water is lowered to about 3.5 by adding acid before chlorination. This decreases the amount of bromine present as bromate(I) ions, which is necessary because bromine cannot be liberated from these. It also prevents hydrolysis of chlorine.

$$Br_2(aq) + H_2O(l) \rightleftharpoons 2H^+(aq) + Br^-(aq) + BrO^-(aq)$$
$$Cl_2(aq) + H_2O(l) \rightleftharpoons 2H^+(aq) + Cl^-(aq) + ClO^-(aq)$$

> **KEY FACT**
>
> A halogen will displace other halogens above it in the electrochemical series from solutions of their halide ions.

2 Why are these reactions described as disproportionations?

Bromine vapour is 'blown out' by a stream of air, and mixed with sulphur dioxide and water as a fine mist. This produces an aqueous solution which has 1500 times the concentration of bromide ions than in the original sea water.

$$Br_2(l) + 2H_2O(l) + SO_2(g) \longrightarrow 4H^+(aq) + 2Br^-(aq) + SO_4^{2-}(aq)$$

This solution is now fed into a tower against a counter-current of steam and chlorine. Oxidation of bromide ions by chlorine again occurs and the reaction product consists of two layers: an aqueous layer (hydrochloric acid) and a layer of impure bromine. The bromine is purified by distillation.

A similar process can be used to obtain iodine from brine deposits, although it also occurs in nitrate deposits in Chile as iodate(V), IO_3^-. The important redox reaction involved in extraction from the nitrate deposits is:

$$2IO_3^-(aq) + 5SO_2(g) + 4H_2O(l) \longrightarrow 2I^-(aq) + 8H^+(aq) + 5SO_4^{2-}(aq)$$

3 **What element is being reduced in this equation?**

Extracting metals from the Earth's crust

Few metals occur uncombined with other elements in nature. Only the very unreactive ones (those at the foot of the electrochemical series such as copper, silver and gold) occur in the free state. We call minerals which are uncombined elements native elements and it is likely that these were the first to be discovered. Far more commonly, metals exist as chemical compounds such as oxides, sulphides, carbonates, sulphates and halides.

4 **What is the oxidation number of a metal present in the Earth's crust as a native element?**

Metals invariably have positive oxidation numbers in their compounds. If the oxidation number is +1, +2 or +3, the compound is predominately ionic (the covalent character of bonds within the compound increases with increasing oxidation number of the metal). Therefore, metals must be extracted from their compounds by reduction. Two techniques are available:

- chemical reduction;
- electrolytic reduction.

Chemical reduction

The metal oxide is heated with an element which has a greater affinity for oxygen (is more easily oxidised). Usually carbon, silicon or aluminium are used. Carbon monoxide is also often used as the reducing agent, being oxidised to carbon dioxide in the reaction. The apparently simple nature of these chemical processes hides a great deal of complexity and sophistication in plant design and operation.

CONNECTION: Extraction of metals by chemical reduction, pages 546–548.

Iron is, perhaps, the most widely used metal in use today. In the extraction of iron, both carbon monoxide and carbon act as reducing agents in the furnace:

$$Fe_2O_3(s) + 3CO(g) \longrightarrow 2Fe(l) + 3CO_2(g)$$
$$Fe_2O_3(s) + 3C(s) \longrightarrow 2Fe(l) + 3CO(g)$$

The extraction of titanium, a metal central to the aerospace industry, presents more of a problem. Titanium(IV) oxide cannot be reduced by carbon, carbon monoxide or hydrogen. Even reduction with very reactive metals such as magnesium and aluminium is incomplete. The answer is to convert the oxide to titanium tetrachloride, $TiCl_4$, and reduce this to titanium. In the UK, sodium is used as the reducing agent but outside the UK magnesium is usually used.

$$TiCl_4(g) + 4Na(l) \longrightarrow Ti(s) + 4NaCl(s)$$

This reaction provides us with another example of the usefulness of oxidation numbers. Titanium tetrachloride is a covalent, molecular compound. Reduction is not a simple electron transfer process, although it is clear that sodium loses electrons to form sodium ions. It is not possible, therefore, to write oxidation and reduction half equations. Instead, it is more useful to describe the reaction as a redox in which the changes in oxidation numbers are:

Reactants: $TiCl_4(g)$: $Ox(Ti) = +4$; $Ox(Cl) = -1$. $Na(l)$: $Ox(Na) = 0$.
Products: $Ti(s)$: $Ox(Ti) = 0$. $NaCl(s)$: $Ox(Na) = +1$; $Ox(Cl) = -1$.

Sodium has been oxidised (change in oxidation number from 0 to +1) and titanium has been reduced (change in oxidation number from +4 to 0).

5 Write an equation for the reduction of titanium tetrachloride with magnesium and state the change in oxidation number of magnesium.

Electrolytic reduction

Electricity is passed through a molten salt and provides electrons for the reduction of metal ions at the cathode:

$$M^{x+}(l) + xe^- \longrightarrow M(l)$$

The method is expensive because of the high temperatures needed to melt the salt and keep it liquid, and the demand for electricity (an expensive form of energy). For this reason, its use is restricted to the extraction of the more reactive metals (those at the top of the electrochemical series), such as, aluminium and sodium. For these elements, electrolytic production is the only practical method. However, electrolysis is a common purification technique for many metals.

CONNECTION: Extraction of metals by electrolysis, page 417.

QUESTIONS

16.1 'Tin cans' are steel cans lined with a tin plating. In an intact 'tin' this provides excellent protection against corrosion but rusting of the steel is rapid once the protective surface has been broken. A zinc coating on steel (as used on a galvanised bucket) provides resistance to corrosion even after the surface has been broken. Explain these findings in the light of the standard reduction potentials:

$$Zn^{2+}(aq) + 2e^- \longrightarrow Zn(s); E^{\ominus}(298) = -0.76 \text{ V}$$
$$Fe^{2+}(aq) + 2e^- \longrightarrow Fe(s); E^{\ominus}(298) = -0.44 \text{ V}$$
$$Sn^{2+}(aq) + 2e^- \longrightarrow Sn(s); E^{\ominus}(298) = -0.14 \text{ V}$$

16.2 On the basis of your reply to question 16.1, why do you think that zinc is not used to line food cans?

16.3 When sodium is added to a solution of calcium nitrate a strong reaction begins in which hydrogen is released in large quantity. After the reaction, careful examination reveals that no calcium metal has been produced. Use the standard electrode potentials in the Database to explain why this is the case.

16.4 In terms of the migration of ions, explain what would take place if two inert electrodes are dipped into a solution containing a mixture of sodium sulphate and potassium nitrate and connected to an external power supply.

16.5 **(a)** Describe, with the aid of an equation, what happens when copper is added to a solution of silver nitrate.
(b) What would you observe during this slow reaction?

ELECTROCHEMISTRY IN USE

16.6   As part of an analytical scheme for bromide and iodide ions, chlorine gas may be bubbled into a solution of the unknown halide. A water-immiscible organic solvent such as 1,1,1-trichloroethane is usually added. If bromide ions are present in the sample, a red-brown organic layer is observed. If iodide ions are present, the organic layer often turns a strong violet colour.

(a) Explain the basis of this test.

(b) Write an ionic equation for the reaction which takes place when chlorine is bubbled into a solution of sodium iodide.

(c) Show how the reaction in (b) could be predicted using the standard electrode potentials shown in the Database.

16.2 ELECTROLYSIS

Electricity at work

Electricity is a flow of electrons which can be used to perform useful work. In our homes, electricity is available at the flick of a switch. Chemical industries use this convenient form of energy for lighting, heating and running machinery. However, they also use it directly in some processes to bring about chemical change.

In our search to obtain useful materials from minerals present in the Earth's crust, electrical energy has played a major part. Faraday postulated his laws of electrolysis in 1834 and this was followed by numerous patents being taken out on the electrolysis of salts. By the 1890s, the many practical problems were beginning to be solved and the electrochemical industry was born.

The major electrochemical industries today are:

* aluminium manufacture and the extraction of other reactive metals;
* sodium hydroxide and chlorine manufacture (part of the chloralkali industry);
* fluorine production.

All use electrical energy to convert raw materials into the desired products. However, electrolysis is very energy-consuming and, therefore, costly. Aluminium production makes the heaviest demands on electrical energy amongst the electrochemical industries and so aluminium is expensive even though it is the most abundant metal in the Earth's crust. The chloralkali industry has the next heaviest demand for electricity.

CONNECTION: Chloralkali and other chemical industries, page 531–536.

The electrolytic cell

When electrodes are connected to opposite poles of an electrical supply, a positively-charged electrode and a negatively-charged electrode are established. Such a combination of electrodes, together with an **electrolyte**, constitutes an **electrolytic cell**. This arrangement can be used successfully on a small scale to produce chemical changes in the laboratory (Fig 16.2). Electrochemical industries use the same principle on a massive scale. The change brought about by electrical energy in an electrolytic cell is known as **electrolysis**.

In addition to its use for manufacturing, electrolysis is used to purify some metals, such as copper and zinc. High levels of purity can be achieved in this way. Electrolysis is also used in such diverse areas as forensic science, medicine and waste disposal.

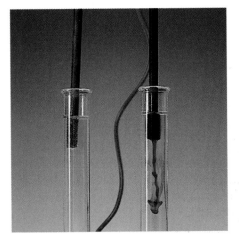

Fig 16.2   The yellow-brown colour around the positive electrode is due to iodine being formed during the electrolysis of aqueous potassium iodide.

Explaining electrolysis

Electrolysis concerns the processes which occur in an electrolyte when an electrical current passes through it. The electrolyte may be a melt (a molten salt) or a solution containing a salt. These processes are:

- the migration of ions towards electrodes placed in the electrolyte;
- the redox reactions which occur at the electrodes (cathode and anode).

In an electrolytic cell, an electrical supply creates a cathode and an anode. Two electrodes placed in an electrolyte connected to an electrical supply become charged according to the polarity of the power source (for example, a battery or a power pack). The **cathode** is the electrode attached to the negative pole of the power source and may be considered as the 'electron source'. The **anode** is attached to the positive pole of the power source and acts as an 'electron remover'. The circuit is completed by migration of ions through the electrolyte, each ion migrating towards the electrode with the opposite charge to itself:

- cations (positively charged ions) migrate towards the cathode;
- anions (negatively charged ions) migrate towards the anode.

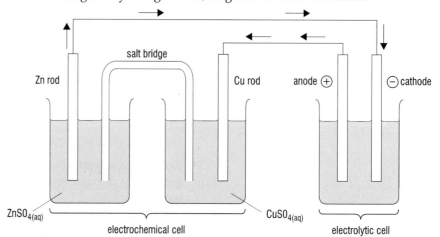

Fig 16.3   Electrical energy produced in an electrochemical cell can be used to bring about chemical change in an electrolytic cell. If insufficient energy is available from one electrochemical cell, two or more may be used in series. The voltage is the sum of the e.m.f.s of all the cells in series.

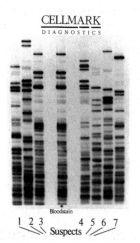

Fig 16.4   'DNA fingerprinting' is now performed routinely in forensic science as it can prove that two samples are identical. It is a reliable technique and a 'positive result' is accepted as strong evidence. This photograph shows suspect 3 matches the bloodstain.

ELECTROPHORESIS AND FORENSIC SCIENCE

Electrophoresis is one of the most powerful tools of the forensic scientist. It involves the migration of charged particles through a solution or a gel in an electric field. The gel is a stationary phase and has a pH gradient (a gradually changing pH level). The charged particles move towards the oppositely charged electrode until they reach a pH level from which they no longer migrate.

Electrophoresis is increasingly used to compare samples of DNA in body fluids. This is useful in crime detection. Electrophoresis of a sample of blood left at the scene of a crime and a sample from the suspect can be run simultaneously. The gel is treated with a stain to show up the DNA samples. If two samples perform identically under electrophoresis (show up in the same positions) then they come from the same source – the suspect is guilty. The technique has become known as **DNA fingerprinting** since, like fingerprints, no two human beings (or animals) have the same DNA sequence. Two individuals who are related may have a similar (but not identical) DNA sequence. In the United States electrophoresis has been used in paternity suits to establish the identity of the father of a child.

An **electrolytic cell** contains two electrodes. Ideally the electrodes should neither react with the electrolyte nor undergo redox reactions within the cell. Such electrodes are referred to as inert electrodes.

An **electrolyte** is a substance which conducts electricity and is simultaneously decomposed by it. The process of decomposition is called **electrolysis**.

Electrons enter the electrolyte through the **cathode** and this is the site of reduction during electrolysis. The **anode** accepts electrons from anions and is, therefore, the site of oxidation.

Cations are reduced at the cathode as they gain electrons:

$$M^{x+} + xe^- \longrightarrow M$$

Anions are oxidised at the anode as they lose electrons:

$$A^{y-} \longrightarrow A + ye^-$$

We will now consider the electrolysis of molten lead(II) bromide, $PbBr_2$ in a laboratory situation. Lead(II) bromide is a white, crystalline ionic solid with a relatively low melting point. An electrical circuit is set up with a battery, an ammeter, a small light bulb and two graphite electrodes (Fig 16.5). When solid lead(II) bromide is placed in a crucible and two graphite electrodes inserted into it, no current passes. However, as the lead(II) bromide is heated strongly and melts, the bulb lights and the ammeter shows a steady reading. The circuit has been completed.

6 **Why was the circuit incomplete before the lead(II) bromide was melted?**

The reactions taking place during the electrolysis of molten lead(II) bromide are:

Anode: Bromide ions are *oxidised* to bromine

$$2Br^-(l) \longrightarrow Br_2(g) + 2e^-$$

Cathode: Lead ions are *reduced* to lead metal

$$Pb^{2+}(l) + 2e^- \longrightarrow Pb(l)$$

Overall reaction: $PbBr_2(l) \longrightarrow Pb(l) + Br_2(g)$

Predicting the products of electrolysis

The chemical changes which take place during electrolysis depend upon the relative ease of reduction and oxidation of the ions concerned. The ions which undergo chemical change at the electrodes are said to be **discharged** and the products of electrolysis may be solids, liquids or gases depending upon the temperature of the electrolyte.

Predicting the products of electrolysis of a melt is straightforward since, usually, only one cation and one anion is present. For example, electrolysis of a sodium chloride melt gives sodium metal at the cathode and chlorine at the anode. Sodium ions are positively charged and migrate to the cathode where they are reduced. Chloride ions move to the anode, where oxidation gives chlorine molecules:

Anode: Chloride ions are *oxidised* to chlorine

$$2Cl^-(l) \longrightarrow Cl_2(g) + 2e^-$$

Cathode: Sodium ions are *reduced* to sodium

$$Na^+(l) + e^- \longrightarrow Na(l)$$

Overall reaction: $2NaCl(l) \longrightarrow 2Na(l) + Cl_2(g)$

7 **Write electrode reactions for the electrolysis of molten lithium iodide.**

Electrolysis of aqueous solutions using inert electrodes

Electrolysis of aqueous solutions is more complicated than electrolysis of a melt. There are often several different ions in solution. Each will migrate towards the oppositely-charged electrode and so two or more ions may accumulate at both cathode and anode (Fig 16.6). In most cases, however, only one product is obtained at each electrode. How can we predict which ions will be transformed during electrolysis?

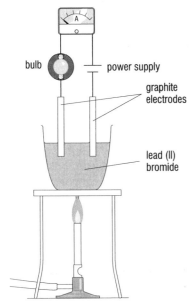

Fig 16.5 Electrolysis of lead(II) bromide. No current is registered by the ammeter until the lead(II) bromide begins to melt. At the anode, bromine gas bubbles off and at the cathode a bead of molten lead forms.

bulb

power supply

graphite electrodes

lead (II) bromide

Fig 16.6 During electrolysis of aqueous solutions in which several different ions may be present, each ion will migrate to the electrode with the opposite charge to itself. Usually, one type of ion only is reduced at the cathode whilst only one type of ion is oxidised at the anode.

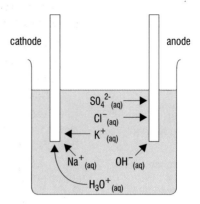

CONNECTION: Electrochemical series of elements, page 404.

At the cathode, the ease of discharge of ions depends on their position in the electrochemical series:

$K^+(aq)$ $Ca^{2+}(aq)$ $Na^+(aq)$ $Mg^{2+}(aq)$ $Al^{3+}(aq)$ $Zn^{2+}(aq)$ $Fe^{2+}(aq)$ $Sn^{2+}(aq)$ $Pb^{2+}(aq)$ $H^+(aq)$ $Cu^{2+}(aq)$ $Ag^+(aq)$

——————————— increasing ease of discharge ————————————————→

This is not surprising, perhaps, since we would expect the cation which is most easily reduced to be discharged first. However, two other factors make prediction more difficult:

- If a cation is in very high concentration it may be discharged in preference to one below it in the electrochemical series but present in lower concentrations. This is because the ease of discharge depends on the electrode potential, and this in turn depends on concentration. Standard electrode potentials are measured for 1 mol dm^{-3} solutions.
- Hydrogen is discharged as predicted if shiny platinum electrodes are used. However, it is less easily discharged if other electrodes, such as lead, aluminium or mercury, are used and this can result in a metal above hydrogen in the electrochemical series being discharged preferentially. This is called the **overvoltage effect** since a higher potential difference is needed to discharge the hydrogen.

Hydrogen is discharged from neutral solutions of salts of those metals which come above it in the electrochemical series. However, the mechanism of this process is not fully understood.

Water undergoes autoionisation:

$$2H_2O(l) \rightleftharpoons H_3O^+(aq) + OH^-(aq)$$

and even though the concentration of oxonium ions is only about 1×10^{-7} mol dm^{-3}, it is usually suggested that hydrogen is formed according to the electrode reaction:

$$2H_3O^+(aq) + 2e^- \longrightarrow H_2(g) + 2H_2O(l)$$

However, it has also been suggested that hydrogen is produced directly from water molecules:

$$2H_2O(l) + 2e^- \longrightarrow H_2(g) + 2OH^-(aq)$$

At the anode, we usually observe that either oxygen is evolved (when the dissolved salt is a nitrate or sulphate) or a halogen is discharged when the dissolved salt is a halide. From experiment, we find:

$SO_4^{2-}(aq)$ $NO_3^-(aq)$ $Cl^-(aq)$ $OH^-(aq)$ $Br^-(aq)$ $I^-(aq)$

————————— increasing ease of discharge ——————→

> **KEY FACT**
>
> $H^+(aq)$ and $H_3O^+(aq)$ are both used to represent the hydrated hydrogen ion, also called an oxonium ion.

CONNECTION: Autoionisation of water, page 472.

As with reactions at the cathode, other factors are important and make predictions more difficult:

- If an anion is in very high concentration it may be discharged in preference to one which would normally be discharged first if their concentrations were similar. For example, the electrolysis of aqueous sodium chloride gives a mixture of chlorine and oxygen and the proportion of oxygen increases with more dilute solutions.
- Oxygen is less easily liberated than might be expected because of an overvoltage effect (similar to the effect described above for hydrogen).

In the same way that the electrode reactions leading to the discharge of hydrogen are not fully understood, there are conflicting views about the evolution of oxygen at the anode. The most popular explanation is:

$$4OH^-(aq) \longrightarrow 2H_2O(l) + O_2(g) + 4e^-$$

but direct formation from water has also been suggested:

$$6H_2O(l) \longrightarrow 4H_3O^+(aq) + O_2(g) + 4e^-$$

Consider the electrolysis of aqueous silver nitrate, $AgNO_3(aq)$. Four aqueous ions are present:

	Cations	Anions
From the silver nitrate	$Ag^+(aq)$	$NO_3^-(aq)$
From the autoionisation of water	$H_3O^+(aq)$	$OH^-(aq)$

The electrode reactions are:

Cathode: Silver ions are reduced to silver (since silver is below hydrogen in the electrochemical series)

$$Ag^+(aq) + e^- \longrightarrow Ag(s)$$

Anode: Hydroxide ions are oxidised to water and oxygen (since hydroxide ions are more easily discharged than nitrate ions)

$$4OH^-(aq) \longrightarrow 2H_2O(l) + O_2(g) + 4e^-$$

Overall reaction: The products of electrolysis of aqueous silver nitrate are silver (at the cathode) and oxygen gas (at the anode).

8 **Explain why the silver nitrate solution becomes acidic during electrolysis.**

Electrolysis of aqueous solutions using reactive electrodes

The material from which the electrode is made is important during electrolysis. If the electrodes are to play no part in the electrolytic processes, then a very unreactive electrode such as graphite or an inert electrode like platinum is chosen. However, sometimes it is essential that the electrodes do take part in the redox reactions. For example, in chromium, nickel or copper electroplating or in the purification of copper. In some cases both electrodes may participate in electrolysis.

Consider, for example, the electrolysis of copper sulphate solution using (a) copper electrodes and (b) platinum electrodes.

When copper electrodes are used copper ions are reduced at the copper cathode, increasing its weight. The copper anode dissolves and gradually the mass of the anode decreases:

KEY FACT

In the electrolysis of aqueous solutions, using inert electrodes:

At the anode: Oxygen is evolved from aqueous solutions of hydroxides, sulphates and nitrates. Halogens are discharged from aqueous solutions of halide ions (though, if their concentrations are low, oxygen is also discharged).

At the cathode: Metals are deposited if they come below hydrogen in the electrochemical series, otherwise hydrogen is evolved.

ELECTROLYSIS AND THE SWIMMING POOL

Electrolysis is used to generate chlorine to sterilise swimming pools 'on site'. If common salt or other source of chloride ions is present in the swimming pool electrolysis can produce chlorine when required. The chlorine is produced at the anode:

$$2Cl^-(aq) \longrightarrow Cl_2(g) + 2e^-$$

The chorine produced first dissolves then quickly reacts with water, forming hydrochloric acid and chloric(I) acid (hypochlorous acid). An equilibrium mixture is obtained:

$$Cl_2(aq) + H_2O(l) \rightleftharpoons HCl(aq) + HOCl(aq)$$

It is the chloric(I) acid which is active against bacteria, penetrating the bacterial cell wall. The bacteria are killed due to its strongly oxidising properties.

As an alternative to electrolysis of sodium chloride, solutions of liquid bleach, sodium chlorate(I), may be added to the pool. This substance is produced 'off-site' by reaction of chlorine with sodium hydroxide, themselves both products of electrolysis:

$$Cl_2 + NaOH(aq) \longrightarrow NaCl(aq) + NaOCl(aq)$$

Fig 16.7 The chloralkali industry produces chlorine by electrolysis on a massive scale.

Anode: Copper is oxidised to copper ions

$$Cu(s) \longrightarrow Cu^{2+}(aq) + 2e^-$$

Cathode: Copper ions are reduced to copper

$$Cu^{2+}(aq) + 2e^- \longrightarrow Cu(s)$$

With inert electrodes such as platinum electrodes the platinum cathode becomes copper plated (we say that it has been **electroplated**) whilst at the anode oxygen gas is observed. Obviously the choice of electrodes depends upon the purpose of the electrolytic process.

9 **Write an equation for the reactions which occur during the electrolysis of aqueous copper(II) sulphate when platinum electrodes are used.**

Fig 16.8 Electrolysis of copper(II) sulphate solution. The larger (stainless steel) electrode is the cathode and becomes electroplated. The smaller cathode is made of copper and is gradually eroded. This simple technique is the basis of the electroplating industries.

ELECTROCHEMISTRY IN USE

ELECTROPLATING

Whether for protection or decoration, electroplating is an important electrolytic process which covers (or **plates**) an object with a thin layer of a metal. Of course, the object itself must be a conductor of electricity. The object is made the cathode in an electrolytic cell containing ions of the plating metal. These ions migrate to the cathode where they are reduced to the metal and deposited on the surface of the cathode. The depth of the layer can be increased by:

- increasing the concentration of the electrolyte;
- raising the electrolysis current;
- exposing the object to the electrolyte for a longer period.

Electroplating is carried out either to protect the base metal against corrosion or simply for decorative purposes. Copper, nickel and chromium plating are all common industrial processes.

Fig 16.9 Chromium plating is used for decorative as well as protective purposes.

QUESTIONS

16.7 Explain carefully, with the aid of suitable equations, the electrode processes taking place at *(i)* the anode, and *(ii)* the cathode, when a lead(II) bromide melt undergoes electrolysis.

16.8 If an aqueous solution of sodium sulphate is electrolysed:
   (a) describe what you would see at each electrode;
   (b) identify and represent, with half-equations, the reduction and oxidation processes taking place;
   (c) explain what would happen if the electrolytic cell were left to run for a very long time at room temperature.

16.9 Electrolysis of dilute sodium chloride produces mainly oxygen at the anode. This oxygen is not, however, suitable for use for medical purposes.
   (a) Explain why this is so.
   (b) How is oxygen normally obtained in large quantities?

16.10 *'When a melt undergoes electrolysis it is usually very easy to predict the products at cathode and anode. Electrolysis of aqueous solutions is more complicated.'* Explain what is meant by this statement.

16.11 Electrolysis of dilute sulphuric acid produces only one product at the anode and one at the cathode. Electrolysis of dilute hydrochloric acid produces one product at the cathode and two at the anode. Using suitable equations, identify the electrolysis products for both of these acids.

16.3 ELECTROLYSIS AND AMOUNT OF SUBSTANCE

Faraday's laws of electrolysis

If we are to harness the power of electricity and use it to bring about chemical change, we need to know how much change we can produce with a given amount of electricity. This important relationship was studied by Michael Faraday, an English chemist and physicist who made major contributions to the study of electrolysis. Faraday was also the first to introduce the terms anode, cathode, cation, anion, electrode and electrolyte.

Faraday's first law

Faraday's **first law**: the mass of a substance liberated at an electrode during electrolysis is proportional to the amount of electricity passed during its formation.

During electrolysis, oxidation at the anode requires that an anion gives up electrons. Reduction at the cathode requires that a cation gains electrons. The greater the number of electrons removed at the anode or passed on at the cathode, the greater the number of ions oxidised or reduced and the greater the mass of products formed. In hindsight, the truth of Faraday's first law seems obvious. At the time, his model of electrolysis was an inspired leap forward.

Faraday's second law

Faraday's **second law**: when an equal amount of electricity passes through different electrolytes, equivalent quantities of product are formed at each electrode.

The meaning of Faraday's second law is less obvious. To help explain the use of the word **equivalent** let us compare electrolysis of two aqueous solutions, one containing silver nitrate and the other copper(II) sulphate. Table 16.2 shows the amounts of each product formed during the electrolysis of these two solutions when equal amounts of electricity (the same number of moles of electrons) are passed:

Table 16.2

	Silver nitrate solution	Copper(II) sulphate solution
Ions present	$Ag^+(aq)$	$Cu^{2+}(aq)$
	$H_3O^+(aq)*$	$H_3O^+(aq)*$
	$NO_3^-(aq)$	$SO_4^{2-}(aq)$
	$OH^-(aq)*$	$OH^-(aq)*$
	*$H_3O^+(aq)$ and $OH^-(aq)$ only present at very low concentrations, about 1×10^{-7} mol dm^{-3}	
Cathode product	1.1 g silver	0.32 g copper
Anode product	57 cm^3 oxygen and 0.091 g water	57 cm^3 oxygen and 0.091 g water

10 Why is silver the cathode product during electrolysis of silver nitrate solution?

However, we are interested in amounts of substances rather than their masses. In the experiment described, we can calculate the amounts of silver and copper produced:

- 1.1 g silver produced. Since silver has molar mass 108 g mol^{-1}, 0.1 mol silver atoms produced;
- 0.32 g copper produced. Since copper has molar mass 63.5 g mol^{-1}, 0.05 mol copper atoms produced.

To explain these amounts we must examine the half-reactions taking place at the cathode in each case. During the electrolysis of silver nitrate solution, every silver ion which is reduced receives one electron. The reduction of one mole of silver ions would require one mole of electrons to be passed around the circuit:

$$Ag^+(aq) + e^- \longrightarrow Ag(s)$$

Electrolysis of aqueous copper(II) sulphate requires two electrons to be passed around the circuit for each copper(II) ion reduced:

$$Cu^{2+}(aq) + 2e^- \longrightarrow Cu(s)$$

ELECTROCHEMISTRY IN USE

The same amount of electricity will only produce half as many moles of copper compared to silver. The amount of copper and silver produced are equivalent in the sense that they are obtained by passage of the same amount of electricity (Faraday's second law). We can now give a modern version of Faraday's second law:

To liberate one mole of ions carrying a charge of either $x+$ or $x-$, x moles of electrons are required.

In each case the same volume of oxygen and the same mass of water are produced. This is because the same reaction takes place at the anode in both cases:

$$4OH^-(aq) \longrightarrow 2H_2O(l) + O_2(g) + 4e^-$$

The transfer of charge during electrolysis

An electrical current is a flow of electrons due to the migration of electrons from a region of high electron density to one of lower electron density. The total charge transferred by an electrical current is the sum of the charges on all the electrons which are passed around the circuit.

During studies of electrolysis a convenient unit of charge known as a Faraday is often used to express the amount of electrical charge which is transferred around an electrolytic circuit. One Faraday is the charge carried by one mole of electrons or singly charged ions and has a value of 96 500 coulombs per mole (96 500 C mol^{-1}). It is given the symbol F. The relationship between Avogadro's constant, L, the charge on the electron, e, and the Faraday is an important one for the study of electrolysis and may be conveniently expressed as:

1 Faraday = Avogadro's constant × the charge on 1 electron, or $F = Le$

$$96\,500 \text{ C mol}^{-1} = 6.022 \times 10^{23} \text{mol}^{-1} \times 1.60 \times 10^{-19}\text{C}$$

Predicting the quantity of products

The relationship between the amount of substance liberated at an electrode and the number of electrons passed around the circuit can be used to calculate the expected mass or volume of substance produced at the cathode and anode. Similarly, it is possible to determine the amount of electricity required to produce a particular mass or volume of product.

For example, suppose we want to deposit 2.54 g copper onto a steel plate. The molar mass of copper = 63.5 g mol^{-1},

therefore, 2.54 g copper contains $\frac{2.54}{63.5}$ moles of copper atoms = 0.04 mol

If the steel plate is the cathode, the reduction which occurs is:

$$Cu^{2+}(aq) + 2e^- \longrightarrow Cu(s)$$

From the equation, one mole of Cu requires two moles of electrons, that is, two Faradays of electricity.

Therefore, 0.04 mole of Cu require 0.080 F or $0.08 \times 96\,500$ C = 7720 C.

We can achieve this by an appropriate combination of time and current. Any of the following would be suitable:

Time/s	Current/A	Time/s	Current/A
10 000	0.772	100	77.2
1000	7.72	10	772

11 How long would you need to pass a current of 500 A in order to deposit 2.54 g copper?

TUTORIAL

Problem: During an electrolysis of molten lead(II) bromide, lead formed at the cathode and bromine at the anode. Calculate the mass of lead deposited and the volume of bromine vapour produced in 15 minutes if a current of 1.5 amps was passed through molten lead(II) bromide. Assume that the volume of bromine was measured at s.t.p.

Molar masses: $Pb = 207$ g mol^{-1}; $Br = 79.9$ g mol^{-1}.
One mole of any gas at s.t.p. occupies 22.4 dm^3.

Solution: A current of 1.5 amps passed for 15 minutes would transfer a charge of $1.5 \times 15 \times 60 = 1350$ coulombs.

This is carried by $\frac{1350}{96\,500} = 0.014$ moles of electrons.

$$Pb^{2+}(aq) + 2e^- \longrightarrow Pb(s)$$

From the equation, two moles of electrons deposit one mole of lead.

Therefore, 0.014 mole of electrons deposit 0.007 mole lead.

Since one mole of lead has a mass of 207 g, 0.007 mole of lead has a mass of 207×0.007 g lead = 1.449 g

$$2Br^-(aq) \longrightarrow Br_2(g) + 2e^-$$

From the equation, two moles of electrons are released when one mole of bromine vapour, $Br_2(g)$, is produced.

Therefore, 0.014 mole of electrons is released when 0.007 mole of bromine molecules form.

Since one mole of bromine molecules occupy 22.4 dm^3, 0.007 mole bromine molecules occupy 22.4×0.007 $dm^3 = 0.1568$ dm^3

Therefore, a mass of 1.45 g of lead and a volume of 157 cm^3 of bromine vapour will have been produced.

QUESTIONS

16.12 In terms of the migration of ions explain what would take place if two inert electrodes are dipped into a solution containing a mixture of sodium sulphate and potassium nitrate and connected to an external power supply.

16.13 Explain carefully with the aid of suitable equations the electrode processes taking place at: **(a)** the anode and **(b)** the cathode when molten potassium iodide is electrolysed.

16.14 To copper plate an electrical switch it was made the cathode in an electrolytic cell containing aqueous copper(II) sulphate as electrolyte. The anode consisted of a piece of copper metal.
(a) Calculate the increase in the mass of the switch if a current of two amps was maintained for 15 minutes.
(b) What product would be expected at the anode during the electrolysis of this solution if the anode had been made of platinum?
(c) Why would platinum be a less suitable choice for the anode?

ELECTROCHEMISTRY IN USE

(d) Why was the switch made the cathode and not the anode during the electrolysis?

A_r [Cu] = 63.5; 1 Faraday = 96 500 C mol^{-1}.

16.15 (a) State Faraday's first and second laws.
   (b) Explain how the laws may be used to predict the quantities of products formed at cathode and anode.

16.16 (a) A current of 0.04 amps is passed through a solution of silver nitrate, $AgNO_3$, for one hour. What mass of silver will be deposited at the cathode?
   (b) Why is the amount of silver deposited independent of the concentration of silver nitrate solution?

A_r[Ag] = 107.87; 1 Faraday = 96 500 C mol^{-1}.

16.17 How long must a current of 0.05 amps be passed through a solution of sodium chloride to produce 1 dm^3 of chlorine gas, the volume being measured at room temperature and $1.01\,325 \times 10^5$ N m^{-2}?
[One mole of any gas occupies about 24 dm^3 at room temperature and at $1.01\,325 \times 10^5$ N m^{-2}]

16.18 Two electrolytic cells are connected in series to a power supply. The first cell contains dilute sulphuric acid. After a period of time, 50 cm^3 of oxygen are collected at the anode at room temperature and at $1.01\,325 \times 10^5$ N m^{-2} pressure. What mass of silver would be accumulated at the cathode of the second cell, containing silver nitrate solution, in the same period of time?

A_r[Ag] = 107.87; 1 Faraday = 96 500 C mol^{-1}.

16.4 ELECTROLYTIC PROCESSES IN INDUSTRY

Metals at the top of the electrochemical series

Aluminium

The world production of this important element is estimated at 15 million tonnes (272 thousand tonnes in the UK). It is obtained from bauxite (an impure form of aluminium oxide).

The first stage of the process is to convert bauxite to pure aluminium oxide, Al_2O_3. The main impurities are Fe_2O_3 (3–25%), SiO_2 (1–7%) and TiO_2 (2–3%). The bauxite is heated at about 150°C, under a pressure of 4 atmospheres, with 2.5 mol dm^{-3} sodium hydroxide solution for about 1–2 hours. Al_2O_3 dissolves (remember the amphoteric nature of this oxide) but the others remain undissolved. These are allowed to settle and are filtered off.

12 **Write an equation for Al_2O_3 dissolving in NaOH(aq).**

Purified Al_2O_3 is obtained by seeding the solution with pure Al_2O_3 in huge precipitator tanks, over 24 m high and with a capacity of 10^3 m^3. It is dried in rotary kilns at about 1000°C.

$$2[Al(OH)_4]^-(aq) \longrightarrow Al_2O_3(s) + 2OH^-(aq) + 3H_2O(l)$$

The melting point of Al_2O_3 is too high to make electrolysis of the molten oxide a viable proposition. Instead, Al_2O_3 is dissolved in molten sodium hexafluoroaluminate(III) (also called cryolite), with a little calcium fluoride added to lower the melting point. The electrolyte is kept at 1000°C.

13 **Write the formula for sodium hexafluoroaluminate(III), and write an equation for its synthesis from Al_2O_3, HF and NaOH.**

In a continuous process, Al_2O_3 is continually added to the melt as molten aluminium is tapped off (Fig 16.10). A concentration of 5% dissolved Al_2O_3 in the electrolyte is maintained. Electrode reactions:

Cathode: $Al^{3+}(l) + 3e^- \longrightarrow Al(l)$ reduction
Anode: $2O^{2-}(l) \longrightarrow O_2(g) + 4e^-$ oxidation

The carbon anodes are attacked by the oxygen which forms, being oxidised to carbon monoxide and carbon dioxide. They are expendable.

These equations are used to summarise a more complex series of reactions. The exact electrode reactions are not known for certain.

14 **Write an overall equation for the electrolytic decomposition of Al_2O_3.**

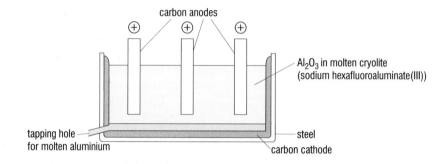

Fig 16.10 Aluminium is produced by electrolysis of a solution of Al_2O_3 in sodium hexafluoroaluminate(III).

carbon anodes

Al_2O_3 in molten cryolite (sodium hexafluoroaluminate(III))

tapping hole for molten aluminium

steel
carbon cathode

ANODISATION

Aluminium may be protected against corrosion by **anodisation**. This process is commonly used with aluminium sheet and involves making the aluminium the anode in an electrolytic process in which the electrolyte is usually sulphuric acid. During the process aluminium is oxidised and the metal surface acquires a deeper 'oxide coat'. This gives increased protection against corrosion. Anodising aluminium in this way also allows the hydrated and porous oxide coat to accept dyes more easily for decorative purposes.

Fig 16.11 Aluminium has a wide range of applications. Here it can be seen used for yacht masts. It is easily worked, has a low density and can be recycled cheaply. The surface oxide layer protects it from corrosion and this layer can be thickened by anodisation.

Other reactive metals

Sodium is a reactive metal and is difficult to extract from its salts by any method other than electrolysis. Molten sodium chloride is electrolysed in a ceramic-lined steel cell, operating at 600°C with graphite anodes and a steel cathode. A voltage of about 7 V is used. Other chlorides, such as calcium chloride, are added to lower the melting point of the electrolyte.

Electrode reactions:

Anode: $2Cl^-(l) \longrightarrow Cl_2(g) + 2e^-$; **Cathode:** $Na^+(l) + e^- \longrightarrow Na(l)$

Magnesium and calcium are reactive metals and like sodium are difficult to extract from their salts. They are obtained by electrolytic decomposition of their molten chlorides. The anode reactions are the same as for molten sodium chloride and the cathode reactions are:

$Ca^{2+}(l) + 2e^- \longrightarrow Ca(l)$
$Mg^{2+}(l) + 2e^- \longrightarrow Mg(l)$

CONNECTION: *Sodium carbonate manufacture, page 533.*

Chlorine and sodium hydroxide

The chloralkali industry is a major branch of the chemical industry and is concerned with the manufacture of sodium hydroxide, chlorine and sodium carbonate. World production is estimated at 33 million tonnes of chlorine (1.5 million in the UK) and about 31 million tonnes of sodium hydroxide (1.2 million in the UK).

Electrolysis is used to obtain chlorine and sodium hydroxide from the readily obtainable and cheap raw materials – sodium chloride and water. Although three different types of electrolysis cells are used, the principle is the same for all processes: an aqueous solution of sodium chloride is electrolysed and at the electrodes

Cathode: $2H_3O^+(aq) + 2e^- \longrightarrow H_2(g) + 2H_2O(l)$

Anode: $2Cl^-(aq) \longrightarrow Cl_2(g) + e^-$

Overall cell reaction

$$2Na^+(aq) + 2Cl^-(aq) + 2H_2O(l) \longrightarrow 2Na^+(aq) + 2OH^-(aq) + H_2(g) + Cl_2(g)$$

| aqueous sodium chloride | aqueous sodium hydroxide |

15 **In all three processes, the electrolysis products, chlorine and sodium hydroxide, must be kept apart. Write an equation for the reaction which would occur if they were not.**

Electrolysis is very energy-consuming and the chloralkali industry makes huge demands on electrical energy – second only to aluminium production.

MERCURY CELL

Cell: 2.5 m × 15 m rubber or PVC-lined, gas-tight steel cell
Anodes: titanium with a rare earth oxide surface layer, typically 100 in each cell
Cathode: flowing mercury
Voltage and current: 4.5 V, 300 kA
Feed: purified, saturated aqueous solution of sodium chloride (brine)
The mercury cell is being phased out because of the hazards associated with using mercury.

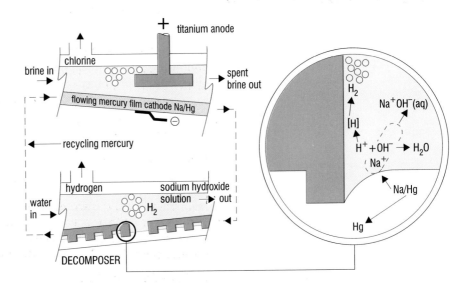

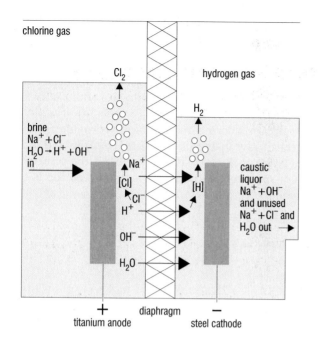

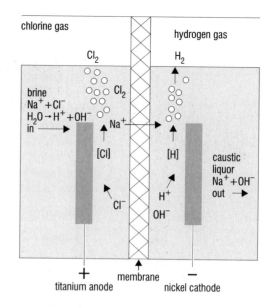

DIAPHRAGM CELL

Cell: large number of cells are connected in series; in each cell the anodes and cathodes are separated by a permeable asbestos diaphragm

Anodes: titanium

Cathodes: steel

Voltage and current: 3.8 V, up to 150 kA

Feed: purified, saturated aqueous solution of sodium chloride (brine)

MEMBRANE CELL

Cell: typically 1 m × 0.5 m × 625 m; anodes and cathodes are separated by an ion-exchange membrane permeable to cations only (a co-polymer of tetrafluoroethene and a similar fluorinated alkene)

Anodes: titanium

Cathodes: nickel

Voltage and current: similar to Diaphragm cell

Feed: highly purified, saturated aqueous solution of sodium chloride (brine)

Fluorine

Fluorine is the most reactive non-metal of all the elements. It is found naturally in fluoride minerals such as calcium fluoride (fluorspar), CaF_2. As an element fluorine is a toxic, corrosive, oxidising agent and very difficult to work with. Despite these problems, there is a demand for elemental fluorine for synthesis of many useful fluorine compounds. Production of fluorine is by an electrolytic method. The difficulty of finding suitable oxidising agents (capable of oxidising fluorides to fluorine) means that straightforward chemical methods are impractical.

Fluorine is produced commercially by passing electricity through an electrolyte of potassium hydrogen difluoride, KHF_2, dissolved in anhydrous hydrogen fluoride. Anhydrous conditions are required because in the presence of water oxygen is produced at the anode in place of fluorine. The electrolyte is contained within a steel cell, maintained at a temperature of 200°C. A voltage of about 8–12 V is used. Fluorine, reacts explosively with the cathode product, hydrogen so that great care must be taken to keep them apart.

The electrode reactions are:

Anode: $2F^- \longrightarrow F_2(g)$

Cathode: $2H^+ + 2e^- \longrightarrow H_2(g)$

(Note that the states of matter of the ions are not given, but they are in a solution of anhydrous hydrofluoric acid).

FLUORINE, TEFLON AND ARTIFICIAL BLOOD

Fluorine is used for producing uranium fuels for use in nuclear power stations and the polymer poly(tetrafluoroethene), better known as 'Teflon'.

'Teflon' is the most slippery substance in the world, according to the Guinness Book of World Records. It was discovered accidentally in 1938 by Roy J. Plunkett, while searching for compounds which would be suitable as refrigerants.

A wide range of compounds containing fluorine are used as anaesthetics, refrigerants, sterilising agents and electrical insulators. One interesting use of a specialised fluorocarbon is as artificial blood (Fig 16.12).

Fig 16.12 An artificial blood can be made using a specialised fluorocarbon. It is inert, non-toxic, has an excellent oxygen carrying capacity and can replace natural blood temporarily following blood loss. As the body's normal red-blood cell levels recover, the artificial blood is lost naturally from the body. This artificial blood is acceptable to Jehova's witnesses, who will not accept transfusions of natural blood.

Metal purification

Many metals can be **refine**d (purified) by electrolysis. This method has been used for many metals but is nowadays used mostly for copper and zinc. The impure metal is usually made the anode in an electrolytic cell and, as the metal oxidises, it dissolves to give ions in solution. Impurities sink to the bottom of the electrolytic cell where they are be removed as a sludge. The metal ions in solution migrate to the cathode. The metal is deposited by reduction and the mass of the cathode increases. The metal deposited has a higher purity relative to the anode. Very high levels of purity can be achieved by making this cathode the anode in another cell and repeating the electrolytic purification.

For zinc the half-reactions are:

Anode: $Zn(s) \longrightarrow Zn^{2+}(aq) + 2e^-$

Cathode: $Zn^{2+}(aq) + 2e^- \longrightarrow Zn(s)$

The ores of metals such as zinc and copper are often found in association with other metals such as silver and gold. The 'sludge' which forms underneath the anode during purification of copper may, for example, contain valuable amounts of these metals.

16 **Why does the electrolyte concentration remain the same during the purification process?**

QUESTIONS

16.19 As the managing director of a small but successful electrolytic production plant, you are hoping to expand onto a new site which will require the construction of a new plant and the appointment of an additional workforce.
  (a) What steps would you take to ensure that the proposed site was an appropriate choice for the company?
  (b) What benefits will a new plant offer to the community?
  (c) What concerns do you think the local community may have about the proposed plant?

16.20 Describe the electrolytic production of aluminium. Your answer should include:
  (a) a brief description of the method used;
  (b) an explanation of the use of molten cryolite as solvent;
  (c) a labelled diagram of the cell used for aluminium production,
  (d) equations for the electrode reactions taking place at (i) the anode; and (ii) the cathode.

16.21 (a) What is meant by the term *anodisation*?
  (b) How can anodisation help to protect an aluminium sheet against corrosion?

16.22 (a) Why is potassium difficult to produce by non-electrolytic methods?
  (b) Describe, with the aid of a simple diagram, an electrolytic cell in which potassium could be produced and the electrolyte which would be employed.
  (c) Give the cathode and anode reactions which take place in your chosen cell.

16.23 (a) How is chlorine produced commercially?
  (b) What two other important co-products are produced with chlorine?
  (c) Give electrode reactions for anode and cathode.
  (d) Give the overall cell reaction.
  (e) State one use for *each* of the three electrolysis products.

16.24 Explain how copper could be purified by an electrolytic method. Your answer should include:
  (a) the nature of the electrolyte;
  (b) the composition of the anode and the cathode;
  (c) an explanation of how this technique improves the purity of the copper.

16.25 From a typical electrolytic plant producing aluminium, 50 000 tonnes of aluminium are produced annually.
  (a) What mass of aluminium oxide is required for this?
  (b) What volume of oxygen would be produced in total at the anodes?
  (c) How much electricity is required for production assuming no loss by earth leakage.

ACTIVITY

You have been asked to estimate the time needed to sterilise electrolytically a swimming pool whose dimensions are 25m long × 10m wide and whose average depth is 2m. It has been suggested that a chlorine concentration of 2×10^{-5} mol dm^{-3} would provide a level of chloric(I) acid in solution adequate to kill 99% of all bacteria (the sterilisation target level). Assume that chloride ions are present in excess and chlorine production takes place with 75% efficiency. Assume also that no loss of chlorine takes place during the sterilisation period. Twelve electrolysis cells are available and each draws 10 amps of current when operational. Roughly how long would sterilisation take? What problems would you expect with this system of sterilisation?

ELECTROCHEMISTRY IN USE

Answers to Quick Questions: Chapter 16

1 The blue solution would slowly lose its colour. There would be slight temperature rise and a brown precipitation would form, replacing the grey zinc powder.

2 Because the halogens undergo simultaneous oxidation (to the halate(I) ion) and reduction (to the halide ion).

3 Iodine

4 Zero

5 $TiCl_4(g) + 2Mg(l) \longrightarrow Ti(l) + 2MgCl_2(l)$
Change in oxidation number for magnesium = 0 to +2

6 There were no free moving ions to carry the charge

7 **Cathode:** $Li^+(l) + e^- \longrightarrow Li(l)$
Anode: $2Cl^-(l) \longrightarrow Cl_2(g) + 2e^-$

8 Hydroxide ions are liberated, leaving an excess of $H^+(aq)$ in solution.

9 Cathode: $Cu^{2+}(aq) + 2e^- \longrightarrow Cu(s)$
Anode: $4OH^-(aq) \longrightarrow 2H_2O(l) + O_2(g) + 4e^-$

10 Because silver comes below hydrogen in the electrochemical series.

11 15.445

12 $Al_2O_3(s) + 2OH^-(aq) + 3H_2O(l) \longrightarrow 2[Al(OH)_4]^-(aq)$

13 Na_3AlF_6
$Al_2O_3(s) + 12HF(aq) + 6NaOH(aq) \longrightarrow 2Na_3AlF_6(aq) + 9H_2O(l)$

14 $2Al_2O_3(l) \longrightarrow 4Al(l) + 3O_2(g)$

15 $Cl_2(aq) + 2NaOH(aq) \longrightarrow NaCl(aq) + NaOCl(aq) + H_2O(l)$

16 Because as quickly as metal ions are deposited at the cathode, an equal amount of ions pass into solution at the anode.

Theme 5

CONTROL

Knowledge of how fast and far chemical reactions proceed is of great importance. Chemists need to understand factors affecting reactions so that they can control them, make them go faster or slower and give high yields of products.

Chapter 17

THE RATES OF CHEMICAL REACTIONS

LEARNING OBJECTIVES

When you have studied this chapter you should be able to:

1. explain the terms reaction kinetics and rate of reaction;

2. explain the importance of investigating rates of chemical change;

3. describe methods used to measure the rates of chemical reactions;

4. explain the term rate curve;

5. describe and explain the effect of changing concentration on the rate of a chemical reaction;

6. define rate equation, rate constant, order of reaction, and half-life for chemical reactions;

7. explain the terms rate determining step and molecularity;

8. use experimental data to determine order of reaction and to calculate the rate constant for a reaction;

9. calculate the half-life of a simple first order reaction;

10. explain the effect of changing temperature on the rate of a chemical reaction;

11. describe in outline the collision and transition state theories;

12. explain what is meant by activation energy and transition state;

13. explain how a catalyst speeds up the rate of a chemical reaction;

14. classify catalysts as homogeneous or heterogeneous;

15. explain the terms autocatalysis, promoter, inhibitor and catalyst poison.

17.1 REACTION KINETICS

The rate at which chemical change occurs depends on the nature of the reaction and the conditions under which it occurs. The study of rates of chemical reactions and the factors which influence them is known as **reaction kinetics** or, simply, **kinetics**.

Several factors affect the rate of chemical change. The most important are:

- concentration of reactants;
- temperature of reaction mixture;
- surface area of solid reactants;
- presence of a catalyst or inhibitor;
- pressure (for those reactions involving gases);
- exposure to electromagnetic radiation, e.g. visible light.

1 **Why is the surface area of a solid reactant important?**

Reaction rates can be changed, often dramatically, by altering one or more factors. For example, the reaction between hydrogen and chlorine proceeds extremely slowly in the dark, but is a fast reaction (often explosive) in the presence of light.

$$H_2(g) + Cl_2(g) \xrightarrow[\text{uv light}]{} 2HCl(g)$$

The importance of rates of reaction

Chemical processes, whether they occur in a beaker, industrial reactor or living cell, are carefully controlled in order to produce the right products at the right time, both efficiently and safely. To achieve this we need to understand how and why the rate at which a reaction takes place can be changed.

Rates and the chemical industry

When a manufacturer decides to make a chemical product, several factors must be considered, such as the demand for the product, the cost of starting materials, the cost of production and the time taken for production. Two major considerations for the company, therefore, are the amount of material produced (the yield) and production time.

Often the optimum conditions for a process may be predicted by studying the kinetics of the reaction and this can save development time and money. However, it is usually necessary to test these conditions experimentally and perhaps further modify the process.

2 **Why do chemical manufacturers prefer processes to take place at normal environmental temperatures to those which require very low or high temperatures?**

The safety of a chemical plant also requires that chemical engineers consider kinetics. If a reaction is exothermic, the energy it produces may accelerate the reaction, thus leading to more rapid production of energy. A cycle of increasing reaction rate can follow, producing a dangerous 'runaway' reaction. The emphasis on safety within the chemical industry ensures that kinetic studies are carried out to check that operating conditions do not exceed the design tolerances of equipment used for production.

Fig 17.1 Establishing optimum operating temperature and pressure and, perhaps, the use of a catalyst, reduces operating costs and may mean the difference between success or failure of a product in the commercial market.

Kinetics and the chemistry of the body

Pharmacology is the study of drugs and drug action and a pharmacologist is concerned with all aspects of the behaviour of drugs in living systems. These include

- the uptake of drugs;
- their subsequent biochemical transformation;
- excretion of any by-products.

The kinetics of each of these significantly affects a drug's effectiveness. A large number of interrelated chemical reactions take place in living cells. Often, the product of one reaction is the starting material of another. The rate of each reaction is controlled by factors such as concentration of starting material, the build-up of a product, cell temperature and the presence of catalysts.

Many illnesses arise from an imbalance due to either excessive or inadequate production of one or more chemical substances. It is difficult to assess the cause of such a problem within the complicated and inaccessible environment of a living cell (**in vivo**). Individual reactions can, however, be duplicated to some extent within the test-tube (**in vitro**), and the factors which influence their rate investigated. This may provide valuable information about cellular processes, but the experimenter must be aware of the many differences between the cell and test-tube environments.

3   Give one important difference between a reaction taking place within a cell and one taking place within a test-tube.

Timed chemical change

We determine **rate of reaction** by measuring the change in concentration of either reactants or products over a period of time:

$$\text{rate of reaction} = \frac{\text{change in concentration}}{\text{time taken for change}}$$

Some reactions occur so quickly that they appear to take no time at all. Such reactions are not instantaneous but they are certainly extremely fast. The precipitations of insoluble salts from aqueous solution are reactions of this type. For example, the formation of silver chloride by addition of aqueous silver nitrate to a solution containing hydrated chloride ions.

$$Ag^+(aq) + Cl^-(aq) \longrightarrow AgCl(s)$$

4 **Why is the precipitation of silver chloride so rapid?**

Other reactions take place very slowly and it is often difficult to detect any signs that reaction is taking place. Rusting is an example of a chemical reaction between iron, water and oxygen which proceeds very slowly to give flaky orange/brown hydrated iron oxide (rust).

$$4Fe(s) + 3O_2(g) \longrightarrow 2Fe_2O_3(s)$$

5 **How could you monitor the rate of rusting of an iron nail?**

For most chemical reactions, the rate changes dramatically during the course of reaction. Usually the rate is more rapid at the beginning of a reaction, gradually decreasing as reactants are used up. This gives rise to a graph of concentration of reactant against time which is initially very steep but whose gradient gradually decreases (Fig 17.8).

Fig 17.3   The bright yellow precipitate of silver iodide seems to form instantly when aqueous silver nitrate is added to aqueous potassium iodide. The reaction is not instantaneous – but it is very fast:
$Ag^+(aq) + I^-(aq) \rightarrow AgI(s)$

QUESTIONS

17.1   Explain what is meant by the term *rate of reaction*.

17.2   Give one example of a very fast reaction and one example of a reaction which takes place very slowly.

17.3   The SI units for the rate of a chemical reaction are mol m^{-3} s^{-1}. In practice, such units are often inconvenient. What units are usually used for expressing the rate of a reaction?

17.4   Name four factors which may influence the rate of a chemical reaction.

17.5   In your own words explain why chemists study the kinetics of chemical reactions on a small scale before scaling the process up for production.

17.2 MEASUREMENT OF REACTION RATES

Monitoring reaction progress

To find out how the rate of a chemical reaction changes as the reaction proceeds and to identify factors which affect it, we need methods to measure reaction rates. Many are available, including:

- chemical analysis;
- spectroscopic analysis;
- measuring changes in pressure or volume;
- measurement of conductivity;
- measurement of optical rotation.

Chemical analysis

The reaction mixture is sampled at regular intervals and the concentration of one of the reaction components in the sample is measured using a chemical test. For example, a sample may be titrated against a solution of known concentration with which it reacts (volumetric analysis).

Consider the acid-catalysed hydrolysis of ethyl ethanoate.

$$CH_3COOC_2H_5(aq) + H_2O(l) \longrightarrow CH_3COOH(aq) + C_2H_5OH(aq)$$

The reaction mixture can be titrated against a standard sodium hydroxide solution, using a suitable indicator. This allows the amount of ethanoic acid to be calculated. The greater the volume of sodium hydroxide solution required the further the reaction has progressed.

CONNECTION: Analysis by chemical tests, page 119.

CONTROLLING BIODEGRADATION

Designing biodegradable polymers is an exercise in the art of compromise. Biodegradable polymers for bottles and packaging must survive long enough for normal use of the product yet must ultimately degrade to harmless products. Polymer scientists are interested in the rate of formation of a polymer from its monomer units and the rate of formation of cross-links between polymer chains. Of equal interest is the rate at which the polymer degrades on exposure to light, heat, oxygen, water and bacterial action.

Biopol, a recent product from ICI, is made by a fermentation process using one bacterium yet is biodegradable due to the action of soil bacteria when buried, forming carbon dioxide and water.

Fig 17.4 Plastic milk bottles, made of a stable poly(ethene) material were introduced in some cities in Britain in the 1970s. Calculations showed that a typical city would be buried under several millions of milk bottles within a few years. Modern biodegradable plastics such as ICI's Biopol, shown in the photograph, break down after a few weeks burial. The rate of biodegradation is critical to the success of such materials.

6 What factors would you vary to carry out a comprehensive study of the above reaction?

7 How could account be taken of the acid present as a catalyst in the hydrolysis of ethyl ethanoate?

Spectroscopic analysis

Spectrophotometers measure changes in the absorption of electromagnetic radiation (usually visible or ultraviolet) and so can be used to monitor reactions. A sample is removed from the reaction mixture and the absorption at a particular frequency measured. This would normally be a frequency at which just one component of the mixture (reactant or product) absorbs.

For example, the iodination of propanone, catalysed by dilute acid.

$$CH_3COCH_3(aq) + I_2(aq) \longrightarrow CH_3COCH_2I(aq) + HI(aq)$$

In this reaction iodine is the only coloured component and its colour varies from a pale straw colour to a darker orange/brown depending on its concentration. In this case, the lower the colour intensity, the further the reaction has progressed.

CONNECTION: Spectrophotometers, pages 194–197.

8 How could you decide when the above reaction was complete?

Fig 17.5 The rate of reaction between bromine and methanoic acid can be followed by the loss of colour of the reaction mixture as bromine is used up. These samples were taken at regular time intervals. It can be seen that the reaction slows down (colour change is less dramatic) as the reaction proceeds.

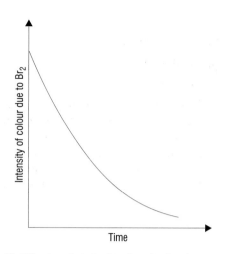

Fig 17.6 A graph of colour intensity against time shows how the concentration of bromine decreases as the reaction proceeds.

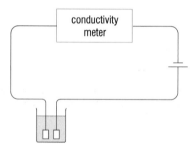

Fig 17.7 This simple circuit can be used to monitor changes in electrical conductivity as a reaction proceeds.

CONNECTION: Optical isomers and plane polarised light, pages 275–277.

Measuring changes in pressure or volume

If a reaction results in a change in pressure or volume, this change can be measured to determine the rate of reaction. It is straightforward to identify reactions which may be followed by this method since in the equation there will always be a different number of moles of gaseous reactants compared to gaseous products.

For example, benzenediazonium chloride decomposes when warmed to give phenol, nitrogen and hydrochloric acid:

$$C_6H_5N \equiv N^+Cl^-(aq) + H_2O(l) \longrightarrow C_6H_5OH(aq) + N_2(g) + HCl(aq)$$

Nitrogen is insoluble in water and so any increase in pressure or volume is a measure of nitrogen formation. The hydrogen chloride produced dissolves too readily in water to produce any appreciable change in pressure or volume. Phenol is not very soluble in weakly acidic solutions and would precipitate, but this does not affect the measurements.

9 **Can you suggest an alternative method for monitoring this reaction?**

Measurement of conductivity

The ease with which a solution conducts electricity is determined by the nature of the ions present, as well as the number present. Solutions containing fast moving ions have higher electrical conductivity. Electrical conductivity may be measured by inserting two platinum electrodes, connected to a conductivity meter, into the reaction mixture (Fig 17.7). The conductivity of the solution is recorded at convenient times during the reaction and used to follow the reaction and determine its rate.

For example, the hydrolysis of bromoethane in alkaline solution:

$$C_2H_5Br(l) + OH^-(aq) \longrightarrow C_2H_5OH(aq) + Br^-(aq)$$

The small and fast-moving hydroxide ions are used up and replaced by the slower bromide ion. Therefore, the electrical conductivity of the reaction mixture decreases as the reaction proceeds.

10 **Would the electrical resistance of this reaction mixture increase or decrease as the hydrolysis proceeds?**

Measurement of optical rotation

An optically active substance will rotate plane-polarised light. Any change in the optical rotation of a reaction mixture can be followed using a polarimeter. This method is often used by chemists and biochemists to monitor the reaction rates of compounds which are chiral. For example, most sugars are optically active and their reactions often involve a measurable change in the optical rotation of a solution.

Consider the hydrolysis of sucrose in acid solution:

$$C_{12}H_{22}O_{11}(aq) + H_2O(l) \longrightarrow C_6H_{12}O_6(aq) + C_6H_{12}O_6(aq)$$
sucrose   glucose   fructose

Sucrose may be hydrolysed to glucose and fructose by mineral acid or by the use of enzymes. During reaction a change in the optical rotation of the solution occurs. Although both reactants and products are optically active, the extent to which and the direction in which they rotate plane-polarised light differs. Therefore, the extent of hydrolysis can be determined by measuring changes in optical rotation of the reaction solution.

Rate curves

Kinetic data are usually presented graphically by either:

- a graph of reactant concentration against time (Fig 17.8);
- a graph of product concentration against time (Fig 17.9).

These graphs are called **rate curves**. The rate at any particular stage of reaction may be found by measuring the gradient of the rate curve at that stage. For example, suppose we want to know the rate of reaction after a particular time. The gradient of the tangent to the rate curve at this time is measured (Fig 17.8). This gives the rate of disappearance of the reactant. A tangent to the concentration of product against time graph gives the rate of appearance of product (Fig 17.9).

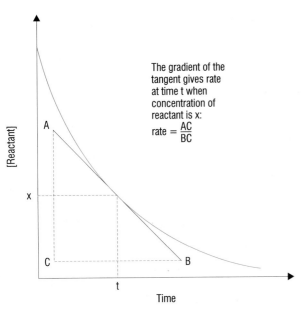

The gradient of the tangent gives rate at time t when concentration of reactant is x:

$$\text{rate} = \frac{AC}{BC}$$

Fig 17.8 A typical rate curve: concentration of reactant is decreasing with time. The gradient of the tangent gives the rate of reaction in terms of the rate of disappearance (therefore, negative gradient) of reactant.

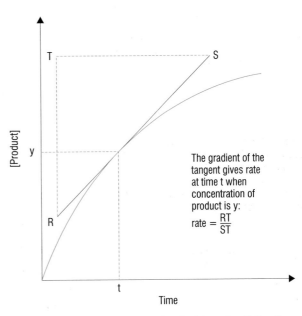

The gradient of the tangent gives rate at time t when concentration of product is y:

$$\text{rate} = \frac{RT}{ST}$$

Fig 17.9 A typical rate curve: concentration of product is increasing with time. The gradient of the tangent gives the rate of reaction in terms of the rate of appearance (therefore, positive gradient) of product.

11 **For some reactions, the rate curve is a straight line. What does this tell you about how the rate changes during the course of the reaction?**

SPOTLIGHT

EXPLOSIVE CHEMISTRY

Explosions are examples of fast reactions. The effectiveness of an explosive relies on the rapid rate of production of a large volume of hot, expanding gases. Four moles of nitroglycerine, occupying about 500 cm^3 decompose to 29 moles of gases. In a typical nitroglycerine explosion the volume of gas produced may be up to 20 000 times the original volume of nitroglycerine. It is the rate of increase in volume which is critical to the success of an explosive.

Fig 17.10 A modern explosive is being used to demolish this bridge. Dynamite was developed by Alfred Nobel. It contains nitroglycerine, mixed with silica to improve its shock-resistance and safe handling. It produces a massive volume of hot gases when exploded. The success of dynamite made its inventor, Alfred Nobel, a wealthy man. In his will he specified that his fortune should be invested and used to provide annual prizes in physics, chemistry, physiology or medicine, literature, economics and peace.

17.6 Suggest methods for investigating the rates of the following reactions:

(a) The alkaline hydrolysis of the ester ethyl ethanoate:

$$CH_3COOCH_2CH_3(aq) + NaOH(aq) \longrightarrow CH_3COONa(aq) + CH_3CH_2OH(aq)$$

(b) The thermal decomposition of calcium carbonate in a sealed system:

$$CaCO_3(s) \rightleftharpoons CaO(s) + CO_2(g)$$

(c) The reaction of zinc powder, shaken with copper(II) sulphate solution:

$$CuSO_4(aq) + Zn(s) \longrightarrow Cu(s) + ZnSO_4(aq)$$

(d) The bromination of ethene, an unsaturated hydrocarbon:

$$\underset{\text{ethene}}{C_2H_4(g)} + \underset{\text{bromine}}{Br_2(g)} \longrightarrow \underset{\text{dibromoethane}}{C_2H_4Br_2(g)}$$

17.7 From your knowledge of the kinetic theory and a study of the states of matter, explain in your own words why the effect of an increase in concentration and an increase in pressure might be expected to have a similar effect on rate of reaction.

17.3 THE EFFECT OF CONCENTRATION ON RATE

The concentration of reactants is one of the most important influences on the rate of a chemical reaction. Vast amounts of kinetic data have been obtained by experiment and, for each reaction, analysis of data has established the relationship between rate and reactant concentration. Before we generalise, it may be helpful to consider two examples:

- the thermal decomposition of dinitrogen pentoxide;
- the reaction between thiosulphate ions and an acid.

The thermal decomposition of dinitrogen pentoxide

A simple reaction whose kinetics have been studied extensively is the thermal decomposition of dinitrogen pentoxide:

$$2N_2O_5(g) \longrightarrow 2N_2O_4(g) + O_2(g)$$

If the concentration of dinitrogen pentoxide is measured over a period of time and the results plotted on a graph, a characteristic rate curve is obtained (Fig 17.11). It is also possible to follow the increasing concentration of dinitrogen tetroxide, one of the products (Fig 17.12)

12 **What technique could be used to follow this reaction?**

The rate of reaction at any particular reactant concentration can be obtained from the gradient of the rate curve at that concentration. The rate of reaction is greatest at the start of reaction (steep gradient) and gradually declines to zero as the reaction comes to an end (Fig 17.11).

Similarly for the graph of dinitrogen tetroxide concentration against time, the gradient of a tangent to the curve at any point in the reaction gives the rate of reaction at that point. We can see that rate of production

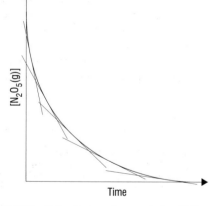

Fig 17.11 Plotting the decreasing concentration of N_2O_5 against time gives a rate curve for the thermal decomposition of N_2O_5 to N_2O_4. Gradients of tangents to the curve give the rate of reaction at different reactant concentrations.

of dinitrogen tetroxide starts at a maximum (steep gradient) and falls away to zero as the reaction is completed (Fig 17.12).

The way in which rate depends on reactant concentration can be established in three ways:

- From inspection of the experimental data and, in particular, seeing what happens to the rate as concentration is halved. For example, we find that rate of dinitrogen pentoxide decomposition is halved when the concentration reaches half its original value.
- For convenience, the initial rate method is often used. A series of experiments is set up in which the concentration of the reactant we are interested in is different for each experiment. The initial rate of reaction for each experiment is determined. This allows the relationship between the concentration of that reactant and the rate of reaction to be established. We find that doubling the initial concentration of dinitrogen pentoxide causes the initial rate of decomposition to double.
- By plotting a graph of rate of reaction against reactant concentration (data obtained from a series of tangents drawn to a rate curve). For the decomposition of dinitrogen pentoxide, for example, a straight line graph is obtained (Fig 17.13).

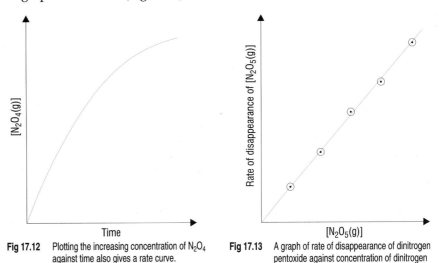

Fig 17.12 Plotting the increasing concentration of N_2O_4 against time also gives a rate curve.

Fig 17.13 A graph of rate of disappearance of dinitrogen pentoxide against concentration of dinitrogen pentoxide. The straight line obtained shows that rate of thermal decomposition is proportional to concentration, i.e. rate $\propto$ [N_2O_5(g)].

Applying any of these methods to the thermal decomposition of dinitrogen pentoxide reveals a simple relationship between rate of reaction and concentration of dinitrogen pentoxide; the rate is proportional to the concentration of dinitrogen pentoxide:

$$\text{rate} \propto [N_2O_5(g)]$$

TUTORIAL

Problem: Calculate the rate of hydrolysis of ethyl ethanoate at 16°C when the concentration of ethanoate ions, CH_3COO^-(aq), is 0.04500 mol dm^{-3}. Kinetic data for the reaction is given in Table 17.1.

Table 17.1 Hydrolysis of ethyl ethanoate at 16°C
$$CH_3COOC_2H_5(aq) + OH^-(aq) \longrightarrow CH_3COO^-(aq) + C_2H_5OH(aq)$$

Time / s	[CH$_3$COO$^-$(aq)] / mol dm^{-3}
0	0.00000
300	0.02304
900	0.03948
1500	0.04672
2100	0.05036
3300	0.05472

Solution:

(a) Plot a graph of [CH$_3$COO$^-$(aq)] against time:

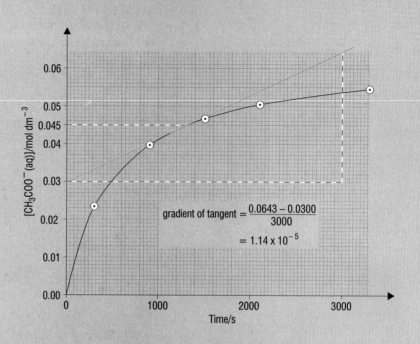

gradient of tangent = $\dfrac{0.0643 - 0.0300}{3000}$

= 1.14×10^{-5}

(b) Draw a tangent to the curve where [CH$_3$COO$^-$(aq)] = 0.04500 mol dm^{-3} (see above).

(c) Determine the gradient of the tangent:

$$\text{gradient} = \frac{0.0643 - 0.0300}{3000}$$

$$= 1.14 \times 10^{-5}$$

(d) Therefore, the rate of reaction is 1.14×10^{-5} mol dm^{-3} s^{-1}

The reaction between thiosulphate ions and an acid

Thiosulphate ions react with a dilute acid (such as nitric acid) to give sulphur dioxide, sulphur and water:

$$S_2O_3^{2-}(aq) + 2H^+(aq) \longrightarrow SO_2(g) + S(s) + H_2O(l)$$

13 **How would the appearance of the reaction mixture change during the course of reaction?**

Several experiments may be set up in which the concentration of sodium thiosulphate is varied but the concentration of nitric acid is kept constant. The time taken for a precipitate of sulphur to become clearly visible can be measured. The time for the first appearance of sulphur in each experiment is different. A graph of 1/time against concentration of sodium thiosulphate is found to be a straight line through the origin (Fig 17.14). Since 1/time is a measure of the rate of reaction, we can see that, in this case, the rate is directly proportional to the concentration of sodium thiosulphate:

rate $\propto$ [Na$_2$S$_2$O$_3$(aq)]

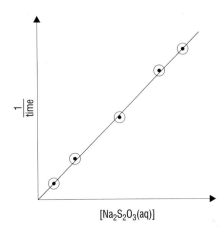

[Na₂S₂O₃(aq)]

Fig 17.14 Plotting 1/time (a measure of rate) against concentration shows these two quantities are proportional for this reaction.
Rate = $k \times$ concentration, where k is a constant.

Fig 17.15 In a modern medical research laboratory, automated sampling at pre-chosen intervals is often used to monitor the progress of a chemical reaction. Many similar samples can be treated in a relatively short time. The results may be stored using a computer and interpreted when required.

14 What would happen to the rate of reaction if [Na₂S₂O₃(aq)] is doubled but other concentrations remain unchanged?

When the experiment is repeated, keeping the concentration of sodium thiosulphate constant and varying the concentration of acid, another straight line graph for 1/time against concentration of acid is obtained. This suggests that the rate of reaction is also directly proportional to the concentration of nitric acid :

$$rate \propto [HNO_3(aq)]$$

In this reaction we can see that rate depends on the concentrations of both reactants. These findings are summarised by the rate equation:

$$rate \propto [Na_2S_2O_3(aq)][HNO_3(aq)]$$

The rate equation

The relationship between concentrations of reactant and rate of reaction depends on the particular reaction under investigation. It can only be established by experiment. It cannot be determined from inspection of the balanced equation for the reaction. For example, we have seen that for the thermal decomposition of dinitrogen pentoxide, the relationship between the rate of the reaction and the concentration of dinitrogen pentoxide (found by experiment) is:

$$rate \propto [N_2O_5(g)]$$

which we can also write in the form:

$$rate = k[N_2O_5(g)] \quad \text{where } k = \text{a constant}$$

This is called the **rate equation** and gives the rate of a chemical reaction in terms of the concentrations of reactants.

Table 17.2 Some experimentally determined rate equations

Reaction	Rate equation
$H_2(g) + I_2(g) \longrightarrow 2HI(g)$	rate = $k[H_2(g)][I_2(g)]$
$2NO(g) + O_2(g) \longrightarrow 2NO_2(g)$	rate = $k[NO(g)]^2[O_2(g)]$
$BrO_3^-(aq) + 5Br^-(aq) + 6H^+(aq) \longrightarrow 3Br_2(aq) + 3H_2O(l)$	rate = $k[BrO_3^-(aq)][Br^-(aq)][H^+(aq)]^2$
$CH_3CHO(g) \longrightarrow CH_4(g) + CO(g)$	rate = $k[CH_3CHO(g)]^{1.5}$

For a general reaction of the type:

$$a\text{A} + b\text{B} + c\text{C} \longrightarrow \text{products}$$

(where A, B and C are the reactants and a, b and c represent the number of moles of each reactant as shown in the balanced equation for the reaction) the rate equation would be:

$$\text{rate} = k[\text{A}]^x[\text{B}]^y[\text{C}]^z$$

where x, y and z are the powers (also called indices) to which the concentration of each reactant must be raised in the rate equation.

The values of a, b and c in the reaction equation have no bearing on the values of x, y and z, which are fixed for a particular reaction. The symbol k represents a constant called the **rate constant**. It always has units but the units can only be decided upon once the values for x, y and z have been established. In the case of dinitrogen pentoxide decomposition the units are easily established:

$$\text{rate} = k \times [\text{N}_2\text{O}_5(\text{g})]$$

units: $\text{mol dm}^{-3}\,\text{s}^{-1} = ? \times \text{mol dm}^{-3}$

therefore, the units for k in this instance are $\dfrac{\text{mol dm}^{-3}\,\text{s}^{-1}}{\text{mol dm}^{-3}} = \text{s}^{-1}$

TUTORIAL

Problem: Experiments show that the rate equation for the reaction:

$$2\text{H}_2(\text{g}) + 2\text{NO}(\text{g}) \longrightarrow 2\text{H}_2\text{O}(\text{g}) + \text{N}_2(\text{g})$$

is

$$\text{rate} = k[\text{H}_2(\text{g})][\text{NO}(\text{g})]^2$$

What are the units for the rate constant, k?

Solution: rate $= k \times [\text{H}_2(\text{g})] \times [\text{NO}(\text{g})]^2$

units: $\text{mol dm}^{-3}\,\text{s}^{-1} = ? \times \text{mol dm}^{-3} \times (\text{mol dm}^{-3})^2$

therefore, the units for k are $\dfrac{\text{mol dm}^{-3}\,\text{s}^{-1}}{(\text{mol dm}^{-3})^3} = \text{mol}^{-2}\,\text{dm}^6\,\text{s}^{-1}$

Order of reaction

Finding, by experiment, the dependence of reaction rate upon the concentration of each reactant allows predictions about reaction rates to be made. The rate equation for a process is a convenient way of communicating the results of kinetic experiments. To describe the effect of the concentration of a particular reactant we use the idea of **order**. The **order of a reaction** with respect to a particular reactant is the power to which the concentration of that reactant is raised in the rate equation. Again, for the general reaction:

$$a\text{A} + b\text{B} + c\text{C} \longrightarrow \text{products}$$

$$\text{rate of reaction} = k[\text{A}]^x[\text{B}]^y[\text{C}]^z$$

The order of reaction with respect to: A is x; B is y; C is z
(since the concentrations are raised to these powers in the rate equation).

The overall order of a reaction is the sum of the powers to which the concentrations of the reactants are raised in the rate equation. So the overall order of reaction is:

$$\text{overall order} = x + y + z$$

The most common values for order with respect to a particular reactant are 0, 1 or 2.

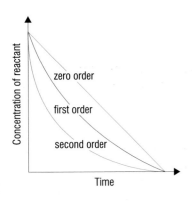

Fig 17.16 Rate curves for zero, first and second order reactions.

- If experiment shows the power for the concentration of a reactant in the rate equation is 0 then the reaction is said to be *zero order* with respect to that reactant.
- If the power is found to be 1 then the reaction is *first order* with respect to that reactant.
- If the power is found to be 2, then the reaction is *second order* with respect to the reactant concerned.

If the reaction $aA + bB + cC \longrightarrow$ products is studied experimentally, for each reactant a graph of its concentration against time enables us to establish the order with respect to this reactant (Fig 17.16).

The reaction is zero order with respect to a reactant if a change in its concentration does not affect the rate. A graph of rate of reaction against concentration is a straight horizontal line (Fig 17.17).

The reaction is first order if a two-fold increase in concentration of A results in a doubling of the reaction rate, that is, the rate is directly proportional to concentration. The units for the rate constant are s^{-1}. When the rate of reaction is plotted against concentration, a straight line through the origin is obtained indicating the simple dependence of rate of reaction on concentration (Fig 17.17).

KEY FACT

A zero order is obtained for a reactant when its concentration does not influence the rate of the reaction at all. The concentration of this reactant need not, therefore, appear in the rate equation and the units of concentration for this substance are ignored when units for the rate constant are established.

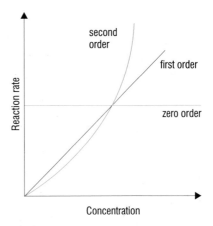

Fig 17.17 Variation of reaction rates with concentration for zero, first and second order reactions.

The reaction is second order if a two-fold increase in concentration of reactant results in the rate of reaction increasing four-fold. When the rate of reaction is plotted against concentration, a curve is obtained (Fig 17.17).

TUTORIAL

Problem: From the kinetic data in Table 17.3, measured at 100°C, determine the order with respect to each reactant for the reaction:

$$2HgCl_2(aq) + K_2C_2O_4(aq) \longrightarrow 2KCl(aq) + 2CO_2(g) + Hg_2Cl_2(s)$$

Calculate the value for the rate constant, stating its units.

Table 17.3

Experiment	[HgCl$_2$(aq)] / mol dm^{-3}	[K$_2$C$_2$O$_4$(aq)] / mol dm^{-3}	Rate / mol dm^{-3} min^{-1}
1	0.0836	0.202	0.52×10^{-4}
2	0.0836	0.404	2.08×10^{-4}
3	0.0418	0.404	1.06×10^{-4}

Solution: (a) By comparing experiments 1 and 2, we see the concentration of mercury(II) chloride remains the same and so the difference in rate must be due to the change in concentration of potassium ethanedioate. Doubling [K$_2$C$_2$O$_4$(aq)] results in a four-fold increase in rate. Therefore, the reaction is second order with respect to K$_2$C$_2$O$_4$(aq).

(b) By comparing experiments 2 and 3, we see the concentration of potassium ethanedioate remains the same and so the difference in rate must be due to the change in concentration of mercury(II) chloride. Halving $[HgCl_2(aq)]$ causes the rate to halve. Therefore, the reaction is first order with respect to $HgCl_2(aq)$.

(c) Therefore, the rate equation is

rate = $k[HgCl_2(aq)][K_2C_2O_4(aq)]^2$

(d) Using the data from experiment 1:

$0.52 \times 10^{-4} = k \times 0.0836 \times (0.202)^2$

Therefore, $k = \dfrac{0.52 \times 10^{-4}}{0.0836 \times (0.202)^2} = 0.0152 \text{ mol}^{-2} \text{ dm}^6 \text{ min}^{-1}$

The half-life of a reaction

The time taken for the concentration of a reactant to fall to half of its original value is called the **half-life** ($t_{\frac{1}{2}}$). For a first order reaction, $\boldsymbol{t_{\frac{1}{2}}}$ is a constant at a particular temperature. The rate curve may be used to demonstrate that a reaction shows first order kinetics (Fig 17.18). To obtain a value for the half-life of a reaction we can use the rate curve to measure the time taken for concentration to fall to half of a given value.

The half-life of a first order reaction can also be obtained from the rate constant, using the relationship:

$$t_{\frac{1}{2}} = \frac{0.693}{k}$$

15 **What are the units of half-life?**

Radioactive decay is a first order process as it exhibits a constant half-life but is, of course, really a special case since, unlike straightforward chemical changes, the rate of radioactive decay is unaffected by temperature.

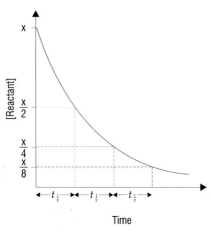

Fig 17.18 The half-life of a first order reaction can be determined from the rate curve.

Rate determining step

Most reactions occur in a series of steps and only the concentrations of those reactants which take part in the slowest step appear in the rate equation. It is the rate of this slowest step which we measure during a kinetic study. This rate is the limiting factor for the rate of the reaction as a whole and is known, therefore, as the rate-determining step. In a single-step reaction obviously the only step is the rate-determining step.

Bromine and propanone react in the presence of hydroxide ions:

$CH_3COCH_3(aq) + Br_2(aq) + OH^-(aq) \longrightarrow CH_3COCH_2Br(aq) + H_2O(l) +$
$Br^-(aq)$

From experiment we find that the rate equation takes the form:

rate = $k[CH_3COCH_3(aq)][OH^-(aq)]$

Changing the concentration of bromine does not affect the reaction rate and so bromine does not appear in the rate equation. Our interpretation is that only propanone and hydroxide ions are involved in the rate-determining step. Bromine must be involved in a faster step which does not limit the rate of the reaction in any way (Fig 17.19). This reaction is therefore zero order with respect to bromine.

CONNECTION: *Reaction mechanisms, Chapter 13.*

16 What effect would doubling the concentration of hydroxide ions have on the rate of bromination of propanone?

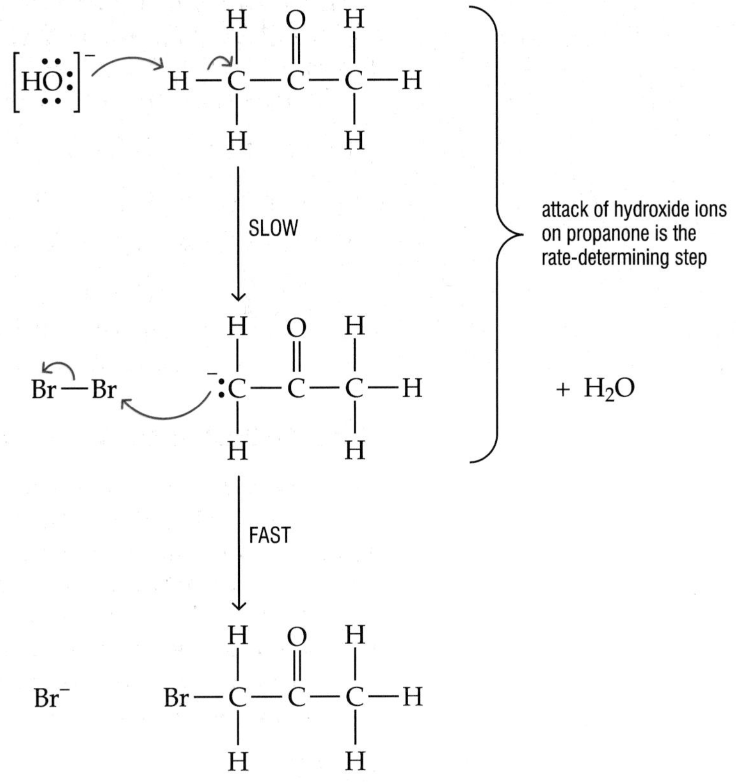

Fig 17.19 The mechanism for bromination of propanone in the presence of hydroxide ions. Initial attack of hydroxide ions on propanone molecules is the slow, rate-determining step.

Occasionally non-integral orders are observed. These are orders which have values between 0 and 1 or 1 and 2. Such reactions are uncommon and are usually either very complex reactions which proceed in many steps, more than one of which affects the overall rate of reaction, or chain-reactions. An example is the thermal decomposition of ethanal which takes place in the gas phase at a temperature of about 450°C.

$$CH_3CHO(g) \longrightarrow CH_4(g) + CO(g)$$

The rate equation is found to be:

$$rate = k[CH_3CHO(g)]^{1.5}$$

Molecularity

Kinetic data for a reaction allow chemists to speculate about the mechanism by which the reaction takes place. The mechanism of a reaction is the sequence of steps by which the reaction is believed to occur, but a proposed mechanism must be consistent with observed kinetic data. A speculation may also be supported by other experimental evidence such as spectroscopic evidence.

The molecularity of a reaction cannot be found experimentally. A reaction step in a mechanism in which only a single species participates is known as a **unimolecular** reaction step. If two species (which may or may not be the same) are involved the step is said to be **bimolecular**. Mechanisms which involve a **trimolecular** step are rare as they involve a simultaneous collision of three particles, and this is statistically unlikely.

KEY FACT

The molecularity of a reaction is defined as the number of species (atoms, molecules, radicals or ions) involved in the rate-determining step.

17.8 In acid solution, bromate(V) ions slowly oxidise bromide ions to bromine:

$$BrO_3^-(aq) + 5Br^-(aq) + 6H^+(aq) \longrightarrow 3Br_2(aq) + 3H_2O(l)$$

The following experimental data was obtained in an investigation to determine how the initial rate depends on the concentration of bromate, bromide and hydrogen ions. The solutions of bromate, bromide and hydrogen ions all have a concentration of 1 mol dm^{-3} and all volumes are measured in cm^3.

	Volumes of solutions used/cm^3				Relative rate
Expt.	BrO_3^-	Br^-	H^+	water	
A	5	25	30	40	1
B	5	25	60	10	4
C	10	25	60	5	8
D	5	12.5	60	22.5	2

(a) What is the order of the reaction with respect to bromate ions, bromide ions and hydrogen ions?

(b) Determine the overall order of reaction.

(c) Do you have enough information to determine a value for the rate constant?

(d) What would the units for the rate constant be?

(e) What would be a suitable method for following reaction progress?

17.9 Propanone reacts with bromine in acid solution to yield bromopropanone:

$$CH_3COCH_3(aq) + Br_2(aq) \longrightarrow CH_3COCH_2Br(aq) + HBr(aq)$$

The initial rate for the disappearance of bromine was measured colorimetrically in six different experiments. In each the concentration of reactants and catalyst was varied as shown below:

	Initial concentrations/mol dm^{-3}			Initial rate/mol dm^{-3} s^{-1}
Expt.	CH_3COCH_3	Br_2	H^+	
1.	0.2	0.1	0.05	4×10^{-5}
2.	0.4	0.1	0.1	1.6×10^{-4}
3.	0.2	0.05	0.05	4×10^{-5}
4.	0.2	0.1	0.1	8×10^{-5}
5.	0.1	0.1	0.1	4×10^{-5}
6.	0.2	0.2	0.05	4×10^{-5}

(a) Identify two experiments which may be compared to determine the order of reaction for bromine.

(b) Write a rate equation for the reaction.

(c) Give the overall order of reaction.

(d) Calculate a value for k, the rate constant, including its units.

(e) Predict the rate of reaction for an experiment in which initial concentrations of reactants are all 0.2 mol dm^{-3}.

17.10 (a) Explain with the aid of a diagram what is meant by the term *constant half life*.

(b) Why is it often useful to use the 'half-life test' to examine a rate curve?

17.11 State what is meant by the terms a) *rate determining step*; b) *the molecularity* of a reaction; c) *non-integral order*.

17.4 TEMPERATURE AND REACTION RATES

Like simple physical processes such as melting or evaporation, chemical processes are faster at higher temperatures. Generally (though certainly not always) a temperature rise of 10°C approximately doubles the rate of a chemical reaction. The rate constant for a reaction is only a constant at a particular temperature.

> **17** What is likely to happen to the rate of reaction if the temperature at which it takes place is increased by 15°C?

The complex series of biochemical reactions known as metabolism are affected by temperature and many animals maintain a constant body temperature to ensure their metabolic rate also remains constant. Cold blooded animals are particularly influenced by the temperature of the environment and are more lively when warm. Their metabolism slows down dramatically in cold conditions. Some surgical operations, particularly those on the heart or brain, are often performed at lower temperatures to deliberately lower the metabolic rate of the patient.

Fig 17.20 Food is stored in a refrigerator to reduce the rate of its decay.

Modelling the influence of temperature on reaction rates

The **collision theory** has developed from the kinetic theory of matter and is based on the idea of particles colliding. The collision frequency (the number of collisions taking place in a particular time interval) can be calculated and theoretical rates of reaction can therefore be worked out. However, reaction rates are usually much slower than predicted by simple calculation, suggesting that a collision does not always result in reaction. To explain the difference between expected and experimentally determined rates of reaction, the collision theory makes certain assumptions:

- reaction only occurs when particles (atoms, molecules, radicals or ions) collide;
- a minimum total energy, known as the **activation energy**, must be possessed by colliding particles if reaction is to take place (that is, for it to be a 'successful' collision);
- some reactions require that particles collide with the correct orientation to one another, that is, the colliding particles must possess the correct **collision geometry**.

The effect of concentration of reactants on reaction rates can be easily explained by the collision theory. In a more concentrated reaction mixture the number of collisions taking place in a particular period of time is greater than in a reaction mixture which is more dilute. The collision frequency is higher. Simply, the more particles that are present, the greater the probability of collisions.

18 What would happen to the collision frequency if the pressure of a gas was lowered?

SPOTLIGHT

SLOWING LIFE DOWN – HIBERNATION

When food is short many animals hibernate. The body temperature of the animal drops so that it may be only a few degrees above the temperature of its surroundings. The chemical reactions in its body (the animal's metabolic processes) take place very slowly using food stored as fat or glycogen. The metabolic rate is dependent on the temperature of the animal's body. An increase in food availability normally coincides with a return of warmer weather. This triggers an increase in the body temperature of the animal and its metabolic rate returns to normal.

Fig 17.21 For some animals, hibernation allows them to survive periods of poor availability of food. As their body temperature drops the rate of chemical reactions taking place inside their cells decreases, thus extending the life of food stored in their bodies. This dormouse can look forward to a long winter sleep.

The collision theory also explains the affect of temperature. A higher temperature for the reaction mixture means that the average kinetic energies of the particles present is higher. Therefore, they move more quickly and collide more often. Although quite valid, this explanation does not account for the enormous increase in reaction rate when the temperature is raised. The Maxwell-Boltzmann distribution of kinetic energies at two different temperatures provides an answer (Fig 17.22). At the higher temperature, T_2, far more particles possess a kinetic energy greater than the activation energy for the reaction. Therefore, not only are there more collisions but many more of these are successful collisions.

Fig 17.22 The effect of temperature on the energies of particles. Raising the temperature moves the most probable energy to a higher value and also flattens the curve. This means that the proportion of particles with high energy increases greatly with a small increase in temperature.

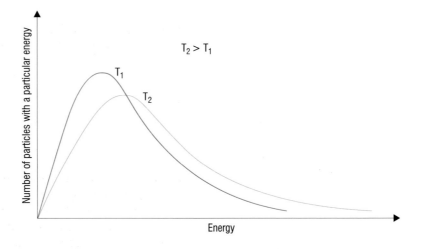

Although the idea of a minimum energy for reaction is used in the collision theory, we have not explained what we mean by a successful collision, other than, that is, one which results in reaction. The **transition state theory** suggests that reaction can only succeed if reacting particles collide with sufficient energy to overcome the energy barrier (the activation energy) to form the **transition state** or **activated complex** (Fig 17.23).

CONNECTION: Activated complexes, intermediates and reaction mechanisms, page 302.

442   THE RATES OF CHEMICAL REACTIONS

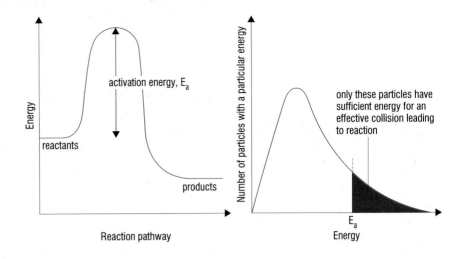

Fig 17.23 Colliding particles must together have sufficient energy to form a transition state and so overcome the energy barrier to reaction (the activation energy).

QUESTIONS

17.12 Sketch a graph with labelled axes to show how kinetic energies of gas molecules are distributed in a sample of a gas. Draw a second curve on the graph to show how the distribution changes at a higher temperature.

17.13 Draw an energy profile diagram to show the energy changes which take place for reactants and products in an endothermic reaction. Show on your diagram the activation energy for the forward reaction and the overall enthalpy change which takes place.

17.14 Theoretical calculations have shown that the increase in the rate of chemical reactions produced by an increase in temperature is far greater than would be expected from an increase in the collision rate alone. How can this effect be explained?

17.15 What is meant by the term *transition state*?

17.5 CATALYSTS

Although we may not be aware of them, catalysts affect our lives in many ways. Life itself depends upon a large number of enzymes, biological catalysts, which control chemical reactions in living cells. Without catalysts, many commercial chemical processes would be so slow that they would either be unworkable or uneconomical. Catalysts are important and, worldwide, around five billion kilograms are used annually in commercial processes.

Characteristics of catalysts

Catalysts are chemical substances which increase the rate of a chemical reaction. They always take part in the reaction but, importantly, can be recovered chemically unchanged at the end of the reaction. The same amount of product is formed but in a shorter time without the need for heating.

THE RATES OF CHEMICAL REACTIONS

Catalysts provide an alternative pathway of lower activation energy than in the absence of a catalyst (Fig 17.24). The reaction proceeds via an intermediate substance (which is not usually isolated) formed between one or more of the reactants and the catalyst. The catalyst appears to work by bringing together two reactants and binding them temporarily while the bonds are rearranged.

19   **Explain why providing an alternative pathway of lower activation energy causes a reaction to speed up.**

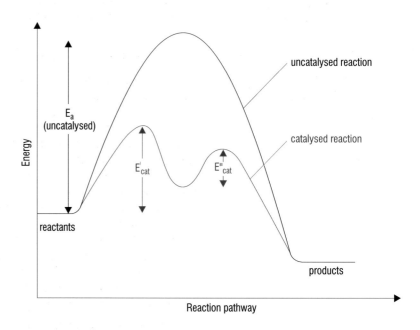

E_a is the activation energy of the uncatalysed reaction

E'_{cat} is the activation energy of the first step in the catalysed reaction

E''_{cat} is the activation energy of the second step in the catalysed reaction

Because $E'_{cat} > E''_{cat}$, in this example, the first step will be the rate-determining step

Fig 17.24   A catalysed reaction takes place via an alternative pathway involving a lower activation energy. The overall enthalpy change for the reaction is unchanged.

It is common to classify catalysts as one of two types:

- **heterogeneous**, when the catalyst is in a different phase to the reaction mixture;
- **homogeneous**, when the catalyst is in the same phase as the reaction mixture.

Enzymes are biological catalysts and do not fit neatly into either category since their sizes are intermediate between those of particles in solution and those regarded as being in suspension. This situation is often described as a **colloidal suspension**. Because of their special biological roles enzymes are regarded as a category of catalyst.

Fig 17.25 The decomposition of hydrogen peroxide is rapid in the presence of manganese(IV) oxide, a heterogeneous catalyst. Here a 30% aqueous solution of hydrogen peroxide is allowed to drip onto solid manganese(IV) oxide. Note the oxygen gas evolved.

The decomposition of hydrogen peroxide illustrates the effect of a chemical catalyst or an enzyme. Hydrogen peroxide is thermally unstable, decomposing to water and oxygen, but without a catalyst the reaction takes place only very slowly. With the addition of a small quantity of manganese(IV) oxide, the reaction is extremely rapid, taking place at the surface of the solid (Fig 17.25).

$$2H_2O_2(aq) \longrightarrow 2H_2O(l) + O_2(g)$$

In this reaction, the manganese(IV) oxide is acting as a heterogeneous catalyst.

Homogeneous catalysis of hydrogen peroxide decomposition is also possible. Addition of a little bromine makes the reaction occur very rapidly. The bromine is believed to oxidise hydrogen peroxide to hydrogen ions and oxygen.

$$H_2O_2(aq) + Br_2(aq) \longrightarrow 2Br^-(aq) + 2H^+(aq) + O_2(g)$$

In a second step bromide ions are oxidised back to bromine by more hydrogen peroxide:

$$2Br^-(aq) + H_2O_2(aq) + 2H^+(aq) \longrightarrow Br_2(aq) + 2H_2O(l)$$

The bromine reacts in the first reaction, but is regenerated in the second. It is **not**, therefore, used up in the reaction.

Hydrogen peroxide is a natural metabolic product which is extremely damaging and must be decomposed rapidly into harmless water and oxygen. The enzyme **catalase** provides a more gentle reaction pathway for its decomposition in the body. A single catalase molecule may decompose 40 000 molecules of hydrogen peroxide in just one second at 40°C within the human liver, and is far more efficient than chemical catalysts such as manganese(IV) oxide. The toxic hydrogen peroxide is safely decomposed. Catalase allows reaction to take place very rapidly at body temperature by lowering the activation energy for the process.

20 **Which catalysed decomposition of hydrogen peroxide has the highest activation energy, with catalase or manganese(IV) oxide?**

CATALYTIC CONVERTERS AT WORK

Catalysts have an important role to play in improving air quality. Increasingly, catalytic converters are fitted to most cars as standard equipment. This has been shown to decrease environmental levels of carbon monoxide, CO and nitrogen dioxide, NO_2, – two toxic pollutants released from car exhausts. The catalytic converter serves two purposes. It oxidises carbon monoxide (and unburnt petrol) to carbon dioxide and reduces nitrogen dioxide to nitrogen gas. Many converters use a platinum-rhodium alloy although transition metal oxides such as copper(II) oxide or chromium(III) oxide have been used.

Fig 17.26 A typical catalytic converter shown in cross-section. The catalyst is in the form of beads containing mixtures of platinum and rhodium. The two chambers contain different catalysts and have different operating temperatures to maximise conversion of gases. Catalytic converters of different design are used in factory chimneys to improve the safety of gaseous emissions.

THE RATES OF CHEMICAL REACTIONS

Autocatalysis

Sometimes one of the reaction products can itself catalyse a reaction. This type of catalysis is called autocatalysis. In this situation, the reaction begins slowly but the rate of reaction increases as the catalytic product is formed. A good example of autocatalysis is in the oxidation of ethanedioate ions by potassium manganate(VII) at room temperature in the presence of acid.

$$2MnO_4^-(aq) + 5C_2O_4^{2-}(aq) + 16H^+(aq) \longrightarrow 10CO_2(g) + 2Mn^{2+}(aq) + 8H_2O(l)$$

Here manganese(II) ions, $Mn^{2+}(aq)$, once formed, catalyse the forward reaction which produced them.

21 How could the rate of this reaction be followed?

Fig 17.27 Evolution of carbon dioxide is a measure of the rate of the autocatalysed reaction between ethanedioate ions and manganate(VII) ions. Manganese(II) ions formed during the reaction act as a catalyst.

Promoters, inhibitors and catalyst poisons

The mode of action of a catalyst may be modified by the presence of other substances, often in trace quantities. These substances are

- promoters;
- inhibitors;
- catalyst poisons.

Promoters are substances which may improve the performance of a catalyst. They do not increase rates of reaction when used alone but support the activity of a catalyst. For example, in the Haber process a promoter is used to increase the effectiveness of the iron catalyst. Traces of molybdenum are used as a promoter in some ammonia production plants whilst alternative promoters for this reaction include potassium or aluminium oxide and, increasingly, potassium hydroxide.

CONNECTION: The Haber process, page 466.

Inhibitors are sometimes called negative catalysts since they slow down chemical reactions. They are believed to operate by reacting with some of the intermediates of the reaction, so reducing the formation of products. An example of an inhibitor is propan-1,2,3-triol (glycerol) which can slow down the decomposition of hydrogen peroxide.

Catalyst poisons such as hydrogen sulphide, mercury(II) salts, arsenic, carbon monoxide and hydrogen cyanide greatly reduce the effectiveness of catalysts. One catalyst which has been used as an alternative to vanadium(V) oxide in the Contact process is platinum. This is an effective catalyst and is usually used at temperatures around 450°C but the quality of starting materials must be carefully controlled since it is poisoned by even very small quantities of arsenic. The effectiveness of catalyst poisons is thought to be due to their capacity to bind to the catalytic site of the catalyst. They are also poisons within biological systems because they bind to proteins – particularly the active sites of enzymes thus preventing their normal activity.

CONNECTION: The Contact process, page 466.

17.16 (a) Explain carefully the meaning of the term catalyst;
  (b) How does the use of a suitable catalyst increase the rate of a reaction?
  (c) Why do catalysts not affect the overall enthalpy change for a reaction?

17.17 For a catalysed reaction of your choice, describe an experiment to follow the progress of a reaction and to determine the effectiveness of the catalyst.

17.18 Give two examples of important industrial processes which employ a catalyst, and state why the use of a catalyst is important.

17.19 Explain, giving an example, what is meant by the term *homogeneous catalyst*.

17.20 Use (a) an energy profile diagram and (b) a Maxwell-Boltzmann distribution curve, to explain the influence of a suitable catalyst on the rate of a chemical reaction.

ACTIVITY

Benzenediazonium chloride is used in organic synthesis. It cannot be isolated but may be prepared *in situ* by the action of nitrous acid on phenylamine at around 0–5°C. At these temperatures it is stable for a reasonable time and may be used in further reactions.

At temperatures above 5°C benzenediazonium chloride becomes increasingly unstable, decomposing to phenol, nitrogen and hydrochloric acid. The following data is for the decomposition of benzenediazonium chloride at 40°C, monitored by measuring the pressure increase during the reaction. The reaction flask was connected to a manometer containing methylbenzene. The height of methylbenzene in the manometer is proportional to the pressure in the reaction flask.

Time from start t/s	Height of methylbenzene h/mm
0	0
300	26
600	51
900	73
1200	90
1500	105
1800	120
2100	138
2400	145
2700	153
3000	161
3300	170
3600	177

When the reaction was left for a very long time the maximum height achieved in the manometer was 225 mm.

Use the above data to show that the reaction is first order with respect to benzenediazonium chloride. Is it possible to find a value for k, the rate constant?

Hint: The concentration at a given time, t, is proportional to:

maximum height of methylbenzene – height of methylbenzene at time t

Answers to Quick Questions: Chapter 17

1   The higher the surface area, the greater contact with the other reactant and, therefore, the faster the reaction.

2   To achieve low or high temperatures requires expensive equipment and makes high energy demands.

3   A reaction in a cell occurs in something close to a closed system. In a test-tube, the reaction occurs in an open system.

4   Hydrated silver ions and chloride ions move relatively freely in aqueous solution and collide with one another frequently, leading to reaction.

5   The rate of increase in mass could be measured as the iron becomes oxidised.

6   Concentrations of components and catalyst; temperature.

7   By titration of a 'blank', that is, a mixture made up with the catalyst but no reactants.

8   The solution would be colourless.

9   By an acid-base titration since the reaction mixture will become increasingly acidic as the reaction proceeds.

10   Decrease

11   The rate is constant throughout the course of the reaction.

12   By monitoring changes in pressure.

13   The reaction would become cloudy (opaque) and, eventually, a pale yellow precipitate of sulphur would settle out.

14   The rate would double.

15   The units of time – commonly, seconds or minutes.

16   The rate would double.

17   The reaction will probably occur about three times as quickly.

18   It would also decrease.

19   A greater number of reacting species will have sufficient energy to overcome the energy barrier to reaction (the activation energy).

20   Catalase

21   Change in colour of the solution: MnO_4^-(aq) is an intense purple, Mn^{2+}(aq) is a very pale pink (colourless when dilute).

Chapter 18

CHEMICAL EQUILIBRIA

LEARNING OBJECTIVES

When you have studied this chapter you should be able to:

1. explain, using suitable examples, what is meant by the terms reversible reaction and dynamic chemical equilibrium;

2. explain the terms homogeneous and heterogeneous equilibria, and describe the characteristics of a chemical equilibrium;

3. explain equilibrium constant (both in terms of partial pressures and concentrations), and write expressions for the equilibrium constants of given reactions;

4. explain the relationship between the equilibrium constant and the position of equilibrium;

5. calculate the value of equilibrium constants or the quantities present in an equilibrium system from appropriate data;

6. explain Le Chatelier's principle and describe its usefulness;

7. describe the influence of concentration, pressure, temperature and a suitable catalyst upon a chemical system at equilibrium;

8. explain why position of equilibria are of importance to industrial chemists, using the Haber process and the Contact process as examples.

18.1 EQUILIBRIA: A QUESTION OF BALANCE

Reversible reactions and dynamic equilibrium

Many chemical reactions appear to simply convert reactants to products, until one of the reactants is used up. Then they stop. For example, magnesium burns in air to form magnesium oxide. Once alight, the magnesium ribbon flares up and in a matter of a few seconds is converted to magnesium oxide:

$$2Mg(s) + O_2(g) \longrightarrow 2MgO(s)$$

Fig 18.1 Although it takes a high temperature to begin the reaction, magnesium burns in excess air to give a white ash of magnesium oxide. The reaction continues until the magnesium has been completely used up.

However, in a great many chemical reactions the reactants are not used up completely even when they are in plentiful supply. Some product is formed but the theoretical yield is not obtained. Why does the reaction fail to use up all the available starting materials? The answer lies in the reversible nature of such reactions.

In reversible reactions, products are formed and undergo further reaction to reform starting materials. Newly-formed starting materials then react to form more products and so on. In this type of reaction, there comes a point when no matter how long we wait (provided the reaction conditions are not altered), the proportions of reactants and products remain unchanged. The forward and reverse reactions are occurring at the same rate and the reaction is said to be in a position of **dynamic equilibrium**.

For example, dinitrogen tetroxide decomposes to form nitrogen dioxide but the reaction does not go to completion. Pure dinitrogen tetroxide dissociates into nitrogen dioxide:

$$N_2O_4(g) \longrightarrow 2NO_2(g)$$

But nitrogen dioxide molecules also recombine to form dinitrogen tetroxide:

$$2NO_2(g) \longrightarrow N_2O_4(g)$$

When the rates of these two reactions are the same, a position of dynamic equilibrium is established and a mixture of $NO_2(g)$ and $N_2O_4(g)$ is formed. Although the molecules are constantly combining and dissociating, the amounts of each type of molecule remains constant, provided conditions are not altered. We show the existence of a dynamic chemical equilibrium by the use of $\rightleftharpoons$ in place of a simple arrow in the equation for the reaction:

$$N_2O_4(g) \rightleftharpoons 2NO_2(g)$$

Because $N_2O_4(g)$ is colourless and $NO_2(g)$ is brown, the intensity of colour of the reaction can be used to determine the relative amounts of the two compounds in a mixture.

Fig 18.2 Equilibrium is reached when the forward and reverse reactions are taking place at the same rate.

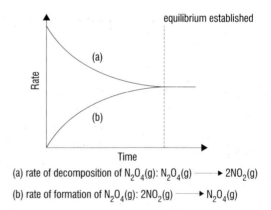

(a) rate of decomposition of $N_2O_4(g)$: $N_2O_4(g) \longrightarrow 2NO_2(g)$

(b) rate of formation of $N_2O_4(g)$: $2NO_2(g) \longrightarrow N_2O_4(g)$

1 **Increasing the pressure causes the mixture to lighten in colour. What has happened to the relative amounts of $N_2O_4(g)$ and $NO_2(g)$ in the mixture?**

A simple analogy might help to understand the idea of dynamic equilibrium. When two people sit balanced on a see-saw, the forces are equal and we say that the system is in **static equilibrium** – nothing is moving (Fig 18.3). When people exercise on a tread-mill they must run fast enough not to fall off backwards, but not so fast that the moving belt carries them forward. At equilibrium, the rates at which the belt and the person are moving are the same, and the system is in a state of **dynamic equilibrium** – it is balanced but in constant motion (Fig 18.4).

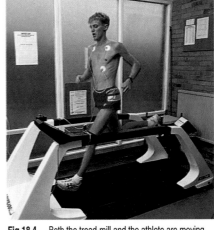

CONNECTION: *Dynamic equilibrium and fractional distillation, page 144.*

Fig 18.3 These children are balancing the see-saw. There is no movement and the system is in static equilibrium.

Fig 18.4 Both the tread-mill and the athlete are moving. When the movement of one balances the movement of the other, athlete and tread-mill are in dynamic equilibrium. Many of the chemical reactions taking place in the athlete's body are also in dynamic equilibrium.

The concept of reversible reactions which establish a position of dynamic equilibrium under a particular set of conditions has enormous implications across chemistry, from large scale manufacturing industries to the chemical changes which occur in our bodies.

SPOTLIGHT

Fig 18.5a The production of ammonia is a key process in making nitrate fertilisers. Feeding the world's growing population relies increasingly on the use of fertilisers to maximise crop yields.

AMMONIA AND FOOD PRODUCTION

In many parts of the world, crop yields are limited by the availability of water and of nitrates in the soil. Modern agriculture depends to a large and increasing extent on the use of industrially manufactured fertilisers, particularly nitrates. The Haber process uses the reaction between nitrogen and hydrogen to make ammonia, a key starting material for nitrate production. At the heart of this process is the reversible reaction:

$$N_2(g) + 3H_2(g) \rightleftharpoons 2NH_3(g)$$

Ammonia production takes place on a massive scale. World-wide production currently exceeds 100 million tonnes. The modern ammonia plant aims to achieve the highest production of ammonia at the lowest cost. As is so often the case, the design solution requires compromise. A yield of only 15% of ammonia is achieved but unreacted nitrogen and hydrogen may be recycled.

Fig 18.5b Changing conditions of temperature and pressure alters the composition of an equilibrium mixture of nitrogen, hydrogen and ammonia. An iron catalyst increases the rate of attainment of this equilibrium. However, complete conversion of nitrogen and hydrogen to ammonia is not achieved under any conditions.

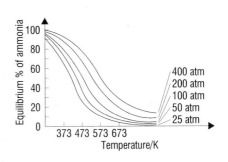

Chemical equilibria

Before we generalise about chemical equilibria, it will be helpful to consider two examples:

- the formation of an ester;
- the hydrogen, iodine, hydrogen iodide equilibrium.

The formation of an ester

Ethanoic acid and ethanol react to form ethyl ethanoate and water. The reaction is reversible:

$$CH_3COOH(aq) + C_2H_5OH(aq) \rightleftharpoons CH_3COOC_2H_5(aq) + H_2O(l)$$

Suppose a mixture of ethanoic acid and ethanol is prepared and divided equally between test-tubes, each of which is kept under the same conditions. The reaction can be monitored by acid-base titrations. At specific times, the contents of one test-tube are titrated against a standard alkali using a suitable indicator. The amount of alkali required allows us to establish the amount of acid present when the reaction mixture is sampled.

2 On the same graph, sketch curves to show how the rates of esterification (forward reaction) and hydrolysis (reverse reaction) change with time.

Initially, the amount of alkali needed decreases each time a titration is performed. However, eventually, it is the same in each successive titration, showing that the amount of ethanoic acid present in each test-tube has reached a constant value. Therefore, the amounts of ethanol, ethyl ethanoate and water must also have reached constant values. It is tempting to believe that chemical reaction has ceased – at least without any evidence that reaction continues. This appearance of calm is misleading, however. Modern spectroscopic methods have allowed us to take this and many similar investigations much further.

Suppose now that we prepare the same mixture of ethanoic acid and ethanol in a small test-tube and place it in a nmr spectrometer. Initially we see peaks in the ^{1}H NMR spectrum for ethanoic acid and ethanol. After a time, new peaks appear due to ethyl ethanoate and water. Eventually all the peaks reach constant values, confirming the finding of titrimetric analysis that the reaction does not go to completion. If deuterium oxide (D_2O) is added to the mixture, within a short time the peak due to water (H_2O) increases in intensity. This is strong evidence that, despite appearances, reaction continues. It has reached a point where the rate of the forward and reverse reactions are equal. A **chemical equilibrium** has been achieved.

3 Write an equation for the formation of deuterated ethanoic acid from $CH_3COOC_2H_5$ and D_2O and explain why the intensity of the peak due to H_2O increases.

Hydrogen and iodine

A typical reversible reaction is that which takes place between gaseous hydrogen and iodine at about 450°C:

$$H_2(g) + I_2(g) \rightleftharpoons 2HI(g)$$

The equation tells us that one mole of iodine molecules and one mole of hydrogen molecules can produce two moles of hydrogen iodide. The reaction may be studied by mixing equal amounts of hydrogen and iodine in a reaction flask, raising the temperature to 450°C and observing any changes. Iodine is an intensely-coloured element and this allows us to monitor changes in the amount of iodine present in the mixture by spectrophotometry. We can use the equation to calculate the amounts of the other two components.

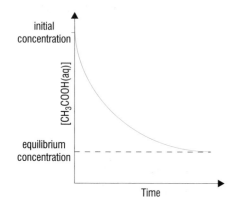

Fig 18.6 In the formation of ethyl ethanoate, the concentration of ethanoic acid decreases until it reaches a constant value. This is its equilibrium concentration.

CONNECTION: NMR spectroscopy, page 189.

CHEMICAL EQUILIBRIA

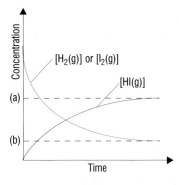

(a) Equilibrium concentration of HI(g)

(b) Equilibrium concentration of H₂(g)

Fig 18.7 As reaction proceeds both hydrogen and iodine are used at the same rate. They both achieve the same final concentration. The concentration of hydrogen iodide, initially zero, increases to a constant value.

The intensity of the iodine colour decreases with time. Eventually a point is reached where no further change in the intensity of the iodine colour is observed. Since intensity of absorption is proportional to concentration we can use this information, together with the equation, to determine the changes in concentration of all three components (Fig 18.7). At equilibrium, the rate at which hydrogen and iodine molecules react to give hydrogen iodide equals the rate at which hydrogen iodide molecules dissociate to give hydrogen and iodine.

4   **If one mole of hydrogen molecules and one mole of iodine molecules are placed in a reactor, we find that one mole of hydrogen iodide is present when equilibrium has been established. What is the percentage conversion of reactants to products?**

At equilibrium the mixture of reactants and products is called the **equilibrium mixture**. The same equilibrium mixture is formed by combining appropriate amounts of either products or reactants. For example, we could establish the same equilibrium concentrations for hydrogen, iodine and hydrogen iodide by starting either with the two elements or with hydrogen iodide (Fig 18.8).

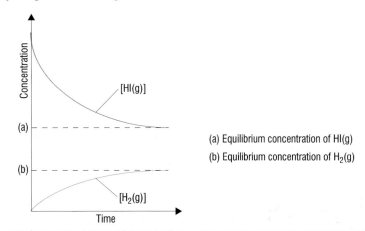

(a) Equilibrium concentration of HI(g)

(b) Equilibrium concentration of H₂(g)

Fig 18.8 Hydrogen iodide dissociates to form hydrogen and iodine. Eventually an equilibrium mixture is formed and the concentration of each equilibrium component becomes constant.

Fig 18.9 Stalagmites and stalactites are formed by the reaction of limestone (calcium carbonate) with rain water containing dissolved carbon dioxide. As the solution of calcium hydrogencarbonate which forms evaporates, calcium carbonate is deposited. This reversible reaction is an example of a heterogeneous equilibrium:

$$CaCO_3(s) + H_2O(l) + CO_2(aq) \rightleftharpoons Ca(HCO_3)_2(aq)$$

5   **Is the equilibrium between hydrogen, iodine and hydrogen iodide an example of homogeneous or heterogeneous equilibrium?**

The equilibrium constant

An important characteristic of any chemical equilibria is the composition of the equilibrium mixture. The amount of each component present in the equilibrium mixture may be expressed as a concentration or, when gases are involved, as a partial pressure. From studies of reactions at equilibrium, general principles have emerged which apply to all equilibria. For a generalised equilibrium with four participants A,B,C and D:

$$aA + bB \rightleftharpoons cC + dD$$

$$\frac{[C]^c[D]^d}{[A]^a[B]^b} = K_c$$

where K_c is a constant at a particular temperature regardless of the amounts of A,B,C and D used initially.

In this expression, the square brackets [ ] around a substance indicate its concentration in the equilibrium mixture measured in mol dm^{-3}. K_c is the **equilibrium constant** expressed in terms of equilibrium concentrations. For example:

$$PCl_5(g) \rightleftharpoons PCl_3(g) + Cl_2(g)$$

$$K_c = \frac{[PCl_3][Cl_2]}{[PCl_5]} \qquad \text{Units} = \frac{\text{mol dm}^{-3} \times \text{mol dm}^{-3}}{\text{mol dm}^{-3}} = \text{mol dm}^{-3}$$

$$CH_3COOH(aq) + C_2H_5OH(aq) \rightleftharpoons CH_3COOC_2H_5(aq) + H_2O(l)$$

$$K_c = \frac{[CH_3COOC_2H_5][H_2O]}{[CH_3COOH][C_2H_5OH]} \qquad \text{Units} = \frac{\text{mol dm}^{-3} \times \text{mol dm}^{-3}}{\text{mol dm}^{-3} \times \text{mol dm}^{-3}} = \text{no units}$$

6 **Write an equation for the equilibrium constant for the reaction: $2SO_2(g) + O_2(g) \rightleftharpoons 2SO_3(g)$ and give its units.**

For homogenous gaseous systems the amount of each component is usually expressed as its partial pressure in the mixture:

$$\frac{(p_C)^c(p_D)^d}{(p_A)^a(p_B)^b} = K_p$$

where K_p is the equilibrium constant expressed in terms of partial pressures (p_C, p_D, p_A and p_B) of the gases present at equilibrium. For example,

$$2NO_2(g) \rightleftharpoons N_2O_4(g)$$

$$K_p = \frac{(p_{N_2O_4})}{(p_{NO_2})^2} \qquad \text{Units} = \frac{\text{Pa}}{\text{Pa} \times \text{Pa}} = \frac{1}{\text{Pa}} = \text{Pa}^{-1}$$

KEY FACT

When the equilibrium constant has no units, K_c is numerically equal to K_p. When this is not the case the values of K_c and K_p are quite different.

Position of equilibrium

The equilibrium constant is a measure of how far a reaction proceeds until equilibrium is reached. For example, an explosive reaction occurs between hydrogen and fluorine:

$$H_2(g) + F_2(g) \rightleftharpoons 2HF(g)$$

The equilibrium constant for this process is about 10^{47} at typical reaction temperatures.

$$\frac{(p_{HF})^2}{(p_{H_2})(p_{F_2})} = 10^{47}$$

The term **position of equilibrium** is often used to describe the balance of reactants and products for a reaction at equilibrium. The position of equilibrium is not related to the rate at which equilibrium is achieved. Some reactions proceed to only a very limited extent and achieve equilibrium very rapidly. Others go virtually to completion but proceed very slowly.

Table 18.1

Reaction	Equilibrium constant, K_p
$H_2(g) + F_2(g) \longrightarrow 2HF(g)$	10^{47}
$H_2(g) + Cl_2(g) \longrightarrow 2HCl(g)$	10^{17}
$H_2(g) + Br_2(g) \longrightarrow 2HBr(g)$	10^9
$H_2(g) + I_2(g) \longrightarrow 2HI(g)$	1

This number is very large because the amount of product (top line) is vastly greater than the amount of the reactants (bottom line). In other words, the amount of reactants remaining is negligible. This reaction is normally regarded as one which goes to completion and is best represented using a simple equation:

$H_2(g) + F_2(g) \longrightarrow 2HF(g)$

Equilibrium constants for the reactions of halogens with hydrogen are shown in Table 18.1. The first three members of the halogen series have very large values for K_p and can be considered to be reactions which go to completion. They are not normally considered as equilibria. However, the value of K_p for the reaction of iodine with hydrogen is only 1. This is a reaction in a state of equilibrium because appreciable amounts of both reactants and products are present. The magnitude of the equilibrium constant indicates how far a reaction has proceeded.

7 Why do the equilibrium constants in Table 18.1 have no units?

QUESTIONS

18.1 What is meant by the term *dynamic chemical equilibrium*?

18.2 Write expressions for the equilibrium constant for each of the reversible reactions shown below expressing the equilibrium constant in terms of the equilibrium partial pressures of each component:
(a) $Cl_2(g) + 3F_2(g) \rightleftharpoons 2ClF_3(g)$
(b) $N_2O_4(g) \rightleftharpoons 2NO_2$
(c) $2SO_2(g) + O_2(g) \rightleftharpoons 2SO_3(g)$
(d) $COCl_2(g) \rightleftharpoons CO(g) + Cl_2(g)$
(e) $CO(g) + 2H_2(g) \rightleftharpoons CH_3OH(g)$

18.3 Write expressions for the equilibrium constant for each of the reversible reactions shown below expressing the equilibrium constant in terms of the equilibrium concentrations of each component:
(a) $CH_3COOCH_2CH_3(aq) + H_2O(l) \rightleftharpoons CH_3COOH(aq) + CH_3CH_2OH(aq)$
(b) $H_2O(l) + H_2O(l) \rightleftharpoons H_3O^+(aq) + OH^-(aq)$
(c) $HSO_4^-(aq) + H_2O(l) \rightleftharpoons H_3O^+(aq) + SO_4^{2-}(aq)$

18.4 Explain with the aid of two suitable examples the meaning of the terms (a) *homogeneous equilibria*; (b) *heterogeneous equilibria*.

18.5 The equilibrium constant for the dissociation of chlorine molecules to chlorine atoms at room temperature has a numerical value of about 1×10^{-38}. (a) Write an equilibrium expression for the dissociation; (b) what does the size of the equilibrium constant tell us about the extent of dissociation at this temperature?

18.2 QUANTIFYING EQUILIBRIA

Calculating K_p

Table 18.2 gives data for the reaction between hydrogen, iodine and hydrogen iodide at 500°C. The partial pressures of hydrogen and of iodine at the start of the reaction and of each of the components of the mixture once the equilibrium has been achieved are given.

Table 18.2 Equilibrium data for the reaction: $H_2(g) + I_2(g) \rightleftharpoons 2HI(g)$

Experiment	Partial pressures of reactants / 10^{-6} Pa		Partial pressures of equilibrium components / 10^{-6} Pa		
	$H_2(g)$	$I_2(g)$	$H_2(g)$	$I_2(g)$	$HI(g)$
1	7.27	4.22	3.41	0.361	7.72
2	7.45	6.04	2.33	0.925	10.2
3	7.47	5.33	2.77	0.643	9.38

The equilibrium constant for this reaction, K_p, may be calculated using the results from Experiment 1 in Table 18.2. Substituting the partial pressures from the table into the equilibrium expression shown below, we get the following:

$$K_p = \frac{(p_{HI})^2}{(p_{H_2})(p_{I_2})} = \frac{(7.72 \times 10^{-6})^2}{(3.41 \times 10^{-6})(0.361 \times 10^{-6})} = 48.41$$

Similar calculations can be performed for Experiments 2 and 3. The values obtained for K_p are 48.27 and 49.40 respectively. It is clear that K_p is indeed a constant and independent of the initial amounts of reactants used.

Calculating K_c

If the concentration of one component is accurately known at equilibrium and the initial amounts of components in the mixture are known, it is possible to determine the concentrations of all the equilibrium components. This enables us to determine a value for the equilibrium constant.

For example, consider an experiment in which 50 g of ethanoic acid and 20 g of ethanol are mixed and left to reach equilibrium at a constant temperature. The reaction mixture is analysed at equilibrium and 0.469 moles of ethanoic acid are found to be present at equilibrium. The equilibrium constant at this temperature may be calculated as follows:

$$CH_3COOH(aq) + C_2H_5OH(aq) \rightleftharpoons CH_3COOC_2H_5(aq) + H_2O(l)$$

mass of each component/g	50	20	0	0
molar mass /g mol^{-1}	60	46		
number of moles initially	$\frac{50}{60}$	$\frac{20}{46}$	0	0
number of moles at equilibrium	$0.833 - x$	$0.435 - x$	x	x

(where x is the number of moles of each reactant undergoing reaction)

But $0.833 - x = 0.469$

Therefore, $x = 0.833 - 0.469 = 0.364$, and so

number of moles at equilibrium	0.469	0.071	0.364	0.364

If we assume that the volume of the reaction mixture is V dm^3

equilibrium concentrations /mol dm^{-3}	$\frac{0.469}{V}$	$\frac{0.071}{V}$	$\frac{0.364}{V}$	$\frac{0.364}{V}$

$$K_c = \frac{[CH_3COOC_2H_5(aq)][H_2O(l)]}{[CH_3COOH(aq)][C_2H_5OH(aq)]}$$

$$K_c = \frac{(0.364/V)(0.364/V)}{(0.469/V)(0.071/V)} = 3.98$$

Therefore, the value for K_c at the temperature of the experiment is 3.98.

8 What would the units be for K_c for the formation of ethanol from ethene and steam:

$C_2H_4(g) + H_2O(g) \rightleftharpoons C_2H_5OH(g)$

Using equilibrium constants to calculate concentrations

It is possible to use the value of the equilibrium constant to calculate the composition of an equilibrium mixture at a particular temperature providing the balanced equation for the reaction is known.

For example, the reaction between nitrogen and oxygen takes place in the gaseous state to give nitrogen monoxide.

$N_2(g) + O_2(g) \rightleftharpoons 2NO(g)$

The equilibrium constant, K_p, determined experimentally at a temperature of 2960°C, is found to be 4×10^{-3}. If the total pressure is 100 kPa, let us see how to calculate the partial pressures of the equilibrium components in the equilibrium mixture at this temperature if equal amounts of nitrogen and oxygen are used.

Since nitrogen and oxygen are present initially in equal amounts (equal numbers of moles), the partial pressure of each initially is 50 kPa.

	$N_2(g)$	$O_2(g)$	$2NO(g)$
partial pressures initially / kPa	50	50	0
partial pressures at equilibrium / kPa	$(50 - x)$	$(50 - x)$	$2x$

(where x is the decrease in partial pressure for nitrogen or oxygen in kPa)

Therefore, $K_p = \frac{(2x)^2}{(50 - x)(50 - x)} = 4 \times 10^{-3}$

$4x^2 = 4 \times 10^{-3} \times (x^2 - 100x + 2500)$

$3.996x^2 + 0.4x - 10 = 0$

This is a quadratic equation of the type $ax^2 + bx + c = 0$, where $a = 3.996$, $b = 0.4$ and $c = -10$. Solving this equation for x gives a value of 1.53 kPa.

Therefore, the partial pressures at equilibrium are

$p_{N_2}(g) = 48.47$ kPa
$p_{O_2}(g) = 48.47$ kPa
$p_{NO}(g) = 3.06$ kPa

These figures show that reaction does not proceed very far in the forward direction at 2960°C. Little nitrogen monoxide is formed.

KEY FACT

For a quadratic equation of the type:

$ax^2 + bx + cx = 0$

the value for x can be calculated from the formula:

$$x = \frac{-b \pm \sqrt{b^2 - 4ac}}{2a}$$

Problem: Calculate the equilibrium concentration of triiodide ions, $I_3^-(aq)$, produced when 0.01 mole of iodine and 0.01 mole of potassium iodide are mixed to form 1 dm^3 of aqueous solution at 25°C. The equilibrium constant, K_c, for the reaction is 1.5×10^{-3} mol dm^{-3} at 25°C.

In 1dm^3 of solution:

$$I_3^-(aq) \rightleftharpoons I_2(aq) + I^-(aq)$$

number of moles initially	0	0.01	0.01
number of moles at equilibrium	x	$(0.01 - x)$	$(0.01 - x)$

(where x is the number of moles of triiodide ions formed at equilibrium)

Therefore, $K_c = \dfrac{[I_2(aq)][I^-(aq)]}{[I_3^-(aq)]} = 1.5 \times 10^{-3}$ mol dm^{-3}

$$\frac{(0.01 - x)(0.01 - x)}{x} = 1.5 \times 10^{-3}$$

$$\frac{x^2 - 0.02x + 0.0001}{x} = 1.5 \times 10^{-3}$$

$$x^2 - 0.02x + 0.0001 = 1.5 \times 10^{-3}x$$

$$x^2 - 0.0215x + 0.0001 = 0$$

Solving this quadratic equation for x, we find that

$x = 0.0147$ or $x = 0.0068$

$x = 0.0068$ is the only sensible answer since the maximum amount of triiodide ions which could be formed from 0.01 moles of iodine and 0.01 moles of iodide ions cannot possibly exceed 0.01 moles.

Therefore, the equilibrium concentrations are

$[I_2(aq)] = 0.0032$ mol dm^{-3}
$[I^-(aq)] = 0.0032$ mol dm^{-3}
$[I_3^-(aq)] = 0.0068$ mol dm^{-3}

Heterogeneous equilibria

Heterogeneous equilibria involve components in more than one phase and, strictly speaking, the idea of equilibrium constants does not apply to them. There are, however, some reactions which involve both gases and solids for which the concept of an equilibrium constant is relevant.

Solids exert a vapour pressure, contributing to the pressure of the gaseous phase. The vapour pressure of a solid, sometimes referred to as its dissociation pressure, is constant regardless of the mass of solid present. When an equilibrium is established the vapour pressure of a solid reactant does not change, and its numerical value may be combined with the equilibrium constant. Often simple expressions for such gas-solid equilibria may be obtained.

Consider the thermal decomposition of calcium carbonate:

$$CaCO_3(s) \rightleftharpoons CaO(s) + CO_2(g)$$

The equilibrium constant may be expressed in terms of the partial pressures of each component:

$$\frac{(p_{CO_2})(p_{CaO})}{(p_{CaCO_3})} = \text{a constant}$$

Fig 18.10 This piece of chalk (calcium carbonate) decomposes to calcium oxide and carbon dioxide when heated strongly. A position of equilibrium is not attained because this is an open system.

However, the values of (p_{CaCO_3}) and (p_{CaO}) do not change, whatever happens to the position of equilibrium. Therefore, they may be combined with the constant and we may write a simplified expression:

$$K_p = (p_{CO_2})$$

The partial pressure of carbon dioxide is independent of the amount of each solid present and is a constant (at a particular temperature). For gas-solid equilibria generally, the equilibrium expression is expressed in terms of the gaseous components only.

Another example is that of an equilibrium established when iron reacts reversibly with steam and we find a similar situation exists:

$$3Fe(s) + 4H_2O(g) \rightleftharpoons Fe_3O_4(s) + 4H_2(g)$$

For this equilibrium we can write an expression involving all four participants:

$$K_p = \frac{(p_{Fe_3O_4})(p_{H_2})^4}{(p_{Fe})^3(p_{H_2O})^4}$$

The vapour pressures of the solid components have constant values at a particular temperature and so the equation simplifies to:

$$K_p = \frac{(p_{H_2})^4}{(p_{H_2O})^4}$$

9 What are the units for K_p for this reaction?

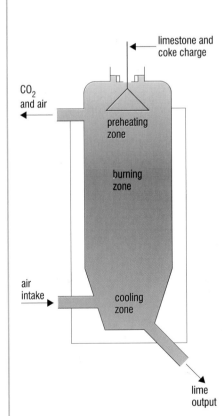

Fig 18.11b Cross-section of a simplified coke-fired vertical lime-kiln. Limestone and coke is added to the top of the kiln. Air is driven into the kiln using fans and a mixture of carbon dioxide and air escapes at the top.

LIME PRODUCTION – OLD AND NEW

Lime or calcium oxide is added to soils to increase their pH and optimise crop production. Early attempts to grow crops successfully in Exmoor in south-west England failed due to the acidic nature of the soil. To remedy this, imported limestone, mainly calcium carbonate, was roasted in primitive lime kilns to produce lime. The lime was then spread on the soil.

Current UK production of lime is in excess of 62 million tonnes – far beyond the capabilities of primitive kilns. Limestone is heated to temperatures of around 1470 K at which temperature very rapid dissociation of calcium carbonate to calcium oxide and carbon dioxide takes place:

$$CaCO_3(s) \longrightarrow CaO(s) + CO_2(g)$$

To drive the reaction in the forward direction great care is taken to ensure that a build-up of carbon dioxide is prevented.

Fig 18.11a Limestone and coal used to be transported by sea from South Wales to Heddon's Mount on the North coast of Devon, shown on this photograph. The limestone was converted into calcium oxide and used to combat the acid soils on Exmoor.

18.6 In an experiment conducted at 300°C, gaseous phosphorus(V) chloride was allowed to reach equilibrium with phosphorus(III) chloride and chlorine:

$$PCl_5(g) \rightleftharpoons PCl_3(g) + Cl_2(g)$$

K_p for the dissociation is known to be 1.16×10^6 N m^{-2} at this temperature. The partial pressures of phosphorus(V) chloride and phosphorus(III) chloride at equilibrium were found to be 1.52×10^5 N m^{-2} and 4.72×10^5 N m^{-2} respectively. Calculate the partial pressure of chlorine in this equilibrium.

18.7 Three moles of iodine and two moles of hydrogen are allowed to react and reach equilibrium with hydrogen iodide at a particular temperature according to the equation below:

$$H_2(g) + I_2(g) \rightleftharpoons 2HI(g)$$

Calculate the number of moles of each component present at equilibrium and the percentage composition of the equilibrium mixture if the equilibrium constant, K_c, has a value of 40 at that temperature.

18.8 Heating equal volumes of nitrogen and hydrogen at a constant temperature of 685 K in the presence of an iron catalyst eventually produced a mixture of ammonia, nitrogen and hydrogen which appeared to undergo no further change. Tests indicated that the percentage of ammonia in the equilibrium mixture was 15% under these conditions. The total pressure at the end of reaction was measured at 200 atmospheres. Calculate the equilibrium constant in terms of partial pressures.

18.9 Equal volumes of carbon monoxide and steam were combined at a temperature of 1000 K. The following equilibrium was established:

$$CO(g) + H_2O(g) \rightleftharpoons CO_2(g) + H_2(g)$$

The equilibrium constant for this reaction was found to have a numerical value of 1.4.
(a) Determine the fraction of each gas present at equilibrium;
(b) Explain why the equilibrium constant in this example has no units.

18.10 At room temperature solid iodine is in equilibrium with its vapour. Express the equilibrium constant for this dissociation in terms of the partial pressures of the equilibrium components.

18.11 A sample of the solid ammonium hydrogensulphide, NH$_4$HS, is allowed to partly dissociate into the gases ammonia and hydrogen sulphide within an evacuated vessel. When the pressure had reached a constant value of 0.645 atmospheres the system was assumed to have come to equilibrium. Calculate a value for the equilibrium constant expressed in terms of partial pressures, K_p.

18.12 During the baking of soda bread, sodium hydrogencarbonate decomposes according to the equation:

$$2NaHCO_3(s) \rightleftharpoons Na_2CO_3(s) + H_2O(g) + CO_2(g)$$

With a sample of sodium hydrogencarbonate at a particular temperature the total gas pressure due to this dissociation is 0.345 atmospheres. Calculate a value for the equilibrium constant at this temperature.

18.3 CONTROLLING EQUILIBRIA

Industries that manufacture chemical products need to control

- the yield of product;
- the time taken to achieve this (the 'throughput').

In reversible chemical processes the position of equilibrium is a major consideration in determining yield and, of equal importance, is the time taken to achieve equilibrium. Often people settle for lower yields but a short throughput time, and material is recycled to avoid wastage and keep the process economical.

Aspects of control

An understanding of the factors which affect equilibria is essential for economic survival of an industrial concern. Important processes such as the Haber process and the Contact process would be uneconomic without careful consideration of plant operating conditions.

Henri Le Chatelier, a French chemist, studied the factors which influence equilibria and in 1888 he summarised his findings as a general principle. This principle predicts the outcome of any change in the conditions under which an equilibrium is established. The principle is known as **Le Chatelier's principle**:

'If the conditions of a system at equilibrium are altered then the position of equilibrium moves in the direction which reduces the effect of the imposed change.'

This principle is quite general in its application and predicts the outcome of changes in:

- the concentration of an equilibrium component;
- the pressure to which the equilibrium is subjected;
- the temperature of the equilibrium system.

The influence of concentration

When the concentration of an equilibrium component is changed the position of equilibrium also changes. If we increase the concentration of a reactant, the equilibrium will be re-established at a new position but the new position of the equilibrium is such that the equilibrium constant remains unchanged.

Consider a reaction with the general equation:

$$A + B \rightleftharpoons C$$

$$K_c = \frac{[C]}{[A][B]}$$

If [A] increases, then [C] must increase and [B] must decrease in order that the ratio remains constant (the equilibrium constant, K_c). Now let us examine a typical equilibrium reaction in greater detail.

One of the simplest but most reliable tests for the presence of iron(III) ions which may be performed in the laboratory involves the addition of potassium thiocyanate to the solution of the suspected iron(III) compound. If the sample does contain hydrated iron(III) ions in solution a blood-red colouration is observed due to the formation of the pentaaquathiocyanato-iron(III) complex ion (Fig 18.12):

$$\{Fe(H_2O)_6\}^{3+}(aq) + SCN^-(aq) \rightleftharpoons \{Fe(H_2O)_5SCN\}^{2+}(aq) + H_2O(l)$$

pale yellow   no colour   blood-red

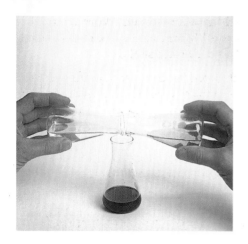

$$K_c = \frac{[\{Fe(H_2O)_5SCN\}^{2+}(aq)][H_2O(l)]}{[\{Fe(H_2O)_6\}^{3+}(aq)][SCN^-(aq)]}$$

Addition of a soluble iron(III) salt, or any thiocyanate salt, leads to the solution becoming a much darker red. The equilibrium position has moved to the right following an increase in concentration of a reactant. If, however, some of the iron(III) is precipitated as an insoluble salt and filtered off, the solution is found to be a much paler colour as the equilibrium moves to the left and the thiocyanato complex dissociates. Once again the equilibrium system is opposing the applied change by a shift in the equilibrium position.

Fig 18.12 The solution being formed when aqueous iron(III) chloride is mixed with aqueous potassium thiocyanate contains a mixture of ions in equilibrium. The blood-red colour is due to the [Fe(H$_2$O)$_5$SCN]$^{2+}$(aq) present.

OXYGEN TRANSPORT AND EQUILIBRIA

The transport of oxgyen around the body relies on a molecule known as haemoglobin, contained in red blood cells. Haemoglobin consists of four peptide chains each containing a haem group, responsible for binding oxygen. Each haem group binds one oxygen molecule, allowing each haemoglobin molecule to transport a maximum of four oxygen molecules at a time. The binding of oxygen is a reversible reaction in which the position of equilibrium is governed by the oxygen concentration in the blood:

deoxyhaemoglobin + oxygen $\rightleftharpoons$ oxyhaemoglobin

Blood returning to the lungs is exposed to high oxygen levels. Under these conditions the binding of oxygen to the haem group is favoured. The oxyhaemoglobin then returns to the tissues where the lower concentration of oxygen favours the release of oxygen. The high levels of carbon dioxide in metabolically active tissues also promotes a shedding of oxygen by haemoglobin.

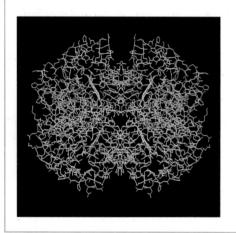

Fig 18.13 Deoxyhaemoglobin contains four haem groups each of which may reversibly bind an oxygen molecule forming oxyhaemoglobin. Where oxygen is required in the body tissues the binding process is reversed releasing essential oxygen to the cells. This computer model of haemoglobin shows haemoglobin's tetrameric structure.

The influence of pressure

The effect of a change in pressure is only really important for equilibria which involve gases. Increasing the pressure of a gaseous chemical equilibrium mixture will cause the position of equilibrium to shift to favour the side of the reaction with least gaseous molecules. This is predicted by Le Chatelier's principle, since the fewer molecules there are, the less pressure is exerted by the gas. So the position of equilibrium moves to oppose the increase in pressure.

For example, consider the gaseous equilibrium reaction:

$PCl_5(g) \rightleftharpoons PCl_3(g) + Cl_2(g)$

An increase in pressure will change the composition of the equilibrium mixture, causing an increase in the concentration of PCl_5 and a corresponding decrease in the concentrations of PCl_3 and Cl_2. The value for the equilibrium constant remains unchanged.

10 PCl₅ and PCl₃ are colourless, Cl₂ is yellow-green. What would be observed if the pressure of an equilibrium mixture was decreased and the mixture left to establish equilibrium again?

Le Chatelier's principle can be used to predict the best operating conditions for an industrial process. For example, consider the hydration of ethene, now used as the main industrial route to ethanol:

$$C_2H_4(g) + H_2O(g) \rightleftharpoons C_2H_5OH(g)$$

From Le Chatelier's principle, high yields (that is, the position of equilibrium lying to the right of the equation) would be favoured at high pressures. In practice, the reaction is conducted by passing ethene and steam over a phosphoric(V) acid catalyst at a temperature of 300°C and at a pressure of about 7×10^6 Pa (some seventy times normal atmospheric pressure).

The dissociation of dinitrogen tetroxide into nitrogen dioxide provides us with an example of a homogeneous gaseous equilibrium that can be studied in the laboratory. Nitrogen dioxide can be made by the action of concentrated nitric acid on copper turnings and collected in a gas syringe. Once the gas has cooled to room temperature (the reaction used to prepare it is exothermic), equilibrium is established and the gas itself will be a mixture of N_2O_4 and NO_2. Since $NO_2(g)$ is brown and $N_2O_4(g)$ is colourless, the colour of the mixture is an indication of the amount of each component.

$$N_2O_4(g) \rightleftharpoons 2NO_2(g)$$

11 Would this equilibrium mixture occupy more or less volume if the position of equilibrium moved to the left?

The pressure can now be increased by simply pushing the syringe plunger. Initially the gas becomes darker, simply because it is now more concentrated. However, it immediately begins to get lighter in colour and eventually settles to a fixed intensity. The position of equilibrium has shifted under the influence of higher pressure. Further nitrogen dioxide molecules have dimerised to form dinitrogen tetroxide. The sequence is shown in Fig 18.14.

Fig 18.14 The effect of increasing pressure on the equilibrium $N_2O_4(g) \rightleftharpoons 2NO_2(g)$ can be seen by this simple experiment in the laboratory. Remember, the darker brown the mixture is, the more nitrogen dioxide that is present.

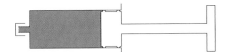

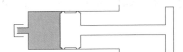

(a) Glass syringe containing an equilibrium mixture of $N_2O_4(g)$ and $NO_2(g)$

(b) Immediately after the pressure has been increased the gaseous mixture appears darker (a more concentrated mixture)

(c) After leaving for a few minutes the mixture lightens in colour and eventually, when equilibrium is re-established, the mixture is a lighter brown than the original (situation (a))

12 For what type of gaseous equilibrium is the position of equilibrium not affected by changes in pressures?

The influence of temperature

The effect of a temperature change upon an equilibrium reaction depends upon the enthalpy change for the reaction. Both the position of equilibrium and the value of the equilibrium constant are altered by a change in temperature. The rate of attainment of equilibrium is also greater at a higher temperature.

Position of equilibrium

Consider the reaction:

$$N_2(g) + O_2(g) \rightleftharpoons 2NO(g); \quad \Delta H = +180 \text{ kJ mol}^{-1}$$

An increase in temperature will favour the forward reaction. This is expected from Le Chatelier's principle, since the forward reaction is endothermic. An endothermic reaction absorbs energy and so offsets an increase in temperature. The system is opposing the imposed change.

In general, the effect of an increase in temperature on an equilibrium is to shift the position of equilibrium in the direction which leads to an absorption of heat, that is the endothermic reaction. The converse is also true. A decrease in temperature leads to a shift in the position of equilibrium in the exothermic reaction, that is the direction which leads to a generation of heat.

Fig 18.15 The decomposition of dinitrogen tetroxide into nitrogen dioxide is endothermic:

$$N_2O_4(g) \rightleftharpoons 2NO_2(g); \Delta H = +57.2 \text{ kJ mol}^{-1}$$

Therefore, the position of equilibrium lies further to the right at higher temperatures, and this can be seen from the change in the intensity of the brown colour of the gaseous mixture at different temperatures (keeping the pressure the same).

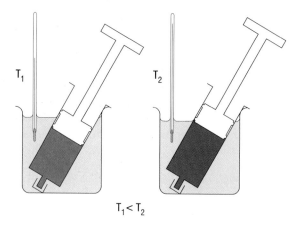

$T_1 < T_2$

Let us look again at the industrial manufacture of ethanol by the hydration of ethene:

$$C_2H_4(g) + H_2O(g) \rightleftharpoons C_2H_5OH(g)$$

This is an exothermic reaction. An increase in temperature drives this equilibrium to the left, that is, it favours the reverse reaction. Consequently, highest yields would be obtained at low temperatures. The problem, of course, is that the lower the temperature, the more slowly the reaction will proceed. Once again, a compromise must be reached. Use of a phosphoric(V) acid catalyst helps by increasing the rate even at lower temperatures.

Equilibrium constant

The change in the value of the equilibrium constant also depends upon the enthalpy change for the reaction under consideration. The equilibrium constant, in terms of partial pressures, for the reaction between nitrogen and oxygen to give nitrogen oxide is:

$$K_P = \frac{(p_{NO})^2}{(p_{N_2})(p_{O_2})}$$

KEY FACT

Changing temperature has a greater affect on an equilibrium system than changing concentration or pressure because it influences the position of equilibrium and the value of the equilibrium constant. An increase in temperature increases the rate of both the forward and the reverse reactions but does not affect them both equally.

An increase in temperature will increase the partial pressure of the products since formation of products is an endothermic process. The value of $(p_{NO})^2$ will increase, as will the value of K_p. A decrease in temperature will favour the reverse reaction and will increase (p_{N_2}) and (p_{O_2}).

13 **What will be the affect on the value of K_p for the reaction between nitrogen and oxygen if the temperature of the reaction mixture is decreased?**

The influence of catalysts

CONNECTION: Catalysts in industrial and biological processes, pages 518, 520 and 567.

Catalysts influence the rate of forward and reverse reactions to the same extent. Consequently, catalysts have no effect on the position of equilibrium or on the value of the equilibrium constant. A catalyst may, however, greatly influence the rate at which equilibrium is achieved and this is why a vast number of chemical processes employ catalysts in the search for the most efficient production method.

QUESTIONS

18.13 Consider the following equilibrium system:

$$Cu^{2+}(aq) + H_2S(aq) \rightleftharpoons CuS(s) + 2H^+(aq)$$

What effect will passing hydrogen sulphide gas into the equilibrium mixture be expected to have upon:
(a) the position of equilibrium;
(b) the mass of copper(II) sulphide precipitated;
(c) the value of the equilibrium constant;
(d) the acidity of the solution?
Assume that the solution is *not* saturated with hydrogen sulphide.

18.14 What effect is the addition of a small amount of sodium hydroxide likely to have upon the factors **(a)** to **(d)** in question 18.13?

18.15 A mixture of PCl_5, PCl_3 and Cl_2 was allowed to reach an equilibrium mixture at a particular temperature. If the applied external pressure is decreased and the system given time to regain equilibrium what would be the influence of this upon:
(a) the equilibrium constant;
(b) the proportion of chlorine in the reaction mixture;
(c) the partial pressure of phosphorus trichloride;
(d) the overall pressure of the system?

18.16 What effect does the addition of a suitable catalyst have upon a reversible reaction with respect to:
(a) the time taken to achieve equilibrium;
(b) the position of equilibrium;
(c) the composition of the equilibrium mixture;
(d) the equilibrium constant?

18.17 The decomposition of dinitrogen tetroxide is a reversible reaction which takes place according to the equation:

$$\underset{\text{colourless}}{N_2O_4(g)} \rightleftharpoons \underset{\text{brown}}{2NO_2(g)};\ \Delta H = +57.2\ \text{kJ mol}^{-1}$$

What would you expect to observe if this reaction was allowed to reach equilibrium and was then cooled?

The Haber process

It is said that Fritz Haber applied Le Chatelier's principle to deduce the optimum conditions for the synthesis of ammonia. The equation for the reaction is:

$$N_2(g) + 3H_2(g) \rightleftharpoons 2NH_3(g); \quad \Delta H = -92 \text{ kJ mol}^{-1}$$

The equation shows a ratio of three moles of hydrogen reacting with one mole of nitrogen to form two moles of ammonia, and so an increase in pressure should promote the forward reaction. The reaction is also exothermic. Therefore, a low temperature should promote the forward reaction. The effect of pressure and temperature on the equilibrium mixture is shown in Fig 18.5(b).

The equilibrium expression for the reaction is shown as:

$$K_p = \frac{(p_{NH_3})^2}{(p_{N_2})(p_{H_2})^3}$$

The value of K_p decreases quite dramatically with increasing temperature (Table 18.3). An increase in temperature favours the reverse reaction and the partial pressures of the reactants on the bottom line of the equilibrium expression increase, decreasing the value of K_p. This has obvious implications for the chemical industry which must choose a temperature which maximises the yield of ammonia while maintaining a satisfactory rate of production.

In practice the reaction takes place at 380–450°C and 200 atmospheres. These conditions are a compromise between the need to establish an equilibrium position as far to the right as possible (high pressure and low temperature) and the requirement to maximise the rate of reaction (high pressure and high temperature required). The removal of the ammonia by condensation also promotes the forward reaction (as predicted by Le Chatelier's principle). Catalysts are used to increase reaction rates.

Table 18.3 The variation of an equilibrium constant K_p for the equilibrium

Temperature / K	Equilibrium constant, K_p / atm^{-2}
625	0.0270
675	0.0129
725	0.0066
775	0.0038
825	0.0014

CONNECTION: Haber process and Contact process, pages 537 and 543.

The Contact process

A central reaction in the production of sulphuric acid by the Contact process is the oxidation of sulphur dioxide to sulphur trioxide:

$$2SO_2(g) + O_2(g) \rightleftharpoons 2SO_3(g); \quad \Delta H = -196 \text{ kJ mol}^{-1}$$

Le Chatelier's principle can be used to select the optimum reaction conditions. The forward reaction (production of sulphur trioxide) is promoted by high pressure since an increase in pressure will drive the reaction in the direction which leads to a reduction in volume, that is the forward direction in this case. Since the forward reaction is exothermic, it will be favoured by low temperatures. The forward reaction will also be promoted by removal of sulphur trioxide.

In practice the pressure used is slightly above atmospheric pressure as this is all that is required for a satisfactory yield of sulphur trioxide. The temperature employed is 420°C since this is an operating temperature at which the catalyst is effective. Below this temperature throughput of the reaction becomes unacceptably low.

QUESTION

18.18 'In industry, Le Chatelier's principle may be used to predict the influence of factors such as concentration, pressure and temperature on the yield of a reversible reaction. The predicted optimum conditions for such reactions are sometimes impractical'. By detailed reference to the Haber and Contact processes explain what is meant by the above statement.

Examine the following table of data for the equilibrium established between hydrogen, carbon dioxide, steam and carbon monoxide at 986°C:

$$H_2(g) + CO_2(g) \rightleftharpoons H_2O(g) + CO(g)$$

Expt.	Initial concentrations				Equilibrium concentrations				K_c
	$[H_2]$	$[CO_2]$	$[H_2O]$	$[CO]$	$[H_2]$	$[CO_2]$	$[H_2O]$	$[CO]$	
1	1.0	1.0	0	0	0.44	0.44	0.56	0.56	–
2	0	0	1.0	1.0	–	–	–	–	1.60
3	1.0	–	0	0	0.27	1.27	0.73	0.73	1.61
4	2.0	2.0	0	0	0.88	–	–	–	1.60
5	0	0	–	–	0.88	0.88	1.12	1.12	–
6	1.0	–	–	1.0	0.883	0.883	1.117	1.117	1.59
7	0.2	0.4	0.6	0.8	–	–	0.456	–	1.60

Make a copy of this table and determine the missing data to complete the table.

Answers to Quick Question: Chapter 18

1   The amount of $NO_2(g)$ has decreased and the amount of $N_2O_4(g)$ has increased.

2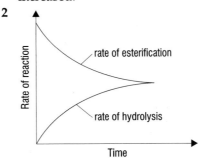

3   $CH_3COOC_2H_5 + D_2O \longrightarrow CH_3COOD + C_2H_5OD$
The deuterated ethanoic acid will then react with ethanol, and the deuterated ethanol will react with ethanoic acid – both reactions resulting in the formation of more water.

4   50%

5   Homogeneous

6   $K_c = \dfrac{[SO_3(g)]^2}{[SO_2(g)]^2[O_2(g)]}$
Units: $mol^{-1}\, dm^3$

7   There are the same number of moles on both sides of the equilibrium equation.

8   $mol^{-1}\, dm^3$

9   No units

10   The mixture would become more intensely yellow-green.

11   Less

12   Those where the number of moles of gas on either side of the equation are the same.

13   K_p will decrease.

Chapter 19

IONIC EQUILIBRIA

LEARNING OBJECTIVES

When you have studied this chapter you should be able to:

1. explain the term electrolyte (stating the difference between a strong and a weak electrolyte);

2. describe how ionisation occurs in aqueous solution;

3. explain the terms degree of ionisation and ionic product of water;

4. use the approximate form of Ostwald's dilution law to determine the degree of ionisation of a weak electrolyte;

5. describe the Brönsted-Lowry definitions of an acid and a base and show what is meant by the terms conjugate acid and conjugate base;

6. describe the Lewis theory of acids and bases;

7. explain the use of the pH scale;

8. explain what is meant by the relative strength of acids and bases;

9. calculate pH values of solutions given suitable data;

10. explain what is meant by acid and base dissociation constants;

11. calculate the pH of weak acids or bases and acid dissociation constants;

12. describe how pH may be measured in aqueous solutions;

13. state how pH changes during the course of acid-base titrations;

14. explain salt hydrolysis;

15. describe with examples the action of buffer solutions;

16. use solubility products to calculate the solubility of a sparingly soluble salt given suitable data;

19.1 IONISATION IN AQUEOUS SOLUTION

Many chemical reactions of interest to human beings, including those which take place within our own bodies, occur in aqueous solution. Of these, a high proportion involve substances which dissociate either partly or completely to give ions. They are said to **ionise** in solution. Hydrated ions in water behave very differently to unionised species. For example, white anhydrous copper(II) sulphate, $CuSO_4(s)$, consists of Cu^{2+} and SO_4^{2-} ions. It does not conduct electricity. When dissolved in water, it forms a blue solution which does conduct electricity. Hydrated ions are formed, $Cu^{2+}(aq)$ and $SO_4^{2-}(aq)$, which are free to move in solution, explaining the conduction of electricity. The blue colour is due to $Cu^{2+}(aq)$.

Ions in solution

In the 1830s, Michael Faraday showed that certain compounds in the molten state or in aqueous solution were able to conduct electricity. Faraday explained his results in terms of charged particles or ions present in solution. In 1887, the Swedish chemist Svante August Arrhenius suggested that all soluble ionic compounds and some covalent compounds ionise or 'come apart' to give aqueous ions. The Arrhenius theory was known as the 'theory of electrolytes' and dealt with the formation, number and speed of movement of ions in solution. This work earned Arrhenius the Nobel prize in 1903.

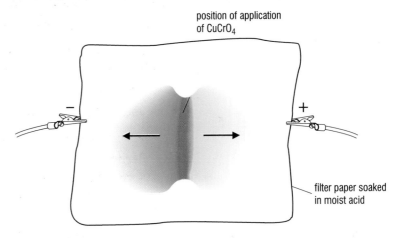

position of application of CuCrO₄

filter paper soaked in moist acid

Fig 19.1 Copper(II) chromate(VI), a brown solid, ionises in aqueous solution to give Cu²⁺(aq) (blue) and CrO₄²⁻(aq) (yellow) in solution. In an electric field two distinct colours are observed as these ions migrate and are seen separately, providing good evidence for ionisation.

Many chemical substances dissociate in aqueous solution to give hydrated ions. The solutions which form are called **electrolytes** and conduct electricity while, at the same time, being chemically changed by the passage of the electrical current.

Solubility in water is not limited to those substances which form ions by dissociation. Many substances, for example sugars, dissolve in water but are non-electrolytes. These substances are water-soluble not because they consist of ions but because they consist of polar molecules which form strong intermolecular bonds with water molecules. It is also true that many ionic compounds do not dissolve in water.

> **KEY FACT**
>
> Ionic compounds dissociate to give ions when melted. The molten compounds conduct electricity while being simultaneously decomposed. The melts are electrolytes.

CONNECTION: Solubility in water, page 86.

1 **What is responsible for the passage of electricity through a copper wire and how does this differ from the conductivity of a solution of copper(II) sulphate?**

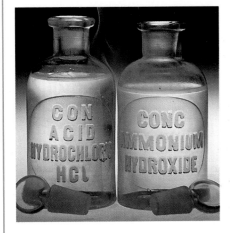

REACTIONS IN LIQUID AMMONIA

Ionic compounds are generally soluble in water but very insoluble in non-polar solvents such as hexane and benzene. By far the majority of reactions involving ionic compounds are performed in aqueous solutions. For some reactions, however, non-aqueous solvents such as liquid ammonia may be used. Like water, ammonia undergoes auto-ionisation to produce ions:

$2NH_3(l) \rightleftharpoons NH_2^-(am) + NH_4^+(am)$.

Despite the difficulty of working with ammonia (boiling point −34.5°C) it has some useful properties which make it an attractive alternative to water for some reactions.

Fig 19.2 Gaseous ammonia and hydrogen chloride react spontaneously, forming white clouds of ammonium chloride.

The process of ionisation

Sometimes it is essential to know the concentration of an ion in solution. This might be, for example, because a drug ionises in water and it is the ion which is responsible for the activity of the drug. Or it may be that we are interested in the concentration of lead or tin ions in water leaking from an old mine. Whatever the reason, the ability to predict the concentration of ions formed in solution is very useful. To do this we need to know the amount of substance present in solution and the extent to which it ionises.

Often the number of ions formed is precisely what would be expected on the basis of the compound's formula. For example, when magnesium chloride dissolves in water it completely ionises to give hydrated magnesium ions and hydrated chloride ions:

$$MgCl_2(s) + aq \longrightarrow Mg^{2+}(aq) + 2Cl^-(aq)$$

2 **How many moles of chloride ions are formed by the ionisation of one mole of magnesium chloride?**

If one mole of magnesium chloride is dissolved in water, three moles of hydrated ions are formed since the salt completely ionises in water. Electrolyte solutions of fully ionised substances in solution are called **strong electrolytes**. They include soluble ionic compounds and strong acids. For example,

$$Na_2SO_4(s) + aq \longrightarrow 2Na^+(aq) + SO_4^{2-}(aq)$$

$$HCl(g) + aq \longrightarrow H^+(aq) + Cl^-(aq)$$

CONNECTION: *Strong and weak acids, page 480.*

The number of ions formed in strong electrolytes is directly related to the formula of the compound and, therefore, can be accurately predicted.

However, some substances only partly ionise when they dissolve in water. In these cases there is no simple relationship between the number of ions formed and the formula of the compound. For example, if ethanoic acid is dissolved in water, it partly ionises. The majority of ethanoic acid molecules remain unionised and a dynamic equilibrium exists:

$$CH_3COOH(l) + aq \longrightarrow CH_3COOH(aq);$$
and then
$$CH_3COOH(aq) + H_2O(l) \rightleftharpoons CH_3COO^-(aq) + H_3O^+(aq)$$

3 **What happens to this equilibrium if a little hydrochloric acid is added?**

For each ethanoic acid molecule which ionises, two ions form. However, once equilibrium has been established, at any moment in time most of the ethanoic acid exists as molecules and very few are ionised. So the total number of particles in solution is only slightly increased by ionisation. Electrolytes which only partly dissociate in water are **weak electrolytes**. They include organic acids, organic bases and some inorganic acids. For example,

$$H_2S(aq) + H_2O(l) \rightleftharpoons HS^-(aq) + H_3O^+(aq)$$

$$HS^-(aq) + H_2O(l) \rightleftharpoons S^{2-}(aq) + H_3O^+(aq)$$

$$CH_3NH_2(aq) + H_2O(l) \rightleftharpoons CH_3NH_3^+(aq) + OH^-(aq)$$

Biologically important substances such as amino acids and proteins (including enzymes and hormones) are also weak electrolytes. Understanding their behaviour in living systems and in the laboratory has led to many improvements in health standards. Pharmaceutical and agrochemical companies make careful studies of the properties in solution of new products that they are developing.

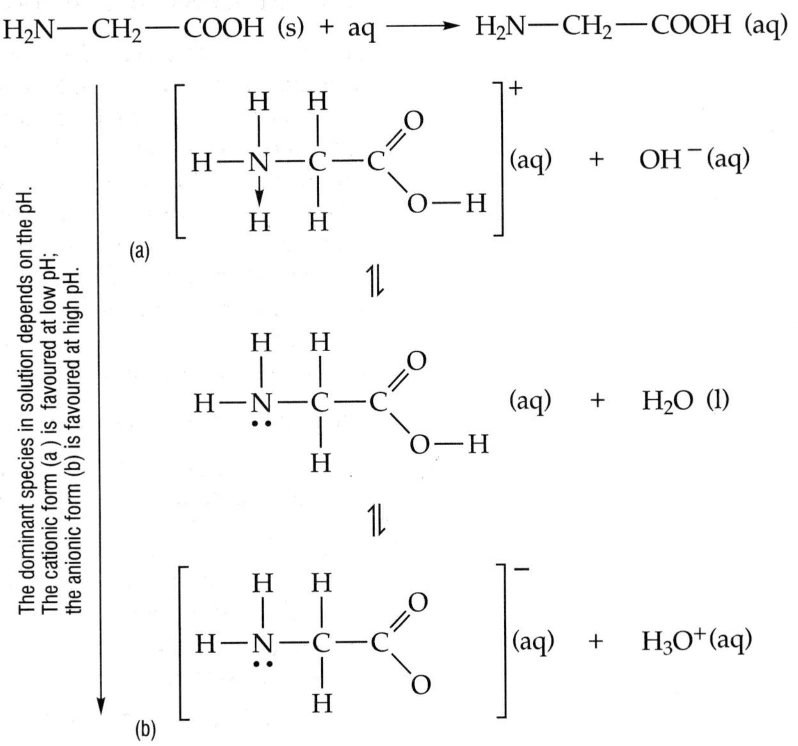

Fig 19.3 2-aminoethanoic acid (glycine) is the simplest amino acid. It is a weak electrolyte and the unionised molecules exist in aqueous solution in dynamic equilibrium with either the cationic form (a) or the anionic form (b), dependent upon pH.

$$H_2N—CH_2—COOH \ (s) + aq \longrightarrow H_2N—CH_2—COOH \ (aq)$$

The dominant species in solution depends on the pH. The cationic form (a) is favoured at low pH; the anionic form (b) is favoured at high pH.

The degree of ionisation

To describe ionisation it is helpful to consider the **degree of ionisation** taking place. This expresses the ionisation of a dissolved substance as a proportion of the amount present. For example, a degree of ionisation of 0.25 for a substance dissolved in water implies that $\frac{1}{4}$ (25%) of the amount of substance has ionised. The degree of ionisation of a weak electrolyte depends on its concentration and the relationship is given by **Ostwald's dilution law** (named after Friedrich Ostwald, a Russian-born, German chemist):

'For a weak binary electrolyte, the degree of ionisation is proportional to the square root of the reciprocal of the concentration of the electrolyte'.

Consider a weak electrolyte, AB (a binary electrolyte), which dissolves in water to give $A^+(aq)$ and $B^-(aq)$. At a concentration of c mol dm^{-3}, a fraction (α) of the electrolyte ionises. This fraction is the **degree of ionisation**. It has no units.

$$AB(aq) \rightleftharpoons A^+(aq) + B^-(aq)$$

initial concentration /mol dm^{-3}	c	0	0
equilibrium concentration /mol dm^{-3}	$(1-\alpha)c$	αc	αc

$$K_c = \frac{[A^+(aq)][B^-(aq)]}{[AB(aq)]} = \frac{(\alpha c)^2}{(1-\alpha)c} = \frac{\alpha^2 c}{(1-\alpha)}$$

◆ **4** **What are the units for K_c in this case?**

A large value of the degree of
ionisation α, implies an electrolyte
which is almost completely ionised.
A small value means little ionisation
has taken place.

For weak electrolytes, the degree of ionisation, α, is very small and the quantity $(1 - \alpha)$ is approximately equal to 1. Therefore, $K_c \approx \alpha^2 c$

5 **What would be the value of α for a strong electrolyte?**

This equation holds only for dilute solutions of weak electrolytes but it is valuable within these limits. Knowing the concentration of an electrolyte and the equilibrium constant for its ionisation we can estimate the degree of ionisation.

TUTORIAL

Problem: Aspirin is an organic acid, a weak electrolyte which dissociates in water to give ions:

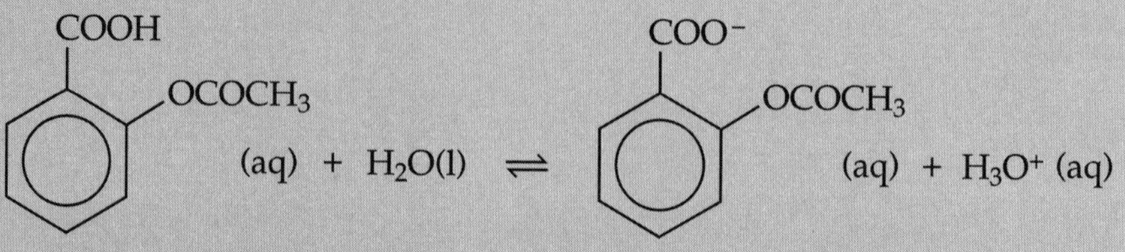

COOH
OCOCH$_3$
(aq) $+$ H$_2$O(l) $\rightleftharpoons$
COO$^-$
OCOCH$_3$
(aq) $+$ H$_3$O$^+$ (aq)

The equilibrium constant for the ionisation of aspirin in water is 3.3×10^{-4} mol dm^{-3}. Calculate its degree of ionisation in a solution of concentration 4×10^{-3} mol dm^{-3}.

Solution: $K_c = \alpha^2 c$

Therefore, $3.3 \times 10^{-4} = \alpha^2 \times 4 \times 10^{-3}$

$$\alpha^2 = \frac{3.3 \times 10^{-4}}{4 \times 10^{-3}} = 0.0825$$

Therefore, $\alpha = 0.29$ (or 29%)

The ionic product of water

Depending on its source, water may contain a large number of dissolved substances, both covalent and ionic. One way of purifying water is to distil it but, even if repeatedly distilled, water still retains a slight electrical conductivity. Therefore, ions must be present. These arise from the **auto-ionisation** of water molecules. Molecules of water react with one another to form ions:

H$_2$O(l) + H$_2$O(l) $\rightleftharpoons$ H$_3$O$^+$(aq) + OH$^-$(aq)
                    oxonium   hydroxide
                    ion   ion

$$K_c = \frac{[H_3O^+(aq)][OH^-(aq)]}{[H_2O(l)]^2}$$

A proton (H$^+$) is transferred from one water molecule to another and a position of dynamic equilibrium is established:

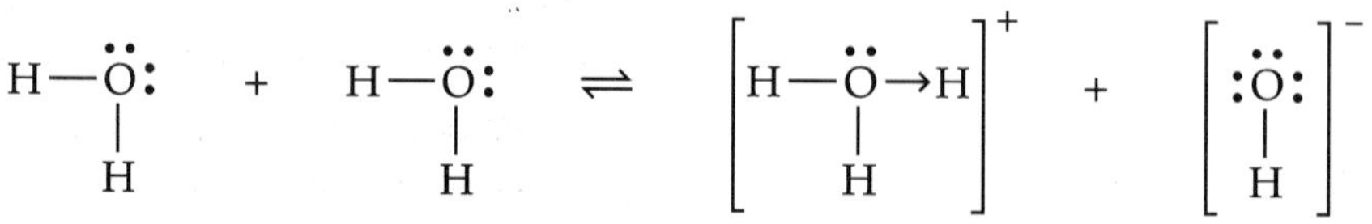

H—O: + H—O: $\rightleftharpoons$ $\left[\text{H—O} \rightarrow \text{H} \right]^+$ + $\left[:O: \right]^-$
    | | | |
    H   H   H   H

At equilibrium the concentration of water molecules in solution far outweighs the concentration of ions and the value of $[H_2O(l)]^2$ may be regarded as constant, whatever the temperature. Shifts in equilibrium might change the ion concentrations but, because they are present in such a large excess, the concentration of water molecules remains essentially constant. The 'approximately constant' value of $[H_2O(l)]^2$ may be incorporated into the equilibrium constant and the expression rewritten:

$$K_w = [H_3O^+(aq)][OH^-(aq)]$$

where K_w is the ionic product of water.

Fig 19.4 The auto-ionisation of water is increased at higher temperatures as indicated by the marked increase in the value of K_w.

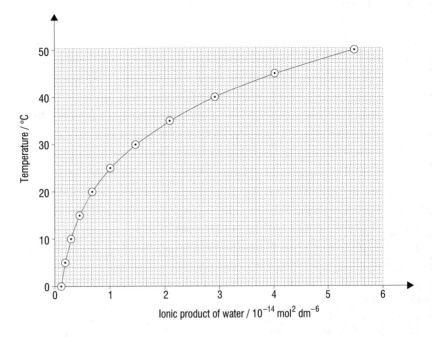

The ionic product of water has a value of 1×10^{-14} mol^2 dm^{-6} at 25°C, but it is very temperature-dependent (Fig 19.4). It has units of (concentration)^2, as expected from the expression for K_w. Individual concentrations of the ions are easily found from the ionic product of water:

$$K_w = [H_3O^+(aq)][OH^-(aq)] = 1 \times 10^{-14} \text{ mol}^2 \text{ dm}^{-6} \text{ at 25°C}$$

but $[H_3O^+(aq)] = [OH^-(aq)]$.

Therefore, $[H_3O^+(aq)]^2 = 1 \times 10^{-14}$

and $[H_3O^+(aq)] = [OH^-(aq)] = \sqrt{(1 \times 10^{-14})}$

$= 1 \times 10^{-7}$ mol dm^{-3}.

6 **From inspection of Fig 19.4, do you think the ionic product of water at its boiling point will be:**
   (a) 5.1×10^{-14} mol^2 dm^{-6};
   (b) 5.1×10^{-13} mol^2 dm^{-6};
   (c) 5.1×10^{-12} mol^2 dm^{-6}?

The ionic product of water, K_w, has a constant value at a particular temperature regardless of what is dissolved in it. This information may be used to calculate the concentration of either hydroxide ions or oxonium ions if the concentration of the other ion is known.

7 **Why are the concentrations of oxonium ions and hydroxide ions equal in pure water?**

Problem: Calculate the oxonium ion concentration in aqueous sodium hydroxide at 25°C in which the hydroxide ion concentration is 1×10^{-2} mol dm^{-3}. The ionic product of water at this temperature is 1×10^{-14} mol^2 dm^{-6}.

Solution: $K_w = [H_3O^+(aq)][OH^-(aq)] = 1 \times 10^{-14}$ mol^2 dm^{-6} at 25°C.

Therefore, $[H_3O^+(aq)] \times 1 \times 10^{-2} = 1 \times 10^{-14}$ mol dm^{-3}

$$[H_3O^+(aq)] = \frac{1 \times 10^{-14}}{1 \times 10^{-2}} \text{ mol dm}^{-3}$$

$= 1 \times 10^{-12}$ mol dm^{-3}.

Therefore, the oxonium ion concentration in 1×10^{-2} mol dm^{-3} sodium hydroxide at 25°C is 1×10^{-12} mol dm^{-3}.

19.1 Explain, giving suitable examples, what is meant by the terms *electrolyte* and *non-electrolyte*.

19.2 The ionic product of water, K_w, is normally quoted as 1×10^{-14} mol^2 dm^{-6} at 25°C. If the value of K_w at 30°C is 1.62×10^{-14} mol^2 dm^{-6} what does this imply about the enthalpy change for the auto-ionisation of water? Explain your answer.

19.3 Calculate the concentration of hydroxide ions in a solution containing oxonium ions at a concentration of 0.01 mol dm^{-3} given that the ionic product for water at the solution temperature is 1.00×10^{-14} mol^2 dm^{-6}.

19.4 A weak binary electrolyte, AB, is dissolved in water at 20°C to give an initial concentration of 0.95 mol dm^{-3}. If the dissociation constant of the electrolyte is 2.46×10^{-4} mol dm^{-3} at 20°C calculate:
(a) the degree of dissociation which takes place;
(b) the percentage dissociation which takes place;
(c) the concentration of B$^-$ ions present after dissociation at 20°C.

19.2 ACIDS AND BASES

Since the earliest days of chemistry there have been many attempts to understand the properties of the substances we call acids and bases. Much of the earliest work was concerned with their behaviour in aqueous solution. Acids dissolve in water to produce **acidic solutions** and their properties in solution are quite different to those in the absence of water. Bases which are soluble in water are known as **alkalis**. Their solutions in water are said to be **alkaline solutions**.

Characteristic reactions of acids and bases

Acids were first classified as substances having a characteristic sour taste. Those bases which dissolved in water on the other hand, produced solutions which were found to have an obvious, bitter taste. The use of taste to identify acids and bases has, predictably, a limited appeal for the modern chemist!

Another characteristic of acids and bases is their ability to change the colour of certain dyes, either natural (for example, vegetable dyes) or artificial. Certain plant extracts, including litmus, were used by Robert Boyle in 1663 to detect the presence of acids and bases. These substances are known as **indicators**.

Fig 19.5   Litmus is a coloured compound extracted from certain plants. It is red in acidic solutions and blue in alkalis.

Acids and bases react together to remove each other's characteristic properties in a process called **neutralisation**, to produce a salt and water. Acids also react with certain metals to form hydrogen and a salt, and with carbonates and hydrogencarbonates to give a salt, carbon dioxide and water.

8 Write equations for the reaction of dilute sulphuric acid with:
(a) potassium hydroxide; (b) magnesium; (c) sodium carbonate.
(Use the Database if necessary.)

Arrhenius, in 1884, showed that acids ionised in aqueous solution to give hydrogen ions (H^+), while bases dissociate to give hydroxide ions (OH^-). Both these ions are hydrated: $H^+(aq)$ and $OH^-(aq)$. For example,

CONNECTION: The hydrated hydrogen ion (the oxonium ion), page 472.

$$H_2SO_4(l) + aq \longrightarrow 2H^+(aq) + SO_4^{2-}(aq)$$
$$KOH(s) + aq \longrightarrow K^+(aq) + OH^-(aq)$$

SPOTLIGHT

SULPHURIC ACID AND THE 'EVENING STAR'

The beautiful clouds of Venus are probably droplets of concentrated sulphuric acid and are suspended in an atmosphere of carbon dioxide. The early Russian-attempted landings on Venus were unsuccessful. The probes were crushed by an atmosphere almost 100 times more dense than that of Earth. Later probes sent in 1975 survived for about an hour and sent back the first pictures of the surface of Venus. Their destruction was probably caused by a combination of high pressure, high temperature and a corrosive atmosphere.

At ground level the temperature is around 500°C on Venus. At this temperature droplets of sulphuric acid falling as 'rain' would decompose to sulphur trioxide and water, reforming sulphuric acid at higher altitude where the lower temperature is cooler.

Fig 19.6 The clouds of Venus as photographed by the space probe Magellan in 1990. NASA plans to send new probes to Venus to investigate the interaction between the planet's surface, atmosphere and sulphuric acid clouds. How long will they last out against such hostile conditions?

---

9 Why are solutions of acids and bases in water electrolytes?

Brönsted-Lowry theory of acids and bases

The study of acids and bases took a great step forward in 1923 with the simultaneous publication by the Dane, Johannes Brönsted and the British chemist Thomas Lowry of a new theory of acidity. It defined an acid as a **proton donor** and a base as a **proton acceptor** and became known as the Brönsted-Lowry theory. A proton, of course, is another name for a hydrogen ion. The Brönsted-Lowry theory explains why pure acids behave differently from their aqueous solutions since, for an acid to behave as a proton donor, there must be a substance present to accept the proton. Water is such a substance and it acts as a base (a proton acceptor) in an aqueous solution of an acid.

> **KEY FACT**
>
> The Brönsted-Lowry definitions of acids and bases are:
>
> an acid is a proton donor;
> a base is a proton acceptor.

$$H_2SO_4(l) + 2H_2O(l) \longrightarrow 2H_3O^+(aq) + SO_4^{2-}(aq)$$

Note in both nitric acid molecules and nitrate ions, delocalization of electrons occurs:

An acid may be represented as HA (H for hydrogen and A for the remaining part of the acid). In the following equation, $A^-(aq)$ is formed from the acid HA by transfer of a proton to a water molecule. The water is acting as a Brönsted-Lowry base – a proton acceptor.

$$HA(aq) + H_2O(l) \longrightarrow H_3O^+(aq) + A^-(aq)$$

We shall see later that some acids do not fully ionise, and an equilibrium is established between the unionised and ionised forms.

10 **In many cases the reaction between HA and H$_2$O is reversible. Explain, therefore, why A$^-$(aq) may be regarded as a base.**

Fig 19.8 Acids do not behave as acids in the absence of a suitable proton acceptor. Here, aqueous ethanoic acid reacts with magnesium ribbon readily, unlike the 'glacial' ethanoic acid which is much more concentrated and contains very little water.

When a base dissolves in water, the water acts as an acid (a proton donor) and the base accepts the proton:

$$B(aq) + H_2O(l) \longrightarrow BH^+(aq) + OH^-(aq)$$

The species BH^+ is formed when B gains a proton from a water molecule. In the reverse reaction, B is reformed by loss of a proton from BH^+. Therefore, B and BH^+ are linked by the transfer of a proton. We say that BH^+ is the **conjugate acid** of B, and B is the **conjugate base** of BH^+.

11 **What is the conjugate base of the acid HA(aq)?**

Water, therefore, can behave as a base and as an acid. It is said to be **amphoteric**. The ability of a substance to behave as acid or base is called **amphoteric character** and this property is not restricted to water. For example, liquid ammonia, can be used as a solvent for some reactions. It may behave as an acid or a base towards other substances dissolved in it. In Fig 19.9 some typical acid-base conjugate pairs are shown.

12 **Write a simple equation for the auto-ionisation of ammonia.**

KEY FACT

A base which is soluble in water is called an alkali.

CONNECTION: The amphoteric character of some metal oxides and hydroxides, Key Fact box page 120.

Fig 19.9

Structural formulae of some typical acid-base conjugate pairs.

Conjugate acid		Conjugate base	
sulphuric acid	H_2SO_4	hydrogensulphate ion	HSO_4^-
hydrogensulphate ion	HSO_4^-	sulphate ion	SO_4^{2-}
ammonium ion	NH_4^+	ammonia	NH_3

protonated (cationic) form of 2-aminopropanoic acid

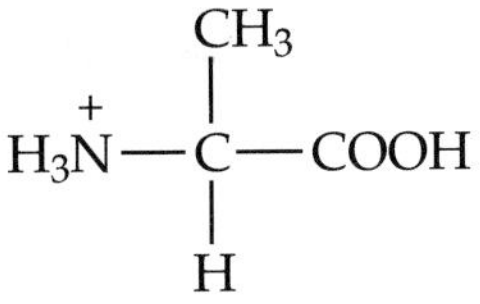

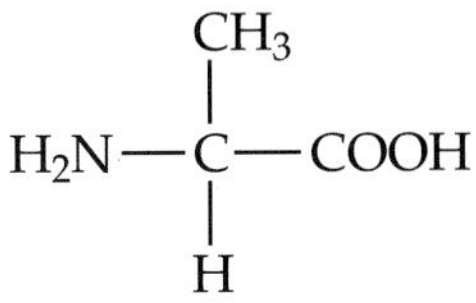

$$H_3\overset{+}{N}-\overset{\overset{\displaystyle CH_3}{|}}{\underset{\underset{\displaystyle H}{|}}{C}}-COOH$$

2-aminopropanoic acid

$$H_2N-\overset{\overset{\displaystyle CH_3}{|}}{\underset{\underset{\displaystyle H}{|}}{C}}-COOH$$

2-aminopropanoic acid

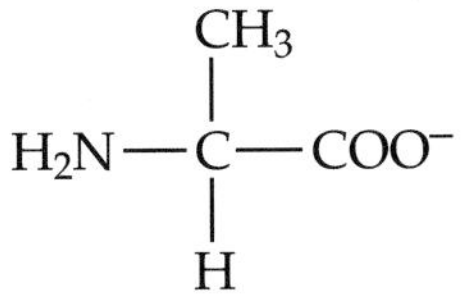

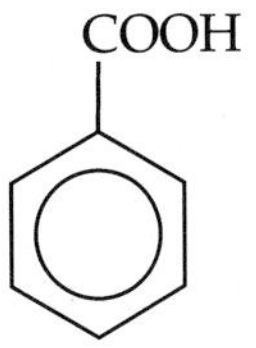

$$H_2N-\overset{\overset{\displaystyle CH_3}{|}}{\underset{\underset{\displaystyle H}{|}}{C}}-COOH$$

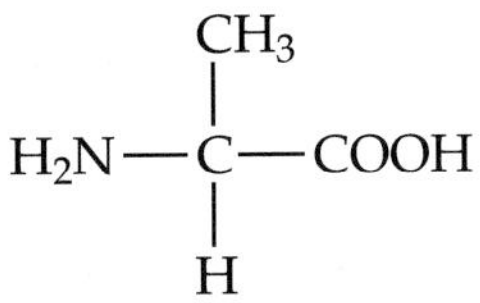

deprotonated (anionic) form of 2-aminopropanoic acid

$$H_2N-\overset{\overset{\displaystyle CH_3}{|}}{\underset{\underset{\displaystyle H}{|}}{C}}-COO^-$$

benzoic acid

COOH (on benzene ring)

benzoate anion

COO⁻ (on benzene ring)

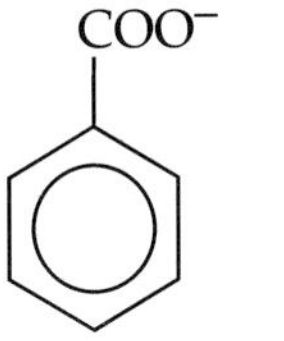

The Lewis theory of acids and bases – extending the model

We have met several examples of a model being developed and refined, for example the model of the atom and the concept of redox reactions. Acid/base theories provide us with yet another example of this refining process.

The American chemist Gilbert Lewis took the definition of acids and bases a step further. He defined an acid as a substance which accepts an electron pair from a base forming a new covalent bond. The base would donate a pair of electrons for the bond. The covalent bond formed is a coordinate bond. This definition included all the compounds covered by earlier definitions as well as other substances not previously considered to be acids or bases. For example, the formation of complex ions. Ligands provide electron pairs for the formation of coordinate bonds to the cation. The ligands are Lewis bases and the cation is a Lewis acid.

13 **Why is the covalent bond formed described as coordinate?**

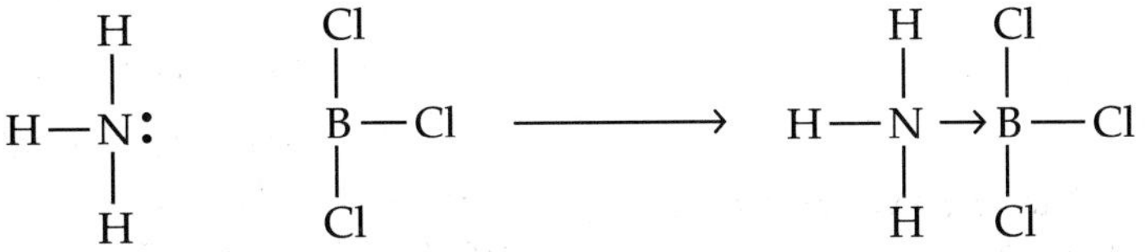

$$H-\overset{\overset{\displaystyle H}{|}}{\underset{\underset{\displaystyle H}{|}}{N}}: \qquad \overset{\overset{\displaystyle Cl}{|}}{\underset{\underset{\displaystyle Cl}{|}}{B}}-Cl \longrightarrow H-\overset{\overset{\displaystyle H}{|}}{\underset{\underset{\displaystyle H}{|}}{N}}\rightarrow\overset{\overset{\displaystyle Cl}{|}}{\underset{\underset{\displaystyle Cl}{|}}{B}}-Cl$$

a Lewis base (electron pair donor)   a Lewis acid (electron pair acceptor)   molecules combine through the formation of a coordinate bond

Fig 19.10   Ammonia acts as a Lewis base when it combines with boron trichloride, a Lewis acid, to form $BCl_3.NH_3$.

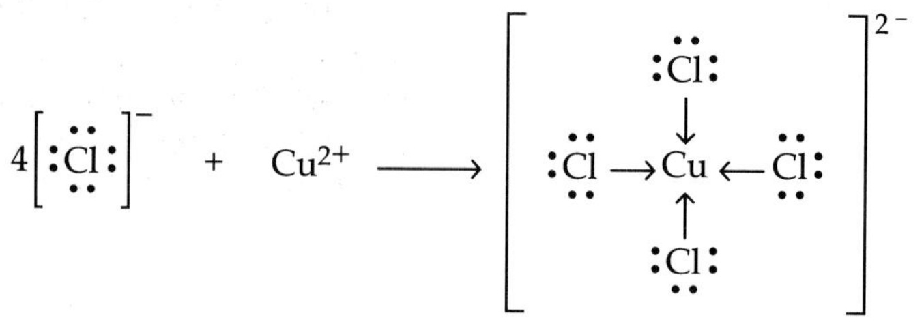

$$4\left[:\overset{\displaystyle ..}{\underset{\displaystyle ..}{Cl}}:\right]^{-} + Cu^{2+} \longrightarrow \left[\begin{array}{c} :\overset{..}{\underset{..}{Cl}}: \\ \downarrow \\ :\overset{..}{\underset{..}{Cl}} \rightarrow Cu \leftarrow \overset{..}{\underset{..}{Cl}}: \\ \uparrow \\ :\overset{..}{\underset{..}{Cl}}: \end{array}\right]^{2-}$$

a tetrahedral complex ion:

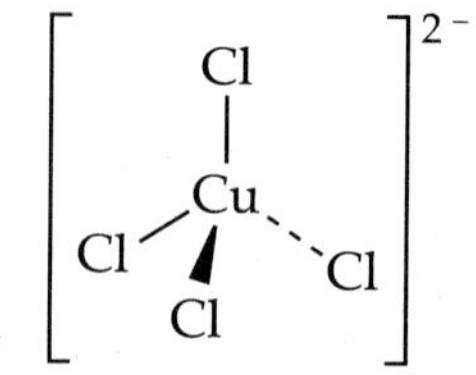

$$\left[\begin{array}{c} Cl \\ | \\ Cl \diagup \overset{\displaystyle Cu}{\underset{\displaystyle Cl}{\quad}} \diagdown Cl \end{array}\right]^{2-}$$

Fig 19.11 Four chloride ions (all acting as electron pair donors and, therefore, Lewis bases) combine with a copper(II) ion (an electron pair acceptor and a Lewis acid) to form a complex ion, the tetrachlorocuprate(II) ion.

The formation of the ammonium ion when an ammonia molecule accepts a proton from a water molecule

$$NH_3(aq) + H_2O(l) \rightleftharpoons NH_4^+(aq) + OH^-(aq)$$

provides an example of substances which can be defined as acids and bases by either the Brönsted-Lowry or the Lewis concepts. Ammonia acts as a base because:

(a) it is a proton acceptor (Brönsted-Lowry); and
(b) it is an electron pair donor (Lewis).

Water acts as an acid in this reaction because:

(a) it is a proton donor (Brönsted Lowry); and
(b) the proton it is transferring is an electron pair acceptor (Lewis).

The two definitions of acids and bases are used differently. The Lewis definition encompasses a wider range of materials, including substances whose reactions do not involve a transfer of protons. It has not replaced the earlier theory but is a valuable means of extending it to explain certain types of reactions involving coordinate bond formation. Chemists usually have the Brönsted-Lowry definition in mind when referring to 'an acid'.

QUESTIONS	**19.5** Give the name and formulae of the conjugate bases for the acids: **(a)** H_2O **(b)** HF **(c)** HCN **(d)** HSO_4^- **(e)** HCOOH
	19.6 Give the name and formulae of the conjugate acids for the bases: **(a)** HCO_3^- **(b)** SO_4^{2-} **(c)** CN^- **(d)** Cl^- **(e)** HPO_4^{2-}
	19.7 Describe what takes place in terms of bond formation when ammonia is bubbled into a solution of hydrochloric acid.
	19.8 Explain, using suitable examples, what is meant by the term amphoteric.
	19.9 Write an equation showing the auto-ionisation of water.
	19.10 Why in capturing a proton to form an oxonium ion, is water behaving as a Brönsted-Lowry base and a Lewis base?

19.3 THE CONCEPT OF pH

Pure water at 25°C contains oxonium ions at a concentration of 1×10^{-7} mol dm^{-3}. But these numbers are hardly convenient for daily use and so, often, oxonium ion concentration is expressed on the **pH scale** instead. This scale was introduced by the Danish biochemist SPL Sørensen, in 1909, to describe simply the concentration of oxonium ions in beer whilst working for the Carlsberg company.

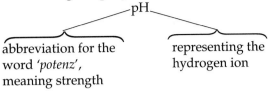

abbreviation for the word *'potenz'*, meaning strength

representing the hydrogen ion

Fig 19.12 Universal indicator is a mixture of three coloured compounds whose colours are sensitive to changes in pH. The colour when one or two drops of this mixture are added to a solution can be used to estimate an approximate pH value.

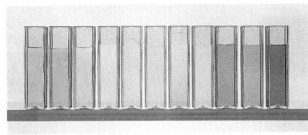

Fig 19.13 The pH scale.

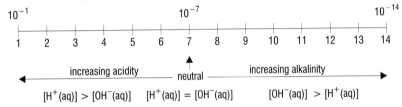

increasing acidity — neutral — increasing alkalinity

$[H^+(aq)] > [OH^-(aq)]$ $[H^+(aq)] = [OH^-(aq)]$ $[OH^-(aq)] > [H^+(aq)]$

KEY FACT

For convenience, the pH of a solution may be defined as the negative logarithm to the base ten of the concentration of oxonium ions:

pH = $-\log[H_3O^+(aq)]$ (usually abbreviated to $-\log[H^+]$)

Concentration is expressed in units of mol dm^{-3}.

However, pH is the negative logarithm to the base ten of the concentration divided by its units. Strictly, we cannot take logarithms of a quantitiy that has units.

For neutral aqueous solutions, the oxonium ion concentration is the same as for pure water (1×10^{-7} mol dm^{-3}). The pH of any neutral aqueous solution is 7 since:

$$pH = -\log[H^+] = -\log(1 \times 10^{-7}) = 7.$$

The vast majority of aqueous solutions do not have an oxonium ion concentration of precisely 1×10^{-7} mol dm^{-3}. The following example shows how oxonium ion concentration is converted to a pH value.

For a solution of oxonium ion concentration 2.5×10^{-3} mol dm^{-3},

$$pH = -\log[H^+] = -\log(2.5 \times 10^{-3}) = 2.6.$$

Most calculators allow you to enter a number, press the log button and then change the sign. Try it yourself with the above example.

SPOTLIGHT

Fig 19.14 Compost is a site of complex bacterial activity in which a number of acid-base reactions take place.

pH AND THE COMPOST HEAP

A typical compost heap contains grass cuttings, discarded food and vegetable matter in the early days of its life. The sap from plant cells is slightly acidic and gives the heap a pH slightly below 7 initially. The fermentation process initially produces an increase in acidity and pH drops further. Bacterial action causes an increase in the temperature of the compost heap and breaks down plant proteins to ammonia. The ammonia dissolves in water to give an alkaline environment and pushes the pH to well above 7. As other bacteria utilise the ammonia pH drops back towards a neutral pH 7, helped by the natural buffering capacity of humus.

If excessive quantities of grass cuttings are used in a compost heap, oxygen access to the heap is restricted and anaerobic bacteria dominate. These bacteria produce only a limited fermentation and the pH remains acidic at around pH 4–5. This material is much less useful as compost.

Problem: Calculate the oxonium ion concentration of a solution whose pH is 4.5.

Solution: $pH = \log[H^+] = 4.5$.

Therefore,
$$\log[H^+] = -4.5$$
$$[H^+] = \text{antilog}\ (-4.5)$$
$$= 3.16 \times 10^{-5}\ mol\ dm^{-3}.$$

Note: On most calculators this may easily be achieved by entering the pH, changing the sign and then pressing inverse log (or antilog button). Check it yourself with the above example.

Concentrations of hydroxide ions may be expressed on a scale similar to the pH scale, referred to as the pOH scale. The pOH value of a solution is the negative logarithm to the base ten of its hydroxide ion concentration in units of $mol\ dm^{-3}$:

$$pOH = -\log[OH^-] \quad \text{(strictly speaking, } -\log[OH^-(aq)]\text{)}$$

The pOH value of neutral solutions and of pure water can be simply calculated. In neutral solutions and pure water, hydroxide ion concentration is $1 \times 10^{-7}\ mol\ dm^{-3}$, and so the pOH value is given by:

$$pOH = -\log[OH^-] = -\log\ (1 \times 10^{-7}) = 7$$

Both the pH and pOH of neutral solutions are equal to 7. Consider again the expression for the ionic product of water:

$$K_w = [H_3O^+(aq)]\ [OH^-(aq)] = 1 \times 10^{-14}\ mol^2\ dm^{-6} \text{ at } 25°C.$$

Taking the negative logarithm of every term here gives:

$$pK_w = pH + pOH$$

This relationship can be useful as it applies to all aqueous solutions whether neutral, acidic or alkaline. We can calculate either pH or pOH provided one value is known. For example, consider a solution with oxonium ion concentration of $1 \times 10^{-1}\ mol\ dm^{-3}$:

$$pH = -\log[H^+] = -\log(1 \times 10^{-1}) = 1.$$

Once the pH value is known, the value of pOH can be calculated since:

$$pOH = 14 - 1 = 13.$$

14 **What is the pH value of a solution for which the concentration of hydroxide ions is $2 \times 10^{-8}\ mol\ dm^{-3}$?**

The relative strengths of acids and bases

Acidity is relative. The term is used to describe the property of one substance compared to another. A solution of an acid in water is a solution with a higher oxonium ion concentration and a lower pH value than that found in pure water. The alkalinity of a solution is also relative to water. When a base dissolves in water to give an alkaline solution, the solution has a lower oxonium ion concentration and a higher pH value than that found in pure water.

The degree of ionisation of an acid determines the concentration of oxonium ions in an aqueous solution of an acid. Solutions of acids and bases in water are electrolytes and may be strongly or weakly ionised. A **strong acid** is an acid which is completely ionised in water. This may be represented by the equation:

$$HA(aq) + H_2O(l) \longrightarrow H_3O^+(aq) + A^-(aq)$$

where HA = the acid (a proton donor as described by the Brönsted-Lowry theory).

Strong acids include hydrochloric acid, sulphuric acid and nitric acid.

A **weak acid** is an acid which only partly ionises in water. This is represented by an equilibrium equation:

$$HA(aq) + H_2O(l) \rightleftharpoons H_3O^+(aq) + A^-(aq)$$

Weak acids include acids such as hydrogen sulphide, hydrogen cyanide and organic carboxylic acids.

15 **Write an equation for the ionisation of hydrogen cyanide in water.**

When a **strong base** dissolves in water it ionises completely. The concentration of hydroxide ions which results is easily calculated from the number of moles of base used. Sodium hydroxide and potassium hydroxide are examples of strong bases:

$$MOH(s) + aq \longrightarrow M^+(aq) + OH^-(aq); \text{ where M = a Group I metal.}$$

A **weak base** only partly ionises in water. Weak bases include ammonia, the carbonate ion, hydrogencarbonate ion and the organic amines. The interaction of a weak base with water is represented as:

$$B(aq) + H_2O(l) \rightleftharpoons BH^+(aq) + OH^-(aq)$$

16 **Write an equation showing the equilibrium established when methylamine, CH_3NH_2, is added to water.**

pH of solutions of strong acids and bases

For strong acids, we can calculate the oxonium ion concentration and pH of its solution from the amount of acid present in a given volume of solution. For example, suppose we want to calculate (i) the oxonium ion concentration and (ii) the pH of a solution formed by dissolving 0.15 mol hydrogen chloride in 1 dm^3 of water.

Hydrochloric acid is a strong acid, undergoing complete ionisation in water. A solution of concentration 0.15 mol dm^{-3} gives rise to a solution of oxonium ions and chloride ions, both at a concentration of 0.15 mol dm^{-3}.

(i) $[H_3O^+(aq)] = 0.15$ mol dm^{-3};
(ii) pH $= -\log(0.15) = 0.82$.

Calculations involving strong bases, such as alkali metal hydroxides, are equally straightforward. It may be assumed that such substances ionise fully into cation and hydroxide ions. This allows straightforward calculation of the hydroxide ion concentration and the pH if the amount of base added is known.

TUTORIAL

Problem: Calculate **(a)** the hydroxide concentration; **(b)** the pOH value; **(c)** the oxonium ion concentration; **(d)** the pH value of 1 dm^3 of solution formed by dissolving 2.5 g of sodium hydroxide pellets in distilled water at a temperature of 25°C.

Solution: The formula of sodium hydroxide is NaOH. The molar mass is 40 g mol^{-1} (Na = 23; O = 16; H = 1). The number of moles dissolved in 1 dm^3 = 2.5/40 = 0.0625 mol.

(a) $[OH^-(aq)]$ = 0.0625 mol dm^{-3};

(b) pOH = $-\log[OH^-]$ = $-\log[0.0625]$ = 1.2;

(c) the oxonium ion concentration is given by:

$K_w = [H_3O^+(aq)][OH^-(aq)] = 1 \times 10^{-14}$ mol^2 dm^{-6} at 25°C.

$$[H_3O^+(aq)] = \frac{1 \times 10^{-14}}{[OH^-]} = \frac{1 \times 10^{-14}}{0.0625} = 1.6 \times 10^{-13} \text{ mol dm}^{-3}$$

(d) the pH value of the solution is given by

$$pH = -\log[H^+] = -\log(1.6 \times 10^{-13}) = 12.8$$

Note The answer in **(d)** could also be obtained using the expression

$$pH + pOH = 14$$

then pH = 14 − 1.2 = 12.8.

Weak acids and bases

Weak acids and bases are far more common in nature than strong acids and bases. Strong acids such as hydrochloric acid have specific roles in digestion but there are thousands of weak acids and bases involved in subtle reactions within all living cells. Calculation of the pH of solutions of weak acids or bases in water is more difficult than with strong electrolytes. Much of the acid or base remains unionised and so we cannot calculate oxonium or hydroxide ion concentrations using the same method as used with strong acids and bases. The interaction of a weak acid or a weak base with water are both equilibrium processes. We can represent each process by the equations and equilibrium expressions shown below.

For weak acids, represented by HA

$$HA(aq) + H_2O(l) \rightleftharpoons H_3O^+(aq) + A^-(aq)$$

$$K_c = \frac{[H_3O^+(aq)][A^-(aq)]}{[HA(aq)][H_2O(l)]}$$

For weak bases, represented by B

$$B(aq) + H_2O(l) \rightleftharpoons BH^+(aq) + OH^-(aq)$$

$$K_c = \frac{[BH^+(aq)][OH^-(aq)]}{[B(aq)][H_2O(l)]}$$

In each case the concentration of water is virtually unchanged by the ionisation process and we may rewrite the two equilibrium expressions by combining the water term in each case with the constant to give:

$$K_a = \frac{[H_3O^+(aq)][A^-(aq)]}{[HA(aq)]} \quad \text{and} \quad K_b = \frac{[BH^+(aq)][OH^-(aq)]}{[B(aq)]}$$

where K_a is the **acid dissociation constant** for HA

K_b is the **base dissociation constant** for B.

17 Why is the concentration of water 'virtually unchanged' by the ionisation of the base?

KEY FACT

Weak acids and bases partly ionise in water, dynamic equilibria being established. We sometimes use the term dissociation rather than ionisation, and the equilibrium constants are most commonly called **acid dissociation constants** and **base dissociation constants** rather than ionisation constants.

The magnitude of acid and base dissociation constants

Dissociation constants are a measure of the strength of an acid or base. They give an immediate idea of the extent of ionisation. Large values for the dissociation constant indicate a high degree of ionisation and a small value suggests a weak acid or base.

Table 19.1 Dissociation constants for a range of acids and bases.

Acid	K_a/mol dm^{-3}	pK_a
methanoic acid, HCOOH	1.78×10^{-4}	3.75
ethanoic acid, CH_3COOH	1.74×10^{-5}	4.76
propanoic acid, C_2H_5COOH	1.35×10^{-5}	4.87
chloroethanoic acid, $CH_2ClCOOH$	1.38×10^{-3}	2.86
dichloroethanoic acid, $CHCl_2COOH$	5.13×10^{-2}	1.29
trichloroethanoic acid, CCl_3COOH	2.24×10^{-1}	0.65
bromoethanoic acid, $CH_2BrCOOH$	1.26×10^{-3}	2.90
iodoethanoic acid, CH_2ICOOH	6.76×10^{-4}	3.17
benzoic acid, C_6H_5COOH	6.31×10^{-5}	4.20
ethanedioic acid, HOOCCOOH	4.68×10^{-2}	1.33 (pK_{a1})
	5.25×10^{-5}	4.28 (pK_{a2})
phenol, C_6H_5OH	1×10^{-10}	10.00
phosphoric(V) acid, H_3PO_4	7.08×10^{-3}	2.15 (pK_{a1})
	6.17×10^{-8}	7.21 (pK_{a2})
	4.37×10^{-13}	12.36 (pK_{a2})
hydrocyanic acid, HCN	3.98×10^{-10}	9.40
hydrogen sulphide, H_2S	8.91×10^{-14}	7.05 (pK_{a1})
	1.20×10^{-14}	13.92 (pK_{a2})
hexaaquaaluminium(III) ion,$[Al(H_2O)_6]^{3+}$	1.26×10^{-5}	4.9 (pK_{a1})

Base	K_b/mol dm^{-3}	pK_b
ammonia, NH_3	1.78×10^{-5}	4.75
methylamine, CH_3NH_2	4.37×10^{-4}	3.36
dimethylamine, $(CH_3)_2NH$	5.25×10^{-4}	3.28
trimethylamine, $(CH_3)_3N$	6.31×10^{-5}	4.20
ethylamine, $C_2H_5NH_2$	5.37×10^{-4}	3.27
phenylamine, $C_6H_5NH_2$	4.17×10^{-10}	9.38

18 **Explain what happens to the basicity of ammonia as its hydrogen atoms are progressively substituted by methyl groups.**

The terms pK_a and pK_b used in Table 19.1 refer to the negative logarithm to the base ten of the dissociation constant in each case. Just as, for convenience, pH values are often used instead of oxonium ion concentrations, pK_a and pK_b values are often used in place of dissociation constants:

$$pK_a = -\log K_a; \quad pK_b = -\log K_b.$$

Notice that the higher the dissociation constant for an acid or base the lower the pK_a or pK_b values.

Polyprotic acids or bases have more than one dissociation constant associated with them. For example, phosphoric(V) acid undergoes a three step ionisation, and each step has an associated dissociation constant:

$$H_3PO_4(aq) + H_2O(l) \rightleftharpoons H_2PO_4^-(aq) + H_3O^+(aq)$$

$$K_{a1} = \frac{[H_2PO_4^-(aq)][H_3O^+(aq)]}{[H_3PO_4(aq)]}$$

$$H_2PO_4^-(aq) + H_2O(l) \rightleftharpoons HPO_4^{2-}(aq) + H_3O^+(aq)$$

$$K_{a2} = \frac{[HPO_4^{2-}(aq)][H_3O^+(aq)]}{[H_2PO_4^-(aq)]}$$

$$HPO_4^{2-}(aq) + H_2O(l) \rightleftharpoons PO_4^{3-}(aq) + H_3O^+(aq)$$

$$pK_{a3} = \frac{[PO_4^{3-}(aq)][H_3O^+(aq)]}{[HPO_4^{2-}(aq)]}$$

The relative sizes of the three acid dissociation constants indicate that the first dissociation is more extensive than the second, which is itself more extensive than the third. Therefore, the relative abundance of phosphorus-containing species in an aqueous solution of phosphoric(V) acid is:

$$[H_3PO_4(aq)] > [H_2PO_4^-(aq)] > [HPO_4^{2-}(aq)] > [PO_4^{3-}(aq)]$$

—————— decreasing relative concentration ——————→

19 Write expressions for the two dissociation constants associated with ethanedioic acid, HOOCCOOH.

Calculating acid dissociation constants and pH

If the pH and the concentration of an acid in water is known, the value of its acid dissociation constant may be calculated. Consider the following example.

The pH of a solution of 0.01 mol dm^{-3} methanoic acid is 2.87 at 25°C. How can we calculate its acid dissociation constant at this temperature?

$$HCOOH(aq) + H_2O(l) \rightleftharpoons HCOO^-(aq) + H_3O^+(aq)$$

$$K_a = \frac{[HCOO^-(aq)][H_3O^+(aq)]}{[HCOOH(aq)]}$$

To calculate the acid dissociation constant we must make some assumptions.

Assumption 1 $[HCOO^-(aq)] = [H_3O^+(aq)]$

In this assumption, we are ignoring any $H_3O^+(aq)$ which come from the autoionisation of water. This is reasonable since the contribution relative to that from the methanoic acid is negligible.

Therefore, $K_a = \dfrac{[H_3O^+(aq)]^2}{[HCOOH(aq)]}$

Assumption 2 $[HCOOH(aq)] = 0.01$ mol dm^{-3}

Here, we are assuming that ionisation of methanoic acid is so small that the concentration of undissociated molecules is approximately the same as that of the original acid. This assumption is only acceptable for weak acids because in these cases the large majority of acid molecules remain undissociated.

Therefore, $K_a = \dfrac{[H_3O^+(aq)]^2}{0.01}$

But pH of the solution = 2.87 = $-\log[H_3O^+(aq)]$

Therefore $[H_3O^+(aq)] = $ antilog$(-2.87) = 1.35 \times 10^{-3}$ mol dm^{-3}

$$K_a = \frac{(1.34 \times 10^{-3})^2}{0.01} = 1.82 \times 10^{-4} \text{ mol dm}^{-3}$$

20 Why does the second assumption become less valid for acids with higher acid dissociation constants?

The calculation of acid dissociation constants shown above may be performed in reverse to predict the pH arising from the dissociation of a known concentration of a weak acid.

TUTORIAL

Problem: Calculate the acid dissociation constant of phenol and the pH of a 0.001 mol dm^{-3} solution of phenol in water given that the degree of ionisation of the acid is 3.6×10^{-4} at the temperature of the solution.

Solution:

$\alpha = 3.6 \times 10^{-4}$

Since phenol is a very weak electrolyte

$$K_a = \alpha^2 c$$

$$= (3.6 \times 10.^{-4})^2 \times 0.001 \text{ mol dm}^{-3}$$

$$= 1.30 \times 10^{-10} \text{ mol dm}^{-3}$$

But $K_a = \dfrac{[H_3O^+(aq)]^2}{[C_6H_5OH(aq)]}$

Therefore $[H_3O^+(aq)]^2 = 1.30 \times 10^{-10} \times 0.001$

$$[H_3O^+(aq)] = 3.61 \times 10^{-7} \text{ mol dm}^{-3}$$

$$pH = -\log(3.61 \times 10^{-7})$$

$$= 6.44$$

Therefore, the pH of a 0.001 mol dm^{-3} solution of phenol is 6.44.

QUESTIONS

19.11 Calculate the pH of solutions containing oxonium ions at the following concentrations:
(a) 0.05 mol dm^{-3} **(b)** 1.3×10^{-8} mol dm^{-3} **(c)** 0.5 mol dm^{-3}

19.12 Calculate an approximate value for the oxonium ion concentration present in solutions with the following pH values:
(a) 6.5 **(b)** 0.95 **(c)** 13.1

19.13 Calculate the pH of solutions with the following hydroxide ion concentrations:
(a) 1.0×10^{-3} mol dm^{-3} **(b)** 1.3×10^{-11} mol dm^{-3}
(c) 2.7×10^{-8} mol dm^{-3}

19.14 The pH of a 0.001 mol dm^{-3} solution of 4-aminobenzoic acid is 3.92 at room temperature. Calculate:
(a) the dissociation constant of the acid;
(b) the pK_a value of the acid;
(c) the degree of dissociation under the conditions stated;
(d) the pOH value of the solution;
(e) the percentage dissociation of the acid;
(f) the pK_b value of the acid.

19.15 Use Table 19.1 to obtain equations and equilibrium expressions for the three stages of the dissociation of phosphoric(V) acid.

19.16 Use Table 19.1 to decide which of the two acids ethanoic acid or phenol is the strongest.

19.17 Examine pK_b values shown in Table 19.1 and decide whether methylamine or dimethylamine is the stronger base. What is the effect on base strength of replacing a methyl group by an ethyl group?

CONNECTION: *Cell e.m.f. and its measurement, pages 389–391.*

The pH of a solution may be accurately measured using a pH meter. There are several different forms of pH meters but most contain a **glass electrode**. This measures the potential difference between a thin glass membrane and the solution under test. The potential difference measured depends upon the concentration of oxonium ions in the solution. This electrode is coupled with a **reference electrode** to complete a cell in which **cell e.m.f.** is proportional to oxonium ion concentration.

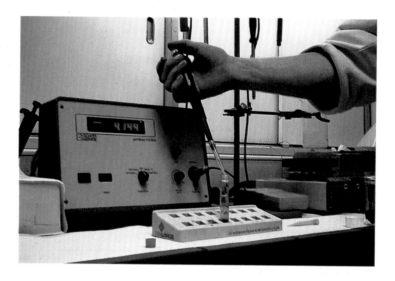

Fig 19.15 A modern pH meter allows the accurate determination of pH, even if the solution is coloured.

Indicators

An important property of acids and alkalis is their ability to change the colour of **indicators**. These are coloured substances whose colour may change at a certain pH.

Many artificial indicators have now been synthesised and have largely replaced the natural dyes. Some indicators, known as **screened indicators**, are available and these contain an additional dye which improves the colour change of the indicator.

Indicators are often themselves weak acids or bases. When placed in a solution, they gain or lose a proton depending upon the pH of the solution. The protonated and deprotonated forms absorb different parts of the visible light falling on the solution and therefore have different characteristic colours.

SPOTLIGHT

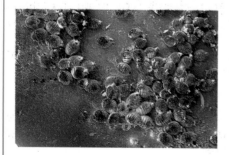

AN UNUSUAL INDICATOR – DACTYLOPIUS COCCUS

Cochineal solution, a scarlet dyestuff containing carminic acid, is made from the ground up bodies of female insects (Dactylopius coccus). Used primarily as a food colouring it is also an acid-base indicator.

Fig 19.16 The dried body of the female insect of the species Dactylopius coccus is a common food colouring and also changes colour with the pH of a solution. The insect is collected from a cactus which grows in Mexico and in the Caribbean.

For a weak acid indicator, HIn unionised indicator molecules exist in equilibrium with their deprotonated form in aqueous solution. This equilibrium may be represented by the equation:

$HIn(aq) + H_2O(l) \rightleftharpoons H_3O^+(aq) +$	$In^-(aq)$
ACID	CONJUGATE BASE
unionised indicator molecules	deprotonated form of indicator
COLOUR 1	COLOUR 2

21 **What important properties should a good indicator have?**

The dissociation of an indicator may be represented, like any other equilibrium, by an equilibrium expression:

$$K_{In} = \frac{[H_3O^+(aq)][In^-(aq)]}{[HIn]}$$

where K_{In} is the acid dissociation constant of the indicator.

22 **Write an equation to represent the equilibrium which exists when a weak base indicator, In, is dissolved in water.**

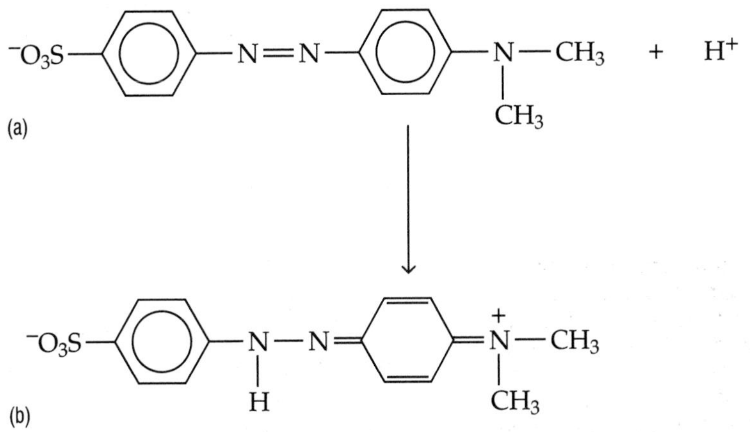

Fig 19.17 Methyl orange is an example of an indicator of the weak base type. Addition of a proton to the orange form **(a)** causes a change of colour to the conjugate acid, which is red **(b)**.

Choosing a suitable indicator

A weak acid or a weak base indicator changes from one colour to another within a narrow pH range. If an indicator of the weak acid type is added to a strongly acidic solution, it will be present almost entirely in its unionised form. This would, of course, be predicted by Le Chatelier's principle. If an alkali is added then the position of equilibrium shifts. When the solution is alkaline, the deprotonated form of the indicator is the most abundant form and so the colour of the solution is different.

At the point where the concentration of each form of an indicator is equal, that is when $[HIn(aq)] = [In^-(aq)]$, the indicator is at the change-over point between its two colours.

Therefore, at the change-over point, the acid dissociation constant for the indicator is equal to the oxonium ion concentration:

$K_{In} = [H_3O^+(aq)]$

Most indicators change colour over a range of pH and the range for some common indicators is shown in Table 19.2. The pK_{In} values indicate the pH of the end-point for the indicator.

KEY FACT

The numerical values for K_{In} for an indicator are no more convenient than those of oxonium ion concentrations or acid dissociation constants and so the extent of ionisation of an indicator is often represented by its pK_{In}, where $pK_{In} = -\log K_{In}$.

The pK_{In} for the indicator is thus equal to the pH of the solution at the end-point. At the end-point, pK_{In} = pH.

Table 19.2

Indicator	pK_{In}	Range	Colour change unionised/ionised form
bromocresol green	4.7	3.8-5.4	yellow/blue
bromophenol blue	4.0	2.8-4.6	yellow/blue
bromothymol blue	7.1	6.0-7.6	yellow/blue
congo red	4.0	3.0-5.0	blue/red
cresol red	8.2	7.2-8.8	yellow/red
methyl orange	3.6	3.2-4.2	red/yellow
methyl red	5.0	4.2-6.3	yellow/red
phenolphthalein	9.4	8.2-10.0	colourless/pink
phenol red	7.9	6.8-8.4	yellow/red
thymol blue (1st)	1.7	1.2-2.8	red/yellow
(2nd)	8.9	8.0-9.6	yellow/blue
thymolphthalein	9.7	8.3-10.5	colourless/blue

Usually an indicator changes colour over a pH range of $pK_{In} \pm 1$. We can understand this by assuming that the colour of the unionised form dominates if it is in a ten-fold excess, that is

$$\frac{[In^-(aq)]}{[HIn(aq)]} = 0.1$$

Similarly the colour of the ionised form dominates if it is in a ten-fold excess, that is

$$\frac{[In^-(aq)]}{[HIn(aq)]} = 10$$

Looking again at the equation for the acid dissociation constant of the indicator HIn

$$K_{In} = \frac{[H_3O^+(aq)][In^-(aq)]}{[HIn(aq)]}$$

we can see that the colour of HIn(aq) dominates when

$$K_{In} = [H_3O^+(aq)]\,0.1$$

Taking logarithms:

$$\log K_{In} = \log[H_3O^+(aq)] + \log(0.1)$$

Therefore, $-\log[H_3O^+(aq)] = -\log K_{In} + \log(0.1)$

$$pH = pK_{In} + 1$$

Similarly, the colour of $In^-(aq)$ dominates when $pH = pK_{In} - 1$.

Acid-base titrations

Acid-base titrations are used to establish the concentration of an acid or alkali by determining the volume of a standard alkali or acid required to neutralise it. A suitable indicator must change colour close to the end-point of the titration, and this may not be at pH 7. The correct choice of indicator is one in which the change-over point for the indicator coincides with the end-point of the titration.

KEY FACT

Within this textbook the term **end-point** for a reaction denotes the point at which the correct amounts of acid and base have been combined. The **change-over point** for an indicator is the point at which the colour change is observed.

CONNECTION: Acid-base titrations, page 129.

Fig 19.18 These four graphs represent the changes in pH recorded during the titration of strong and weak acids and bases.

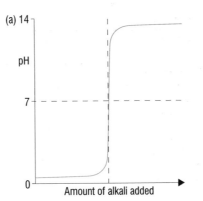
(a)

Strong acid/strong base titrations:
Example: hydrochloric acid and sodium hydroxide
Note here that the vertical portion of the graph over which pH changes most rapidly lies between pH 4 and pH 10. This is a sharp change and suggests that any indicator which has its change-over point in this range will be suitable. Phenolphthalein (9.4) and methyl red (5.0) amongst other indicators, are suitable for this kind of titration

Strong acid/weak base titrations:
Example: hydrochloric acid and ammonia solution
In this graph the maximum rate of pH change lies between pH 3.5 and pH 6.5. A suitable indicator will therefore change colour within this range. Table 19.2 suggests that a suitable indicator would be methyl orange. Note here that at the end-point in the titration the solution has a pH of approximately 5 and is therefore distinctly acid.

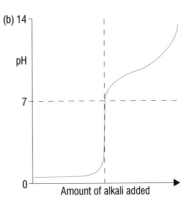
(b)

23 **Why is phenolphthalein unsuitable for this titration?**

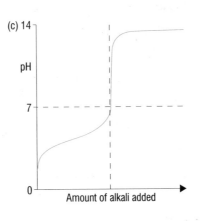
(c)

Weak acid/strong base titrations:
In this titration, as can be seen from the graph, the pH changes rapidly between pH 7.5 and pH 10.5. The end-point for the titration is about pH 9 and gives rise to an alkaline solution. A suitable indicator here would be phenolphthalein since it has a pK_{In} of 9.4 and, therefore, a change-over point which is quite close to the end-point for this titration. Any indicator with a pK_{In} of less than 7.5 including methyl orange would be quite unsuitable since it would suggest a false end-point.

24 **Would methyl orange give an early or late end-point?**

Weak acid/weak base titrations:
With titrations of this kind there is no sharp change in pH and it is not possible to detect the end-point using an indicator. The end-point here is best detected by performing the titration with the use of a pH meter.

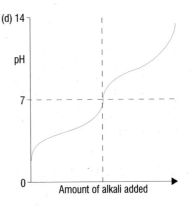
(d)

19.18 A strong base is to be titrated with a standard solution of a weak acid. The pH at the end-point is known to be approximately 7.9. Using Table 19.2, suggest two suitable indicators and one indicator which would *not* be suitable for this titration. Give reasons for the choices you have made.

19.19 Thymolphthalein was used as indicator for the titration of a solution of ethanoic acid with sodium hydroxide solution. Explain **(a)** why the colour change does not take place at pH 7; **(b)** why methyl orange gives misleading and inaccurate results when used for the same reaction; **(c)** at what pH the colour change for thymolphthalein is observed.

19.20 'When selecting an indicator for use in a titration care must be taken to ensure that the change-over point for the indicator should occur close to the end-point of the reaction.' Explain what is meant by this statement.

19.21 List two suitable indicators in each case for reactions between:
(a) a strong base and a weak acid;
(b) a strong base and a strong acid;
(c) a weak base and a strong acid.

19.22 Why is a pH meter usually used to show the end-point in titrations between a weak acid and a weak base?

19.5 BUFFER SOLUTIONS

It is important that the environment inside a living cell remains essentially the same, regardless of what reactions are occurring within it. This includes maintaining an almost constant pH. A constant pH within cells is essential in order to maintain optimum pH for the action of enzymes which control metabolism. The pH of blood is maintained between 7.35 and 7.45 despite changes in the concentration of dissolved nutrients and the levels of carbon dioxide. If the pH of blood extends above pH 7.45 or below pH 7.35 this alone is sufficient to induce a coma. A constant pH is achieved by substances known as **buffers**. Buffer solutions resist changes in pH when small amounts of acids or alkalis are added. The buffer is usually a weak acid combined with a salt of that acid, or a weak base combined with a salt of that weak base.

A buffer can only prevent pH changes caused by the addition of small amounts of acids or bases. Larger quantities overwhelm the buffer and dramatically change the pH. Buffers have many and widespread applications.

Fig 19.19 This man is receiving an intravenous saline drip. Apart from drugs a saline drip infusion may contain amino acids, dextrose, fat, vitamins and minerals as a source of body nutrients. Physiological saline is used to replace blood plasma following an accident or during hospital treatment and must be buffered within a very narrow pH range.

- Many medicines are buffered to ensure their stability.
- Agar plates prepared for growth of bacterial cultures use buffers to optimise bacterial growth.
- In industry buffers are used widely to control chemical processes including electroplating, dyeing and the preparation of photographic films and paper.
- Analytical biochemists use buffers to determine the number and type of amino acids present in proteins by **electrophoresis**, a technique which involves the migration of charged particles in solution through an electric field.
- Chemists use buffers in analysis, for example, in the calibration of pH meters.

Buffer solutions can only maintain the pH of a solution within a particular value and so a buffer solution is prepared with that pH value in mind. Both the substances used and their concentrations are important in determining the effective pH range of a buffer. Pre-prepared packs of buffer solutions are available commercially for a number of pH ranges.

Weak acid buffers

A weak acid buffer is used to maintain the pH of a solution between 4 and 7 and may be prepared by combining, in solution, a weak acid and one of its salts. For example, ethanoic acid and sodium ethanoate. The ethanoic acid is a weak acid and, therefore, only partly dissociates in aqueous solution:

$$CH_3COOH(aq) + H_2O(l) \rightleftharpoons CH_3COO^-(aq) + H_3O^+(aq)$$

Sodium ethanoate, on the other hand, is a soluble ionic compound and completely dissociates in aqueous solution (it is a strong electrolyte):

$$CH_3COONa(s) + (aq) \longrightarrow CH_3COO^-(aq) + Na^+(aq)$$

25 **What is the major source of CH_3COO^-(aq) in a buffer solution consisting of ethanoic acid and sodium ethanoate?**

If an acid is added to this buffer solution, the oxonium ions tend to transfer protons to the ethanoate ions (from the sodium ethanoate, predominantly) to form ethanoic acid. This occurs because ethanoic acid is a weak acid:

$$CH_3COO^-(aq) + H_3O^+(aq) \longrightarrow CH_3COOH(aq) + H_2O(l)$$

ethanoate ions   oxonium ions
from the buffer from the added acid

If an alkali is added to the buffer solution, hydroxide ions combine with oxonium ions (from the ethanoic acid). The position of equilibrium now shifts so as to oppose this change (as we would predict by Le Chatelier's principle). The overall effect is a negligible change in oxonium ion concentration. We can represent the overall reaction as follows:

$$CH_3COOH(aq) + OH^-(aq) \longrightarrow CH_3COO^-(aq) + H_2O(l)$$

Looking back at this explanation we can see why a buffer solution contains the two components:

- the weak acid buffers against the addition of OH^-(aq);
- the salt of the weak acid buffers against the addition of H_3O^+(aq).

Weak base buffers

If a weak base and one of its salts are mixed together in water then a weak base buffer forms. Buffers of this kind are used to maintain the pH of a solution between 7 and 10. A typical weak base buffer is formed by adding ammonia and ammonium chloride to water. In this system, both substances present provide ammonium ions but the vast majority of these ions are provided by the ammonium chloride since it is fully ionised in aqueous solution:

$$NH_4Cl(s) + aq \longrightarrow NH_4^+(aq) + Cl^-(aq)$$

Ammonia is only partly ionised in solution since it is a weak base:

$$NH_3(aq) + H_2O(l) \rightleftharpoons NH_4^+(aq) + OH^-(aq).$$

Addition of an acid to the buffer produces a sudden increase in $[H_3O^+(aq)]$. Reaction between oxonium ions and hydroxide ions present then occurs, decreasing the concentration of oxonium ions. More hydroxide ions are provided by a shift in the position of equilibrium for ammonia ionisation and so the system resists a decrease in pH.

26 **In which direction must the equilibrium for the ionisation of ammonia move when an acid is added?**

Addition of an alkali to a buffered solution causes an increase in $[OH^-(aq)]$. A reaction occurs between the ammonium ions present and hydroxide ions added and the position of equilibrium for the ionisation of ammonia moves from right to left. The system has thus prevented a large increase in the pH of the solution. Again, looking back, we can see why a weak base buffer contains two components:

- the weak base buffers against the addition of $H_3O^+(aq)$;
- the salt of the weak base buffers against the addition of $OH^-(aq)$.

The effective pH range of a buffer

An approximate value for the pH of a buffer system may be calculated if the dissociation constant of the weak acid or base it contains and the concentration of the buffer components are known. Suppose we want to calculate the pH of a solution made by mixing 500 cm^3 of a 0.2 mol dm^{-3} solution of propanoic acid and 500 cm^3 of a 0.02 mol dm^{-3} solution of sodium propanoate at 25°C. The acid dissociation constant of propanoic acid is 1.35×10^{-5} mol dm^{-3} at this temperature.

$$K_a = \frac{[H_3O^+(aq)][C_2H_5COO^-(aq)]}{[C_2H_5COOH(aq)]}$$

To carry out the calculation we must make three assumptions:

- the concentration of oxonium ions arises from the dissociation of the acid (we ignore any $H_3O^+(aq)$ arising from autoionisation of water);
- the concentration of propanoate ions is due entirely to the sodium propanoate (we ignore any contribution from the dissociation of propanoic acid);
- the concentration of the propanoic acid is unchanged by the small amount of dissociation which takes place.

Now, rearranging the expression for K_a

$$[H_3O^+(aq)] = \frac{K_a \times [C_2H_5COOH(aq)]}{[C_2H_5COO^-(aq)]}$$

The concentration of both acid and salt are halved as two solutions of volume 500 cm^3 combined to form 1 dm^3 of solution.

Therefore:

$$[H_3O^+(aq)] = \frac{(1.35 \times 10^{-5}) \times (0.1)}{(0.01)} = 1.35 \times 10^{-4} \text{ mol dm}^{-3}$$

$$pH = -\log(1.35 \times 10^{-4})$$

$$= 3.87.$$

So, a buffer consisting of equal volumes of 0.02 mol dm^{-3} propanoic acid and 0.2 mol dm^{-3} sodium propanoate has a pH of 3.87 and this value does not change significantly when small amounts of either an acid or an alkali are added.

AGAR PLATES AND pH CONTROL

Bacterial cultures may be grown on agar, an extract prepared from seaweeds. The optimum growth of bacteria is dependent upon the pH of the agar plate which may need to be buffered to produce satisfactory conditions for the bacteria to grow.

Fig 19.20 This agar plate has been used as a medium for bacterial growth. The optimum pH is maintained by a buffer solution in which the plate is prepared.

QUESTIONS

19.23 (a) Calculate the pH of a buffer solution made by mixing 50 cm^3 of aqueous 0.100 mol dm^{-3} ethanoic acid solution with a solution of 50 cm^3 of 0.2 mol dm^{-3} sodium ethanoate solution.

(b) Explain how this buffer solution resists changes in pH due to the addition of small quantities of: i) a dilute aqueous acid ii) a dilute alkali.

K_a (CH$_3$COOH) = 1.74 × 10^{-5} mol dm^{-3}

19.24 Why are buffer solutions unable to resist changes in pH due to the addition of large quantities of acids or bases?

19.25 A buffer solution is prepared by combining methylamine, CH$_3$NH$_2$, with methylammonium chloride, CH$_3$NH$_3$Cl. Explain how such a buffer would react to addition of (a) a small quantity of hydrochloric acid; (b) a small quantity of sodium hydroxide solution.

19.26 To prepare 1 dm^3 of a buffer solution which will maintain a pH of 5.5, 0.6 g of ethanoic acid was used. What mass of sodium ethanoate should the solution contain?

The pH of salt solutions

It is often assumed, wrongly, that all salts dissolve in water to give neutral solutions, that is with pH 7. However, many salts produce acidic or alkaline solutions in water. For example, a solution of ammonium chloride in water has a pH of less than 7. By contrast, an aqueous sodium ethanoate solution has a pH of slightly more than 7. How can we explain this?

The 'unexpected' acidity or alkalinity arises when the ions of which the salt is composed dissolve and then react with water molecules, a reaction known as **salt hydrolysis.** Salts may form from any of the following four combinations:

1 **Strong acids and strong bases**
   For example, potassium nitrate, formed from nitric acid and potassium hydroxide.
   Salts of strong acids and strong bases produce neutral solutions (pH 7) and no salt hydrolysis takes place.

2 **Weak acids and strong bases**
   For example, sodium ethanoate, formed from ethanoic acid and sodium hydroxide.
   Salts of weak acids and strong bases form slightly alkaline solutions (pH > 7) as the conjugate base of the acid has a tendency to react with water to leave an excess of hydroxide ions and create an alkaline solution:

   $$CH_3COO^-(aq) + H_2O(l) \longrightarrow CH_3COOH(aq) + OH^-(aq)$$

3 **Strong acids and weak bases**
   For example, ammonium chloride, formed from hydrochloric acid and ammonia.
   Salts formed from the reaction of a strong acid and a weak base form acidic solutions (pH < 7). This is due to a tendency for the conjugate acid of the base to react with water, producing oxonium ions and a low pH.

   $$NH_4^+(aq) + H_2O(l) \longrightarrow NH_3(aq) + H_3O^+(aq)$$

4 **Weak acids and weak bases**
   For example, ammonium ethanoate, formed from ammonia and ethanoic acid.
   Salts formed by the reaction between a weak acid and a weak base have a pH which is close to 7. The exact value depends upon the relative strengths of the acids and bases concerned. The solutions may be slightly acidic, neutral or slightly alkaline.

Solubility and the solubility product

All chemical substances have a maximum solubility in water, depending upon the substance and the temperature of the solution. The **solubility** of a salt is usually expressed in terms of the mass of a solute which will dissolve in 100 g of the solvent at a quoted temperature. A highly soluble salt such as sodium chloride will dissolve to produce a saturated solution of concentration 36 g per 100 g of water at room temperature, while lead(II) chloride, usually regarded as an insoluble compound, has a solubility of 0.99 g per 100 g of water. It should be noted that the term solubility is a relative term and that all metal salts have a measurable solubility in water. Substances which are only very slightly soluble in a solvent are often said to be **sparingly soluble.**

If an undissolved ionic compound is in contact with a *saturated* solution of its ions a **solubility equilibrium** is established. The solid dissolves at the same rate that ions combine to precipitate fresh solid.

An understanding of solubility equilibria is essential to industrial chemists. For example, the conditions under which a particular ionic substance precipitates from solution are important. If a product precipitates at the wrong stage in an industrial process it might block piping or damage pumping equipment with environmentally disastrous consequences. Knowledge of solubility equilibria is also of use to waste disposal specialists involved in removing toxic ions from industrial effluent. Analytical chemists use solubility equilibria to determine the concentration of ions present in trace quantities.

Consider a saturated solution of the sparingly soluble compound, silver chloride in contact with an excess of solid silver chloride:

$$AgCl(s) + aq \longrightarrow Ag^+(aq) + Cl^-(aq)$$

We can write an expression for the equilibrium constant

$$K_c = \frac{[Ag^+(aq)][Cl^-(aq)]}{[AgCl(s)]}$$

The equilibrium is a heterogeneous equilibrium and, as with equilibria of this type that we have met before, solid phase components do not appear in the equilibrium expression since their masses may be taken as constant at a particular temperature. The equilibrium expression, therefore simplifies to:

$$K_{sp} = [Ag^+(aq)][Cl^-(aq)]$$

where K_{sp} is the **solubility product** for silver chloride. Like all equilibrium constants, it is temperature dependent.

The concept of solubility product is only useful for sparingly soluble salts. The solubility product is a constant for a given solute in water and provides another way of comparing the solubilities of sparingly soluble ionic substances. A very small value for the solubility product of a substance implies that the substance is poorly soluble.

In general, the solubility product of any sparingly soluble electrolyte, $AxBy$, can be represented by the expression:

$$K_{sp} = [A]^x[B]^y$$

The units of the solubility product are $(mol\ dm^{-3})^{(x+y)}$.

27 **What would be the units for the solubility product of lead iodide, PbI_2?**

Three situations can exist:

- $[A]^x[B]^y < K_{sp}$: the solution is **not** saturated and more solute (A_xB_y) can dissolve;
- $[A]^x[B]^y = K_{sp}$: the solution is saturated;
- $[A]^x[B]^y > K_{sp}$: precipitation of A_xB_y will occur.

TUTORIAL

Problem: Given that the solubility of silver chloride in water is $1.34 \times 10^{-5}\ mol\ dm^{-3}$ 25°C, calculate the solubility product for silver chloride at this temperature.

Solution: Since the concentration of a saturated solution of silver chloride is $1.34 \times 10^{-5}\ mol\ dm^{-3}$ at 25°C and ionisation gives two ions:

$$AgCl(aq) \longrightarrow Ag^+(aq) + Cl^-(aq)$$

The concentration of both silver ions and chloride ions must equal $1.34 \times 10^{-5}\ mol\ dm^{-3}$ at 25°C.

$$K_{sp} = [Ag^+(aq)] [Cl^-(aq)]$$
$$= (1.34 \times 10^{-5})^2 \text{ mol}^2 \text{ dm}^{-6}$$
$$= 1.8 \times 10^{-10} \text{ mol}^2 \text{ dm}^{-6}$$

Problem: If the solubility of antimony sulphide, Sb_2S_3 is 3.73×10^{-17} g in 100 g of water at 25°C, calculate the value of the solubility product at this temperature.
$Ar[Sb] = 122$, $Ar[S] = 32.1$.

Solution: The solubility of $Sb_2S_3 = 3.73 \times 10^{-17}$ g in 100 g of water

$= 3.73 \times 10^{-16}$ g in 1 dm³ water

Molar mass of $Sb_2S_3 = (2 \times 122) + (3 \times 32.1) = 340.3$ g mol⁻¹

The concentration of saturated $Sb_2S_3 = \dfrac{3.73 \times 10^{-17}}{340.3} = 1.1 \times 10^{-19}$ mol dm⁻³

Since $Sb_2S_3(s) \longrightarrow 2Sb^{3+}(aq) + 3S^{2-}(aq)$

$[Sb^{3+}(aq)] = (1.1 \times 10^{-19}) \times 2$ mol dm⁻³

$= 2.2 \times 10^{-19}$ mol dm⁻³

$[S^{2-}(aq)] = (1.1 \times 10^{-19}) \times 3$ mol dm⁻³

$= 3.3 \times 10^{-19}$ mol dm⁻³

$K_{sp} = [Sb^{3+}(aq)]^2 [S^{2-}(aq)]^3$

$= (2.2 \times 10^{-19})^2 \times (3.3 \times 10^{-19})^3$ mol⁵ dm⁻¹⁵

$= 1.7 \times 10^{-93}$ mol⁵ dm⁻¹⁵

SPOTLIGHT

PRECIPITATION AND THE GREAT BARRIER REEF

Sea creatures such as shellfish or those which grow in colonies, such as coral, need a solid coat for protection. The creature must produce a high localised concentration of calcium and carbonate ions to precipitate a protective shell. The coral is produced in a controlled way and takes a very long time to achieve significant size. The coral of the Great Barrier Reef is largely calcium carbonate produced in this way.

Fig 19.21   Coral is composed of calcium carbonate precipitated for protective purposes by a number of different sea-creatures. To do this the creature must cause the product of the concentration of calcium ions and carbonate ions in its immediate environment to exceed the solubility product of calcium carbonate.

The common ion effect

If we reconsider the general solubility product expression for a sparingly soluble electrolyte, A_xB_y:

$$K_{sp} = [A]^x[B]^y$$

We can see that the expression means that the concentrations $[A]^x$ and $[B]^y$ when multiplied together are a constant for a saturated solution of an electrolyte at a particular temperature. This expression remains true *whatever* the individual concentrations of the ions. We should now examine the situation which arises when *the solution itself* provides a source of either of the two ions, A^{n+} or B^{m-}. Here the solution has an ion *in common* with the electrolyte and both electrolyte and solution contribute to the total concentration of that ion. This contribution reduces the solubility of the electrolyte in the solution and this effect is known as the **common ion effect**.

The common ion effect can be used to determine the solubility of electrolytes in any solution if the solubility product of the electrolyte is known. In a chemical plant working with a sparingly soluble electrolyte an understanding of the solubility product and the common ion effect allows scientists to calculate the maximum concentration of the product which may be held in solution before solid precipitates.

QUESTIONS

19.27 Calculate the solubility product of calcium fluoride given that the concentration of a saturated solution of this substance is 2.15×10^{-4} mol dm^{-3}.

19.28 Express the solubility product for aluminium hydroxide, $Al(OH)_3$, in terms of the concentrations of aluminium and hydroxide ions.

19.29 Explain what is meant by the term **common ion effect**.

19.30 Explain, in terms of solubility products, why silver nitrate solution added to a solution of sodium chloride precipitates silver chloride and **not** sodium nitrate.

19.31 Explain why silver chromate(VI) is precipitated when its saturated solution is treated with one drop of sodium chromate solution.

19.32 If the solubility product of barium sulphate in aqueous solution at 25°C is 1×10^{-10} mol^2 dm^{-6}, determine:
   (a) the maximum solubility in grams of solute per 100 cm^3 of aqueous solution which may be achieved at this temperature; $A_r[Ba] = 137$, $A_r[S] = 32$, $A_r[O] = 16$
   (b) the maximum concentration of barium sulphate in a solution of sulphuric acid in which $[SO_4^{2-}]$ is equal to 0.01 mol dm^{-3}.

ACTIVITY

The carbonates of Group II are generally regarded as insoluble in water. The following table gives the solubility products of three Group II carbonates:

Compound	K_{sp} at 25°C/mol^2 dm^{-6}
CaCO$_3$	3.8×10^{-9}
SrCO$_3$	9.4×10^{-10}
BaCO$_3$	8.1×10^{-9}

1 dm^3 of a solution is prepared in which calcium nitrate, strontium nitrate and barium nitrate are all present at a concentration of 0.005 mol dm^{-3}. How many micrograms (millionths of a gram) of sodium carbonate could be stirred into this solution before the first permanent precipitate of a metal carbonate appeared? Which metal carbonate would precipitate first?

Answers to Quick Questions: Chapter 19

1 Delocalised electrons; no decomposition of the metal occurs. In aqueous copper(II) sulphate, it is the free moving ions which carry the current from one electrode to the other; the solution is decomposed by the passage of electricity.

2 Two

3 The position of equilibrium would move to the left, that is, the concentration of undissociated ethanoic acid would increase.

4 $mol\ dm^{-3}$

5 1

6 $5.1 \times 10^{-13}\ mol^2\ dm^{-6}$

7 Each pair of water molecules gives rise to one oxonium ion and one hydroxide ion.

8 $2KOH(aq) + H_2SO_4(aq) \longrightarrow K_2SO_4(aq) + 2H_2O(l)$
$Mg(s)\ H_2SO_4(aq) \longrightarrow MgSO_4(aq) + H_2(g)$
$Na_2CO_3(aq) + H_2SO_4(aq) \longrightarrow Na_2SO_4(aq) + CO_2(g) + H_2O(l)$

9 They consist of solutions of hydrated ions and, therefore, can conduct electricity.

10 $A^-(aq)$ acts as a proton acceptor from $H_3O^+(aq)$ and, therefore, is behaving as a base.

11 $A^-(aq)$

12 $2NH_3 \rightleftharpoons NH_4^+ + NH_2^-$

13 The two electrons shared in the bond both originate from the same atom.

14 6.30

15 $HCN(aq) + H_2O(l) \rightleftharpoons H_3O^+(aq) + CN^-(aq)$

16 $CH_3NH_2(aq) + H_2O(l) \rightleftharpoons CH_3NH_3^+(aq) + OH^-(aq)$

17 It is due to the weakness of the base and so the amount of water which donates protons to the base is negligible compared with the total amount of water present.

18 It increases until all three hydrogens are replaced whereupon it decreases.

19 $K_{a1} = \dfrac{[HOOCCOO^-(aq)][H_3O^+(aq)]}{[HOOCCOOH(aq)]}$

$K_{a2} = \dfrac{[^-OOCCOO^-(aq)][H_3O^+(aq)]}{[HOOCCOO^-(aq)]}$

20 The concentration of undissociated acid molecules does not approximate to the original concentration of the acid.

21 The two colours should be distinctly different and the colour change from one form to the other should occur sharply.

22 $In(aq) + H_2O(l) \rightleftharpoons InH^+(aq) + OH^-(aq)$

23 It does not change colour within this pH range.

24 Early – and it would be very indistinct, with the change occurring slowly with the addition of several cm^3 of base.

25 The complete ionisation of sodium ethanoate in aqueous solution.

26 To the right.

27 $mol^3\ dm^{-9}$

Theme 6

CHEMISTRY AT WORK

Throughout this book you have met many chemical ideas. In this final theme these ideas are used in a number of important areas of industrial chemistry and the chemistry of the body. Whether we are considering industrial plants or reactions which occur in living cells, the chemical concepts are the same.

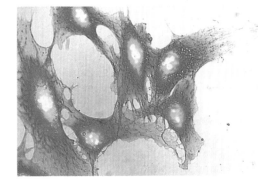

Chapter 20

THE CHEMICAL INDUSTRY

20.1 THE IMPORTANCE OF THE CHEMICAL INDUSTRY

Chemists convert materials that occur naturally – organic material (coal, oil, natural gas and biomass), rocks and minerals – into substances which are of use to us. We live in an Aladdin's cave of natural resources and it is the chemical industry's business to transform nature's materials into more useful products on a large scale. These products include: medicines, plastics, agrochemicals, fibres, detergents, dyes, fragrances, fuels, explosives, fertilisers, metals and alloys, glass and ceramics, superconductors, other advanced materials.

We rely on modern drugs, antiseptics, disinfectants, detergents and other cleaners to maintain a healthy population and demand materials for building, clothing and transport. All these products derive from the chemical industry. Fertilisers manufactured by the chemical industry are used to supply nutrients to the plants we grow. Controlled and sensible use of carefully tested agrochemicals such as herbicides and pesticides protect our crops. But food deteriorates rapidly and must be preserved on its journey from where it is grown to where it is consumed and also during storage. Preservatives, such as antioxidants, are important.

1 **Food may also be preserved by freezing or by freeze-drying. What advantages might preservation using compounds such as ethanoic acid (present in vinegar), benzoic acid and ascorbic acid (vitamin C) have over these methods?**

Fig 20.1 Hygiene is important in our lives. The toiletries in this bathroom were made by the chemical industry from natural materials such as crude oil and natural gas. Also, the ceramic tiles were manufactured from sand and the towels are brightly coloured thanks to the use of synthetic dyestuffs.

In the chemical industry, chemistry and technology unite to find ways of producing these materials upon which we depend. The industry manufactures a vast range of compounds. Some are used directly by the consumer while others are synthesised into other useful compounds. Importantly, the chemical industry makes a large contribution to a nation's balance of trade, often more than any other manufacturing industry.

However, it is not easy to define the 'chemical industry'. We would certainly classify the manufacture of sulphuric acid as part of the chemical industry but other areas, such as the formulation of a paint, are less clear-cut. The pigments and polymers found in paint are made by the chemical industry, but is the mixing and blending of these to make paint still part of the industry?

2 **How does the chemical industry contribute to car manufacture?**

Sectors of the industry

All chemical industries change raw materials into new products. The raw materials may be organic or inorganic. Primary industries are concerned with manufacturing the compounds from which other industries make products. Products from these primary industries are rarely bought by the public and, perhaps, their importance is not immediately appreciated.

For example, sulphuric acid is found in car batteries but that hardly accounts for the vast quantities manufactured annually. The importance of sulphuric acid lies in its use by other industries to make products such as fertilisers, paints, pigments, detergents, fibres and plastics. The primary industries are summarised in Table 20.1 and discussed in more detail later in this Theme. Four of the main industries which consume products from the primary chemical industries are summarised in Table 20.2. They are major sectors of the chemical industry.

Table 20.1 The main primary chemical industries

	Raw materials	Manufactures	Supplies
Petrochemical industry	Crude oil and natural gas (essentially methane) mainly; also coal and biomass.	Simple unsaturated hydrocarbons such as ethene, propene and benzene. These are converted into simple organic compounds containing functional groups, such as alcohols, aldehydes, ketones, carboxylic acids. Although closely linked to the petrochemical industry, oil refining and the manufacture of petrol is not considered part of the petrochemical industry.	Other industries which use the simple organic compounds to synthesise more complex compounds such as polymers, pharmaceuticals, agrochemicals and dyestuffs.
Chloralkali industry	Common salt (sodium chloride); limestone (calcium carbonate).	Chlorine (hence '**chlor**'), sodium hydroxide (hence '**alkali**') and sodium carbonate.	Chlorine to industries which make organic chlorine compounds (and hence solvents, some polymers and other more complex organic compounds), bleaches and disinfectants. Inorganic chlorine-containing compounds (for example, hydrochloric acid); sodium hydroxide to a vast range of industries making inorganic and organic compounds; sodium carbonate to the glass and paper industries and for the manufacture of detergents.
Sulphur industry	Sulphur and air (to supply oxygen).	Sulphuric acid (the largest volume compound in production).	Industries concerned with the manufacture of fertilisers, paints, pigments, detergents, fibres and plastics.
Nitrogen industry	Air (to supply nitrogen) and crude oil (to supply hydrogen).	Ammonia, nitric acid and carbamide.	The fertiliser industry (major user), also industries concerned with making explosives, dyestuffs, textiles, food, pharmaceuticals, plastics and nitrogen-containing organic compounds.
Phosphorus industry	Phosphate rock and sulphuric acid (from the sulphur industry).	Phosphoric(V) acids and phosphates	The fertiliser industry (major users), also industries concerned with making detergents, food, flameproofing materials, lubricants, catalysts and industrial phosphates.

Table 20.2 Some important consumers of products from the petrochemical industry

Industry	Products
Polymers industry	Manufactures plastics, synthetic fibres, rubbers and adhesives (it is the major user of the organic building blocks produced by the petrochemical industry). Modern polymers extend beyond organic polymers and include the silicones and other inorganic polymers.
Pharmaceutical industry	Manufactures drugs and medicines. It is a high-profit industry but with very high research and development costs – in excess of £100 million to discover, test and get on to market a single drug. It does, however, have an exceptional record of innovation.
Agrochemical industry	Manufactures pesticides (herbicide to control weeds, insecticides to control insects, and fungicides to control fungal growths). Very similar to the pharmaceutical industry in terms of profits, costs and innovation.
Dyestuffs industry	Manufactures the chemicals used to colour anything from clothing to paint, from hair to plastics.

The size of different sectors can be compared in two ways:

- the output of chemical compounds by their mass (Fig 20.2);
- the gross value added of the end-products manufactured from these compounds, that is the difference between a product's saleable value and the cost of the raw materials (Fig 20.3).

3 **Which three organic compounds (in order) are manufactured in the largest quantities?**

4 **Explain why the data shown in Fig 20.3 confirm the assertion that the pharmaceutical and agrochemical industries are high profit industries (Table 20.2).**

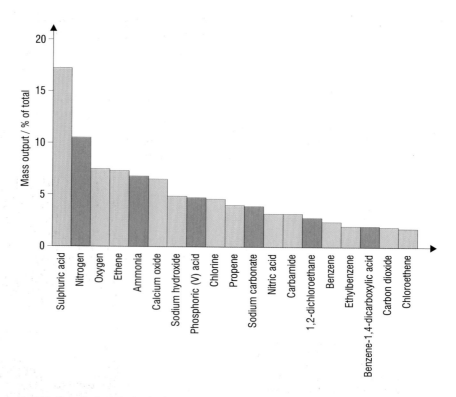

Fig 20.2 Output by mass of major chemical compounds (based on 1988 data from USA, where total production was around 225 millions of tonnes). A similar distribution is found in all developed countries with a chemical industry.

THE CHEMICAL INDUSTRY

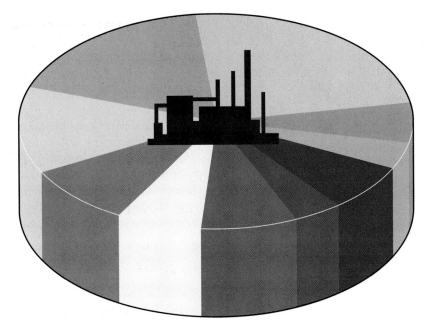

Fig 20.3 The main sectors of the UK chemical industry by percentage share of the gross value added. In the UK, total sales of the chemical industry are well in excess of £20 million each year.

- Pharmaceuticals 24%
- Specialised chemical products for industrial and agricultural use 18%
- Organic compounds 13%
- Detergents, soaps and toiletries 9%
- Synthetic plastics, rubber and resins 7%
- Paints, varnishes and printing inks 8%
- Dyestuffs and pigments 4%
- Inorganic compounds 7%
- Specialised chemical products for household and office use 4%
- Fertiliser 6%

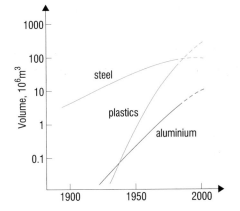

Fig 20.4 World production of plastics has grown at a faster rate than that of either steel or aluminium.

Expansion and development of the industry

Polymers are just one example of the expanding chemical industry. Over 60 million tonnes of polymers are manufactured each year and it is hard to imagine a world without plastics and synthetic fibres. Yet the first truly artificial plastic was only made in 1907 when Bakeland synthesised 'Bakelite' from phenol and methanal. Nowadays, world production of plastics rivals that of steel (Fig 20.4). Similarly, many other areas of the industry are growing to meet the demands for their products.

5 **Make a list of items you would find in a kitchen which are made of plastic.**

Fig 20.5 'Procom' is an advanced material which is a composite of a polymer mixed with a mineral filler. The polymer is poly(propene) and the presence of very small particles of mineral distributed throughout it (only visible under an electron microscope) gives a strong material which may be used to replace metals in some applications, for example in the one shown here.

The chemical industry is growing more rapidly than most other manufacturing industries. Investment in research and development (R&D) and well planned research portfolios are vital to the success of large chemical companies. However, a problem facing countries wishing to develop a chemical industry is the enormous capital outlay (the money needed to set up the chemical plant). The manufacture of bulk chemicals, those produced on a very large scale, requires expensive chemical plant perhaps costing £10 million to £300 million.

Modern chemical industry takes full advantage of developments in computer technology and advances in engineering and electronics. Computers control the operation of continuous-process chemical plants. Analytical methods used to monitor production as well as to test products are often fully automated.

6 **Explain why the modern chemical industry tends to be capital intensive rather than labour intensive.**

20.2 THE SCALE OF MANUFACTURING

Chemical plant is the equipment associated with manufacturing a compound. It might include the reactors, separation units (for example distillation columns), pumps, compressors, pipes, storage vessels, heat exchange units, control facilities and measuring instruments.

Fig 20.6   Advances in technology enabled ICI to win the Pollution Abatement Award in 1989. In the LCA (Leading Concept Ammonia) Process, ammonia is manufactured using less feed gas, less land and materials, and is much quieter than conventional plants. These photographs show part of the interior of a conventional plant (left) and a view of a LCA plant (right)

Compounds are manufactured in varying amounts but they fall into two broad categories, **bulk chemicals** and **fine chemicals**. Bulk chemicals are produced on the scale of thousands of tonnes per year. Fine chemicals, however, are produced on a much smaller scale, in the order of hundreds of tonnes per year. There is no sharp dividing line.

Continuous processes

Bulk chemicals are manufactured invariably by a **continuous process**. Reactors are fed continuously with reactants as the products are drawn off. The plant operates 24 hours a day. Although the method of choice, continuous processing requires a plant which is dedicated to the manufacture of a single product. Closing the plant down for repair or maintenance is costly. Examples include the production of sodium hydroxide and chlorine by electrolysis, sulphuric acid, ammonia, and nearly all the processes used in the petrochemical industry for making the organic compounds which are the building blocks for a vast range of chemical compounds.

7 In the manufacture of phenylethene:
(a) Name and give formulae for the compounds which are recycled.
(b) The key reaction during the second stage of manufacture is dehydrogenation (Fig 20.7). Explain this term and write a suitable equation for this reaction.

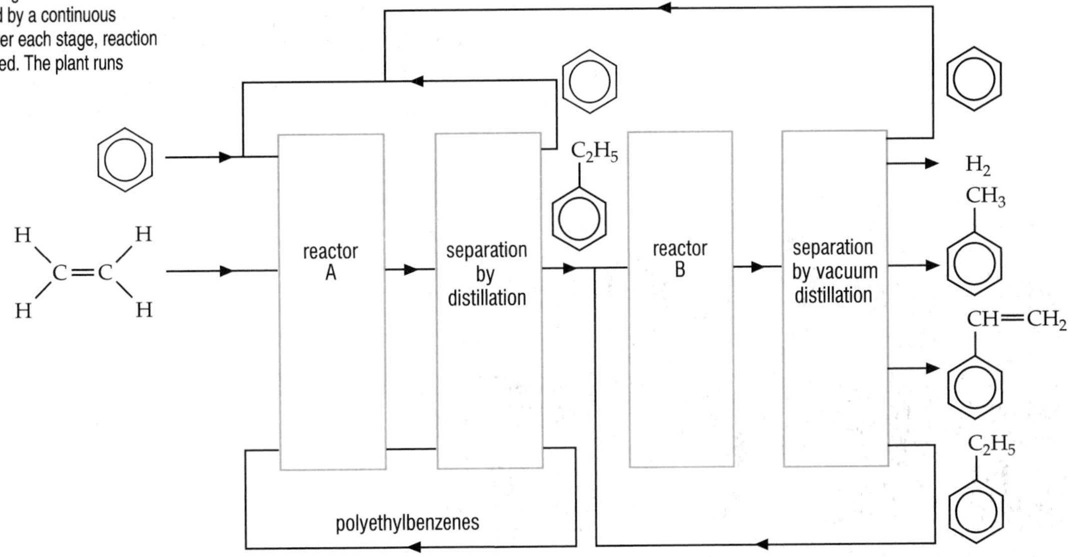

Fig 20.7 Poly(phenylethene), or polystyrene as it is more commonly known, is made by polymerising phenylethene. This flow diagram shows how the monomer is manufactured by a continuous process in two stages. After each stage, reaction products must be separated. The plant runs 24 hours a day.

Batch processes

Compounds which are produced on less than 100 tonnes per year are made by a **batch process**. The syntheses of many pharmaceuticals and agrochemicals are carried out in apparatus which resembles that used in the laboratory, with the exception, often, that glass tubing and reaction vessels are replaced by steel versions. The aim of a batch process is to produce a specific quantity of product in a single operation. The compound is separated and collected from the reaction mixture. The apparatus is dismantled, cleaned and made ready for use again. An advantage of batch processing is that the apparatus need not be dedicated to the production of a single compound. It can be re-assembled easily to serve a number of purposes. The disadvantage is that it is more labour intensive than continuous processing.

The choice between continuous and batch processing usually comes down to a question of scale. For the manufacture of compounds where the volume of production is sufficiently large, continuous processing is likely to be the economical solution.

Fig 20.8 An industrial poly(phenylethene) plant

8 Which is likely to require the greatest capital outlay, a continuous process or a batch process?

9 Explain why a batch process is more labour intensive than a continuous process.

Fig 20.9 Inside a perfume factory. The compounds which give expensive perfumes their odour are required in small quantities only. They are ideally suited to batch processing.

20.3 DESIGNING A CHEMICAL PLANT

The design of a chemical plant requires consideration of many factors and, as is often the case in design problems, compromises must be reached. Decisions must be made concerning, for example, the location of the plant, resource requirements, construction materials, reactor and control system design and, of course, costs.

Locating the plant

Siting a chemical plant is not straightforward. For example, in the UK the location of chemical industry has depended upon the historical availability of a number of factors:

- raw materials;
- transport infra-structure;
- sales outlets;
- human resources (skilled and unskilled);
- energy;
- water;
- effluent disposal facilities.

Other factors which influence choice include:

- social and environmental issues (for example, the social benefit of siting a chemical plant in a particular place even if this is not the most economically viable choice);
- the possibility of government inducements through tax and rates exemption for a specified period;
- possible assistance from government with capital investment.

Government may intervene with financial incentives to encourage industrial development in priority areas, for example regions of high local unemployment.

10 **Comment briefly on the location of British chemical industries, areas of population and available transport systems. Compare the map in Fig 20.10 with a map of the UK which shows major towns and cities, rivers, ports, motorways and railways.**

Fig 20.10 The distribution of chemical industries in the UK.

THE CHEMICAL INDUSTRY

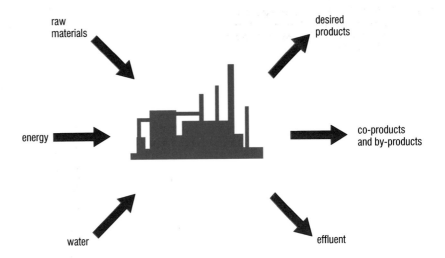

raw materials

desired products

energy

co-products and by-products

water

effluent

Resource requirements

A suitable supply of raw materials is essential, for example, metallic ores, minerals or hydrocarbons. What resources are needed depends on the nature of the chemical process and the plant used to house it. Water is usually required for at least one of the following purposes: raw material, solvent, coolant or heating medium (often as steam). Commonly it needs to be purified both before and after use.

> **11** Many nuclear power stations are situated near the sea. Explain why, and in what way, sea water must be purified before it can be used for steam generation in power stations.

Supplies of energy are another essential resource. Efficient use of energy is the key to economically successful processes. Much chemical plant is complex as a result of making best use of resources: recycling unreacted materials, separating and using co-products, removing or minimising the formation of by-products (the terms co- and by-product are explained shortly). Importantly, plant complexity is often due to the need to balance energy-consuming processes with energy-producing processes. The demand for energy is particularly high in the separation units such as distillation columns.

Energy-producing processes

Exothermic reactions: these are usually cooled and the hot coolant can be put to work providing heat somewhere else in the plant where it is needed.

Pressure changes: when pressure is lowered, for example a gas is allowed to expand as it leaves a pressurised reactor, heat is produced.

Energy-consuming processes

Endothermic reactions: these require heat to maintain their reaction temperature.

Pressure changes: the compression of gases, necessary to provide the high pressures needed for some processes, requires a considerable amount of energy.

> **12** Explain why an endothermic reaction requires heat to maintain a reaction temperature.

> **13** What types of gaseous reactions are likely to require high pressures?

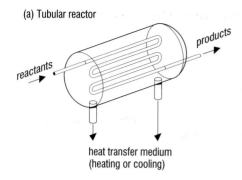

(a) Tubular reactor

reactants

products

heat transfer medium
(heating or cooling)

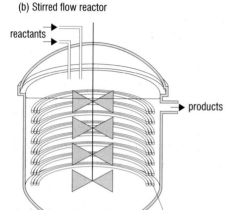

(b) Stirred flow reactor

reactants

products

coil for heat
transfer medium
(heating or cooling)

Fig 20.12 Two types of reactors used in continuous processing (a) a tubular reactor, (b) a stirred flow reactor.

Industrial reaction vessels

Batch reactors are scaled-up versions of laboratory equipment. The reaction takes place, the apparatus is dismantled and the product isolated. Continuous processes require carefully designed reactors since reactants are constantly being fed in and products removed as they form.

The design of reactors for continuous processes depends on the phase in which reaction takes place: liquid or gaseous. In the liquid phase, tubular reactors or stirred-flow reactors are used (Fig 20.12). Tubular reactors consist of coiled tubing through which reactants are passed, with the tubing kept at the desired temperature by passing a heat transfer medium around its outside. In a stirred flow reactor, the reactants are stirred into the reaction mixture by rotating paddles and the product tapped from the top of the reactor.

14 **What determines the length of tubing in a tubular reactor?**

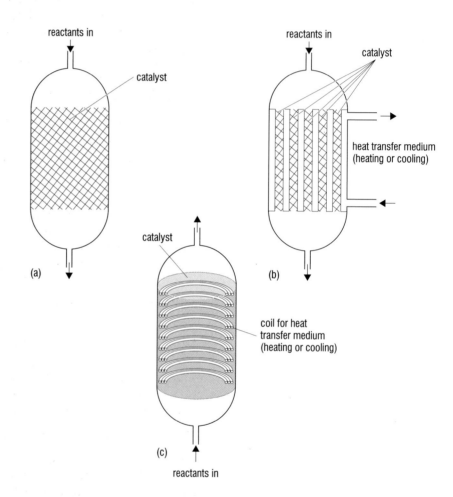

Fig 20.13 Catalytic reactors for reactions between gases which require a catalyst: (a) adiabatic reactor; (b) tubular reactor; (c) fluidised-bed reactor.

For gaseous reactions catalytic reactors are nearly always employed. These may be adiabatic, tubular or fluidised-bed reactors (Fig 20.13). In an adiabatic reactor no attempt is made to control the temperature of the reaction. Tubular reactors are similar to those for liquid phase reactions with the tubing packed with catalyst, while in fluidised-bed reactors the gaseous reactants are passed through the catalyst at such a speed that the solid catalyst is lifted and the catalyst bed is said to have been fluidised.

15 **What is the major disadvantage of an adiabatic reactor?**

Materials used to construct the reactors and other parts of the plant must meet certain requirements:

- inert to reactants, intermediates and products;
- cheap and long-lasting;
- capable of withstanding high temperatures and pressures.

16 **Why is it desirable to keep the number of stages in a synthetic route to a minimum if a continuous process is to be used?**

Costs

Costs of manufacturing chemical compounds, as in all other manufacturing industries, come under two headings: **capital cost** and **production cost**.

Money required to set-up the plant – to buy the reactors and other facilities and to build it – is the capital cost. Once established there is a regular input of money required to buy raw materials, pay for services (water, electricity, gas, steam and many others), meet salaries and wages of employees, and pay for maintenance work. The factory also has a number of overheads to meet (rates, rental charges and so on). Finally, the chemical plant will deteriorate with age and use, eventually needing to be replaced. This 'hidden' cost in running the plant is called **depreciation**.

17 **Explain why capital costs are fixed but production costs are variable.**

The economics of a process are usually complex, being dependent on several interacting factors. However, we can make two important generalisations.

- It is desirable to restrict a process to a small number of stages, since each stage requires a reactor and, usually, a separation unit.
- It is desirable to use conditions which do not make heavy demands on energy (extremes of high or low temperatures and pressures) or which require specialised, expensive equipment, such as high pressure reactors or reactors which are resistant to corrosive compounds.

18 **Why are catalysts so important in the chemical industry?**

Co-product or by-product?

Chemical reactions usually give more than one product, even if the manufacturer is only interested in one of them for sale or use in later processes. For example, sulphur dioxide to be used in the manufacture of sulphuric acid is sometimes made by roasting zinc sulphide in air:

$$2ZnS(s) + 3O_2(g) \longrightarrow 2ZnO(s) + 2SO_2(g)$$

The sulphur dioxide is the desired product but zinc oxide is also produced. This is called a **co-product.** Co-products are the reaction products formed in the same chemical reaction as that which gives the desired compound.

The nature of co-products formed in a chemical process affect the economics significantly. Useful co-products may have saleable value and therefore make an important contribution to the production cost of a process. Co-products with no commercial value may involve expensive disposal costs.

Sometimes other reactions (often called **side reactions**) occur alongside the main reaction. These give compounds, called **by-products**, other than those formed by the main reaction. They might arise, for example, by thermal decomposition of one of the products of the main reaction. Alternatively, one of the reactants may undergo reaction with the solvent. For example,

Main reaction: Compound C is made by reacting compounds A and B. Compound D also forms and is a co-product.

$$A + B \longrightarrow \quad C \quad + \quad D$$

$$\text{desired} \qquad \text{co-product}$$
$$\text{product}$$

Side reaction: Compound A reacts with the solvent to give an unwanted by-product E:

$$A + S \longrightarrow E$$

$$\text{by-product}$$

Side reaction: As well as reacting with compound A in the main reaction, compound B undergoes an undesirable thermal decomposition to give unwanted by-products:

$$B \longrightarrow F$$

$$\text{by-product}$$

ACTIVITY

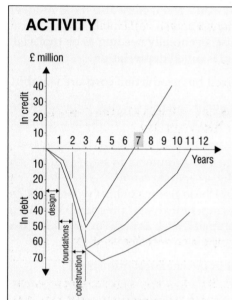

Fig 20.14 Cash flow during the design, construction and early years of production of a typical ammonia plant.

THE ECONOMICS OF INDUSTRIAL PROCESSES

A typical continuous process

A capital outlay is required to build a chemical plant even for an established process. It takes a number of years before the bank balance for such a plant is in credit. Fig 20.14 shows how the bank balance varies during the plant design phase, as the foundations are laid, the plant is constructed, and the early years of production for a typical ammonia plant producing over 103 tonnes of ammonia daily.

A typical batch process for the production of a fine chemical

The pharmaceutical industry invests heavily in research. Although pharmaceuticals can be big profit makers, investment is high and many projects are unsuccessful. Fig 20.15 shows the cash flow for a typical R&D programme in the pharmaceutical or agrochemical industries. The UK market is relatively small and so the financial benefits of the initial UK launch are far less than those gained by a launch in the much bigger USA market. Patents are vital to the economic success of the pharmaceutical industry. After filing, a patent has a lifetime of twenty years. When it expires, competitors may market the same drug under a different name; this is called the generic competition phase.

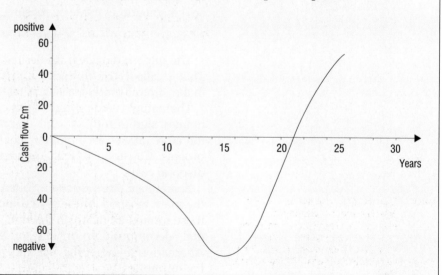

Fig 20.15 Cash flow for a typical R&D programme in the pharmaceutical or agrochemical industries.

(a) This part relates to the economics of ammonia production.

(i) Which of the three graphs on Fig 20.14 represents the cash flow in each of the following situations:

- interest is charged on the debt at a rate of 15%;
- no interest is charged on the debt;
- there is a delay of one year in start-up?

(ii) Sketch two graphs for the cash flow for the same operation assuming:

- the interest charged on the debt is 10%;
- the interest charged on the debt is 15% for the first six years but then falls to 10% for the remainder of the period of debt.

(iii) The plant is designed to operate for 365 days each year but doesn't.

For what reasons would the plant be closed down?

(b) This part relates to the economics of producing a drug.

Place the following labels on the appropriate parts of the cash flow diagram:

positive cash flow (programme in credit)
negative cash flow (programme in debt)
discovery of drug
research phase (searching for candidate drugs for an identified disease)
development phase (testing the most likely candidate)
break-even point
launch in the UK
launch in the USA
patent filed
patent expiry
profit-making period
other companies produce drugs in competition with the product

20.4 CLASSICAL SYNTHESIS COMPARED WITH INDUSTRIAL PROCESSING

Laboratory preparations are often limited to those requiring equipment and conditions that can be easily obtained. For example, most preparations are carried out at room pressure and within the temperature range 0 – 100°C. More extreme conditions can be used but require specialised equipment. In industry, simpler synthetic routes are often found by utilising extremes of temperature or pressure. The scale of manufacture makes investment in the more expensive apparatus worthwhile. The manufacture of epoxyethane provides an example.

Fig 20.16 Epoxyethane has many important uses. For example, it reacts with water to give ethan-1,2-diol, an industrial process that uses about 60% of the epoxyethane produced. Ethan-1,2-diol is used as anti-freeze and in the manufacture of polyesters. Will it help this car to start?

In the laboratory, epoxyethane is made by treating ethene with chlorine dissolved in aqueous alkali, followed by treatment with concentrated alkali.

$$C_2H_4(g) + Cl_2(aq) \xrightarrow{\text{Cl}_2 \text{ in dilute NaOH(aq); rt}} CH_2ClCH_2OH(aq)$$

$$CH_2ClCH_2OH(aq) \xrightarrow{\text{conc NaOH(aq); warm}} \underset{O}{CH_2 — CH_2}(aq) + Cl^-(aq) + H_2O(l)$$

19 **Explain why CH$_2$ClCH$_2$OH is formed in the first stage and name a likely by-product (one that is formed in a side reaction) of this stage.**

Before the 1930s this route was used to manufacture epoxyethane. Ethene was passed into an aqueous solution of chlorine at about 50°C to give 2-chloroethanol and hydrochloric acid. Treatment of this reaction mixture with calcium hydroxide, at 100°C, gave epoxyethane and calcium chloride.

$$C_2H_4(g) + Cl_2(aq) \xrightarrow{\text{Cl}_2 \text{ in H}_2\text{O; 50°C}} CH_2ClCH_2OH(aq) + HCl(aq)$$

$$CH_2ClCH_2OH(aq) + HCl(aq) + Ca(OH)_2(s) \xrightarrow{100°C} \underset{O}{CH_2 — CH_2}(aq) +$$

$$CaCl_2(aq) + 2H_2O(l)$$

20 **Why is calcium hydroxide used in the industrial process whilst sodium hydroxide is the preferred alkali for use in the laboratory?**

The problem was how to dispose of the co-product of the second stage – calcium chloride. It has no saleable value.

The introduction of a single stage reaction solved the problem. This came about when, in the 1930s, chemists found that ethene could be oxidised directly to epoxyethane by reaction with oxygen in the presence of a silver catalyst. Ethene and oxygen are passed over a silver catalyst at 250–350°C and 10–15 atmospheres pressure. The silver is supported on aluminium oxide and the reaction mixture is in the reactor for only 1–4 seconds!

$$2C_2H_4(g) + O_2(g) \xrightarrow{\text{cat = Ag (on Al}_2\text{O}_3\text{); 250–350°C, 10–15 atm}} \underset{O}{2CH_2 — CH_2}$$

21 **Why is the silver finely divided on aluminium oxide?**

This method has the enormous economic advantage of being a single stage reaction, as well as not producing the unwanted calcium chloride.

ACTIVITY

Fig 20.17a Variation in (a) acidity of rainwater and (b) barren lakes in Norway, and (c) the consumption of fossil fuels in Europe.

ACID RAIN

Between 1955 and 1975, statistics were gathered on the number of barren lakes in Norway, the acidity of the rainwater and the fossil consumption in Europe during the same time period (Fig 20.17a). Data is also available for the emissions of sulphur dioxide in Sweden for the period 1950 to 1983 (Fig 20.17b).

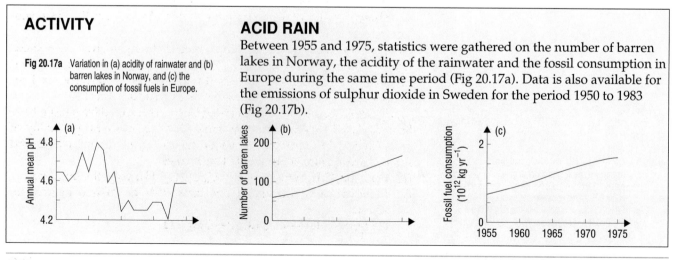

THE CHEMICAL INDUSTRY

ACTIVITY *continued*

Fig 20.17b Sulphur dioxide emssions in Sweden.

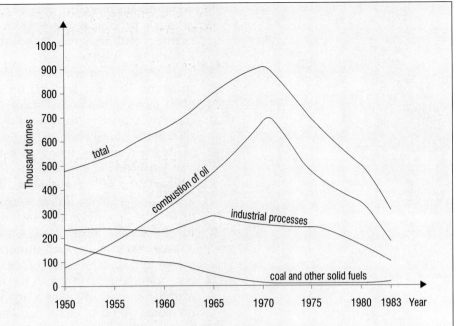

(a) Plot a graph to show how oxonium ion concentration, $[H_3O^+(aq)]$, in rainwater varied over the twenty-year period.

(b) Suggest reasons for the changes in pH of the rainwater from 1955 to 1975.

(c) What other information do you need before you can explain the cause of changes in pH?

(d) Assume you are working for an environmental group. The group wishes to release the findings with an informed interpretation of the data. Draft a briefing paper in which issues raised by the data are discussed.

Answers to Quick Questions: Chapter 20

1 Freezing and dry-freezing require a continuous energy input; the use of chemical preservatives does not.

2 Plastics, rubber, high performance advanced materials, oils and lubricants, paint and other materials used in the manufacture of cars are produced by the chemical industry.

3 Ethene, propene, carbamide (urea).

4 Pharmaceutical (about 24%) and agrochemical products (about 18%) have biggest share of the gross value added of sectors of the UK chemical industry.

5 Storage containers, some cooking utensils, lining of non-stick cookware, containers for microwave cooking, work surfaces, floor tiles, possibly cups, plates and saucers (though these are usually used for picnics etc). There may be others that you can add to this list.

6 Many operations which were carried out or controlled by people have been replaced by automated systems. This requires a larger outlay of cash (capital investment) but lower running costs (since wages and salaries are a major component of such costs).

7 (a) Benzene, C_6H_6; ethylbenzene, $C_6H_5CH_2CH_3$; polyethylbenzenes.

   (b) Removal of hydrogen, in this case from an alkane to give an alkene.

   $$C_6H_5CH_2CH_3 \longrightarrow C_6H_5CH = CH_2 + H_2$$

8   A continuous process.

9   A batch process requires labour to set up the reactor apparatus, charge it with reactants, isolate the products, dismantle and clean the apparatus.

10 The sites of chemical industry are associated with good transport infrastructure. They are serviced by good road and rail systems, and are often close to sea-ports. Major towns grew up around the sites of chemical industry – it is not the case that the chemical industry was sited at the centre of a highly populated area!

11 Sea-water contains many dissolved salts and, therefore, must be purified before use. Several techniques are available. In the nuclear industry, the salts are removed by ion exchange prior to use of the water to raise steam. If they were not removed, there would be a build-up of solid deposits in the boilers.

12 An endothermic reaction has a positive enthalpy change, which means it takes heat from the surroundings. Therefore, such a reaction must be heated to maintain the reaction temperature.

13 Gaseous equilibria in which the number of moles of products are less than the number of moles of reactants (according to Le Chatelier, high pressures would cause the equilibrium to favour products in this case).

14 The time the reaction mixture must be in the tubing in order that the desired conversion from reactants to products has been achieved.

15 Reactions which require heating or cooling cannot be carried out.

16 More stages – more equipment – more expense!

17 Capital costs require one initial outlay of money. Production costs are on-going, and wages and salaries, costs of materials, energy, transport etc will vary over a period of time.

18 They allow milder reaction conditions (lower temperatures and pressures), which are more economic, to be employed.

19 Cl_2 in dilute NaOH(aq) gives a solution of HOCl(aq):
$Cl_2(g) + H_2O(l) \rightleftharpoons HCl(aq) + HOCl(aq)$
The dilute NaOH(aq) moves the position of equilibrium to the right. Therefore, the addition of HOCl to C_2H_4 takes place.
$CH_2CH_2 + HOCl \longrightarrow CH_2ClCH_2OH$
A likely by-product is 1,2-dichloroethane, CH_2ClCH_2Cl:
$CH_2CH_2 + Cl_2 \longrightarrow CH_2ClCH_2Cl$

20 Calcium hydroxide is cheap and readily available. However, it is relatively insoluble if water. Therefore, in the laboratory, aqueous solutions of sodium hydroxide are preferred because of the ease of handling them (in particular, controlled addition to a reaction mixture).

21 To provide a larger surface area upon which the reaction can take place.

Chapter 21

THE PETROCHEMICAL INDUSTRY

21.1 CRUDE OIL AND NATURAL GAS

Crude oil, also called **petroleum**, is an invaluable source of organic compounds from which most of the synthetic parts of our world have been created. Around 95% of all synthesised organic compounds are made from a number of key compounds which are produced in bulk from crude oil or natural gas. These are the compounds we call **petrochemicals** and the petrochemical industry is one of the world's major industries. But, however important these compounds may be, we still use over 90% of our crude oil and natural gas as fuel.

Fig 21.1 A modern petrochemical plant is an example of sophisticated design and construction, bringing together science, engineering and technology.

The first stage in the petrochemical industry is the processing of crude oil. Natural gas does not require similar processing since it is essentially methane. However, it is purified before further conversion into other organic compounds.

Distillation of crude oil

Crude oil is a complex mixture of hydrocarbons. These are mainly saturated hydrocarbons – alkanes and cycloalkanes. Smaller amounts of unsaturated hydrocarbons are present (alkenes and aromatics), along with small amounts of compounds containing nitrogen, oxygen or sulphur.

The first stage in processing crude oil, whether in preparation for its use as a fuel or as a source of petrochemicals, is **fractional distillation** (Fig 21.2). This is carried out as a continuous process, with crude oil being constantly supplied to the distillation vessel. The crude oil is separated into **fractions**, each of which has a characteristic boiling point range and is itself a mixture of hydrocarbons (Table 21.1). These fractions are used to make either fuels or petrochemical feedstocks.

Fig 21.2 Fractionating towers are designed for maximum efficiency in the shortest possible distance (otherwise the towers would be unacceptably tall – both aesthetically and economically). Bubble caps are used to achieve this efficiency.

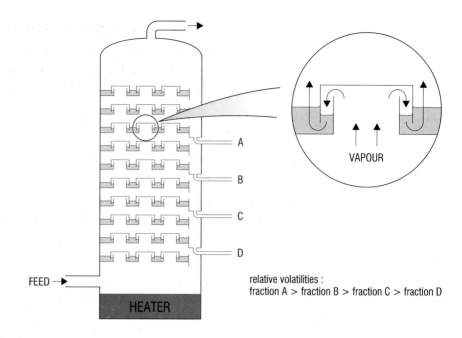

FEED →

HEATER

relative volatilities :
fraction A > fraction B > fraction C > fraction D

Table 21.1 The fractions obtained from the fractional distillation of crude oil

Boiling range / °C	Name of fraction	Volume / %	Number of carbon atoms in typical hydrocarbon	Nature of fraction
<0	gases	1–2	1–4	unbranched alkanes
0–65	gasolines	20–40	5–6	unbranched and branched alkanes
65–170	naphtha		6–10	unbranched, branched alkanes cycloalkanes, and aromatics
170–250	kerosine (paraffin)	10–15	10–14	complex alkanes, cycloalkanes and aromatics
250–340	gas oil (diesel)	15–20	14–19	complex alkanes, cycloalkanes, aromatics and alkylaromatics
340–500	lubricating oil and wax	40–45	19–35	high relative molecular mass complex alkanes, complex cycloalkanes and aromatics, alkylaromatics
>500	bitumen		>35	very high relative molecular mass hydrocarbons of all types

1 Why does volatility decrease with increasing number of carbon atoms in a typical hydrocarbon?

2 Why do you think that no aromatic compounds are found in the gases or gasolines fractions?

The refining processes for fuel production

The growth of the petrochemical industry owes much to the motor car. The most common internal combustion engine runs on petrol. Liquid fuel is vaporised in the carburettor and mixed with air. This gaseous mixture is fed into the piston chamber where it is compressed and then ignited by a spark. Expansion of the gases occurs, forcing the piston down and providing the mechanical energy to drive the car.

3 **Explain why the hydrocarbon/air mixture expands when it is ignited.**

The fuel must have two properties if it is to work efficiently and effectively:

• It must be sufficiently volatile to vaporise in the carburettor but not vaporise so easily that it causes 'vapour lock' in the fuel pipes. A fuel which has a boiling range of 20–200°C is the best compromise.
• It must not ignite too soon as this causes 'knocking' or 'pinking', with a resulting loss of power. Octane number is a measure of a fuel's burning characteristics (the octane numbers of various hydrocarbons are given in Table 11.2). Most modern cars run on 90–98 octane petrol.

Fractions from crude oil distillation which have the right volatility for car engines are gasolines and naphtha. However, the octane number of these combined fractions is about 50 – totally unsuitable for modern cars. Early cars ran on gasoline, with an octane number of about 65, and it was increasing car sales which gave rise to huge expansion of the refining industry. Not only were more cars made but increased performance required higher octane fuels. Fractions obtained by distillation of crude oil could not meet the demand. The crude oil needed to be refined.

There are two processes which can be used to refine crude oil:

• catalytic cracking;
• catalytic reforming.

Both involve treating the lower volatility (high boiling range) fractions in order to break down the large hydrocarbon molecules into smaller ones.

Fig 21.3 Petrol must have a suitable octane rating if it is to be effective in an internal combustion engine. An engine is designed to have a precise ignition point and additives are used to ensure that petrol has the correct octane number for the engine.

Fig 21.4 A Boeing 747 aircraft being re-fuelled. High performance engines demand high performance fuels with high octane numbers. This allows increased compression of the hydrocarbon/air mixture before ignition and improved performance.

Catalytic cracking

The residue from distillation at atmospheric pressure is distilled under reduced pressure and passed over a strongly acidic catalyst (nowadays zeolites are used) at about 500°C and 2 atmospheres. This produces gasoline with an octane number of about 90, together with higher boiling range fractions and catalytic cracker gas (10–20% of the feed). The gas contains substantial quantities of propene and isomers of butene.

4 **Why does distillation under reduced pressure allow the residue to be vaporised at lower temperatures?**

Fig 21.5 Modern catalytic crackers are fluidised bed reactors. During cracking, carbon is deposited on the catalyst and must be burnt off to regenerate it. The catalyst, in particle form, is moved continuously through alternating zones: reaction zones and catalyst-regeneration zones.

Table 21.2 Typical composition of catalytic cracker

Gas	% by mass
H_2	0.6
H_2S	3.4
CH_4	10.3
C_2H_4	3.4
C_2H_6	7.5
C_3H_6	13.8
C_3H_8	12.1
C_4H_8	18.4
C_4H_{10}	30.5

The low relative molecular mass alkanes and alkenes which form are used either in other refining processes or as raw materials for the petrochemical industry.

5 Classify the compounds in Table 21.2 as saturated hydrocarbons, unsaturated hydrocarbons or other compounds.

6 Why are unsaturated hydrocarbons more important than saturated hydrocarbons as the starting materials in synthesis?

7 Name the compounds which have the molecular formula 'C_4H_8'.

Catalytic reforming

Naphtha has an octane number of about 40. To raise its octane number it is passed over a platinum on aluminium oxide catalyst at about 500°C and 7–30 atmospheres. Most importantly, alkanes are simultaneously cyclised and dehydrogenated to give aromatics, though many other reactions also occur.

Table 21.3 Conversions in a catalytic reformer

Hydrocarbons	% of feed	% of products
alkanes	60	32
cycloalkanes	25	2
aromatics	15	66

8 What other gas is produced in a catalytic reformer?

ACTIVITY

ADDITIVES FOR PETROL

An important form of scientific communication at conferences is the poster presentation. Conference organisers invite contributors to provide a poster (usually 'A1' size, about 120 cm × 80 cm) summarising the work they have carried out and its implications.

Read the article below and design a poster to convey the importance of this work to the delegates attending a scientific conference on fuels for transport systems. Note that some non-standard chemical names are used.

Cheaper MTBE feed in sight

US petrochemical producer Lyondell has announced a new process that could cut the cost of setting up production for petrol additives. Methyl *tert*-butyl ether (MTBE) is the leading oxygenate compound for making high-octane fuel blends without using poisonous lead alkyls or aromatics. MTBE plants are springing up all over the world and in turn these are driving the construction of plants to make the key feedstocks: methanol and but-2-ene.

Isobutene supply has been a limiting factor for some MTBE plants. Conventionally, but-2-ene can be made from butane by a two-step process of isomerisation and dehydrogenation (1). Lyondell, however, has developed a process to convert but-1-ene to

but-2-ene in a single step (2). n-Butenes are abundant, and Lyondell has significant supplies from its refining and alkene production. There are a number of catalysts for but-1-ene → but-2-ene conversion in the literature, based on alumina or silica and operating at high temperature. For instance, Air Products patented a silica catalysed process operating at 520°C.

Lyondell has patents pending on the new process, which it has successfully tested at lab and pilot plant scales. The company plans a large-scale demonstration this year. 'Once we have demonstrated the success of this technology on a large scale basis, we expect to build a commercial unit at our Channelview (Texas) Petrochemical

Complex. We also plan to license the technology to other producers', said Bob Gower, Lyondell's president. Early indications are that the capital cost of building a plant with the new technology will be less than half that for the conventional process.

The US Clean Air Act has been a boost for Lyondell, which already produces the low-aromatic petrol needed to reformulate fuels under the act (recovering the aromatics for the chemical market), as well as abundant methanol and butene feedstocks for MTBE.

The need to refine crude oil fractions to produce a suitable fuel for the motor car led to the birth of the modern petrochemical industry. There were two important side-products from the refining processes – alkenes from catalytic cracking and aromatics from catalytic reforming. These are the basis for thousands of useful compounds.

The petrochemical industry now uses naphtha and natural gas as the major feedstocks and processes have been developed, based on the refining processes, to convert these fractions into the basic building blocks, often called **petrochemical feedstocks**, for the synthesis of organic compounds. There are three main petrochemical processes for naphtha:

- steam cracking;
- catalytic reforming;
- steam reforming.

9 **What types of hydrocarbon are present in the naphtha fraction obtained from the fractional distillation of crude oil?**

Steam cracking

Steam cracking is also called thermal cracking. A mixture of naphtha and pre-heated steam is heated to about 800–1000°C. It is in the reactor (called a cracking furnace, Fig 21.6) for less than 0.2 seconds. The cracking process may be represented by the typical equation:

$$2C_6H_{14}(g) \longrightarrow CH_4(g) + 3C_2H_4(g) + C_2H_6(g) + C_3H_6(g)$$

But remember that this is greatly simplified. The reactions occurring in the cracking furnace are numerous and complex.

Fig 21.6 Steam cracking is carried out in a cracking furnace.

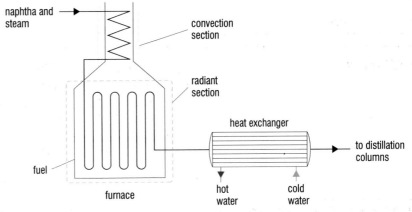

10 **How do the cracking processes used in refining and in the manufacture of petrochemical feedstocks differ?**

Cracking involves free radical chain reactions. This explains the mixture of compounds obtained, for example, when ethane is used as the feedstock (Table 21.4). A suggested mechanism for the steam cracking of ethane is:

initiation $\quad H_3C - CH_3 \longrightarrow 2 \cdot CH_3$

propagation
$$\begin{cases} \cdot CH_3 + CH_3CH_3 \longrightarrow CH_4 + \cdot CH_2CH_3 \\ \cdot CH_2 - CH_2 - H \longrightarrow H\cdot + CH_2 = CH_2 \\ H\cdot + CH_3CH_3 \longrightarrow H_2 + \cdot CH_2CH_3 \end{cases}$$

termination
$$\begin{cases} H\cdot + \cdot CH_2CH_3 \longrightarrow CH_3CH_3 \\ \cdot CH_3 + \cdot CH_2CH_3 \longrightarrow CH_3CH_2CH_3 \\ \cdot CH_2CH_3 + \cdot CH_2CH_3 \longrightarrow CH_3CH_2CH_2CH_3 \end{cases}$$

Remember that a dot (·) represents a single, unpaired electron.

11 **Suggest an initiation step for the cracking of propane.**

Steam cracking is used with various feedstocks other than naphtha, for example ethane, propane and gas oil. The compositions vary considerably with precise reaction conditions and nature of the feed and so the percentages given in Table 21.4 are only approximate.

Table 21.4 Percentage composition of the products of steam cracking

| Product | Feedstock | | | |
	ethane	propane	naphtha	gas oil
H_2	5	2	1	1
CH_4	9	27	15	8
C_2H_4	78	42	20–30	15–23
C_2H_6	3	19	16	14
C_3H_6	2	3	5	5
$CH_2{=}CH{-}CH{=}CH_2$	2	3	5	6
petrol	3	7	23–33	20
fuel oil	3	7	4	22–30

Catalytic reforming

The process is the same catalytic reforming of crude oil fractions used to produce higher octane fuels. When naphtha is used as the feed, a range of aromatic hydrocarbons are obtained, including benzene, methylbenzene and the three isomers of dimethylbenzene. Three main reactions occur:

i) Dehydrogenation

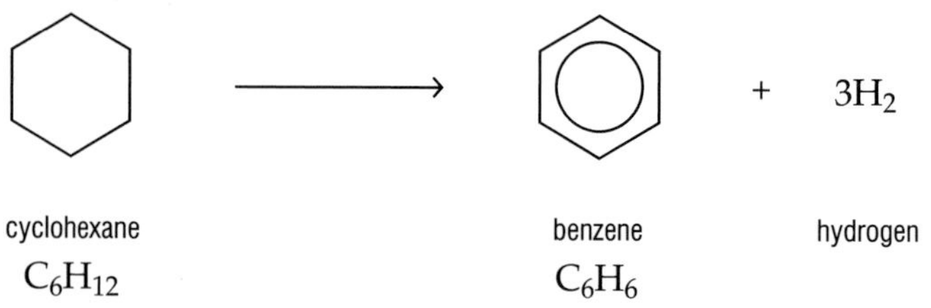

| cyclohexane | benzene | hydrogen |
| C_6H_{12} | C_6H_6 | |

ii) Dehydroisomerisation

| methylcyclopentane | benzene | hydrogen |
| $C_5H_9CH_3$ | C_6H_6 | |

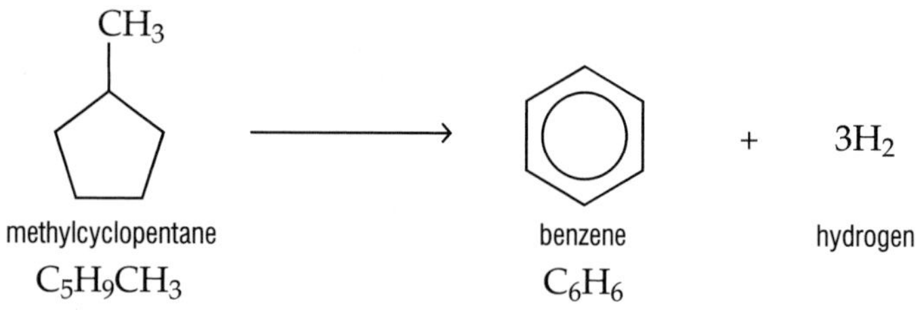

iii) Dehydrocyclisation

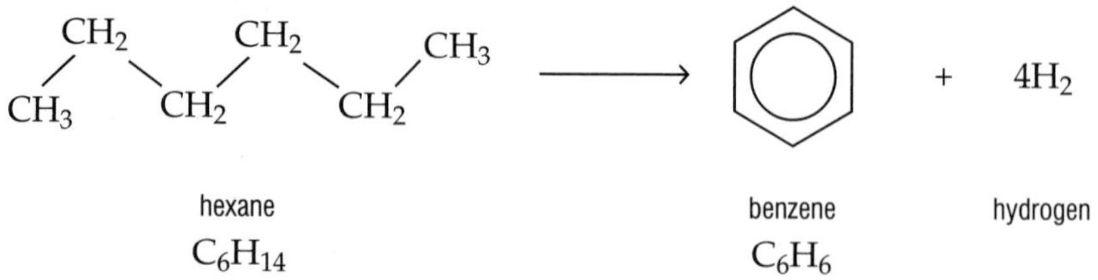

| hexane | benzene | hydrogen |
| C_6H_{14} | C_6H_6 | |

12　'Dehydroisomerisation' and 'dehydrocyclisation' are terms we have not met before. From the examples given try to explain their meaning.

13 Write equations to represent (a) the dehydroisomerism of 1,3-dimethylpentane, (b) the dehydrocyclisation of heptane.

Steam reforming

The normal feed is natural gas (mainly methane) but where this is not available naphtha is used. The key reaction is

$$CH_4(g) + H_2O(g) \rightleftharpoons CO(g) + 3H_2(g)$$

The reaction is endothermic and reversible. From a consideration of equilibria, reaction conditions for maximising yield would be high temperature and low pressure. In practice a steam/methane molar ratio of about 3 : 1 is used and the mixture passes over a nickel catalyst at about 800°C. It is important that sulphur-containing impurities are removed from the feed since they would poison the catalyst. Poisoning by sulphur-containing impurities is a common problem in the petrochemical industry.

Steam reforming of naphtha takes place under similar conditions and may be represented as

$$(CH_2)_n(g) + nH_2O(g) \longrightarrow nCO(g) + 2nH_2(g)$$

14 Write equations for the steam reforming of (a) pentane, (b) heptane.

Nowadays the major feedstocks for the petrochemical industry are naphtha and natural gas. Looking to the future, it is likely that natural gas will increase in importance and that the cracking of higher boiling range fractions from crude oil will also gain in importance. Natural gas liquids (NGLs – mostly ethane, propane and butane) may be used more, particularly as they are relatively pure mixtures and gas cracking is technically easier and cheaper than cracking liquids.

The building blocks for petrochemical manufacturing processes, and their vast array of important and useful products, come from four major processes. These are given in Table 21.5.

Table 21.5 Petrochemical manufacturing processes

Type of building block	Source
Alkenes	Refining crude oil by catalytic cracking.
Alkenes	Steam cracking of ethane, propane or naphtha.
Aromatics	Catalytic reforming of naphtha.
Carbon monoxide	Steam reforming of natural gas or naphtha.

ACTIVITY

ETHANOL BY FERMENTATION

Ethanol may prove to be an important feedstock for the chemical industries of some developing countries. Monosaccharides are fermented to ethanol and carbon dioxide with the help of suitable enzyme catalysts.

Monosaccharides are not the usual starting point, however, for the manufacture of ethanol by fermentation. A by-product of sugar refining, molasses, is commonly used. Molasses contains about 35–40% of sucrose and 15–20% of glucose and fructose (percentages are by mass). The molasses is diluted to 150–200 g dm^{-3} total sugar, the pH adjusted to 4–5 and yeast added. After about 36 hours at 30°C, the mixture contains 60–90 g dm^{-3} ethanol.

(a) What reaction must occur before the sucrose present in the molasses can be metabolised to ethanol?

(b) Write an equation for the fermentation process.

(c) How can the pH be adjusted and what method might be used to ensure that the desired pH had been reached?

(d) Calculate the approximate theoretical yield of ethanol from a diluted molasses solution containing 200 g dm^{-3} total sugar. Explain any assumptions that you make. Compare the theoretical yield with the actual yield obtained after 36 hours.

(e) The yield of ethanol does not increase significantly after 36 hours. Offer an explanation for this.

(f) Design and plan an investigation to study the factors which affect the fermentation of a molasses solution.

(g) Explain how organic molecules containing more than two carbon atoms could be synthesised from ethanol.

21.3 FROM PETRO-CHEMICALS TO BULK ORGANIC COMPOUNDS

There is not enough room here to give details of all the industrial processes used to convert the basic building blocks of the petrochemical industry into the bulk organic compounds which serve the needs of industries such as pharmaceuticals, agrochemicals, polymers and dyestuffs. Instead summary charts are given for chemicals obtained from ethene (Fig 21.7), propene (Fig 21.8) and benzene (Fig 21.9). We will discuss some of the processes involving ethene or propene where they illustrate important ideas we have met earlier in the book. A discussion of the detailed reactions of benzene are beyond the scope of this book.

Fig 21.7   Compounds obtained from ethene

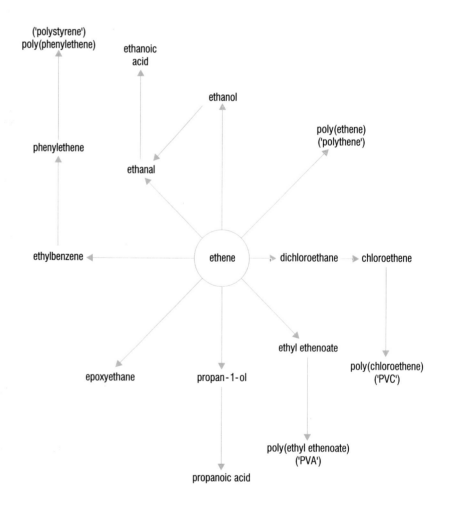

Fig 21.8 Compounds obtained from propene

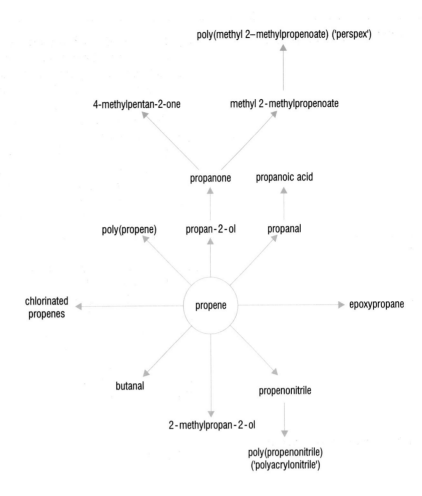

Fig 21.9 Compounds obtained from benzene

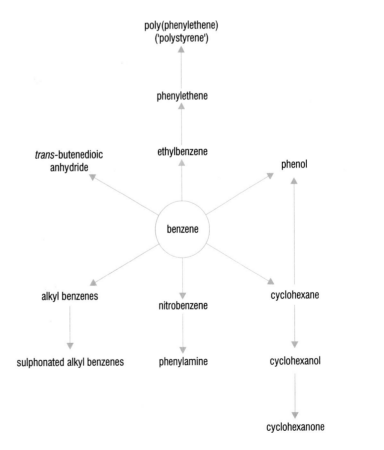

Chloroethene

Chloroethene is not easily prepared from ethene in the laboratory. However, oxychlorination processes are used industrially and give yields of around 90–95%.

$$CH_2\!=\!CH_2(g) + 2HCl(g) + \tfrac{1}{2}O_2(g) \xrightarrow[250\text{--}350^\circ C]{cat = CuCl_2/KCl} CH_2ClCH_2Cl(g) + H_2O(g)$$

$$CH_2ClCH_2Cl(g) \xrightarrow{\text{heat at about } 500^\circ C} CH_2\!=\!CHCl(g) + HCl(g)$$

Hydrogen chloride is recycled and used for the first reaction.

Epoxyethane and epoxypropane

The manufacture of epoxyethane has been discussed as an example of how an industrial process differs from a classical synthesis in Chapter 20. Epoxypropane is manufactured by an equivalent process.

15 Write equations for (a) the classical synthesis and (b) the industrial production of epoxypropane from propene.

An alternative process involves a reaction between propene and a hydroperoxide, for example

$$(CH_3)_3COOH + CH_2\!=\!CHCH_3 \rightarrow (CH_3)_3COH + CH_2\!-\!CHCH_3$$
$$\overset{\diagdown}{}O\overset{\diagup}{}$$

This is known as the Halcon process, named after the American company where it was invented in 1968.

16 Suggest a structure for $(CH_3)_3COOH$, and explain why it cannot represent the formula of a carboxylic acid.

The Halcon process is economically viable because the co-product 2-methylpropan-2-ol has value. It may be sold as an additive to petrol, to raise the octane number, or it can be converted into 2-methylpropene.

17 Under what conditions would 2-methylpropan-2-ol be converted into 2-methylpropene? What type of reaction is this?

Fig 21.10 Washing powders and liquids contain non-ionic surfactants which lower the surface tension of water and allow the detergent to penetrate more deeply into the fabric, improving washing power. These surfactants are manufactured from epoxyethane or epoxypropane.

THE PETROCHEMICAL INDUSTRY

Ethanol and propan-2-ol

The laboratory syntheses of these alcohols involve dissolving the alkene in concentrated sulphuric acid and diluting the resulting solution with water.

Industrially, however, ethanol is manufactured from ethene by the reaction

$$C_2H_4(g) + H_2O(g) \xrightarrow[\text{300°C}]{\text{conc H}_3\text{PO}_4 \text{ adsorbed on a solid, inert support}} C_2H_5OH(g)$$

The reaction is exothermic and reversible. High yields are favoured by low temperatures and high pressures but at low temperatures the reaction is too slow even in the presence of a catalyst. The reaction proceeds at an acceptable rate at about 300°C, but at this temperature K_p is very small and an equimolar mixture of steam and ethene gives a poor yield.

18 **Explain why K_p decreases with increasing temperature for this reaction.**

There are two possible solutions:

- increase the proportion of steam (use a large excess);
- increase the pressure.

Problems arise, however. The catalyst absorbs water and is rendered ineffective and so needs to be replaced periodically and high pressures require expensive equipment. Running costs, therefore, are high.

19 **Explain why increasing the proportion of water would improve yield.**

Inevitably a compromise is reached. The actual reaction conditions are 300°C, 70 atmospheres and a water : ethene mole ratio of 0.6 : 1. This only gives 5% conversion and so unreacted ethene is recycled.

In contrast to the hydration of ethene, a variety of catalysts can be used to increase the rate of reaction between propene and water. A 'solid' H_3PO_4 catalyst can be used (as for ethanol production) at 200–250°C and about 40 atmosphere. The yield is about 4%. However, recently two important processes have been developed:

- ICI process. Uses a tungsten oxide catalyst, 250–300°C, 200–300 atmosphere pressure and a water : propene molar ratio of 2.5 : 1.
- Deutsche Texaco process. Uses an acidic ion exchange resin as catalyst, 130–160°C, 80–100 atmosphere pressure and a water : propene molar ratio of 14 : 1.

In both processes water is in the liquid phase (because of the high pressures employed) and the propan-2-ol dissolves in it as it forms. This displaces the equilibrium in favour of products.

20 **Explain the advantage of the ICI and Deutsche Texaco processes over a process using 'solid' H_3PO_4 catalyst.**

Fig 21.11 Solvents play an important role in paints, dyes, adhesives, pharmaceuticals and cosmetics. Hydrocarbons, alcohols, ketones, esters, ethers and halogenated hydrocarbons are all used commonly. For example, gloss paints leave a hard film on a treated surface as the blend of solvents evaporates.

Ethanal and propanone

21 How can ethanal and propanone be obtained from ethene and propene respectively using classical synthetic methods?

Ethanal can be manufactured by a one-stage process. Oxygen and an excess of ethene are passed into a solution of palladium(II) chloride and copper(II) chloride in dilute hydrochloric acid at about 125°C and 3 atmosphere pressure. This is called the Wacker-Chemie process.

$$CH_2{=}CH_2 + \tfrac{1}{2}O_2 \longrightarrow CH_3CHO$$

Ethanal is removed from the reaction mixture by scrubbing with water. Unreacted ethene is recycled and the ethanal is recovered from the aqueous solution by distillation.

A two-stage process is also used which at first might seem unattractive in view of our criterion that chemical plants should have the minimum number of reactors and separation units. However, a closer look at the flow diagram for the plant (Figure 21.12) shows that pure oxygen is not required, air will suffice as the source of oxygen. Therefore, an oxygen separation plant is not needed as it is for the one-stage process (and an oxygen separation unit is very costly).

Fig 21.12 The Wacker-Chemie process for manufacturing ethanal from ethene can be a one or two-stage process. The one-stage process requires pure oxygen but the two-stage process can operate with air.

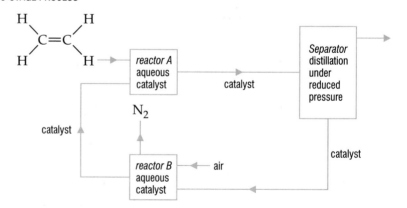

Although propanone could be produced directly from propene by the Wacker-Chemie process, this is not used significantly. The reason is that there has been no need to build plants specifically for propanone production since it is formed as a co-product in a number of other industrial processes. Instead, propene is hydrated to give propan-2-ol which can be dehydrogenated to propanone if required.

$$CH_3CH(OH)CH_3(g) \xrightarrow[\text{or cat = ZnO; about 380°C}]{\text{cat = Cu/Zn ; 400–500°C}} CH_3COCH_3(g) + H_2(g)$$

Propanone is a co-product in the cumene process for manufacturing phenol. The process involves three stages:

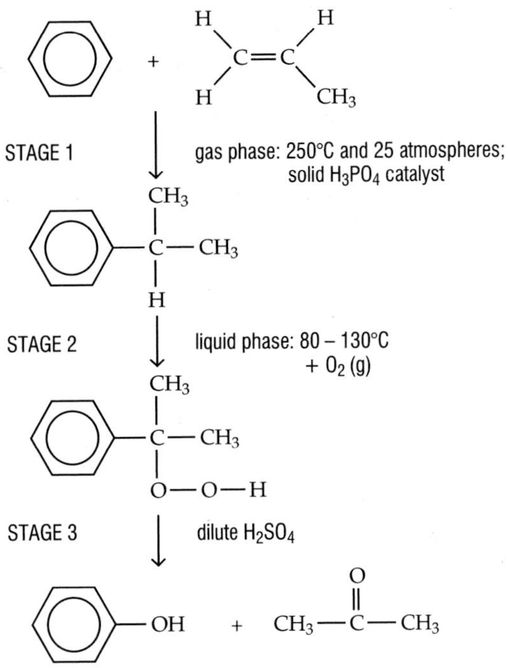
STAGE 1
gas phase: 250°C and 25 atmospheres; solid H3PO4 catalyst
STAGE 2
liquid phase: 80 – 130°C + O2 (g)
STAGE 3
dilute H2SO4

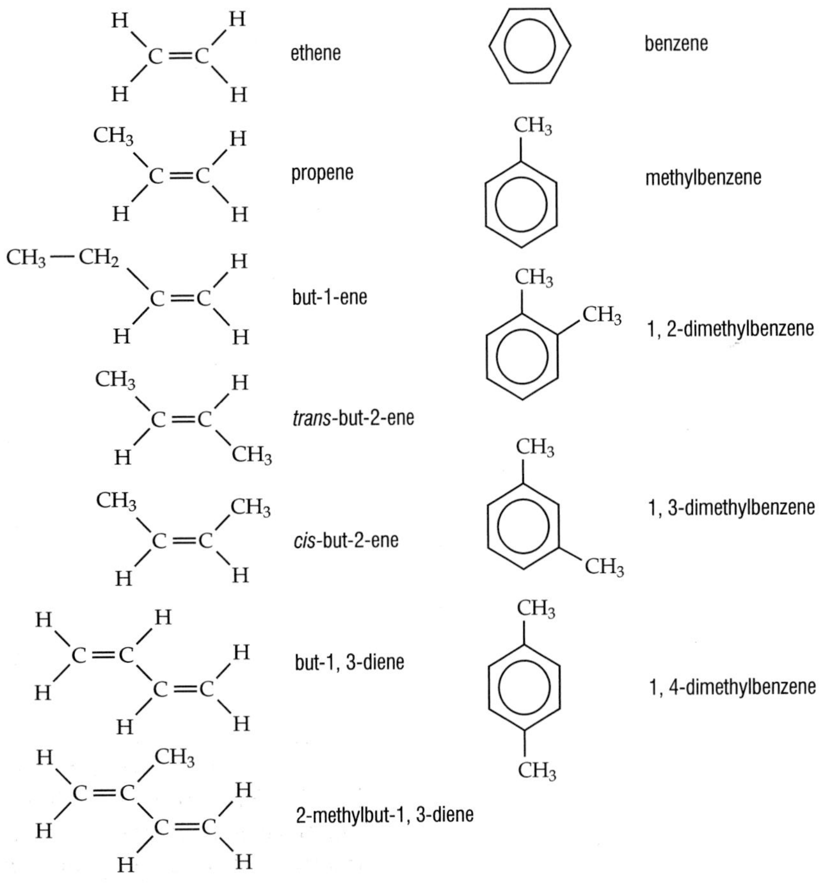
ethene
propene
but-1-ene
trans-but-2-ene
cis-but-2-ene
but-1, 3-diene
2-methylbut-1, 3-diene
benzene
methylbenzene
1, 2-dimethylbenzene
1, 3-dimethylbenzene
1, 4-dimethylbenzene

Fig 21.13 Petrochemical feedstocks. These are the basic building blocks produced by the petrochemical industry, used to make over 1,000,000 different organic compounds, of which around 30,000 are listed as being of commercial importance.

The manufacture of ketones

Read the article below and then answer the questions.

Ketones

One-step technology

Catalytica, of California, US, has developed a new technology to convert alkenes to ketones in a simple one-step process. Catalytica claims both economic and safety benefits over existing industrial processes.

The Catalytica process consists of the direct oxidation of the alkene with oxygen under mild conditions using a proprietary catalyst system. A variety of alkenes will undergo the reaction (*see* Scheme 1); propene produces propanone; cyclohexene produces cyclohexanone; all linear butenes produce methyl ethyl ketone (MEK) and ethene gives ethanal. But most of Catalytica's efforts so far have been directed towards the commercialisation of the MEK process.

The existing commercial processes for MEK production mainly involve vapour phase dehydrogenation of butan-2-ol: first butenes are hydrated to butan-2-ol using sulphuric acid, and then butan-2-ol is dehydrogenated at 400–500°C over a catalyst. This process is costly in energy – the reaction proceeds only at elevated temperature – and it involves sulphuric acid and sometimes Pd/ CuCl in the catalyst system, which are corrosive and an environmental hazard.

The Catalytica process proceeds at less than 100°C, without the use of sulphuric acid, and without halide in the catalyst. Catalytica claims a fivefold cost saving on energy alone, not to mention the benefit to safety and the environment. In the laboratory, yields of more than 95 per cent overall have been achieved, and Catalytica sees no problem in scaling up.

Catalytica is at present in negotiations with four firms that may build new MEK plants using Catalytica technology, or convert existing technology. A further potential application of this technology is the conversion of existing Wacker oxidation plants used for ethanol or propanone production, to the production of MEK.

(a) From the information supplied work out the systematic name for 'MEK' and name the class of organic compounds to which it belongs.

(b) Explain the advantages of the new process over the existing technology.

(c) What do you think the term 'proprietary catalyst system' means?

(d) Write equations for the two stages in the production of MEK from but-1-ene by the existing commercial process. Explain why the use of but-2-ene as the starting material would also give MEK.

(e) Write equations for the reactions of ethene, propene, pent-1-ene, pent-2-ene and cyclohexene with oxygen using the new technology.

(f) Design and plan a series of experiments which could be used in the laboratory to investigate the claims for the new catalyst.

THE MANUFACTURE OF A PESTICIDE

A selective weed killer, 2,4-D is made by a four step process (Fig 21.14).

(a) What reagents could be used for step 1 and what other compounds might be formed at this stage?

(b) Explain why careful control of pH allows 2,4-dichlorophenol to be separated from 2,4-D in steps 3 and 4.

(c) How could the two products formed in step 3 be separated?

(d) Describe how this reaction sequence could be carried out in the laboratory on a small scale, for example starting with 5 g phenol.

(e) What problems might be encountered when 'working up' the process to an industrial scale?

(f) Devise an analytical procedure that could be used in the control laboratory of the company manufacturing 2,4-D to assess the purity of the product.

Fig 21.14

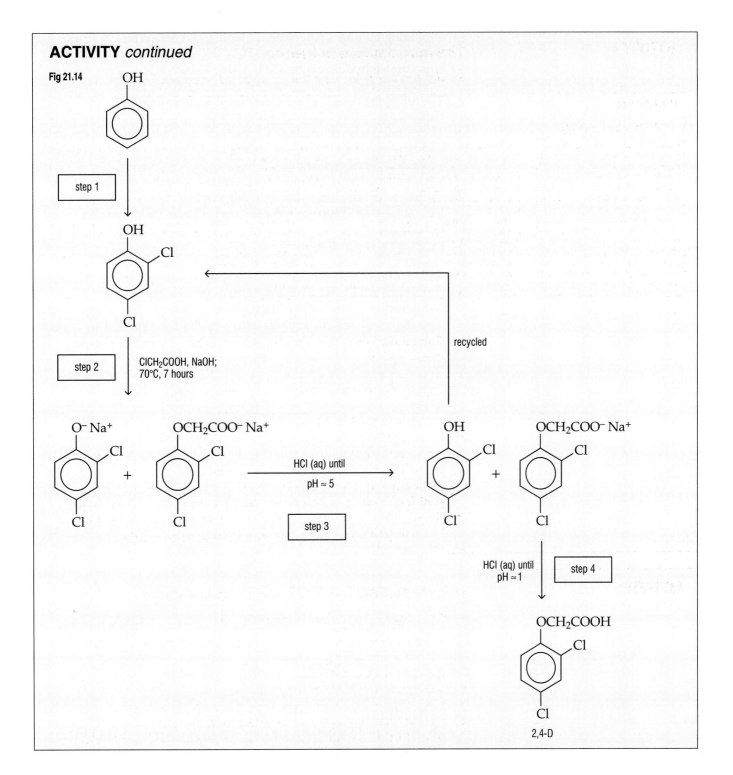

Answers to Quick Questions: Chapter 21

1 Dispersion forces are the intermolecular forces between these molecules, and therefore determine volatility. Dispersion forces decrease with decreasing molar mass.

2 They are not sufficiently volatile to distil over at these temperatures.

3 It produces carbon dioxide and water vapour. The volume produced is greater than the hydrocarbon/air mixture because one mole of a hydrocarbon with x carbon atoms will give x moles of carbon dioxide.

4 Boiling point decreases with decreasing pressure.

5   Assuming they are open-chain molecules and not cyclic:
Saturated hydrocarbons: CH_4, C_2H_6, C_3H_8, C_4H_{10}
Unsaturated hydrocarbons: C_2H_4, C_3H_6, C_4H_8
Others: H_2, H_2S
The following could be cyclic, saturated hydrocarbons:
C_3H_6, C_4H_8

6   They undergo a wider range of reactions (predominately electrophilic addition reactions with, for example, halogens, hydrogen halides, water, hydrogen cyanide).

7   Cyclobutane, but-1-ene, but-2-ene, 2-methylpropene.

8   Hydrogen (since alkanes are dehydrogenated).

9   Unbranched and branched alkanes, cycloalkanes and aromatics.

10 Cracking used to make petrochemical feedstocks employs higher temperatures and no catalyst.

11 $H_3C - CH_2 - CH_3 \longrightarrow H_3C\cdot + \cdot CH_2CH_3$

12 Dehydroisomerisation: dehydrogenation followed by a molecular rearrangement to give a different structural isomer.
Dehydrocyclisation: dehydrogenation followed by a molecular rearrangement which converts an open-chain compound to a cyclic compound.

13 (a)

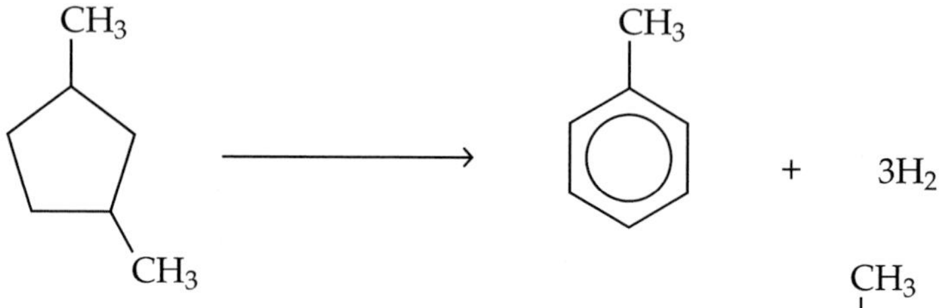

(b)

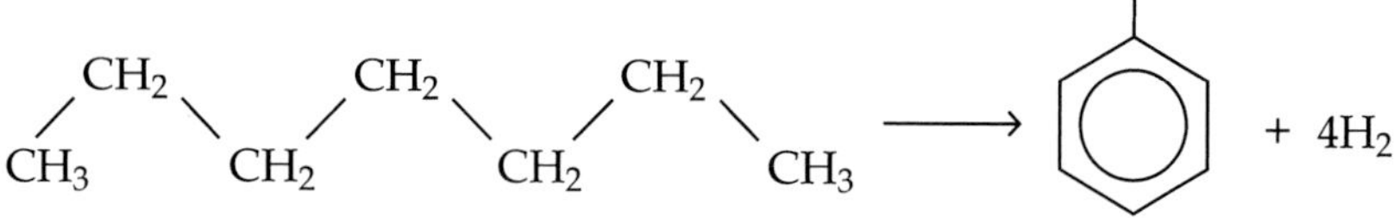

14 (a) $C_5H_{12}(g) + 5H_2O(g) \longrightarrow 5CO(g) + 11H_2(g)$
    (b) $C_7H_{16}(g) + 7H_2O(g) \longrightarrow 7CO(g) + 15H_2(g)$

15 (a) $C_3H_6 + Cl_2 \xrightarrow{\text{in dil. NaOH(aq); rt}} CH_3CH(OH)CH_2Cl$

$CH_3CH(OH)CH_2Cl \xrightarrow{\text{conc. NaOH(aq); warm}} CH_3 - CH - CH_2 + Cl^- + H_2O$ (with O bridging)

    (b) $C_3H_6 + O_2 \xrightarrow[\text{high temp. and press}]{\text{Ag catalyst}} CH_3 - CH - CH_2$ (with O bridging)

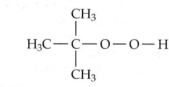

$$H_3C - \underset{\underset{CH_3}{|}}{\overset{\overset{CH_3}{|}}{C}} - O - O - H$$

Fig 21.14

$$- C \overset{\displaystyle O}{\underset{\displaystyle O - H}{<}}$$

Fig 21.15

16 See Fig 21.14
The structure does not contain the carboxylic acid group: Fig 21.15.

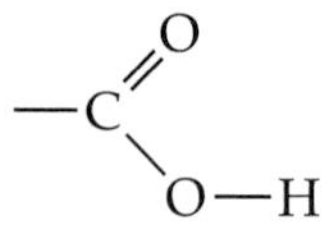

17 Pass over an aluminium oxide catalyst at about 250°C. This is a dehydration.

18 The reaction is exothermic. Therefore, increasing temperature favours the right to left reaction (ethanol to ethene and water) and decreases the value of K_p.

19 Increasing the concentration of a component on the left-hand side of the equation will result in the position of equilibrium changing to increase the amounts of products present at equilibrium.

20 The catalysts used do not absorb water and, therefore, do not have to be replaced as regularly as 'solid' H_3PO_4.

21 Convert the alkene to the alcohol (ethene to ethanol; propene to propan-2-ol) and then oxidise to give a carbonyl compound (ethanol to ethanal; propan-2-ol to propanone). See Database for conditions.

Chapter 22

INDUSTRIES USING MINERAL RESOURCES

22.1 THE CHLORALKALI INDUSTRY

The raw material for the chloralkali industry is sodium chloride (common salt). From this chlorine, sodium hydroxide and sodium carbonate are manufactured. It is an expanding industry with world production of chlorine increasing more than ten-fold between 1950 and 1990 (Fig 22.1).

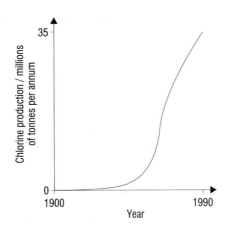

Fig 22.1 World production of chlorine during the 20th Century.

The salt is obtained by dissolving underground deposits in water, pumping the solution to the surface and evaporating it. In the UK, the solution is piped 20 miles from where it is pumped out of the ground (Holford, near Northwich) to the salt works where it is evaporated (Runcorn). Rapid evaporation produces small crystals of 'fine salt', suitable for foods, while slower evaporation gives 'common salt' which is used by chemical industries. A third form, 'rock salt', is obtained by mining, but is used primarily to de-ice roads.

1 **Why do you think that common salt is used rather than rock salt by the chemical industries?**

Chlorine and sodium hydroxide

Electrolysis is used to obtain chlorine and sodium hydroxide from common salt. The overall process may be summarised by the equation:

$$2NaCl(aq) + H_2O(l) \longrightarrow 2NaOH(aq) + Cl_2(g) + H_2(g)$$

where the energy supplied is in the form of electricity. The uses of chlorine and sodium hydroxide are summarised over the page. The process was described in detail in Theme 4.

Fig 22.2 Rock salt, used for de-icing roads, is mined from huge underground deposits in Cheshire. The major operation in the UK is at Winsford in Cheshire.

2 Write equations for the electrode reactions which occur during the electrolysis of sodium chloride solution.

Chlorine and the manufacture of organic compounds

The manufacture of chlorinated organic compounds accounts for about 70% of chlorine produced. The most important compound produced from chlorine is chloroethene, the monomer for PVC manufacture.

$$nCH_2 = CHCl \longrightarrow -[CH_2CHCl]_n$$

3 How is chloroethene manufactured from ethene in the petrochemical industry?

Many chlorinated hydrocarbons are used as solvents, with the chlorinated methanes being produced in the largest amounts. Chlorinated organic compounds are also used in antiseptics, for example TCP (trichlorophenol), and agrochemical products such as DDT and 2,4-D.

Chlorine used in bleaches and disinfectants

This accounts for about 20% of the chlorine manufactured. Chlorine is used to bleach, for example, paper and wood pulp, to disinfect domestic water supplies and to treat sewage.

Chlorine itself is not a bleach, nor does it display properties as a disinfectant. Instead, it is the chlorate(I) ion, formed when chlorine dissolves in water, which is the active species.

Fig 22.3 Bleach is used in the production of paper

4 Write an equation for the reaction of chlorine with (a) water, (b) cold, dilute sodium hydroxide solution.

Chlorine and the manufacture of inorganic compounds

Many inorganic compounds are produced by direct reactions involving chlorine. For example, hydrogen chloride, chlorine(I) oxide, chloric(I) acid, sodium chlorate(V), and numerous anhydrous metal chlorides. Apart from their end uses, for example as catalysts, metal chlorides are also intermediates in the extraction of some metals from their ores. An example is described later in this chapter, the extraction of titanium – a metal of enormous importance in the aerospace industry.

5 How can aluminium chloride be prepared and for what reactions is it an important catalyst?

Sodium hydroxide

Sodium hydroxide is manufactured on a massive scale and is the most important alkali used in industry. About half the sodium hydroxide produced is used to make chemical compounds, important amongst these are various phenols, sodium chlorate(I), sodium cyanide and sodium peroxide. Sodium hydroxide is also used for neutralisation reactions, paper processing, manufacturing Rayon and making soaps and detergents.

6 Sodium hydroxide is used industrially to manufacture 'sodium phosphates' by reaction with phosphoric(V) acid. Three 'phosphates' are manufactured. Write equations for their formation and name them.

Sodium carbonate

The reaction used to manufacture sodium carbonate can be represented as

'$CaCO_3 + 2NaCl \longrightarrow Na_2CO_3 + CaCl_2$'

The equation is given in inverted commas and no states of matter are shown because the reaction does not actually occur directly. The reasons are:

- it is energetically unfavourable;
- calcium carbonate is too insoluble in water.

7 Write an ionic equation for the reaction between aqueous sodium carbonate and aqueous calcium chloride.

To achieve this 'simple' conversion a complex industrial process has been designed. It is called the **Solvay process**.

Stage 1 Purifying the sodium chloride solution
The sodium chloride feed must be purified. The addition of sodium carbonate and sodium hydroxide removes calcium ions and magnesium ions which are present as impurities.

8 Write ionic equations for the reactions which result in the removal of $Ca^{2+}(aq)$ and $Mg^{2+}(aq)$ from the sodium chloride solution.

Stage 2 Production of carbon dioxide
Carbon dioxide is produced by feeding a 13 : 1 (by mass) mixture of limestone and coke into the top of a kiln with a counter-current of pre-heated air fed into the bottom. The process is continuous.

$$C(s) + O_2(g) \longrightarrow CO_2(g); \quad \Delta H = -393 \text{ kJ mol}^{-1}$$

This exothermic reaction provides sufficient energy to raise the temperature in the kiln to around 1200°C, at which temperature the limestone decomposes.

$$CaCO_3(s) \rightleftharpoons CaO(s) + CO_2(g); \quad \Delta H = +180 \text{ kJ mol}^{-1}$$

The emerging gas is about 40% by volume of carbon dioxide. After removal of any dust particles, it is used for the carbonating towers in the Solvay process.

9 What is the probable composition of the remaining 60% by volume of gases coming from the calcining kiln?

The calcium oxide which is formed is converted to calcium hydroxide and used for ammonia recovery later in the process (stage 6).

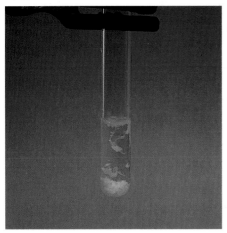

Fig 22.4 Group II carbonates are relatively insoluble in water and so ions such as $Mg^{2+}(aq)$ and $Ca^{2+}(aq)$ can be removed from solution by addition of sodium carbonate. Here, addition of $Na_2CO_3(aq)$ to $CaCl_2(aq)$ results in precipitation of $CaCO_3$.

10 The emerging gases are cooled in a heat exchanger by air. The coolant (air) becomes hot. To what use can this air be put in the Solvay process?

Stage 3 Ammoniation

The pH of the purified brine must be raised to about 9.5 by adding ammonia. The sodium chloride solution produced in stage 1 is passed down an absorption tower (some 35 m tall) with ammonia being fed in from the other end.

$$NH_3(g) + aq \longrightarrow NH_3(aq)$$
$$NH_3(aq) + H_2O(l) \rightleftharpoons NH_4^+(aq) + OH^-(aq)$$

$\left.\right\}$ combined reactions $\Delta H = -30 \text{ kJ mol}^{-1}$

11 Why does the solution need to be cooled as it emerges from the absorption tower?

12 What lessons could be learned from the petroleum industry when designing the absorption tower to ensure thorough mixing of the gaseous ammonia and the sodium chloride solution?

Fig 22.5 A flow diagram for the modern Solvay process. Many by-products are recycled to make best use of the materials – a characteristic of a modern chemical plant.

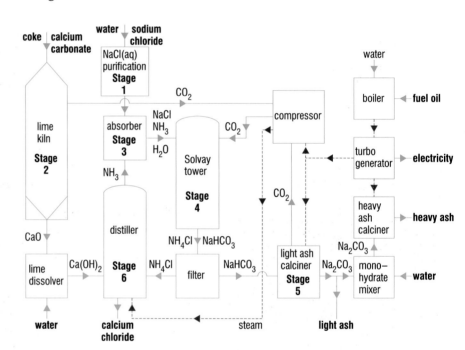

Stage 4 Carbonation

Ammoniated brine is passed down Solvay towers (around 25 m tall) with gases from the kilns and with carbon dioxide recycled from the calciners (called secheurs) used in stage 5.

$$CO_2(g) + 2OH^-(aq) \rightleftharpoons CO_3^{2-}(aq) + H_2O(l)$$

$$H_2O(l) + CO_3^{2-}(aq) + CO_2(g) \rightleftharpoons 2HCO_3^-(aq)$$

13 Explain how the yield of the desired product, $HCO_3^-(aq)$, can be increased by controlling the carbon dioxide pressure.

As hydroxide ions are removed, so more ammonium ions form

$$NH_3(aq) + H_2O(l) \rightleftharpoons NH_4^+(aq) + OH^-(aq)$$

The solution now contains $Na^+(aq)$, $Cl^-(aq)$, $NH_4^+(aq)$ and $HCO_3^-(aq)$. The solution is cooled until the sodium hydrogencarbonate precipitates. The rate of cooling controls the size of crystals obtained.

Table 22.1

Compound	Solubility / g in 100 g water
sodium hydrogencarbonate	9.6
ammonium hydrogencarbonate	12
sodium chloride	36
ammonium chloride	37

14 Use Table 22.1 to explain why sodium hydrogencarbonate precipitates first.

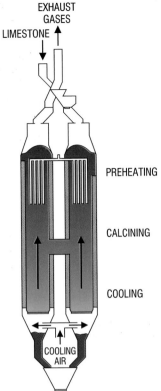

Fig 22.6 Sodium hydrogencarbonate is decomposed by heating in rotating calciners (30 m long and 3 m in diameter) during stage 5. The products are sodium carbonate, carbon dioxide and water.

Fig 22.7 NaHCO$_3$(s) decomposes upon heating to give CO$_2$(g) (identified because it turns lime water milky) and water, which can be seen to condense on the cool part of the tube. The residue in the test tube is Na$_2$CO$_3$(s).

Stage 5 Calcining

Solid sodium hydrogencarbonate is heated in a rotating calciner (Fig 22.6) at 180°C for 35–40 minutes.

$$2NaHCO_3(s) \longrightarrow Na_2CO_3(s) + CO_2(g) + H_2O(g)$$

The carbon dioxide is recycled.

The product is called **light ash**. By adding water to form Na$_2$CO$_3$.H$_2$O(s) and dehydrating this compound, anhydrous sodium carbonate of increased particle size is formed, called **heavy ash**.

Stage 6 Ammonia recovery

Ammonia is too valuable to be wasted (the Haber process for its manufacture is complex and expensive). Therefore, a recovery process is included in the plant. The filtrate from the Solvay tower (stage 4) is heated to decompose the hydrogencarbonates.

15 Write an ionic equation to represent the thermal decomposition of the hydrogencarbonates.

At the same time, the following reaction occurs

$$2NH_4^+(aq) + CO_3^{2-}(aq) \longrightarrow 2NH_3(g) + CO_2(g) + H_2O(l)$$

Some aqueous ammonia is added to make up for any losses during the processes and the solution is mixed with calcium hydroxide in a **distiller tower**.

$$Ca(OH)_2(aq) + 2NH_4Cl(aq) \longrightarrow 2NH_3(aq) + CaCl_2(aq) + 2H_2O(l)$$

16 Write an ionic equation for this reaction.

The ammonia gas is then recovered:

$$NH_3(aq) \rightleftharpoons NH_3(g) + aq$$

What happens to the calcium chloride? Some is recovered but most is used to fill the holes in the Earth's crust where the salt used to be!

Uses of sodium carbonate

Sodium carbonate can be used industrially as a replacement for sodium hydroxide, for example in paper processing, or the manufacture of soaps and detergents. Most of the heavy ash produced is used by the glass manufacturing industry. Light ash is used in the production of chemical compounds. An increasing use is to remove the pollutant sulphur dioxide from the stack gases of fossil fuel power stations.

17 Write an equation for the reaction between sodium carbonate and sulphur dioxide.

Fig 22.8 Around two-thirds of sodium carbonate made in the UK is in the form of heavy ash. By far the dominant use of this is in the glass manufacturing industry. This photograph shows the raw material for glass – which includes sand, dolomite, salt cake, soda ash and limestone, being pushed into the float glass furnace.

ACTIVITY

BURNING METAL CARBONATES

In 1992 Japanese chemists at the Tokyo Institute of Technology found that metal carbonates could be converted into methane by reacting them with hydrogen. A remarkable and unexpected reaction. High temperatures (200 – 400°C) and a metal catalyst were needed.

The following results were obtained for calcium carbonate. In each case 20 g $CaCO_3$ with 2% catalyst was used at 400°C:

Catalyst	Rate of methane production / cm^3 $hour^{-1}$
Nickel	2
Cobalt	1
Platinum	0.5
Copper	0.08

After an initial induction period when no reaction occurred, complete conversion took place and all the calcium carbonate was used up. Other metal carbonates were tried and, in some instances, a higher rate of methane production was obtained.

(a) Write an equation for the reaction between calcium carbonate and hydrogen, given that the only products are methane, water and calcium oxide.

(b) How long would it take to complete the conversion of calcium carbonate to methane, at 400°C, in the presence of (i) Ni, (ii) Co, (iii) Pt, (iv) Cu?

(c) A number of other metal carbonates showed greater reactivity than calcium carbonate towards hydrogen. How might other metal carbonates be manufactured on a large scale? [*Hint:* The chloralkali industry produces vast tonnages of sodium carbonate annually.]

INDUSTRIES USING MINERAL RESOURCES

(d) Natural gas (methane) is the starting point for manufacturing many petrochemicals. However, fossil fuel reserves are finite. Discuss the feasibility of using metal carbonates to provide methane for the petrochemicals industry when the fossil fuels run out.

(e) Design and plan a series of experiments which could be used to confirm and extend the Japanese findings. Think carefully about the variables that must be controlled and safety precautions.

22.2 NITROGEN COMPOUNDS

Both oxygen and nitrogen are major bulk chemicals but they are rarely used in industrial processes in their pure form. Air is nearly always used instead as the source.

18 **Name one petrochemical process where pure oxygen rather than air is used.**

Pure nitrogen, obtained by fractional distillation of liquid air, is itself used in some industries. For example, it is used for annealing steel. It is also used in industries where certain operations need to be carried out in an inert atmosphere, such as food packaging, glass making, chemical processes and silicon chip production. Liquid nitrogen is used in refrigeration when very cold temperatures are required, for example in food transport and for medical purposes.

Ammonia

There are two main stages involved in the manufacture of ammonia:

* the manufacture and purification of synthesis gas ('syngas') – a mixture composed mainly of hydrogen and nitrogen but also containing a little argon;
* the synthesis of ammonia.

Ninety percent of the ammonia produced is used to make fertilisers. It is also used in the manufacture of nylon and for the preparation of other chemical compounds.

Fig 22.9 (a) The manufacture of ammonia takes place by a highly automated, continuous production process. The complexity of the plant reflects the number of stages involved.
(b) A flow diagram for the manufacture of ammonia.

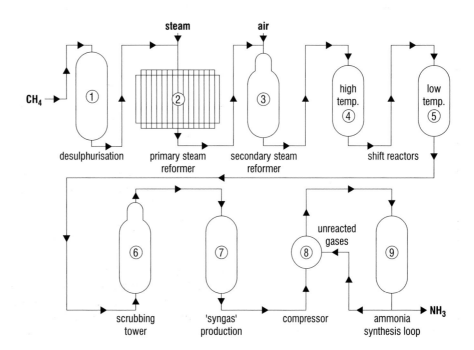

The manufacture and purification of syngas

The first stage in the production of syngas is the removal of sulphur-containing compounds from methane. Their presence in natural gas is because the decaying matter from which it formed comprised amino acids containing sulphur. The **desulphurisation** step is necessary because sulphur compounds poison some of the catalysts used later in the process. The methane feedstock is mixed with hydrogen and passed over a cobalt/nickel/alumina catalyst. This converts sulphur compounds to hydrogen sulphide which is then removed by reaction with zinc oxide:

$$RSH(g) + H_2(g) \xrightarrow{\text{Co/Ni/Al}_2\text{O}_3} RH(g) + H_2S(g)$$
$$ZnO(s) + H_2S(g) \longrightarrow ZnS(s) + H_2O(g)$$

Next, steam is reduced by methane, in a process called **primary steam reforming**. A temperature of about 850°C, a pressure of 30 atmospheres and a nickel catalyst are used. About 90% of the methane reacts under these conditions.

$$CH_4(g) + H_2O(g) \underset{}{\overset{\text{Ni}}{\rightleftharpoons}} CO(g) + 3H_2(g)$$

19 Would a higher overall pressure help to increase the yield of hydrogen? Explain your answer.

In the third stage of the process, air (a mixture of approximately 78% nitrogen, 21% oxygen and 1% argon) is introduced into the gas and some of the hydrogen is burnt to give steam. This needs to be controlled in order that the emerging gases contain sufficient hydrogen for the ammonia synthesis later. The steam reacts with most of the remaining methane to form carbon monoxide and hydrogen, as in the primary steam reforming stage.

$$2H_2(g) + O_2(g) \longrightarrow 2H_2O(g)$$

$$CH_4(g) + H_2O(g) \underset{}{\overset{\text{Ni}}{\rightleftharpoons}} CO(g) + 3H_2(g)$$

At this point the gas contains hydrogen, nitrogen, carbon dioxide, carbon monoxide and about 0.25% methane.

The fourth stage in the process is called the **shift reaction**. Carbon monoxide (another poison for the catalysts used) is converted to carbon dioxide by reaction with steam over an iron(III) oxide catalyst at 400°C. This reduces the carbon monoxide content from 11% to 3%. The mixture is then passed over a copper catalyst at 220°C, reducing carbon monoxide content to only 0.3%.

$$CO(g) + H_2O(g) \rightleftharpoons CO_2(g) + H_2(g)$$

20 Look at the equation above. What is the added advantage of this stage?

Unwanted carbon dioxide (present at about 18% by this stage) is removed by **scrubbing**, a process used to dissolve unwanted gases in a suitable solvent. A concentrated solution of potassium carbonate reacts with carbon dioxide to give potassium hydrogencarbonate. This is steam-treated later to regenerate the carbonate which can then be used again.

$$K_2CO_3(aq) + CO_2(g) + H_2O(l) \longrightarrow 2KHCO_3(aq)$$

The sixth and final stage of synthesis gas production converts any remaining carbon monoxide to methane by passing the gas over a nickel catalyst at 600°C.

$$CO(g) + 3H_2(g) \longrightarrow CH_4(g) + H_2O(g)$$
$$CO_2(g) + 4H_2(g) \longrightarrow CH_4(g) + 2H_2O(g)$$

INDUSTRIES USING MINERAL RESOURCES

Ammonia synthesis

Ammonia is synthesised from syngas according to the equation:

$$N_2(g) + 3H_2(g) \rightleftharpoons 2NH_3(g); \quad \Delta H = -92 \text{ kJ mol}^{-1}$$

This is called the **Haber process**. The syngas has a typical composition by volume:

74.3% hydrogen; 24.7% nitrogen; 0.8% methane; 0.3% argon

21 **How does the stoichiometry of the equation compare with the composition of syngas?**

Syngas is compressed to a pressure of 200 atmospheres and passed over an iron catalyst (containing potassium hydroxide as a promoter) at 380–450°C. With each pass there is about 15% conversion. On leaving, ammonia is condensed out by refrigeration and unchanged reactants replenished and passed over the catalyst again (the ammonia synthesis loop). After several passes, yields of up to 98% can be obtained.

22 **Ammonia can be condensed from the mixture because of its relatively low volatility. Explain the range of boiling points of the components emerging from the ammonia synthesis plant (Table 22.2).**

Table 22.2 Boiling points of components emerging from the ammonia synthesis plant

Component	Boiling point / °C
Hydrogen	–252
Nitrogen	–196
Methane	–161
Argon	–186
Ammonia	–33

Fig 22.10 A nitrate fertiliser being applied to barley. Liquid ammonia can be used directly as a fertiliser. It is pumped through drills into the ground. In the soil it dissolves in water and is absorbed by plants through their roots.

Nitric acid

Nitric acid is used for the manufacture of fertilisers (accounting for some 90% of the nitric acid produced). It is also used for the manufacture of explosives. The first stage in manufacturing nitric acid involves oxidising ammonia

$$4NH_3(g) + 5O_2(g) \rightleftharpoons 4NO(g) + 6H_2O(g); \quad \Delta H = -909 \text{ kJ mol}^{-1}$$

Theoretically, high yields and an acceptable reaction rate would be favoured by:

- low pressure;
- excess oxygen (air);
- a catalyst;
- as low a temperature as possible while maintaining practicable reaction rates, catalyst efficiency and operating pressure.

23 **Comment on the theoretical justification for these conditions.**

Fig 22.11 This man is holding up a large circular gauze, woven from rhodium-platinum wire for use as a catalyst in the commercial production of nitric acid by the oxidation of ammonia

In practice, liquid ammonia is allowed to evaporate and is mixed with air. Commonly, a 12% ammonia and 88% air mixture is used, with exact amounts dependent on the operating conditions of the plant.

24 Assuming a feed of 12% ammonia and 88% air, calculate the approximate mole ratio of $NH_3:O_2$ in the feed. How does this compare to the stoichiometry of the equation?

Typically, several reactors are used in parallel, each containing catalyst gauzes (90% platinum/10% rhodium) at about 900°C. They are operated at a pressure of 4–10 atmospheres and give a minimum of 96% conversion of ammonia. The hot gases are cooled in a heat exchanger and can be used to generate steam, which in turn may be used, via a turbine, to produce electricity. Air is now added and the pressure raised to 7–12 atmospheres. In doing so, the gases heat up and so must be further cooled.

25 Why does the temperature of the gas mixture rise when pressure is increased?

Initially two reactions occur:

$$2NO(g) + O_2(g) \rightleftharpoons 2NO_2(g); \ \Delta H = -115 \text{ kJ mol}^{-1}$$
$$2NO_2(g) \rightleftharpoons N_2O_4(g); \ \Delta H = -58 \text{ kJ mol}^{-1}$$

Raising pressure and lowering temperature would favour the formation of products in both these reactions.

26 How would the value of the equilibrium constant, K_p, vary with temperature for each of these reactions?

The partially oxidised mixture is finally passed counter-current to a stream of water.

$$N_2O_4(g) + H_2O(l) \rightleftharpoons HNO_2(aq) + HNO_3(aq) \quad \text{(SLOW)}$$
$$4HNO_2(aq) \rightleftharpoons 2NO(g) + N_2O_4(aq) + 2H_2O(l) \quad \text{(FAST)}$$

Overall, these final two reactions can be summarised by the equation

$$3N_2O_4(g) + 2H_2O(l) \rightleftharpoons 4HNO_3(aq) + 2NO(g); \ \Delta H = -103 \text{ kJ mol}^{-1}$$

27 Chart the changes in oxidation number of nitrogen from ammonia through to the product in the manufacture of nitric acid and classify any reactions as oxidation, reduction or disproportionation.

28 How can metal nitrates be prepared from nitric acid?

Carbamide (urea)

Carbamide is used as a fertiliser and animal feed, for the manufacture of some resins and some pharmaceutical products, and in paper-processing. Carbon dioxide and ammonia are brought into contact at 180–210°C and 200–400 atmospheres (60% conversion).

$$CO_2(g) + 2NH_3(l) \rightleftharpoons NH_2COONH_4(s)$$

This is passed into a low pressure vessel for the decomposition into carbamide

$$NH_2COONH_4(s) \rightleftharpoons NH_2CONH_2(s) + H_2O(g)$$

29 Why is it necessary to transfer the reaction mixture to a low pressure reactor for the decomposition stage?

Fertilisers: the major use of nitrogen compounds

The food we need to sustain the world's increasing population can only be grown if the land is able to supply the nutrients necessary for healthy plant growth. However, growing plants and harvesting crops strips the soil of nutrients. Natural fertilisers, such as manure and composted materials, supply these nutrients but only at low concentrations. Intensive farming requires an intensive supply of nutrients. The primary nutrients are nitrogen, phosphorus and potassium (often abbreviated to 'NPK'). Their respective roles are:

- nitrogen: to promote root development and leaf growth (it determines the protein content of a plant);
- phosphorus: to regulate leaf growth and plant size;
- potassium: to promote the growth rate of plants.

Artificial fertilisers are often used to add particular nutrients to the soil and the most common ones are those which supply the primary nutrients. 'Straight' fertilisers are a source of nitrogen only. As well as liquid ammonia, ammonium nitrate and carbamide (urea) are used as straight fertilisers. The ammonium nitrate is prepared by reaction between aqueous ammonia and nitric acid:

$$NH_3(aq) + HNO_3(aq) \longrightarrow NH_4NO_3(aq)$$

30 **Which of the following is the richest source of nitrogen: 1 g ammonia, 1 g ammonium nitrate or 1 g carbamide? Explain your answer.**

'Compound' fertilisers are combination fertilisers which supply nitrogen, phosphorus and potassium in a single application. Their compositions vary depending on the precise purpose of the fertiliser.

Fig 22.12 Different NPK ratios are used for particular purposes. Compound fertilisers can be bought with the desired NPK ratio.

- Nitrogen is usually provided in the form of ammonium nitrate.
- Phosphorus is usually provided in the form of 'ammonium phosphates' manufactured from phosphoric(V) acid and anhydrous ammonia. For example:

$$2NH_3 + H_3PO_4 \longrightarrow (NH_4)_2HPO_4$$

'Ammonium phosphates' is a term used to describe the mixture of $NH_4H_2PO_4$ and, predominately, $(NH_4)_2HPO_4$ obtained in the reaction.
- Potassium is added in the form of potassium chloride, which is mined directly from the earth's crust.

31 Explain why $(NH_4)_3PO_4$ cannot be obtained by reaction between phosphoric(V) acid and ammonia. *Hint:* both H_3PO_4 and NH_3 are weak electrolytes.

Fig 22.13 NPK granules are manufactured by slurrying a finely ground mixture of ammonium nitrate, diammonium hydrogenphosphate and potassium chloride with the minimum amount of water and allowing it to evaporate under controlled conditions.

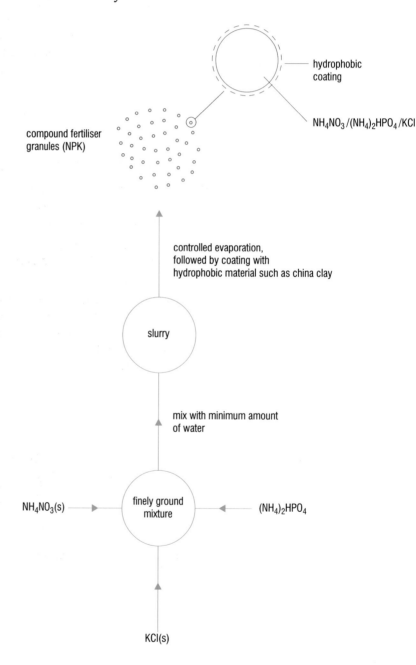

22.3 SULPHUR COMPOUNDS

The German chemist Justis von Liebig once suggested that the level of production of sulphuric acid by a country could be used as a satisfactory measure of its industrial activity. To some extent this simple indicator is still true. Certainly, sulphuric acid is produced on a larger scale than any other chemical compound.

Uses of sulphuric acid

The actual uses of sulphuric acid varies with country. For example, the major use in the USA is for fertiliser manufacture, some 65%. In the UK, use is more evenly distributed (the following percentages are approximate and vary from year to year):

30 – 32%	fertilisers (predominately to manufacture 'superphosphate' from phosphate rock);
18 – 20%	paints, pigments and dyestuff intermediates;
16 – 18%	manufacturing other chemical compounds;
10 – 12%	soaps and detergents;
10 – 14%	fibres;
8 – 10%	plastics;
about 3%	processing metals.

The manufacture of sulphuric acid

In Britain, annual production of sulphuric acid is currently around 3 million tonnes. It is manufactured almost exclusively by the Contact process which may be considered as taking place in three stages (Fig 22.14):

- the production of sulphur dioxide;
- the oxidation of sulphur dioxide to sulphur trioxide;
- the absorption of sulphur trioxide.

There are two major sources of sulphur: sulphur extracted from the Earth's crust by the Frasch process and sulphur recovered from oil refining and petrochemical processing. The sulphur obtained by both processes is very pure and ideal for burning in the first stage of the sulphuric acid manufacturing process.

Fig 22.14 A flow diagram for the Contact process.

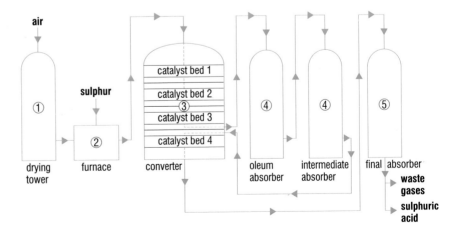

Fig 22.15 Natural sulphur deposits flowing from a blow hole halfway up Mt Subyak, a volcano near Medan, Sumatra. Burning sulphur to produce sulphur dioxide is the first stage in the manufacture of sulphuric acid.

The production of sulphur dioxide

Molten sulphur is sprayed into a current of air at 950–1000°C. The gases emerging from the reactor contain about 10% by volume sulphur dioxide and 10% by volume oxygen.

$$S(l) + O_2(g) \longrightarrow SO_2(g); \quad \Delta H = -297 \text{ kJ mol}^{-1}$$

Alternatively, sulphur dioxide can be produced by roasting (heating very strongly) sulphide ores such as zinc sulphide (zinc blende) or iron sulphide (iron pyrites) in air:

$$4FeS_2(s) + 11O_2(g) \longrightarrow 2Fe_2O_3(s) + 8SO_2(g)$$
$$2ZnS(s) + 3O_2(g) \longrightarrow 2ZnO(s) + 2SO_2(g)$$

32 Which of the two sulphur ores mentioned above contains the highest percentage of sulphur by mass?

Oxidation of sulphur dioxide

In the Contact process, sulphur dioxide and oxygen are passed over catalyst beds of vanadium(V) oxide in a series of reactors at about 420°C. Although the equilibrium would be favoured by a lower temperature, the catalyst becomes inactive below about 380°C. Each reactor has four catalyst beds and the gases are cooled as they pass through one bed to the next.

$$2SO_2(g) + O_2(g) \rightleftharpoons 2SO_3(g); \quad \Delta H = -196 \text{ kJ mol}^{-1}$$

33 Why is it necessary to cool the gases after each pass over a heated catalyst?

After three passes the emerging gases contain about 95% by volume sulphur trioxide. The sulphur trioxide is removed but the remaining gas cannot be allowed to escape into the atmosphere because the sulphur dioxide content is environmentally damaging. Therefore, sulphur trioxide is removed after three beds and the remaining gases passed over a fourth bed which not only removes the sulphur dioxide but converts it also to sulphur trioxide. A final conversion of greater than 99.5% is achieved.

Recently, a single fluidised bed has replaced the first three static beds in the reactor of modern plants. This avoids the need for cooling between beds and is an example of chemists taking full advantage of advances in technology – in this case the development of fluidised bed reactors.

Fig 22.16 A series of reaction beds, each containing vanadium(V) oxide catalyst, are used to convert sulphur dioxide and oxygen into sulphur trioxide. The reaction mixture is cooled between beds.

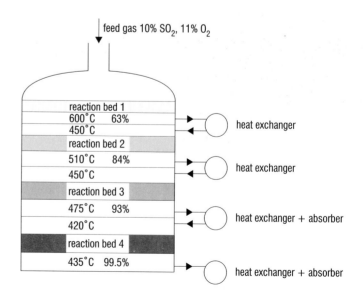

The absorption process

The final stage is to dissolve sulphur trioxide in water to give sulphuric acid. However, using water produces a fine acid mist which cannot be condensed and which would pass through the absorber and out into the atmosphere. This is unacceptable, both commercially and environmentally. The answer is to use 98% sulphuric acid ('oleum') to dissolve the sulphur trioxide and to obtain sulphuric acid of the desired concentration by dilution. However, we use the following simplified equation to represent the reaction:

$$SO_3(g) + H_2O(l) \longrightarrow H_2SO_4(aq); \quad \Delta H = -130 \text{ kJ mol}^{-1}$$

In practice, two absorbers are used in series to remove the sulphur trioxide from the mixture obtained after passage through the third catalyst bed. The gases from the fourth bed are passed through a final absorber (Fig 22.16). All the absorbers contain oleum.

34 Why is a ceramic material used to pack the absorbers?

22.4 PHOSPHORUS COMPOUNDS

The compound of phosphorus which is produced in by far the largest quantities is phosphoric(V) acid. It is the starting point for the manufacture of nearly all the useful compounds of phosphorus, in particular the phosphates.

Uses of phosphoric(V) acid

Nine-tenths of the phosphoric(V) acid manufactured is used to make fertilisers. 'Compound' fertilisers (see earlier in this chapter) are combination fertilisers which supply nitrogen, phosphorus and potassium in a single application. Their compositions vary depending on the precise purpose of the fertiliser. Phosphoric(V) acid is also used for metal treatments and for making other phosphorus compounds but the quantity used is much less than that for fertilisers.

Salts of phosphoric(V) acid are important. Ammonium salts are used as fertilisers. Sodium salts are used in detergents, foods and water treatment.

35 Mono, di and trisodium phosphate(V) salts have many commercial applications. How could they be prepared from phosphoric(V) acid?

Fig 22.17 Phosphoric(V) acid is present in some food products. The labels on cans and bottles of Pepsi-Cola confirm its presence in this popular drink.

The manufacture of phosphoric(V) acid

The source of phosphorus is phosphate rock, $3Ca_3(PO_4)_2.CaF_2$. This is crushed, heated strongly to remove carbonates and mixed with sulphuric acid and recycled phosphoric(V) acid in a series of stirred reactors.

36 Write an equation to show how the carbonates are removed from the phosphate rock. Are they actually 'removed'?

A slurry of precipitated calcium sulphate in aqueous phosphoric(V) acid is formed. The slurry cannot be filtered very easily and water is added to aid the filtration.

$$Ca_3(PO_4)_2(s) + 3H_2SO_4(aq) \longrightarrow 2H_3PO_4(aq) + 3CaSO_4(s)$$

Two side reactions occur.

$$3CaF_2(s) + SiO_2(s) + 3H_2SO_4(aq) \longrightarrow H_2SiF_6(aq) + 3CaSO_4(s) + 2H_2O(l)$$
$$CaCO_3(s) + H_2SO_4(aq) \longrightarrow CO_2(g) + CaSO_4(s) + H_2O(l)$$

The process produces about 46% P_2O_5 as phosphoric(V) acid, which is usually concentrated to a solution equivalent to 56% P_2O_5. The acid produced is very impure.

37 It is common to give the phosphorus content of fertilisers in terms of 'P_2O_5'. If the phosphoric(V) acid is equivalent to 46% by mass P_2O_5, what is the concentration, in mol dm^{-3}, of the acid?

Pure phosphoric(V) acid can be made by the thermal process which involves burning pure phosphorus and dissolving the phosphorus(V) oxide which forms in water.

$$P_4(l) + 5O_2(g) \longrightarrow 2P_2O_5(g)$$
$$P_2O_5(s) + 3H_2O(l) \longrightarrow 2H_3PO_4(aq)$$

ACTIVITY

COMPOUNDS OF ZINC

One of the major ores of zinc is zinc blende, ZnS. It is used as a source of sulphur dioxide for the manufacture of sulphuric acid:

$$2ZnS(s) + 3O_2(g) \longrightarrow 2ZnO(s) + 2SO_2(g)$$

The economics of this process for making sulphur dioxide would be better if there was a saleable use for the co-product in this reaction.

Imagine you are working for an industrial company which produces sulphur dioxide in this way. You realise that zinc oxide could provide a starting point for the synthesis of a range of zinc compounds.

Assuming an adequate supply of the inorganic compounds produced by the major industries described in this chapter, plan a series of reactions which could be used by your company to manufacture compounds of zinc.

22.5 METAL EXTRACTION BY CHEMICAL REDUCTION

The use of electricity to extract metals from their ores has been described in Theme 4. However, electricity is expensive and for metals of lower reactivity (lower down in the electrochemical series) chemical reduction is used. Two examples are examined in some detail, iron and titanium.

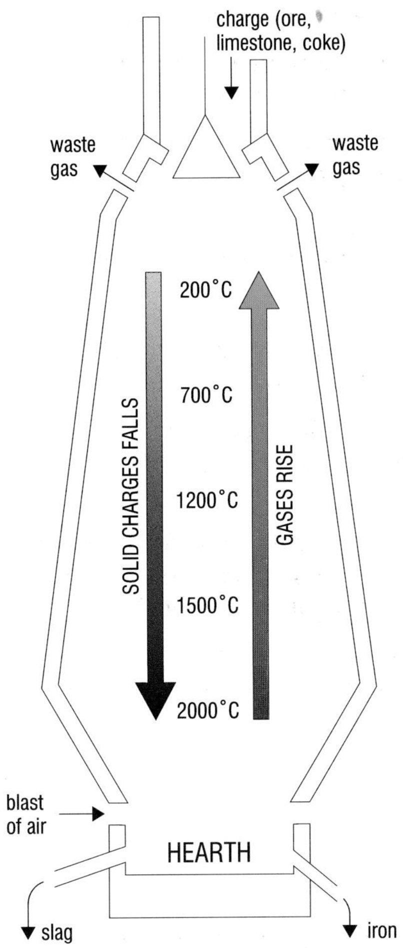

Iron and steel

No metal has had more impact on the material development of the human race than iron. Most of the metal we see around us will be iron or steel, and further vast quantities are buried in concrete, reinforcing it and improving its properties as a construction material. The major iron ores are oxides, haematite (Fe_2O_3) and magnetite (Fe_3O_4). Iron ore, coke (carbon) and limestone (calcium carbonate) are roasted together to produce sinter. This is mixed with further ore, coke and limestone and fed into a blast furnace (Fig 22.18) against a counter current of hot air. The furnace operates continuously and produces around 7000–8000 tonnes a day. A series of reactions occur which can be summarised:

$$2C(s) + O_2(g) \longrightarrow 2CO(g)$$

In the hotter part of the furnace: $Fe_2O_3(s) + 3CO(g) \longrightarrow 2Fe(l) + 3CO_2(g)$

In the cooler part of the furnace: $Fe_2O_3(s) + 3C(s) \longrightarrow 2Fe(l) + 3CO(g)$

38 **Describe the redox reactions which occur in a blast furnace in terms of oxidation numbers.**

The ore contains silicon(IV) oxide as an impurity and the purpose of the limestone is to remove this as a liquid slag:

$$CaCO_3(s) \longrightarrow CaO(s) + CO_2(g)$$
$$CaO(s) + SiO_2(s) \longrightarrow CaSiO_3(l)$$

When cooled, the molten iron solidifies to form 'pig iron' (also called cast iron). Most of this is transferred immediately to the steel furnaces.

Iron forms alloys, called **steels**, with other metals and non-metals. Steels are usually made by the **basic oxygen process**. Scrap iron (30%) and molten iron (70%) are put into a furnace, together with the small quantities of alloying elements required for the particular steel being made. High purity oxygen is passed into the melt through a retractable oxygen lance (a pipe made of steel and water cooled) and calcium oxide is added at this stage. Slag forms as it does in iron manufacture and the steel is tapped off.

39 **Our environment is hostile to metals and oxidation occurs more rapidly in acidic conditions. Which gases in the atmosphere are likely to accelerate the rusting of iron?**

charge (ore, limestone, coke)

waste gas

waste gas

200°C

700°C

1200°C

1500°C

2000°C

SOLID CHARGES FALLS

GASES RISE

blast of air

HEARTH

slag

iron

Fig 22.18 The blast furnace.

Fig 22.19 Titanium has a low density but good mechanical strength (especially when alloyed with small amounts of other metals such as aluminium and tin). It finds extensive use in the aerospace industry.

Titanium

Titanium occurs naturally in its ores, rutile (TiO_2) and ilmenite ($FeTiO_3$). Rutile is the ore preferred for titanium manufacture but ilmenite is used to make titanium(IV) oxide for use as a white pigment.

Carbon is cheap and plentiful and so would be the element of choice to reduce a metal oxide to the metal. However, TiO_2 cannot be reduced by carbon, carbon monoxide or hydrogen. Even reduction with very reactive metals such as magnesium and aluminium is incomplete. The problem has been solved by converting the oxide to titanium tetrachloride, $TiCl_4$, and reducing this to titanium. In the UK, sodium is used as the reducing agent; outside the UK, magnesium is usually used.

Chlorination of TiO_2 is usually carried out in a continuous process in a fluidised bed reactor at about 1000°C in the presence of coke. The principal reactions are:

$$TiO_2(s) + 2Cl_2(g) + 2C(s) \longrightarrow TiCl_4(l) + 2CO(g)$$
$$TiO_2(s) + 2Cl_2(g) + C(s) \longrightarrow TiCl_4(l) + CO_2(g) \text{ (highly exothermic)}$$
$$TiO_2(s) + 4Cl_2(g) + 2C(s) \longrightarrow TiCl_4(l) + 2COCl_2(g)$$

Under these conditions, impurities are also converted to chlorides. Involatile chlorides solidify on cooling but the volatile chlorides (those with molecular structures) are removed and separated by fractional distillation in an inert atmosphere (nitrogen or argon).

40 Why must the storage tanks for TiCl₄ be completely dry?

Finally, TiCl₄ (which has a purity greater than 99.5%) is passed into a reactor containing molten, high purity sodium. An argon atmosphere is required. A series of exothermic reactions result in the formation of $TiCl_3$ and $TiCl_2$, causing the temperature to rise from about 500°C to about 800°C. This is insufficient, however, to effect complete reduction and so the reactor is heated to about 1000°C.

$$TiCl_4(g) + 4Na(l) \longrightarrow Ti(s) + 4NaCl(s)$$

41 Write an equation for the reduction of TiCl₄ with magnesium.

The reactor is maintained at operating temperature for 3–4 days. It is then cooled for about 4 days before the sodium chloride/titanium mixture is recovered. The mixture is crushed and sodium chloride removed by dissolving it in dilute hydrochloric acid.

ACTIVITY

The production of tin(II) sulphate

Read the article below and answer the following questions.

Electrosynthesis

Membranes help to force corrosion

Corrosion is a nuisance. It happens to some metals when you don't want it, and it won't happen to others when you do–to dissolve them, for instance. Take the manufacture of stannous sulphate (Sn(II) sulphate, $SnSO_4$). This compound is used widely for tin plating baths in the electronic industries, and for manufacturing printed circuit boards *etc.*

Yet its manufacture is decidedly low-tech. According to the *Kirk-Othmer Encyclopedia of Chemical Technology* (3rd edn, Vol 23, New York: Wiley, 1983) it takes several days at 100°C to dissolve granulated tin in sulphuric acid. The process currently used by UK manufacturers has a seven-step flow chart, which uses a sequence of aggressive chemicals and is still energy intensive:)

1. Tin ingot
   ↓ energy
2. Tin granules
   ↓ HCl
3. Tin chloride
   ↓ NaOH
4. Tin hydroxide
   ↓ H₂SO₄
5. Filter coke
   ↓ H₂SO₄
6. Tin sulphate slurry
   ↓ energy
7. Dry tin sulphate product

Chemists at the Electricity Council Research Centre, Capenhurst, Cheshire, have reduced this to a single step electrochemical process, which is energy-efficient and produces high-quality product from low-grade materials. The tin ingot is forced to dissolve by making it the anode in an electrochemical cell using sulphuric acid as the electrolyte. Only a low DC voltage is required.

Normally metal ions would deposit at the cathode in this arrangement; the trick is to separate the cell into two sub-cells by means of a membrane. This allows electrons to pass but excludes cations and restricts them to the anolyte. When the tin sulphate anolyte has reached a sufficiently high concentration it can be drained off as product.

Other arrangements of the cell could be employed using cationic or anionic exclusion membranes to manufacture a wide range of metal salts. It will find uses in the scrap metal recycling industries Capenhurst scientists are discussing the process, which they like to be called 'forced corrosion', with a number of possible licensees.

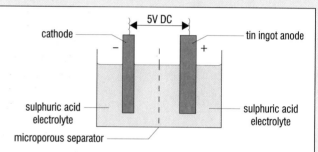

(a) Consider the process currently used in the UK.
   (i) Draw a flow diagram for the process, giving equations for each step.
   (ii) Why is the tin ingot converted to granules?
   (iii) Which two steps make the greatest demands on energy?
   (iv) Which step involves a redox process?
   (v) The process 'uses a sequence of aggressive chemicals'. Identify these compounds and explain the implications that they have for plant design.
   (vi) Explain why hydrochloric acid is used rather than sulphuric acid or nitric acid.

(vii) Describe a series of experiments that could be carried out in the laboratory in order to determine the optimum conditions for the process.

(b) Now consider the electrosynthetic method.
  (i) Explain the function of the membrane.
  (ii) Write equations for the electrode reactions that would occur if no membrane were present and those which would occur with the membrane in place.
  (iii) Explain the term *'tin sulphate anolyte'*.

(c) You have been asked to compare the energy demands of the two processes. Design and plan a series of experiments that could be carried out in the laboratory to establish the relative energy demands.

Answers to Quick Questions: Chapter 22

1 Common salt is less contaminated with impurities than is rock salt.

2 Cathode reaction: $2H^+(aq) + 2e^- \longrightarrow H_2(g)$
Anode reaction: $2Cl^-(aq) \longrightarrow Cl_2(g) + 2e^-$

3 $CH_2 = CH_2 + 2HCl + \frac{1}{2}O_2 \longrightarrow CH_2ClCH_2Cl + H_2O$

$CH_2ClCH_2Cl \longrightarrow CH_2 = CHCl + HCl$
(see Page 524 for reaction conditions)

4 (a) $Cl_2(aq) + H_2O(l) \rightleftharpoons HCl(aq) + HOCl(aq)$
(b) $Cl_2(aq) + 2OH^-(aq) \longrightarrow Cl^-(aq) + OCl^-(aq) + H_2O(l)$

5 Passing dry, oxygen-free chlorine over aluminium, heated to initiate the reaction. Precautions taken to exclude air and moisture. $AlCl_3$ is used as a catalyst for halogenation, alkylation and alkanoylation reactions of aromatic hydrocarbons.

6 $H_3PO_4 + NaOH \longrightarrow NaH_2PO_4 + H_2O$; sodium dihydrogen phosphate(V)
$H_3PO_4 + 2NaOH \longrightarrow Na_2HPO_4 + 2H_2O$; disodium hydrogen phosphate(V)
$H_3PO_4 + 3NaOH \longrightarrow Na_3PO_4 + 3H_2O$; trisodium phosphate(V)

7 $Ca^{2+}(aq) + CO_3^{2-}(aq) \longrightarrow CaCO_3(s)$

8 $Ca^{2+}(aq) + 2OH^-(aq) \longrightarrow Ca(OH)_2(s)$
$Mg^{2+}(aq) + 2OH^-(aq) \longrightarrow Mg(OH)_2(s)$

9 It is largely nitrogen (from the air fed into the kiln).

10 It may be used to feed the kiln for the production of carbon dioxide.

11 An exothermic reaction has taken place. Therefore, the temperature will rise, but the solubility of ammonia in water decreases with increasing temperature.

12 Might use bubble caps similar to those used in distillation columns.

13 Increased pressure of carbon dioxide will, according to Le Chatelier's principle, favour the right to left reaction.

14 It is the least soluble.

15 $2HCO_3^-(aq) \longrightarrow CO_3^{2-}(aq) + CO_2(g) + H_2O(l)$

16 $2OH^-(aq) + 2NH_4^+(aq) \longrightarrow 2NH_3(g) + 2H_2O(l)$

17 $Na_2CO_3(s) + SO_2(g) \longrightarrow Na_2SO_3(s) + CO_2(g)$
However, this reaction only occurs if moisture is present. Sulphurous acid is first produced from SO_2 and H_2O, and it is this that reacts with Na_2CO_3.

18 Manufacture of ethanal by the one-stage Wacker-Chemie process.

19 No. There are a greater number of moles of gas on the right hand side of the equation. Therefore, from Le Chatelier's principle, increasing pressure would favour the right to left reaction.

20 As well as reducing the amount of carbon monoxide, the reaction also produces hydrogen.

21 The stoichiometry of the reaction is 1 mole N_2 to 3 mole H_2. This is the same mole ratio in which these two gases are present in syngas.

22 Ammonia is far less volatile than might be expected from molar mass alone due to the presence of hydrogen-bonding. The only intermolecular forces between particles of the other components are dispersion forces (they are all non-polar molecules).

23 Le Chatelier's principle predicts that a position of equilibrium where concentrations of NO and H_2O are high is favoured by low pressure, low temperature and increasing the concentration of one or more of the reactants. Catalysts do not affect the position of equilibrium but do enable the position of equilibrium to be attained more rapidly.

24 Since air is about 21% oxygen by volume, of the 88% of air in the feed, 18.5% is oxygen. Therefore the mole ratio of ammonia to oxygen is 12:18.5 (about 2 : 3). The stoichiometry of the equation gives a 4:5 mole ratio of ammonia to oxygen.

25 Assuming that the volume does not change, the ideal gas equation must be obeyed ($pV = nRT$). Therefore, an increase in pressure must result in an increase in temperature.

26 An increase in temperature would result in a decrease in K_p for both reactions.

27

$$4NH_3 + 5O_2 \rightleftharpoons 4NO + 6H_2O$$

oxidation number of nitrogen -3 $+2$

oxidation

$$2NO + O_2 \rightleftharpoons 2NO_2$$

oxidation number of nitrogen $+2$ $+4$

oxidation

$$2NO_2 \rightleftharpoons N_2O_4$$

oxidation number of nitrogen $+4$ $+4$

no change

$$3N_2O_4 + 2H_2O \rightleftharpoons 4HNO_3 + 2NO$$

oxidation number of nitrogen $+4$ $+5$ $+2$

disproportionation

28 Reaction of cold, dilute nitric acid on a reactive metal, oxide, hydroxide or carbonate. Warming may be required in some cases.

29 To shift the position of equilibrium in favour of products (greater number of moles of gas).

30 1 g ammonia.

Explanation: percentage by mass of nitrogen present in each of the samples:

1 g NH_3 (molar mass 17 g mol^{-1}), 82%;
1 g NH_4NO_3 (molar mass 80 g mol^{-1}), 35%;
1 g NH_2CONH_2 (molar mass 60 g mol^{-1}), 47%.

31 There are two competing equilibria:

$HPO_4^{2-}(aq) \rightleftharpoons H^+(aq) + PO_4^{3-}(aq)$ [reaction A]
$NH_4^+(aq) \rightleftharpoons H^+(aq) + NH_3(aq)$ [reaction B]

It turns out that ammonia has a stronger attraction for the hydrated hydrogen ions that does the phosphate(V) ions (the equilibrium constant for reaction A is smaller than that for reaction B).

32 FeS_2

33 The reaction is exothermic, therefore the temperature rises during reaction. However, the position of equilibrium which favours the products is promoted by low temperatures (Le Chatelier's principle).

34 Ceramics are highly resistant to corrosive compounds.

35 Reaction between $H_3PO_4(aq)$ and $NaOH(aq)$ or $Na_2CO_3(aq)$.

36 $CO_3^{2-}(s) \longrightarrow O^{2-}(s) + CO_2(g)$

They are not really 'removed'. Rather, they are converted to oxides.

37 Assume 100 g of the acid solution.

46 g P_2O_5 is equivalent to $46 \times \dfrac{98}{142}$ g H_3PO_4 = 32 g H_3PO_4

Therefore, 32 g in about 100 g solution

320 g in about 1000 g solution

Therefore, concentration of solution is about 320 g dm^{-3}

Molar mass H_3PO_4 is 98 g mol^{-1}

Therefore, concentration of the acid is about 3.3 mol dm^{-3}

38 $2C + O_2 \longrightarrow 2CO$;

$Ox[C]$ changes from 0 to +2
$Ox[O]$ changes from 0 to −2

$Fe_2O_3 + 3CO \longrightarrow 2Fe + 3CO_2$

$Ox[Fe]$ changes from +3 to 0
$Ox[C]$ changes from +2 to +4

$Fe_2O_3 + 3C \longrightarrow 2Fe + 3CO$

$Ox[Fe]$ changes from +3 to 0
$Ox[C]$ changes from 0 to +2

39 CO_2 and the 'acid rain' gases, SO_2 and NO_2.

40 $TiCl_4$ is a molecular chloride which hydrolyses very easily:

$TiCl_4(l) + 2H_2O(l) \longrightarrow TiO_2(s) + 4HCl(aq)$

41 $TiCl_4(g) + 2Mg(s) \longrightarrow Ti(s) + 2MgCl_2(s)$

Chapter 23

CHEMISTRY IN THE BODY

23.1 COMPOUNDS IN THE BODY

The human body is a highly sophisticated and beautifully designed chemical plant. At one level we are no more than a collection of chemical compounds which are being interconverted constantly, including exchange with the environment. Every action we take, every perception and process – smell, taste, sight, movement, thought, recognition, memory and so on – are the result of a series of interconnected chemical changes.

Despite all their knowledge and access to modern techniques, using the latest advances in computers, engineering and technology, chemists cannot synthesise most of the molecules found in the body given the same starting compounds that the body uses. Equally fascinating is the ability of the body to store these compounds. They are often very sensitive to subtle changes in pH, temperature and other factors, yet our bodies manage to buffer them against such environmental changes and preserve them in it.

The chemical industry is characterised by large scale industrial plant, operating under extreme conditions of temperature and pressure. Even then, catalysts are often required to speed up reactions and ensure they become economical. Yet the human body carries out a multitude of complex and highly selective syntheses at a fairly low temperature and pressure.

Analysis of living tissue

When a sample of muscle, or other living tissue, is reduced to a pulp in cold, dilute acid, it is possible to separate soluble and insoluble materials. The soluble material contains inorganic ions and small molecules. The insoluble material contains polymeric molecules with average molecular masses up to several millions (Table 23.1).

Table 23.1 Compounds found in living tissue

Inorganic ions:	The most common ions found in our bodies and in the soluble extract from living tissue are potassium, sodium, calcium, chloride and phosphate(V) (although the latter might also be found as hydrogenphosphate(V) and dihydrogenphosphate(V)).
Small molecules:	For convenience these may be considered to fall into five categories:
	• carboxylic acids;
	• sugars;
	• aminoacids;
	• organic phosphates;
	• purines and pyrimidines.
Polymeric molecules:	These are sometimes called **biopolymers**. For convenience these may be considered under four headings:
	• polysaccharides;
	• proteins;
	• lipids;
	• nucleic acids.

Carboxylic acids

Carboxylic acids play a vital role in enabling chemical reactions to occur in the living cell easily and rapidly. Without them, many polymeric molecules found in the body (called biopolymers) would not be able to function. Some of these carboxylic acids are given in Table 23.2. Ethanoic acid is, of course, vinegar. Others are found in fruits, for example, citric acid is found in citrous fruits such as lemons and limes.

Table 23.2 Important carboxylic acids that occur in nature

Carboxylic acid	Structural formula	Functional groups present	Description
ethanoic acid (acetic acid, vinegar)	CH_3COOH	—COOH carboxylic acid	monocarboxylic acid
butanedioic acid (succinic acid)	CH_2COOH | CH_2COOH	—COOH carboxylic acid	dicarboxylic acid
butenedioic acid (fumaric acid)	$CHCOOH$ || $CHCOOH$	—COOH carboxylic acid $>C=C<$ alkene	unsaturated dicarboxylic acid
2-hydroxypropanoic acid (lactic acid)	$CH_3CH(OH)COOH$	—COOH carboxylic acid —OH alcohol	hydroxycarboxylic acid
2-hydroxybutanedioic acid (malic acid)	$COOH$ | $CHOH$ | CH_2COOH	—COOH carboxylic acid —OH alcohol	hydroxydicarboxylic acid
3-hydroxy-3-carboxy-pentanedioic acid (citric acid)	H | H—C—COOH | HO—C—COOH | H—C—COOH | H	—COOH carboxylic acid —OH alcohol	hydroxytricarboxylic acid
2-ketopropanoic acid (pyruvic acid)	O || CH_3—C—COOH	—COOH carboxylic acid $>C=O$ ketone	ketocarboxylic acid
2-ketobutanedioic acid (oxaloacetic acid)	O || C—COOH | CH_2—COOH	—COOH carboxylic acid $>C=O$ ketone	ketocarboxylic acid

1 **What type of isomerism is shown by (a) butenedioic acid and (b) 2-hydroxypropanoic acid?**

One of the reaction sequences which enables energy from food to be released, is the so-called 'Krebs cycle'. Although a discussion of the full cycle is beyond the scope of this book, some of the chemical changes within this sequence illustrate the complexity and subtlety of body chemistry. A section of the cycle is given in Fig 23.1. Think about the conditions necessary to achieve these conversions in the laboratory and remember that our bodies carry them out in a highly sensitive environment (the cell) which can be destroyed by all sorts of changes, such as temperature, pH and ion concentration.

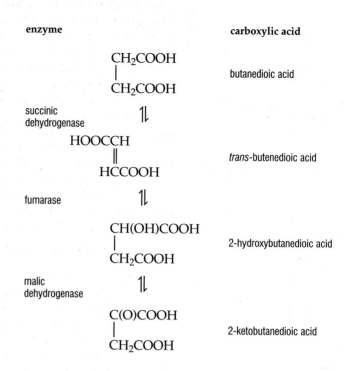

Fig 23.1 A section of the Krebs cycle

enzyme   carboxylic acid

CH_2COOH
|
CH_2COOH butanedioic acid

succinic
dehydrogenase   ⇅

HOOCCH
‖
HCCOOH   *trans*-butenedioic acid

fumarase   ⇅

$CH(OH)COOH$
|
CH_2COOH 2-hydroxybutanedioic acid

malic
dehydrogenase   ⇅

$C(O)COOH$
|
CH_2COOH 2-ketobutanedioic acid

2 How could butanedioic acid be converted into 2-ketobutanedioic acid in the laboratory? Comment on these reaction conditions compared with those found in the living cell.

Sugars

Sugars are made up of carbon, hydrogen and oxygen and nearly always have the molecular formula $(CH_2O)_n$. They have two roles in nature. They are:

- fuels (providing energy for the body);
- building blocks for the biopolymers called polysaccharides (giant carbohydrates), such as cellulose in plants and glycogen in the human body.

The simplest sugars are called **monosaccharides**. They have a ring structure and the molecular formula $C_6H_{12}O_6$. The three most important monosaccharides are glucose, fructose and galactose (Fig 23.2).

3 What functional groups are present in (a) glucose, (b) fructose and (c) galactose?

4 What type of isomerism do the following pairs display: (a) galactose and glucose, (b) fructose and glucose, (c) galactose and fructose?

CHEMISTRY IN THE BODY

(a)

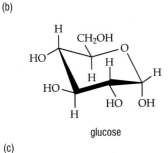

galactose

(b)

glucose

(c)

fructose

Fig 23.2 Structural formulae of galactose, glucose and fructose.

Fig 23.3 Glucose solution is used when an immediate source of energy is required.

The concentration of glucose in the blood is about 8×10^{-3} mol dm^{-3}. It is the most important carbohydrate fuel in human cells, providing a ready source of energy. This is because of its small size and solubility in water which, together, allow it to pass through the cell membrane into the cell where it is metabolised (chemically broken down) to give smaller molecules, releasing energy in the process.

In nature, most sugars are **disaccharides**. As this term indicates, they are made up of two simple sugar molecules. The three most important disaccharides are shown in Fig 23.4. All have the same molecular formula, $C_{12}H_{22}O_{11}$. Monosaccharides are rare in nature and are formed in the cell by metabolism of disaccharides.

Although soluble in water, the disaccharides are too large to pass through the cell membrane. They are broken down by hydrolysis in the small intestine during digestion to give the smaller monosaccharides which pass into the blood and then through cell membranes into the cell.

$$C_{12}H_{22}O_{11} + H_2O \longrightarrow C_6H_{12}O_6 + C_6H_{12}O_6$$

Monosaccharides are stored in cells by undergoing **condensation reactions**, initially reforming disaccharides and then polymerising further to form polysaccharides. These are too large to escape from the cell and provide a means of storing energy inside it. Hydrolysis back to monosaccharides and their subsequent metabolism to carbon dioxide and water are the energy-producing reactions which help meet the body's energy demands. This is because they are exothermic reactions.

5 Explain why the formation of a disaccharide from two monosaccharide molecules is called a condensation reaction.

(a)

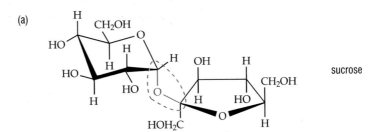

sucrose

(b)

lactose

(c)

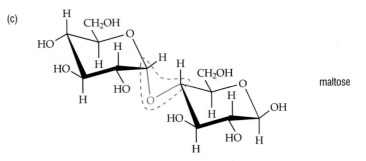

maltose

Fig 23.4 Structural formulae of the three most important disaccharides found in nature (a) sucrose (common sugar), (b) lactose (found in milk), and (c) maltose (used, for example, in brewing). In each molecule the glycosidic link is highlighted.

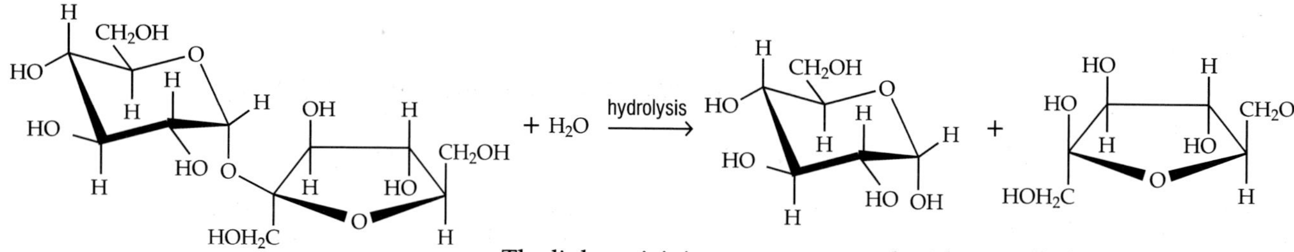

Fig 23.5 The formation of glucose and fructose from the hydrolysis of sucrose.

The linkage joining two monosaccharides is called a **glycosidic bond** (Fig 23.4). It is formed when two hydroxyl functional groups undergo a condensation reaction. When two simple alcohol molecules undergo a condensation reaction, an ether molecule is formed. For example, two ethanol molecules condense in particular conditions to give ethoxyethane:

$$CH_3CH_2OH + CH_3CH_2OH \longrightarrow CH_3CH_2OCH_2CH_3 + H_2O$$

6   **What reaction conditions are needed to bring about a condensation reaction between two simple alcohols?**

Ethers are not hydrolysed even under forcing conditions such as reflux in acidic or alkaline media, yet saccharides hydrolyse in the mild conditions inside a living cell. The formation of disaccharides and their hydrolysis compared with the chemistry of ethers demonstrate how effective enzymes are as catalysts.

Amino acids

Amino acids are the building blocks of the body's most important class of biopolymers – the proteins. Twenty different amino acids are needed to make the body's proteins. Of these, eleven are called **non-essential** and nine are **essential**. The essential amino acids can only be obtained through diet whereas the others may be synthesised in the body, usually by modification of an essential amino acid, although they are usually taken in as food.

All amino acids have an amine group and a carboxylic acid group. Their reactions, therefore, are characteristic of these two functional groups. The side group, –R, determines the precise nature of the amino acid. All enzymes are proteins and, therefore, are made from amino acids.

Table 23.3 The essential and non-essential amino acids

Non-essential	R	Essential	R
arginine	$-(CH_2)_3-NH-\underset{\underset{NH}{\|\|}}{C}-NH_2$	histidine	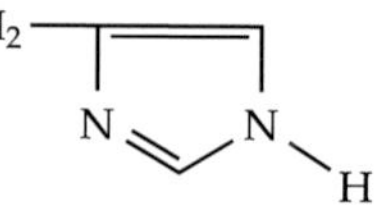
asparagine	$-CH_2-\underset{\underset{O}{\|\|}}{C}-NH_2$		
		isoleucine	$-CH(CH_3)-CH_2-CH_3$
aspartic acid	$-CH_2-COOH$	leucine	$-CH_2-CH(CH_3)_2$
cysteine	$-CH_2-SH$	lysine	$-(CH_2)_4-NH_2$
glutamic acid	$-(CH_2)_2-COOH$	methionine	$-(CH_2)_2-S-CH_3$
glutamine	$-(CH_2)_2-\underset{\underset{O}{\|\|}}{C}-NH_2$		
glycine	$-H$	phenylalanine	
*proline			
serine	$-CH_2-OH$	tryptophan	
threonine	$-CH(OH)-CH_3$		
tyrosine		valine	$-CH(CH_3)_2$

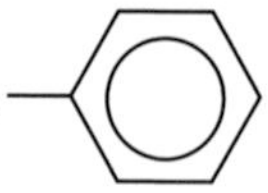

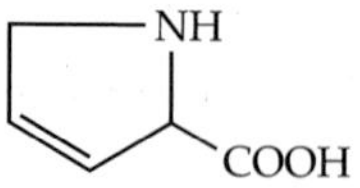

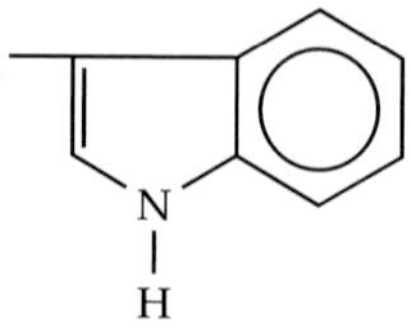

*structure of whole amino acid

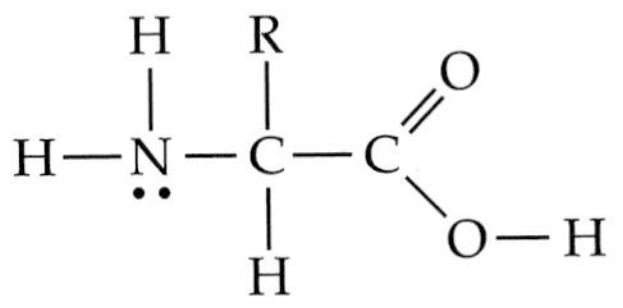

(a) Structural formula for the general amino acid $H_2NCH(R)COOH$

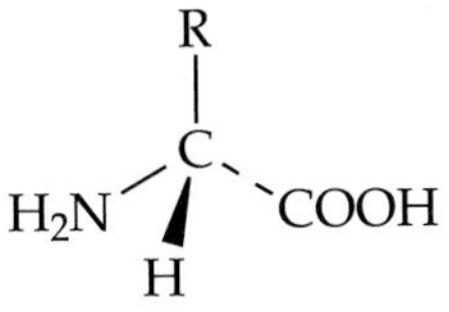

(b) Shape of the amino acid molecule

Fig 23.6 General structural formula and displayed formula for amino acids.

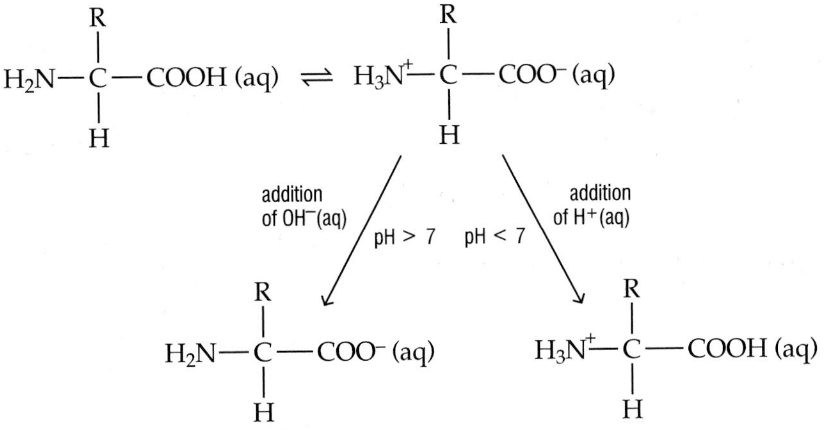

Fig 23.7 The form in which an amino acid exists depends upon pH. The exact pH at which the Zwitterion ($^+NH_3CH(R)COO^-$) exists depends on the nature of the side-group, R.

7 What type of isomerism do nineteen amino acids used in protein synthesis display? Which simple amino acid does not display this form of isomerism?

8 What type of isomerism is displayed by the pair of amino acids, isoleucine and leucine?

9 Write structural formulae for the species present when alanine is dissolved in (a) dilute hydrochloric acid, (b) dilute sodium hydroxide.

In addition to their role as protein building blocks, amino acids also help buffer against sharp changes in pH. Amino acids can regulate pH in a cell by reacting with either oxonium ions (hydrated hydrogen ions) or hydrated hydroxide ions which enter or are suddenly formed in the delicately balanced cell. In this way, sudden fluctuations in pH which would destroy the cell are avoided. The buffering reactions are:

$$H_3N^+\!\!-\!CH(R)\!-\!COO^- (aq) + H^+ (aq) \longrightarrow H_3N^+\!\!-\!CH(R)\!-\!COOH (aq) \\ + H_2O(l)$$

$$H_3N^+\!\!-\!CH(R)\!-\!COO^- (aq) + OH^- (aq) \longrightarrow H_2N\!-\!CH(R)\!-\!COO^- (aq) \\ + H_2O(l)$$

Phosphates

Phosphoric(V) acid is an inorganic, molecular compound. It is manufactured by the chemical industry in huge quantities from phosphate rock. Yet this inorganic compound is central to vast numbers of biochemical reactions, in particular those associated with the energy-producing reactions which body cells depend upon.

Phosphoric(V) acid can form salts by reaction with bases, and esters by reaction with any organic compound containing the $-CH_2OH$ group.

10 What is the shape of a phosphoric(V) acid molecule, H_3PO_4?

11 Write equations to show the ionisation of phosphoric(V) acid in water.

Of particular interest in our study of body chemistry is the formation of phosphate esters. This often makes the organic molecule which forms the ester biochemically more reactive. Indeed, cell chemistry frequently involves phosphate ester formation as the first stage in a reaction sequence.

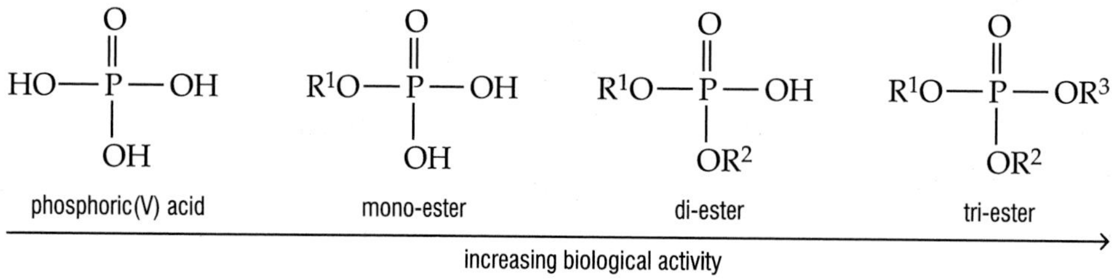

phosphoric(V) acid   mono-ester   di-ester   tri-ester

increasing biological activity

Fig 23.8 Esters of phosphoric(V) acid. In these esters, the organic groups R^1, R^2 and R^3 may or may not be the same as one another.

12 Write structural formulae of two examples in each case of mono-esters formed by reacting phosphoric(V) acid with (a) primary alcohols, (b) sugars, (c) amino acids, (d) hydroxyacids.

One further important reaction of phosphoric(V) acid and phosphate(V) ions is the condensation reaction that results in the formation of diphosphates and triphosphates, such as adenosine diphosphate (ADP) and adenosine triphosphate (ATP).

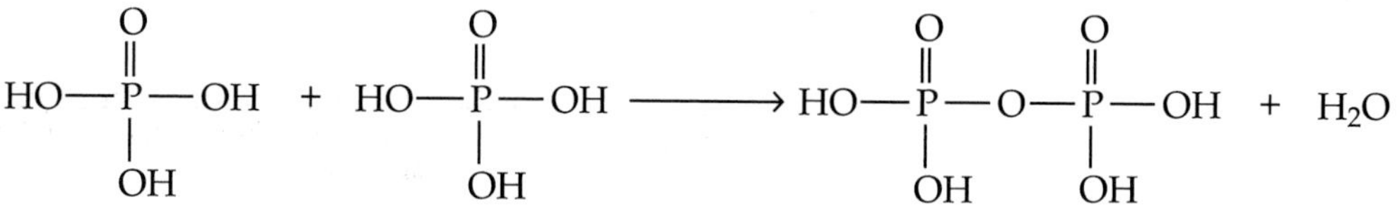

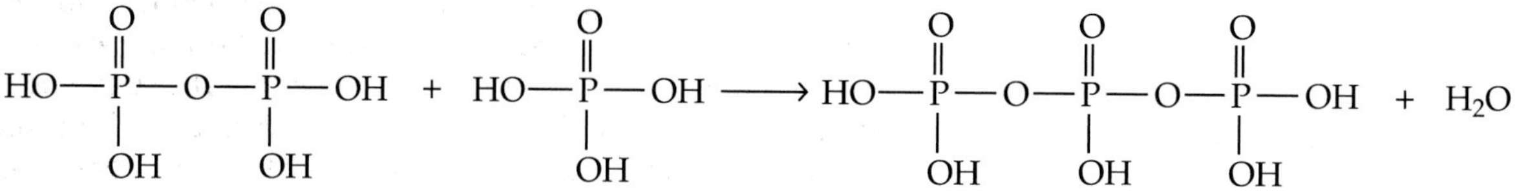

The reverse reaction (a hydrolysis reaction) releases energy and plays a significant role in meeting the body's energy demands.

13 Explain why the formation of di- and triphosphates may be described as condensation reactions.

Purines and pyrimidines

Adenine and guanine are natural bases derived from purine. Cytosine, uracil and thymine are natural bases derived from pyrimidine. Collectively, these small molecules are often referred to as nitrogenous bases.

Most do not occur freely, however. They are usually combined with the sugar, ribose, and one or more phosphates. Molecules containing this combination are called **nucleotides** and are important in the formation of **nucleic acids**. Adenosine is the nucleotide formed from adenine and ribose.

14 Explain why pyrimidine does not have discrete single and double bonds in the ring but, instead, is an aromatic ring structure.

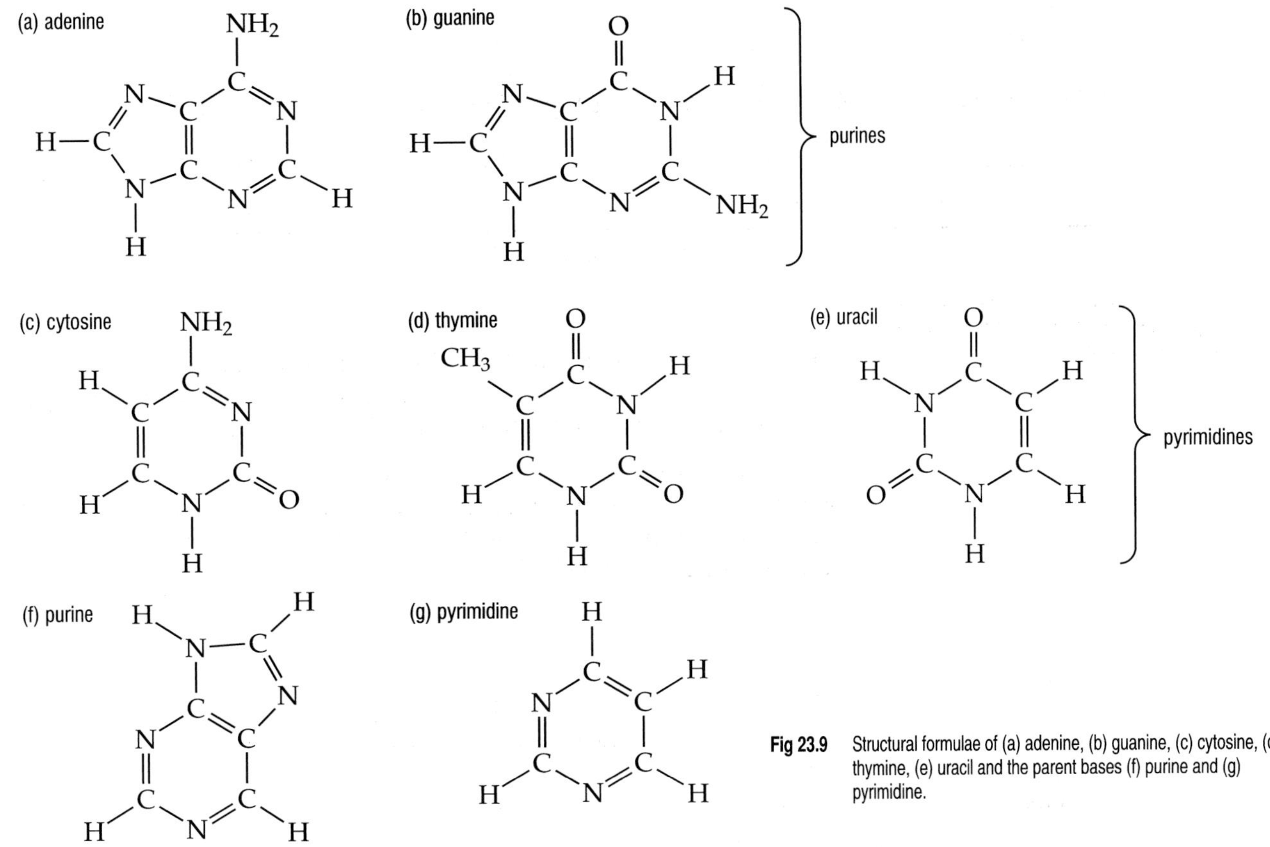

(a) adenine
(b) guanine
} purines

(c) cytosine
(d) thymine
(e) uracil
} pyrimidines

(f) purine
(g) pyrimidine

Fig 23.9 Structural formulae of (a) adenine, (b) guanine, (c) cytosine, (d) thymine, (e) uracil and the parent bases (f) purine and (g) pyrimidine.

Adenosine triphosphate (ATP) is the major compound involved in the body's energy transfer reactions (see later in this chapter). Although a complex molecule, it is easy to understand both its structure and its activity by applying basic chemical principles. One molecule of ATP is made from five smaller molecules by a series of condensation reactions (Fig 23.10). Each condensation is accompanied by the elimination of a water molecule.

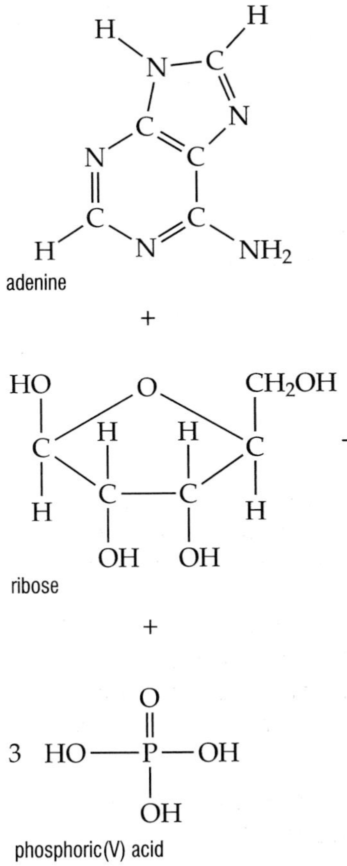

adenine

+

ribose

+

$3 \quad$ phosphoric(V) acid

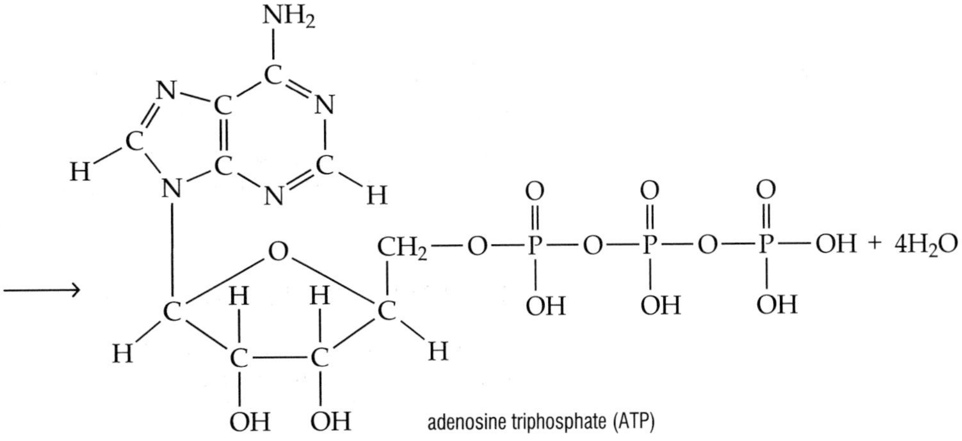

adenosine triphosphate (ATP)

$+ 4H_2O$

Fig 23.10 It takes one adenine molecule, one ribose molecule and three phosphoric(V) acid molecules to form a molecule of ATP. The overall process is shown here.

15 **What type of reactions would produce ADP (adenosine *di*phosphate) and AMP (adenosine *mono*phosphate) from ATP?**

23.3 BIOPOLYMERS

A detailed account of the chemistry of biopolymers is beyond the scope of this book. We shall, however, look briefly at their synthesis and degradation, and the role they play in our bodies.

Polysaccharides

Starch is often produced as a way of storing energy in plants. It is a polymer of glucose, not very soluble in water but hydrolysed in the presence of a suitable catalyst (dilute mineral acid or appropriate enzyme) to give glucose as the only product.

Hydrolysis of starch is the key reaction which enables us to tap this energy store. Starch molecules are too large to enter the living cell and so enzymes catalyse its hydrolysis in the digestive system to glucose. The small glucose molecule can penetrate the cell membrane and, once inside, is polymerised again. There is a subtle change, however. Instead of starch being reformed a different biopolymer, **glycogen**, is synthesised.

Glycogen molecules are too big to escape from the cell, (just as starch molecules would have been). However, glycogen is soluble in water whereas starch is not (to any appreciable extent). When the body demands energy, glycogen is hydrolysed in the cell to glucose which can escape into the bloodstream and be transported to the energy-demanding site in the body. We store glycogen primarily in our liver and muscles.

Fig 23.11   Depolymerisation and polymerisation are the key to storing and utilising energy in the body.

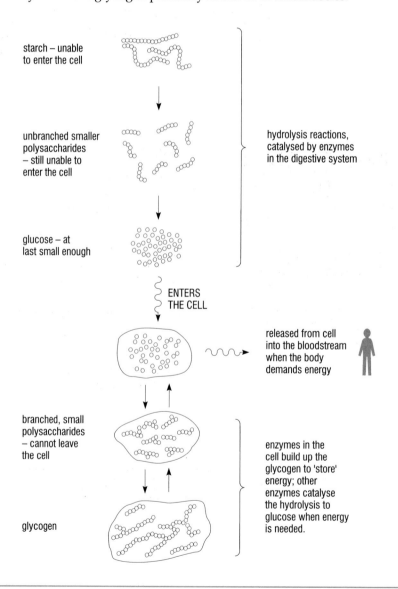

starch – unable to enter the cell

unbranched smaller polysaccharides – still unable to enter the cell

glucose – at last small enough

hydrolysis reactions, catalysed by enzymes in the digestive system

ENTERS THE CELL

released from cell into the bloodstream when the body demands energy

branched, small polysaccharides – cannot leave the cell

glycogen

enzymes in the cell build up the glycogen to 'store' energy; other enzymes catalyse the hydrolysis to glucose when energy is needed.

16 Explain why, in principle, the number of polysaccharides that could be formed from monosaccharides (simple sugars) is so great.

Despite the vast number of possible polysaccharide molecules, in nature the combinations of monosaccharides are limited. Starch and glycogen are two polysaccharides which are synthesised in nature. The third important one is **cellulose**. We will meet this later.

Fig 23.12 Structural formulae of (a) starch, (b) cellulose.

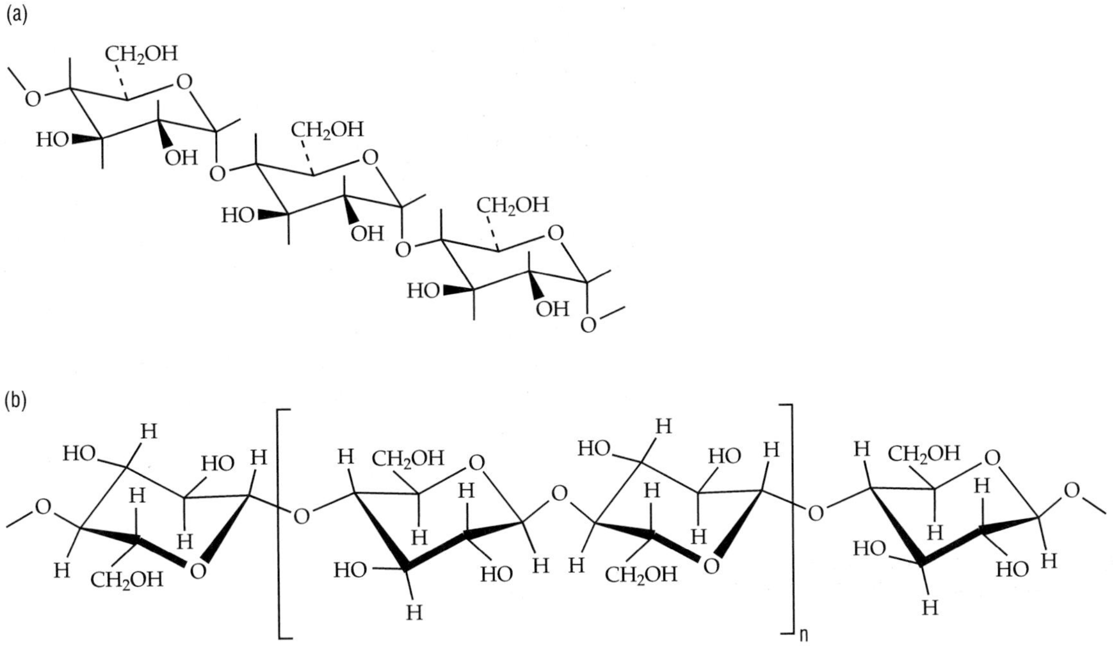

(a)

(b)

Fig 23.13 We can buy monosaccharides, disaccharides and polysaccharides at the local supermarket. Glucose is a monosaccharide, sucrose is a disaccharide and starch is a polysaccharide.

Proteins

Proteins make up some 15% of our body weight. They are the most abundant 'solid' substances in our bodies, (which are about 65% water), and perform many functions in the body. They

- contribute to our mechanical structure;
- enable us to move;
- facilitate the transport of smaller molecules and ions around the body;
- control the types and rates of chemical reactions (these proteins are called enzymes);
- form the immunological system.

This range of tasks suggests the need for many and varied proteins, and, indeed, this is the case.

Proteins are biopolymers formed when amino acids condense, with the elimination of water:

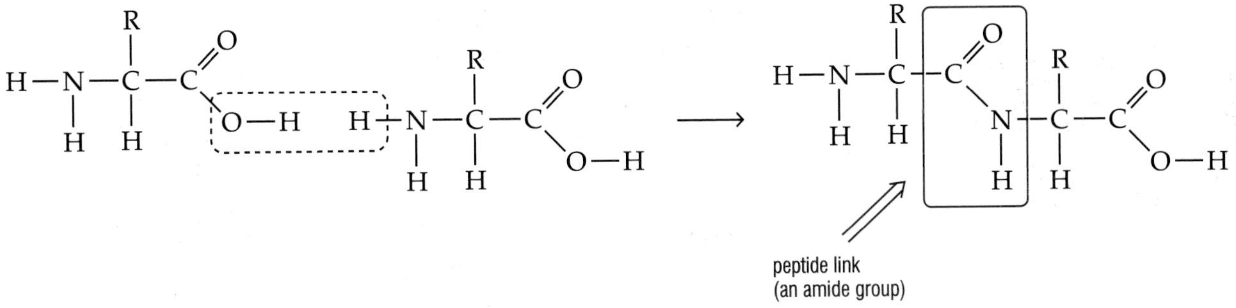

peptide link
(an amide group)

Fig 23.14 Proteins are broken down into their constituent amino acids by hydrolysis. Enzymes catalyse this reaction in the body.

Carboxylic acids react with primary amines to give amides (a reaction used to synthesise polyamides such as the nylons). The remarkable feature of protein formation in the body is the conditions under which reaction occurs. In aqueous solutions of pH about 7, amino acids exist as zwitterions. In this form, a condensation reaction is extremely unlikely. It is made possible by catalysts in the cell. These catalysts are in ribosomes – tiny, insoluble particles which are distributed throughout the cell. In the ribosomes, RNA (discussed later in this chapter) is bonded to protein. So, protein is needed to catalyse the synthesis of proteins.

17 Write the zwitterion structure for leucine.

18 What conditions are needed to form a substituted amide from a carboxylic acid and a primary amine?

The nature of the peptide link

The peptide link has the structure:

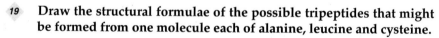

At first sight we might imagine the C—N bond is free to rotate. Surprisingly, this is not the case. Electron density within the peptide link is delocalised, giving the C—N bond some multiple covalent bond character and restricting its rotation. The origin of the delocalisation might be pictured as being the result of two extreme arrangements of electrons, shown above and below:

19 Draw the structural formulae of the possible tripeptides that might be formed from one molecule each of alanine, leucine and cysteine.

Vast numbers of proteins are theoretically possible. If a protein were composed of 100 amino acids, each of which might be one of the twenty amino acids necessary in the body, there are 20^{100} possible arrangements! And many proteins contain considerably more than 100 amino acids. While there are indeed a very large number of proteins, they do not even begin to approach the number of theoretical possibilities.

Haemoglobin found in red blood cells provides an example of how a minute change in structure can result in a devasting difference in function. The prime function of haemoglobin is to carry oxygen around the body in the blood. Sickle cell anaemia is caused by the presence of sickle-cell haemoglobin, a molecule which cannot transport oxygen efficiently. Yet, remarkably, sickle-cell haemoglobin and normal haemoglobin differ by just one amino acid in 287.

Fig 23.15 Chromatographic analysis of an amino acids mixture obtained when a protein is hydrolysed can help to identify the protein.

CHEMISTRY IN THE BODY

Primary, secondary and tertiary structures of proteins

The sequence of amino acids in a protein is called its **primary structure**.

Three types of bonding can occur both within (intramolecular) and between (intermolecular) protein molecules.

- hydrogen bonds

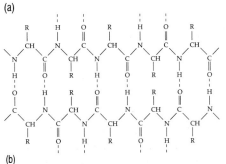

- sulphur-sulphur bonds (also called 'cystine bonds'. These are formed between the cysteine amino acid units by reduction. See table 23.3)

$$-SH + HS- \longrightarrow -S-S- + 2[H]$$

- 'salt links', due to the amino acids units which have amine or carboxylic acid side groups

(a)

(b)

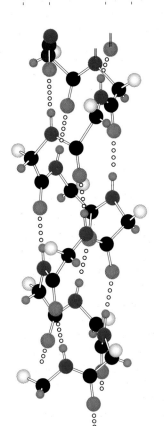

Single covalent bonds are free to rotate and so protein molecules can bend and flex, adopting a range of conformations. In certain conformations, intramolecular bonding is at its greatest. The protein is locked into a particular shape by this intramolecular bonding to give the protein its **secondary structure**. Proteins are either sheet-like (for example, fibroin, the protein found in silk) or helical (for example, keratin, the protein found in hair) (Fig 23.16).

Pauling and Corey, in 1951, were the first to propose an α-helical structure for proteins. Their model was deduced from X-ray diffraction measurements. The loosely coiled spring makes five complete turns for every 18 amino acids. In this way, the amino acids which fall directly above and below one another in the spring are able to form hydrogen-bonds. The same two scientists refined their model later by suggesting that the coil itself was not straight. Indeed, they proposed that in keratin, six helical protein molecules wrapped themselves around a seventh into a platted bundle. In 1954, Linus Pauling received the Nobel prize for his contributions to protein structure determination and, in particular, discovering the α-helical structure.

The third level of protein structure is called its **tertiary structure**. This occurs when the helix bends or folds. It also describes the shape taken up by globular proteins. As with the secondary structure, this third level of structure is held in place by intramolecular bonding.

20 **Explain why the secondary and tertiary structures of proteins can be largely destroyed by water.**

Fig 23.16 (a) The sheet-like structure of a protein, and (b) the helical structure of a protein.

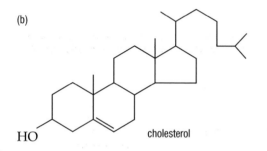

ester linkage

R¹, R², R³ may or may not be the same

Fig 23.17 The molecular structure of a fat or oil, with the ester group highlighted.

(a)

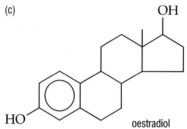

(b)

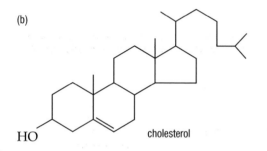

cholesterol

HO

Fig 23.18 (a) The basic steroid ring structure, (b) cholesterol, (c) oestrogen (female hormone), (d) testosterone (male hormone).

Lipids

Lipids is the collective name for compounds fats, oils and fat-like molecules such as steroids. Their role in the body includes being:

- components in cell membranes;
- fuels;
- chemical messengers (some 'hormones').

Fats and oils are esters of fatty acids and propan-1,2,3-triol. They are hydrolysed in the intestine where the enzyme lipase catalyses the reaction.

21 **How can the constituent fatty acids be obtained from a fat or oil?**

Steroids have a complex structure. They are not fats even though they are classed as lipids. Perhaps the most familiar steroid is cholesterol, high concentrations of which lead to heart disorders.

The sex of a person is determined by the sex hormone present. Subtle differences in two steroid molecules, oestrogen and testosterone, manifest themselves in striking differences. It would be a dull world indeed if it were not for molecular subtleties.

(c)

(d)

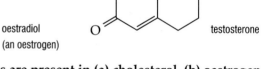

OH

OH

HO

oestradiol
(an oestrogen)

O

testosterone

22 **What functional groups are present in (a) cholesterol, (b) oestrogen, (c) testosterone?**

23 **Write the structural formulae of the products obtained when cholesterol is reacted with (a) ethanoic acid, CH₃COOH, in the presence of a little concentrated sulphuric acid, (b) a solution of bromine, Br₂, in hexane.**

24 **If you were given a sample of a white powder and told that it was either oestrogen or testosterone, how would you identify it? Give examples of chemical tests and spectroscopic methods.**

Nucleic acids

The final group of biopolymers are the nucleic acids, with relative molecular masses of up to 1 million. Nucleic acids serve two functions in the cell. They:

- store and record all the inherited characteristics of the person, animal or plant;
- translate this information so that chemical reactions in the cell are directed towards pre-determined goals.

CHEMISTRY IN THE BODY

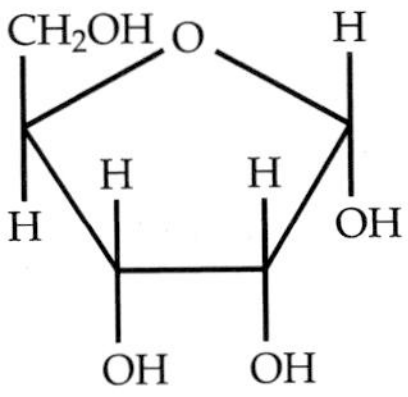

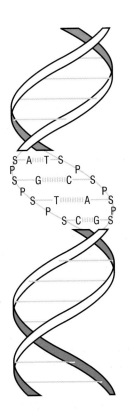

Nucleic acids are made from phosphoric(V) acid, either ribose or deoxyribose (Fig 23.19), and a nitrogen compound which has basic properties through a series of condensation reactions (Fig 23.10). The nitrogen compounds are the purines and pyrimidines we met earlier, often called nitrogenous bases.

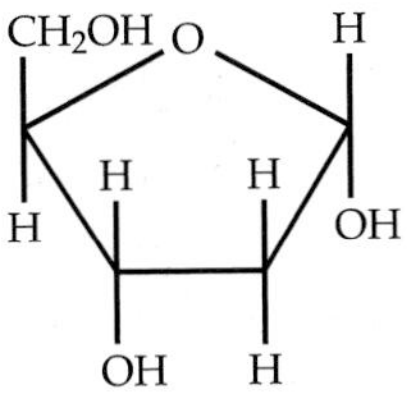

- Ribonucleic acid (RNA) is formed from phosphoric(V) acid, ribose and a nitrogenous base.
- Deoxyribonucleic acid (DNA) is formed from phosphoric(V) acid, deoxyribose and a nitrogenous base.

Polymeric chains form because each phosphoric(V) acid molecule has three —OH groups and so can form mono-, di-, and tri-esters. In both DNA and RNA, the phosphate(V) forms di-esters with two sugars, giving a backbone of alternating phosphate-sugar (either ribose or deoxyribose). The nitrogenous bases are attached to this polymeric backbone as pendant-like side groups. This is the primary structure.

Fig 23.19 Ribose is the monosaccharide from which RNA is made and deoxyribose is the monosaccharide from which DNA is made.

25 Explain why adenine has basic properties. Write structural formulae for species present in aqueous solution.

The secondary structure of nucleic acids is due to hydrogen-bonding between base pairs (Fig 23.20). This is similar to the situation for proteins, where the secondary structure is held in place by intramolecular bonds. In the nucleic acids, it is the geometry of the bases that determines which pairing can take place and, therefore, the overall shape of the double helix.

Fig 23.20 Part of the double helix of DNA. Hydrogen-bonding between base pairs hold the molecule in shape. The key to the shape of the DNA double helix was an understanding of which base pairs had just the right geometry to hold the coiled molecules in place.

In every cell there is some 1.4 m of DNA with a mass of just 6×10^{-12} g. Clearly, intramolecular bonding is essential to keep the DNA tightly wound and contained within a cell. When cells divide and multiply, the DNA is replicated in the new cells, carrying with it all the genetic information from the original cell. The sequence of bases in the DNA molecule differs from person to person and it is this that is responsible for giving each of us our own individual characteristics.

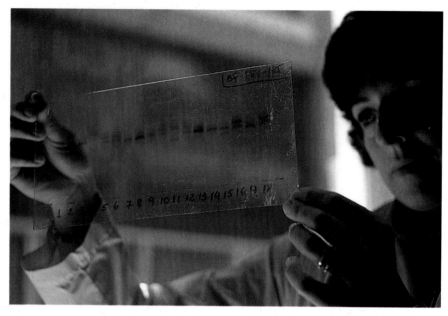

Fig 23.21 Analysis of DNA extracted from a cell can be used to identify a person. Each of us has a unique DNA fingerprint and the technique is finding increasing application in forensic science.

ACTIVITY

CORROSIVE POISONS

Corrosive poisons are toxic substances that destroy living tissue. Acids and bases catalyse the hydrolysis of protein molecules, initially degrading them into smaller peptide molecules and eventually into amino acids. Other corrosive poisons act because of their ability either to oxidise or to reduce compounds which occur in the body.

We can categorise corrosive poisons as having one of the following four toxic actions:

- acid hydrolysis;
- base hydrolysis;
- oxidising agent;
- reducing agent.

(a) Classify the following compounds according to the four categories of corrosive poisons and suggest where you might come into contact with these species in everyday life:

chlorate(I) ion; hydrochloric acid; hydrogen peroxide; iodine; nitrogen dioxide; ozone; sodium hydroxide; sulphite ion; sulphuric acid; trisodium phosphate(V).

(b) Identify some compounds in the body that might be oxidised or reduced by corrosive poisons.

(c) Make a poster to show how (i) hydrochloric acid and (ii) sodium hydroxide act as corrosive poisons towards proteins in the body. Sketch a design on a sheet of A4 paper (30 cm × 21 cm) and then scale up to make a poster which could be displayed in the laboratory.

23.4 FROM MOLECULES TO LIFE

Despite the enormous advances in synthetic chemistry and our successes as designer chemists, the artificial synthesis of living systems by human beings remains in the realms of science fiction.

In purely chemical terms, a body that has just died is indistinguishable from a live one. The same ions, small molecules and biopolymers are present. So what is the difference? The major function of the living cell is to constantly recreate itself from within. In other words, the body is able to maintain chemical substances at their present working levels during the unimaginable number of changes that are constantly occurring. There are two things in particular that the living body does:

- synthesise the precise chemicals needed from a range of building blocks and avoid the production of unwanted materials;
- store, transport and release energy to bring about the required chemical changes.

Compounds in the body are assembled and dis-assembled (synthesised and degraded) all the time.

26 **Explain why hydrolysis and condensation reactions are both involved in these assembly/dis-assembley processes.**

27 **Name another pair of reaction types which are related in a similar way and take place inside the body and in the laboratory.**

Enzymes: controlling speed and selectivity

The remarkable precision of biochemical synthesis is due to enzymes. These catalysts mediate virtually all the chemical reactions in our bodies, controlling the rate of reactions and the nature of reaction products. Despite immense advances in our understanding of catalysis, we have so far been unable to mimic the unique properties of enzymes.

Enzymes are proteins. They are central both to synthetic routes (for example, building a protein from amino acids) and degradation routes (for example, multi-stage route from glucose to carbon dioxide and water). Perhaps the most important feature of enzymes, and one which separates them from the catalysts we normally use in the laboratory, is their selectivity. These catalysts are not scatter-gun catalysts. Instead they are highly specific – often one enzyme for one chemical reaction. This is illustrated by the following reaction:

Fig 23.22 Biological washing powders contain enzymes which catalyse the breakdown of dirt, grease and fats, improving the washing efficiency of the detergent.

$$
\begin{array}{c}
CH_3 \\
| \\
H-C-OH \\
| \\
COOH
\end{array}
\xrightarrow[\text{by lactic dehydrogenase}]{\text{oxidation, catalysed}}
\begin{array}{c}
CH_3 \\
| \\
C=O \\
| \\
COOH
\end{array}
$$

$$
\begin{array}{c}
CH_3 \\
| \\
CH_2 \\
| \\
H-C-OH \\
| \\
COOH
\end{array}
\xrightarrow[\text{catalysed by lactic dehydrogenase}]{\text{oxidation is } not}
$$

The oxidation of 2-hydroxypropanoic acid (lactic acid) to 2-keto-propanoic acid (pyruvic acid) is catalysed by the enzyme, lactic dehydrogenase. However, this enzyme will not catalyse the oxidation of 2-hydroxy-butanoic acid.

It is possible to characterise enzymes in several ways. A convenient one is according to the type of reaction they catalyse.

- hydrolyses, for example proteins to amino acids, polysaccharides to monosaccharides;
- adding enzymes, for example,
  $CH_3COCOOH + CO_2 \longrightarrow HOOCCH_2COCOOH$;
- transferring enzymes, for example,
  glucose-glucose + glucose-glucose $\longrightarrow$ glucose-glucose-glucose + glucose;
- isomerases, for example moving a phosphate group from one part of a glucose molecule to another;
- oxidising enzymes, for example, $C_2H_5OH \longrightarrow CH_3CHO + H_2$

Enzymes work, like all catalysts, by providing an alternative reaction pathway – one which has a lower activation energy than the uncatalysed pathway.

28 **Sketch reaction profiles for (a) an uncatalysed reaction and (b) an enzyme catalysed reaction.**

The enzyme forms a reactive intermediate with one or more of the reactants (we call these **substrates** in the context of an enzyme catalysed reaction). The uniqueness of enzymes in catalysing specific reactions can be explained by our understanding of their structure, bonding and shape.

We often picture the catalysis as having a 'lock and key' mechanism. The enzyme is the lock and the substrate is the key. In order to undergo the necessary reaction (unlock the door), the substrate key must fit the enzyme lock (Fig 23.23).

Fig 23.23 The lock and key mechanism of enzyme catalysed reaction. With some locks only the right key will unlock the door. Others may be less specific.

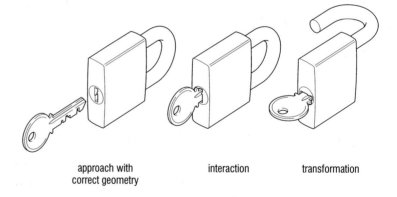

approach with   interaction   transformation
correct geometry

In chemical terms, the substrate must have the correct size and shape to fit a site in the enzyme, and the necessary functional groups to form some type of bond to hold it in place. This bonding is usually intermolecular – dipole-dipole or hydrogen-bonding. The enzyme may already be in the necessary conformation, or this shape may be induced by the presence of the substrate. In order for the enzyme to fit, it must also approach the substrate with the correct alignment (or orientation).

It is often the case that the enzyme is unable to create the necessary shape. Another component is needed. These components are called activators or co-enzymes and are frequently metal ions or compounds such as nucleotides. The secret is that the enzyme is tailored to a specific reaction.

29 Explain why an enzyme containing a metal ion may be considered to be a coordination compound.

Fig 23.24 Geometrical shape and functional groups enable a complex fit to be made which facilitates reaction of the substrate.

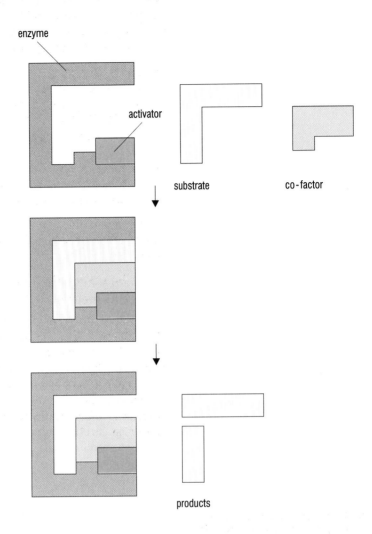

The conservation of energy in our bodies

Energy production distinguishes a live body from a dead one. The key reaction which provides energy is:

$$C_6H_{12}O_6 + 6O_2 \longrightarrow 6CO_2 + 6H_2O; \quad \Delta H = -2816 \text{ kJ mol}^{-1}$$

Importantly, this reaction does not occur in one step. To do so would release so much energy that the cell would 'burn-up' and be destroyed. Instead the body contrives a sequence of reactions in the cell, about 30 altogether. Each releases energy in small, more useful quantities.

30 Why must the total energy released by the various steps equal 2816 kJ for 1 mole of glucose as the starting point?

What happens to this energy? It is used, along with energy from other energy-producing reactions, to synthesise adenosine triphosphate (ATP). This molecule hydrolyses in the body, under enzyme control, to release energy as required (Fig 23.25).

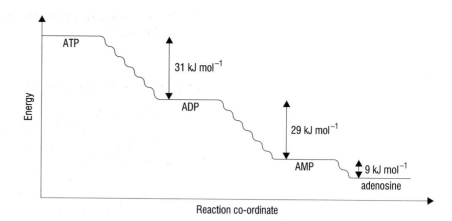

Fig 23.25 An energy diagram for the stepwise hydrolysis of ATP to adenosine. The ATP ⇌ ADP + P cycle provides a major means of 'storing' energy from food and using it to perform work in the body.

ACTIVITY

OPTIMUM CONDITIONS FOR ENZYME ACTIVITY

Saliva contains the enzyme amylase which catalyses the initial hydrolysis of starch. The disappearance of starch can be monitored taking samples from the reaction mixture and adding them to a standard solution of iodine in potassium iodide solution. The starch gives a characteristic deep blue colour, the intensity being proportional to the amount of starch present in solution.

Design and plan a series of experiments that you could perform in the laboratory in order to establish the optimum conditions for amylase activity.

23.5 FOOD FOR THE BODY

To stay healthy we need a sensible and balanced diet. Food is made up of proteins, carbohydrates, fats and water, together with minerals and vitamins. A good diet is one in which sufficient amounts of these substances are made available to the chemical factories inside our bodies.

'Real' chemical plants

Plants are a major source of food and, for some people the only source of food. They perform three major chemical functions:

- concentration of cations and anions in the soil to provide the minerals essential to life;
- synthesis of complex carbohydrates and other molecules from the simple molecules, carbon dioxide and water;
- building protein molecules from nitrogen. Atmospheric nitrogen is converted to ammonium compounds and then proteins by leguminous plants such as peas and beans. Bacteria in the soil convert atmospheric nitrogen into nitrates and these are synthesised into proteins in the roots of other plants. When animals eat plants they then convert them into their own types of proteins.

However, the most common substance found in plants, other than water, is not a food which humans can use. Structurally, plants are made of cellulose – a polysaccharide built from glucose molecules. It is the most abundant organic compound on earth. Cellulose is an enormous energy resource but humans are unable to use this supply directly because we lack the necessary enzymes to hydrolyse cellulose and release its 'stored' energy. Other carnivorous animals are also unable to use cellulose as an energy source. However, herbivores can and cows will eat grass. Enzymes in their gut hydrolyse cellulose to glucose which is then used to provide energy.

Fig 23.26 Nodules on the roots of these pea plants 'fix' nitrogen. Cells in the nodules convert molecular nitrogen into plant proteins.

31 What is the structural difference between cellulose and starch (look at Fig 23.12)?

The elements required for healthy plant growth

Many plants require at least 18 elements if they are to grow healthily. These may be classified in at least two ways. The first is to crudely divide them into those required in reasonably substantial quantities, we call these the **essential elements**, and others needed in only very small amounts, the **trace elements**. Plants have developed elegant ways of capturing the necessary elements from their surrounding environment.

The essential and trace elements can be further divided.

Table 23.4 Elements necessary for the healthy growth of green plants

Essential elements	Trace elements
Carbon	Boron
Hydrogen	Chlorine
Oxygen	Copper
Nitrogen	Iron
Phosphorus	Manganese
Potassium	Molybdenum
Calcium	Sodium
Magnesium	Vanadium
Sulphur	Zinc

- Carbon, hydrogen and oxygen are 'non-mineral' elements. They are obtained from the atmosphere and water and captured by plants by the process known as **photosynthesis**.
- All the other elements are 'minerals' and obtained from chemical compounds found in the soil. Do not get confused with the more accurate meaning of mineral as it is used to describe compounds which occur in the Earth's crust. Hydrated ions are absorbed through the plant's root system. These 15 elements fall into three categories according to the amounts required to sustain healthy growth:

Primary nutrients: N, P, K;
Secondary nutrients: Ca, Mg, S;
Micronutrients: B, Cl, Cu, Fe, Mn, Mo, Na, V, Zn.

32 Suggest the formulae of hydrated ions which are absorbed by a plant's roots to provide the following elements: N, P, K, Ca, Mg, S.

33 Which mineral provides a source of the micronutrient vanadium.

Photosynthesis – trapping the Sun's energy

In the final analysis, all the fuel reserves on our planet have their origins in the Sun. Perhaps the most remarkable chemical synthesis is **photosynthesis**, the reactions by which green plants convert the very simple molecular compounds, carbon dioxide and water, into more complex carbohydrates.

$$6CO_2(g) + 6H_2O(l) \longrightarrow C_6H_{12}O_6(aq) + 6O_2(g); \quad \Delta H = + 2816 \text{ kJ}$$

This equation summarises a very complicated reaction sequence. Even now the entire sequence is not fully understood.

The glucose is polymerised to starch and stored in plants in this form. This provides humans and other animals with an edible energy supply. Our bodies are able to reverse this reaction in a multi-stage process, releasing small packets of energy upon demand.

34 Iron is a micronutrient which helps to catalyse the formation of chlorophyll in green plants. A lack of iron causes plants to yellow. Explain why the addition of lime to soil, to raise the pH, lowers the availability of iron.

The key to photosynthesis is the ability of certain pigments to absorb visible light. The most important are chlorophyll A and chlorophyll B. Both are coordination compounds, with magnesium ions at the centre of a complex ring structure. Electrons in the compounds are excited by absorbed visible light and are promoted to higher energy levels. As they return to their original levels, the energy released is captured and used by the plant for chemical syntheses. An important example is the synthesis of ATP and we have seen how the energy stored in ATP can be released in small amounts when the body requires it, through a series of hydrolysis reactions.

(a)

(b)

The structure of (a) chlorophyll A and (b) chlorophyll B. Differences in their molecular structures mean that each absorbs in a slightly different region of the visible spectrum from the other and, therefore, have slightly different colours.

chlorophyll A

chlorophyll B

35 **What is the structural difference between chlorophyll A and chlorophyll B?**

36 **Describe the similarities between chlorophylls A and B and haemoglobin.**

37 **Green plants which have an abnormally low chlorophyll content are said to suffer from chlorosis. Explain why magnesium deficiencies in the soil result in chlorosis.**

Nitrogen fixation

Atmospheric nitrogen is very unreactive. This is due largely to the strength of bonding within the nitrogen molecule. The enthalpy of atomisation of nitrogen is 472 kJ mol^{-1}. A triple covalent bond (six electrons being shared, three from each nitrogen atom) binds the two nitrogen atoms in a molecule.

38 **Explain why the bond energy of a nitrogen molecule is 944 kJ mol^{-1}.**

Fortunately, this lack of reactivity has been overcome in nature in order to convert nitrogen into usable forms. Atmospheric nitrogen is said to be 'fixed'. Lightning provides sufficient energy to cause a reaction between nitrogen and other gases in the atmosphere to give nitrate ions. Certain bacteria in the soil also convert nitrogen to nitrate ions. Others convert nitrate ions, taken up by a plant from the soil, into protein through a series of stages:

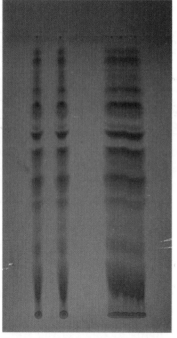

Fig 23.28 The two chlorophylls can be isolated from green plants by solvent extraction and separated by chromatography.

$$NO_3^- \longrightarrow NO_2^- \longrightarrow NH_3 \longrightarrow \quad \text{amino acids} \quad \longrightarrow \quad \text{proteins}$$

Other bacteria, found in the root nodules of legumes such as peas, beans and clover, take nitrogen directly from the atmosphere and fix it to produce proteins.

39 **What is the oxidation number of nitrogen in each of the species from nitrate ion to protein?**

In a lightning storm, unreactive nitrogen molecules are forced to react with oxygen:

$$N_2(g) + O_2(g) \longrightarrow 2NO(g)$$

Nitrogen oxide, NO, is oxidised readily in air to nitrogen dioxide:

$$NO(g) + \tfrac{1}{2}O_2(g) \longrightarrow NO_2(g)$$

This dissolves in water to give a mixture of nitric acid, HNO_3, and nitrous acid, HNO_2.

$$2NO_2(g) + H_2O(l) \longrightarrow HNO_3(aq) + HNO_2(aq)$$

40 **Explain why the formation of nitric acid and nitrous acid from nitrogen dioxide is a disproportionation reaction.**

Fig 23.29   Lightning initiates a sequence of reactions which result in atmospheric nitrogen being converted into nitric acid and nitrous acid.

ACTIVITY

TRACE ELEMENTS

Imagine you are a teacher in a school and want to devise an experiment for your pupils to carry out in order demonstrate the importance of trace elements on plant growth.

You want the experiment to show the uptake of each trace element by a plant and the effect of the element on the growth of the plant.

Describe how you would set about this task.

Answers to Quick Questions: Chapter 23

1 (a) geometrical isomerism, *cis-trans* about the $C=C$ double bond;
(b) optical isomerism (it is a chiral compound).
2 Step 1: butanedioic acid to *trans*-butenedioic acid; although dehydrogenation of alkanes to alkenes can be achieved industrially, this method is not available in the laboratory.

Step 2: *trans*-butenedioic acid to 2-hydroxybutanedioic acid; add conc H_2SO_4 at room temperature, followed by water.

Step 3: 2-hydroxybutanedioic acid to 2-ketobutanedioic acid; reflux with an acidic solution of $K_2Cr_2O_7$.

All the conditions required to achieve these conversions in the laboratory require vigorous reaction conditions and corrosive compounds. In contrast, such reactions are achieved in cells under considerably milder conditions.

3   **(a)** primary alcohol (1), secondary alcohol (4), ether (1);
    (b) primary alcohol (2), secondary alcohol (3), ether (1);
    (c) primary alcohol (1), secondary alcohol (4), ether (1).

4   **(a)** geometrical (*cis-trans*);
    (b) structural isomerism;
    (c) structural isomerism.

5   Two simple molecules combine with the elimination of a simple molecule (water).

6   Conc H_2SO_4 at 140°C (but if excess acid is used elimation of water can occur to give the alkene).

7   Optical isomerism (they are all chiral compounds, except for glycine).

8   Structural isomerism.

9   See Fig 23.30

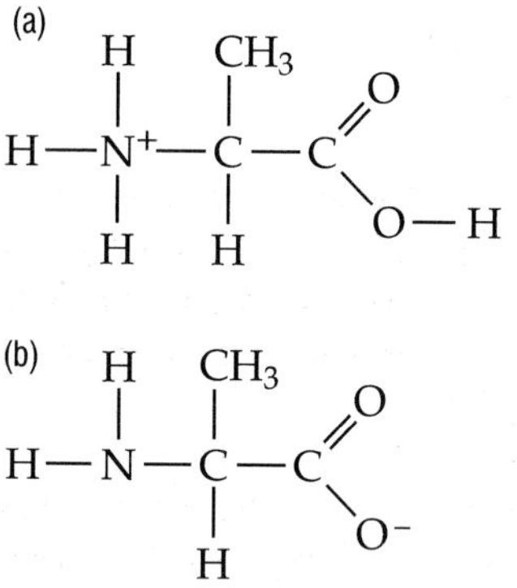

10 Tetrahedral

11 $H_3PO_4(aq) + H_2O(l) \rightleftharpoons H_2PO_4^-(aq) + H_3O^+(aq)$
    $H_2PO_4^-(aq) + H_2O(l) \rightleftharpoons HPO_4^{2-}(aq) + H_3O^+(aq)$
    $HPO_4^{2-}(aq) + H_2O(l) \rightleftharpoons PO_4^{3-}(aq) + H_3O^+(aq)$

12 There are numerous examples in each category. The general formula for the ester is shown in Fig 23.31.
    where R is the remainder of the organic molecule (whether it is a primary alcohol, sugar, amino acid or hydroxyacid) containing the hydroxyl group.

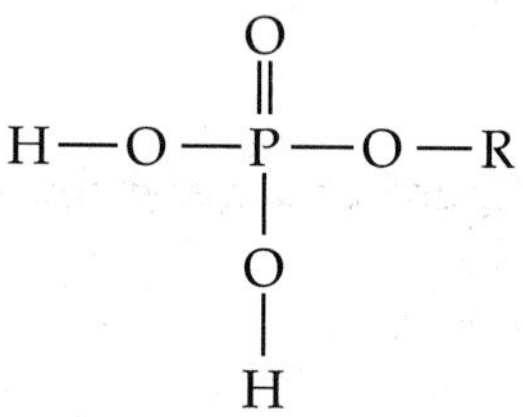

13 Their formation involves two molecules combining with the elimination of a simple molecule (water).

14 Pyrimidine is isoelectronic with benzene. Electrons are delocalised around the pyrimidine ring in a similar way to the benzene system arrangement in the molecule.

15 Hydrolysis in both cases.

16 Each monosaccharide has six hydroxyl groups. Since a condensation reaction could occur between any pair of hydroxyl groups in reacting sugar molecules, countless combinations are possible as the polysaccharide chain is built up.

17 See Fig 23.32

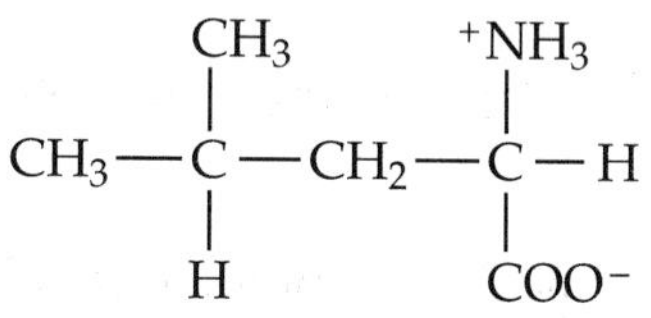

18 The carboxylic acid and the amine react at room temperature to form a salt. The solid salt is then heated to produce the substituted amide.

19 The general formula is:

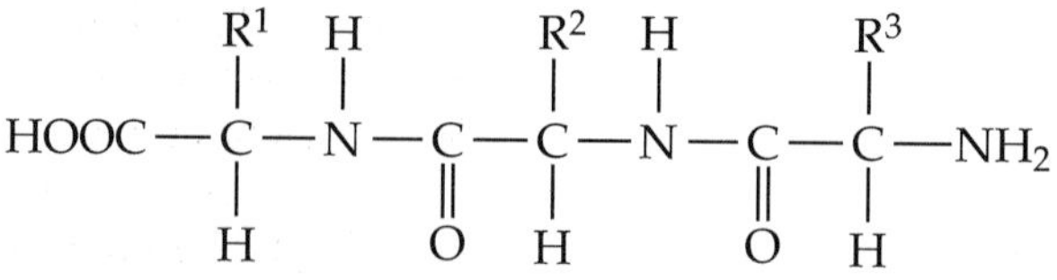

$$\underset{H}{\overset{R^1}{HOOC-C}}-N-\underset{H}{\overset{H}{C}}\overset{}{\underset{O}{\underset{\parallel}{C}}}-N-\underset{H}{\overset{H}{C}}\overset{}{\underset{O}{\underset{\parallel}{C}}}-\overset{R^3}{C}-NH_2$$

For three amino acids there are six possible combinations which give tripeptides with the general formula given above.

20 The hydrogen bonds which hold them in the secondary and tertiary structures may be broken by interaction with water.

21 By a hydrolysis reaction, for example, warming with dilute sodium hydroxide.

22 **(a)** cholersterol: secondary alcohol, alkene
    (b) oestrogen: phenol, secondary alcohol
    (c) testosterone: ketone, alkene, secondary alcohol

(a) Species present in dilute hydrochloric acid, (b) species present in dilute sodium hydroxide

Fig 23.30

Fig 23.31

Fig 23.32

(a)

(chemical structure of cholesterol acetate)

$$CH_3-\overset{\overset{\displaystyle}{}}{\underset{\underset{\displaystyle O}{}}{C}}-O-\text{(steroid structure)}$$

(b)

(chemical structure of dibromo steroid)

Br
Br

24 There are several tests that might be used. The task is to identify:
the phenol group if the sample is oestrogen (neutral iron(III) chloride, for example); the alkene group if the sample is testosterone (bromine water, for example); the ketone group if the sample is testosterone (2,4-dinitrophenylhydrazine, for example).
Infrared spectroscopy or NMR could be used.

25 It contains a secondary amine group. There is also a primary amine group but this is attached to an aromatic ring and so is likely to be less basic.

(chemical equation showing adenine protonation)

$$+ \quad 2H_2O \rightleftharpoons \quad + \quad 2OH^-$$

26 The reverse of a condensation reaction which eliminates a water molecule is hydrolysis.

27 Oxidation and reduction.

28 See Fig 23.33

29 The protein acts as a polydentate ligand, bonding to the central metal ion to form a coordination compound.

30 Hess's law: the total energy change accompanying a chemical reaction is independent of the route taken.

31 In starch, all the C—O—C linkages are *cis*, resulting in a helical structure. In cellulose, the C—O—C linkages alternate between *cis* and *trans*, giving a linear chain.

32 $NO_3^-(aq)$; $PO_4^{3-}(aq)$, $HPO_4^{2-}(aq)$, $H_2PO_4^-(aq)$; $K^+(aq)$; $Ca^{2+}(aq)$; $Mg^{2+}(aq)$, $SO_4^{2-}(aq)$.

33 Vanadinite.

34 Raising the pH of the soil increases the hydroxide ion concentration in the moisture present. These can react with $Fe^{3+}(aq)$ to give the insoluble $Fe(OH)_3$ and so decrease the concentration of $Fe^{3+}(aq)$ in solution.

35 Chlorophyll A has a CH_3 group where chlorophyll B has a CHO group.

36 Both contain a metal ion coordinated to four nitrogen atoms at the centre of a pyrrole ring.

37 Magnesium deficiency would result in smaller amounts of chlorophyll, since Mg^{2+} ions are central to its formation.

38 It contains a triple covalent bond and, therefore, the two nitrogen atoms are held together very strongly.

39 Oxidation numbers of nitrogen:
NO_3^-, +5; NO_2^-, +3; NH_3, –3; amino acids, –3; proteins, –3

40 The nitrogen undergoes a simultaneous oxidation and reduction:
$NO_2 \longrightarrow HNO_3$; $Ox[N]$ changes from +4 to +5 (oxidation)
$NO_2 \longrightarrow HNO_2$; $Ox[N]$ changes from +4 to +3 (reduction)

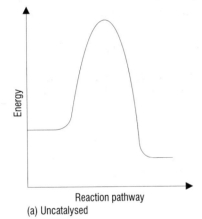

Energy

Reaction pathway
(a) Uncatalysed

Energy

Reaction pathway
(b) Catalysed

Fig 23.33

EXAMINATION QUESTIONS

1 (a) State the meaning of the term *activation energy* for a chemical reactoin. (2)

(b) Explain how the following factors influence the rate of a chemical reaction, referring to the activation energy in each case.
(i) concentration of reactants
(ii) temperature (12)

(c) A reaction sequence which may contribute to the formation of acid rain is given below.

$$NO(g) + \tfrac{1}{2}O_2(g) \rightarrow NO_2(g)$$

$$NO_2(g) + SO_2(g) \rightarrow NO(g) + SO_3(g)$$

Discuss the role of the NO in this reaction sequence. (5)

(d) The following data were obtained from studies of the reaction between NO and O_2 in a vessel at constant temperature.

Experiment	1	2	3
Initial total pressure of NO and O_2/atm	1.00	1.60	2.00
Initial partial pressure of NO/atm	0.40	0.40	0.80
Initial rate of reaction/atm s^{-1}	1.08	2.16	8.64

Use the information given to determine the order of the reaction, the rate equation and the value of the rate constant at that temperature. Show how you derive your answers. (11)

(JMB 1989)

2 (a) What is meant by dynamic equilibrium? (4)

(b) Using the Contact Process for the exothermic conversion of SO_2 and O_2 into SO_3 as an example, discuss the factors which influence both the position of equilibrium and the value of the equilibrium constant. (12)

(c) Ethanoic (*acetic*) acid reacts with ethanol as shown.

$$CH_3COOH(aq) + C_2H_5OH(aq) \rightleftharpoons CH_3COOC_2H_5(aq) + H_2O(l)$$

0.15 mol of ethanoic acid and 0.13 mol of ethanol were added to 0.04 mol of ethyl ethanoate (*acetate*), and the mixture was allowed to reach equilibrium at a fixed temperature. Analysis showed that 0.05 mol of the acid was present in the equilibrium mixture. Calculate the equilibrium constant for the reaction. (5)

(d) Why does the apparently simlar reaction

$$CH_3COOH(aq) + NaOH(aq) \rightleftharpoons CH_3COONa(aq) + H_2O(l)$$

proceed almost to completion in aqueous solution? (3)

(e) $K_p = 0.403$ at 1000°C for the equilibrium
$$FeO(s) + CO(g) \rightleftharpoons Fe(s) + CO_2(g)$$
If CO at a pressure of 1.00 atm and excess solid FeO are placed in a closed container at 1000°C, what are the partial pressures of CO and CO_2 when equilibrium is reached? (6)

(JMB 1989)

(3) (a) (i) Draw a detailed labelled sketch of the apparatus which you would use to measure the standard (reduction) electrode potential of the electrode $Fe^{3+}(aq), Fe^{2+}(aq) \mid Pt$.

(ii) State the purpose of each of the main parts of the apparatus. (8)

The following standard (reduction) eletrode portentials in acid solution will be needed to answer parts of the remainder of this question.

Electrode	$I_2(s), I^-(aq) \mid Pt$	$Cr_2O_7^{2-}(aq), Cr^{3+}(aq) \mid Pt$	$Fe^{3+}(aq), Fe^{2+}(aq) \mid Pt$
$E^{\circ}(298\ K)/V$	0.54	1.33	0.77

(b) (i) Write balanced half-equations for the chemical reactions which occur in the half-cells $Cr_2O_7^{2-}(aq), Cr^{3+}(aq) \mid Pt$ and $Fe^{3+}(aq), Fe^{2+}(aq) \mid Pt$.

(ii) Write an overall balanced equation for the oxidation of $Fe^{2+}(aq)$ by $Cr_2O_7^{2-}(aq)$ in acid solution.

(iii) Calculate the standard emf of the cell obtained by combining the two half-cells in **(b)**(i). (5)

(c) State whether the emf of this cell in **(b)**(iii) will increase, decrease, or remain unchanged if
(i) the concentration of $Fe^{3+}(aq)$ in the iron half-cell is increased,
(ii) the concentration of hydrogen ions is increased equally in both half-cells. (2)

(d) State, giving your reason, whether or not iodine will oxidise $Fe^{2+}(aq)$ ions in acid solution. (2)

(e) Aluminium is extracted by an electrolytic process. Write chemical equations for reactions occurring at (i) the cathode, and (ii) the anode. (3)

(WJEC 1989)

4 (a) Three types of isomerism shown by organic compounds are: structural, *cis-trans* and optical. The following two formulae together involve all of these types of isomerism. The two part-structural formulae **A** and **B** below are ambiguous in that each represents more than one compound. Draw **full** structural formulae to illustrate the **four** isomers that formula **A** represents and the two isomers that formula **B** represents. In each case, identify the type of isomerism involved.
(i) $CH_3CH(OH)C_3H_7$ (ii) $CH_3CH=CHC_2H_5$
　　　　A　　　　　　　　　　**B** (6)

(b) What class of compound, and how many isomers of the product, will be formed when
(i) **A** reacts with acidified potassium dichromate(VI),
(ii) **B** reacts with hydrogen in the presence of nickel? (4)

(UCLES 1990)

5   **(a)** Describe the bonding, molecular shapes and reactions with water of the tetrachlorides of the elements in Group IV. Explain how their thermal stabilities vary down the group.   (7)

  (b) Some cases of lead poisoning have been traced to the reaction of lead in water pipes with oxygen dissolved in the water to give lead(II) hydroxide. Write a balanced equation for this reaction. Use the *Database* to find the two redox potentials involved and briefly discuss why lead(II) hydroxide is formed.   (3)

(UCLES 1990)

6   Consider the following reactions:

	Reaction	$\Delta H/\text{kJmol}^{-1}$
P	$C(s) \rightarrow C(g)$	+715
Q	$C_3H_8(g) + 5O_2(g) \rightarrow 3CO_2(g) + 4H_2O(l)$	−2220
R	$3C(s) + 4H_2(g) \rightarrow C_3H_8(g)$	−104

  (a) Name enthalpy change R.   (1)

  (b) Select information from the above table to calculate the enthalpy change for the reactoin:
$3C(g) + 4H_2(g) \rightarrow C_3H_8(g)$ (2)

  (c) Use information from the table to calculate the enthalpy change during the complete combusion of 1120 cm^3 of propane (measured at s.t.p.).   (2)

(SCE 1991)

7   **(a)** Use the electron-pair repulsion theory to explain the shapes of the following species:
    (i) SF_6 and ICl_4^- (4)
    (ii) PCl_5 and SF_4 (4)
    (iii) NH_4^+ and PF_3.   (4)

  (b) Predict the shapes of the molecules

        FNO   NSF

  where the atoms are bonded in the orders shown, and the oxidation number of F is -1, of O is -2, and of S is +4.   (6)

  (c) Give the shape of and the bonding in the ion NO_2^+.   (2)

(UODLE 1991)

8   **(a)** Describe the structure of and nature of bonding in benzene. State and explain the thermochemical and other evidence in support of these.   (5)

  (b) State **two** typical examples of the chemical reactivity and **two** of the lack of reactivity of benzene and explain these in terms of your answer to **(a)** above.   (4)

  (c) Compare and contrast the chemical behaviour of
    (i)   hydroxyl groups, and
    (ii)   chlorine atoms
  when attached directly to the benzene nucleus with their behaviour when attached to alkyl groups.   (5)

  (d) (i)   The cyclic unsaturated hydrocarbon, naphthalene ($C_{10}H_8$), which may be written as ⬡⬡, absorbs 5 mol of hydrogen per mole of hydrocarbon on complete hydrogenation, the accompanying enthalpy change (ΔH°(298K)) being −284 kJ per mole of naphthalene.
The average enthalpy of hydrogenation of a $C{=}C$ double bond in a ring is −120 kJ mol^{-1}. Use this information to calculate the delocalisation or resonance energy in naphthalene and explain the basis of your calculation.

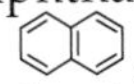

    (ii)   Bearing in mind your result in **(d)**(i) above and your answers in **(a)** and **(b)** above, state what you would expect to be the characteristic chemical behaviour of naphthalene. Give a reason.

    (iii)   A simple substituted naphthalene is used in the manufacture of a range of useful chemical compounds. Name this naphthalene derivative and state the way in which it is used.   (6)

(WJEC 1991)

9   **(a)** Discuss the oxidation and reduction reactions of aliphatic compounds containing the carbonyl group. Indicate the reagents used, the products formed and give equations where appropriate. Show, in your answer, how these reactions can be used to distinguish between aldehydes and ketones.   (17)

  (b) Alkenes are another class of compounds that undergo reduction reactions. Using propene as your example indicate what reagents and conditions you would use for its reduction and give a balanced equation. Name one important commercial process which makes use of the reduction of alkenes.   (5)

  (c) Suggest how propene may be converted into poly(propene). Explain with the aid of a mechanism how propene reacts with bromine. Explain why poly(propene) does not react with bromine in the same way.   (8)

(JMB 1990)

10   **(a)** Describe, giving detailed procedures and conditions, how ethyl ethanoate may be prepared from ethanoic acid.   (4)

  (b) State, giving equations but not practical details, how carboxylic acids may be converted into
    (i)   acid anhydrides,
    (ii)   hydrocarbons,
    (iii)   acid amides.   (6)

  (c) Account for the following facts.
    (i)   Esters are much less soluble in water than carboxylic acids of similar relative molecular mass.

(ii) A carboxylic acid of high relative molecular mass is much less soluble in water than one of low relative molecular mass.

(iii) Sodium salts of carboxylic acids of high relative molecular mass are however, soluble in water. (4)

(d) (i) Describe, giving equations, how polyesters such as Terylene are made. Manufacturing details are **not** required.

(ii) State the main type of use for Terylene and explain how the structure of the polymer makes it suitable for such use. (4)

(e) Ethyl ethanoate ($CH_3COOCH_2CH_3$) and methyl propanoate ($CH_3CH_2COOCH_3$) have the same relative molecular mass and almost the same boiling point. Given an unlabelled sample of each, state how you could distinguish between them using methods available in a school laboroatory. **No** practical details are required. (2)

(WJEC 1989)

11 The anions in each of the sodium salts below can be identified by investigating the reactions of the salts with sulphuric acid. Each salt is first tested with dilute (1.0 M) sulphuric acid and then, if no gas is evolved, with concentrated acid. Describe what happens in these tests, and give balanced equations for the reactions occurring. Give a chemical test for any gas evolved if it cannot be identified by its appearance.

(a) Na_2CO_3 (5)
(b) $NaCl$ (4)
(c) NaI (6)
(d) $NaNO_2$ (5)
(e) $NaNO_3$ (5)
(f) Na_2SO_3 (5)

(JMB 1989)

12

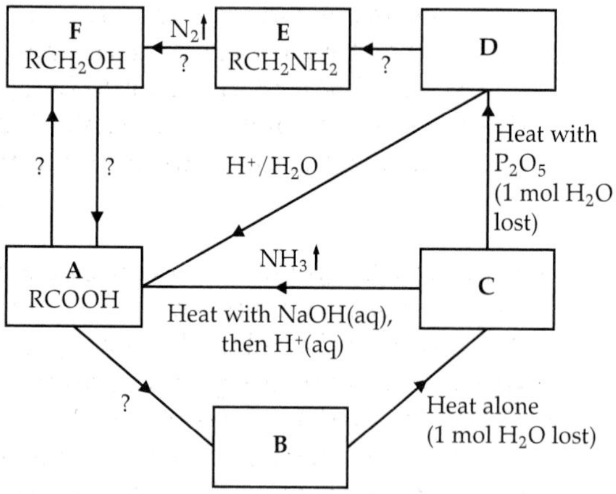

(a) Consider the reaction scheme shown above. The compounds, **A–F**, each contain only one functional group; R represents an alkyl group.

(i) Identify the functional groups present in compounds **B**, **C** and **D** respectively.

(ii) Identify the reagents whereby the conversions A → B, D → E, E → F, F → A, and **A → F** may be brought about, giving *brief* reasoning in **each** case. (10)

(b) Use **all** the quantitative information given below to identify fully the compounds **A–F** above.

(i) When 0.2500 g of **A**, a monobasic carboxylic acid, was titrated against 0.1000 mol dm^{-3} sodium hydroxide solution, 41.63 cm^3 of the latter was required for complete reaction.

(ii) When **C** is heated, as shown, with sodium hydroxide solution, ammonia is evolved in a 1:1 mole ratio to **C**. When 0.3000 g of **C** was thus treated, and the ammonia released absorbed in water, the resulting solution required 25.39 cm^3 of 0.2000 mol dm^{-3} hydrochloric acid solution for complete reaction.

(iii) When **E** reacts to form **F**, as shown above, nitrogen is evolved in a 1:1 mole ratio. When 0.2000 g of **E** was thus treated, 108.44 cm^3 of nitrogen gas, measured at 298K and 1.01×10^5 Pa, was evolved.

[$A_r(C) = 12.01$; $A_r(H) = 1.01$; $A_r(N) = 14.01$; $A_r(O) = 16.00$; 1 mol of a gas occupies 2.24×10^4 cm^3 at 273 K and 1.01×10^5 Pa.] (10)

(WJEC 1991)

13 (a) The equation below represents one of the reactions which may occur when the alkanes present in the gas oil fraction from petroleum are heated with steam.
$$C_{16}H_{34} \rightarrow C_7H_{16} + 3CH_2{=}CH_2 + CH_3CH{=}C_{H2}$$
heptane   ethene   propene

(i) Give the name of this type of process.

(ii) Explain why such processes are of economic importance. (3)

(b) If propene is passed under pressure into an inert solvent containing a Zieglar-Natta catalyst poly(propene) is produced.

(i) Write an equation for the reactoin.

(ii) What type of reaction occurs?

(iii) Explain how the catalyst influences the structure of the polymer produced.

(iv) Give **one** advantage and **one** disadvantage of poly(propene) as a packaging material compared to traditional materials. (6)

(AEB 1993)

14 (a) The atomic radii and melting points of elements in Period 3 are given below.

	Na	Mg	Al	Si	P	S	Cl
Atomic radius/nm	0.157	0.136	0.125	0.117	0.110	0.104	0.099
Melting point/°C	98	651	660	1410	44	114	−101

    (i) State and explain the trend shown by the values for the atomic radii across Period 3.

    (ii) Account for the melting points of these elements in terms of structure and bonding.

(b) Discuss, writing equations where appropriate, how the following properties vary across Period 3.

    (i) the acid-base character of the oxides

    (ii) the reaction with water of the chlorides of the elements sodium to phosphorus

                                     (JMB 1991)

15 Read the following paragraph, and answer the questions below it.

The Preparation of Chloropentaamminecobalt(III) Chloride

Dissolve 12 g of cobalt(II) chloride-6-water in $12 \, cm^3$ of hot water; cool; add to it a suspension of 25 g of NH_4Cl in $60 \, cm^3$ of concentrated aqueous ammonia. Then add $6 \, cm^3$ of hydrogen peroxide and a further $6 \, cm^3$ in small portions. When the oxidation is complete, neutralise the solution with concentrated hydrochloric acid; the aquo complex is precipitated; add a further $12 \, cm^3$ of acid. Heat the solution to boiling, cool, filter off the precipitated product.

(a) Write down the formula of the product, showing the ions.   (2)

(b) Give an ion-electron equation (half-equation) for the oxidation reaction occurring.   (1)

(c) What is meant by the term *suspension*?   (1)

(d) Explain what is being neutralised.   (2)

(e) Suggest a formula for the aquo complex ion. (1)

(f) What reaction is occurring during the final boiling?   (1)

                                 (UODLE 1990)

16 (a) Explain the term *electronegativity*.   (20)

State, and explain, how the electronegativity of the elements changes:

    (i) across a row of the Periodic Table;   (2)

    (ii) down a group of the Periodic Table.   (2)

Illustrate the use of electronegativity in predicting chemical properties.   (2)

(b) Illustrate, with the aid of a suitable example, what is meant by a hydrogen bond.   (4)

Comment on the following observations.

    (i) The boiling points of CH_4, NH_3, and H_2O are −161°C, −33°C and 100°C respectively.   (2)

    (ii) The boiling points of HF and H_2O are 20°C and 100°C respectively.   (2)

(c) State, and explain briefly, whether or not each of the following molecules possesses a dipole moment:

O_2 O_3 CO_2 H_2O (4)

                                 (UODLE 1990)

17 (a) Calculate the pH at 25°C of

    (i) aqueous hydrochloric acid of concentration $0.1 \, mol \, dm^{-3}$;

    (ii) aqueous sodium hydroxide of concentration $0.1 \, mol \, dm^{-3}$;

    (iii) aqueous ethanoic acid of concentration $0.1 \, mol \, dm^{-3}$.

(The value for K_a for ethanoic acid at 25°C is $1.8 \times 10^{-5} \, mol \, dm^{-3}$, and the value for the ionic product of water, K_w, is $1.0 \times 10^{-14} \, mol^2 \, dm^{-6}$.)   (8)

(b) Draw sketches of the curves you would expect of the change in pH of the solution as $50.0 \, cm^3$ aqueous sodium hydroxide of concentration $0.1 \, mol \, dm^3$ is added to

    (i) $25.0 \, cm^3$ of hydrochloric acid of concentration $0.1 \, mol \, dm^{-3}$;

    (ii) $25.0 \, cm^3$ of ethanoic acid of concentration $0.1 \, mol \, dm^{-3}$.   (5)

(c) The table below gives the pH ranges of some acid-base indicators.

Indicator	pH range
A	3.0 – 4.6
B	4.2 – 6.3
C	8.0 – 9.6

    (i) What properties make a compound suitable for use as an acid-base indicator?

    (ii) Explain why care must be taken when choosing an indicator for a particular titration. Illustrate your answer by choosing an indicator from the above table suitable for use in titrations involving the two acid-base reactions given in (b).

    (iii) Suggest why universal indicator is not suitable for use in acid-base titrations.   (12)

                                 (AEB 1992)

18 (a) Show how the physical properties of sodium chloride can be explained by a consideration of its structure and bonding. Illustrate your answer with a diagram of its crystal structure.   (8)

(b) The following data were obtained by recording the pressure when a sample of pentane was placed in a sealed container and then heated. Pentane was the only substance in the container.

Temperature/K	250	300	350	400	450	480	500	550	600
Pressure/atm	0.1	0.7	3.5	11.5	27.0	33.9	35.4	39.0	42.5

(i) Plot a graph of pressure against temperature on the graph paper provided.

(ii) Use the graph to estimate the normal boiling point of pentane and explain how you deduce your answer.

(iii) Explain, by reference to the kinetic theory, the shape of the curve from 250 K to 450K.

(iv) Explain the shape of the curve from 480 K to 600 K. (22)

(JMB 1991)

19 (a) State what you would observe and give the formulae of the metal-containing species involved when aqueous solutions of iron(III) nitrate and cobalt(II) nitrate react separately with
  (i) concentrated hydrochloric acid,
  (ii) aqueous sodium carbonate,
  (iii) concentrated aqueous ammonia. (18)

(b) The tests and observations in the following reaction scheme were recorded. Use this information to identify the metal ions which can be present in **A**. Give the formulae for **all** the metal-containing species involved in **B** to **I**. (12)

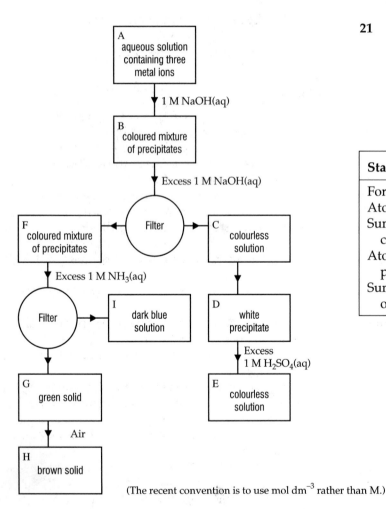

(The recent convention is to use mol dm^{-3} rather than M.)

(JMB 1991)

20 (a) Epoxyethane (CH_2—CH_2, with O bridging) is obtained commercially by the oxidation of ethene (ethyelene) with air. It reacts with hydrogen cyanide to give a compound **A**, C_3H_5NO. Treatment of A with sodium hydroxide, followed by acidification, yields **B**, $C_3H_6O_3$, which on heating gives a liquid, **C**, $C_3H_4O_2$. **C** effervesces with aqueous sodium carbonate, and it polymerises on standing.

Treatment of **A** with sulphuric acid and methanol yields **D**, $C_4H_6O_2$, which also forms a polymer. Warm aqueous sulphuric acid converts **D** into **C**.

Giving your reasoning in full, deduce the structures of the compounds **A** to **D**, and account for the ability of **C** and **D** to polymerise. (14)

(b) Suggest a reason why epoxyethane reacts with hydrogen cyanide, whereas ethoxyethane (diethyl ether) does not. **Predict** the structure of the product obtained by treating epoxyethane with (i) ammonia, and (ii) sodium ethoxide followed by aqueous acid. (6)

(OCSEB 1989)

21 (a) Set out, in the form of a diagram, the complete Born-Haber cycle for the formation of sodium chloride from metallic sodium and gaseous chlorine. Clearly show all the species involved and all the energy/enthalpy changes. (6)

(b) Use the following data to calculate the lattice energy of calcium oxide. (5)

Standard molar enthalpy change	kJ mol^{-1}
Formation of calcium oxide	-635
Atomisation (sublimation) of calcium	+193
Sum of first two ionisation energies of calcium	+1740
Atomisation (dissociation) of oxygen, per mole of atoms formed	+250
Sum of first two electron affinities of oxygen	+702

(c) Although the standard molar enthalpy change of formation of magnesium oxide is almost the same as that of calcium oxide, its lattice energy is about 300 kJ mol^{-1} greater in magnitude. Which enthalpy changes in the Born-Haber cycle for magnesium oxide will differ significantly from those in the corresponding cycle for calcium oxide? Briefly account for any differences. (5)

(d) The compound $CaCl_2$ is well-known but neither $CaCl_3$ nor $CaCl$ has yet been made as a stable compound. Calculations show that:
(i) $CaCl_3$ would have a *large positive* standard molar enthalpy change of formation;
(ii) $CaCl$ would have a *small negative* standard molar enthalpy change of formation.
Account for each of (i) and (ii) in terms of enthalpy change of atomisation, ionisation energy, electron affinity, and lattic energy. (9)

(AEB 1992)

22 (a) When discussing the **mechanisms** of organic reactions, **three** different types of reagent are considered. Give the names of these **three** main types of reagent, and say what characterises each of them. (6)

(b) For each of the following types of compound, pick from your list of three a reagent type which is commonly associated with attacking
(i) alkenes;
(ii) alkanes;
(iii) arenes (aromatic hydrocarbons);
(iv) halogenoalkanes (alkyl halides). (4)

(c) Give, in the form of a balanced equation, a specific example of a reaction between a halogenoalkane and the type of reagent which you have selected in **(b)** as commonly attacking halogenoalkanes. (2)

(d) Contrast the reactions of chlorine with all four types of compound [(i) to (iv) in **(b)**], both in terms of reaction mechanism and in terms of products. (8)

(OCSEB 1989)

23 (a) Explain what is meant by the terms **structural isomerism** and **stereoisomerism**. Illustrate your answer by discussing the isomerism shown by the following examples:
(i) compounds of formula $C_3H_6O_2$ which belong to the same homologous series;
(ii) compounds of formula C_3H_6O which belong to different homologous series;
(iii) carboxylic acids of formula $C_3H_6O_3$;
(iv) ions of formula $[Co(NH_3)_4 Cl_2]^+$. (20)

(b) Give **one** chemical test that could be used to distinguish between the isomers given for **(a)** (ii). (5)

(AEB 1989)

24 (a) An alcohol has the molecular formula C_4H_7OH. Oxidation of this alcohol gives a ketone. Ozonolysis of this alcohol gives methanal as one of the products.
Giving your reasoning in full, deduce the structural formula for this alcohol and predict whether it will be optically active. (8)

(b) Outline the stages by which ethanal may be converted into 2-oxopropanoic acid (pyruvic acid) CH_3COCO_2H. Full practical details are not required, although you should specify reagents and any important conditions. (6)

(c) A gaseous hydrocarbon **X** can exist as geometric isomers. 30 cm^3 of gaseous **X** were exploded with 200 cm^3 of oxygen. The residual gases were cooled to 25°C and found to occupy 140 cm^3. After treatment with aqueous sodium hydroxide, the final volume of gas remaining was 20 cm^3. Assuming that all volumes were measured at 25°C and the same pressure, deduce the molecular and structural formulae of hydrocarbon **X**. (6)

25 (a) Distinguish between *order* and *molecularity* of a reaction. (3)

(b) For the gas phase isomerisation reaction
$$A \rightarrow B$$
the following data were obtained at 700 K.

Time/hour	0	2	5	10	20	30
Fraction of A unreacted	1.0	0.91	0.79	0.63	0.40	0.25

(i) Plot an appropriate graph to show that this is a first-order reaction. (6)
(ii) Use your graph to find the *half-life* and *first-order rate constant* for the reaction. (4)
(iii) How would trebling the initial concentration of A affect (1) the half-life, (2) the initial rate of this reaction? (2)

(c) Discuss briefly the factors which are responsible for the increase of a reaction rate with temperature and indicate their relative importance. (5)

(UODLE 1991)

26 The element strontium (*atomic number:* 38) has four *isotopes* of *mass numbers* 84, 86, 87, 88 having abundances of 0.5%, 9.9%, 7.0% and 82.6%, respectively.
(a) Explain the meaning of the terms in italics. (3)
(b) Describe the principles underlying the use of the mass spectrometer. (8)
(c) Define *relative atomic mass*. Use the information given above
(i) to draw a labelled sketch of the mass spectrum trace for strontium;
(ii) to calculate its relative atomic mass. (8)

(d) Give the electronic structure of an atom of strontium. Draw a sketch of the graph you would expect for the logarithm of the first eight successive molar ionisation energies against the number of electrons removed and briefly explain its shape. (6)

(AEB 1989)

27 (a) Discuss changes in two physical and two chemical properties of the elements of Group VII, fluorine to iodine, (8)

(b) Explain the following observations.
  (i) Sulphur hexafluoride exists, and is stable, but sulphur hexachloride is unknown. (3)
  (ii) Hydrogen chloride can be prepared by the action of concentrated sulphuric acid on an ionic chloride, but similar methods cannot be used to prepare pure hydrogen bromide. (3)

(c) In the presence of sodium hydrogencarbonate, sodium sulphite is oxidised quantitatively by iodine.
A sample of 25.0 cm^3 of a solution containing 5.00 g dm^{-3} of impure Na_2SO_3 required 22.5 cm^3 of $0.042 \text{ mol dm}^{-3}$ I_2 solution for complete reaction. Calculate the purity of the Na_2SO_3. (6)

(UODLE 1990)

28 A naturally occurring compound **A**, $CH_3CH(OH)COOH$, exhibits optical activity. However when compound **A** is prepared from ethanal in the laboratory, the product is found to be optically inactive.

(a) (i) Explain what is mean by 'optical activity' and why compound **A** shows this property.
  (ii) Give reagents, conditions and equations to show how **A** can be made from ethanal. By considering the mechanism of the initial step, explain whey the sample of **A** made this way is optically inactive. (14)

(b) Write a mechanism for the reaction of **A** with ethanoyl chloride. Draw structures for the organic compounds formed from the reactions of **A** with
  (i) sodium hydroxide,
  (ii) acidified potassium dichromate(VI). (7)

(c) A second optically active compound **B**, which has the molecular formula $C_3H_7NO_2$, reacts with nitrous acid to form **A**. Deduce the structure of **B** and write a balanced equation for its reaction with nitrous acid to form **A**. Explain why **B** has a much higher melting point than **A**. (9)

(JMB 1989)

29 (a) Indicate, by stating the necessary reagents and observations, how you would distinguish between sodium bromides and sodium iodide using a chemical test. (3)

(b) A number of oxoanions of chlorine are known; examples include ClO^-, ClO_3^- and ClO_4^-. (i) What is the oxidation number of chlorine in
ClO_3^-
ClO^-
  (ii) ClO^- is formed when chlorine reacts with aqueous alkali. Write an ionic equation for this reaction.
  (iii) When $KClO_3$ is heated just above its melting point it forms KCl and $KClO_4$. Write an equation for this reaction. (4)

(c) (i) What is meant by the term *electron affinity*?
  (ii) How is the electron affinity of fluorine anomalous?
  (iii) Suggest an explanation for this anomaly. (3)

(ULSEB 1991)

30 (a) α-Alanine is an amino acid of formula
$CH_3CHCOOH$.
$\quad\quad |$
$\quad NH_2$
  (i) Give the systematic name of α-alanine and write equations for the reactions it has separately with ethanol and with concentrated hydrochloric acid.
  (ii) Draw the structural formula of the product formed when two α-alanine molecules react together. Use this to illustrate a peptide link and briefly explain the importance of such links. (6)

(b) α-Alanine can act as a bidentate *ligand* in the formation of *complex ions*. Explain the words in italics and from your knowledge of other ligands identify the potential donor atoms which α-alanine could use when acting as a ligand. (6)

(c) $CrCl_3(H_2O)_6$, represents the formula of three possible isomers. 1 mole of each in aqueous solution, when treated with excess aqueous silver nitrate, gave 1, 2, and 3 moles of silver chloride precipitate respectively.
  (i) Draw the structures of the complex ions present in the three isomers.
  (ii) Explain why the isomers have different colours.
  (iii) Explain why different amounts of AgCl precipitate are obtained. (13)

(AEB 1992)

31 In the stratosphere 30 km above the Earth's surface, ozone is being made continuously by the following reaction:

$$O + O_2 \rightarrow O_3; \quad \text{rate} = k[O][O_2].$$

The free oxygen atoms arise from the splitting of oxygen molecules by ultraviolet light from the Sun:

$$O_2 \rightarrow 2O.$$

(a) Calculate the rate at which ozone forms, given the following values:

$k = 3.9 \times 10^{-5}$ dm^3 mol^{-1}s^{-1},
$[O] = 3 \times 10^{-14}$ mol dm^{-3},
$[O_2] = 1.3 \times 10^{-4}$ mol dm^{-3}. (2)

(b) The temperature of the stratosphere is –50°C.
  (i) State a typical value of air temperature at sea level.
  (ii) Calculate the molar concentration of oxygen molecules (in mol dm^{-3}).
  [One mole of gas occupies 24 dm^3 under the conditions at sea level.]
  (iii) By using the collision theory of reaction kinetics and your answers to **(b)** (i) and **(b)** (ii), discuss qualitatively how the rate of ozone formation at sea level would compare with that in the stratosphere.

  In fact, practically no ozone is formed in the lower atmosphere. Suggest a reason for this. (6)

(c) It is assumed that the concentration of ozone in the stratosphere has remained roughly constant for many thousands of years but there is now some evidence that chlorofluorocarbons, (CFCs – used as refrigerants and aerosol propellants) are causing the ozone concentration to decrease. It is not thought that their presence affects the rate of ozone formation, however.
What does this tell you about the role of CFCs in the other reactions involving ozone that must be occurring in the stratosphere? (2)
(UCLES 1990)

32 (a) (i) What are the main factors responsible for the observed trends in the physical and chemical properties of the Group 2 elements?
  (ii) Give reasons for the atypical properties of beryllium. (6)

(b) In the light of the trends demonstrated in the sequence magnesium - calcium - strontium - barium, predict the following behaviour of radium and its compounds. Give brief reasons for your prediction.
  (i) The reaction of radium with water.
  (ii) The solubility in water of radium sulphate.
  (iii) The solubility in water of radium hydroxide.
  (iv) The type of bonding in radium compounds, and the typical physical properties of these compounds.
  (v) A method for obtaining radium from its compounds. (12)

(c) (i) Part of the decay series of naturally-occurring thorium is as follows. The half-lives are shown under the arrows.

$$^{228}\text{Ra} \xrightarrow[\text{6.67 years}]{\beta} W \xrightarrow[\text{6.13 hours}]{\beta} X \xrightarrow[\text{1.91 years}]{\alpha} Y \xrightarrow[\text{3.64 days}]{\alpha} Z$$

Use the Periodic Table to identify **W, X, Y** and **Z**, and give the correct mass number and atomic number for each.
  (ii) Why is the decay of a radioisotope described as a *first-order* process?
  (iii) A sample of ^{228}Ra with an initial mass of 100 mg is left in a safe container. What mass of ^{228}Ra would remain after 20 years? (7)
(AEB 1992)

33 Fore each of the following reactions of carbonyl compounds
(a) draw the full structural formula(e) of the organic compound(s) produced,
(b) comment on the use to which the reaction is put, either in identifying, synthesising or distinguishing between compounds.

  (i) $CH_3CH_2COCH_3 \xrightarrow{I_2/NaOH}$ (3)

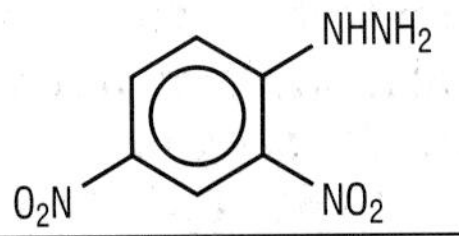

  (ii) $C_6H_5CHO \xrightarrow{} $ (3)

  (iii) $CH_3CHO \xrightarrow[\text{Tollens' reagent}]{\text{Fehling's } or}$ (2)

  (iv) $C_6H_5COCH_3 \xrightarrow{HCN}$ (2)
(UCLES 1990)

34 40 cm³ of aqueous sodium hydroxide of concentration 0.50 mol dm⁻³ were placed in a polystyrene cup and, using a burette, 40 cm³ of aqueous sulphuric acid were added in 5.0 cm³ portions. The mixture was stirred with a thermometer after each addition and the temperature recorded. The results were plotted as shown in the graph below.

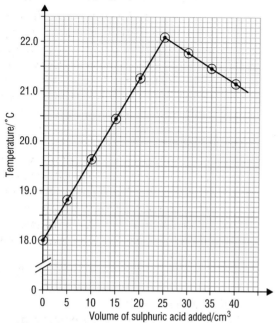

(a) Define *enthalpy change of neutralisation*. (2)

(b) How do you account for the shape of the graph? (2)

(c) (i) What is the maximum temperature rise observed?

(ii) What volume of sulphuric acid exactly neutralised the 40 cm³ of aqueous sodium hydroxide?

(iii) How many moles of sodium hydroxide were neutralised?

(iv) Calculate a value for the enthalpy of neutralisation of sodium hydroxide with sulphuric acid. (Assume that the specific heat capacity of the solutions is 4.2 J K⁻¹ g⁻¹, that the heat capacity of the cup is zero and that the density of the solutions is 1 g cm⁻³.) (6)

(d) Calculate the concentration in mol dm⁻³ of the sulphuric acid used. (3)

(e) (i) Define pH.

(ii) Calculate the pH of the sulphuric acid used in this experiment.

(iii) Given that $K_w = 1 \times 10^{-14}$ mol² dm⁻⁶ at the temperature of the experiment, calculate the pH of the aqueous sodium hydroxide. (5)

(AEB 1993)

35 This question is concerned with the extraction of caffeine from tea leaves. Tea leaves contain between 3% and 5% by mass of caffeine. The caffeine can be extracted initially with hot water, in which it is fairly soluble (18 g/100 g water at 80°C; 2.2 g/100 g water at 20°C). Coloured impurities such as tannic acids can be removed as calcium salts by adding calcium carbonate. After filtering, the caffeine in the filtrate is extracted by shaking with a number of successive portions of dichloromethane to produce a caffeine solution in dichloromethane. Most of the solvent is distilled off to produce a concentrated solution of caffeine in the organic solvent, and this solution is then evaporated over a water bath to yield the crude product.

The product is purified by dissolving it in hot methylbenzene (10 cm³ is correct for about 30 g of tea as starting material), adding 15–20 cm³ of hexane, filtering hot and allowing the filtrate to cool. The crystals are often greenish in colour.

(a) Using the basic description set out above, describe the extraction of caffeine from tea leaves in sufficient detail to allow an A-level student to carry out the experiment. Begin with 30 g or so of tea leaves. Safety considerations should be stressed where appropriate. (15)

(b) What name is given to the process where methylbenzene and hexane are used? Explain briefly how this process removes both soluble and insoluble impurities. (5)

(c) When a substance X, dissolved in a particular solvent, is shaken up with another solvent which is immiscible with the first, an equilibrium of the form

$$[X]_{solvent\ 1} \rightleftharpoons [X]_{solvent\ 2}$$

is established. Use the equilibrium law to derive a relationship between the concentration of **X** in solvent 1 and solvent 2 when equilibrium is attained.

Starting with 100 cm³ of an aqueous solution containing 10 g of **X**, and given that the equilibrium constant is 1, show that it is more efficient to extract the caffeine, **X**, with two portions of dichloromethane of volume 50 cm³ rather than with one portion of volume 100 cm³. (Note that one 100 cm³ portoin will distribute the 10 g as 5 g into each solvent.) (5)

(d) (i) Tea leaves contain many other organic compounds. Explain why this method is suitable for the isolation of the one specific compound.

(ii) Explain how the calcium carbonate allows the removal of tannic acids. (5)

(ULSEB 1991)

36 (a) Ethylamine is a base. Explain why this is so, illustrating your answer with one example of its reactions.
Put the compounds ethylamine, ethanamide and phenylamine (aniline) into order of increasing basicity (basic strength). (5)

(b) Give the mechanism for the reaction between dimethylamine and iodomethane. What is the name of this **type** of mechanism? If equimolar quantities of dimethylamine and iodomethane are allowed to react, what nitrogen-containing **products** will be present in the final mixture (6)

(c) Explain
(i) what is meant by acylation of amines,
and (ii) why a tertiary amine cannot be acylated.
Give an equation for the reaction between benzoyl chloride (C_6H_5COCl) and phenylamine ($C_6H_5NH_2$). (6)

(d) An amide may be hydrolysed by warming it with either aqueous acid or alkali. State the products formed in each case. (3)
(OCSEB 1990)

37 (a) By reference to the F_2 molecule, explain the meaning of the term *bond energy*. (1)
Sulphur hexafluoride can be made by reacting sulphur tetrafluoride with fluorine in the gas phase:

$$SF_4(g) + F_2(g) \rightarrow SF_6(g); \Delta H = -434 \text{ kJ mol}^{-1}$$

By considering the bonds broken and bonds formed during this reaction, calculate an average value for the S–F bond energy. State any assumptions you have made. (3)

(b) (i) Construct dot-and-cross diagrams to illustrate the bonding in the molecules of sulphur difluoride, SF_2, and sulphur hexafluoride, and predict their shapes. (3)
(ii) Caesium fluoride has a similar formula mass to sulphur hexafluoride. Describe three major differences you would expect to find in the physical properties of the two compounds. (3)
(UCLES 1992)

38 (a) Consider the following oxides:
Al_2O_3, CaO, CO, MgO, Na_2O, P_4O_{10}, SO_2, SiO_2.
State which of these oxides are:
(i) **basic** and give **two** further examples,
(ii) **acidic** and give **two** further examples,
(iii) **neutral** and give **one** further example,
(iv) **amphoteric** and give **one** further example. (5)

(b) (i) Describe, *in outline*, tests which you would carry out to demonstrate whether a given oxide was basic, acidic or neutral.

(ii) Describe tests which you would carry out to show how amphoteric character is exhibited chemically, *stating clearly the experimental observations expected*. Write equations for the reactions which occur. (5)

(c) Consider the following chlorides:
CCl_4, HCl, $MgCl_2$, NaCl, PCl_5, $PbCl_2$, $PbCl_4$, $SiCl_4$.
(i) State with **brief** explanation, which of the above chlorides contain dominantly (1) ionic bonding, and (2) covalent bonding. For compounds which you describe as ionic, state whether this is for the anhydrous compound or for aqueous solution or both.
(ii) Explain why you may be unable to assign clearly certain of these chlorides to either category.
(iii) Discuss with particular (but not necessarily exclusive) reference to the above listed chlorides, the characteristic differences in behaviour (e.g. physical properties, interaction with water, electrolytic properties) between ionic and covalent compounds. (10)
(WJEC 1990)

39 (a) Define the term *mole*.
Why is the phrase "the mass of one mole of oxygen" ambiguous? (2)

(b) Briefly describe how a value for the Faraday constant can be determined experimentally. Assuming the value for the charge on the electron, how can the Avogadro constant then be obtained? (4)

(c) A meteorological balloon of 2 m diameter has a volume of 4.19 m^3 or 4190 dm^3. It floats since it is given an upthrust equal to the mass of air it displaces.
Calculate (i) the mass of hydrogen in the balloon, (ii) the mass of air it displaces, (iii) the load the balloon can carry for it just to lift off from the ground. (4)
[Average M_r of air = 29. Under the conditions of inflation, 1 mol of hydrogen occupies a volume of 23 dm^3.]
(UCLES 1990)

40 (a) Outline the principles behind the use of each of the following methods of purification:
(i) solvent extraction,
(ii) steam distillation,
(iii) fractional distillation,
(iv) ion exchange. (12)

(b) Describe briefly one example of the use of each method in the school laboratory. (8)
(OCSEB 1989)

41 (a) Describe the components in any one fuel cell and the reactions taking place at the electrodes. Suggest one advantage of fuel cells as sources of energy. (5)

(b) Hydrogen and oxygen can be used in a fuel cell. By choosing suitable electrode reactions from the *Database*, calculate the e.m.f. of a cell consisting of an oxygen electrode and a hydrogen electrode, each under standard acidic conditions. (2)

(c) One type of rechargeable battery makes use of the nickel-cadmium cell, in which nickel and cadmium electrodes, coated with their respective hydroxides, are immersed in potassium hydroxide solution. In normal use, the cadmium electrode is the one that releases electrons to the external circuit.
The relevant electrode reactions are:

$Cd(OH)_2(s) + 2e^- \rightleftharpoons Cd(s) + 2OH^-(aq)$,
$NiO(OH)(s) + H_2O(l) + e^- \rightleftharpoons$
$Ni(OH)_2(s) + OH^-(aq)$.

Draw a cell diagram for this arrangement, showing the polarity of the electrodes, and construct a balanced equation for the reaction that occurs during discharge. (3)
(UCLES 1990)

42 (a) What are the main factors which account for the variation in character of the elements across the period sodium to argon? (6)

(b) Draw, or describe, the structure and give **one** important chemical property of a hydride which is formed by each of the elements Na, Si, P and Cl.

(c) Compare the structure, type of chemical bonding, and properties of NaCl with those which you expect the compound AlP to have. Explain your reasoning. (6)
(OCSEB 1989)

43 Hexane and heptane have the following properties

	Relative molar mass	Boiling point /°C	Vapour pressure at 25°C /Pa
hexane	86.2	68.7	20180
heptane	100.2	98.1	6081

(a) State Raoult's law as it applies to hexane in an ideal mixture of hexane and heptane. (2)

(b) Suggest an experiment which you could use to investigate the variation in boiling point with liquid composition for a mixture of two miscible, flammable liquids such as hexane and heptane. You should give a digram of the apparatus and the quantities of liquids you would use. State how the mixtures would be made up, give a careful description of how accurate measurements would be taken, and give any safety procedures involved. (6)

(c) In a liquid mixture containing 23.0 g hexane and 42.0 g of heptane at 25°C calculate
(i) the mole fraction of hexane in the liquid. (2)
(ii) the partial vapour pressure of each component in the vapour, (2)
(iii) the total vapour pressure above the mixture, (1)
(iv) the mole fraction of hexane in the vapour, (2)
(v) the difference between your answers in (i) and (iv) and state what technique depends on this difference. (2)

(d) Mixtures of ethoxyethane, $(C_2H_5)_2O$ and trichloromethane, $CHCl_3$, do not obey Raoult's law; they have higher boiling points than expected. Explain this observation in terms of the intermolecular forces present in the pure liquids and in the mixture. (3)
(OCSEB 1991)

44 Some properties of four compounds are given in the Table:

Property	Compounds			
	NaCl sodium chloride	SiO_2 silicon dioxide	CH_4 methane	C_2H_5OH ethanol
melting point/K	1074	2043	91	156
solubility in water	good	insoluble	poor	very good
electrical conductivity of liquid	good	poor	poor	poor

For **each** of these 4 compounds explain these properties in terms of both the bonding and the structure involved. (20)
(OCSEB 1991)

45 (a) Suggest reasons why the strength of a nylon rope is severely impaired by prolonged contact with aqueous sodium hydroxide, whereas the strength of a poly(propene) rope is unaffected. (4)

(b) Explain why white wine changes the colour of acidified potassium dichromate(VI) on warming but has no effect on Fehling's (or Tollens') reagent. Some other wines, such as sherry, give positive results with all of these reagents. Suggest an explanation. (6)
(UCLES 1991)

46 (a) State the partition law for the distribution of a solute between two immisible solvents. (2)

(b) The following experimental results were obtained for a number of mixtures of ammonia distributed between water and 1,1,1-trichloroethane.

Mixture	Total ammonia concentration	
	in water / mol dm^{-3}	in 1,1,1-trichloroethane / mol dm^{-3}
1	3.89	0.0335
2	3.24	0.0284
3	2.61	0.0221

(i) Describe how such results could be obtained. You should include essential steps in the method, the approximate quantities or concentrations of any substances used and the safety procedures involved. (8)

(ii) Calculate the ratio of the ammonia concentrations in the two solvents for each of the mixtures. Comment on the values obtained. (4)

(c) Explain how partition is involved in chromatographic separation techniques. Illustrate your answer by reference to one particular technique giving one example of a mixture that could be separated by this technique. (6)

(OCSEB 1990)

47 A certain industrial cleaner and paint solvent was distilled to produce a single compound **D**. When **D** reacted with 2,4-dinitrophenylhydrazine, an orange precipitate was produced. With alkaline aqueous iodine, **D** gave a pale yellow precipitate. **D** did *not* react either with warm acidified potassium dichromate(VI) or with aqueous bromine. Reduction of **D** with hydrogen over a catalyst produced an equimolecular mixture of two isomers, **E** and **F**, with the molecular formula $C_4H_{10}O$.
Suggest structural formulae for **D**, **E** and **F** and explain the reactions involved. (10)

(UCLES 1991)

48 (a) Describe and explain the variation in volatility of the elements in Group VII. (3)

(b) How does a study of bond energies help to explain the trends observed in the relative thermal stabilities of the hydrides of the Group VII elements? (4)

(c) The carbon monoxide in a sample of polluted air can readily be determined by passing it over solid iodine(V) oxide, I_2O_5, to give carbon dioxide and iodine.

(i) Write a balanced equation for the reaction between carbon monoxide and iodine(V) oxide. (1)

The iodine produced is removed and titrated with aqueous sodium thiosulphate:

$$2Na_2S_2O_3 + I_2 \rightarrow Na_2S_4O_6 + 2NaI.$$

A 1.0 dm3 sample of air produced iodine that required 20.0 cm^3 of 0.10 mol dm^{-3}

sodium thiosulphate to discharge the iodine colour.

(ii) Calculate the mass of carbon monoxide in this sample of polluted air. (2)

(UCLES 1992)

49 Three methods of production of ethanol are as follows.

Method 1 Ethene is passed into 64% sulphuric acid and ethanol is formed when the resulting solution is diluted with water and warmed. This method is used in the laboratory and has also been used on an industrial scale.

Method 2 A variety of foodstuffs and other materials can be fermented to give ethanol on a medium or large industrial scale.

Method 3 Ethanol is produced on a large scale by the reaction of ethene and steam in a catalytic reactor.

$$C_2H_4(g) + H_2O(g) \rightleftharpoons C_2H_5OH(g) \quad \Delta H^{\ominus} = -46 \text{ kJ mol}^{-1}$$

An inert supporting material is impregnated with the phosphoric(V) acid catalyst. Phosphoric(V) acid is a strong, involatile acid. The reactor operates at 300°C and 70 atm pressure. Unreacted ethene is separated from the product and recycled.

(a) Write balanced chemical equations for *Method 1*. (3)

(b) An employee is considering entering a recently emptied fermentation tank in order to clean it after use in *Method 2*. What safety hazard(s) and precautions would be involved? (4)

(c) Considering the equilibrium involved in *Method 3*, explain with reasons the choice of the reactor temperature and pressure used. (8)

(d) (i) Why is phosphoric(V) acid chosen as a suitable catalyst for *Method 3*?

(ii) Suggest the mechanism of the reaction between ethene, steam and the catalyst to produce ethanol in *Method 3*.

(iii) The phosphoric(V) acid catalyst in *Method 3* is used and re-used in the continuous process. Why would this be so much less easy for the sulphuric acid in *Method 1*? (7)

(e) What advantages does *Method 3* offer over *Method 2* as an industrial process? (6)

(f) If 8% of the ethene feedstock forms by-products in *Method 3*, what mass of ethanol is produced per tonne of ethene? (4)

(g) The production of a product of 100% purity is not always the goal of an industrial process. Illustrate this statement, referring to the production of ethanol by *Method 2*. (3)

(JMB 1990)

50 **(a)** Describe (by drawing suitable dot-and-cross diagrams) the type of bonding in
   (i) solid sodium chloride;
   (ii) gaseous hydrogen chloride;
   (iii) the addition compound formed between boron trichloride, BCl3, and ammonia. (4)

(b) Write equations to show the processes that occur when
   (i) sodium chloride and (ii) hydrogen chloride are dissoved in water. (2)

(c) Explain why aqueous hydrogen chloride is strongly acidic, aqueous sodium chloride is neutral and aqueous ammonium chloride is slightly acidic. (6)
   (AEB 1991)

51 By carrying out appropriate calculations, predict what will happen at each stage of the following experiment and describe what you expect to see.
(a) 1 g of calcium hydroxide was mixed with 100 cm^3 of water at room temperature.
(b) 10 cm^3 of 2 mol dm^{-3} sulphuric acid was then added to the mixture.
(c) 5 drops of a full range universal indicator was then added.
Describe an experimental procedure you could use on your resulting mixture to obtain a value for the solubility of calcium sulphate in water.
   (ULSEB 1991)

52 **(a)** Describe the reaction of phenylamine with
   (i) dilute hydrochloric acid,
   (ii) aqueous bromine,
   (iii) ethanoyl chloride.

   For each reaction, write a balanced equation and draw the full structural formula of the product. (6)
(b) Sunset Yellow is used as a food colourant (code number E110) in sweets, orange squash and jams. It is produced by the following reactions:

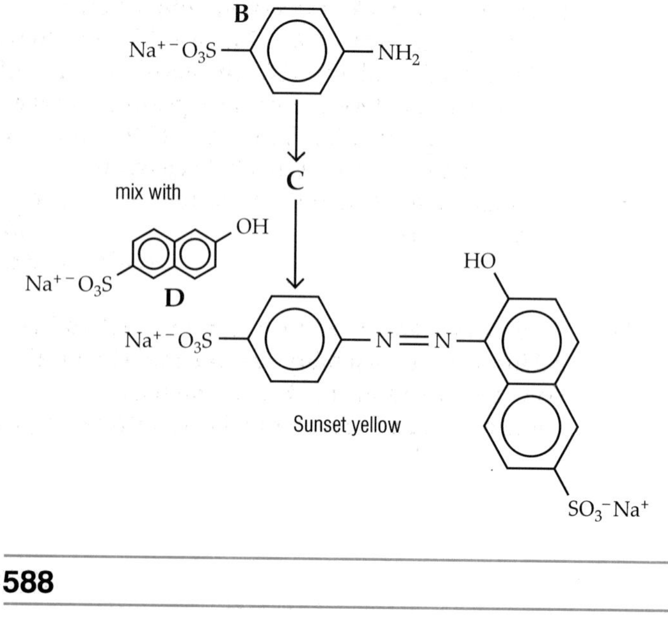

(i) What type of compound is Sunset Yellow?
(ii) What reagent must compound **B** be treated with and under what conditions to make compound **C**, before **C** is added to **D** to produce Sunset Yellow?
(iii) The sodium sulphate ($SO_3^-Na^+$) groups have little effect on the colour of Sunset Yellow. Suggest a reason why these groups are needed in the molecule. (4)
   (UCLES 1990)

53 Identify, as fully as possible, the substances **W**, **X**, **Y** and **Z**. Explain each of the reactions described and give equations where possible.
(a) Substance **W** is a white powder which dissolves in water to form a blue solution. When concentrated hydrochloric acid is added to aqueous **W** the solution turns green but the blue colour returns when the solution is diluted with water. A white precipitate forms when aqueous barium chloride is added to aqueous **W**. The precipitate is insoluble in dilute hydrochloric acid. (6)
(b) Substance **X** is a yellow solid which dissolves in water to give a yellow solution. On the addition of dilute sulphuric acid the solution turns orange. When the acidified aqueous **X** is warmed with ethanol the solutions turns green. A flame test carried out on solid **X** produced a lilac flame. (6)
(c) Substance **Y** is a pale pink solid which dissolves in water to give an almost colourless solution. A white precipitate forms when aqueous sodium hydroxide is added but this precipitate turns brown on standing in an open test tube. When aqueous silver nitrate is added to aqueous **Y** a white precipitate forms which is insoluble in dilute nitric acid. (6)
(d) Substance **Z** is a white solid which dissolves in water to form a colourless solution. When aqueous ammonia is added a white precipitate forms which redissolves in excess of the reagent. When dilute sulphuric acid is added to **Z** and the mixture is warmed, a vinegar like smell is observed and the vapour turns blue litmus red. (7)
   (AEB 1990)

54 **(a)** Explain the following terms
   (i) Order of reaction;
   (ii) rate constant;
   (iii) half-life of a reaction. (6)
(b) The half-life, $t_{\frac{1}{2}}$, of a reaction may be related to the order, n, with respect to a particular reactant by:
$$t_{\frac{1}{2}} = ka_0^{(1-n)}$$
where k is a constant for the reaction and a_0 is

the initial concentration of the reactant. For a first order reaction, show that $t_{\frac{1}{2}}$ is constant and sketch a graph to show how the concentration of the reactant changes with time. (4)

(c) State and explain how the rate constants for the forward and reverse reactions and the equilibrium constant for the reaction

$$H_2(g) + I_2(g) \rightleftharpoons 2HI(g) \quad \Delta H \text{ positive}$$

would change if at all.
(i) in the presence of a catalyst,
(ii) with an increase of temperature. (6)

(d) Explain how a small increase in temperature can cause a large increase in the rate of a chemical reaction. (4)

(ULSEB 1991)

55 (a) Describe the reactions that occur when chlorine is bubbled through (i) cold, (ii) hot, aqueous sodium hydroxide.
Write equations and explain how the oxidation number of chlorine changes during these reactions. What observation would make it clear that the chlorine had reacted? (5)

(b) When chlorine is bubbled through a solution of iodine in hot aqueous sodium hydroxide, the two halogens react in the $Cl_2 : I_2$ ratio of 7 : 1, forming a white precipitate **A** and a solution of sodium chloride. **A** has the following composition by mass:

Na, 16.9%;
H, 1.1%;
I, 46.7%;
O, 35.3%.

Calculate the empirical formula of **A** and thus deduce the balanced equation for the reaction. What is the oxidation number of iodine in **A**? (5)

(UCLES 1990)

56 (a) What is meant by the terms *activation energy* and *catalyst*?
Substances **A** and **B** react together exothermically to give **C** and **D**. Draw sketches of the reaction profile for this reaction when it takes place both with and without a catalyst. (7)

(b) Explain what is meant by the term *heterogeneous catalysis*. Write an equation for an industrial process which involves a heterogeneous catalyst and give the name of the catalyst used. (5)

(c) Peroxodisulphate(VI) ions are capable of oxidising iodide ions to iodine according to the equation
$$S_2O_8^{2-}(aq) + 2I^-(aq) \rightarrow 2SO_4^{2-}(aq) + I_2(aq).$$
The reaction can be catalysed by some d-block metal ions. Describe in outline an experiment to investigate the catalytic effect of $Fe^{2+}(aq)$ and $Zn^{2+}(aq)$ on the reaction. (7)

(d) $Fe^{2+}(aq)$ ions are found to catalyse the reaction. Given the standard reduction potentials below, suggest, with relevant equations, a possible explanation for this.

Electrode reaction	$E^{\ominus}/V$
$Fe^{3+}(aq) + e \rightarrow Fe^{2+}(aq)$	+0.77
$I_2(aq) + 2e \rightarrow 2I(aq)$	+0.54
$S_2O_8^{2-}(aq) + 2e \rightarrow 2SO_4^{2-}(aq)$	+2.01

Why is it unlikely that the reaction would be catalysed by $Zn^{2+}(aq)$? (6)

(AEB 1991)

57 Describe what happens in each of the following, writing balanced equations, naming products and including reaction mechanisms wherever possible.
(a) Ethyl ethanoate is gently refluxed with aqueous sodium hydroxide. (4)
(b) Ethene is bubbled through a solution of bromine in hexane. (6)
(c) Propan-1-ol is warmed with acidified potassium dichromate(VI). (5)
(d) A potential difference is applied to aqueous aminoethanoic acid
(i) at pH 4 (ii) at pH 10.
Products need not be named in this part. (5)

(ULSEB 1991)

58 When chlorine is bubbled through a concentrated aqueous solution of ammonium chloride, a yellow oily liquid, nitrogen trichloride, NCl_3, is formed, together with a solution of hydrochloric acid. Nitrogen trichloride is hydrolysed by aqueous sodium hydroxide, producing ammonia gas and a solution of sodium chlorate(I).
(a) Write balanced equations for the formation and hydrolysis of nitrogen trichloride. (2)
(b) Draw the shape of the nitrogen trichloride molecule. (2)
(c) Apart from peaks associated with solitary nitrogen atoms (at $m/e = 14$) and chlorine atoms (at $m/e = 35$ and $m/e = 37$), the mass spectrum of nitrogen trichloride contains 9 peaks arranged in 3 groups, ranging from $m/e = 49$ to $m/e = 125$. Predict the m/e values of all 9 peaks, and suggest a formula for the species responsible for each one. (6)

(UCLES 1990)

59 Describe how the major factors, which determine and influence the viability of industrial chemical processes, are applied to the commercial production of EITHER ammonia OR nitric acid. (20)

60 Sulphuric acid is manufactured by the catalytic oxidation of sulphur dioxide with purified air over a vanadium(V) oxide catalyst in a four-stage process at 500 °C and atmospheric pressure to give sulphur trioxide. The sulphure trioxide is then absorbed in 98% sulphuric acid and the 98.5% acid is then diluted with water to give the commercial 98% concentrated acid. The sulphur used is imported from Europe or America.

The reaction between sulphur dioxide and oxygen is an equilibrium:

$$2SO_2(g) + O_2(g) \rightleftharpoons 2SO_3(g)$$
$$\Delta H = -94..6 \text{ kJ mol}^{-1} \text{ of } SO_3$$

(a) Describe and explain the effect on the yield of SO_3 of
  (i) increasing the pressure;
  (ii) raising the temperature.
  Comment on the actual operating conditions in the light of your answers. (6)
(b) Discuss and explain the effect of the catalyst on
  (i) the yield of the reaction;
  (ii) the rate of attainment of equilibrium. (4)
(c) Describe two likely environmental consequences of a substantial leakage of sulphur dioxide from the plant. (2)
(d) About one-third of the world production of sulphuric acid is used in the manufacture of chemicals for use in agriculture.
  (i) Name three types of chemical widely used in modern agriculture. (3)
  (ii) Outline the environmental impact of any two of these chemicals. (4)
(e) Describe and give equations for reactions which show that sulphuric acid behaves as
  (i) an oxidising agent;
  (ii) a dehydrating agent. (6)
  (AEB 1990)

61 (a) Use the following substances to illustrate the relationship between physical properties and bonding: iron, argon, ammonia, sodium chloride, silicon(IV) oxide and poly(chloroethene) (PVC). (18)
(b) Summarise the main ideas of the electron pair repulsion theory. Deduce the shapes of the following: PF_3, CS_2, XeF_4 and PBr_4^-, giving your reasons. (12)
  (ULSEB 1991)

62 (a) (i) State the conditions under which magnesium and calcium will react with water, and write balanced equations for the reactions. (5)
  (ii) Explain any differences between the two reactions in terms of the atomic properties of the two metals. (3)
(b) Compare the chemistries of magnesium and calcium with reference to the following:
  (i) the solubilities of their sulphates in water;

(ii) the thermal stabilities of their carbonates;
(iii) the reaction of their oxides with water. (9)
(c) A mineral, which can be represented by the formula $Mg_xBa_y(CO_3)_z$, was analysed as described below.
From the results, calculate the formula of the mineral.
A sample of the mineral was dissolved in excess hydrochloric acid and the solution made up to 100 cm³ with water. During the process 48 cm³ of carbon dioxide, measured at 25°C and 1 atmosphere pressure, were evolved.
A 25.0 cm³ portion of the resulting solution required 25.0 cm³ of EDTA solution of concentration 0.02 mol dm⁻³ to reach an end-point. A further 25.0 cm³ portion gave a precipitate of barium sulphate of mass 0.058 g on treatment with excess dilute sulphuric acid. You may assume that Group 2 metal ions form 1 : 1 complexes with EDTA. (8)
(Molar volume of any gas at 25°C and 1 atmosphere pressure = 24 dm³).
  (AEB 1990)

63 For **both** part **(a)** and part **(b)**, draw structural formulae for the lettered compounds **C, D, E, F, G, H, J, K, L**. Give your reasoning.
(a) A liquid **C**, $C_5H_{10}O_2$, reacts with LiAlH₄ to give a mixture of two alcohols, **D** and **E**. Both **D** and **C** give a pale yellow crystalline product **F** when treated with iodine in alkaline solution. The liquid **C** is insoluble in cold, dilute aqueous NaOH; but on boiling, the mixture gradually becomes homogeneous. (8)
(b) A hydrocarbon, **G**, does not react with chlorine in the dark. When a mixture of chlorine and **G** is irradiated with ultraviolet light, only two monochlorinated products, **H** and **J**, are formed. When **H** and **J** are each treated with warm aqueous NaOH solution, and then with KMnO₄ solution, **H** gives an acid, **K**, whereas **J** gives a ketone **L**. (12)
  (UODLE 1991)

64 (a) (i) State Hess's law.
  (ii) Define standard enthalpy change of formation.
  (iii) Propene and benzene can react together according to the equation

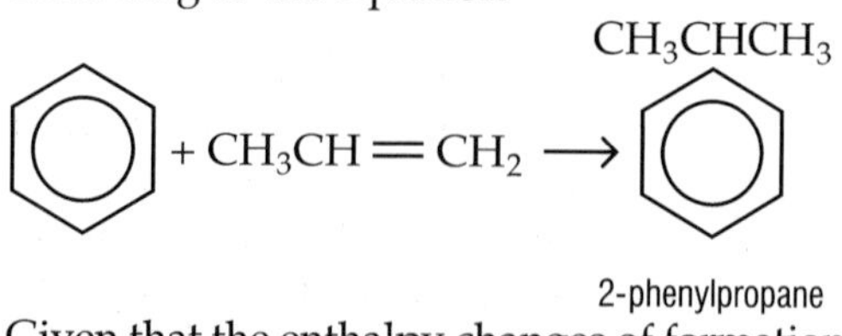

$$\text{C}_6\text{H}_6 + CH_3CH=CH_2 \longrightarrow \text{2-phenylpropane (CH}_3\text{CHCH}_3)$$

2-phenylpropane
Given that the enthalpy changes of formation of benzene, propene and 2-phenylpropane are +49.0, +20.4, and –30.1 kJ mol⁻¹ respectively, calculate the standard enthalpy change for the above reaction. (7)

(b) What would be observed if bromine was added separately to samples of benzene and propene? Write an equation for any reaction(s) occurring and explain any differences between the behaviour of the two compounds. (6)

(c) Propene reacts with hydrogen bromide according to the equation

$$CH_3CH\!=\!CH + HBr \rightarrow CH_3CHBrCH_3$$

  (i) What is the name of the product of the reaction?

  (ii) Give the name of, and outline the mechanism for, the reaction. (6)

(d) One important use of propene is the manufacture of poly(propene).

  (i) Write an equation to represent the process and give the name of the catalyst used.

  (ii) State **two** different properties of poly(propene) and a use associated with **each** property. (6)

(AEB 1993)

65 Butan-1-ol can be converted into other compounds as shown below.

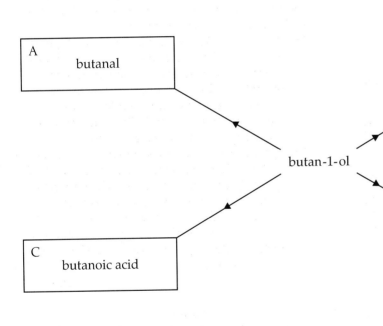

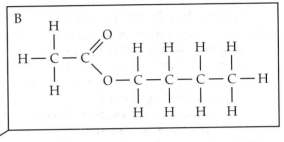

(a) Name compound B. (1)

(b) Which organic chemical, not shown in the diagram, would be required in the conversion of butan-1-ol to B? (1)

(c) Draw structural formulae for

  (i) an isomer of B which belongs to the **same** homologous series as B;

  (ii) an isomer of B which belongs to a **different** homologous series from B. (2)

(d) Which two compounds in the flow chart can be formed from butan-1-ol by the same type of chemical reaction? (1)

(e) A, B and C are colourless liquids.

Describe **two** tests which together could be used to distinguish them. (2)

(f) Which reagent could be used to convert butan-1-ol to D? (1)

(g) Compound D undergoes addition polymerisation.

  (i) Name the polymer formed. (1)

  (ii) Draw part of this polymer structure showing at least three monomer units linked together. (1)

(h) In the conversion of butan-1-ol to A, the product can be separated from the reaction mixture by distillation.

Draw a diagram of apparatus which could be used. (2)

(SCE 1992)

66 **(a)** Compare and contrast the chemistry of benzene and alkenes by commenting on the following observations:

  (i) The carbon-carbon bond length in benzene is 139 pm and that of ethene is 133 pm. (5)

  (ii) Both benzene and ethene react with bromine, but the conditions and type of reaction are different. (5)

  (iii) Benzene can be used as the solvent in some reactions involving $KMnO_4$ but the liquid alkene, cyclohexene, cannot. (3)

(b) Define the term *electrophile*. (1)
Predict the products, if any, obtained by treating ethene with each of the following reagents. Give your reasoning.
(i) HCN; (2)
(ii) HBr; (2)
(iii) NaBr. (2)
(UODLE 1991)

67 **(a)** (i) Explain what is meant by the term *acid dissociation constant*, K_a. (4)
(ii) The pH of a solution of methanoic acid of concentration 0.1 mol dm^{-3} is 2.38. Calculate K_a for methanoic acid. (5)
(iii) Explain why the dissociation constants of ethanoic acid and chloroethanoic acid differ from one another and from the dissociation constant of methanoic acid. (6)
(b) (i) Explain why an aqueous solution containing one mole of methanoic acid and one mole of sodium methanoate shows little change in pH when small amounts of hydrochloric acid or sodium hydroxide are added. (6)
(ii) What is the name given to solutions such as the one described in (b)(i)? State **two** uses of such solutions and outline their importance in nature. (4)
(AEB 1993)

68 **(a)** For the industrially important reaction
$2SO_2(g) + O_2(g) \rightleftharpoons 2SO_3(g)$
$\Delta H(298\text{ K}) = -94.5\text{ kJ mol}^{-1}$.
Describe, giving reasons, the effect on the position of equilibrium of:
(i) increase of temperature; (3)
(ii) decrease of pressure; (2)
(iii) a platinum catalyst; (2)
(iv) excess oxygen. (2)
At 1300 K and a total pressure of 1 atm, the partial pressures at equilibrium are 0.27 atm for SO_2 and 0.41 atm for O_2.
Calculate the equilibrium constant K_p. Be careful to give the units of K_p. (6)
(b) Discuss the application of the Equilibrium law to the equilibrium
$CaCO_3(s) \rightleftharpoons CaO(s) + CO_2(g)$
at 700 K, and explain what would happen if carbon dioxide were added to the system. (5)
(UODLE 1991)

69 The drug Salvarsan has been used in the treatment of parasitic infections for over eighty years. It has the formula

Suggest the physical and chemical properties the drug is likely to have, giving equations and formulae where appropriate.
Properties you might like to consider, include:
(i) solubility in water
(ii) reactions with acids and alkalis
(iii) features of the infrared spectrum
(iv) reactions with halogens
(v) combustion
(vi) reaction with copper(II) ions, Cu^{2+}
(vii) reaction with nitrous acid, $HNO_2(aq)$
This is not an exhaustive list and other properties can be considered instead.
(ULSEB 1991)

70 **(a)** (i) Draw a labelled diagram of a blast-furnace to show how iron ore, mainly Fe_2O_3, is converted to iron on the commercial scale. (7)
(ii) Give an equation in each case for:
the reduction of the ore
the formation of the reducing agent
the removal of impurities. (3)
(b) Suggest explanations for the following observations.
(i) When an iron bar is immersed in aqueous copper(II) sulphate an immediate deposit of copper appears on the surface of the iron. If the bar is first immersed in concentrated nitric acid, removed carefully, and then placed in aqueous copper(II) sulphate, no reaction occurs. When the iron bar is scratched, a deposit of copper appears immediately. (4)
(ii) When an aqueous solution containing Fe^{3+} ions is added to an aqueous solution of an iodide it gives a blue coloration with starch If a soluble fluoride is then added, the blue coloration disappears. (6)
(OCSEB 1991)

Theme 1

Chapter 1

1.1 the four components are the atmosphere, the lithosphere, the hydrosphere and the biosphere. Their characteristics may be found in the text;

1.2 (a) neon is the second most abundant noble gas in air;
(b) 0.018 cm^3;
(c) the difference in the two values is due to global variations in the water content of air. Other variations are small by comparison;

1.3 (a) atmosphere 20.95% by volume, Earth's crust 46.4% by mass.
In the atmosphere much of the oxygen is free, elemental oxygen (O_2). In the Earth's crust the oxygen is largely chemically combined with other elements;
(b) the figures indicate that most of the iron content of the Earth is not concentrated in the Earth's crust. It is in fact contained largely in the core and mantle;

1.4 (a) 12.38 g;
(b) 5.5 g;

1.5 the water in rivers is derived from rainfall and from melting snow. This water contains little salt. In the sea, water is lost by evaporation and any dissolved salts become more concentrated;

1.6 (a) iron;
(b) by eating plants and animals, which concentrate potassium compounds within them and by developing systems to retain potassium within their bodies;

1.7 the chemical compounds described are produced by human activity and include carbon dioxide, sulphur dioxide and other pollutants;

1.8 an element, (example, iron) is a substance composed of one kind of atom only. When iron combines with water and oxygen a chemical reaction takes place forming hydrated iron oxide. This is an example of a compound - a substance formed from more than one element. Sea-water is a physical combination of water and other compounds. A purely physical combination of substances is known as a mixture;

1.9 using a magnet will separate iron filings from the sand. This is a simple physical method of separation indicating that the sand and iron filings are a simple mixture and not chemically combined;

1.10 the chemical compound formed, sodium chloride, has very different properties to the individual elements sodium and chlorine;

1.11 iron, oxygen, silicon and magnesium are abundant on Earth but rare in the universe as a whole. By contrast, the universe as a whole contains a very large porportion of hydrogen and helium;

1.12 silicon, aluminium and potassium;

1.13 make your own list!

1.14 the melting of butter is a physical change, reversed by cooling. Frying an egg involves an irreversible chemical change for the proteins of which the egg is composed;

1.15 this is a process by which plants use the energy of the Sun to bring about reaction between carbon dioxide and water, forming sugars;

1.16 it undergoes corrosion in the presence of oxygen and water and the rusty surface which results causes poor electrical contact;

1.17 sodium and magnesium are examples of metals, chlorine and neon are non-metals whilst good examples of metalloids are silicon and germanium;

Chapter 2

2.1 models may show the structure of something, on a scale which is of use to human beings or show how a process takes place;

2.2 Dalton's model of the atom argued that atoms were the smallest particles of matter but did not examine the structure of atoms. The nuclear model of atomic structure recognised that atoms consisted of subatomic particles - the neutrons, protons and electrons;

2.3 they consist of different types of atoms;

2.4 the subatomic particles (neutrons, protons and electrons) are the building blocks from which atoms are constructed. Any differences between atoms are entirely due to a difference in the numbers of these particles they contain;

2.5

Subatomic particle	Charge	Location
neutron	0	nucleus
proton	+1	nucleus
electron	−1	outside the nucleus

2.6 about 10^5 or 100 000 times larger;

2.7 this is a much more powerful tool than the conventional microscope and can examine much smaller objects. The image it produces may be further improved by using a computer. It can help to show that a model of a chemical structure, developed as a result of chemical tests is accurate (or inaccurate);

2.8 some alpha particles would be deflected towards the gold nuclei whilst others might be 'captured';

2.9 $^{20}_{10}$Ne 10p, 10n; $^{132}_{54}$Xe 54p, 78n; $^{195}_{78}$Pt 78p, 117n; $^{79}_{35}$Br 35p, 44n;

2.10 $^{12}_{6}$C, $^{28}_{14}$Si, $^{73}_{32}$Ge, $^{118}_{50}$Sn and $^{208}_{82}$Pb. The neutron/proton ratio increases as the Group is descended;

2.11 $^{1}_{1}$H, $^{2}_{1}$H and $^{3}_{1}$H;

2.12 $^{235}_{92}$U contains 92 protons and 143 neutrons;
$^{238}_{92}$U contains 92 protons and 146 neutrons;

2.13 the argon isotope contains 18 protons and 22 neutrons; the potassium isotope contains 19 protons and 20 neutrons. They appear in this order because the Periodic Table lists elements in order of increasing atomic number not increasing mass number;

2.14 the relative atomic mass of an element is the average relative mass of its atoms, based on a scale in which the $^{12}_{6}$C isotope is assigned a relative mass of 12 and taking into account the relative abundance of isotopes of the element. The concept allows comparison of the masses of particles by determining the mass of the same number of atoms of each element rather than by weighing individual atoms;

2.15 approximately −1.922 × 10^{-18} coulombs;

2.16 approximately +4.806 × 10^{-19} coulombs;

2.17 (a) 28.09;
(b) 467.45 g;

2.18 W = $^{25}_{12}$Mg^{2+};
X = $^{16}_{8}$O;
Y = $^{127}_{53}$I$^-$;
Z = $^{15}_{8}$O^{2-};

2.19 63.6;

2.20 (a) 4; (b) 5; (c) 9; (d) 4; (e) 2;

2.21 the heaviest relative molecular mass is 74 and the lowest 70;

2.22 atomic spectra and a study of ionisation energies both provide evidence that electrons exist within energy levels;

2.23 2.8.5;

2.24 (a) it has one electron in its outer energy level;
(b) it is 'one electron short' of an inert gas configuration;

2.25 these are electrons, usually in the outer energy level of an atom only, which are available for bonding;

2.26 each of these elements has five electrons in its outer energy level;

2.27 H 1; H_2 2; H^+ 0; H^- 2;

2.28 2.8.4; 2.8.6; 2.8.8; 2.8.8.2;

2.29 2; 2; 2.8; 2.8.8; 2.8.8;

2.30 Na;

2.31 nuclear changes alter the number of protons and neutrons in the nuclei of the atoms concerned. Simple chemical changes involve a reorganisation of electrons only;

2.32 ^6_3Li is the more stable isotope. For lighter elements the neutron/proton ratio is close to 1 for stable isotopes;

2.33 an electron must be produced; $^{198}_{80}Hg \longrightarrow ^{198}_{79}Au + ^0_{-1}e$

2.34 P is a neutron. It has a relative mass of 1 but no charge. Neutrons combine with an atom of $^{235}_{92}U$ to give two smaller nuclides, $^{94}_{36}Kr$ and $^{139}_{56}Ba$ plus three neutrons;

2.35 The amount of carbon-14 in non-living material gradually decreases with age. The high value obtained for the shroud suggests that it is not as old as previously thought;

2.36 **(a)** n/p ratio increases from 1.543 to 1.555;
(b) n/p ratio decreases from 1.598 to 1.570;

2.37 X is a proton;

2.38 Z must be a member of Group VIII;

2.39 this is important since equal masses of different substances do not contain the same number of particles. To calculate the amounts of materials required for reactions chemists must compare the *amount of substance* which is a measure of the number of particles involved;

2.40 **(a)** 0.15 mol;
(b) 56 g;

2.41 approximately 6.1×10^{-3} mol;

2.42 664.2 g;

2.43 the relative molecular mass of a substance is the ratio of the mass of a molecule of the substance to the mass of an atom of the $^{12}_6C$ isotope of carbon. It has no units. The molar mass of a substance is the mass of one mole of that substance. It has units of g mol^{-1};

2.44 2.01×10^{23};

2.45 1.42×10^6 mol;

2.46 about 55.56 mol;

2.47 it would be better to buy 1 mol at £9.50 since 100 g is only 0.7 mol. At this price it would be £12.78 for 1 mol;

2.48 **(a)** 1.67×10^{-3} mol aspirin;
(b) approximately 600;

2.49 **(a)** 45 g;
(b) 1.81×10^{-5} mol;

Chapter 3

3.1 it is an inert gas and helps to prevent the crisps going stale;

3.2 **(a)** plants use carbon dioxide and act as a 'living reservoir' for carbon containing compounds. They also produce oxygen gas for animals to breathe;
(b) like lungs, the rainforests are a site for gaseous exchange and help to maintain a suitable atmosphere for humans and other animals;

3.3 the carbon dioxide concentration is at a minimum around noon when sunlight is at a maximum and plants are actively photosynthesising. During the hours of darkness plants use oxygen and liberate carbon dioxide, causing a maximum CO_2 concentration just before sunrise;

3.4 primary pollutants are directly produced by processes taking place on Earth such as burning coal. Secondary pollutants result from further reactions of primary pollutants. Using these definitions, tobacco smoke is clearly a primary pollutant;

3.5 reduce the use of cars and other motorised transport or improve their design to minimise emission of unburnt hydrocarbons and nitrogen oxides. Better still, use a bicycle!

3.6 the difference is due to their relative distances from the Sun.

3.7 **(a)** it suggests that the C—Br bond must be less strong than the C—Cl bond;

(b) Datasheet 4 confirms this. The average bond enthalpy of the C—Cl bond is 330 kJ mol^{-1} whilst that of the C—Br bond is 276 kJ mol^{-1};

3.8 a covalent bond results when two or more atoms share a number of electrons, usually two. The bond is an attractive force which arises due to the attraction of the atoms nuclei for the electrons they share;

3.9 **(a)**

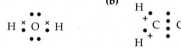

(b) neon;

3.10 **(a)** 3; **(b)** 2; **(c)** 1; **(d)** 2; **(e)** 0;

3.11 H—Cl single bond between hydrogen and chlorine atoms;
O=O double bond between two oxygen atoms in an oxygen molecule;
H—C≡N triple bond between carbon and nitrogen atoms in the hydrogen cyanide molecule;
In general, the strength of bonds increases in the series: single->double->triple. Detailed data for bond strengths are available in Datasheet 4;

3.12 **(a)**   **(b)**

(c)

3.13 **(a)**   **(b)**   **(c)**

3.14 **(a)**

(b) a chlorine atom in one $AlCl_3$ molecule can provide an electron pair to form a co-ordinate bond with aluminium. This arrangement is repeated for the other $AlCl_3$ molecule giving the structure shown in **(c)**;

(c)

594

3.15 **(a)** $CF_3Cl \longrightarrow CF_3 + Cl$

(b)

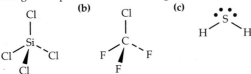

3.16 **(a)** you may well have predicted that the bond energies gradually decrease in the series $F{-}F \longrightarrow Cl{-}Cl \longrightarrow Br{-}Br \longrightarrow I{-}I$ as the size of the halogens increases (giving longer bonds);

(b) the Datasheet reveals a flaw in this prediction. The unexpectedly low bond energy for fluorine is probably due to the closeness of the fluorine atoms in F_2 giving rise to some repulsion between non-bonding electron pairs;

3.17 see page 47;

3.18 there are six electrons arranged around the central aluminium atom (three from aluminium and one from each chlorine atom). These are organised in three pairs which adopt the position of minimum repulsion between them. This results in a trigonal planar structure with bond angles of 120°;

3.19 **(a)** there are four pairs of electrons around nitrogen, three bonding pairs and one lone pair. This gives a pyramidal distribution of bonding pairs;

(b) around carbon there are three regions of negative charge as the $C{=}O$ double bond behaves as a single region of negative charge. The distribution here is trigonal planar;

3.20 in a water molecule the central oxygen atom carries two lone pairs of electrons. Lone pairs give rise to greater repulsive effects than bonding pairs. This forces the bonding pairs together, giving a smaller bond angle than in ammonia, where only a single lone pair is centred on nitrogen;

3.21 **(a)** **(b)** **(c)**

(d) **(e)**

associated bond angles are **(a)** 109.5°; **(b)** approximately 109.5°; **(c)** approximately 104.5°; **(d)** approximately 108°; **(e)** 120°;

3.22 **(a)** 44 g mol^{-1}; **(b)** 98 g mol^{-1}; **(c)** 78 g mol^{-1}; **(d)** 58 g mol^{-1}; **(e)** 228 g mol^{-1}; **(f)** 180 g mol^{-1};

Assuming 1 mol of gas occupies 24 dm^3 at room temperature and pressure (**3.23**, **3.24**, **3.25**, **3.28**, **3.29**):

3.23 **(a)** 12 dm^3; **(b)** 0.8 dm^3; **(c)** 0.34 dm^3;

3.24 **(a)** 0.17 g; **(b)** 1.33 g; **(c)** 1.67 g;

3.25 1 dm^3 of nitrogen contains 0.04 mol nitrogen molecules therefore 0.08 mol nitrogen. 1 g of nitrogen will contain 0.036 mol nitrogen molecules. Thus 1 dm^3 contains the larger amount of nitrogen and hence the larger number of atoms;

3.26 **(a)** 0.0005 mol; **(b)** 0.004 mol; **(c)** 0.004 mol;

3.27 **(a)** 1.20×10^{23}; **(b)** 1.80×10^{25}; **(c)** 3.01×10^{21}; **(d)** 6.02×10^{17};

3.28 **(a)** 0.255 g;

(b) the mass would be 1.67 kg;

3.29 assume that 24 dm^3 of air contains one mole of gas molecules at room temperature and pressure. The number of molecules of each gas is:
N_2 4.70×10^{23}; O_2 1.26×10^{23}; Ar 5.60×10^{21}; CO_2 1.81×10^{20};

3.30 28.95;

3.31 **(a)** the average relative molecular mass of air is 28.95. The volume of 12 kg of air will be 12 000/28.95 × 24 dm^3 = 9948 dm^3. An average female uses 5% of this volume of oxygen daily = 497.4 dm^3 of oxygen;

(b) since 1 mol O_2 occupies 24 dm^3, this volume of oxygen contains 497.4/24 = 20.72 mol O_2. This would contain $20.72 \times 6.023 \times 10^{23}$ molecules of oxygen, = 1.248×10^{25} molecules oxygen;

3.32 **(a)** N_2 mole fraction = 0.78, O_2 mole fraction = 0.21, Ar mole fraction = 0.01;

(b) N_2 partial pressure = 89.7 kPa, O_2 partial pressure = 24.2 kPa, Ar partial pressure = 1.15 kPa;

(c) normal air contains some water vapour. This will lower the mole fraction and hence the partial pressure of all other components;

3.33 the ideal gas equation shows clearly that a large increase in temperature must, if volume is constant, lead to a great increase in pressure inside the tube. Thin-walled tubes will not survive this increase in pressure and may explode. The precise increase in pressure expected may be calculated using the ideal gas equation if required;

3.34 **(a)** the concept of an ideal gas assumes no forces of attraction between particles of the gas and assumes that individual particles occupy zero volume. For real gases these assumptions are never true;

(b) under many conditions, particularly those of low pressure and high temperature, the above two assumptions are approximately true;

3.35 hydrogen 6.66×10^4;
oxygen 3.33×10^4;

3.36 O_2 mole fraction = 0.2095;
O_2 partial pressure = 2.123×10^4 Pa;

3.37 **(a)** 6.12 cm^3; **(b)** 9.40 cm^3.

Chapter 4

4.1 most elements form compounds with other elements in the environment. These compounds are more stable than the free elements;

4.2 with the exception of simple oxides of carbon, carbonates, hydrogencarbonates and cyanides, inorganic compounds do not contain the element carbon. Organic compounds all contain carbon;

4.3 **(a)** **(d)** and **(i)** are organic. The other examples are all inorganic compounds;

4.4 these data may be found in Table 1.2 in Chapter 1. There are several comparisons which can be made. Notably much less iron is found in the Earth's crust than in the Earth as a whole whilst much of the oxygen is contained in the Earth's crust. The Earth's crust consists largely of aluminosilicate rocks reflected in the high proportion of silicon and aluminium in the crust. Calcium, present in limestone rock is present in a higher proportion in the Earth's crust. Sodium and potassium are present as salts in the crust but form only a small proportion of the Earth as a whole;

4.5 **(a)** +1; **(b)** −1; **(c)** +1; **(d)** +6; **(e)** +5; **(f)** −1; **(g)** 0; **(h)** +7; **(i)** +5; **(j)** +2; **(k)** +5; **(l)** +5;

4.6 manganese +7 $\longrightarrow$ +2; carbon +3 $\longrightarrow$ +4;

4.7 **(a)** Na_2S; **(b)** Fe_2O_3; **(c)** HgI; **(d)** CuCN; **(e)** $PbBr_2$; **(f)** FeO; **(g)** IF_7;

4.8 **(a)** potassium dihydrogenphosphate(V);
(b) strontium sulphate;
(c) silver carbonate;
(d) potassium oxide;
(e) magnesium sulphate-6-water;
(f) chromium(III) chloride;
(g) aluminium hydroxide;
(h) ammonium nitrite;
(i) ammonium nitrate;
(j) potassium hydrogensulphate;

4.9 UCl_3 = +3, UBr_4 = +4, UF_6 = +6, UO_2Cl_2 = +6, UO^+ = +3, UO_2^{2+} = +6, UF_8^{3-} +5;

4.10   boron trifluoride; carbon disulphide; diiodine hexachloride; dinitrogen oxide; xenon difluoride; xenon tetrafluoride;

4.11   bismuth(III) chloride; iodine(V) oxide; lead(IV) oxide; molybdenum(IV) sulphide; titanium(IV) oxide; vanadium(IV) oxide;

4.12   a crystal lattice is a regular, three-dimensional array of closely-packed particles in the solid state. The four types of crystal lattice are: molecular lattice, giant covalent lattice, ionic lattice and metallic lattice;

4.13   this is a solid which does not have a regular lattice;

4.14   allotropes are different physical forms of the same element;

4.15   the answer does not say specifically which bonds are referred to. The covalent bonds between iodine atoms in iodine are strong bonds. The intermolecular forces between iodine molecules in the iodine lattice are weak bonds;

4.16   intermolecular bonds arise as a result of attractions between particles. They may be due to permanent or temporary dipoles present in the particles. The greater the intermolecular forces between particles, the higher its melting and boiling points are likely to be;

4.17   diamond is a much harder form of carbon and may cause physical damage to the motor. It is also a very poor electrical conductor and would be a poor choice for the brushes;

4.18

(figure: silicon bonded to four oxygen atoms)

each silicon atom is bonded to four oxygen atoms but each of these is bonded to a further silicon atom. Each silicon atom is thus considered to have a half share in four oxygen atoms - the equivalent of SiO_2;

4.19   in the solid state ionic compounds consist of ions arranged in fixed positions within a lattice. The ions are not free and cannot carry a current. When molten or in aqueous solution the ions can move and will conduct electricity;

4.20   KCl, NaCl, KF, NaF, $CaCl_2$, K_2O, CaF_2, MgF_2, CaO, MgO;

4.21   metals adopt a structure in which their valence electrons are delocalised. When a potential is applied across a piece of metal the electrons can move and carry a current;

4.22   **(a)**

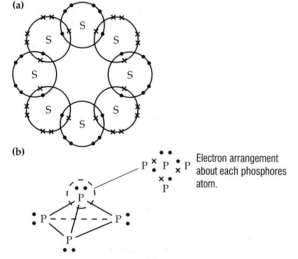

(b)

P × P × P   Electron arrangement about each phosphores atom.

4.23   each silicon atom contributes an electron to each of four bonds with its neighbours. The four bonding pairs repel one another and take up positions of minimum repulsion – the tetrahedral geometry;

4.24   both elements have four electrons in their outer shell. The most stable lattice, which satisfies the electronic requirements of each element is a giant covalent lattice in which each carbon atom is tetrahedrally bonded to four silicon atoms and vice-versa. It would therefore have a diamond structure;

4.25   **(a)**

Ca O $\longrightarrow$ $[Ca]^{2+}$ $\begin{bmatrix} O \end{bmatrix}^{2-}$

(b)

F Mg F $\longrightarrow$ $[Mg]^{2+}$ $\begin{bmatrix} F \end{bmatrix}^{-}$ $\begin{bmatrix} F \end{bmatrix}^{-}$

(c)

Na S Na $\longrightarrow$ $[Na]^{+}$ $[Na]^{+}$ $\begin{bmatrix} S \end{bmatrix}^{2-}$

4.26   the melting points are dependent upon the lattice energies of the ionic compounds concerned. Lattice energies in each series of halides in Table 4.5 decrease because the increasing size of the anion means that cation and anion cannot approach each other so closely. Melting points thus decrease down each series for a given halide ion, the strongest lattice is when $r^+ \simeq r^-$;

4.27   the atoms in a metallic lattice may be viewed as cations arranged in a regular three-dimensional structure, and surrounded by delocalised electrons. The strength of attraction between the cations and the electrons dictates the strength of the metallic bonding. The lithium cation has a high charge density due to its very small size. It has a strong attraction for the electrons which surround it and a stronger metallic lattice than the other members of the series. The trend is for progressively weaker lattices as Group 1 is descended;

4.28   2.70%;

4.29 $C_6H_{12}O_6$;

4.30   41.49;

4.31   **(a)** 5.12g; **(b)** 0.62 g; **(c)** 9.00 g; **(d)** 28.8 g; **(e)** 7.485 g;

4.32 $SrSO_4$; celestite is a sulphate mineral;

4.33 $MgSiO_3$;

4.34   184.5 kg;

4.35 $(NH_4)_2SO_4$ 21.2%; NH_4NO_3 35%, contains a higher proportion of nitrogen;

4.36   79.07 g;

Chapter 5

5.1   **(a)** water is used as a reaction medium and as a reactant in biological processes. It also functions as a transport medium for nutrients and for waste products;

   (b) it is an effective solvent for a very wide range of chemical substances despite the diversity in their bonding;

5.2   in most aqueous solutions the number of water molecules vastly outweighs the number of dissolved particles, which have little effect on the density of the solution;

5.3   a knowledge of the solubility of carbon dioxide in water under various conditions would be needed;

5.4

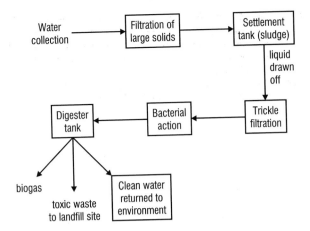

5.5 it would be necessary to ensure that those substances which are expensive to remove are prevented from entering the water to be purified;

5.6 these ions are difficult to remove because most of their compounds have a very high solubility in water;

5.7 this term is used to describe any change of state which takes place; eg, a change from solid state to liquid or to solution states;

5.8 after a time the amount of water present in the liquid state and in the vapour state (above the liquid) reach constant, equilibrium values. However, there is a constant exchange of water molecules between the two phases. The equilibrium is thus *dynamic*;

5.9 (a) the line T——→A on the diagram represents the variation in boiling point of water with changing pressure. Clearly at higher pressures the boiling point of water increases. This is expected since water requires more energy to overcome the higher external pressure and enter the vapour state;

(b) if the sample is cooled some water will be condensed to the liquid state, leaving a sample of air at the lower temperature which is still saturated with water vapour;

5.10 the ammonium chloride dissociates at the high temperature at the bottom of the test-tube to give ammonia and hydrogen chloride gases. This is an endothermic dissociation. In the cooler region near the mouth of the tube an exothermic process takes place in which the gases recombine to form ammonium chloride. The overall process is one of *sublimation*;

5.11 2.618×10^3 kJ is released;

5.12 (a) gas; (b) solid;

5.13 when a gas condenses to a liquid a great deal of energy is released as the attractive forces between particles overcome the kinetic energy of the gas particles and trap the substance in the liquid state. This confirms the existence of intermolecular forces and reminds us that the notion of an ideal gas is a theoretical one;

5.14 (a) water 2.28 kJ g^{-1}; ethanol 0.95 kJ g^{-1}; tetrachloromethane 0.20 kJ g^{-1};

(b) 228 kJ;

5.15 (a)

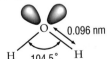

(b) the two lone pairs of electrons centred on oxygen have a greater repulsive effect than the bonding pairs and distort the distribution of electron pairs to a non-regular tetrahedral geometry;

5.16 if it were not for the existence of hydrogen bonding the predicted melting points and boiling points for water would be approximately –90°C and –70°C respectively. There would be no solid or liquid water on Earth! All life forms use water as a reaction medium for biological processes and thus life as we know it could not exist. In addition the vast expansion of the

atmosphere that would take place would have dramatic effects on the temperature of the Earth. If life were to evolve on Earth it would certainly take a very different form;

5.17 the trichloromethane has a net dipole and will be electrostatically attracted to the charged rod;

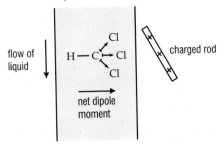

5.18 boiling point decreases with altitude since pressure decreases with altitude; if a series of measurements are made for the boiling point of the chosen liquid at a wide range of altitudes a calibration curve may be produced. Note: different calibration curves would be necessary for different pressures. If the boiling point of the liquid is measured at an unknown altitude the graph can be used to determine altitude;

5.19 water expands as it freezes because the open structure of ice contains water molecules which are less densely packed than in the liquid state;

5.20 the other hydrides of Group VI do not have a sufficiently powerful dipole to give rise to hydrogen bonding because the elements they contain are less electronegative than oxygen;

5.21 the water interrupts existing hydrogen bonds between keratin molecules, causing the hair to lose its shape. On drying, hydrogen bonds reform between keratin molecules in new positions and the hair adopts a new overall shape;

5.22 (a) $Ca^{2+}(aq)$, $Cl^-(aq)$ (b) $Pb^{2+}(aq)$, $NO_3^-(aq)$
(c) $Na^+(aq)$, $CrO_4^{2-}(aq)$ (d) $K^+(aq)$, $HSO_4^-(aq)$
(e) $Al^{3+}(aq)$, $NO_3^-(aq)$

5.23 silver and chloride ions are continuously coming together to precipitate as solid silver chloride, and solid silver chloride is continuously dissolving to give hydrated silver and chloride ions.

5.24 ammonia solution contains hydrated hydroxide ions; Hydrochloric acid contains hydrated oxonium ions. During the reaction a proton is transferred from an oxonium to a hydroxide ion creating two molecules of water.

5.25 (a) the high solubility of ammonia is largely explained by the formation of hydrogen bonds between water and ammonia molecules. Hydrogen sulphide cannot form hydrogen bonds with water molecules;

(b)

$$H-\overset{..}{\underset{H}{S}}-H \; + \; H-\overset{..}{\underset{H}{\overset{..}{O}}}-H \; \rightleftharpoons \; \left[H-\overset{..}{\underset{H}{\overset{..}{S}}}-H\right]^- + \left[H-\overset{..}{\underset{H}{O}}-H\right]^+$$

5.26 (a) 0.300 mol dm^{-3} (b) 0.126 mol dm^{-3}; (c) 0.070 mol dm^{-3};
(d) 0.129 mol dm^{-3}; (e) 0.027 mol dm^{-3};

5.27 33.8 g of tartaric acid and 43.2 g of citric acid should be made up to a total volume of 4.5 dm^3 of aqueous solution;

5.28 (a) $H_2SO_4(aq) + 2KOH(aq) \longrightarrow K_2SO_4(aq) + 2H_2O(l)$;
(b) 0.0025 mol KOH(aq);
(c) 0.01 mol dm^{-3};
(d) potassium sulphate;
(e) 0.4 mol dm^{-3};

5.29 (a) 0.0625 mol dm^{-3}; (b) 0.0067 mol dm^{-3};

5.30 4.15 g;

5.31 (a) 0.02 mol; (b) 0.0001 mol; (c) 0.050 mol;

5.32 0.16 mol dm^{-3};

Chapter 6

6.1 **(a)** metabolism refers to the sum of all reactions taking place within living organisms;
(b) enzymes control metabolic rate within cells;

6.2 the fossil fuels include coal, natural gas and petroleum. They are formed from dead plants and animals whose remains are subjected to heat and high pressures;

6.3 this was the first synthesis of an organic compound in the laboratory (the first synthetic organic compound);

6.4 2 structural isomers exist;

6.5 **(a)** the lone pairs of electrons centred on nitrogen, oxygen and fluorine would render the chain too reactive and cause cleavage;
(b) the silicon-oxygen bond is stronger than the silicon-silicon bond. Chains based on silicon are unstable in an oxygen atmosphere;

6.6 there are several examples;

6.7 **(a)**

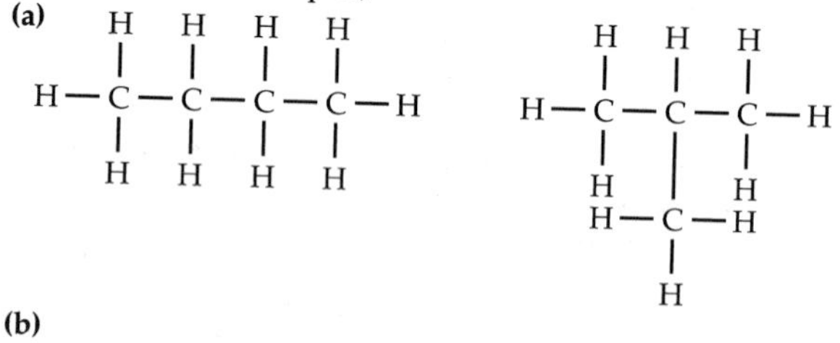

(b)

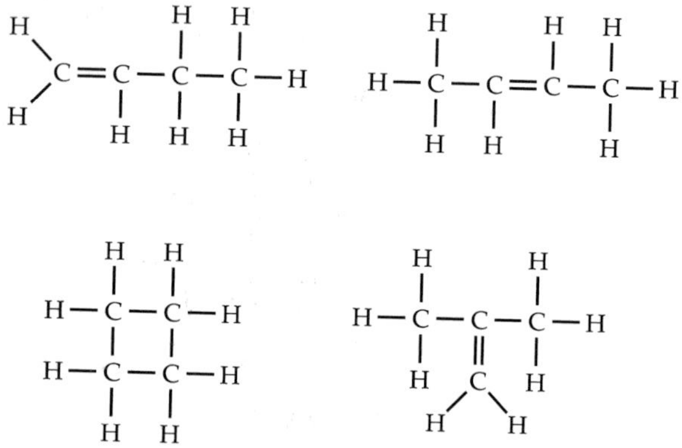

(c)

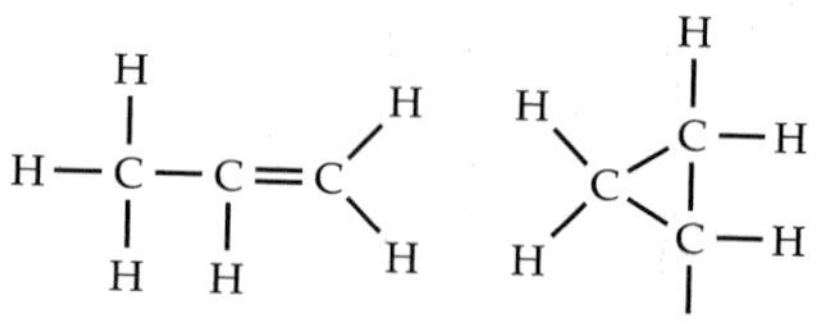

6.8 there are no 'terminal' carbon atoms in a cyclic structure. Hence every carbon atom will have two hydrogen atoms attached. This explains the general formula;

6.9 **(a)** **(b)** $H—C≡C—H$ **(c)**

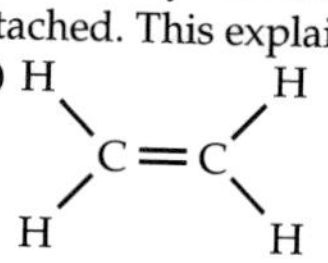

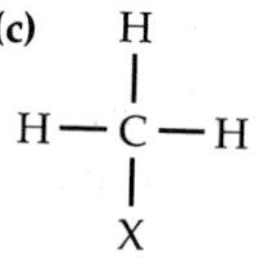

(d) 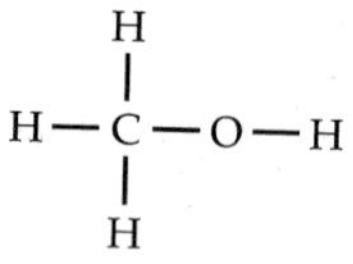 **(e)**

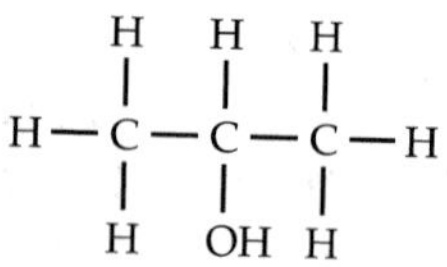

(f) 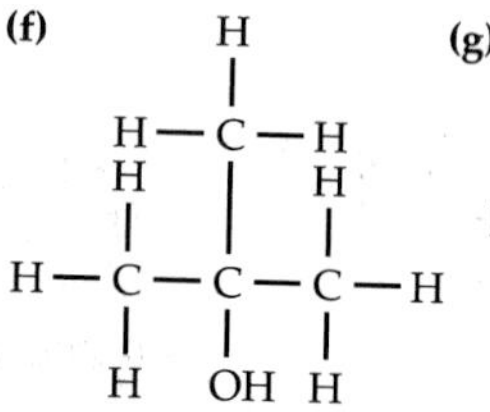 **(g)** 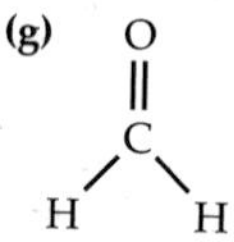 **(h)**

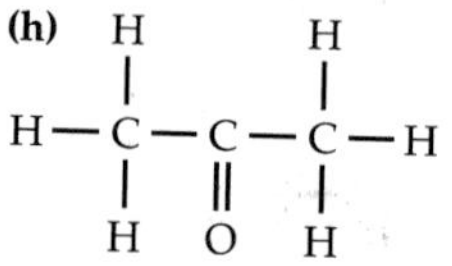

(i) 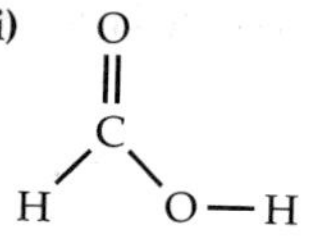 **(j)**

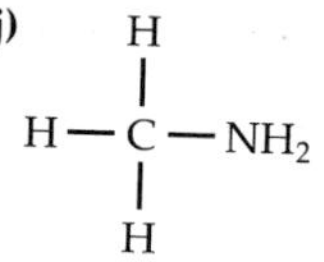

6.10 **(a)**

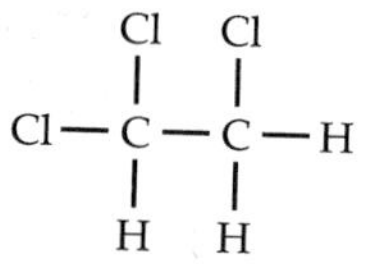

(b) it would be the same compound as 1,1,2-trichloroethane.

6.11

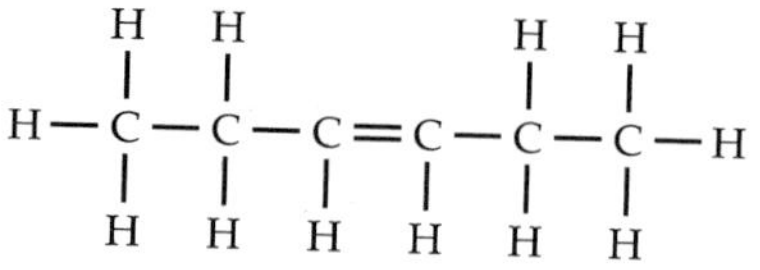

other isomers also possible

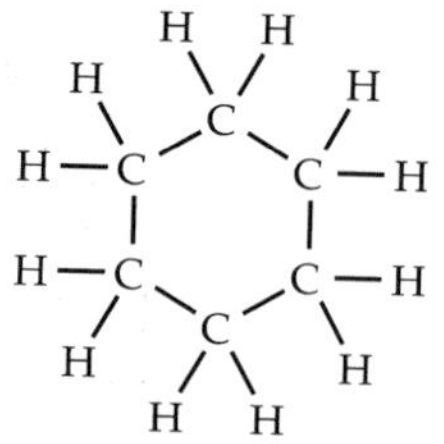

6.12

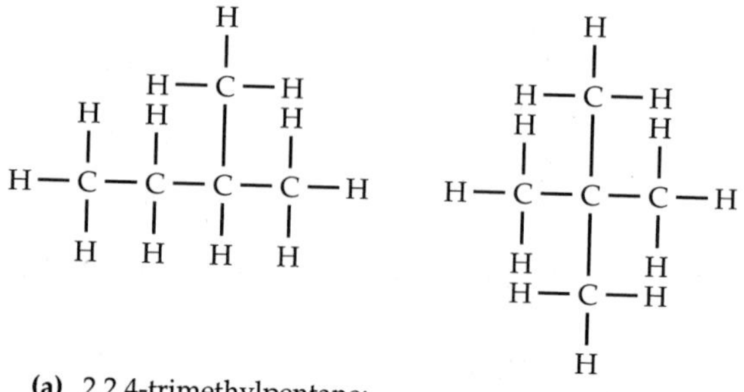

6.13 **(a)** 2,2,4-trimethylpentane;
(b) **(i)** it contains hydrogen and carbon only;
(ii) it contains no multiple bonds of any kind;
(iii) it is both saturated and a hydrocarbon;
(iv) not all the carbon atoms in the compound are arranged in sequence. Three are attached to the chain, projecting from it;
(v) the compound does not contain any delocalised ring systems;

6.14 the molecular formula of propene is C_3H_6; the structural formula is

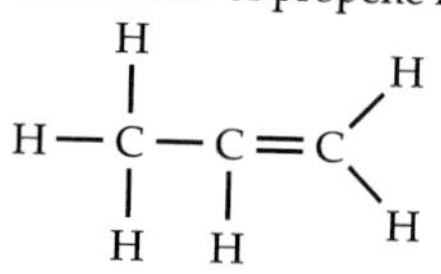

Propene is an alkene;

6.15

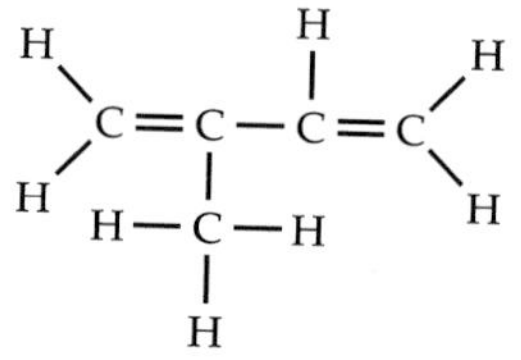

6.16

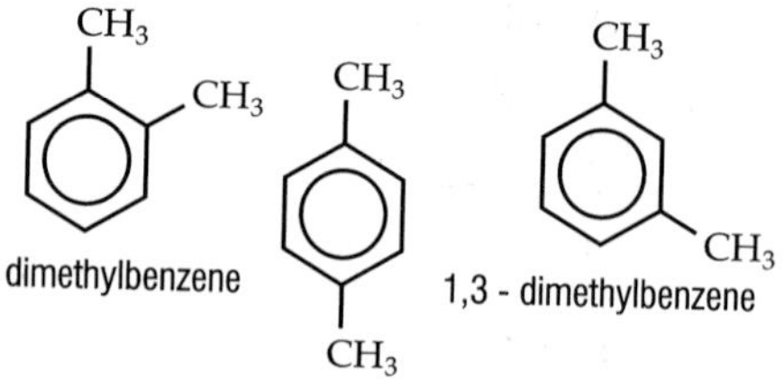

1,2 - dimethylbenzene

1,4 - dimethylbenzene

1,3 - dimethylbenzene

6.17

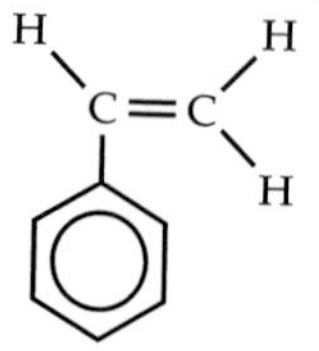

6.18 methanol, ethanol, propan-1-ol, butan-1-ol, pentan-1-ol and hexan-1-ol;

6.19 **(a)** there are ten;

(b)

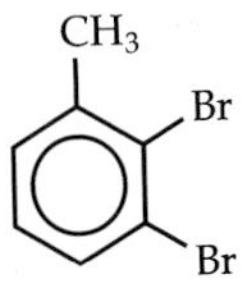

2,3-dibromo-1-methylbenzene

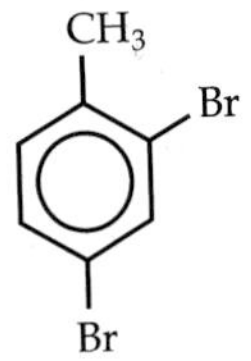

2,4-dibromo-1-methylbenzene

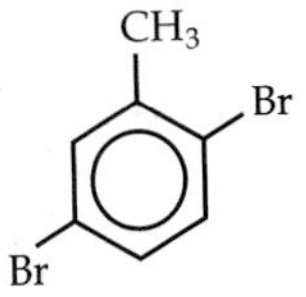

2,5-dibromo-1-methylbenzene

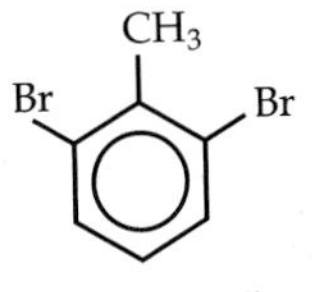

2,6-dibromo-1-methylbenzene

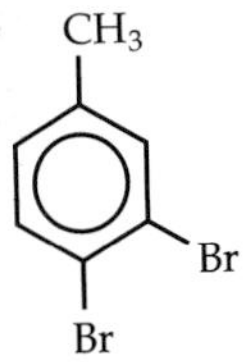

3,4-dibromo-1-methylbenzene

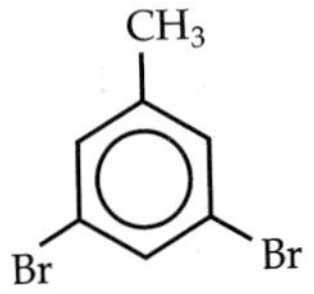

3,5-dibromo-1-methylbenzene

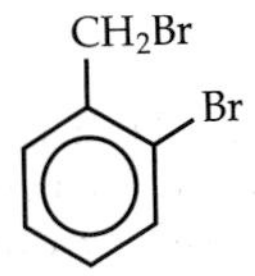

2-bromo-1-
bromomethylbenzene

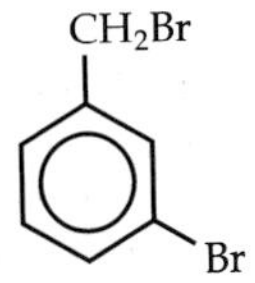

3-bromo-1-
bromomethylbenzene

4-bromo-1-
bromomethylbenzene

dibromomethylbenzene

6.20 **(a)**

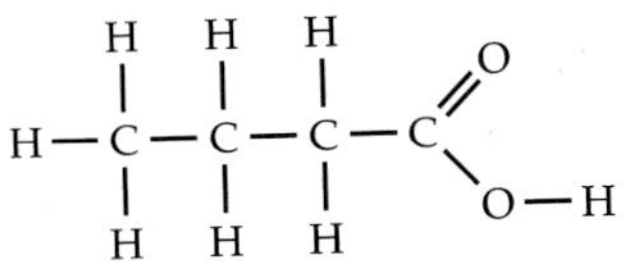

carboxylic acid

(b)

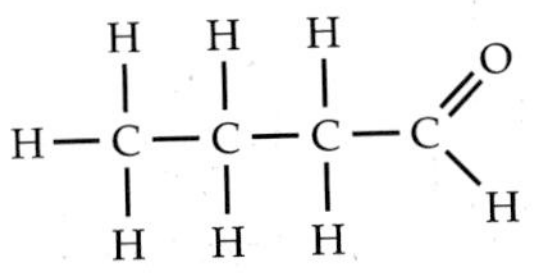

aldehyde

(c) CH₂CH₂CH₂CH₃

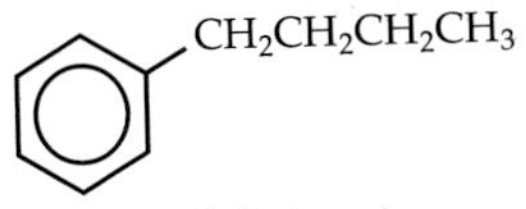

aromatic hydrocarbon

(d)

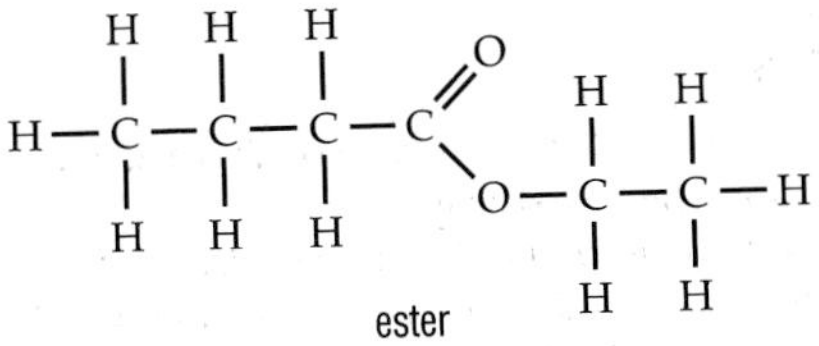

ester

(e)

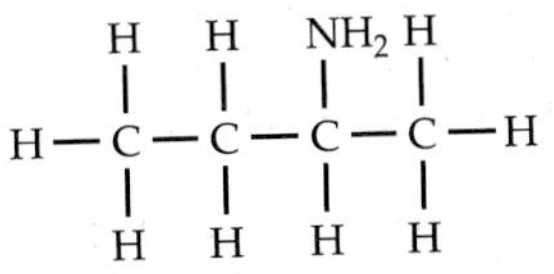

primary amine

(f)

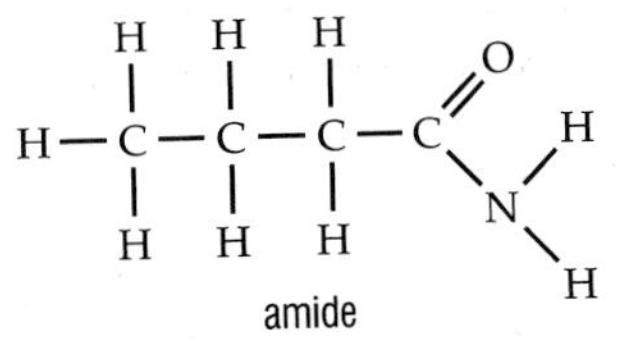

amide

6.21 **(a)** it contains the alkene and nitrile functional groups;

(b) propenenitrile;

6.22 **(a)** aldehydes;

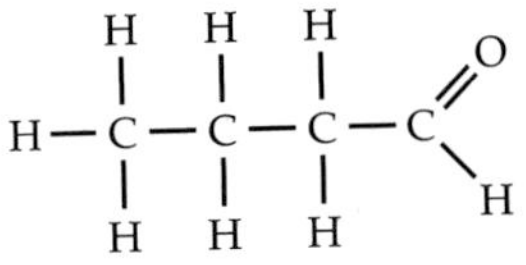

butanal

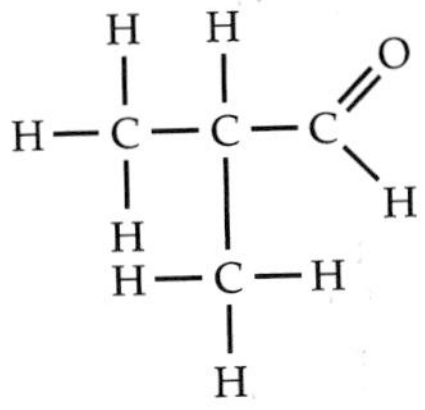

2 - methylpropanal

(b) ketone;

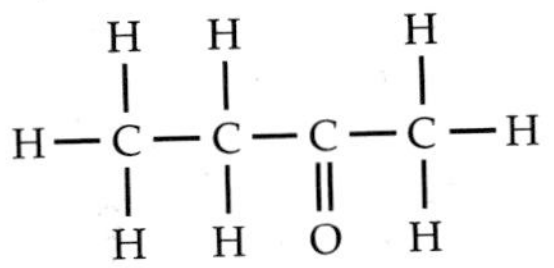

butanone

6.23

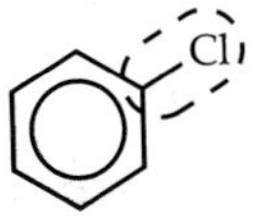

halogenoarene

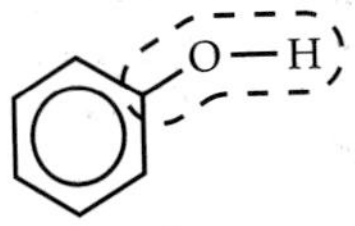

hydroxyl group

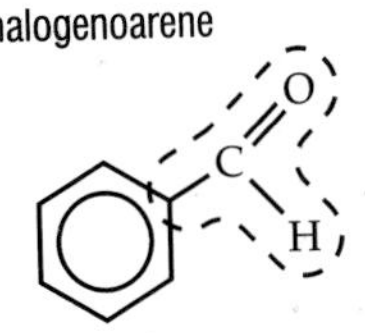

aldehyde

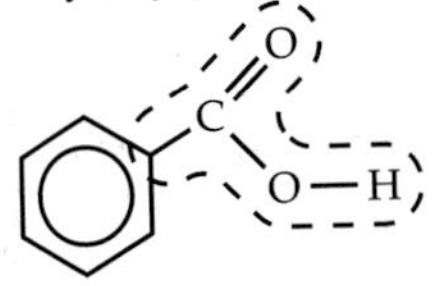

carboxylic acid

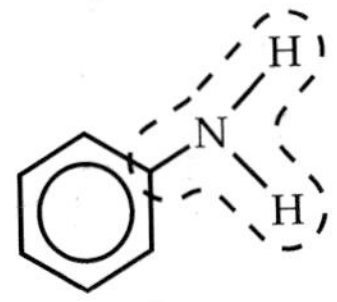

primary amine

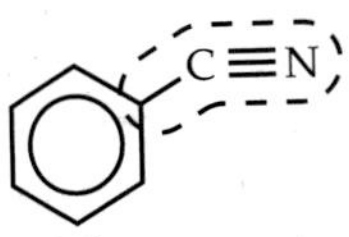

nitrile

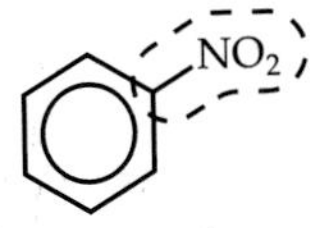

nitro compound

Theme 2

Chapter 7

7.1 qualitative analysis is concerned with investigation of the *type* of chemical substances present in a sample. Quantitative analysis aims to establish how much of a substance is present in the sample;

7.2 drinking water may contain a mixture of many different substances present in trace or ultratrace quantities. Water scientists need to ensure that the concentration of each substance does not exceed the legally permitted maximum;

7.3 **(a) (i)** modern instruments detect the presence and amount of chemical substances in very small samples with a very high degree of consistency;

    (ii) many instrumental techniques measure a characteristic of a sample (such as frequency of absorption) very quickly. Much bench analysis ('by hand') is time-consuming;

  (b) (i) automation has ensured that a rapid throughput of analytical samples can be achieved even during the absence of the analyst – eg during breaks or at night;

7.4 **(a)** $CaCl_2(aq) + Na_2CO_3(aq) \longrightarrow CaCO_3(s) + 2NaCl(aq)$;

  (b) sodium and chloride ions are spectator ions here;

  (c) $Ca^{2+}(aq) + CO_3^{2-}(aq) \longrightarrow CaCO_3(s)$;

7.5 **(a)** addition of a soluble sulphate salt such as sodium sulphate to a solution of barium chloride would precipitate insoluble barium sulphate. This may be filtered off and dried;

  (b) for the example in **(a)**:
$Na_2SO_4(aq) + BaCl_2(aq) \longrightarrow BaSO_4(s) + 2NaCl(aq)$;

  (c) sodium and chloride ions;

  (d) barium nitrate;

7.6 **(a)** one mole of aqueous calcium hydroxide is neutralised by two moles of hydrochloric acid to yield one mole of solid aqueous calcium chloride and two moles of water;

  (b) two moles of aluminium chloride react with three moles of aqueous sodium carbonate and three moles of water to yield two moles of aluminium hydroxide, six moles of aqueous sodium chloride and three moles of carbon dioxide gas;

  (c) one mole of aqueous lead(II) nitrate reacts with one mole of aqueous magnesium chloride to yield one mole of lead(II) chloride and one mole of aqueous magnesium nitrate;

  (d) one mole of aqueous iron(III) chloride reacts with three moles of aqueous potassium hydroxide to yield one mole of solid iron(III) hydroxide and three moles of aqueous potassium chloride;

  (e) one mole of solid zinc hydroxide reacts with two moles of aqueous hydroxide ions to yield one mole of tetrahydroxyzincate ions;

7.7 **(a)** the formula unit of magnesium nitrate contains two nitrate ions. To indicate this the nitrate ion is enclosed in brackets;

  (b) a '2' after the brackets shows that only the contents of the preceeding brackets must be doubled. A '2' in front of the entire formula unit would double the magnesium also;

7.8 **(a)** potassium iodide $\longrightarrow$ 1 mol K^+(aq) + 1 mol I^-(aq);

  (b) sodium chromate(VI) $\longrightarrow$
2 mol Na^+(aq) + 1 mol CrO_4^{2-}(aq);

  (c) potassium manganate(VII) $\longrightarrow$
1 mol K^+(aq) + 1 mol MnO_4^-(aq);

  (d) chromium(III) sulphate $\longrightarrow$
2 mol Cr^{3+}(aq) + 3 mol SO_4^{2-}(aq);

  (e) ammonium sulphate $\longrightarrow$
2 mol NH_4^+(aq) + 1 mol SO_4^{2-}(aq);

  (f) zinc nitrate $\longrightarrow$ 1 mol Zn^{2+}(aq) + 2 mol NO_3^-(aq);

7.9 **(a)** 0.156 mol; **(b)** 22.78 g; **(c)** 21.95 g;

7.10 **(a)** oxygen is present in excess; **(b)** 217.58 g; **(c)** 4.32 g;

7.11 2.56 g;

7.12 **(a)** 3.91 g; **(b)** 1.41 g;

7.13 **(a)** 1.31 g;

  (b) 1 mol of any gas occupies approximately 24 dm^3 at room temperature and pressure. The volume of carbon dioxide produced will be about 0.71 dm^3;

  (c) 3.15 g;

7.14 $[Al(H_2O)_6]^{3+}$(aq) + OH^-(aq) $\longrightarrow$
$[Al(H_2O)_5(OH)]^{2+}$(aq) + H_2O(l);
$[Al(H_2O)_5(OH)]^{2+}$(aq) + OH^-(aq) $\longrightarrow$
$[Al(H_2O)_4(OH)_2]^+$(aq) + H_2O(l);
$[Al(H_2O)_4(OH)_2]^+$(aq) + OH^-(aq) $\longrightarrow$
$[Al(H_2O)_3(OH)_3]$(s) + H_2O(l);
$[Al(H_2O)_3(OH)_3]$(s) + OH^-(aq) $\longrightarrow [Al(OH)_4]^-$(aq) + $3H_2O$(l);

7.15 $(NH_4)_2SO_4$(aq) + $2NaOH$(aq) $\longrightarrow$
$2NH_3$(g) + Na_2SO_4(aq) + $2H_2O$(l);

7.16 if a solution of the mixture is treated with sodium hydroxide a precipitate will be produced which is a mixture of blue copper(II) hydroxide and tin hydroxide. On addition of excess sodium hydroxide, tin(II) hydroxide (amphoteric) redissolves as $[Sn(OH)_4]^{2-}$. The copper(II) hydroxide may then be filtered off, washed with distilled water and dried. Addition of a little aqueous hydrochloric acid to the filtrate precipitates hydrated tin(II) hydroxide which may be washed and dried to yield $Sn(OH)_2$;

7.17 aluminium hydroxide: $Al(OH)_3$;
zinc hydroxide: $Zn(OH)_2$;
lead hydroxide: $Pb(OH)_2$;

7.18 each of the water molecules in the hexaaquonickel(II) complex ion may be replaced, one at a time, by ammonia ligands, leading to the hexaamminenickel(II) complex ion;

7.19 addition of ammonium thiocyanate solution to the iron(III) chloride solution gives a blood-red complex,
$[Fe(H_2O)_6]^{3+}$(aq) + SCN^-(aq) $\longrightarrow$
$[Fe(H_2O)_5SCN]^{2+}$(aq) + H_2O(l);

7.20 **(a)** magnesium hydroxide, $Mg(OH)_2$(s);

  (b) the cloudiness was due to formation of $Zn(OH)_2$(s);

  (c) addition of acid forms $[Zn(H_2O)_6]^{2+}$(aq) which is soluble;

7.21 $[Cu(NH_3)_4(H_2O)_2]Br_2$, containing the complex $[Cu(NH_3)_4(H_2O)_2]^{2+}$;

7.22 **(a)** the $[Cr(H_2O)_6]^{3+}$ ion, which forms initially, is acidic, forming the $[Cr(H_2O)_5(OH)]^{2+}$ ion and oxonium ions;

7.23 **(a)** Na_2CO_3(s) + H_2SO_4(aq) $\longrightarrow Na_2SO_4$(aq) + H_2O(l) + CO_2(g);

  (b) $CaCO_3$(s) + $2HCl$(aq) $\longrightarrow CaCl_2$(aq) + H_2O(l) + CO_2(g);

  (c) $Mg(HCO_3)_2$(s) + $2HNO_3$(aq) $\longrightarrow$
$Mg(NO_3)_2$(aq) + $2CO_2$(g) + $2H_2O$(l);

7.24 copper(II) nitrate;

7.25 it has a creamy-white colour and is less soluble in NH_3(aq);

7.26 $2Br^-$(aq) + Cl_2(aq) $\longrightarrow Br_2$(aq) + $2Cl^-$(aq);

7.27 **(a)** barium sulphate, $BaSO_4$, is a very insoluble white solid;

  (b) $Ba(CH_3COO)_2$(aq) + H_2SO_4(aq) $\longrightarrow$
$BaSO_4$(s) + $2CH_3COOH$(aq);

7.28 **(a)** the bromine has been used up in an addition reaction adding across the cyclohexene double bond;

  (b)

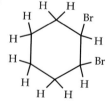

1,2-dibromocyclohexane

  (c) the bromine colour would then remain;

7.29 **(a)**

butan-1-ol

butan-2-ol

(b)

Butan-2-ol reacts similarly

7.30 **(a)**

(b) the phenol is precipitated as a milky suspension;

(c) phenols react with a neutral solution of iron(III) chloride to give intense violet/blue colours;

7.31 $2CH_3CH_2COOH(aq) + K_2CO_3(aq) \longrightarrow$
$2CH_3CH_2COO^-K^+(aq) + CO_2(g) + H_2O(l)$;

7.32 the aromatic ring is hydrophobic in character and the acids are thus water-insoluble. In alkaline solution salts are formed which are much more hydrophilic due to interactions between the ions and the polar water molecules;

7.33 **(a)**

propanal

(b)

(c) reduction of the diamminesilver ions to give metallic silver;

7.34 **(a)** warming with sodium hydroxide liberates the free amine. In acid solution a quaternary ammonium salt remains, producing no ammonia;

(b)

butylammonium chloride

(c) an equilibrium is established between protonated and unprotonated amine molecules. Oxonium ions are responsible for acidity;

7.35 **(a)** salt formation takes place yielding ethylammonium methanoate;

(b) $CH_3CH_2NH_3^+\ HCOO^-$;

7.36 **(a)** wash first with distilled water and then with standard sodium carbonate solution;

(b) wash first with distilled water and then with the HCl(aq) to be used;

(c) wash with distilled water only;

7.37 **(a)** $0.056\ mol\ dm^{-3}$; **(b)** $0.045\ g$;

7.38 **(a)** $0.194\ mol\ dm^{-3}$; **(b)** $3.30\ g$;

7.39 **(a)** $20\ cm^3$; **(b)** 99.5%;

7.40 $0.039\ mol\ dm^{-3}$;

7.41 5.80% solution;

7.42 **(a)** a titration involving the formation of a complex;

(b) $2.5 \times 10^{-3}\ mol$;

7.43 tap-water contains chloride ions;

7.44 **(a)** $Cu^{2+}(aq) + S^{2-}(aq) \longrightarrow CuS(s)$; **(b)** $0.012\ g$;

(c) $1.93 \times 10^{-4}\ mol\ dm^{-3}$;

7.45 **(a)** 0.0024; **(b)** $0.096\ g$; **(c)** $0.024\ mol\ dm^{-3}$;

7.46 your words not ours! The information on pages 134 and 135 will be helpful;

7.47 **(a)** C_2H_4NO; **(b)** $C_4H_8N_2O_2$;

7.48 **(a)** NO_2; **(b)** NO_2 or N_2O_4;

7.49 **(a)** C_5H_7N; **(b)** $C_{10}H_{14}N$;

Chapter 8

8.1 **(a)** see Fig 8.1(a). The sample is placed in a capillary tube, sealed at one end and heated in oil. The melting point is taken to be the point at which the first complete meniscus is formed;

(b) the 'literature value' is the quoted melting point for the pure substance. If the experimental value is lower this is likely to be due to impurities;

8.2 **(a)** the term impurity refers to any substance present in a sample (often in very small quantity) other than the named substance;

(b) impurities must be carefully monitored to ensure that a product is safe to use and performs satisfactorily;

8.3 the curve you add should be the same shape as the existing curve but should be below this, since the solute raises the boiling point;

8.4 simple distillation is often unsuccessful at separating two liquids of closely similar boiling points. Fractional distillation is a much more efficient method for such mixtures;

8.5 **(a)** an homogenous mixture is a physical combination of two or more chemical substances forming a single phase, eg solid, liquid or gas;

(b) ethanol and water both contain hydrogen atoms bonded to oxygen atoms. This gives rise to the possibility of hydrogen bonding between water and ethanol molecules and ensures a homogenous mixture;

8.6 **(a)** $76\,440\ Pa$;

(b) $21\,560\ Pa$;

8.7 an ideal mixture is one in which every component of the mixture exerts a vapour pressure proportional to its mole fraction. Such a mixture obeys Raoult's law precisely;

8.8 **(i)** ideal mixture;

(ii) react together;

(iii) non-ideal mixture;

(iv) immiscible liquids;

8.9

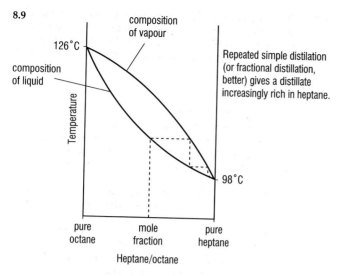

126°C

composition of vapour

composition of liquid

Temperature

Repeated simple distilation (or fractional distillation, better) gives a distillate increasingly rich in heptane.

98°C

pure octane

mole fraction

pure heptane

Heptane/octane

8.10 0.833g;

8.11 X = 0.39; Y = 0.8;

8.12 the two isomers of butenedioic acid are geometrical isomers and hence show significantly different physical properties. They can thus be separated by HPLC. The isomers of 2-hydroxypropanoic acid are optical isomers and (apart from their effect on plane polarised light) have identical physical properties. HPLC cannot separate these;

8.13 oxygen reacts with some substances (oxidation taking place on the surface of the stationary phase), chemically changing them. It is thus unsuitable as a carrier gas for GLC;

8.14 a TLC of three samples may be run; i) pure caffeine ii) regular coffee and iii) decaffeinated coffee. Regular coffee should show a 'spot' due to caffeine (confirmed by the caffeine sample), decaffeinated should not give a caffeine 'spot';

8.15 2.86 g;

8.16 **(a)** a cationic exchanger contains cations which may be substituted for those in the sample - thus effecting an exchange of cations. An anionic exchanger substitutes anions in the sample for those held on the exchange medium. A cationic exchanger is necessary to remove cadmium ions from water;

(b) $2RSO_3^-Na^+(s) + Cd^{2+}(aq) \rightleftharpoons (RSO_3)_2Cd(s) + 2Na^+(aq)$;

(c) the last traces of cadmium ions cannot easily be removed from an aqueous sample by this method since an equilibrium is established between cadmium ions in solution and cadmium ions retained by the exchange resin. Thus a finite quantitiy of cadmium ions will always pass through the column and will not be removed;

8.17 the cost of the resin required to deal with water on a commercial scale is prohibitive. In addition, used resin, when regenerated, would release nitrate ions. This nitrate waste would need to be dealt with in a suitable manner;

Chapter 9

9.1 each element produces a characteristic emission spectrum as a consequence of its excited electrons returning to lower energy levels. When the frequency of radiation is in the visible region the colours observed may be used to identify the element concerned;

9.2 Bohr suggested that electrons of certain allowed energies did not radiate energy. This idea was used to account for the stability of atoms;

9.3 **(a)** the visible region of an emission spectrum appears as a set of coloured lines on a dark background. These lines appear in series and in each series as the frequency increases the lines get closer together and finally coalesce;

(b) it shows us that electrons in atoms possess different energies but only certain energies are allowed. Also, the higher the energy associated with an electron the smaller the difference in energy between successive energy levels;

(c) an absorption spectrum of an element has lines corresponding to the same frequencies as the lines in an emission spectrum but they appear as dark lines on a white background;

9.4 **(a)** the atom loses control over its electron and becomes an ion;

(b) if we divide the energy requirement per mole of hydrogen atoms by the number of hydrogen atoms contained in one mole then this will give a value for a single hydrogen atom. [2.18×10^{-21} kJ];

9.5 **(a)** the electrical supply is used to excite the gaseous atoms;

(b) the pattern of excitation of electrons is different for each element. Consequently, on relaxation, atoms of different elements have their strongest emission lines in different parts of the electromagnetic spectrum, giving rise to different colours;

9.6 **(a)** this refers to the energy needed to form a specified gaseous ion by the removal of an electron from the appropriate gaseous atom or ion. It is usually quoted for formation of one mole of the gaseous ion;

(b) **(i)** $Ca(g) \longrightarrow Ca^+(g) + e^-$;

(ii) $Li^+(g) \longrightarrow Li^{2+}(g) + e^-$;

9.7 **(a)** see Key Facts box on page 165;

(b) the first three electron affinities of nitrogen refer to the successive capture of three electrons by a mole of gaseous nitrogen atoms as shown in the following equations:
$N(g) + e^- \longrightarrow N^-(g)$
$N^-(g) + e^- \longrightarrow N^{2-}(g)$;
$N^{2-}(g) + e^- \longrightarrow N^{3-}(g)$;

9.8 **(a)** energy is always required to remove electrons from the attractive power of an atom's nucleus – hence the ionisation energies always have positive values. **(b)** The sign of the electron affinity for an atom depends upon two factors; i) energy released when a nucleus captures an electron and ii) energy required to overcome the repulsion between the existing electrons and the electron to be added;

9.9 the second electron to be removed is usually closer to the nucleus and must be removed from a positively charged ion. Hence the second ionisation energy is larger;

9.10 **(a)** lithium has the electron structure $1s^2 2s^1$. The second electron is much more difficult to remove since it resides in the 1s orbital which is very close to the nucleus;

(b) for neon ($1s^2 2s^2 2p^6$) the first and second ionisation energies are similar since they correspond to the removal of electrons from the same $2p^6$ subshell;

9.11 briefly, your answer should emphasise that the data show clearly that the fourth electron removed from an aluminium atom is in a significantly different environment to the first three. This could then be used to establish a model for the aluminium atom in which the outer energy level contains three electrons;

9.12 **(a)** true. As a Group is descended the outer electrons are further from the nucleus and are shielded by inner electrons;

(b) false. Here the effect of increasing nuclear charge dominates and leads to an increase in first ionisation energy;

(c) false. Chlorine is below fluorine in Group VII and would have a smaller first ionisation energy for reasons described in **(a)**;

(d) false. Nuclear charge is an influence on first ionisation energy but, for example, on descending a Group, the increase in nuclear charge is outweighed by the increase in atomic radius and the shielding effect of electrons close to the nucleus;

9.13 in potassium, the electron being removed is in the 4s subshell; in argon it is in a 3p subshell. The 4s subshell is further from the nucleus and more effectively shielded from it by other electrons than the 3p is;

9.14 your diagram should show a trend of increasing first ionisation energy from left to right across the Periodic Table and a general decrease on descending;

9.15 (a) increase in nuclear charge;
(b) once again, the explanation is that on descending a Group, the increase in nuclear charge is outweighed by the increase in atomic radius and the shielding effect of electrons closer to the nucleus;

9.16 (a) it implies that electrons 'orbit' the nucleus like planets around the Sun;
(b) electrons in atoms exist in three-dimensional regions of space known as orbitals. The term orbit is inaccurate since it suggests the electrons move only on the circumference of a circle;

9.17 an atomic orbital is a mathematically defined region of space which may hold up to a maximum of two electrons;

9.18 (a) see Fig 9.14;
(b) 8. The 2s orbital can accomodate two electrons and the three 2p orbitals can accomodate two electrons each;

9.19 these orbitals have the same energy associated with them;

9.20 it explains that within each energy level different environments are available to electrons, corresponding to the different orbitals available. Thus a greater number of transitions are possible than predicted by the Bohr model of the atom;

9.21 (a) $1s^2 2s^2 2p^6 3s^2 3p^3$;
(b) $1s^2 2s^2 2p^6 3s^2 3p^6$;

9.22 (a) He $1s^2$;
(b) Na^+ $1s^2 2s^2 2p^6$
(c) S $1s^2 2s^2 2p^6 3s^2 3p^4$
(d) S^{2-} $1s^2 2s^2 2p^6 3s^2 3p^6$;
(e) Ca^{2+} $1s^2 2s^2 2p^6 3s^2 3p^6$;
(f) H^- $1s^2$;
(g) K $1s^2 2s^2 2p^6 3s^2 3p^6 4s^1$;
(h) Cl^- $1s^2 2s^2 2p^6 3s^2 3p^6$;
(i) Br $1s^2 2s^2 2p^6 3s^2 3p^6 3d^{10} 4s^2 4p^5$;

9.23 the electronic structure of beryllium is $1s^2 2s^2$. When the first two electrons are removed from a beryllium atom these are removed from the 2s orbital. The third electron must be removed from the 1s orbital which is closer to the nucleus and not shielded by other electrons. It demands much more energy. The fourth electron also comes from the 1s orbital;

9.24 (a) the ionisation energies increase as the electrons associated with them are being removed from ions with increasing positive charge;
(b) Group VI
(c) the seventh ionisation energy must involve removal of an electron from an orbital nearer to the nucleus and less well shielded;
(d) this increase means that the fifth electron is removed from a new subshell;

9.25 your own words!

9.26 N $1s^2 2s^2 2p^3$; O $1s^2 2s^2 2p^4$.
The electronic structure of nitrogen is actually the more stable due to the half-filled 2p subshell. In oxygen we see the addition of a further electron to one of the 2p orbitals which already contains an electron, leading to a repulsion. This electron is therefore easily removed;

9.27 on descending a Group, the effect of increasing nuclear charge is outweighed by the increase in atomic radius and the shielding effect of electrons closer to the nucleus;

9.28

S S^{2-}
(a) $1s^2 2s^2 2p^6 3s^2 3p^4$ $1s^2 2s^2 2p^6 3s^2 3p^6$
(b) $[Ne]3s^2 3p^4$ $[Ne]3s^2 3p^6$
(c)

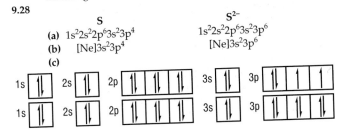

9.29 They are unusual in that the 4s subshell is normally filled before the 3d subshell (lower energy). For chromium and copper, however, the stability associated with a half-filled (Cr) and a full (Cu) 3d subshell means that only one electron is in the 4s orbital;

Chapter 10

10.1 (a) the type of elements present;
(b) the ratios of each type of atom or ion present in the sample;

10.2 (a) formation of positive ions from the sample and separation of these according to their mass/charge ratios;
(b) examination of the environments of certain nuclei, notably ^1H and ^{13}C;
(c) idenfication of functional groups present in the sample by examination of the characteristic absorption of infrared radiation;

10.3 X-ray diffraction is used to establish the relative positions in space of the atoms or ions in a sample;

10.4 your sketch should show the three peaks, at 24, 25 and 26, each with the correct height (relative abundance); $A_r(Mg) = 24.33$;

10.5 $M_r = 74$; butan-1-ol: most abundant peak at 31, probably due to CH_2OH^+ 2-methylpropan-2-ol: most abundant peak at 59, probably due to $(CH_3)_2COH^+$

10.6 m/e 46, 45, 43, 42;

10.7 (a) C_4H_8 (b)

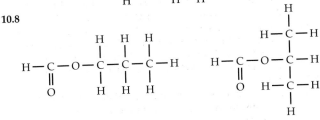

10.8

H H H   H
| | | |
H—C—O—C—C—C—H   H—C—H
‖ | | | |
O   H H H   H—C—O—C—H
                               ‖ |
                               O   H—C—H
                                   |
                                   H

Propyl methanoate   2-Propyl methanoate

10.9 121.86;

10.10 (a) both substances give a molecular ion of m/e 58;
(b) high resolution mass spectrometry will distinguish between the m/e ratios of the molecular ions for each substance;

10.11 one advantage of NMR spectroscopy is that it does not damage the sample, which may be recovered after investigation;

10.12 in methanol there are two types of environment for hydrogen atoms and hence peaks. Tetramethylsilane has twelve hydrogen atoms, each in identical environments – hence one (large) peaks;

10.13 (a) 3
(b) the ratio of the areas under the peaks will be 3 : 2 : 3;

10.14 (a) one large peak, corresponding to six hydrogen atoms, all in an identical environment;
(b) there are three environments for hydrogen atoms:
(i) attached to the same carbon atom as the chlorine atom;
(ii) attached to an adjacent carbon atom on the same side of the ring;
(iii) attached to an adjacent carbon atom on the opposite side of the ring;
[If you are still not sure – make a model!!]

10.15 see pages 189–191;

10.16 the absence of a peak at m/e 15 in the mass spectrum suggests that there are no methyl groups present. The NMR spectrum shows that all hydrogen atoms are in the same environment. The compound must be cyclohexane;

10.17 your sketch should show the following peaks/chemical shifts:
methyl hydrogens	approx 7.8	3 hydrogen atoms
2 and 6 positions	approx 2.7	2 hydrogen atoms
3 and 5 positions	approx 2.7	2 hydrogen atoms
4 position	approx 2.7	1 hydrogen atom;

10.18 an absorption spectrum is obtained when a sample selectively absorbs certain frequencies of radiation from a range of frequencies available. The energy absorbed is used to effect electronic changes in the sample. The appearance of such a spectrum (in the visible range) is of a series of dark lines on a coloured background;

10.19 energy absorbed is used for vibration, twisting and bending of molecules. Ammonia has a number of possible vibrations. You should try to draw some of these. Fig 10.20 will be useful here;

10.20 ethanol $1000–1300 \text{ cm}^{-1}$ $3580–3670 \text{ cm}^{-1}$;
methoxymethane $1000–1300 \text{ cm}^{-1}$;

10.21 **(a)** NMR spectroscopy of cyclopropane will show one large peak only, with a chemical shift of about 8.0 (6H). An NMR spectrum of propene should show three peaks;

(b) infrared spectroscopy is less useful but should show a peak for C=C at around 1650 cm^{-1}. Of course the fingerprint region of the IR spectrum of both substances can also be compared with a genuine sample, which allows positive identification;

10.22 a calibration graph of concentration against absorption may be plotted. The concentration of an unknown sample may then be found by measuring its absorption at 800 nm and reading its concentration from the graph;

10.23 **(a)** the conjugated double bond of buta-1,3-diene is a chromophore and absorbs radiation at 220 nm;

(b)

buta-1,3-diene

1,2,3,4-tetrabromobutane

(c) the chromophoric group has undergone an addition reaction and is no longer present;

10.24 **(a)** only the C=C double bond of the side-chain undergoes addition reactions under these conditions, adding 1 mole of hydrogen per C=C;

(b) the C=C bond was part of a larger delocalised system, shared with the ring. When this group reacts the chromophoric group is altered and now absorbs at a different wavelength;

(c) addition of bromine is unlikely to produce chemical reaction (unless reaction takes place in strong ultraviolet light). The chromophoric group remaining (the benzene ring) is unchanged, as is the absorption maximum;

Theme 3

Chapter 11

11.1 a research portfolio is produced after establishing a projected demand for a new product. This follows detailed market research which gives information about the size of the market for a new product, its likely lifespan and preliminary investigation of the cost of production, distribution and marketing;

11.2 the chemical and physical properties of any existing products are considered and the common properties of these products noted. These qualities are used as the basis for a 'shopping list' of candidate compounds;

11.3 a substructure search is a hunt through the chemical literature for a specific arrangement of atoms or groups believed to contribute significantly to the properties of a compound or range of compounds;

11.4 **(a)**

(b) the amino and carboxylic acid functional groups;

(c) $NH_2CH(CH_3)COOH$ $NH_2CH(CH_2CH(CH_3)_2)COOH$;
alanine   leucine
$HOOCCH_2(NH_2)COOH$ $H_2NCH_2CH_2CH_2CH_2CH(NH_2)COOH$
aspartic acid   lysine

11.5 it would need to combine enhanced antiseptic properties, low toxicity and ease of use (including long shelf-life). The structure and properties of existing compounds would be carefully examined and a substructure search carried out to find existing compounds of similar structure if possible. Then, a list of new candidate compounds can be prepared and synthesised;

11.6

11.7 **(a)** both dyes contain an aromatic ring bearing a sulphonate group. The ring is also bridged by a N=N bridge to another aromatic system.

(b) For red dyes it would be wise to search for a substructure in which a sulphonate group was attached to a naphthalene ring whilst yellow dyes are more likely to contain a sulphonate group attached to a benzene ring;

11.8 **(a)** bacterial cells ceased to divide when an electric field was applied yet careful investigation of variables such as pH, temperature and nutrient availability showed that these were not responsible;

(b) platinum was believed to be completely inert under these conditions;

(c) these are less toxic;

11.9 **(a)** a greater chain length for an aliphatic polyamide decreases the frequency with which the amide functional group is encountered along the chain. This weakens the bonding between one polymer chain and another, allowing more flexing and twisting;

(b) the Kevlar polymer chains pack together to give rigid fibres;

(c) a substructure search for starting materials containing a benzene ring and suitable functional groups may identify new staring materials to prepare a related polymer with modified properties;

11.10 **(a)** many drugs are believed to function by virtue of their shape and the interaction of their structure with a receptor molecule or site. Computer modelling allows identification of candidate molecules with suitable shapes;

(b) without computer modelling (or alternative modelling system) a research team may waste valuable time synthesising molecules whose shape is inappropriate;

(c) need to know bond lengths and angles of the molecule to be modelled, as well as its full structural formula;

11.11 mathematical modelling is used to predict the properties of compounds and is based on the use of mathematical equations to describe the arrangements of atoms. Computer modelling allows display of a molecule with bond lengths and angles displayed. The two techniques allow prediction of the properties and likely behaviour of an unknown molecule;

11.12 **(a)**

(b) only the 1,4-dimethylbenzene has a shape which allows it to pass through the pores of the zeolite ZSM-5. It has a (small) methyl group in each of the 1- and 4-positions;

11.13 (a) see Datasheet 1 for data required;
(b) about 0.176 nm;

11.14 (a)

Element	1st IE	2nd IE	3rd IE
Be	215	420	3547
Mg	176	346	1850
Ca	141	274	1180
Sr	131	253	986
Ba	120	231	811

(b) draw the graph. The explanation for these variations has been met in the answer to question 9.12(d);

11.15 (a) the literature value is 130°C;
(b) the strength of intermolecular bonding increases with chain length since longer chains allow more opportunity for intermolecular forces to act;

11.16

Class		Datasheet
(a)	halogenoalkane	18
(b)	ketone	22
(c)	nitrile	23
(d)	alcohol	19
(e)	aldehyde	22
(f)	ester	21
(g)	nitrile	23
(h)	carboxylic acid	21
(i)	alcohol	19
(j)	amine	23
(k)	amide	21
(l)	carboxylic acid	21

11.17

Compound		Class	Semistructural formula	Datasheet
(a)	butanal	aldehyde	$CH_3CH_2CH_2CHO$	22
(b)	butanone	ketone	$CH_3CH_2COCH_3$	22
(c)	but-1-ene	alkene	$CH_2{=}CHCH_2CH_3$	16
(d)	propan-1-ol	alcohol	$CH_3CH_2CH_2OH$	19
(e)	ethyl propanoate	ester	$CH_3CH_2CO_2CH_2CH_3$	21
(f)	ethylamine	amine	$CH_3CH_2NH_2$	23
(g)	ethanenitrile	nitrile	CH_3CN	23
(h)	1-bromopropane	halogenoalkane	$CH_3CH_2CH_2Br$	18

11.18 (a) add water;
(b) add sodium carbonate solution;
(c) use ammoniacal silver nitrate;
(d) reflux with acidifid dichromate (VI). The tertiary alcohol will not oxidise;

11.19 see Database;

11.20 (a)

(b)

(c)

11.21 (a) KCN/ethanol/reflux;
(b) CH_3COOH/conc H_2SO_4 catalyst/reflux;
(c) acidfied potassium dichromate(VI)/reflux;
(d) acidified potassium dichromate(VI)/reflux;
(e) Sn/conc HCl/reflux;
(f) H_2SO_4(aq)/reflux;

11.22 (a) $LiAlH_4$/dry ethoxyethane/reflux;
(b) e.g. $SOCl_2$/room temperature;
(c) H_2SO_4(aq)/reflux;
(d) Br_2/room temperature;
(e) $KMnO_4$(aq)/OH^-/reflux;
(f) $K_2Cr_2O_7$/H^+/distil;

11.23 use the Datasheets and synthetic route maps;

11.24 use the Datasheets and synthetic route maps;

11.25 (a) add CH_3COOH and a few drops of conc sulphuric acid to the acid and reflux;
(b) warm with excess ethanoyl chloride which reacts with both the alcohol and phenol groups to give the desired product;

11.26 use the acid chloride formed from butanedioic acid and warm this with 1,4-dihydroxybenzene; about 104;

11.27 (a)

Repeat unit for polymer

(b) the acid chloride and amine functional groups will react at room temperature to give a polyamide;

(c) ester cross-linkages may be formed by warming the polymer chains with a little sulphuric acid;

(d) the extent of polymerisation may be controlled through control of pH. If acid conditions are allowed to persist the amine functional groups will be become protonated and unreactive. Cross linking may be controlled by controlling the temperature whilst the chains are exposed to the sulphuric acid catalyst;

Chapter 12

12.1   see page 243;
12.2   see page 243;
12.3   Table 12.2, p247 may be helpful here;
12.5   it is likely that the lattice energy of the calcium salt is greater;
12.6   **(b)** $Ca(CH_3(CH_2)_{12}COO)_2$;
12.7   Hint: examine the electronic structure of sulphur!
12.8   examine the argument on page 246;
12.10   Hint: Linus Pauling, the Nobel Prize-winning American chemist estimated the % ionic character of the sodium chloride bond at about 70%. He estimated silver chloride at about 31%;
12.11   page 249 may be useful;
12.13   use the data supplied carefully and you can show that the side-chain double bond is conjugated to the ring system and does not behave like an isolated double bond. Delocalisation!
12.15   you should discuss the nature of the bonding within a carbonate ion and the bonding between calcium *and* carbonate ions;
12.16   Hint: what is the distribution of electron pairs around fluorine?

12.18	H_2S	non-linear		104.5°
	SCl_2	non-linear		104.5°
	PF_3	pyramidal		107°
	PF_5	trigonal bipyramidal		90° and 120°
	IF_5	distorted square pyramidal	approx	90°
	SF_6	octahedral		90°
	XeF_4	square planar		90°
12.19	$SOCl_2$	distorted trigonal planar	approx	120°
	XeF_2	linear		180°
	SO_2	non-linear approx		120
	SO_3	trigonal planar		120°
12.20	NH_4^+	tetrahedral		109.5°
	NH_2^-	non-linear		104.5°
	NH_3	pyramidal		107°

12.22   see page 266;
12.24   octahedral;
12.25   **(b)** $[Au(CN)_2]^-$ dicyanoaurate(I) ion, linear;
12.26   the polar isomer will have greater intermolecular attractions;
12.27   glycine;
12.28   propyl methanoate and methyl propanoate;
12.32   Hint: benzene is planar

Chapter 13

13.2   **(a)** homolytic;
13.4   **(a)** substitution; **(b)** elimination; **(c)** addition; **(d)** addition; **(e)** substitution; **(f)** addition; **(g)** addition; **(h)** substitution;
13.9   **(a)** 1-chloro-2-iodoethane; **(b)** 2-chloro-1-iodobutane.
13.10   Hint: try writing the structures of the intermediate carbocations first;
13.11   **(a)**, **(b)** and **(d)** are chiral;
13.12   the dominant reaction type for alkenes is electrophilic addition. The cyanide ion is a nucleophile;
13.14   transfer of Cl^- takes place;
13.15   chlorine radicals must first be produced (uv light needed);
13.16   combination of radicals in the reaction mixture;
13.17   the nitro group is electron withdrawing;

13.18   **(a)** pentan-1-ol; **(b)** ethanoic acid, **(c)** methyl ethanoate and **(d)** but-1-thiol are the names. Now check your structures;
13.19   2-iodohexane is the most easily hydrolysed, followed by bromomethane then chloroethane. 1-fluorobutane is unreactive;
13.20   bromoethane is water-insoluble;
13.22   Hint: think about the relative electronegativities of chlorine, carbon and silicon;

Theme 4

Chapter 14

14.4   endothermic;
14.5   **(a)** $CH_4(g) + 2O_2(g) \longrightarrow CO_2(g) + 2H_2O(l)$;
    (b) the burning methane is the system, everything else is the surroundings;
    (c) an open system;
    (d) exothermic;
14.7   **(a)** $Sn(s) + 2HNO_3(aq) \longrightarrow Sn(NO_3)_2(aq) + H_2(g)$;
    (b) about 24 cm^3;
    (c) 24 cm^3 is 2.4×10^{-5} m^3. The work done is therefore,
$$w = P\Delta V$$
$$= 10^5 \times 2.4 \times 10^{-5} = 2.4 \text{ J};$$
14.9   see Key Facts box, p353;
14.10   aluminium;
14.12 $\Delta H = mc\Delta T = 750 \times 4.184 \times 75 = 235.35$ kJ
14.14   13.64 kJ;
14.17 $\Delta H = -48.8$ kJ;
14.20   13.5 kJ;
14.23   **(a)** 95.2 J;
    (b) it will decrease;
    (c) 0.023°C, a negligible decrease in temperature;
14.25   at standard conditions, α-sulphur is the more stable. β-sulphur must contain more energy and will have the highest enthalpy change of combustion. They both give sulphur dioxide on burning, of course;
14.26   3.2 g methanol evolves 72.63 kJ; 2 g ethanol evolves 59.43 kJ;
14.27   2712.4 kJ;
14.28   one mole reacts with three moles of potassium hydroxide;
14.30   5141.3 kJ mol^{-1};
14.31   706 kJ mol^{-1};
14.33   −52 kJ mol^{-1};
14.34   **(a)** −238.2 kJ mol^{-1};
    (b) −276.6 kJ mol^{-1};
    (c) approximately −315 kJ mol^{-1};
14.35   −2275 kJ mol^{-1};
14.36   **(a)** −392.9 kJ mol^{-1};
    (b) it differs by only 0.9 kJ mol^{-1} – good agreement;

Chapter 15

15.3   **(a)** Fe oxidised, water reduced;
    (b) Mg oxidised, oxygen reduced;
    (c) iron oxidised, silver ion reduced;
    (d) carbon oxidised, iron ion reduced;
15.12   **(a)** 0.46 V;
    (b) 1.22 V;
    (c) 1.2 V;
15.16   a combination of half-cells (i) and (iv) give the largest e.m.f.;
15.19   **(a)** the oxygen half-cell;
    (b) 1.55 V;
    (c) $N_2H_4(aq) + O_2(g) \longrightarrow N_2(g) + 2H_2O(l)$;
    (d) the size of the cell e.m.f. suggests the cell is feasible;
    (e) hydrazine and oxygen;
    (f) the only products are 'safe' nitrogen and water;

Chapter 16

16.1 the ease of oxidation is Zn(–0.76 V)/Fe(–0.44 V)/Sn(–0.14 V);
16.3 calcium has a more negative electrode potential than H_3O^+;
16.6 **(c)** chlorine has a more positive electrode potential than bromine and will be reduced by bromine;
16.10 Hint: remember the autoionisation of water;
16.11 H_2SO_4: cathode; hydrogen, anode; oxygen;
HCl: cathode; hydrogen, anode; chlorine and oxygen;

16.14 **(a)** 0.59 g
    (b) oxygen;
    (c) it does not liberate any copper(II) ions into the solution, which would soon be exhausted of $Cu^{2+}(aq)$ ions;
    (d) at the cathode reduction of $Cu^{2+}(aq)$ takes place allowing the copper-plating to take place. At the anode oxidation of copper takes place;
6.16 **(a)** 0.16 g;
    (b) it depends only on the number of electrons passed around the circuit. This alone determines the amount of silver deposited;
16.17 44.67 hours (2680 min);
16.18 0.90 g;
16.25 **(a)** 94 477 tonnes;
    (b) 3.34×10^4 dm^3, measured at room temperature and pressure;
    (c) 5.37×10^8 C;

Theme 5

Chapter 17

17.6 **(a)** a conductivity method or pH method;
    (b) a method which monitors change in pressure;
    (c) a colorimetric method;
    (d) a colormetric method;
17.8 **(a)** it is first order with respect to bromate(V) and bromide ions but second order with respect to hydrogen ions;
    (b) the reaction must be fourth order overall;
    (c) no – these are relative rates only;
    (d) mol^{-3} dm^9 s^{-1} are the SI units;
    (e) monitoring the depth of colour of the solution or monitoring the increase in pH;
17.9 **(a)** 1 and 3;
    (b) rate = $k[CH_3COCH_3][H^+]$;
    (c) 2;
    (d) $k = 4 \times 10^{-3}$ mol^{-1} dm^3;
    (e) initial rate = 1.6×10^{-4} mol dm^3 s^{-1};

Chapter 18

18.5 **(a)** $K = \left(\dfrac{[Cl(g)]^2}{[Cl_2(g)]}\right)_{eq}$
    (b) dissociation is *very* limited;
18.6 3.74×10^5 N m^{-2};
18.7 $H_2(g) = 0.247$ mol present = 4.94%;
    $I_2(g) = 1.247$ mol present = 24.94%
    $HI(g) = 3.507$ mol present = 70.12%;
18.8 $K_p = 1.72 \times 10^{-5}$ atm^{-2};
18.9 CO and H_2O are each present as mole fraction 0.23 (23%);
    CO_2 and H_2 are each present as mole fraction 0.27 (27%);
18.10 $K_p = p(I_2(g))$;
18.11 $K_p = 0.104$ atm^2;
18.12 $K_p = 0.0298$ atm^2;
18.13 **(a)** it will move to the right;
    (b) it will be increased;

    (c) no effect;
    (d) it will be increased;
18.14 **(a)** it will move to the right;
    (b) it will be increased;
    (c) no effect;
    (d) it will decrease;
18.15 **(a)** no effect;
    (b) it will increase;
    (c) it will increase;
    (d) it will decrease slightly;
18.16 **(a)** the time taken will decrease;
    (b) – **(d)** no effect;
18.17 the brown colour would become less intense as the equilibrium position moves to the left;

Chapter 19

19.2 it must be an endothermic process;
19.3 1×10^{-12} mol dm^{-3};
19.4 **(a)** 0.0161;
    (b) 1.61%;
    (c) 0.01563 mol dm^{-3};
19.11 **(a)** 1.30;
    (b) 7.89;
    (c) 0.30;
19.12 **(a)** 3×10^{-7} mol dm^{-3};
    (b) 1.1×10^{-1} mol dm^{-3};
    (c) 7.94×10^{-14};
19.13 **(a)** 11;
    (b) 3.1;
    (c) 6.4;
19.14 **(a)** 1.45×10^{-5};
    (b) 4.84;
    (c) 0.12;
    (d) 10.08;
    (e) 12%;
    (f) 9.16;
19.16 ethanoic acid;
19.17 a slight increase in base strength results;
19.18 phenolphthalein and phenol red can be used. Methyl orange should not;
19.19 **(a)** the salt that is formed does not form a neutral solution;
    (b) it changes colour far too early – its pK_{in} value is 3.6;
    (c) 9.7;
19.21 see Table 19.2 on page 487;
19.23 **(a)** 5.06;
    (b) see pages 490 and 491;
19.26 4.51 g CH_3COONa should be added;
19.27 3.98×10^{11} mol^3 dm^{-3};
19.28 $K_{sp} = [Al^{3+}][OH^-]^3$;
19.30 sodium nitrate has a much larger K_{sp} than silver chloride;
19.32 **(a)** 2.43×10^{-4};
    (b) the maximum achievable concentration of $BaSO_4$ is approximately 1×10^{-3} mol dm^{-3};

Index

Page numbers in italics refer to information given only in an illustration or caption